FRP Composite Structures

FRP Composite Structures
Theory, Fundamentals, and Design

Hota V. S. GangaRao and Woraphot Prachasaree

CRC Press
Taylor & Francis Group
Boca Raton London New York

CRC Press is an imprint of the
Taylor & Francis Group, an **informa** business

First edition published 2022
by CRC Press
6000 Broken Sound Parkway NW, Suite 300, Boca Raton, FL 33487-2742

and by CRC Press
2 Park Square, Milton Park, Abingdon, Oxon, OX14 4RN

ISBN: 978-1-032-05251-9 (hbk)
ISBN: 978-1-032-05252-6 (pbk)
ISBN: 978-1-003-19675-4 (ebk)

DOI: 10.1201/9781003196754

Typeset in Times
by codeMantra

Dedication

To my wife, Rajeswari, who brought the meaning of wisdom to fore, to my children, Siva and Lekha, who taught me patience, and in memory of my parents (Siva Rama Sastry and Sodemma HOTA), who equipped me with the art of living.

G. HOTA

To my mother, Nucharee

To my wife and son, Janjira and Theetat

To the memory of my late father, Vichit

W. Prachasaree

"Knowledge trumps stupidity as Sun's rays burn through fog"

Advitamruta Upanishad

Contents

Preface

This textbook deals with the theory and design of structural components and systems made of fiber-reinforced polymer (FRP) composites. The primary focus herein is on the design of multifunctional structures that are commonly encountered in civil and military infrastructure.

FRP composites are made of two or more constituent materials (fibers, resins, fillers, and additives) with different physical and chemical properties. These constituents retain their identity in a finished composite product without blending fully or even losing their microstructural characteristics. The finished product (composite) exhibits superior thermomechanical performance over its constituents.

The need for the theory, design, and implementation of FRP composites in infrastructure systems has been growing in recent years because of superior material durability and higher strength to self-weight ratio over conventional materials. New constituent materials, manufacturing techniques, design approaches, and construction methods are being advanced and introduced in practice by the FRP composites community to cost-effectively build FRP structural systems. The primary focus of the polymer composites industry is to advance the field implementation of structural systems with enhanced durability (~100-year service life) and reduced initial and maintenance costs.

The focus of this textbook is to develop simplified mathematical models representing the behavior of beams and plates under static loads, after introducing generalized Hooke's Law for materials with different properties, i.e., anisotropic, orthotropic, transversely isotropic, and isotropic properties. Subsequently, the simplified models coupled with the design methods including FRP composite material degradation factors have been introduced by solving a wide range of practical design problems. The idea of deferring computer-based analysis and design is to focus fully on the basic mechanics of composite materials before involving in large-scale number crunching. The initial emphasis has been on the mechanics of FRP composite lamina and laminate followed by the analysis of composite beams and plates under a variety of static and hygro-thermal loads, such as axial, bending, shear, and their combined load effects including thermal and moisture-related responses. In later chapters (7–12), this textbook deals with the design of composite members under a variety of load conditions including the connections for composite members, resulting in composite structural systems. A snapshot of this textbook's outline and interconnection of various chapters is given in Figure 1.16, Chapter 1. Emphasis is placed on fabric lay-up that is symmetric to the center of gravity of a section to get across the fundamental principles of the mechanics of composite materials without overwhelming a reader with the computational cumbersomeness.

Majority of this work is based on pultruded member responses, wherein pultrusion is a mass-manufacturing technique as described briefly in Section 1.6, to produce high-volume products at high quality in an economical manner. Many other important topics, such as durability, cost–benefit analysis, fire responses, nondestructive evaluation, in-depth manufacturing details, quality control/assurance, and polymer recycling, have been deferred to different compendia, recognizing that numerous research publications are available in the literature. However, summaries on fiber stacking sequences and durability responses are examined cursorily in Appendices A–B of this textbook.

It is recognized that FRP composites are getting accepted to be one of the commodity products such as steel, concrete, and timber. In addition, FRP composites are being hybridized with other materials to optimize the advantages of composites as well as those of commodity products, to attain a longer-lasting hybrid; however, with added advantages of enhanced durability, specific strength and stiffness, and other thermomechanical properties.

Many higher educational institutions across the globe are offering courses in FRP composite constituent material, classifications, properties, and the applications of composite structural components and systems. These courses are offered at both the undergraduate technical elective level and graduate level in the areas of structural mechanics and structural engineering. In addition, many practicing engineers are attending short courses, webinars, and seminars in the areas of manufacturing, analysis and design of composite materials, components, and systems. The contents of this textbook are intended to bring clarity to analysis with emphasis on assumptions and approximations, and design of polymer composite structural systems. However, it is expected that readers of this textbook should possess working knowledge in the area of mechanics of materials of structural coupons and components to attain full grasp of the subject matter provided herein. As stated above, it is impossible to cover all aspects (cradle-to-cradle) of composite components and systems. Hence, this textbook is limited to theory and design of composite component response, and this textbook provides an expose on the design of composite components and systems including joint designs under static loads only. Through good grounding on fundamental aspects of FRP composites, a student will be able to advance to complex topics as the composite component and system behavior under fatigue, sustained (creep) and dynamic loading, buckling, and other modes of structural responses. Without question, additional knowledge and in-depth treatment of constituent material selection, manufacturing, in-situ evaluation of composite component integrity, degradation mechanisms under harsh environments, and other topics would be helpful; hence, some of these aspects are covered in a cursory manner in the appendices of this textbook and underpinned with an extensive list of references.

The authors hope that this textbook will be of great value in terms of understanding the fundamental static behavior of composite laminate, components, and systems; but also a great resource in designing composite components and systems under static loads. Hopefully, students and practicing engineers can use these concepts and design guides as a basis to advance the design knowledge judiciously, especially in field-implementing structural composites for military and civil infrastructure systems.

Acknowledgments

We acknowledge that the seeds for this effort were sown in years 1987–95 when the U.S. National Science Foundation, U.S. Department of Transportation, and West Virginia Department of Transportation along with the polymer composites industry provided initial funding to research, develop, and implement Fiber Reinforced Polymer (FRP) composites for infrastructure systems. We are grateful to West Virginia University for their continued support in advancing the state-of-the-art of FRP composites since 1987, allowing us to develop and teach new courses, field-implement and evaluate composite systems, and encourage teaching short courses and webinars.

We would like to thank many sponsoring agencies including federal, state, and county governments; manufacturing and construction industries; several foreign institutions such as Nanjing Tech, China, University of Southern Queensland, Australia, Prince of Songkla University, Thailand; and others such as professional and trade organizations including ASCE, ACI, ASTM, PIANC, and others. Special thanks to ACMA for providing opportunity to participate in developing pre-standards for pultruded products.

Hota thanks his wife, Rajeswari, and their children and families for their encouragement during the development of this manuscript.

Grateful acknowledgment is made to all the friends, colleagues, and graduate students who contributed to the development of this book. Woraphot would like to express deep gratitude to his academic advisor, Dr. Hota V.S. GangaRao, for providing continuous academic support throughout his academic profession. Finally, the constant encouragement and enormous understanding of his wife and son, Janjira and Theetat, were essential to the completion of this project.

This acknowledgment would be incomplete without profusely thanking many faculty and administrative colleagues at West Virginia University (WVU) and at other great institutions of learning including those (research engineers, graduate and undergraduate students) who worked with us at WVU's Constructed Facilities Center (CFC), US-Army Corps of Engineers (ACE) and other federal agencies, and US- NSF's IUCRC -Center for Integration of Composites into Infrastructure (CICI).

Acknowledgments

Authors

Prof. Hota V.S. GangaRao Ph. D., PE After joining West Virginia University in 1969 as an Assistant Professor, Dr. Hota attained the rank of Maurice & JoAnn Wadsworth Distinguished Professor in the Department of Civil and Environmental Engineering, Statler College of Engineering and Mineral Resources, and became a Fellow of ASCE and SEI. Dr. Hota has been directing the Constructed Facilities Center since 1988 and also the Center for Integration of Composites into Infrastructure, both co-sponsored by the National Science Foundation.

Through many interdisciplinary activities, he has worked to advance the state-of-the-art of fiber-reinforced polymer (FRP) composite materials and their applications to infrastructure systems. Dr. Hota has demonstrated successful use of composites by field-implementing his research findings and technical innovations for construction and rehabilitation of a wide spectrum of engineering systems in West Virginia, Ohio, Pennsylvania, Alaska, and other states of the United States of America. In addition to the application of FRPs for highway, waterway, and railroad structures, he has utilized FRP composites for corrosion-resistant storage buildings and economical modular housing. Recently, Dr. Hota has been involved with innovations of naval vessels, prefabricated pavements, utility poles, high-pressure gas pipes, sheet piling, natural fiber-reinforced composites, and others.

Dr. Hota has published over 400 technical papers on a wide range of subjects in refereed journals and proceedings, in addition to textbooks and book chapters. He has received fifteen patents and many national awards, and advised over 300 M.S. and Ph.D. students. His pioneering accomplishments have been covered by CNN, ABC Evening News, KDKA-Pittsburgh, WV-PBS, and many others.

Woraphot Prachasaree received a B.Eng. degree in civil engineering from Prince of Songkla University, Thailand in 1997 and an M.Eng. degree in the field of structural engineering from Kasetsart University, Thailand in 2000. In 2001, he joined WVU's Constructed Facilities Center (CFC) as a research assistant. He holds an M.S. degree and a Ph.D. degree in civil engineering, both from West Virginia University. In 2007, Woraphot became a Lecturer at the Department of Civil and Environmental Engineering, Prince of Songkla University, where he currently holds the position of assistant to the President. His present research interests cover several areas, including FRP composites, concrete structures, structural durability, and FE modeling. He has authored or coauthored over 40 publications in engineering journals. He is a registered professional engineer in Thailand. He has engaged in consulting work in the field of structural inspection, design, and construction for many engineering and infrastructure projects.

1 Introduction

This textbook deals with the theory and design of structural components and systems made of fiber-reinforced polymer (FRP) composites. The primary focus of this chapter is on the design of multifunctional structures that are commonly encountered in civil and military infrastructure.

FRP composites are made of two or more constituent materials (fibers, resins, fillers, and additives) with different physical and chemical properties. These constituents retain their identity in a finished composite product without blending fully or even losing their microstructural characteristics. The finished product (composite) exhibits superior thermomechanical performance over its constituents. For example, fibers are weak in compression and do not transfer shear from one fiber to another, while the binder (polymer resin) after curing (commonly known as matrix) provides the shear transfer capability and compressive strength; together they form a composite that is made of glass, carbon, or other fibers or fabrics with polymeric resin, such as epoxy or polyurethane. Composites are tailor-made with different fibers (glass, carbon, aramid, or plant-based) providing strength and stiffness, while the polymer matrix provides force transfer between fibers. The idea of such juxtaposition is to enhance strength, durability, cost-effectiveness, or other features. Furthermore, the matrix protects the fiber from environmental degradation and helps in efficiently manufacturing a product on a high-volume basis.

Polymer composites are identified by a specific fiber. For example, composites with glass or carbon fibers are referred to as glass FRP composites or carbon fiber-reinforced composites, respectively. In addition, composite manufacturers and users refer to these products as fiber-reinforced composites (FRP), glass-reinforced plastics (GRP), and polymer matrix composites (PMC). Though several examples of composites occur in nature, the modern synthetic composite field implementation only began in the late 1940s with the advent of quality glass reinforcements and room-temperature-cured resins. Although Leo Hendrik Baekeland invented modern composites with synthetic resins in the early 1900s, composites have been successfully utilized in a diverse range of applications, only after World War II.

In recent years, several large-scale FRP composite structures have been erected to demonstrate the potential of composites in major civil engineering applications. These include a 112 m pedestrian bridge in Scotland (Head, 1994), a 38 m pedestrian bridge in Denmark (McManus, 1997), and several highway bridges in the USA (GangaRao et al., 1999). Yet, despite the strong indications of the potential of these materials, FRP composites continue to be slow in penetrating the mainstream civil engineering marketplace. One of the reasons for this continued lack of market penetration and growth has been that civil engineers have not possessed the technical information and design tools necessary to exploit composite materials cost-effectively.

This textbook addresses the need for an improved understanding of composite materials and their behavior and design within an infrastructure system context. The broad aim is to improve the fundamental understanding of composite material and system response from a structural engineering perspective. It also aims to assist in the development of a more complete understanding of the constituent materials for fiber composite structural elements and how their behavior influences the overall behavior of such elements, especially when they are assembled. The first half of this book covers engineering properties of the constituent materials (fiber and matrix), mechanics and analyses of composite lamina and components, and consideration of hygrothermal and other environmental effects. The second half then covers the design philosophy for composite members under different loading actions, including the design of composite members, connections and structural systems. The Appendices provide supporting information and theories behind the mechanics and design of fiber composites.

DOI: 10.1201/9781003196754-1

1.1 HISTORIC PERSPECTIVE

FRP composites are composed of at least two constituent materials (e.g., glass or carbon fibers combined with resins like epoxy or polyester) differing in terms of thermomechanical properties. However, the idea is to engineer the combined properties to be superior to the constituent materials. The constituent materials remain separate and distinct in a composite, that is, the fiber reinforcement remains distinct from the matrix, which is the cured resin. The resins help bond together fibers, fabrics, or fibrous particulates. The resins primarily dictate the manufacturing process and processing conditions. They partially protect fibers and fabrics from environmental damage such as chemical or thermal exposure. In short, fibers and resins contribute their strengths (i.e., fibers with maximum tension or bending strengths while resin providing shear transfer) to "tailor-make" the end product with a specific use in mind, i.e., optimal strength, stiffness, or durability.

The history of FRP composites can be traced back to Mesopotamia (present-day Iraq). The Mesopotamians used mud and straw in bricks and bonded copper sheets with natural resins. In addition, decorative pieces made of cloisonnes (a kind of enamel) were produced in Asia and Europe. All the items were made with at least two distinctly different materials combined into one final product of improved performance. Organic polymers from tree and plant secretions, fossilized resins, and those from fish and animal offal were in practice, also. Furthermore, cotton, wool, and linen are natural organic fibers that have been decorated with fine gold threads as a part of earlier (old) composites. The inclusions of metallic wires to reinforce nonmetallic applications are many. For example, decorative organic plastic and paper pulp were used to make canes and umbrella handles in the late nineteenth century in Europe (Seymour and Deanin, 1986), as nonstructural composites.

Sophisticated applications of composites have been identified in nature from time immemorial, and man-made composite applications such as straw-reinforced adobe or brick in Tulou houses (Figure 1.1a) (Liang et al., 2011) and beehives (Figure 1.1b) (Chernick, 2009) have been cited in the literature. Typical beehive houses of a high-domed structure with adobe brick were found to be extremely durable with significant heat resistance due to favorable thermal properties for hot countries.

In the ensuing years, more sophisticated composites have appeared, such as Damascus steel (Figure 1.2), which is made from using wootz steel imported from India. Recent discoveries (Inman, 2006) revealed that nanowires and carbon nanotubes in a blade were forged from Damascus steel, which is a composite. This discovery provides a lesson that it is preferable to incorporate nanomaterials as an integral means of mass manufacturing for their maximum thermomechanical properties.

A few additional examples of recent origins of polymer composites are reinforced rubber tires, aerospace components, laminated polymer parts, and others. Similarly, polymer-coated metal hardware appeared in the markets in the last 70–80 years, such as polymer-insulated copper wires. Henry Ford first exhibited a finished soybean car (Figure 1.3) in 1941 (Soybean Car, 2012), which was uneconomical for that era. Other notable examples made of composites have appeared in thousands of household washing machines, refrigerators, drains, and other mass applications.

(a) Tulou house   (b) Beehive house

FIGURE 1.1   (a) Tulou house (Liang et al., 2011) and (b) beehive house.

FIGURE 1.2 Damascus steel (Srinivasan and Ranganathan, 2004).

FIGURE 1.3 Soybean car (Soybean Car, 2012).

The modulus and strength of previous FRP composite materials were mainly designed and produced to sustain water and chemical resistance. More recently, FRP composites are formed as load-bearing structural components for aircraft and marine structures. The surfaces of marine structures are protected through a class of polymer coatings including sacrificial coatings (lead or silver). For example, the protection of ship hulls against marine growth has been achieved through these novel coatings. As recently as 2012, the US Navy and Air Force (SERDP and ESTCP, 2012) are demonstrating a novel protective coating (e.g., ceramic-metal matrix) to reduce wear on compressor airfoils on gas turbine engines that significantly increase fuel efficiency and reduce maintenance requirements.

After World War II, the US manufacturers began producing fiberglass and polyester resin composite boards and the automotive industry introduced the first composite vehicle bodies in the early 1950s (Tang, 2011). The obvious reasons for moving in the direction of polymer composites are their corrosion resistance and magnetic transparency, high strength-to-weight ratio, and ease of shaping. Modern composites are made primarily of glass, carbon, and/or aramid reinforcements. However, natural fiber composites are just beginning to make some inroads into high-volume applications. For example, interior parts of automobiles such as dashboards and door panels are made of natural fiber composites. Similarly, wood plastics are being used extensively in Europe including deck panels. The natural fibers are bonded together with a variety of resin systems such as esters, urethanes, and phenolics. In addition, a variety of additives and fillers are used in composites to improve structural performance such as the bonding characteristics between the fiber and resin, manufacturing process improvement, colors, fire protection, and cost reductions. Additional details can be found on natural fiber composites in a technical paper written by Dittenber and GangaRao (2012).

1.2 FIBER-REINFORCED POLYMER COMPOSITES – GENERAL FEATURES

Composite materials are nonhomogeneous, consisting of two or more constituents including fibers as a reinforcing component and matrix as a shear force transfer component. Other constituents are fillers, pigments, and promoters. Thermomechanical properties of composites can be determined from the properties of reinforcement and matrix with reasonable accuracy. The reinforcing component is normally discontinuous, stiffer, and stronger than the matrix component. Composite materials can be classified into various types, as summarized in Figure 1.4.

Most composite materials with reinforcements in the form of short or chopped fibers provide low performance in strength since the matrix part plays a major role in carrying applied or induced forces in those composite materials. To improve the structural performance of composite materials, continuous fibers with high-specific strength and stiffness were developed in the late 1950s. Composites with high-performance continuous fibers have been customarily used in the aerospace and automotive industries. One of the above-identified advanced composite materials with continuous fiber and fabric reinforcement and polymer matrix has been successfully applied for a wide range of field applications, particularly in infrastructural applications. A few of these examples are illustrated in Section 1.4.

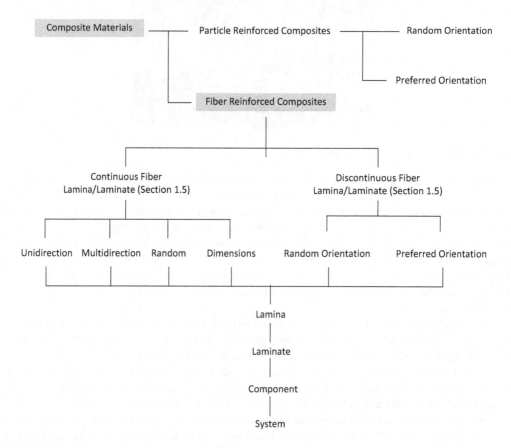

FIGURE 1.4 Classification of composite materials.

For FRP composites, the continuous fiber is embedded into polyester, vinyl ester, or other matrices. Typically, there are three FRPs, commonly referred to as glass fiber-reinforced polymers (GFRP), carbon fiber-reinforced polymers (CFRP), and aramid fiber-reinforced polymers (AFRP). The continuous fiber reinforcement is the main load-bearing part of FRP composite materials. The matrix plays important roles in protecting fibers, transferring shear stresses, and bonding the fibers together.

Applications of FRP composites continue to widely expand (refer to Figures 1.10–1.13) to be used in armor, automotive, aerospace, aircraft, marine, military, medical, energy, chemical, oil and gas industry, recreational, construction, and infrastructural applications. FRP composites have unique advantages over conventional engineering materials (concrete, steel, timber, etc.) such as high strength and stiffness-to-weight ratio, superior resistance to environmental attack, thermal insulation and conductivity, wear resistance, longer fatigue life, better temperature-dependent behavior below glass transition temperature (up to $\approx -40°F$), acoustic insulation, and environmental stability.

1.3 CONSTITUENT MATERIALS

In general, modern composite materials consist of two phases: one as a reinforcement (fibers) and the other as a binder/resin, but more distinct constituents of small quantities are added, which are promoters, fire retardants, and colorants. For a reinforcement phase, glass fibers are commonly used for numerous applications in infrastructural engineering. High-performance fibers such as carbon, boron, aramid, silicon carbide, etc. are widely accepted for advanced applications (i.e., aerospace and military).

As mentioned in Section 1.2, the matrix provides stability, local stress transfer, and protection for fibers. There are four different types of matrices used in modern composite materials: polymeric, ceramic, metallic, and carbon. However, the most extensively used matrices in commercial products and construction are polymeric resins such as thermosets within cross-linking of molecules or thermoplastics without cross-linking.

In general, the most predominant type of polymeric matrices is thermoset polymers such as epoxies, polyesters, vinyl esters, and polyimides. The thermoset polymers are cured by a chemical reaction. Polymerization and cross-linking develop during the curing process. This curing process is irreversible. The thermoset polymers cannot be formed repeatedly by reheating.

Thermoplastics are linear polymers and can be reformed by melting them with heat. Thermoplastics do not use a cure cycle. The cross-linking is not formed, and bonding between the chains in thermoplastics mainly comes from the van der Waals forces. The most widely used thermoplastics as matrices are polypropylene (PP), poly-ether-ether-ketone (PEEK), polysulfone, polyphenylene sulfide (PPS), and thermoplastic polyimides. Typical chemical compositions of thermoplastics and thermosets can be found in any standard textbook on polymer chemistry (GangaRao et al., 2007).

1.3.1 GLASS FIBERS

Glass fibers are the most common fiber type for FRP composites. In general, glass fibers have lower mechanical properties than other fibers (carbon and aramid fibers). However, glass fiber is much lower in cost and significantly less brittle, as reinforcement than carbon in polymer composites. Glass fibers are made from liquid, which is a melted ingredient with combinations of silica sand, limestone, fluorspar, kaolin clay, dolomite, colemanite, and other minerals. The melted glass is extruded through bushing to a typical diameter of 5–25 μm. Individual filaments are treated with a chemical solution to provide a protective coating before being bundled together to form rovings, as shown in Figure 1.5.

The most important properties of glass fibers are combinations of corrosion resistance, low cost, and high tensile strength, but the limitations in advanced applications are relatively low stiffness, fatigue endurance, and degradation with severe hygrothermal environments (i.e., strength loss under load and moisture). Glass fibers are produced in several types: (1) S-type (structural type) with higher strength than other types; (2) E-type (electrical type) with low alkali content, low-cost fiber, low susceptibility to corrosion, and high mechanical properties; and (3) C-type (chemical resistance type) with high corrosion and chemical resistance, etc. Chemical compositions and mechanical properties of glass fiber are summarized in Tables 1.1 and 1.2.

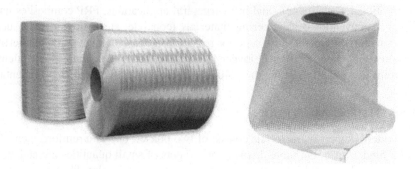

FIGURE 1.5   Glass roving and glass fabrics.

TABLE 1.1
Compositions of Glass Fibers (ASM vol. 21, 2001)

Fiber	Composition (% Weight)							
	SiO_2	Al_2O_3	CaO	MgO	B_2O_3	Fe_2O_3	Na_2O	TiO_2
E (boron containing)	52–56	12–15	21–23	0.4–4	4–6	0.2–0.4	0–1	0.2–0.5
E	59	12.1	22.6	3.4	–	0.2	0.9	–
ECR	58.2	11.6	0.5	2.0	–	0.1	1.0	2.5
D	74.5	0.3	2.8	–	22.0	0–0.1	1.0	–
S, R	60–65.5	23–25	0–9	6–11	–		0–0.1	–

Note: E=electrical, S or R=strength (structural), C=chemical, D=tempered glass with curved edge.

TABLE 1.2
Mechanical Properties of Glass Fibers (ASM vol. 21, 2001)

Fiber	Mechanical Properties					
	Bulk Density (lb/ft³)	Tensile Strength		Young's Modulus		Elongation at Break (%)
		GPa	ksi	GPa	10⁶ psi	
E (boron containing)	158.6–159.2	3.1–3.8	450–551	76–78	11.0–11.3	4.5–4.9
E	163.6	3.1–3.8	450–551	80–81	11.6–11.7	4.6
ECR	166.1–167.3	3.1–3.8	450–551	80–81	11.6–11.7	4.5–4.9
D	134.8	2.4	349	–	–	–
S, R	154.8–155.4	4.38–4.59	635–666	88–91	12.8–13.2	5.4–5.8

Note: E=electrical, S or R=strength (structural), C=chemical, D=tempered glass with curved edge.

1.3.2 CARBON FIBERS

Carbon fibers are most widely used for aerospace and advanced composite applications due to relatively high stiffness/strength to weight. Carbon fibers have high temperature, fatigue, corrosion, and chemical resistance. However, the mechanical properties of carbon fibers have broadly varied depending on precursors used in the manufacturing process. Typically, carbon fiber contains at least 90% of carbon by weight. The most common precursors are polyacrylonitrile (PAN), petroleum or coal tar pitch, phenolic fibers, cellulosic fibers, etc. Carbon fibers are produced from synthetic polymers (such as PAN) spun into filament yarns. Carbon fibers are finished by heating and stretching treatment through chemical and mechanical process. Typically, the filament of carbon fiber has a diameter of 5–10 μm. Carbon fibers are supplied in a continuous tow wound as shown in Figure 1.6.

Advantages of carbon fibers are summarized as (1) high tensile strength to weight, (2) high tensile stiffness to weight, (3) high fatigue, toughness, and stress rupture resistance, (4) high-dimensional stability, (5) low abrasion, (6) low coefficient of thermal expansion, (7) good vibration damping, and (8) high corrosion resistance and chemical inertness. Some of the disadvantages of carbon fibers include high cost and brittleness, electrical conductivity, and electromagnetic properties; such disadvantages may limit their potential use in some applications. Tensile strength and elastic modulus of different fiber grades are provided in Figure 1.7.

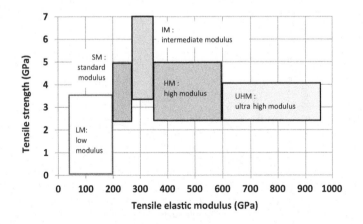

FIGURE 1.6 Carbon roving and carbon fabrics.

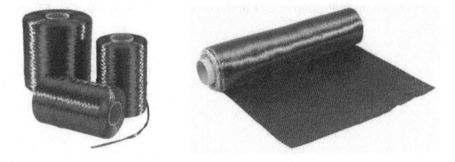

FIGURE 1.7 Tensile strength and elastic modulus of carbon fibers. (Adapted from Carbon types: product types by mechanical: The Japan Carbon Fiber Manufacturers Association, 2014.)

1.3.3 ARAMID FIBERS

Aramid or aromatic polyamide fibers are manufactured as long synthetic polyamide chains. At least 85% of the amide linkages are attached directly to two aromatic rings (Lubin, 1982; Schwartz, 1992). Aramid fibers, commercially known as Kevlar and Twaron, have higher strength and stiffness-to-weight ratios than glass fibers. Aramid fibers offer excellent toughness and impact resistance. In addition, the mechanical response of aramid fibers under impact loading tends to be ductile when compared with carbon fibers.

Kevlar is manufactured from a condensation reaction of para-phenylenediamine and terephthe-loyl chloride. The aromatic ring structure and para configuration are formed, thus contributing to high strength and modulus and high thermal stability (Mallick, 1993). The outstanding characteristics of aramid fibers are summarized as low flammability and no melting point, very high damping coefficient, high strength and modulus with a slightly different molecular structure, good fabric integrity at elevated temperatures, very low thermal conductivity, high impact, and dynamic load resistance. The mechanical properties of aramid fibers are compared to those of other commercial fibers as presented in Table 1.3. In addition, the tensile stress and strain relations of various reinforcing fibers are illustrated in Figure 1.8.

TABLE 1.3
Thermal and Mechanical Properties of Commercially Available Reinforcing Fibers (Kaw, 2006)

Fiber	Diameter (in⁻⁶)	Fiber Density (lb/ft³)	Tensile Strength (ksi)	Tensile Modulus (10⁶ psi)	Thermal Expansion (10⁻⁶ in/in/°F) Axial	Transverse
S-glass	315–551	115	664	12.5	2.778	2.778
E-glass	315–551	159	500	10.5	2.778	2.778
Aramid	472	90	525	18.9	−2.778	+2.278
Basalt	394–787	164	696	33.4	–	–
IM 7 carbon	197	112	785	40	–	–
AS4 carbon	276	112	580	33.1	–	–
GY70 graphite	335	112	270	83	−0.722	+3.889

Note: These properties vary slightly, depending on the manufacturing process.

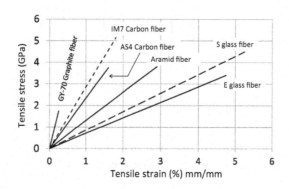

FIGURE 1.8   Stress and strain relations of reinforcing fibers (Daniel and Ishai, 2006).

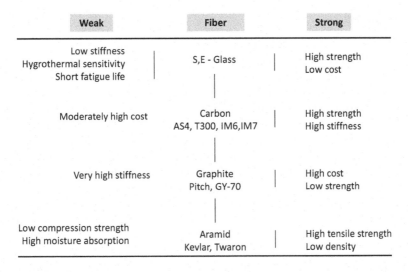

Weak	Fiber	Strong
Low stiffness Hygrothermal sensitivity Short fatigue life	S,E - Glass	High strength Low cost
Moderately high cost	Carbon AS4, T300, IM6,IM7	High strength High stiffness
Very high stiffness	Graphite Pitch, GY-70	High cost Low strength
Low compression strength High moisture absorption	Aramid Kevlar, Twaron	High tensile strength Low density

FIGURE 1.9 Summary of strengths and limitations for reinforcing fibers (Daniel and Ishai, 2006).

However, disadvantages of aramid fibers include high moisture absorption (up to 10% of fiber weight) leading to internal longitudinal splits and cracks at pre-existing micro-voids. In addition, the mechanical properties of aramid fibers deteriorate rapidly under ultraviolet (UV) lights without proper protective coating. The summary of strengths and limitations for select reinforcing fibers is presented in Figure 1.9.

1.3.4 BASALT FIBERS

Basalt fiber is a relative newcomer to FRPs and structural composites. Basalt is produced from natural material found in volcanic rocks. It can be melted at 1,300°C–1,700°C and can be spun into fibers. The finished basalt fibers have a similar chemical composition as glass fiber but have better strength characteristics. Unlike most glass fibers, basalt fiber is highly resistant to alkaline, acidic, and salt attack, making it a good candidate for concrete, bridge, and shoreline structures. Compared to carbon and aramid fibers, basalt fibers have the features of a wider service temperature range, −452°F to 1,200°F (−269°C to +650°C), higher oxidation resistance, higher radiation resistance, higher compression strength, and higher shear strength. However, basalt fibers are more brittle and more expensive than glass fibers.

1.3.5 POLYESTER RESIN

Polyester resins are one of the most common and economical thermosetting resin systems. They are widely used in composite industries, up to almost 75% of the total thermoset resins. Polyester resins as unsaturated synthetic resins are produced by the reaction of polyhydric alcohols and dibasic organic acids. To form a finished polymer, styrene as a reactive monomer is added to obtain low viscosity. Polyester resins are cured exothermically with conventional organic peroxides. Polyester resins are low cost and least toxic. However, the strength and modulus of polyester resins are low compared with epoxy resins. The volumetric shrinkage of polyester resins can be up to 12%, which may leave sink marks in finished products. Moreover, one of the major concerns with respect to polyester resins is their response to moisture due to a high number of ester linkages, which occur along the polyester molecule. These segments of the molecule are susceptible to attack both from water and from a range of chemicals, particularly in an alkaline environment.

1.3.6 Vinyl Ester Resin

Vinyl ester resins are formed by the reaction of epoxy resin and acrylic or methacrylic acid. As in polyester resin, styrene is used to obtain vinyl ester resin. Vinyl ester resins are cured with conventional organic peroxides through an exothermic reaction, similar to polyester resins. The most common form of vinyl ester resin is formed by the reaction of a DGEBA epoxide with methacrylic acid. The advantages of vinyl ester resins are high fracture toughness and resilience, good chemical and corrosion resistance, excellent resistance to acids and caustics, and good hydrolytic stability. In addition, the presence of hydroxyl groups along the polymer chain provides vinyl esters with good fiber wetting properties. Vinyl ester resins do appear to give better strain to failure characteristics than many of the standard polyester-based laminating systems. However, vinyl ester resins provide volume shrinkage of the order of 5%–10%, which can lead to part distortion and stress cracking due to residual stress buildup.

1.3.7 Epoxy

Epoxy resins are formed by reacting epichlorohydrin with bis-phenol wherein the finished product is linear with cross-linking by reacting with amine-type compounds. Epoxy resins have excellent performance compared to other organic resins. They are used in composite products that require superior resistance to corrosive environment and better mechanical properties than polyester resins. The characteristics of epoxy resins are low cure shrinkage, low water absorption, excellent resistance to chemical attack, absence of volatile matters during cure, good electrical properties, excellent adhesion to a wide range of fibers and fillers, etc. However, epoxy resins are more expensive than other thermoset resins and take a long time to cure to attain full strength.

1.3.8 Polyurethane Resins

Polyurethane resins can be classified as thermosetting or thermoplastic resin. Polyurethane resins are produced by reaction injection molding (RIM) process with a low pressure of 2–5 psi. Polyurethane resins produced by the RIM process provide high compressive strengths and toughness, high abrasion resistance, and good flexibility at low temperatures. In addition, they are typically used to increase the stiffness of structural components and better bonding between structural members.

The thermomechanical properties of thermoset resins (polyurethane, polyester, and epoxy) are given in Table 1.4. Since there are many variations in those three polyset resins, the properties in Table 1.4 are approximate and a general range for these properties is suggested. To design a laminate, exact properties of these materials must be obtained from the resin manufacturers.

1.4 FUTURE PERSPECTIVE

FRP composite system demonstration and high-volume implementation efforts recently have been focused on marine, aerospace, civil, and military infrastructure applications such as deckhouses for ships (Figure 1.10), airplane components, wind turbine blades, cold-water pipes for Ocean Thermal Energy Conversion (OTEC) (Figure 1.11a), high-pressure composite pipes for natural gas transmission (Figure 1.11b), carbon-cored aluminum alloy conductor electric transmission lines, utility poles up to 120 ft in height (Figure 1.12), modular housing construction, and so on. Some of the current and future application details were reported by GangaRao (2011b).

The airframe is built with carbon and glass FRP composites with 70% and 15% of all materials, respectively, to deliver a low radar profile. FRP composite materials are important to the design of this aircraft to provide better stability and lower self-weight. FRP composite materials provide higher tensile strength and better aerodynamic performance with less weight and more reliability

TABLE 1.4

Approximate Structural Properties for Representative Thermoset Resins at Room Temperate

Material Property	Polyurethane	Polyester (cast)	Epoxy (cast)
Specific gravity	1.1–1.46	1.1–1.46	1.11–1.40
Tensile strength (ksi)	6.00–10.00	6–13	4–13
% Elongation (rupture)	$\cong 6.00$	2.00–5.00	3.00–8.00
Tensile modulus (ksi)	400–600	300–640	350–450
Compressive strength (ksi)	13–36	13–30	15–25
Impact strength (ft-lb/in)	–	0.2–0.4	0.2–1.0
Bending strength (ksi)	4.39	–	16–20
Bending modulus (ksi)	175–220	–	365–560
Thermal expansion (10^{-6} in/in/°F)	50–60	30.6–55.5	25.0–36.1
Water absorption, 24 h, 1/8″ thickness (%)	0.2–0.9	0.15–0.60	0.08–0.15
Thermal conductivity (Btu-in/hr-ft²-°F)	–	31.16	1.16–1.45

Note: The above are suggested values (guide) only, and vary as a function of the polymer composition. The values suggested herein must not be used in the final design computations for laminates. The matrix properties for final designs must be supplied by the resin manufacturers because final values may vary depending on resin formulations. Adopted and modified from "Reinforced Concrete Design with FRP Composites", by Hota V.S. GangaRao, et.al., CRC Press, 2007.

(a) DDG-1000 Zumwalt   (b) EUROFIGHTER TYPHONE airframe
                                        (Eurofighter Typhoon, 2013)

FIGURE 1.10   Military infrastructure. DDG-1000 Zumwalt.

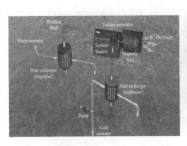

(a) Lockheed OTEC   (b) Pipe or pole section under bending

FIGURE 1.11   Ocean thermal energy conversion and FRP composite pipe: (a) Lockheed OTEC and (b) pipe or pole section under bending (GangaRao, 2011b).

FIGURE 1.12   FRP utility poles.

than conventional materials. The self-weight of an FRP composite airframe can be reduced to 30% compared to the self-weight of traditional engineering materials such as aluminum.

The fabricators of pipeline infrastructural systems are searching for alternatives, including the FRP pipeline for high-pressure natural gas transmission. Moreover, FRP pipeline systems have been successfully implemented for marine, chemical, and sewage applications.

1.4.1   Bridges

For infrastructural and highway systems, FRP composite materials have led to structural modules and members that allow for rapid construction and replacement of infrastructural systems (Figure 1.13) as superstructures, substructures, and structural members of other highway structures. One of the most attractive applications is to replace old deteriorated conventional bridge decks with a lightweight bridge deck system. Since the early 1990s, FRP composite decks have been successfully used to replace deteriorated bridge decks. At present, designs of modern highway bridges to upgrade or build a new bridge with modular and lighter FRP structural members, are being used for longer spans and heavier vehicles.

The FRP composite elements and components are fabricated to arrive at structural systems for hydraulic structures. In addition, FRP composite components are combined with traditional materials to develop hybrid structural systems. A few structural items not subjected to extreme loadings, such as ladders and gratings, exposed to extreme weather are made of FRP composites for loading service life. Many such hydraulic structures have been developed and field-implemented (GangaRao et al., 2020) and are functioning well. Some of these structures are miter blocks, wicket gates, miter

FIGURE 1.13   FRP bridge deck replacement (Market Street Bridge, Wheeling, WV).

gates, recess panels, rehabilitation of corroded steel piles, concrete discharge ports under water, and others. For additional details, readers are encouraged to review the report entitled "Composites for Hydraulic Structures" (GangaRao et al., PIANC WGR191, 2020).

FRP composites can be used as floating bridges, underwater bridges, and even folding/unfolding bridges. Movable bridges of 10 lb/ft^2 self-weight can be developed in a cost-competitive manner. However, some of these bridging systems have to be either anchored or balanced with dampers for stability. For example, a novel rolling bridge system was designed by Thomas Heatherwick Studio, London in 2005. Similar advances made by the US Army in lightweight bridging with rapid deployment capabilities such as the Wolverine Heavy Assault Bridge System (FAS: MAN, 2000) and new developments in composite bridging are being pursued by USDOD (ONR BAA 04–009) in terms of lightweight expeditionary bridging capability.

1.4.2 Smart Materials

The community of advanced composites manufacturers has been constantly assessing self-healing materials, including coatings that serve as sensors and self-cleaning functionality. For example, certain surface coatings can reduce or even eliminate algae growth on marine structures. Similarly, paints containing nanobubbles can heal wall paints that might be damaged due to accidents. Recent developments using carbon fibers as sensing materials and healing damage through polymers housed in nanofibers are some excellent examples for future research. Similarly, electrically conductive coatings with wireless networks are being developed to detect fire and other structural hazards.

Examples of smart materials with shape memory (Figure 1.14) are being developed for novel applications (CRG Technology Overviews, 2011). As fundamental research and development matures in shape memory polymers, a wide range of applications can open up for structural health monitoring and can be expanded into fields of health care, security, and altering/controlling human behavior patterns. For example, a polymer that releases slow doses of insulin (commonly known as humulin) is being marketed extensively to control diabetes.

1.4.3 Fire

Fire-retardant properties of composites have to be quantified with a special emphasis on the chemical reaction (including toxicity) of polymers to fire (ignition, flame spread, and heat release rate). This is especially important when composites are filled with nanoparticles such as nanoclay, carbon

FIGURE 1.14 Shape memory polymers (CRG Technology overviews).

nanotubes or filaments, and even traditional additives such as halogens or phosphorus. Even non-toxic inorganic fillers are used to minimize smoke output and to retard the burn rate through the thickness of a composite laminate. These additives can alter the mechanical and physical properties of polymers and composites. Therefore, protective coatings against fire using innovative materials are being considered. For example, nanofiber sheets in the form of coatings are being researched (Zhao et al., 2009). These coatings have a "bilayer structure", with the top layer being thermally conductive and expected to redirect the heat flux on the surface. Since the in-plane heat flux is of several magnitudes higher than the through-thickness flux, fire-induced heat is dissipated through the outer thin layer. The thermally insulated bottom layer turns into char (silicon oxide) due to an immense heat buildup from the top layer to the bottom; thus, the silicon oxide layer protects the polymer composite. Similarly, researchers are working on the use of graphite to alter thermal properties of composites (Yu et al., 2007). Intertek Testing Services NA, Inc has successfully conducted fire testing (ASTM E119) on a composite wall panel for WVU-CFC. The wall panel consisted of an intumescent coating in addition to gypsum board and natural resin-based FRP sheathings. The fire test data revealed a fire rating of 60 min. For additional details on the fire performance of FRP composites, many excellent reports are available in the literature, and readers are recommended to review "Fire Performance of FRP Utility Poles" by Liang, Agarwal, GangaRao, and Gupta, published by EPRI in December 2020.

1.4.4 Natural Fiber Composites

Natural fibers with excellent specific mechanical properties cost around 40%–60% of the cost of common E-glass fiber, require only 20%–40% of the embodied (production) energy, and are sustainable. Even though several limitations of natural fibers are well documented (Dittenber and GangaRao, 2012), many researchers have been advancing the state-of-the-art natural fiber composites focusing on surface treatments of fibers and improving the fiber–matrix interfacial bond strength properties. Advances in natural fiber composites are being made in such a way that a class of natural fiber composites are yielding better mechanical properties than typical glass fiber composites, albeit these natural fiber composites are more expensive than GFRPs (Netravali et al., 2007; Dittenber, 2013). Similarly, studies on the performance of natural fiber composites under severe environments such as elevated temperature have been conducted (Manalo et al., 2015). However, most research activities are focused on reducing costs and improving performance characteristics so that they can compete directly with GFRPs.

1.5   LEVELS OF ANALYSIS AND DESIGN FOR FRP LAMINATE COMPOSITES

In general, FRP composite structural members are made of thin composite layers called laminae or plies. At the lamina level, fiber reinforcements can be arranged in the same direction or in different directions, depending on the force transfer requirement for a given application. The mechanical properties and behavior of each lamina are determined using *Micromechanics*, which deals with the interactions of the constituents (reinforcement and matrix) at the microscopic level. To obtain desired performances of FRP composites, several laminae are combined into a laminate. At this level, the average properties of laminae are used in the analysis approach called *Macromechanics*. Lamination theory, including of fiber architectures and laminae properties, is applied to this approach. To form FRP composite structural members, many layers of laminates are combined and assembled through various manufacturing processes. To determine structural responses and the design of FRP composite members, several structural analysis methods, classical as well as numerical, can be used. Levels of the analysis and design for FRP composite structural members are summarized as shown in Figure 1.15.

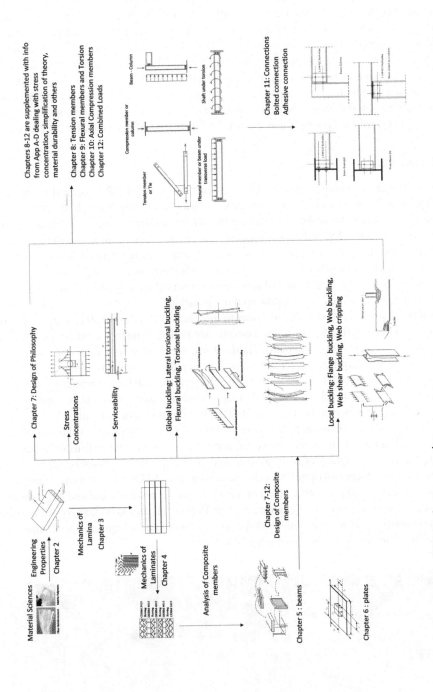

FIGURE 1.15 Levels of analysis for FRP laminate composites.

1.6 MANUFACTURING PROCESS

The analysis and design methodologies provided herein are mainly focused on the FRP composite structural members manufactured by the pultrusion method. However, over 20 different methods of composite component manufacturing are in practice such as pultrusion, filament winding, resin transfer molding (RTM), sheet molding compounds (SMC), etc. (CISPI, 1992). Most of the manufacturing methods are described well in the literature. In general, any FRP structural members can be fabricated either from an automated or from a manual process. Each process has its own advantages and limitations, which is a function of various process parameters. The following four major process parameters are commonly evaluated in any mass manufacturing setting: (1) efficiency and speed of processing composite constituents into required forms, (2) percent cure of resins in the presence of fibers and fabrics, (3) fiber/fabric tension during process to minimize kinks and stress risers and maximize strength and stiffness, and (4) cure temperature and its variation with time during and after the process.

1.6.1 PULTRUSION

To achieve optimized mechanical properties, most of the modern FRP structural members have been manufactured by means of a pultrusion process. Manufacturing through the pultrusion process can be formulated to meet the most demanding chemical, flame retardant, electrical, and/or environmental conditions. The pultrusion process can produce both simple and complex profiles, eliminating the need for extensive post-production assembly of components. Glass fibers and resin (polyester and vinyl ester or other resins) systems are efficiently used to result in low cost for finished pultruded products. However, special precautions have to be taken to minimize stress concentrations at corners, and residual stress build-up during curing.

In the pultrusion process, reinforcements (fiber rovings and/or fabrics) are aligned and continuously pulled into the forming part. The dry reinforcements are pulled through a resin bath to make resin-impregnated reinforcement. The resin bath consists of suitable accelerators, filler materials, catalysts, and wetting agents. In the forming part, reinforcements are pulled together to form a closer configuration of the structural member. Then, the saturated reinforcements will be continuously pulled through a heated die to cure and form a hard composite member when a part comes out of a die. Finally, the continuous structural members are cut into the desired lengths as shown in Figure 1.16.

The advantages of the pultrusion process are as follows: (1) low labor cost, (2) minimal material wastage, and (3) high production rate. However, some limitations of the pultrusion process are as follows: (1) potential inadequacy of resin wet-out and cure, especially at corners, (2) production of constant section, and (3) control of pull speed with minimal voids. Due to lower mechanical properties of a composite part and premature aging of a part, improper resin wet-out and curing will result in shortened durability and initiate premature failure. Pull speed should be kept low to improve the quality of a composite part with a complex shape so that proper and adequate wet-out and curing of resin occur while the part is moving through a heated die.

FIGURE 1.16   Pultrusion process (Bedford Reinforced Plastics, Inc., PA, US).

1.6.2 Pultruded FRP Structural Sections

The FRP structural members manufactured by the pultrusion process can be formed for simple and complex profiles (Fiberline, 2003; Creative Pultrutions, 2004; Strongwell, 2010; Bedford, 2012 and others). The simple cross sections are I-shape, L-shape (angle), C-channel, T-shape, H-shape or WF wide-flange, round, square, plate, hollow square, or round tube. All these shapes are shown in Figure 1.17, and they are widely available on the market.

These shapes can be obtained at any length since the pultrusion process is a continuous process. In addition, complex profiles such as multi hollow cells, hat-shape for sheet piles, Z-shape are shown in Figure 1.18, and such complex shapes are used in various applications such as FRP bridge decks as shown in Figure 1.19. FRP bridge decks (Prachasaree et al., 2006, 2007, 2008a, b, 2009a–c, 2012, 2013, 2015) have been studied widely for decades. For these simple structural profiles (except rods and bars), each element of a pultruded FRP structural member includes reinforcing fibers orientated in two different directions separated by a minimum of 30° (ACMA, 2010a). The minimum fiber volume fraction of each profile element must not be less than 30%. The continuous fiber reinforcement in an element of a profile (web, flange, diagonal, or horizontal stiffener of webs) in the strong axis direction should not be less than about 30% (by volume) of the total fiber reinforcement for any structural shapes and not less than 25% for a plate. For multiple elements sharing a common edge in the pultrusion direction, at least 50% of the non-roving reinforcement in the flange or web element (having the largest percentage of non-roving reinforcement and sharing the common edge) should extend through the junction connecting the elements (ACMA, 2021).

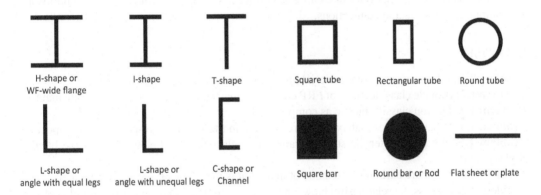

FIGURE 1.17 Simple pultruded FRP structural shapes.

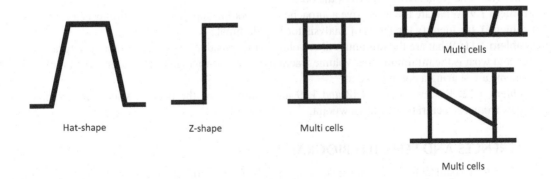

FIGURE 1.18 Complex pultruded FRP structural shapes.

| H-deck and double trapezoid FRP deck | High profile FRP bridge deck (ProDeck 8) | Low profile FRP bridge deck (ProDeck 4) |

FIGURE 1.19 FRP bridge decks and floor sections.

1.7 SUMMARY

Composites are composed of two or more constituent materials that remain distinct in the final form but bond to create a product that has superior properties to the constituent materials. The use of composites dates back to the earliest civilizations, but the use of engineered composites has dramatically increased in the last century. Codes and specifications are rapidly evolving to keep up with the recent growth of composites in civil infrastructure; often the most detailed guides are provided by manufacturers based on their testing and other experiences for specific products. With properties that can be completely customized to fit specific needs, future prospects of FRP composites are limitless. The succeeding chapters provide fundamental understandings of constituent materials, mechanics and analyses of composite lamina and components, and the design philosophy for composite members and connections.

EXERCISES

Problem 1.1: What is a composite material and what are they made of?

Problem 1.2: Provide classifications of FRP composite materials and is wood a composite material?

Problem 1.3: Do you classify plastic as composites?

Problem 1.4: What role does matrix (cured resin) play in the structural response of a composite?

Problem 1.5: What is the tensile strength range of E-glass fiber and elongations at the break of S-glass fiber?

Problem 1.6: What are carbon fibers made of and what is the filament size of carbon fibers?

Problem 1.7: Discuss the relationship between tensile strength and tensile elastic modulus of commercially available carbon fibers of different grades and what is their elongation at failure?

Problem 1.8: What are the common thermoset resins used in manufacturing composite components? How one resin is different from the other?

Problem 1.9: What are the protective approaches one would adopt to improve fire resistance?

Problem 1.10: Identify the levels of analysis for FRP laminate composites.

Problem 1.11: What are the minimum fiber volume fractions that are suggested for structural profiles and what is the minimum fiber volume fraction of the non-roving reinforcement when multiple elements are sharing a common edge?

Problem 1.12: Review Figures 1.18 and 1.19 and comment on the need to select such complex shapes from a structural efficient viewpoint.

REFERENCES AND SELECTED BIOGRAPHY

ACMA, Pre-standard for load & resistance factor design (LRFD) of pultruded fiber reinforced polymer (FRP) structures, Arlington, VA, 2021.

ACMA, Pre-Standard for Load & Resistance Factor Design (LRFD) of Pultruded Fiber Reinforced Polymer (FRP) Structures, Arlington, VA, 2021.

ASCE, In Gray, E.F. Jr. (Ed.), *Structural Plastics Design Manual*, New York, Task Committee on Design of the Structural Plastics Research Council of the Technical Council on Research of the American Society of Civil Engineers, doi: 978-0-87262-391-0, 1984.

ASM International, *ASM Handbook*, Vol. 21: Materials Park, OH, Composites, 2001.

Bank, L.C., *Composites for Construction: Structural Design with FRP Materials*, Hoboken, NJ, John Wiley & Sons, Inc., 2006.

Bedford Reinforced Plastics, Design Guide, Bedford, PA, Bedford Reinforced Plastics, Inc., 2012.

Callister, W.D. Jr., *Fundamentals of Materials Science and Engineering*, 5th edition, New York, NY, John Wiley & Sons, Inc., 2001.

CEN, *Reinforced Plastic Composites: Specifications for Pultruded Profiles Parts 1–3 EN 13706*, Brussels, Belgium, European Committee for Standardization (CEN), 2000.

Chernick, K., Traditional Syrian beehive houses kept heat out the natural way. Retrieved May 11, 2012 from http://www.greenprophet.com/200/07/syrian-beehive-houses/, July 2009.

CISPI, *Introduction to Composites*, Composites Institute of the Society of the Plastics Industry Inc. 1992.

Clark, J.L., *Structural Design of Polymer Composites: Eurocomp Design Code and Background Document*, London, UK, E & FN Spon, 1996.

Creative Pultrutions, *The Pultex Pultrusion Design Manual of Standard and Custom Fiber Reinforced Structural Profiles*, 5, revision (2), Alum Bank, PA, Creative Pultrutions Inc, 2004.

CRG technology overviews, Retrieved May 15, 2012, from http://www.crgrp.com/technology/ overviews/ morphing.shtml CRN DT 205/2007, 2008.

CUR, *CUR Recommendation 96: Fiber Reinforced Plastics in Civil Engineering Structures*, Amsterdam, the Netherlands, CUR, 2003.

Daniel, I.M., and Ishai, O., *Engineering Mechanics of Composite Materials*, 2nd edition, New York, NY, Oxford University Press, Inc., 10016, 2006.

Dittenber, D.B., Natural kenaf fiber reinforced composites as engineered structural materials, PhD dissertation, West Virginia University, Morgantown, WV, 2013.

Dittenber, D.B., and GangaRao, H.V.S., Critical review of recent publications on use of natural composites in infrastructure, *Composites Part A: Applied Science and Manufacturing*, 43(8), 2012, pp. 1419–1429.

Eastes, W.L., Hofman, D.A., and Wingert, J.W., Boron-Free Glass Fibers, U.S. Patent 5,789,329, August 4, 1998.

Eurofighter typhoon, Technical Guide, Eurofighter Jagdflugzeug GmbH, Am Söldnermoos 17, 85399, Hallbergmoos, Germany, 2013.

FAS: MAN., XM 1004 wolverine: Heavy assault bridge system H82510, Retrieved May 18, 2012, from http://www.fas.org/man/dod-101/sys/land/wolverine.htm, 2000.

Fiberline, *Design Manual*, Kolding, Denmark, Fiberline, 2003.

GangaRao, H.V.S., Infrastructure applications of fiber-reinforced polymer composites. In Kutz, M. (Ed.), *Applied Plastics Engineering Handbook—Processing and Materials*, New York, NY, Elsevier, 2011a, Retrieved http://www.knovel.com/web/portal/browse/display?_EXT_KNOVEL_DISPLAY_bookid=4 606

GangaRao, H.V.S., *Infrastructure applications of fiber-reinforced polymer composites, Applied Plastics Engineering Handbook – Processing and Materials*, New York, NY, Elsevier, 2011b.

GangaRao, H.V.S, et al., Composites for hydraulic structures, InCom WG Report 191, PIANC, 2020.

GangaRao, H.V.S., Taly, N.B., and Vijay, P.V., *Reinforced Concrete Design with FRP Composites*, Boca Raton, FL, CRC Press, 2007.

GangaRao, H.V.S., Thippeswamy, H.K., Shekar, V., and Craigo, C., Development of glass fiber reinforced polymer composite bridge deck, *SAMPE Journal*, 35(4), 1999, pp. 12–24.

Hart-Smith, L.J., The key to designing efficient bolted composite joints, *Composites*, 25(8), 1994, pp. 835–837.

Head, P.R., Engineering our infrastructure: The next 60 years, *The Structural Engineer*, 72(9), 1994, p. 143.

Heatherwick Studio, Rolling Bridge, London, 2002, UK. Retrieved http://www.heatherwick.com/ projects/ infrastructure/rolling-bridge/, March 1, 2013.

Inman, M., Legendary swords' sharpness, strength from nanotubes, study says, 2006, Retrieved May 15, 2012, from http://news.nationalgeographic.com/news/2006/11/061116-nanotech-swords.html., 2006, pp. 1–2.

Kaw, A.K., *Mechanics of Composite Materials*, 2nd edition, Boca Raton, FL, CRC Press, Taylor & Francis Group, 2006.

Kutz, M., *Applied Plastics Engineering Handbook – Processing and Materials*, New York, NY, Elsevier, 2011, Retrieved http://www.knovel.com/web/portal/browse/display?_EXT_KNOVEL_DISPLAY_ bookid= 4606.

Liang, R., Agarwal, S., Gangarao, H.V.S., and Gupta, R.K., *Fire Performance of FRP Utility Poles: A Critical Review*, Electric Power Research Institute (EPRI), Washington, DC, December 2020.

Liang, R., GangaRao, H.V.S., and Stanislawski, D., *SGER: Material and Structural Response of Historic Hakka Rammed Earth Structures*. Final Report No. 090819, 9, Morgantown, WV, National Science Foundation, 2011.

Loewenstein, K.L., *The Manufacturing Technology of Continuous Glass Fibers*, 3rd revised ed., Elsevier, Amsterdam, The Netherlands, 1993.

Lubin, G., *Handbook of* Composites, Springer, Boston, MA, 1982.

Mallick, P.K., *Fibre Reinforced Composites: Materials, Manufacturing and Design*. 2nd Edition, Marcel Dekker Inc., New York, NY, 1993.

Manalo, A.C., Wani, E., Zukarnain, N.A., Karunasena, W, and Lau, K.T., Effects of alkali treatment and elevated temperature on the mechanical properties of bamboo fibre-polyester composites, *Composites Part B: Engineering*, 80, 2015, pp. 73–83.

McManus, S, Built by Lego, bridge design and engineering, No.8, August 1997.

Mosallam, A.S., *Design Guide for FRP Composite Connections*, Reston, VA, American Society of Civil Engineers, 2011. Retrieved http://books.google.com/books?id=-M-Yz0FXCeAC/.

National Research Council of Italy (CNR), Guide for the design and construction of structures made of thin FRP pultruded elements, Rome, Italy, CNR-DT 205/2007, 2008.

Netravali, A.N., Huang, X., and Mizuta, K., Advanced 'green' composites, *Advanced Composite Materials: The Official Journal of the Japan Society of Composite Materials*, 16(4), 2007, pp. 269–282.

Prachasaree, W., and Chaiviriyawong, P., State of the art review on FRP composite bridge decks: Development, fabrication and field implementation, *Proceedings of the Sixth Regional Symposium on Infrastructure Development (RSID 6)*, January 12–13, Bangkok, Thailand, 2009, pp. 26.1–26.6.

Prachasaree, W., and GangaRao, H.V.S., Local deflection state limit of light weight FRP bridge deck composites, *Science and Engineering of Composite Materials*, 15(4), 2008a, pp. 273–284.

Prachasaree, W., and GangaRao, H.V.S., Web buckling strength evaluation of multicellular FRP bridge deck module, *PSU-UNS 4th International Conference on Engineering Technologies (ICET 2009)*, Novi-Sad, Serbia, April, 2009b.

Prachasaree, W., GangaRao, H.V.S., Laosiriphong, K., Shekar, V., and Whitlock, J., Theoretical and experimental analysis of FRP bridge deck under cold temperatures, *International Journal of Materials and Product Technology*, 28(1–2), 2007, pp. 103–121.

Prachasaree, W., GangaRao, H.V.S., and Shekar, V., Performance evaluation of FRP bridge deck component under torsion, *Journal of Bridge Engineering, ASCE*, 11(4), 2006, pp. 430–442.

Prachasaree, W., GangaRao, H.V.S., and Shekar, V., Performance evaluation of FRP bridge deck component under shear loads, *Journal of Composite Materials*, 43(4), 2009c, pp. 377–395.

Prachasaree, W., Sangkaew, A., Limkatanyu, S., and GangaRao H.V.S., Parametric study on dynamic response of fiber reinforced polymer composite bridges, *The International Journal of Polymer Science*, Article ID 565301, 2015, pp. 1–13.

Prachasaree, W., and Shekar, V., Experimental evaluation and field implementation of FRP bridge deck modules, *Songklanakarin Journal of Science and Technology*, 30(4), 2008b, pp. 501–508.

Prachasaree, W., and Sookmanee, P., Structural performance of light weight multicellular FRP composite bridge desk using finite element analysis, *Journal Wuhan University of Technology*, Materials Science Edition, 27(5), 2012, pp. 939–943.

Prachasaree, W., Sookmanee, P., Limkatanyu, S., and GangaRao, H.V.S., Simplified load distribution factor of fiber reinforced polymer composite bridge desks, *The Baltic Journal of Road and Bridge Engineering*, 8(4), 2013, pp. 271–280.

Ramkumar, R.L., Bhatia, N.M., Labor, J.D., and Wilkes, J.S., *Handbook: An Engineering Compendium on the Manufacture and Repair of Fiber-Reinforced Composites*, Atlantic City International Airport, Atlantic City, NJ, Prepared for Department of Transportation FAA Technical Center, 1986.

Schwartz, M.M., *Composite Materials Handbook*. 2nd Edition, McGraw-Hill, New York, 1992

SERDP and ESTCP, Erosion resistant coating improves engine efficiency, 2012. Retrieved May 16, 2012, from http://www.serdp.org/News-and-Events/In-the-Spotlight/Erosion-Resistant-CoatingImproves-Engine-Efficiency

Seymour, R.B., and Deanin, R.D., History of polymeric composites, In *Proceedings of the Symposium Held during the 192nd ACS National Meeting*, Anaheim, CA, 1986.

Sinopoli, C.M., *The Political Economy of Craft Production: Crafting Empire in South India*, England, UK, Cambridge University Press, 2003.

Soybean Car, Popular research topics, Benson Ford Research Center, The Henry Ford, Retrieved June 6, 2015.

Sproull, J.F., Fiber Glass Composition, U.S. Patent 4,542,106, 17 Sept 1985.

Srinivasan, S., and Ranganathan, S. India's legendary wootz steel: An advanced material of the ancient world. National Institute of Advanced Studies. OCLC 82439861. Archived from the original on 2019-02-11. Retrieved 2014–08–12, 2004.

Strongwell Corporation, Design manual: EXTERN and other proprietary pultruded products, Bristol, VA, 2010.

Tang, B., Fiber reinforced polymer composites applications in USA, Structural Engineer, HIBT-10, Bridge Specialist Group 1200, 1997, pp. 28–29.

Tang, B., Fiber reinforced polymer composites applications in USA. Retrieved 5/11, 2011, from http://www.fhwa.dot.gov/bridge/frp/frp197.cfm

The Japan Carbon Fiber Manufacturers Association, Carbon types: Product types by mechanical, 2014.

Wallenberger, F.T., *Structural Silicate and Silica Glass Fibers, Advanced Inorganic Fibers*, Boston, MA, Springer, 2000, pp. 129–168.

Yu, A., Ramesh, P., Itkis, M.E., Bekyarova, E., and Haddon, R.C., Graphite nanoplatelet—Epoxy composite thermal interface materials, *The Journal of Physical Chemistry C*, 111(21), 2007, pp. 7565–7569, doi: 10.1021/jp0717615.

Zhao, Z., Gou, J., Bietto, S., Ibeh, C., and Hui, D., Fire retardancy of clay/carbon nanofiber hybrid sheet in fiber reinforced polymer composites. *Composites Science and Technology*, 69(13), 2009, pp. 2081–2087, doi: 10.1018/j.compscitech.2008.11.004.

2 Engineering Properties of Composite Materials

The basic building block of composite structures is the unidirectional lamina. A composite lamina is a thin layer of resin reinforcing multiple fibers or fabrics. Many micromechanics models were developed by researchers for predicting properties of a composite lamina based on the percent of fibers in relation to resin either by weight or by volume. In general, those models can be classified into three main categories: (1) mechanics of materials, (2) theory of elasticity, and (3) numerical (finite element) methods. In numerical predictions, lamina is modeled as though fibers are periodically spaced, infinitely long, and fully surrounded by matrix (no dry patch) in square or hexagonal arrays. Through the theory of elasticity approach, three fundamental laws of physics in a system are applied by satisfying (1) force equilibrium, (2) constitutive relations, and (3) strain compatibility. Likely, the compatibility equations may not be satisfied for models of mechanics of material approach because of approximations based on certain assumptions in developing these approaches. The lamina of the elastic model approach will be represented by the representative volume elements (RVEs). The details of the theory of elasticity approach or the numerical approach will not be discussed herein as they are beyond the scope of this textbook.

2.1 CHARACTERISTICS OF A COMPOSITE LAMINA

The mechanics of materials approach does not account for the stresses at the fiber–matrix interface, the characteristic of representative volume element (RVE) models, fiber packing arrangement, etc. The lamina is modeled in a way that the fibers are orientated parallel and bonded with the matrix as shown in Figure 2.1.

The micromechanics models using the mechanics of materials approach precisely predict the extensional modulus in the fiber direction, but the other moduli of lamina (e.g., transverse modulus) may be inaccurate. However, the predicted results from the micromechanical models are improved by using appropriate experimental results, as incorporated in the semi-empirical model by Halpin and Kardos (1976).

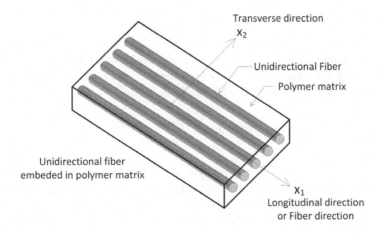

FIGURE 2.1 Model of unidirectional fiber lamina.

DOI: 10.1201/9781003196754-2

Before discussing any details of the abovementioned approaches, the basic terms related to fiber volume fractions are presented and used in micromechanics models. In general, the mechanical and hygrothermal characteristics of composite lamina are presented in terms of fiber volume. Moreover, the fiber and matrix are usually measured in terms of constituent material masses, mass or volume fraction, and their densities. Therefore, the basic terms such as fiber volume fraction, matrix volume fraction, mass fractions, density, and void content are defined in the following sections.

2.2   VOLUME AND MASS FRACTIONS

As mentioned before, the properties of composite lamina mainly depend on the volume ratio of either the fiber or the matrix. The fiber and matrix volume fractions are defined in this section. The composite lamina consisting wholly of fiber and matrix implies that the composite volume is filled by fiber and matrix without any voids or air pockets in the composite. The fiber (V_f) and matrix (V_m) volume fractions are defined as the ratios of fiber volume (v_f) and matrix volume (v_m) to composite volume (v_c) as follows:

$$V_f = \frac{v_f}{v_c} \tag{2.1.1}$$

$$V_m = \frac{v_m}{v_c} \tag{2.1.2}$$

Assume that the total volume of a composite lamina, after neglecting any void content (volume) is the sum of fiber and matrix volumes, i.e., the sum of the fiber and matrix volume fractions is equal to unity.

$$V_f + V_m = \frac{v_f + v_m}{v_c} = 1 \tag{2.2}$$

The fiber mass (W_f) and matrix mass (W_m) (or weight) fraction are defined as the ratios of fiber mass (w_f) and matrix mass (w_m) to composite mass (w_c) as follows:

$$W_f = \frac{w_f}{w_c} \tag{2.3.1}$$

$$W_m = \frac{w_m}{w_c} \tag{2.3.2}$$

From the assumption, the total mass (weight) of a composite lamina is the sum of both the fiber and matrix masses (weight). Therefore, the sum of both the fiber and matrix mass fractions is equal to unity, i.e., the sum of W_f & W_m from Eqs. (2.3.1 and 2.3.2).

$$W_f + W_m = \frac{w_f + w_m}{w_c} = 1 \tag{2.4}$$

In practice, the fiber weight (or volume) fraction of composite materials can be experimentally found by the ASTM "burn test" or other test methods. Then, the fiber weight fraction can be directly calculated from Eq. (2.3.1). By burning the polymer matrix composites, matrix mass including filler material can be removed using the principles given in the burn test (ASTM D 3014), matrix digestion (ASTM D 3171), or solvent extraction (ASTM C 613). However, the resin is mixed with fillers (such as clay,) which does not burn and remains, at least partially, in between the fibers after conducting the burn test. Therefore, after the burn test, fillers must be washed off from the fibers/fabric before weighing to determine the weight fraction.

2.3 MASS DENSITY

The mass density of composite lamina in terms of fiber and matrix fractions is presented herein. The mass density of materials is generally defined as a ratio of material mass to material volume. Neglecting the influence of voids, the fiber, matrix, and composite masses are written in terms of products of density and volume, as:

$$\rho_c = \frac{w_c}{v_c} \quad \rho_f = \frac{w_f}{v_f} \quad \rho_m = \frac{w_m}{v_m} \tag{2.5.1}$$

Note: $w_c = w_f + w_m$

From Eqs. (2.1 and 2.5.1), we get:

$$\rho_c = \rho_f V_f + \rho_m V_m \tag{2.5.2}$$

Noting that the composite volume is a sum of fiber volume and matrix volume, and invoking Eq. (2.3), the density of composite lamina can be also written in terms of mass and density of fibers and matrix as:

$$\frac{1}{\rho_c} = \frac{W_f}{\rho_f} + \frac{W_m}{\rho_m} \tag{2.6}$$

In general, the characteristic of composite lamina is presented in terms of volume fraction for design purposes, while it is convenient to write in terms of weight fraction for easier understanding of fundamental concepts of relations of mass and volume. Thus, the composite mass fraction can also be written in terms of density and volume fraction from Eqs. (2.5 and 2.6) as:

$$W_f = \frac{\rho_f}{\rho_f V_f + \rho_m V_m} V_f \tag{2.7.1}$$

$$W_m = \frac{\rho_m}{\rho_f V_f + \rho_m V_m} V_m \tag{2.7.2}$$

The density of composite materials can be also found by the liquid displacement method (Kaw, 2006). If the composite specimens sink in water, then the density can be given as below:

$$\rho_c = \frac{w_c}{w_c - w_i} \rho_s \tag{2.8.1}$$

If composite specimens float in water, then additional weight (sinker) needs to be attached. The density of composite materials can be expressed as:

$$\rho_c = \frac{w_c}{w_c + w_a - w_{ca}} \rho_s \tag{2.8.2}$$

where
w_c = weight of composite in air,
w_i = weight of composite weighed in water,
ρ_s = density of the liquid solution,

w_a = weight of sinker when immersed in water, and
w_{ca} = weight of the composite specimen plus weight of sinker when immersed in water.

Weights are measured directly, which are products of masses of different constituents multiplied by the acceleration due to gravity, g (~9.84 m/sec^2 or 32.2 ft/sec^2)

2.4   MORE THAN TWO CONSTITUENTS

The constituents of composite lamina, composed of more than two constituents (e.g., other than fiber and resins), are discussed below. The composite lamina may include two or more fiber types and cured resin (matrix). The mass of a composite lamina is the sum of all constituent masses (w_i) as:

$$w_c = w_1 + w_2 + w_3 + \cdots + w_n = \sum_{i=1}^{n} w_i \tag{2.9.1}$$

where n is the number of constituents, such as fabrics, mats, resins, fillers, and other additives.

Using the definition of density (ρ_i) and volume fraction (V_i), Eq. (2.9.1) can be presented in terms of constituent material densities as:

$$\rho_c = \rho_1 V_1 + \rho_2 V_2 + \rho_3 V_3 + \cdots + \rho_n V_n = \sum_{i=1}^{n} \rho_i V_i \tag{2.9.2}$$

A composite lamina volume is the sum of all constituent material volumes (v_i) as given below:

$$v_c = v_1 + v_2 + v_3 + \cdots + v_n = \sum_{i=1}^{n} v_i \tag{2.9.3}$$

Using the definition of density (ρ_i) and mass fraction (W_i), Eq. (2.9.3) can be presented in terms of density as:

$$\frac{1}{\rho_c} = \frac{W_1}{\rho_1} + \frac{W_2}{\rho_2} + \frac{W_3}{\rho_3} + \cdots + \frac{W_n}{\rho_n} = \sum_{i=1}^{n} \frac{W_i}{\rho_i} \tag{2.9.4}$$

In addition, mass fractions of all constituents are presented as:

$$W_i = \frac{\rho_i}{\rho_c} V_i \tag{2.9.5}$$

Note: The above equations will be used in the design part of this textbook, i.e., Chapters 7–12.

2.5   VOID CONTENT

In general, voids inside polymer composite materials are formed during the manufacturing process. The existence of voids leads to the density difference between experimental (actual) and theoretical values. The results in the experimental (actual) density are lesser than the results in the theoretical density. The voids inside a composite may be filled by chemicals, moisture, and/or air. These are important factors that influence the mechanical characteristics such as compressive strength, transverse tensile strength, shear strength, and corresponding stiffnesses, fatigue resistance, and others.

From a theoretical viewpoint, the void volume fraction is derived as a function of composite density, volume fractions of fiber, and matrix.

The void volume fraction (V_v) is defined as the ratio of void volume (v_v) to composite volume (v_c).

$$V_v = \frac{v_v}{v_c} \qquad (2.10.1)$$

The total composite volume (v_c) with voids is given below:

$$v_c = v_f + v_m + v_v \qquad (2.10.2)$$

By the definition of density, the composite volume (2.10.2) is written in terms of the actual composite density $\left(\rho_c^{act}\right)$, which is the actual composite weight over volume, whereas the theoretical density ρ_c^{th} of composite is presented in terms of fiber and matrix volumes without voids.

$$v_c = \frac{w_c}{\rho_c^{act}} \qquad (2.10.3)$$

$$v_f + v_m = \frac{w_c}{\rho_c^{th}} \qquad (2.10.4)$$

By substituting Eqs. (2.10.3 and 2.10.4) into Eq. (2.10.2):

$$v_v = \frac{w_c}{\rho_c^{act}}\left(\frac{\rho_c^{th} - \rho_c^{act}}{\rho_c^{th}}\right) \qquad (2.10.5)$$

By substituting Eqs. (2.10.3 and 2.10.4) into Eq. (2.10.1):

$$V_v = \frac{\rho_c^{th} - \rho_c^{act}}{\rho_c^{th}} \qquad (2.10.6)$$

where ρ_c^{th} is defined in Eq. (2.10.4).

Note: For an in-depth treatment on void volume fraction, readers are directed to Qureshi (2012).

Example 2.1

A 5 kg of the FRP composite with 60% fiber volume fraction where specific gravities of both the fiber and matrix are given as 2.5 and 1.5, respectively.

   a. Determine the density of this FRP composite.
   b. Determine mass fractions of fiber and matrix.
   c. Determine the volume of fiber and matrix.

Solution

   a. Density of FRP composite
      From Eq. (2.5.2):
      Density of fiber (ρ_f) = 2.5(1,000) = 2,500 kg/m³
      Density of matrix (ρ_m) = 1.5(1,000) = 1,500 kg/m³

$$\rho_c = \rho_f V_f + \rho_m V_m = 0.6(2,500) + 0.4(1,500) = 2,100\, kg/m^3$$

b. Mass fractions of fiber and matrix
    From Eqs. (2.1.2 and 2.3.1):

$$W_f = \frac{\rho_f}{\rho_c} V_f = \left(\frac{2,500}{2,100}\right) 0.6 = 0.714$$

$$W_m = \frac{\rho_m}{\rho_c} V_m = \left(\frac{1,500}{2,100}\right) 0.4 = 0.286$$

c. Volume of fiber and matrix
    The volume of the composite is:
    From Eq. (2.5.1):

$$v_c = \frac{w_c}{\rho_c} = \frac{5}{2,100} = \frac{2.38}{1,000} \, m^3$$

From Eqs. (2.1.1 and 2.1.2):

$$v_f = V_f \times v_c = 0.6\left(2.38 \times 10^{-3}\right) = 1.429 \times 10^{-3} \, m^3$$

$$v_m = V_m \times v_c = 0.4\left(2.38 \times 10^{-3}\right) = 9.52 \times 10^{-4} \, m^3$$

Example 2.2

The experimental data revealed that the FRP composite with dimensions (4 × 4 × 0.5 cm) and self-weight of 15 and 12 g for fibers after burn test. The specific gravity of fiber and matrix is 2.5 and 1.5, respectively.

a. Determine the density of this FRP composite.
b. Determine the volume of fiber, matrix, and void.
c. Determine the volume fraction of fiber, matrix, and void.in

Solution

a. Density of the FRP composite
    The density of the composite is:
    From Eq. (2.5.1):

$$\rho_c = \frac{w_c}{v_c} = \frac{15 \times 10^{-3}}{4 \times 4 \times 0.5 \times 10^{-6}} = 1,875 \, kg/m^3$$

b. Volume of fiber, matrix, and void
    Density of fiber (ρ_f) = 2.5 (1,000) = 2,500 kg/m³
    Density of matrix (ρ_m) = 1.5 (1,000) = 1,500 kg/m³
    From Eqs. (2.5.1 and 2.5.2):

$$v_f = \frac{w_f}{\rho_f} = \frac{12\left(10^{-3}\right)}{2,500} = 4.8 \times 10^{-6} \, m^3$$

$$v_m = \frac{w_m}{\rho_m} = \frac{(15-12)\left(10^{-3}\right)}{1,500} = 2 \times 10^{-6} \, m^3$$

$$v_v = v_c - v_f - v_m = (8 - 4.8 - 2) \times 10^{-6} = 1.2 \times 10^{-6} \, m^3$$

c. Volume fraction of fiber, matrix, and void
   From Eqs. (2.1.1, 2.1.2, and 2.10.1):

$$V_f = \frac{v_f}{v_c} = \frac{4.8}{8} = 0.6$$

$$V_m = \frac{v_m}{v_c} = \frac{2}{8} = 0.25$$

$$V_v = \frac{v_v}{v_c} = \frac{1.2}{8} = 0.15$$

2.6   REPRESENTATIVE VOLUME ELEMENT (RVE)

The RVE is defined as the smallest portion of the composite that contains all of its characteristics (Jones, 1975) including different packing arrays. Then, the RVE is utilized to represent a composite material in micromechanics models. The stress and strain of the RVE are not
uniform due to the heterogeneous nature of constituents in composite materials. However, the
RVE can be replaced by an equivalent homogenous material without affecting the state of stress
around the RVE. In general, the RVE includes two different kinds of packing arrays. These are
(1) square packing geometry and (2) hexagonal packing geometry as shown in Figure 2.2a and b,
respectively. The maximum fiber volume fraction that is possible in both packing geometries is
presented in Section 2.6.1 and 2.6.2.

2.6.1   SQUARE PACKING GEOMETRY

The packing geometry has a quarter cross-sectional area of circular fiber for each of the four corners
of the RVE, as shown in Figure 2.2a. The cross-sectional areas of the RVE (A_{RVE}) and circular fiber
(A_f) inside the packing geometry are used to determine the fiber volume fraction. The theoretical
maximum fiber volume fraction of the square packing geometry occurs when the diameter (d) of
circular fiber is equal to the distance between the center points of the circular fibers (s).

$$V_f = \frac{A_f}{A_{RVE}} = \frac{\pi}{4}(d)^2 \left(\frac{1}{s}\right)^2 = \frac{\pi}{4}\left(\frac{s}{s}\right)^2 = 0.785 \qquad (2.11.1)$$

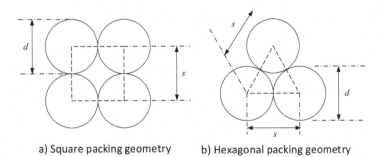

a) Square packing geometry
of RVE

b) Hexagonal packing geometry
of RVE

FIGURE 2.2   Packing arrays of RVE (Jones, 1975): (a) Square packing geometry of RVE and (b) hexagonal
packing geometry of RVE.

2.6.2 Hexagonal Packing Geometry

The packing geometry is considered in the triangular shape of equal sides, as shown in Figure 2.2b. The cross-sectional areas of the RVE and circular fiber inside the packing geometry are used to determine fiber volume fraction. The theoretical maximum fiber volume fraction of the hexagonal packing geometry occurs when the diameter (d) of circular fiber is equal to the distance between the center points of the circular fibers (s).

$$V_f = \frac{A_f}{A_{RVE}} = \frac{\pi \frac{d^2}{4}}{\frac{\sqrt{3}}{2}s^2} = \frac{\pi}{2\sqrt{3}}\left(\frac{s}{s}\right)^2 = 0.907 \tag{2.11.2}$$

2.7 ELASTIC PROPERTIES OF COMPOSITE LAMINA

The packing forms and interface stress–strain are not accounted for in the RVE mode while determining the elastic properties of composite lamina. The RVE model is assumed to be a combination of two constituents (fiber and matrix) with the same thickness throughout a composite lamina. The fibers and matrix are perfectly bonded together, without any voids inside the lamina. The fibers are continuous, and the responses are assumed to be linearly elastic. The shapes of fibers in this model do not have any effect on property calculations. However, the width of each material is still different depending on the material volume fractions in the model. Indeed, the properties of composite lamina are thus influenced by both the fiber and matrix volume ratios.

The above approach is often referred to as the "rule of mixtures" (Jones, 1975). Conveniently, the model is assumed to be a rectangular cross section, and the thickness of both constituents can be taken as unity. The fiber and matrix parts are assumed to be orientated in parallel in the material coordinate system in the longitudinal direction (fiber direction x_1) as shown in Figure 2.3.

From the RVE model in Figure 2.3, the volumes of fiber and matrix are determined from the fiber and matrix areas of A_f and A_m, together having thickness of unity. Thus, the fiber and matrix volume fractions can be determined as:

$$V_f + V_m = \frac{A_f L + A_m L}{A_c L} = \frac{w_f + w_m}{w_c} = 1 \tag{2.12}$$

Notations: Mechanical properties of the FRP composite lamina are presented in the following section. Notations of mechanical properties are defined as follows:

E_{11}^c = longitudinal elastic modulus of composite lamina,
E_{11}^f = longitudinal elastic modulus of fiber,
E_{22}^c = transverse elastic modulus of composite lamina,
E_{22}^f = transverse elastic modulus of fiber,

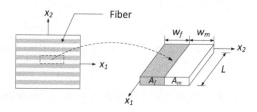

FIGURE 2.3   RVE model based on mechanics of materials.

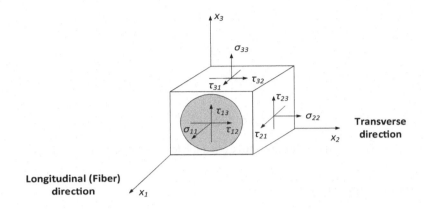

FIGURE 2.4   Stress notations.

G_{12}^c = in-plane shear modulus of composite lamina,
G_{12}^f = in-plane shear modulus of fiber,
G_{23}^c = transverse shear modulus of composite lamina,
G_{23}^f = transverse shear modulus of fiber,
v_{12}^c = major Poisson's ratio of composite lamina,
v_{12}^f = major Poisson's ratio of fiber,
E^m = elastic modulus of matrix,
G^m = shear modulus of matrix, and
v^m = Poisson's ratio of matrix.

Stress notations are shown in Figure 2.4.

2.7.1 Longitudinal Elastic Modulus (E_{11}^c)

To determine the composite's elastic modulus $\left(E_{11}^c\right)$ and Poisson ratio $\left(v_{12}^c\right)$, longitudinal stress $\left(\sigma_{11}^c\right)$ in the material coordinate system is applied on the RVE model as shown in Figure 2.5. The model is uniformly extended through the cross section in the applied stress direction assuming a perfect bond between the constituent materials. In addition, the model is contracted in the direction perpendicular to the applied stress $\left(\sigma_{11}^c\right)$ direction. Thus, both the induced strains of the fiber and matrix are identical $\left(\varepsilon_{11}^f = \varepsilon_{11}^m = \varepsilon_{11}^c\right)$.

The sum of forces of both the fiber and matrix portions is equal to the total applied force on the RVE model.

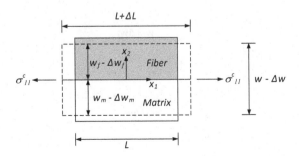

FIGURE 2.5   RVE model under longitudinal stress.

$$F_{11}^c = F_{11}^f + F_{11}^m \tag{2.13.1}$$

$$\sigma_{11}^c A_c = A_f \sigma_{11}^f + A_m \sigma_{11}^m \tag{2.13.2}$$

$$A_c E_{11}^c \varepsilon_{11}^c = \left(A_f E_{11}^f \varepsilon_{11}^f \right) + \left(A_m E^m \varepsilon_{11}^m \right) \tag{2.13.3}$$

$$E_{11}^c = V_f E_{11}^f + V_m E^m \tag{2.14}$$

Equation (2.14) is known as the **rule of mixture** equation for the elastic modulus $\left(E_{11}^c \right)$ of a composite.

2.7.2 Poisson Ratio's $\left(v_{12}^c \right)$

It should be noted that Poisson's ratio of fiber $\left(v_{12}^f \right)$ is different from that of the matrix $\left(v_{12}^m \right)$ resulting in contractions perpendicular to the applied load direction. These reductions are also different for the fiber and matrix, as shown in Figure 2.5. However, the total contraction (Δw) in the direction perpendicular to the stress direction is the sum of both the contracting parts ($\Delta w_f + \Delta w_m$) as mentioned above. The negative sign represents the contraction in the direction perpendicular to the applied load direction.

$$v_{12} = -\frac{\varepsilon_{22}^c}{\varepsilon_{11}^c} = -\frac{\left(\dfrac{\Delta w_f}{w} + \dfrac{\Delta w_m}{w} \right)}{\left(\dfrac{\Delta L}{L} \right)} = -\frac{\left(\dfrac{v_{12}^f w_f}{w}\left(\dfrac{\Delta L}{L} \right) + \dfrac{v_{12}^m w_m}{w}\left(\dfrac{\Delta L}{L} \right) \right)}{\left(\dfrac{\Delta L}{L} \right)} \tag{2.15}$$

$$v_{12} = v_{12}^f V_f + v_{12}^m V_m \tag{2.16}$$

Equation (2.16) is known as the **rule of mixture** equation for major Poisson's ratio.

2.7.3 Transverse Elastic Modulus E_{22}^c

To determine the composite's transverse elastic modulus $\left(E_{22}^c \right)$, the transverse stress $\left(\sigma_{22}^c \right)$ in the material coordinate system is applied to the RVE model. As shown in Figure 2.3, the constituent materials are idealized as rectangular blocks. The axial stresses in constituents and the composite are assumed to be identical with the understanding that fibers and matrix are uniformly positioned during the manufacturing process. Therefore, the transverse stress is uniform in both the constituents under the equilibrium condition $\left(\sigma_{22}^c = \sigma_{22}^f = \sigma_{22}^m \right)$ and the total extension (Δw) of the RVE model is the sum of each constituent part (fiber and matrix $\Delta w_f + \Delta w_m$) as shown in Figure 2.6.

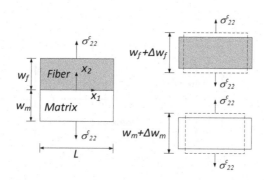

FIGURE 2.6   RVE model under transverse stress.

$$\varepsilon_{22}^c = \frac{\sigma_{22}^c}{E_{22}^c} = \frac{\Delta w_f + \Delta w_m}{w_c} = \left(\frac{w_f}{w_c}\right)\frac{\sigma_{22}^f}{E_{22}^f} + \left(\frac{w_m}{w_c}\right)\frac{\sigma_{22}^m}{E^m} = V_f\frac{\sigma_{22}^f}{E_{22}^f} + V_m\frac{\sigma_{22}^m}{E^m} \tag{2.17}$$

$$\frac{1}{E_{22}^c} = \frac{V_f}{E_{22}^f} + \frac{V_m}{E^m} \tag{2.18}$$

Equation (2.18) is presented in the **inverse form of the rule of mixture** for the elastic modulus $\left(E_{22}^c\right)$. For unidirectional composites, fiber stiffness typically has less contribution to the transverse elastic modulus $\left(E_{22}^c\right)$. However, at high fiber volume fractions, the transverse elastic modulus is dominated generally by the matrix.

2.7.4   IN-PLANE SHEAR MODULUS $\left(G_{12}^c\right)$

To determine the in-plane shear modulus $\left(G_{12}^c\right)$, the in-plane shear stress $\left(\tau_{12}^c\right)$ in the material coordinate system is applied to the RVE model. The in-plane shear stresses exerted on both the fiber and matrix are identical. Shear deformation from the RVE model is the sum of the shear deformation of fibers and that of the matrix, as shown in Figure 2.7.

The in-plane shear strain from the RVE model is the sum of deformations of each constituent material:

$$\gamma_{12}^c = \frac{\Delta_f + \Delta_m}{w_c} = \left(\frac{w_f}{w_c}\right)\frac{\tau_{12}^f}{G_{12}^f} + \left(\frac{w_m}{w_c}\right)\frac{\tau_{12}^m}{G^m} = V_f\frac{\tau_{12}^f}{G_{12}^f} + V_m\frac{\tau_{12}^m}{G^m} \tag{2.19}$$

$$\frac{1}{G_{12}^c} = \frac{V_f}{G_{12}^f} + \frac{V_m}{G^m} \tag{2.20}$$

Equation (2.20) is known as the **rule of mixture** equation for the elastic in-plane shear modulus $\left(G_{12}^c\right)$. As seen from Eq. (2.20), if the in-plane shear modulus of fiber is very high, then the fiber stiffness has a lower contribution to the in-plane shear modulus $\left(G_{12}^c\right)$ than the matrix shear modulus, unless the fiber volume fraction is very high. Thus, one can conclude that the elastic in-plane shear modulus of a composite is usually dominated by matrix properties. On average, the first term on the right hand of Eq. (2.20), contribution of fiber, is around 90% in-plane shear modulus.

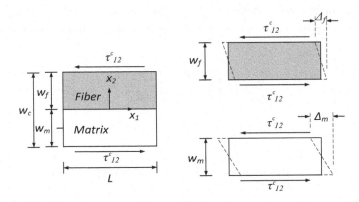

FIGURE 2.7   RVE model under shear stress for $G^c{}_{12}$.

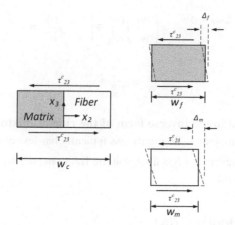

FIGURE 2.8   RVE model under shear stress for G^c_{23}.

2.7.5   TRANSVERSE SHEAR MODULUS $\left(G^c_{23}\right)$

To determine the transverse shear modulus $\left(G^c_{23}\right)$, the transverse shear stress $\left(\tau^c_{23}\right)$ in the material coordinate system is applied to the RVE model. The transverse shear stresses of both the fiber and matrix are equal. The shear deformation of the RVE model is a sum of the shear deformation of the fiber and matrix as shown in Figure 2.8.

The transverse shear strain of the RVE model is obtained from the deformations of each constituent.

$$\gamma^c_{23} = \frac{\Delta_f + \Delta_m}{w_c} = \left(\frac{w_f}{w_c}\right)\frac{\tau^f_{23}}{G^f_{23}} + \left(\frac{w_m}{w_c}\right)\frac{\tau^m_{23}}{G^m} = V_f \frac{\tau^f_{23}}{G^f_{23}} + V_m \frac{\tau^m_{12}}{G^m} \tag{2.21}$$

$$\frac{1}{G^c_{23}} = \frac{V_f}{G^f_{23}} + \frac{V_m}{G^m} \tag{2.22}$$

2.7.6   POISSON'S RATIO $\left(\nu^c_{23}\right)$

For unidirectional fiber composites, each composite lamina is classified as transversely isotropic. The Poisson's ratio (ν_{23}) can be evaluated from relations between longitudinal and shear modulus as those for isotropic materials. For additional information on the relationship between Poisson's ratio and moduli, the reader is referred to Boresi and Schmidt (2003). Invoking Eqs. (2.18 and 2.22) into Poisson's equation corresponding to the transversely isotropic material system, ν^c_{23} can be obtained as:

$$\nu^c_{23} = \frac{E^c_{22}}{2G^c_{23}} - 1 = \left(\frac{E^f_{22}E^m}{2G^f_{23}G^m}\right)\left(\frac{V_f G^m + V_m G^f_{23}}{V_f E^m + V_m E^f_{22}}\right) - 1 \tag{2.23}$$

Example 2.3

For unidirectional FRP composite lamina with fiber volume fraction 0.6, find:

    a. Elastic and shear modulus of FRP composite lamina and
    b. Poisson's ratio of FRP composite lamina

If properties of fiber and matrix are assumed to be isotropic as: $G_{12} = \dfrac{E_{11}}{2(1+\nu_{12})}$

Given: Properties of both the fiber and matrix, i.e., longitudinal E_{11}^f and transverse E_{22}^f elastic modulus of fiber are equal to 90 GPa, elastic modulus of matrix is 4 GPa, and Poisson's ratios of fiber ν_{12}^f and matrix ν^m are 0.2 and 0.3, respectively.

Solution

a. Elastic and shear modulus of FRP composite lamina:

From Eq. (2.14):

$$E_{11}^c = V_f E_{11}^f + V_m E^m = 0.6(90) + 0.4(4) = 55.6\,\text{GPa}$$

From Eq. (2.18):

$$\frac{1}{E_{22}^c} = \frac{V_f}{E_{22}^f} + \frac{V_m}{E^m} = \frac{0.6}{90} + \frac{0.4}{4} = 0.1067$$

$$E_{22}^c = 9.38\,\text{GPa}$$

From Eq. (2.20):

$$G_{12}^f = \frac{E_{11}^f}{2\left(1+\nu_{12}^f\right)} = \frac{90}{2(1+0.2)} = 37.5\,\text{GPa}$$

$$G^m = \frac{E^m}{2\left(1+\nu^m\right)} = \frac{4}{2(1+0.3)} = 1.538\,\text{GPa}$$

$$\frac{1}{G_{12}^c} = \frac{V_f}{G_{12}^f} + \frac{V_m}{G^m} = \frac{0.6}{37.5} + \frac{0.4}{1.538} = 0.276$$

$$G_{12}^c = 3.62\,\text{GPa}$$

b. Poisson's ratio of FRP composite lamina

From Eq. (2.16):

$$\nu_{12} = \nu_{12}^f V_f + \nu^m V_m = (0.2 \times 0.6) + (0.3 \times 0.4) = 0.24$$

2.8 THERMAL EXPANSION COEFFICIENTS

When the RVE model is extended to cases with temperature changes (ΔT), the RVE model is related to a case with boundaries that are free to expand (or contract). This concept is illustrated in Figure 2.9. Only strains are induced in both constituents. We assume that the total stress of the RVE model is still equal to zero. However, the stress of each constituent is not equal to zero. The above information is used to establish thermal expansion coefficients of the RVE model based on the mechanics of materials approach (**rule of mixture**).

2.8.1 LONGITUDINAL THERMAL EXPANSION COEFFICIENTS $\left(\alpha_{11}^c\right)$

Consider the elongation (or contraction) of the RVE model in the fiber direction under thermal changes (ΔT). Since the RVE model is assumed that the fiber and matrix are perfectly bonded with no slip between them, both constituents must elongate (or contract) in the same magnitude as that with the RVE model.

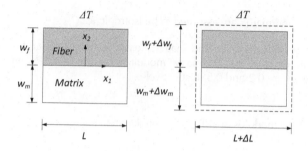

FIGURE 2.9 RVE model under temperature difference (ΔT).

The total thermal stress of the RVE model is equal to be zero.

$$F_{11}^c = F_{11}^f + F^m = 0 \tag{2.24.1}$$

$$\sigma_{11}^c = \sigma_{11}^f \left(\frac{A^f}{A^c} \right) + \sigma^m \left(\frac{A^m}{A^c} \right) = \sigma_{11}^f V_f + \sigma^m V_m = 0 \tag{2.24.2}$$

$$\sigma_{11}^f = E_{11}^f \left(\varepsilon_{11}^f \right)^{mec} = E_{11}^f \left(\varepsilon_{11}^f \right)^{total} - E_{11}^f \alpha_f \Delta T \tag{2.24.3}$$

$$\sigma^m = E^m \left(\varepsilon^m \right)^{mec} = E_{11}^m \left(\varepsilon^m \right)^{total} - E^m \alpha_m \Delta T \tag{2.24.4}$$

where α_f and α_m = thermal coefficient of fiber and matrix, respectively.
ε^{mec} = strain induced by an applied force only.
$V_{f, m}$ = volume fraction of fiber and matrix respectively, in relation to composite volume.

The thermal stresses in the constituents under free thermal condition are determined from the thermo-mechanical strains. The normal strains in the fiber direction are equal $\left(\varepsilon_{11}^f = \varepsilon_{11}^m = \varepsilon_{11}^c \right)$. The total strain of the RVE model is obtained from the induced thermal strain as:

$$E_{11}^f \left(\varepsilon_{11}^f \right)^{total} V_f - E_{11}^f \alpha_f \Delta T V_f + E_{11}^m \left(\varepsilon_{11}^m \right)^{total} V_m - E^m \alpha_m \Delta T V_m = 0 \tag{2.25.1}$$

$$E_{11}^f \left(\varepsilon_{11}^c \right) V_f + E_{11}^m \left(\varepsilon_{11}^c \right) V_m = E_{11}^f \alpha_f \Delta T V_f + E^m \alpha_m \Delta T V_m \tag{2.25.2}$$

$$\varepsilon_{11}^c = \alpha_{11}^c \Delta T = \frac{\left(E_{11}^f \alpha_f V_f + E^m \alpha_m V_m \right) \Delta T}{E_{11}^f V_f + E^m V_m = E_{11}^c} \tag{2.25.3}$$

The denominator relation of E_{11}^c in Eq. (2.25.3) is derived in Eq. (2.14).

$$\alpha_{11}^c = \frac{\alpha_f E_{11}^f}{E_{11}^c} V_f + \frac{\alpha_m E^m}{E_{11}^c} V_m \tag{2.26}$$

2.8.2 Transverse Thermal Expansion Coefficients $\left(\alpha_{22}^c \right)$

Consider the elongation (or contraction) using the RVE model in the direction perpendicular to the fiber direction under thermal change of (ΔT). The total deformation is the sum of deformations ($\Delta w_f + \Delta w_m$) of constituents as shown in Figure 2.9.

The deformation and strain relations of constituents are given as:

$$\Delta w = \varepsilon_{22}^c w_c = \varepsilon_{22}^c \left(w_f + w_m \right) \tag{2.27.1}$$

$$\varepsilon_{22}^c = \varepsilon_{22}^f \left(\frac{w_f}{w_c} \right) + \varepsilon^m \left(\frac{w_m}{w_c} \right) = \varepsilon_{22}^f V_f + \varepsilon^m V_m \tag{2.27.2}$$

The thermomechanical strain can be written in terms of longitudinal stress and Poisson's ratio.

$$\varepsilon_{22}^c = \left(-v_{12}^f \frac{\sigma_{11}^f}{E_{11}^f} + \alpha_f \Delta T \right) V_f + \left(-v^m \frac{\sigma_{11}^m}{E^m} + \alpha_m \Delta T \right) V_m \tag{2.27.3}$$

$$\alpha_{22}^c = \left(\alpha_f - v_{12}^f \left(\alpha_{11}^c - \alpha_f \right) \right) V_f + \left(\alpha_m - v^m \left(\alpha_{11}^c - \alpha_m \right) \right) V_m \tag{2.28}$$

2.9 MOISTURE EXPANSION COEFFICIENTS

When the RVE model absorbs moisture (ΔM) under unrestrained boundaries, the RVE model is free to expand as shown in Figure 2.10. As in unrestrained thermal condition case of composite material response, strains are induced in both the constituents of the RVE model, but the total stress of the RVE model is still equal to zero. However, the stress of each constituent is not equal to zero. The moisture expansion coefficients of the RVE model based on the mechanics of materials approach (**rule of mixture**) are presented in Section 2.9.1 and 2.92.

2.9.1 Longitudinal Moisture Expansion Coefficient $\left(\beta_{11}^c \right)$

As shown in Figure 2.10, fiber and matrix portions in the fiber direction must be elongated in the same magnitude with the RVE model in the moisture concentration or moisture loading (ΔM). Therefore, the normal strains in the fiber direction are equal $\left(\varepsilon_{11}^f = \varepsilon_{11}^m = \varepsilon_{11}^c \right)$. The total strain of the RVE model is obtained from the induced moisture strain as follows:

The total moisture stress (force) of the RVE model is equal to zero.

$$F_{11}^c = F_{11}^f + F^m = 0 \tag{2.29.1}$$

or

$$\sigma_{11}^c = \sigma_{11}^f \left(\frac{A^f}{A^c} \right) + \sigma^m \left(\frac{A^m}{A^c} \right) = \sigma_{11}^f V_f + \sigma^m V_m = 0 \left(\text{recognizing } F_{11}^c = \sigma_{11}^c A_c, \text{etc} \right) \tag{2.29.2}$$

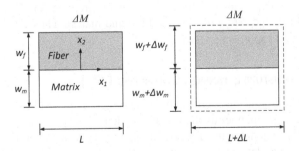

FIGURE 2.10   RVE model under moisture loading (ΔM).

$$\sigma_{11}^{f} = E_{11}^{f}\left(\varepsilon_{11}^{f}\right)^{mec} = E_{11}^{f}\left(\varepsilon_{11}^{f}\right)^{total} - E_{11}^{f}\beta_{f}\Delta M_{f} \qquad (2.29.3)$$

$$\sigma^{m} = E^{m}\left(\varepsilon^{m}\right)^{mec} = E_{11}^{m}\left(\varepsilon^{m}\right)^{total} - E^{m}\beta_{m}\Delta M_{m} \qquad (2.29.4)$$

where β_{f} and β_{m} = moisture coefficient of both the fiber and matrix with units of in/in/lb/lb (strain per unit weight increase in moisture per unit weight of composite), respectively.

ε^{mec} = strain induced by applied forces only.

$\Delta M_{f, m}$ = moisture weight increase in fiber and matrix, respectively, per unit weight of composite.

The stresses in the constituents induced under moisture uptake are determined from the hygro-mechanical strains. The normal strains in the fiber direction are equal $\left(\varepsilon_{11}^{f} = \varepsilon_{11}^{m} = \varepsilon_{11}^{c}\right)$. The total strain of the RVE model is obtained from the induced moisture strain.

By substituting Eqs. (2.29.3 and 2.29.4) into Eq. (2.29.2) and rearranging for $\varepsilon_{11}^{f} = \varepsilon_{11}^{m} = \varepsilon_{11}^{c}$

$$\left(E_{11}^{f}V_{f} + E^{m}V_{m}\right)\varepsilon_{11}^{c} = E_{11}^{f}\beta_{f}\Delta M_{f}V_{f} + E^{m}\beta_{m}\Delta M_{m}V_{m} \qquad (2.29.5)$$

$$\varepsilon_{11}^{c} = \beta_{11}^{c}\Delta M_{c} = \frac{E_{11}^{f}\beta_{f}V_{f}\Delta M_{f} + E^{m}\beta_{m}V_{m}\Delta M_{m}}{\left(E_{11}^{f}V_{f} + E^{m}V_{m}\right) = \left(E_{11}^{c}\right)} \qquad (2.29.6)$$

$$\beta_{11}^{c} = \frac{E_{11}^{f}\beta_{f}V_{f}\Delta M_{f}}{E_{11}^{c}\Delta M_{c}} + \frac{E^{m}\beta_{m}V_{m}\Delta M_{m}}{E_{11}^{c}\Delta M_{c}} \qquad (2.29.7)$$

where $\Delta M_{f, m, c}$ are moisture-induced loadings corresponding to fiber, matrix, and composite, respectively.

In general, the moisture absorption capacity of the constituents is different from one to another. The fibers in the RVE model are often assumed to absorb zero moisture. The matrix is assumed to absorb moisture uniformly. Thus, the moisture expansion coefficient, Eq. (2.29.7), is simplified as Eq. (2.30) after recognizing $V_{m} = 1.0$. Also, $\Delta M_{m}/\Delta M_{c} = \rho_{c}/\rho_{m}$ for negligible increase in fiber weight due to moisture uptake, which comes from the fact that weight increases are measured per unit weights of matrix and composite.

Note: Volume increase in composite is the same as the volume increase in matrix since fiber does not take moisture in the hygrothermal response, i.e., $\Delta V_{c} = \Delta V_{m,}$ and multiplying both sides with ρ_{c} and dividing and multiplying both sides with ρ_{m} we realize that $\Delta M_{m}/\Delta M_{c} = \rho_{c}/\rho_{m}$

$$\beta_{11}^{c} = \frac{E^{m}\beta_{m}\rho_{c}}{E_{11}^{c}\rho_{m}} \qquad (2.30)$$

If the fiber is very stiff when compared with the matrix or the fiber volume fraction is very high (say ~70%), then the moisture expansion coefficient in the fiber direction is independent of the increase in moisture content (ΔM_{f} and ΔM_{m}) of both the fiber and matrix. Thus, the longitudinal moisture expansion coefficient $\left(\beta_{11}^{c}\right)$ approaches zero.

2.9.2 TRANSVERSE MOISTURE EXPANSION COEFFICIENT $\left(\beta_{22}^{c}\right)$

The transverse moisture expansion coefficients in the direction perpendicular to the fiber direction can be derived in the same manner as the transverse thermal expansion coefficient. Consider the elongation using the RVE model in the direction perpendicular to the fiber direction under moisture loading of (ΔM). The total deformation is the sum of deformations ($\Delta w_{f} + \Delta w_{m}$) of constituents as shown in Figure 2.10.

Repeating Eqs. (2.27.1 and 2.27.2):

$$\Delta w = \varepsilon_{22}^c w_c = \varepsilon_{22}^c \left(w_f + w_m \right) \tag{2.31.1}$$

$$\varepsilon_{22}^c = \varepsilon_{22}^f \left(\frac{w_f}{w_c} \right) + \varepsilon^m \left(\frac{w_m}{w_c} \right) = \varepsilon_{22}^f V_f + \varepsilon^m V_m \tag{2.31.2}$$

The hygro-mechanical strain of the composite can be written in terms of longitudinal stress and the Poisson's ratio as given below and converted to fiber and matrix stress.

$$\varepsilon_{22}^c = -v_{12}^c \frac{\sigma_{11}^c}{E_{11}^f} - \beta_{22}^c \Delta M_c = -\left(v_{12}^f \frac{\sigma_{11}^f}{E_{11}^f} + \beta_f \Delta M_f \right) - \left(v^m \frac{\sigma_{11}^m}{E^m} + \beta_m \Delta M_m \right) \tag{2.31.3}$$

$$\beta_{22}^c = \frac{1}{\Delta M_c} \left(\left(v_{12}^f \frac{\sigma_{11}^f}{E_{11}^f} + \beta_f \Delta M_f \right) + \left(v^m \frac{\sigma_{11}^m}{E^m} + \beta_m \Delta M_m \right) - v_{12}^c \frac{\sigma_{11}^c}{E_{11}^f} \right) \tag{2.31.4}$$

If the fibers inside the RVE model do not absorb moisture and the matrix is under uniform moisture distribution, then the transverse moisture expansion coefficient in Eq. (2.31.4) is simplified by considering fiber moisture contents (ΔM_f) as zero. Herein, β_{22}^c is simplified to Eq. (2.31.5) using the same logic that was adopted to simplify Eq. (2.29.7) to Eq. (2.30).

$$\beta_{22}^c = \beta^m \frac{\rho_c}{\rho_m} \left((1 + v^m) - \frac{E^m}{E_{11}^c} \left(v_{12}^f V_f + v^m V_m \right) \right) \tag{2.31.5}$$

If the fiber is very stiff compared to the matrix, then the moisture expansion coefficient perpendicular to the fiber direction is also independent of moisture contents (ΔM_f and ΔM_m) of both the fiber and matrix. Thus, the transverse moisture expansion coefficient $\left(\beta_{22}^c \right)$ will approach the expression in Eq. (2.32).

$$\beta_{22}^c = \beta^m \frac{\rho_c}{\rho_m} \left(1 + v^m \right) \tag{2.32}$$

Example 2.4

For unidirectional FRP composite lamina with fiber volume fraction 0.6, find:

a. Thermal coefficients of FRP composite lamina
b. Moisture expansion coefficients of FRP composite lamina.

Given: properties of fiber and matrix are given as follows: longitudinal E_{11}^f and transverse E_{22}^f elastic modulus of fiber are equal to 90 GPa, Elastic modulus of matrix is 4 GPa, Poisson's of fiber v_{12}^f and matrix $v^m = 0.2$ and 0.3, respectively. The density of fiber ρ_f and matrix ρ_m is 2,400 and 1,200 kg/m³. The coefficient of thermal expansion is 6×10^{-6} and 60×10^{-6} m/m/°C for fiber α_f and matrix α_m, respectively. The moisture expansion of matrix β_m is 0.4 m/m/kg/kg.

Solution

a. Thermal coefficients of FRP composite lamina:
   From Eq. (2.25.3):

$$\alpha_{11}^c = \frac{\alpha_f E_{11}^f V_f + \alpha_m E^m V_m}{E_{11}^f V_f + E^m V_m} = \left(\frac{(6 \times 90 \times 0.6) + (60 \times 4 \times 0.4)}{(90)0.6 + (4)0.4} \right) (10^{-6})$$

$$= 7.55 (10^{-6}) \, \text{m/m/°C}$$

From Eq. (2.28):

$$\alpha_{22}^c = \left(\alpha_f - v_{12}^f\left(\alpha_{11}^c - \alpha_f\right)\right)V_f + \left(\alpha_m - v^m\left(\alpha_{11}^c - \alpha_m\right)\right)V_m$$

$$\left((6 - 0.2(7.55 - 6))0.6 + (60 - 0.3(7.55 - 60))0.4\right)\left(10^{-6}\right) = 21.1\left(10^{-6}\right)\text{m/m/}^\circ\text{C}$$

b. Moisture expansion coefficients of FRP composite lamina
   From Eq. (2.5.2):

$$\rho_c = \rho_f V_f + \rho_m V_m = (2,400 \times 0.6) + (1,200 \times 0.4) = 1,920\,\text{kg/m}^3$$

From Eq. (2.30), E_{11}^c was determined as 55.6 GPa in Ex. 2.3a:

$$\beta_{11}^c = \frac{E_m \beta_m \rho_c}{E_{11}^c \rho_m} = \frac{4 \times 0.4 \times 1,920}{55.6 \times 1,200} = 0.0460\,\text{m/m/kg/kg}$$

From Eq. (2.31.5):

$$\beta_{22}^c = \beta_m \frac{\rho_c}{\rho_m}\left((1 + v^m) - \frac{E_m}{E_{11}^c}\left(v_{12}^f V_f + v^m V_m\right)\right)$$

$$\beta_{22}^c = \frac{0.4 \times 1,920}{1,200}\left((1 + 0.3) - \frac{4}{55.6}(0.2 \times 0.6 + 0.3 \times 0.4)\right) = 0.843\,\text{m/m/kg/kg}$$

2.10 SEMI-EMPIRICAL HALPIN–TSAI APPROACH

The semi-empirical Halpin–Tsai approach is developed by using curve fitting to the results from the theory of elasticity approach. The two empirical factors (reinforcing factor (β) and partitioning factor (η)) are introduced into the original elastic constants (rule of mixtures). The reinforcing factor (β) accounts for fiber-packing geometry and loading conditions that are neglected in the original model of the rule of mixture. The partitioning (η) factor accounts for an appropriate stress ratio of the two constituents (Halpin and Kardos, 1976).

Longitudinal Elastic Modulus $\left(E_{11}^c\right)$ Composite

The longitudinal elastic modulus $\left(E_{11}^c\right)$ of a polymer composite in Halpin–Tsai semi-empirical formula is still the same as one of the mechanics of the material approach (rule of mixtures) as given in Eq. (2.14).

Poisson's Ratio (ν_{12})

Poisson's ratio (ν_{12}) is given in Eq. (2.16), Section 2.7.2.

Transverse Elastic Modulus $\left(E_{22}^c\right)$ of Composite

The transverse elastic modulus $\left(E_{22}^c\right)$ of the Halpin–Tsai semi-empirical formula is:

$$E_{22}^c = \left(\frac{1 + \beta\eta V_f}{1 - \eta V_f}\right)E^m \quad \text{and} \quad \eta = \frac{\left(\dfrac{E_{22}^f}{E^m}\right) - 1}{\left(\dfrac{E_{22}^f}{E^m}\right) + \beta} \tag{2.33}$$

For circular fibers in the RVE of a square array, the reinforcing factor (β) is taken as 2. For rectangular fibers with the cross section (ab) in a hexagonal array as shown in Figure 2.10, the reinforcing factor (β) is given to be ($2a/b$). When the modular ratio of fiber to matrix is equal to unity then, the RVE model is a homogeneous medium where the partitioning factor (η) is zero. If, the modular ratio of fiber to matrix is about zero then, the fibers inside the RVE model are represented by voids due to the partitioning factor ($\eta = -1/\beta$).

In-Plane Shear Modulus $\left(G_{12}^c\right)$

The transverse elastic modulus $\left(G_{12}^c\right)$ of the Halpin–Tsai semi-empirical formula is:

$$G_{12}^c = \left(\frac{1+\beta\eta V_f}{1-\eta V_f}\right)G^m \quad \text{and} \quad \eta = \frac{\left(\dfrac{G_{22}^f}{G^m}\right)-1}{\left(\dfrac{G_{22}^f}{G^m}\right)+\beta} \tag{2.34}$$

where β = reinforcing factor
η = the partitioning factor

For circular fibers in the RVE of a square array, the reinforcing factor (β) is given to be 1. For rectangular fibers with the cross section in a hexagonal array as shown in Figure 2.11, the reinforcing factor (β) is given to be 1.73 ln (a/b). However, the reinforcing factor gives reasonable results up to some limits of fiber volume fraction (more than 50%). Then, another formula of the reinforcing factor that was introduced by Hewitt and Malherbe (1970) may be used as:

$$\beta = 1 + 40 V_f^{10} \tag{2.35}$$

Note: The magnitudes of both the reinforcing and partitioning factors are empirical, and their relationships given in Eqs. (2.33 and 2.35) are based on limited experimental data (Halpin and Kadros, 1976).

Example 2.5

Using the semi-empirical Halpin–Tsai approach, FRP composite lamina with fiber volume fraction 0.6 (assuming circular fibers are arranged in a square array of RVE), find:

    a. Longitudinal and transverse elastic modulus
    b. Poisson's ratio of lamina
    c. In-plane shear modulus for lamina.

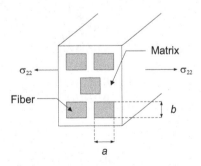

FIGURE 2.11 Rectangular fibers in an array based on the semi-empirical Halpin–Tsai approach (Kaw, 2006; Hyer, 2009).

Properties of fiber and matrix are assumed to be isotropic as: $G_{12} = \dfrac{E_{11}}{2(1+v_{12})}$

Given: Properties of fiber and matrix are given as follows: longitudinal E_{11}^f and transverse E_{22}^f elastic modulus of fiber are equal to 90 GPa, the elastic modulus of matrix is 4 GPa, Poisson's ratio of fiber v_{12}^f and matrix $v^m = 0.2$ and 0.3, respectively.

Solution

a. Longitudinal and transverse elastic modulus
   From Eq. (2.14):

$$E_{11}^c = V_f E_{11}^f + V_m E^m = 0.6(90) + 0.4(4) = 55.6\,\text{GPa}$$

From Eq. (2.33):
   For circular fibers in the RVE of a square array, the reinforcing factor $(\beta) = 2$.

$$\eta = \frac{\left(\dfrac{E_{22}^f}{E^m}\right) - 1}{\left(\dfrac{E_{22}^f}{E^m}\right) + \beta} = \frac{\left(\dfrac{90}{4}\right) - 1}{\left(\dfrac{90}{4}\right) + 2} = 0.878$$

$$E_{22}^c = \left(\frac{1 + \beta\eta V_f}{1 - \eta V_f}\right) E^m = \left(\frac{1 + 2(0.878)0.6}{1 - (0.878)0.6}\right) 4 = 17.36\,\text{GPa}$$

b. Poisson's ratio of lamina
   From Eq. (2.16):

$$v_{12} = v_{12}^f V_f + v_{12}^m V_m = 0.2(0.6) + 0.3(0.4) = 0.24$$

c. In-plane shear modulus

$$G_{12}^f = \frac{E_{11}^f}{2\left(1 + v_{12}^f\right)} = \frac{90}{2(1+0.2)} = 37.5\,\text{GPa}$$

$$G^m = \frac{E^m}{2\left(1 + v^m\right)} = \frac{4}{2(1+0.3)} = 1.538\,\text{GPa}$$

From Eq. (2.34)
   For circular fibers in the RVE of a square array, the reinforcing factor $(\beta) = 1$.

$$\eta = \frac{\left(\dfrac{G_{12}^f}{G^m}\right) - 1}{\left(\dfrac{G_{12}^f}{G^m}\right) + \beta} = \frac{\left(\dfrac{37.5}{1.538}\right) - 1}{\left(\dfrac{37.5}{1.538}\right) + 1} = 0.921$$

$$G_{12}^c = \left(\frac{1 + \beta\eta V_f}{1 - \eta V_f}\right) G^m = \left(\frac{1 + 0.921(0.6)}{1 - 0.921(0.6)}\right)(1.538) = 5.34\,\text{GPa}$$

2.11   SUMMARY

This chapter provides the micromechanics models and mechanics of material-based theories to estimate the engineering properties of a unidirectional lamina, the basic building block of composite structures. The mechanics of the materials-based approach is easy to comprehend but provides lower bound values for lamina stiffness. More advanced theory of elasticity-based approach provides better accuracy in relation to experimental data (Kaw, 2006). However, those advanced theories are deemed beyond the scope of simple design methods. In addition, the hygrothermal characteristics of composite lamina are presented by considering the additional stresses induced by the thermal and moisture expansion in the laminates. In all the models presented, the fibers and matrix are assumed to be uniform and perfectly bonded together, without any voids or fiber kinks inside the lamina. These properties are fundamental to the calculation of the constitutive relations of composite lamina and laminates (combination of several laminae) presented in Chapters 3 and 4.

EXERCISES

Problem 2.1: The weight fraction of carbon in carbon-vinyl ester composite is 0.6. Find the carbon-VE volume fractions and the specific gravity of the composite if carbon-specific gravity is 1.8 and that of vinyl ester is 1.20.

Problem 2.2: The E-glass and carbon fiber volume fraction are 35% and 25% respectively, in a composite made with epoxy matrix. The specific gravity of glass, carbon, and epoxy, respectively, are 2.8, 1.6, and 1.2. Find the weight fractions of each constituent material in the composite.

Problem 2.3: Using the properties of carbon and vinyl ester given in Chapter 1, determine the modulus of elasticity (E) of a unidirectional lamina made of carbon and epoxy with a fiber volume fraction of 55%.

Problem 2.4: Based on the data given in Problem 2.3, calculate the transverse modulus of elasticity and in-plane shear modulus using the Halpin–Tsai empirical adjustments given in Eqs. (2.33 and 2.34). Assume that the circular fibers are arranged in a square array of RVE.

Problem 2.5: Assuming that the ratio of longitudinal to the transverse modulus of elasticity is 0.3, find the values of moduli in these two directions (transverse and longitudinal) for E-glass and vinyl ester unidirectional composite by employing the strength of the material approach.

Problem 2.6: Assuming a square array of arrangement for E-glass fibers, find the spacing in fiber radius (f_r) for the fiber volume fraction (fvf) of 0.50.

Problem 2.7: The standard burn test of an epoxy-glass revealed the weight of the sample before burn-off is 0.0001 lbs (crucible weight is excluded here) and the same sample weighed 0.00009 lbs after burn-off. Find the fiber weight and volume fractions, assuming the fiber weighs 170 lb/ft^3 and resin weighs 80 lb/ft^3.

Problem 2.8: The fvf of the epoxy/E-glass composite of 50 lbs is 30%. Using properties in Chapter 1, find the weight of epoxy and the density of the new hybrid composite where glass weight is replaced by carbon.

Problem 2.9: Find the density of a hybrid beam with 10 layers of carbon of 0.1″ thickness and 12 layers of glass of 0.15″ thickness, which are bonded with epoxy matrix. The fiber volume fractions of carbon and glass are 0.2 and 0.25, respectively.

Problem 2.10: Using properties given in Chapter 1, find the thermal expansion coefficient of epoxy/E-glass composite with 70% fvf.

Problem 2.11: Prove that the thermal expansion coefficient of a unidirectional epoxy/E-glass composite never exceeds that of the epoxy matrix.

Problem 2.12: In a unidirectional vinyl ester/E-glass composite, the transverse thermal expansion coefficient is 8 in/in/°F. Find the longitudinal thermal expansion coefficient based on the material data given in Chapter 1.

Problem 2.13: Find the moisture expansion coefficient of epoxy/E-glass fiber composite with 60% fvf using properties in Chapter 1 and using moisture expansion coefficient of 0.33 (in/in/lb/lb) for epoxy and zero for glass fibers. How much does the moisture expansion coefficient in this problem vary if the epoxy resin absorbs 6% of moisture of its weight?

Problem 2.14: If an epoxy/E-glass laminate ($12'' \times 3'' \times 1/16''$) absorbs 6% moisture, find the water uptake in the laminate by weight and the volume change when 4% moisture is absorbed by epoxy.

REFERENCES AND SELECTED BIOGRAPHY

American Society for Testing and Materials, Standard test method for constituent content of composite prepreg by soxhlet extraction, ASTM C 613–97, Annual Book of ASTM Standards, 1997.

American Society for Testing and Materials, Standard test methods for constituent content of composite materials, ASTM D 3171-11, Annual Book of ASTM Standards, 2011.

Barbero, E.J., *Introduction to Composite Materials Design*, Taylor &Francis, Inc., Philadelphia, PA, 1998.

Boresi, A.P, and Schmidt, R.J., *Advanced Mechanics of Materials*, 6th edition, John Wiley & Sons, Inc., New York, NY, 2003.

Callister, W.D. Jr, *Fundamentals of Materials Science and Engineering*, 5th edition, John Wiley & Sons, Inc., New York, NY, 2000.

Daniel, I.M., and Ishai, O., *Engineering Mechanics of Composite Materials*, 2nd edition, Oxford University Press, Inc., New York, NY, 2006.

Gibson, R.F., *Principles of Composite Material Mechanics*, McGraw-Hill, Inc., New York, NY1994.

Halpin, J.C., and Kardos, J.L., The Halpin-Tsai equations: A review, *Polymer Engineering and Science*, 16(5), 1976, pp. 344–352.

Hewitt, R.L., and Malherbe, M.C.de., An approximation for longitudinal shear modulus of continuous fibre composites, *Journal of Composite Materials*, 4(2), 1970, pp. 280–282.

Hyer, M.W., *Stress Analysis of Fiber-Reinforced Composite Materials*, International edition, WCB/McGraw-Hill, Singapore, 2009.

Jones, R.M., *Mechanics of Composite Materials*, Hemisphere Publishing, New York, NY, 1975.

Kaw, A.K., *Mechanics of Composite Materials*, 2nd edition, CRC Press, Taylor & Francis Group, Boca Raton, FL, 2006.

Kollar, L.P., and Springer, G.S., *Mechanics of Composite Structures*, Cambridge University Press, New York, NY, 2003.

Qureshi, M.A.M., Failure behavior of pultruded GFRP members under combined bending and torrision, PhD Dissertation, submitted to West Virginia University, 2012.

Swanson, S.R., *Introduction to Design and Analysis with Advance Composite Materials*, International edition, Prentice-Hall, Inc., Upper Saddle River, NJ, 1997.

3 Mechanics of FRP Composite Lamina

In this chapter, constitutive relations (stress vs. strain) of laminated composites are introduced in addition to the stress–strain relations of a lamina. The constitutive relations of the generally anisotropic materials under linear elastic responses are presented in a simplified form for monoclinic, orthotropic, transversely isotropic, and isotropic materials because fiber-reinforced polymer (FRP) composite laminates are made from unidirectional, bidirectional, or even multidirectional fabrics that form thin lamina before forming into a laminate. It is assumed that laminate thickness, sum of composite lamina layers, is very small (~10% or less of the least dimension) compared to its other dimensions. Constitutive relations of composite laminates (combination of several laminae) can be reduced to plane stress condition because of disproportionately large dimensions in the other two directions. In other words, the laminate thickness is still very small compared to the other two dimensions. The constitutive relations of plane stress condition are used to establish both laminate stiffness (*ABD*) and compliance (*abd*) matrices. The hygro-thermomechanical responses of laminated composites are functions of the laminate stiffness matrix, and these responses are presented for in-plane stress versus strain relations at the end of this chapter. The main objective of this chapter is to present the stiffness formulation of FRP composite lamina and derive the lamina response under hygro-thermomechanical loads. In addition, stress and strain relationships in global coordinates through transformations are highlighted. It is important to reiterate that the stress–strain relation is independent of any applied loading or external actions induced by temperature or moisture.

3.1 STRESS AND STRAIN RELATIONSHIP

In general, hygro-thermomechanical properties of composite lamina and laminates can be classified as generally anisotropic, monoclinic, orthotropic, transversely isotropic, and isotropic. These classifications are discussed in depth in Section 3.2. The most important assumption while studying FRP composites response is that the constituent material (fiber and matrix) properties are smeared, i.e., equivalent of homogenous materials. The properties of the FRP composites depend on fiber architecture (fiber arrangement and stacking sequence); for example: [0/90], [0/+45/−45/90], as shown in Figure 3.1, fabric type and pattern including weaving, braiding or stitching and even 3-D stitching, and resin type. Cure kinetics of a composite is not directly accounted for while developing these experimental responses, except those composite coupon properties (strength and stiffness in different directions) reflect the percent of cure or the amount of fiber-matrix bond in an indirect sense. For theoretical developments, composites are assumed to be fully cured.

For non-orthogonal random fiber arrangement, the laminate behavior is presented to be generally anisotropic. The stress–strain relationship of generally anisotropic composite materials is discussed under the linear elastic response of a lamina. The stress–strain relationships of monoclinic, orthotropic, transversely isotropic, and isotropic materials are also presented after defining these new technical terms and by simplifying the generally anisotropic stress–strain relation. Under a linear elastic state, stresses under an undeformed state are defined on an infinitesimal element in a Cartesian coordinate system as shown in Figure 3.2.

The sign convention for stress on an infinitesimal face is positive when it (stress) is normal to that infinitesimal face (on the stress side at distance $dx_{1,2,3}$) and stress directions are in the same direction as the positive directions of the axes. The negative stress is defined as the normal line of

DOI: 10.1201/9781003196754-3

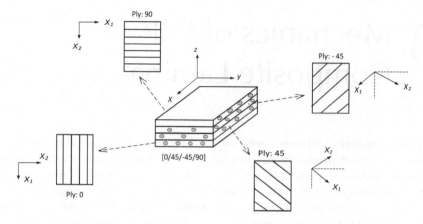

FIGURE 3.1   Fiber architecture and stacking sequence [0/+45/−45/90].

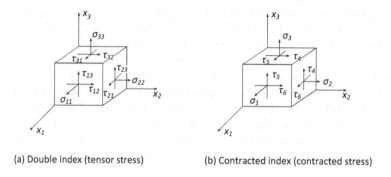

(a) Double index (tensor stress)   (b) Contracted index (contracted stress)

FIGURE 3.2   State of stress (positive stress): (a) double index (tensor stress) and (b) contracted index (contracted stress).

that infinitesimal face (on the stress side) and the stress direction is in the direction opposite to the positive axes, as shown in Figure 3.2.

3.2   GENERALLY ANISOTROPIC STRESS–STRAIN RELATIONSHIP

In the orthogonal coordinate system (x_1–x_2–x_3), the deformational responses of an infinitesimal rectangular parallelepiped can be separated into two parts by considering an infinitesimal rectangular parallelepiped of anisotropic materials, under arbitrary loadings as (1) dilatation and (2) distortion. Dilatation is the change of volume with respect to the original volume affecting normal strains along the orthogonal coordinates while distortion is an angular deformation of an infinitesimal rectangular parallelepiped due to shear strains acting on orthogonal planes.

If an infinitesimal element is under unidirectional tensile stress in direction X_1 (σ_{11} direction), then the element extends in direction (X_1). Thus, axial strain (ε_{11}) in direction X_1 is presented by following the linear elastic response as:

$$\varepsilon_{11} = S_{1111}\sigma_{11} \tag{3.1.1}$$

where S_{1111} = elastic constant under applied load and dilatational response in direction (X_1), while the element simultaneously contracts/expands in directions (X_2 and X_3) normal to direction (X_1) due to the Poisson effect, as shown in Figure 3.3. Thus, both axial strains (ε_{22}) and (ε_{33}) in directions (X_2) and (X_3), respectively, can also be presented as well as the axial strain (ε_{11}) in (X_1) direction.

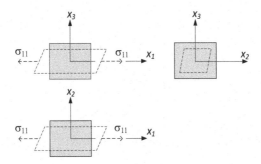

FIGURE 3.3 Infinitesimal element under unidirectional tensile stress (X_1).

$$\varepsilon_{22} = S_{2211}\sigma_{11} \qquad (3.1.2)$$

$$\varepsilon_{33} = S_{3311}\sigma_{11} \qquad (3.1.3)$$

where

S_{2211} = elastic constant under applied axial load in direction (X_1) and dilatational response in direction (X_2)

S_{3311} = elastic constant under applied axial load in direction (X_1) and dilatational response in direction (X_3).

In addition, the element simultaneously distorts on the orthogonal planes as shown in Figure 3.3. Similarly, strains induced under shear in different orthogonal planes are presented below:

$$\gamma_{23} = S_{2311}\sigma_{11} \qquad (3.1.4)$$

$$\gamma_{13} = S_{1311}\sigma_{11} \qquad (3.1.5)$$

$$\gamma_{12} = S_{1211}\sigma_{11} \qquad (3.1.6)$$

where

S_{2311} = elastic constant under applied axial load in direction (X_1) and distortional response in plane (X_2–X_3)

S_{1311} = elastic constant under applied axial load in direction (X_1) and distortional response in plane (X_1–X_3)

S_{1211} = elastic constant under applied axial load in direction (X_1) and distortional response in plane (X_1–X_2)

Notations:

σ_{ii} and ε_{ii} = normal stresses and strains, respectively, in direction (X_i) of the orthogonal coordinate system (i = 1, 2, 3) as shown in Figure 3.2a.

γ_{ij} = shear strains on the plane (X_i–X_j) of the orthogonal coordinates (i, j = 1, 2, 3).

S = elastic constant (dilatational or distortional response) under an applied unit stress.

Similarly, unidirectional tensile stresses (σ_{22}) and (σ_{33}) are applied to directions (X_2) and (X_3), respectively. Strains under unidirectional tensile stresses (σ_{22}) are given as:

$$\varepsilon_{11} = S_{1122}\sigma_{22} \qquad (3.2.1)$$

$$\varepsilon_{22} = S_{2222}\sigma_{22} \qquad (3.2.2)$$

$$\varepsilon_{33} = S_{3322}\sigma_{22} \tag{3.2.3}$$

$$\gamma_{23} = S_{2322}\sigma_{22} \tag{3.2.4}$$

$$\gamma_{13} = S_{1322}\sigma_{22} \tag{3.2.5}$$

$$\gamma_{12} = S_{1222}\sigma_{22} \tag{3.2.6}$$

Strains under unidirectional tensile stresses (σ_{33}) are given as:

$$\varepsilon_{11} = S_{1133}\sigma_{33} \tag{3.3.1}$$

$$\varepsilon_{22} = S_{2233}\sigma_{33} \tag{3.3.2}$$

$$\varepsilon_{33} = S_{3333}\sigma_{33} \tag{3.3.3}$$

$$\gamma_{23} = S_{2333}\sigma_{33} \tag{3.3.4}$$

$$\gamma_{13} = S_{1333}\sigma_{33} \tag{3.3.5}$$

$$\gamma_{12} = S_{1233}\sigma_{33} \tag{3.3.6}.$$

Shear stresses (τ_{23}, τ_{13}, and τ_{12}) are also applied on planes (X_2–X_3), (X_1–X_3), and (X_1–X_2), respectively. The deformation resultants under shear are written in the same form as the deformation resultants under applied normal stresses. Thus, the total normal strain (ε_{11}) in direction (X_1) is obtained from the additional resultants of normal strain (ε_{11}) from all six applied stress conditions. Moreover, strains in other directions are determined in the same manner. Following normal and shear strains (Eqs. 3.4.1–3.4.6) in the orthogonal coordinate system of the generally anisotropic materials can be written as:

$$\varepsilon_{11} = S_{1111}\sigma_{11} + S_{1122}\sigma_{22} + S_{1133}\sigma_{33} + S_{1123}\tau_{23} + S_{1113}\tau_{13} + S_{1112}\tau_{12} \tag{3.4.1}$$

$$\varepsilon_{22} = S_{2211}\sigma_{11} + S_{2222}\sigma_{22} + S_{2233}\sigma_{33} + S_{2223}\tau_{23} + S_{2213}\tau_{13} + S_{2212}\tau_{12} \tag{3.4.2}$$

$$\varepsilon_{33} = S_{3311}\sigma_{11} + S_{3322}\sigma_{22} + S_{3333}\sigma_{33} + S_{3323}\tau_{23} + S_{3313}\tau_{13} + S_{3312}\tau_{12} \tag{3.4.3}$$

$$\gamma_{23} = S_{2311}\sigma_{11} + S_{2322}\sigma_{22} + S_{2333}\sigma_{33} + S_{2323}\tau_{23} + S_{2313}\tau_{13} + S_{2312}\tau_{12} \tag{3.4.4}$$

$$\gamma_{13} = S_{1311}\sigma_{11} + S_{1322}\sigma_{22} + S_{1333}\sigma_{33} + S_{1323}\tau_{23} + S_{1313}\tau_{13} + S_{1312}\tau_{12} \tag{3.4.5}$$

$$\gamma_{12} = S_{1211}\sigma_{11} + S_{1222}\sigma_{22} + S_{1233}\sigma_{33} + S_{1223}\tau_{23} + S_{1213}\tau_{13} + S_{1212}\tau_{12} \tag{3.4.6}$$

where
S_{iikl} = elastic constants or strain responses (ε_{ii} and γ_{ij}) due to applied unit normal stress (σ_{ii})
S_{ijkl} = elastic constants or strain responses (ε_{ii} and γ_{ij}) due to applied unit shear stress (τ_{ij}).

Note: the first two indices of S_{ijkl} represent the strain response plane of an infinitesimal element, whereas the last two indices of S_{ijkl} represent the applied stress plane on an infinitesimal element. In the three-dimensional elasticity approach, S_{ijkl} is the fourth-order tensor resulting in ($3^4 = 81$) elements. However, both stresses and strains are symmetric, i.e., normal stress ($\sigma_{ii} = \sigma_{ii}$ or $\sigma_{jj} = \sigma_{jj}$), normal strain ($\varepsilon_{ii} = \varepsilon_{ii}$ or $\varepsilon_{jj} = \varepsilon_{jj}$), shear stress ($\tau_{ij} = \tau_{ji}$), and shear strain ($\gamma_{ij} = \gamma_{ji}$).

Therefore, the above symmetrical results can reduce the number of elements of S_{ijkl} as follows: Since $S_{ijkl} = S_{jikl}$ and $S_{ijkl} = S_{ijlk}$, only six independent numbers of the first index pair (ij) are needed and are as follows: (11 = 11, 22 = 22, 33 = 33, 23 = 32, 13 = 31, and 12 = 21). Similarly, the independent number of the last index pair (kl) is identical to the first index pair (ij). Hence, it can be concluded that only the 36 (6×6) of 81 elements of S_{ijkl} are independent.

The number of elements of tensor S_{ijkl} is identical to those from the mechanics of materials approach in Eqs. (3.4.1–3.4.6). The generally anisotropic stress–strain relation in Eqs. (3.4.1–3.4.6) can be written in the contracted notation as shown in Figure 3.2b.

The compliance matrix in terms of contracted notation is presented in Eq. (3.5). However, elements of stresses and strains are still presented in tensor notation (double index) since they conveniently represent specific directions and planes.

3.2.1 ANISOTROPIC STRESS–STRAIN RELATIONSHIP

From Eqs. (3.4.1–3.4.6), strain–stress relation of a generally anisotropic element, in the contracted notation and matrix form, is:

$$
\begin{bmatrix} \varepsilon_{11} \\ \varepsilon_{22} \\ \varepsilon_{33} \\ \gamma_{23} \\ \gamma_{13} \\ \gamma_{12} \end{bmatrix} = \begin{bmatrix} S_{11} & S_{12} & S_{13} & S_{14} & S_{15} & S_{16} \\ S_{21} & S_{22} & S_{23} & S_{24} & S_{25} & S_{26} \\ S_{31} & S_{32} & S_{33} & S_{34} & S_{35} & S_{36} \\ S_{41} & S_{42} & S_{43} & S_{44} & S_{45} & S_{46} \\ S_{51} & S_{52} & S_{53} & S_{54} & S_{55} & S_{56} \\ S_{61} & S_{62} & S_{63} & S_{64} & S_{65} & S_{66} \end{bmatrix} \begin{bmatrix} \sigma_{11} \\ \sigma_{22} \\ \sigma_{33} \\ \tau_{23} \\ \tau_{13} \\ \tau_{12} \end{bmatrix}
\tag{3.5}
$$

where S_{ij} = elements of compliance under the orthogonal coordinate system. Through matrix inversion of Eq. (3.5), compliance matrix S_{ij} can be presented in terms of elastic constants (stiffness) as follows: The compliance matrix $[S_{ij}]$ in Eq. (3.5) is symmetric for the linear elastic system. Materials that are symmetric in the compliance matrix $[S_{ij}]$ are known as hyperelastic materials. The hyperelastic materials behave elastically under very large strain. They can be both nonlinear and correspond to large deformation behavior such as rubbery response in polymeric materials, where maximum rupture strains are around 6%–8%, for select polymer types.

The compliance matrix element can be reduced to 21 elements because of symmetry, i.e., $[S_{ij}] = [S_{ji}]$. Inversion of compliance matrix $[S_{ij}]$ is generally denoted as stiffness matrix $[C_{ij}]$. If the compliance matrix $[S_{ij}]$ is not symmetric, then the inversion of the compliance matrix $[S_{ij}]$ has no physical meaning.

Inversion of generally anisotropic strain–stress relations (3.5) is commonly written either in the form of:

$$
\begin{bmatrix} \sigma_{11} \\ \sigma_{22} \\ \sigma_{33} \\ \tau_{23} \\ \tau_{13} \\ \tau_{12} \end{bmatrix} = \begin{bmatrix} S_{11} & S_{12} & S_{13} & S_{14} & S_{15} & S_{16} \\ S_{21} & S_{22} & S_{23} & S_{24} & S_{25} & S_{26} \\ S_{31} & S_{32} & S_{33} & S_{34} & S_{35} & S_{36} \\ S_{41} & S_{42} & S_{43} & S_{44} & S_{45} & S_{46} \\ S_{51} & S_{52} & S_{53} & S_{54} & S_{55} & S_{56} \\ S_{61} & S_{62} & S_{63} & S_{64} & S_{65} & S_{66} \end{bmatrix}^{-1} \begin{bmatrix} \varepsilon_{11} \\ \varepsilon_{22} \\ \varepsilon_{33} \\ \gamma_{23} \\ \gamma_{13} \\ \gamma_{12} \end{bmatrix}
\tag{3.6.1}
$$

or,

$$
\begin{bmatrix} \sigma_{11} \\ \sigma_{22} \\ \sigma_{33} \\ \tau_{23} \\ \tau_{13} \\ \tau_{12} \end{bmatrix} = \begin{bmatrix} C_{11} & C_{12} & C_{13} & C_{14} & C_{15} & C_{16} \\ C_{21} & C_{22} & C_{23} & C_{24} & C_{25} & C_{26} \\ C_{31} & C_{32} & C_{33} & C_{34} & C_{35} & C_{36} \\ C_{41} & C_{42} & C_{43} & C_{44} & C_{45} & C_{46} \\ C_{51} & C_{52} & C_{53} & C_{54} & C_{55} & C_{56} \\ C_{61} & C_{62} & C_{63} & C_{64} & C_{65} & C_{66} \end{bmatrix} \begin{bmatrix} \varepsilon_{11} \\ \varepsilon_{22} \\ \varepsilon_{33} \\ \gamma_{23} \\ \gamma_{13} \\ \gamma_{12} \end{bmatrix}
\tag{3.6.2}
$$

where C_{ij} are the elements of the stiffness matrix under the orthogonal coordinate system.

If materials are hyperelastic, then the independent elements of compliance and stiffness matrices are reduced from 36 elements to 21 elements because of the symmetry of the matrices given in Eqs. (3.6.1 and 3.6.2).

3.2.2 MONOCLINIC STRESS–STRAIN RELATIONSHIP

For the FRP composite materials, typically fibers are aligned perpendicular to a symmetric plane or parallel to a symmetric plane. In addition, they can be both parallel and perpendicular to a symmetric plane. These different configurations are shown in Figure 3.4, where the symmetric plane is identified with dotted lines.

The composite materials are classified as monoclinic (e.g., feldspar rock), wherein the stiffness and compliance can be obtained by simplifying the compliance of generally anisotropic materials in Eqs. (3.4.1–3.4.6). In both stiffness and compliance matrices, elements corresponding to the normal shear effect of an unsymmetrical plane are taken to be zero. In addition, elements corresponding to shear effects between the symmetrical and unsymmetrical planes are also taken to be zero. For example, fibers aligned perpendicular to a symmetrical plane (X_1–X_2) are shown in Figure 3.5.

Elements corresponding to the normal shear effect of an unsymmetrical plane (X_2–X_3) are S_{1123}, S_{2223}, S_{3323}, S_{2311}, S_{2322}, and S_{2333} (contracted notation fourth column = tensor notation 23)

Elements corresponding to normal shear effect of unsymmetrical plane (X_3–X_1) are S_{1113}, S_{2213}, S_{3313}, S_{1311}, S_{1322}, and S_{1333} (contracted notation fifth column = tensor notation 13)

Elements corresponding to the shear effect between symmetrical and unsymmetrical planes are S_{2312}, S_{1312}, S_{12234}, and S_{1213} (contracted notation sixth column = tensor notation 12).

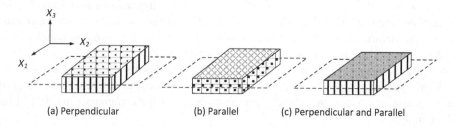

(a) Perpendicular   (b) Parallel   (c) Perpendicular and Parallel

FIGURE 3.4 Monoclinic FRP materials with a symmetrical plane (X_1–X_2): (a) perpendicular, (b) parallel, and (c) perpendicular and parallel.

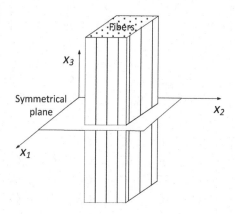

FIGURE 3.5 Fibers aligned perpendicular to symmetrical plane (X_1–X_2).

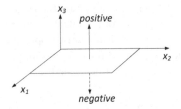

FIGURE 3.6 Normal directions of a symmetrical plane (X_1–X_2).

All the above shear compliance elements are taken as zero, while others are nonzero in the compliance matrix.

It should be noted that normal directions of the symmetric plane (X_1–X_2) are positive and negative directions with reference to the X_3 axis as shown in Figure 3.6. Because of symmetry about the x_1–x_2 plane, any shear compliance element corresponding to shear stress in planes 2–3 and 1–3 must be zero. Derivation of shear compliance corresponding to shear stress in planes 2–3 and 1–3 being zero can be found on pages 90–91 of Kaw's book (Kaw, 2006), which is based on shear strains in the "negative" direction are "negative" compared to the "positive" direction in planes 2–3 and 3–1, and not so in planes 1–2.

$[S_{ij}] = [S_{ji}]$ and $[C_{ij}] = [C_{ji}]$ for linear elastic materials. Monoclinic strain–stress relations are given as:

$$
\begin{bmatrix}
\varepsilon_{11} \\
\varepsilon_{22} \\
\varepsilon_{33} \\
\gamma_{23} \\
\gamma_{13} \\
\gamma_{12}
\end{bmatrix}
=
\begin{bmatrix}
S_{11} & S_{12} & S_{13} & 0 & 0 & S_{16} \\
S_{21} & S_{22} & S_{23} & 0 & 0 & S_{26} \\
S_{31} & S_{32} & S_{33} & 0 & 0 & S_{36} \\
0 & 0 & 0 & S_{44} & S_{45} & 0 \\
0 & 0 & 0 & S_{54} & S_{55} & 0 \\
S_{61} & S_{62} & S_{63} & 0 & 0 & S_{66}
\end{bmatrix}
\begin{bmatrix}
\sigma_{11} \\
\sigma_{22} \\
\sigma_{33} \\
\tau_{23} \\
\tau_{13} \\
\tau_{12}
\end{bmatrix}
\tag{3.7.1}
$$

Similarly, stress versus strain relations are:

$$
\begin{bmatrix}
\sigma_{11} \\
\sigma_{22} \\
\sigma_{33} \\
\tau_{23} \\
\tau_{13} \\
\tau_{12}
\end{bmatrix}
=
\begin{bmatrix}
S_{11} & S_{12} & S_{13} & 0 & 0 & S_{16} \\
S_{21} & S_{22} & S_{23} & 0 & 0 & S_{26} \\
S_{31} & S_{32} & S_{33} & 0 & 0 & S_{36} \\
0 & 0 & 0 & S_{44} & S_{45} & 0 \\
0 & 0 & 0 & S_{54} & S_{55} & 0 \\
S_{61} & S_{62} & S_{63} & 0 & 0 & S_{66}
\end{bmatrix}^{-1}
\begin{bmatrix}
\varepsilon_{11} \\
\varepsilon_{22} \\
\varepsilon_{33} \\
\gamma_{23} \\
\gamma_{13} \\
\gamma_{12}
\end{bmatrix}
\tag{3.7.2}
$$

or

$$
\begin{bmatrix}
\sigma_{11} \\
\sigma_{22} \\
\sigma_{33} \\
\tau_{23} \\
\tau_{13} \\
\tau_{12}
\end{bmatrix}
=
\begin{bmatrix}
C_{11} & C_{12} & C_{13} & 0 & 0 & C_{16} \\
C_{21} & C_{22} & C_{23} & 0 & 0 & C_{26} \\
C_{31} & C_{32} & C_{33} & 0 & 0 & C_{36} \\
0 & 0 & 0 & C_{44} & C_{45} & 0 \\
0 & 0 & 0 & C_{54} & C_{55} & 0 \\
C_{61} & C_{62} & C_{63} & 0 & 0 & C_{66}
\end{bmatrix}
\begin{bmatrix}
\varepsilon_{11} \\
\varepsilon_{22} \\
\varepsilon_{33} \\
\gamma_{23} \\
\gamma_{13} \\
\gamma_{12}
\end{bmatrix}
\tag{3.7.3}
$$

where C_{ij} are the elements of the stiffness matrix under the orthogonal coordinate system. In addition, the 21 constants of hyperelastic anisotropic materials are reduced to 13 constants for hyperelastic monoclinic materials. These hyperelastic anisotropic materials have stress versus strain relations derived from a strain energy density function, commonly found in textbooks on the theory of elasticity, where linear elastic models do not describe the actual material behavior.

3.2.3 Orthotropic Stress–Strain Relationship

For FRP composite materials, fibers are oriented in three mutually perpendicular symmetrical planes $(X_1–X_2)$, $(X_2–X_3)$, and $(X_3–X_1)$ as shown in Figure 3.7. In these materials, if a unidirectional load is applied along an orthogonal axis, then only normal strains exist. In addition, if shear force is applied on an orthogonal plane, then the response of shear strain corresponding to only shear force exists. However, when the unidirectional load is not applied along an orthogonal direction or shear load is not applied on an orthogonal plane, both normal and shear strains exist simultaneously.

To formulate the orthogonal compliance, the compliance matrix of generally anisotropic materials is simplified by considering the symmetrical plane of the materials. The normal directions of three mutually perpendicular symmetry planes $(X_1–X_2)$, $(X_2–X_3)$, and $(X_3–X_1)$ in positive and negative directions are X_3, X_1, and X_2, respectively. These normal directions have both positive and negative directions.

As previously stated for the monoclinic case, any compliance element of Eqs. (3.4.1–3.4.6) that has shear compliance elements corresponding to symmetric planes $(X_1–X_2)$ is zero, and shear compliances corresponding to symmetric planes of orthotropic materials $(X_1–X_2, X_2–X_3,$ and $X_3–X_1)$ are also zero, i.e., $S_{1123}, S_{1113}, S_{2223}, S_{2213}, S_{3323}, S_{3313}, S_{2311}, S_{1311}, S_{2322}, S_{1322}, S_{1333}, S_{1312}, S_{1223},$ $S_{2312}, S_{1312}, S_{1213}, S_{2313}, S_{1323}, S_{1112}, S_{2212}, S_{3312}, S_{2312}, S_{1211}, S_{1222}, S_{1233},$ and S_{1213}. (In contracted notation: $S_{14}, S_{15}, S_{24}, S_{25}, S_{34}, S_{35}, S_{41}, S_{42}, S_{43}, S_{51}, S_{52}, S_{53}, S_{56}, S_{64}, S_{46}, S_{45}, S_{54}, S_{16}, S_{26}, S_{36}, S_{61}, S_{62}, S_{63},$ and S_{65} are zero.) The compliance matrix of an orthotropic material is presented in Eq. (3.8.1). The stress and strain for orthogonal materials is generally written in the double index notation as given in Eq. (3.8.1).

$$\begin{bmatrix} \varepsilon_{11} \\ \varepsilon_{22} \\ \varepsilon_{33} \\ \gamma_{23} \\ \gamma_{13} \\ \gamma_{12} \end{bmatrix} = \begin{bmatrix} S_{11} & S_{12} & S_{13} & 0 & 0 & 0 \\ S_{21} & S_{22} & S_{23} & 0 & 0 & 0 \\ S_{31} & S_{32} & S_{33} & 0 & 0 & 0 \\ 0 & 0 & 0 & S_{44} & 0 & 0 \\ 0 & 0 & 0 & 0 & S_{55} & 0 \\ 0 & 0 & 0 & 0 & 0 & S_{66} \end{bmatrix} \begin{bmatrix} \sigma_{11} \\ \sigma_{22} \\ \sigma_{33} \\ \tau_{23} \\ \tau_{13} \\ \tau_{12} \end{bmatrix}$$

$$= \begin{bmatrix} \dfrac{1}{E_{11}} & -\dfrac{v_{21}}{E_{22}} & -\dfrac{v_{31}}{E_{33}} & 0 & 0 & 0 \\ -\dfrac{v_{12}}{E_{11}} & \dfrac{1}{E_{22}} & -\dfrac{v_{32}}{E_{33}} & 0 & 0 & 0 \\ -\dfrac{v_{13}}{E_{11}} & -\dfrac{v_{23}}{E_{22}} & \dfrac{1}{E_{33}} & 0 & 0 & 0 \\ 0 & 0 & 0 & \dfrac{1}{G_{23}} & 0 & 0 \\ 0 & 0 & 0 & 0 & \dfrac{1}{G_{31}} & 0 \\ 0 & 0 & 0 & 0 & 0 & \dfrac{1}{G_{12}} \end{bmatrix} \begin{bmatrix} \sigma_{11} \\ \sigma_{22} \\ \sigma_{33} \\ \tau_{23} \\ \tau_{13} \\ \tau_{12} \end{bmatrix} \qquad (3.8.1)$$

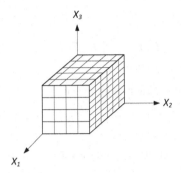

FIGURE 3.7 Orthotropic FRP materials with three mutually symmetrical planes.

where elastic modulus (E_{ii}) is defined as the normal stress (σ_{ii}) divided by the normal strain (ε_{ii}) while other stresses are taken to be zero.

Shear modulus (G_{ij}) is defined as the shear stress (τ_{ij}) divided by the shear strain (γ_{ij}) while other stresses are taken to be zero.

Poisson's ratio (ν_{ij}) is the negative result of the transverse normal strain (ε_{ij}) divided by the axial strain (ε_{ii}) where both strains are induced by normal stress (σ_{ii}) with zero stress in other directions.

Inversion of Eq. (3.8.1) is given for linear elastic orthotropic stress–strain relation as:

$$
\begin{bmatrix}
\sigma_{11} \\
\sigma_{22} \\
\sigma_{33} \\
\tau_{23} \\
\tau_{13} \\
\tau_{12}
\end{bmatrix}
=
\begin{bmatrix}
C_{11} & C_{12} & C_{13} & 0 & 0 & 0 \\
C_{21} & C_{22} & C_{23} & 0 & 0 & 0 \\
C_{31} & C_{32} & C_{33} & 0 & 0 & 0 \\
0 & 0 & 0 & C_{44} & 0 & 0 \\
0 & 0 & 0 & 0 & C_{55} & 0 \\
0 & 0 & 0 & 0 & 0 & C_{66}
\end{bmatrix}
\begin{bmatrix}
\varepsilon_{11} \\
\varepsilon_{22} \\
\varepsilon_{33} \\
\gamma_{23} \\
\gamma_{13} \\
\gamma_{12}
\end{bmatrix}
\tag{3.8.2}
$$

$$
\begin{bmatrix}
\sigma_{11} \\
\sigma_{22} \\
\sigma_{33} \\
\tau_{23} \\
\tau_{13} \\
\tau_{12}
\end{bmatrix}
= \frac{1}{D}
\begin{bmatrix}
E_{11}\left(1-\dfrac{E_{33}}{E_{22}}v_{23}^2\right) & E_{22}\left(v_{12}+\dfrac{E_{33}}{E_{22}}v_{13}v_{23}\right) & E_{33}\left(v_{13}+v_{12}v_{23}\right) & 0 & 0 & 0 \\
E_{22}\left(v_{12}+\dfrac{E_{33}}{E_{22}}v_{13}v_{23}\right) & E_{22}\left(1-\dfrac{E_{33}}{E_{11}}v_{13}^2\right) & E_{33}\left(v_{23}+\dfrac{E_{22}}{E_{11}}v_{12}v_{13}\right) & 0 & 0 & 0 \\
E_{33}\left(v_{13}+v_{12}v_{23}\right) & E_{33}\left(v_{23}+\dfrac{E_{22}}{E_{11}}v_{12}v_{13}\right) & E_{33}\left(1-\dfrac{E_{22}}{E_{11}}v_{12}^2\right) & 0 & 0 & 0 \\
0 & 0 & 0 & G_{23} & 0 & 0 \\
0 & 0 & 0 & 0 & G_{13} & 0 \\
0 & 0 & 0 & 0 & 0 & G_{12}
\end{bmatrix}
$$

$$
\times
\begin{bmatrix}
\varepsilon_{11} \\
\varepsilon_{22} \\
\varepsilon_{33} \\
\gamma_{23} \\
\gamma_{13} \\
\gamma_{12}
\end{bmatrix}
\tag{3.8.3}
$$

where

$$D = \frac{E_{11}E_{22}E_{33} - v_{23}^2 E_{22}E_{33}^2 - v_{12}^2 E_{33}E_{22}^2 - v_{13}^2 E_{22}E_{33}^2 - 2v_{12}v_{13}v_{23}E_{22}E_{33}^2}{E_{11}E_{22}E_{33}}$$

Note: The coefficients in the constituent relations are symmetric, hence: $\dfrac{v_{12}}{E_{11}} = \dfrac{v_{21}}{E_{22}}; \dfrac{v_{13}}{E_{11}} = \dfrac{v_{31}}{E_{33}};$

$\dfrac{v_{32}}{E_{33}} = \dfrac{v_{23}}{E_{22}}$

The 13 elastic monoclinic material constants are reduced to nine constants for hyperelastic-orthotropic materials.

3.2.4 Transversely Isotropic Stress–Strain Relationship

If the materials have three mutually perpendicular symmetrical planes, one of those symmetrical planes is taken to be an isotropic plane as shown in Figure 3.8. Any properties of that isotropic plane are independent of the load direction. These materials are classified as transversely isotropic materials. An example of transversely isotropic material is the unidirectional fiber-reinforced composite. Fiber reinforcements are aligned in one of the orthogonal directions while the plane perpendicular to the fiber direction is considered to be isotropic; hence, properties of the matrix are also assumed to be isotropic.

For example, fibers aligned in the orthogonal direction to plane $(X_2–X_3)$ and in the direction of (X_1) is taken to be isotropic, then the elastic moduli and Poisson's ratio of the perpendicular plane $(X_2–X_3)$ are considered as $E_{22} = E_{33}$, $G_{13} = G_{12}$, $v_{13} = v_{12}$. In addition, the shear modulus of the plane perpendicular to $(X_2–X_3)$ is given as:

$$G_{23} = \frac{E_{22}}{2(1+v_{23})} \tag{3.9.1}$$

Thus, elements of compliance in orthotropic materials (3.8.1) are simplified as $S_{12} = S_{13}$ and $S_{44} = 2(S_{22}–S_{33})$. The compliance matrix of the transversely isotropic material is presented as:

$$
\begin{bmatrix} \varepsilon_{11} \\ \varepsilon_{22} \\ \varepsilon_{33} \\ \gamma_{23} \\ \gamma_{13} \\ \gamma_{12} \end{bmatrix} =
\begin{bmatrix}
S_{11} & S_{12} & S_{12} & 0 & 0 & 0 \\
S_{12} & S_{22} & S_{23} & 0 & 0 & 0 \\
S_{12} & S_{23} & S_{22} & 0 & 0 & 0 \\
0 & 0 & 0 & 2(S_{22}-S_{33}) & 0 & 0 \\
0 & 0 & 0 & 0 & S_{55} & 0 \\
0 & 0 & 0 & 0 & 0 & S_{66}
\end{bmatrix}
\begin{bmatrix} \sigma_{11} \\ \sigma_{22} \\ \sigma_{33} \\ \tau_{23} \\ \tau_{13} \\ \tau_{12} \end{bmatrix} \tag{3.9.2}
$$

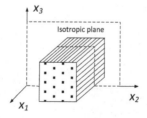

FIGURE 3.8 Transversely isotropic FRP materials with an isotropic plane $(X_2–X_3)$.

$$
\begin{bmatrix}
\varepsilon_{11} \\
\varepsilon_{22} \\
\varepsilon_{33} \\
\gamma_{23} \\
\gamma_{13} \\
\gamma_{12}
\end{bmatrix}
=
\begin{bmatrix}
\dfrac{1}{E_{11}} & -\dfrac{v_{21}}{E_{22}} & -\dfrac{v_{21}}{E_{22}} & 0 & 0 & 0 \\[2mm]
-\dfrac{v_{21}}{E_{22}} & \dfrac{1}{E_{22}} & -\dfrac{v_{32}}{E_{22}} & 0 & 0 & 0 \\[2mm]
-\dfrac{v_{21}}{E_{22}} & -\dfrac{v_{23}}{E_{22}} & \dfrac{1}{E_{22}} & 0 & 0 & 0 \\[2mm]
0 & 0 & 0 & \dfrac{2(1+v_{23})}{E_{22}} & 0 & 0 \\[2mm]
0 & 0 & 0 & 0 & \dfrac{1}{G_{13}} & 0 \\[2mm]
0 & 0 & 0 & 0 & 0 & \dfrac{1}{G_{12}}
\end{bmatrix}
\begin{bmatrix}
\sigma_{11} \\
\sigma_{22} \\
\sigma_{33} \\
\tau_{23} \\
\tau_{13} \\
\tau_{12}
\end{bmatrix}
\qquad (3.9.3)
$$

Inversion of Eq. (3.9.3) is linear elastic in terms of stress–strain of a transversely isotropic element under the orthogonal coordinate system. It is given as:

$$
\begin{bmatrix}
\sigma_{11} \\
\sigma_{22} \\
\sigma_{33} \\
\tau_{23} \\
\tau_{13} \\
\tau_{12}
\end{bmatrix}
=
\begin{bmatrix}
C_{11} & C_{12} & C_{12} & 0 & 0 & 0 \\
C_{12} & C_{22} & C_{23} & 0 & 0 & 0 \\
C_{12} & C_{23} & C_{22} & 0 & 0 & 0 \\
0 & 0 & 0 & \dfrac{C_{22}-C_{23}}{2} & 0 & 0 \\
0 & 0 & 0 & 0 & C_{66} & 0 \\
0 & 0 & 0 & 0 & 0 & C_{66}
\end{bmatrix}
\begin{bmatrix}
\varepsilon_{11} \\
\varepsilon_{22} \\
\varepsilon_{33} \\
\gamma_{23} \\
\gamma_{13} \\
\gamma_{12}
\end{bmatrix}
\qquad (3.9.4)
$$

$$
\begin{bmatrix}
\sigma_{11} \\
\sigma_{22} \\
\sigma_{33} \\
\tau_{23} \\
\tau_{13} \\
\tau_{12}
\end{bmatrix}
=
\frac{1}{D}
\begin{bmatrix}
E_{11}\left(1-v_{23}^{2}\right) & E_{22}v_{12}\left(1+v_{23}\right) & E_{33}v_{12}\left(1+v_{23}\right) & 0 & 0 & 0 \\[2mm]
E_{22}v_{12}\left(1+v_{23}\right) & E_{22}\left(1-\dfrac{E_{33}}{E_{11}}v_{13}^{2}\right) & E_{33}\left(v_{23}+\dfrac{E_{22}}{E_{11}}v_{13}^{2}\right) & 0 & 0 & 0 \\[2mm]
E_{33}v_{12}\left(1+v_{23}\right) & E_{33}\left(v_{23}+\dfrac{E_{22}}{E_{11}}v_{13}^{2}\right) & E_{33}\left(1-\dfrac{E_{22}}{E_{11}}v_{12}^{2}\right) & 0 & 0 & 0 \\[2mm]
0 & 0 & 0 & \dfrac{E_{22}}{2(1+v_{23})} & 0 & 0 \\[2mm]
0 & 0 & 0 & 0 & G_{13} & 0 \\[2mm]
0 & 0 & 0 & 0 & 0 & G_{12}
\end{bmatrix}
\begin{bmatrix}
\varepsilon_{11} \\
\varepsilon_{22} \\
\varepsilon_{33} \\
\gamma_{23} \\
\gamma_{13} \\
\gamma_{12}
\end{bmatrix}
$$

$$(3.9.5)$$

$$
\text{where,} \quad D = 1 - v_{23}^{2} - 2\left(1+v_{23}\right)\frac{E_{22}}{E_{11}}v_{12}^{2}
$$

The nine constants of elastic orthotropic materials are reduced to five constants (say: E_{11}, $E_{22} = E_{13}$, $v_{12} = v_{13}$, v_{23}, $G_{13} = G_{12}$) for transversely isotropic and elastic materials. Equation (3.9.5) is identical to Eq. (3.8.1) where $E_{22} = E_{33}$, $G_{13} = G_{12}$ and $v_{13} = v_{12}$ for transversely isotropic relationship.

3.2.5 ISOTROPIC STRESS–STRAIN RELATIONSHIP

When properties of materials are independent of any direction of applied loading, then every plane of the material is taken as symmetric and the materials are classified as isotropic materials. For example, the randomly oriented short fibers and particle reinforcements in composite materials can be classified as isotropic.

Thus, the elastic constants and Poisson's ratio of generally anisotropic materials are simplified as follows: $(E = E_{11} = E_{22} = E_{33})$, $(G = G_{13} = G_{12} = G_{23})$, and $(\nu = \nu_{13} = \nu_{12} = \nu_{23})$. In addition, the relation between the elastic and shear modulus is given as:

$$G = \frac{E}{2(1+\nu)} \tag{3.10.1}$$

Hence, the compliance matrix of the transversely isotropic material in Eq. (3.9.3) is simplified to be the compliance matrix of isotropic material as:

$$
\begin{bmatrix}
\varepsilon_{11} \\
\varepsilon_{22} \\
\varepsilon_{33} \\
\gamma_{23} \\
\gamma_{13} \\
\gamma_{12}
\end{bmatrix}
=
\begin{bmatrix}
S_{11} & S_{12} & S_{12} & 0 & 0 & 0 \\
S_{12} & S_{11} & S_{12} & 0 & 0 & 0 \\
S_{12} & S_{12} & S_{11} & 0 & 0 & 0 \\
0 & 0 & 0 & 2(S_{11}-S_{12}) & 0 & 0 \\
0 & 0 & 0 & 0 & 2(S_{11}-S_{12}) & 0 \\
0 & 0 & 0 & 0 & 0 & 2(S_{11}-S_{12})
\end{bmatrix}
\begin{bmatrix}
\sigma_{11} \\
\sigma_{22} \\
\sigma_{33} \\
\tau_{23} \\
\tau_{13} \\
\tau_{12}
\end{bmatrix}
\tag{3.10.2}
$$

$$
\begin{bmatrix}
\varepsilon_{11} \\
\varepsilon_{22} \\
\varepsilon_{33} \\
\gamma_{23} \\
\gamma_{13} \\
\gamma_{12}
\end{bmatrix}
=
\begin{bmatrix}
\frac{1}{E} & -\frac{\nu}{E} & -\frac{\nu}{E} & 0 & 0 & 0 \\
-\frac{\nu}{E} & \frac{1}{E} & -\frac{\nu}{E} & 0 & 0 & 0 \\
-\frac{\nu}{E} & -\frac{\nu}{E} & \frac{1}{E} & 0 & 0 & 0 \\
0 & 0 & 0 & \frac{2(1+\nu)}{E} & 0 & 0 \\
0 & 0 & 0 & 0 & \frac{2(1+\nu)}{E} & 0 \\
0 & 0 & 0 & 0 & 0 & \frac{2(1+\nu)}{E}
\end{bmatrix}
\begin{bmatrix}
\sigma_{11} \\
\sigma_{22} \\
\sigma_{33} \\
\tau_{23} \\
\tau_{13} \\
\tau_{12}
\end{bmatrix}
\tag{3.10.3}
$$

For isotropic materials, the compliance and stiffness matrix have only two independent elastic constants (i.e., Young's modulus (E) and Poisson's ratio (ν)). The inversions of Eqs. (3.10.2 and 3.10.3) are linear elastic stress and strain relation of isotropic materials under the orthogonal coordinate system shown in Eqs. (3.10.4 and 3.10.5).

$$
\begin{bmatrix}
\sigma_{11} \\
\sigma_{22} \\
\sigma_{33} \\
\tau_{23} \\
\tau_{13} \\
\tau_{12}
\end{bmatrix}
=
\begin{bmatrix}
C_{11} & C_{12} & C_{12} & 0 & 0 & 0 \\
C_{12} & C_{11} & C_{12} & 0 & 0 & 0 \\
C_{12} & C_{12} & C_{11} & 0 & 0 & 0 \\
0 & 0 & 0 & \frac{C_{11}-C_{12}}{2} & 0 & 0 \\
0 & 0 & 0 & 0 & \frac{C_{11}-C_{12}}{2} & 0 \\
0 & 0 & 0 & 0 & 0 & \frac{C_{11}-C_{12}}{2}
\end{bmatrix}
\begin{bmatrix}
\varepsilon_{11} \\
\varepsilon_{22} \\
\varepsilon_{33} \\
\gamma_{23} \\
\gamma_{13} \\
\gamma_{12}
\end{bmatrix}
\tag{3.10.4}
$$

$$
\begin{bmatrix} \sigma_{11} \\ \sigma_{22} \\ \sigma_{33} \\ \tau_{23} \\ \tau_{13} \\ \tau_{12} \end{bmatrix} = \frac{1}{D} \begin{bmatrix} 1-v & v & v & 0 & 0 & 0 \\ v & 1-v & v & 0 & 0 & 0 \\ v & v & 1-v & 0 & 0 & 0 \\ 0 & 0 & 0 & \frac{(1-v)}{2} & 0 & 0 \\ 0 & 0 & 0 & 0 & \frac{(1-v)}{2} & 0 \\ 0 & 0 & 0 & 0 & 0 & \frac{(1-v)}{2} \end{bmatrix} \begin{bmatrix} \varepsilon_{11} \\ \varepsilon_{22} \\ \varepsilon_{33} \\ \gamma_{23} \\ \gamma_{13} \\ \gamma_{12} \end{bmatrix}
\qquad (3.10.5)
$$

$$
\text{where, } D = \frac{(1+v)(1-2v)}{E}
$$

The compliance and stiffness matrix of the different types of materials is summarized in Table 3.1.

Example 3.1

The properties of orthotropic materials are given in the table below:

Elastic Modulus (GPa)	Shear Modulus (GPa)	Poisson's Ratio
$E_{11} = 150$	$G_{12} = 15$	$v_{12} = 0.25$
$E_{22} = 20$	$G_{13} = 10$	$v_{13} = 0.35$
$E_{33} = 10$	$G_{23} = 5$	$v_{23} = 0.45$

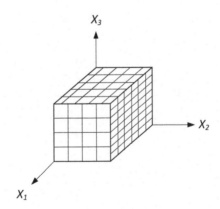

EXAMPLE FIGURE 3.1 Orthotropic materials.

a. Determine compliance and stiffness matrices of the material
b. Determine strain under state of stress: $\sigma_{11} = 250\,\text{MPa}$, $\sigma_{22} = 20\,\text{MPa}$, $\sigma_{33} = 150\,\text{MPa}$, $\tau_{23} = 25\,\text{MPa}$, $\tau_{13} = 10\,\text{MPa}$, and $\tau_{12} = 10\,\text{MPa}$
c. Determine stress under state of strain: $\varepsilon_{11} = -1{,}000\,\mu\text{m/m}$, $\varepsilon_{22} = 1{,}500\,\mu\text{m/m}$, $\varepsilon_{33} = 500\,\mu\text{m/m}$, $\gamma_{23} = -250\,\mu\text{rad}$, $\gamma_{13} = -100\,\mu\text{rad}$ and $\gamma_{12} = 1{,}000\,\mu\text{rad}$.

TABLE 3.1
Compliance and Stiffness Matrix of Materials

Materials	Compliance	Stiffness

Generally anisotropic
Eqs. (3.6.1 and 3.6.2)

Compliance:
$$\begin{bmatrix} S_{11} & S_{12} & S_{13} & S_{14} & S_{15} & S_{16} \\ S_{21} & S_{22} & S_{23} & S_{24} & S_{25} & S_{26} \\ S_{31} & S_{32} & S_{33} & S_{34} & S_{35} & S_{36} \\ S_{41} & S_{42} & S_{43} & S_{44} & S_{45} & S_{46} \\ S_{51} & S_{52} & S_{53} & S_{54} & S_{55} & S_{56} \\ S_{61} & S_{62} & S_{63} & S_{64} & S_{65} & S_{66} \end{bmatrix}$$

Stiffness:
$$\begin{bmatrix} C_{11} & C_{12} & C_{13} & C_{14} & C_{15} & C_{16} \\ C_{21} & C_{22} & C_{23} & C_{24} & C_{25} & C_{26} \\ C_{31} & C_{32} & C_{33} & C_{34} & C_{35} & C_{36} \\ C_{41} & C_{42} & C_{43} & C_{44} & C_{45} & C_{46} \\ C_{51} & C_{52} & C_{53} & C_{54} & C_{55} & C_{56} \\ C_{61} & C_{62} & C_{63} & C_{64} & C_{65} & C_{66} \end{bmatrix}$$

Monoclinic
Eqs. (3.7.1–3.7.3)

Compliance:
$$\begin{bmatrix} S_{11} & S_{12} & S_{13} & 0 & 0 & S_{16} \\ S_{21} & S_{22} & S_{23} & 0 & 0 & S_{26} \\ S_{31} & S_{32} & S_{33} & 0 & 0 & S_{36} \\ 0 & 0 & 0 & S_{44} & S_{45} & 0 \\ 0 & 0 & 0 & S_{54} & S_{55} & 0 \\ S_{61} & S_{62} & S_{63} & 0 & 0 & S_{66} \end{bmatrix}$$

Stiffness:
$$\begin{bmatrix} C_{11} & C_{12} & C_{13} & 0 & 0 & C_{16} \\ C_{21} & C_{22} & C_{23} & 0 & 0 & C_{26} \\ C_{31} & C_{32} & C_{33} & 0 & 0 & C_{36} \\ 0 & 0 & 0 & C_{44} & C_{45} & 0 \\ 0 & 0 & 0 & C_{54} & C_{55} & 0 \\ C_{61} & C_{62} & C_{63} & 0 & 0 & C_{66} \end{bmatrix}$$

Orthotropic
Eqs. (3.8.1–3.8.3)

Compliance:
$$\begin{bmatrix} S_{11} & S_{12} & S_{13} & 0 & 0 & 0 \\ S_{21} & S_{22} & S_{23} & 0 & 0 & 0 \\ S_{31} & S_{32} & S_{33} & 0 & 0 & 0 \\ 0 & 0 & 0 & S_{44} & 0 & 0 \\ 0 & 0 & 0 & 0 & S_{55} & 0 \\ 0 & 0 & 0 & 0 & 0 & S_{66} \end{bmatrix}$$

Stiffness:
$$\begin{bmatrix} C_{11} & C_{12} & C_{13} & 0 & 0 & 0 \\ C_{21} & C_{22} & C_{23} & 0 & 0 & 0 \\ C_{31} & C_{32} & C_{33} & 0 & 0 & 0 \\ 0 & 0 & 0 & C_{44} & 0 & 0 \\ 0 & 0 & 0 & 0 & C_{55} & 0 \\ 0 & 0 & 0 & 0 & 0 & C_{66} \end{bmatrix}$$

(Continued)

TABLE 3.1 (Continued)
Compliance and Stiffness Matrix of Materials

Materials	Compliance	Stiffness

Transversely Isotropic
Eqs. (3.9.1–3.9.5)

Compliance:
$$\begin{bmatrix} S_{11} & S_{12} & S_{12} & 0 & 0 & 0 \\ S_{12} & S_{22} & S_{23} & 0 & 0 & 0 \\ S_{12} & S_{23} & S_{22} & 0 & 0 & 0 \\ 0 & 0 & 0 & 2(S_{22}-S_{33}) & 0 & 0 \\ 0 & 0 & 0 & 0 & S_{55} & 0 \\ 0 & 0 & 0 & 0 & 0 & S_{66} \end{bmatrix}$$

Stiffness:
$$\begin{bmatrix} C_{11} & C_{12} & C_{12} & 0 & 0 & 0 \\ C_{12} & C_{22} & C_{23} & 0 & 0 & 0 \\ C_{12} & C_{23} & C_{22} & 0 & 0 & 0 \\ 0 & 0 & 0 & \dfrac{C_{22}-C_{23}}{2} & 0 & 0 \\ 0 & 0 & 0 & 0 & C_{66} & 0 \\ 0 & 0 & 0 & 0 & 0 & C_{66} \end{bmatrix}$$

Isotropic
Eqs. (3.10.1–3.10.5)

Compliance:
$$\begin{bmatrix} S_{11} & S_{12} & S_{12} & 0 & 0 & 0 \\ S_{12} & S_{11} & S_{12} & 0 & 0 & 0 \\ S_{12} & S_{12} & S_{11} & 0 & 0 & 0 \\ 0 & 0 & 0 & 2(S_{11}-S_{12}) & 0 & 0 \\ 0 & 0 & 0 & 0 & 2(S_{11}-S_{12}) & 0 \\ 0 & 0 & 0 & 0 & 0 & 2(S_{11}-S_{12}) \end{bmatrix}$$

Stiffness:
$$\begin{bmatrix} C_{11} & C_{12} & C_{12} & 0 & 0 & 0 \\ C_{12} & C_{11} & C_{12} & 0 & 0 & 0 \\ C_{12} & C_{12} & C_{11} & 0 & 0 & 0 \\ 0 & 0 & 0 & \dfrac{C_{11}-C_{12}}{2} & 0 & 0 \\ 0 & 0 & 0 & 0 & \dfrac{C_{11}-C_{12}}{2} & 0 \\ 0 & 0 & 0 & 0 & 0 & \dfrac{C_{11}-C_{12}}{2} \end{bmatrix}$$

Solution

a. Compliance and stiffness matrices of the material:

By substituting the orthotropic material properties into Eq. (3.8.1), the compliance and stiffness are found as:

$$
\begin{bmatrix}
S_{11} & S_{12} & S_{13} & 0 & 0 & 0 \\
S_{21} & S_{22} & S_{23} & 0 & 0 & 0 \\
S_{31} & S_{32} & S_{33} & 0 & 0 & 0 \\
0 & 0 & 0 & S_{44} & 0 & 0 \\
0 & 0 & 0 & 0 & S_{55} & 0 \\
0 & 0 & 0 & 0 & 0 & S_{66}
\end{bmatrix}
=
\begin{bmatrix}
\dfrac{1}{150} & -\dfrac{0.25}{150} & -\dfrac{0.35}{150} & 0 & 0 & 0 \\
-\dfrac{0.25}{150} & \dfrac{1}{20} & -\dfrac{0.45}{10} & 0 & 0 & 0 \\
-\dfrac{0.35}{150} & -\dfrac{0.45}{10} & \dfrac{1}{10} & 0 & 0 & 0 \\
0 & 0 & 0 & \dfrac{1}{5} & 0 & 0 \\
0 & 0 & 0 & 0 & \dfrac{1}{10} & 0 \\
0 & 0 & 0 & 0 & 0 & \dfrac{1}{15}
\end{bmatrix}
\dfrac{1}{\text{GPa}}
$$

$$
\begin{bmatrix}
C_{11} & C_{12} & C_{13} & 0 & 0 & 0 \\
C_{21} & C_{22} & C_{23} & 0 & 0 & 0 \\
C_{31} & C_{32} & C_{33} & 0 & 0 & 0 \\
0 & 0 & 0 & C_{44} & 0 & 0 \\
0 & 0 & 0 & 0 & C_{55} & 0 \\
0 & 0 & 0 & 0 & 0 & C_{66}
\end{bmatrix}
=
\begin{bmatrix}
\dfrac{1}{150} & -\dfrac{0.25}{150} & -\dfrac{0.35}{150} & 0 & 0 & 0 \\
-\dfrac{0.25}{150} & \dfrac{1}{20} & -\dfrac{0.45}{10} & 0 & 0 & 0 \\
-\dfrac{0.35}{150} & -\dfrac{0.45}{10} & \dfrac{1}{10} & 0 & 0 & 0 \\
0 & 0 & 0 & \dfrac{1}{5} & 0 & 0 \\
0 & 0 & 0 & 0 & \dfrac{1}{10} & 0 \\
0 & 0 & 0 & 0 & 0 & \dfrac{1}{15}
\end{bmatrix}^{-1}
$$

$$
=
\begin{bmatrix}
157.1 & 14.35 & 10.12 & 0 & 0 & 0 \\
14.35 & 34.9 & 16.05 & 0 & 0 & 0 \\
10.12 & 16.05 & 17.46 & 0 & 0 & 0 \\
0 & 0 & 0 & 5 & 0 & 0 \\
0 & 0 & 0 & 0 & 10 & 0 \\
0 & 0 & 0 & 0 & 0 & 15
\end{bmatrix}
\text{GPa}
$$

b. Strain under the state of stress:

From Eq. (3.8.1), the strain under the state of stress is determined as:

$$
\begin{bmatrix}
\varepsilon_{11} \\
\varepsilon_{22} \\
\varepsilon_{33} \\
\gamma_{23} \\
\gamma_{13} \\
\gamma_{12}
\end{bmatrix}
=
\begin{bmatrix}
\dfrac{1}{150} & -\dfrac{0.25}{150} & -\dfrac{0.35}{150} & 0 & 0 & 0 \\
-\dfrac{0.25}{150} & \dfrac{1}{20} & -\dfrac{0.45}{10} & 0 & 0 & 0 \\
-\dfrac{0.35}{150} & -\dfrac{0.45}{10} & \dfrac{1}{10} & 0 & 0 & 0 \\
0 & 0 & 0 & \dfrac{1}{5} & 0 & 0 \\
0 & 0 & 0 & 0 & \dfrac{1}{10} & 0 \\
0 & 0 & 0 & 0 & 0 & \dfrac{1}{15}
\end{bmatrix}
\dfrac{1}{\text{GPa}}
\begin{bmatrix}
250 \\
20 \\
15 \\
25 \\
10 \\
10
\end{bmatrix}
\text{MPa} =
\begin{bmatrix}
1{,}598 \\
-91.7 \\
16.67 \\
5{,}000 \\
1{,}000 \\
667
\end{bmatrix}
\mu
$$

Note: Unit: normal strain ε_{ii} (m/m) and shear strain (γ_{ij}) rad/rad

c. Stress under the state of strain:
From Eq. (3.8.2), the stress under the state of strain is determined as:

$$
\begin{bmatrix}
\sigma_{11} \\
\sigma_{22} \\
\sigma_{33} \\
\tau_{23} \\
\tau_{13} \\
\tau_{12}
\end{bmatrix}
=
\begin{bmatrix}
157.1 & 14.35 & 10.12 & 0 & 0 & 0 \\
14.35 & 34.9 & 16.05 & 0 & 0 & 0 \\
10.12 & 16.05 & 17.46 & 0 & 0 & 0 \\
0 & 0 & 0 & 5 & 0 & 0 \\
0 & 0 & 0 & 0 & 10 & 0 \\
0 & 0 & 0 & 0 & 0 & 15
\end{bmatrix}
\begin{bmatrix}
-1{,}000 \\
1{,}500 \\
500 \\
-250 \\
-100 \\
1{,}000
\end{bmatrix}
\mu =
\begin{bmatrix}
-130.5 \\
46.1 \\
22.7 \\
-1.25 \\
-1 \\
15
\end{bmatrix}
\text{MPa}
$$

Note: Unit: normal strain ε_{ii} (m/m) and shear strain (γ_{ij}) rad/rad.

3.3 THE PLANE STRESS RELATION

FRP laminated composite structures such as beams, plates, and other structural shapes are generally made from numerous plies or lamina of FRP composite materials. In addition, the thickness of those FRP laminated composite structures is much smaller than the other two dimensions in the plane of lamina. Any stress variation through the thickness is normally much smaller and often taken to be zero. Thus, it is possible to reduce complicated tasks of dealing with a 3-D constitutive relation using plane stress models. If the thickness in direction (X_3) of an FRP lamina is assumed to be much smaller than its length (X_1) and width (X_2) dimensions, stresses in that direction $(\sigma_{33}, \tau_{23},$ and $\tau_{13})$ are considered to be out-of-plane and also assumed to be negligible. Thus, stresses of an infinitesimal element for FRP lamina under the orthogonal coordinate system are presented in Figure 3.9.

The plane stress assumptions can lead to inaccuracies near the edges of laminates because of the absence of out-of-plane stresses in a plane stress model. If plane $(X_1–X_2)$ is defined as the in-plane (of a problem), then the generally anisotropic strain and stress relation in Eq. (3.5) under the plane stress condition can be simplified as: $\sigma_{33} = \tau_{23} = \tau_{13} = 0$

$$
\begin{bmatrix}
\varepsilon_{11} \\
\varepsilon_{22} \\
\varepsilon_{33} \\
\gamma_{23} \\
\gamma_{13} \\
\gamma_{12}
\end{bmatrix}
=
\begin{bmatrix}
S_{11} & S_{12} & S_{13} & S_{14} & S_{15} & S_{16} \\
S_{21} & S_{22} & S_{23} & S_{24} & S_{25} & S_{26} \\
S_{31} & S_{32} & S_{33} & S_{34} & S_{35} & S_{36} \\
S_{41} & S_{42} & S_{43} & S_{44} & S_{45} & S_{46} \\
S_{51} & S_{52} & S_{53} & S_{54} & S_{55} & S_{56} \\
S_{61} & S_{62} & S_{63} & S_{64} & S_{65} & S_{66}
\end{bmatrix}
\begin{bmatrix}
\sigma_{11} \\
\sigma_{22} \\
0 \\
0 \\
0 \\
\tau_{12}
\end{bmatrix}
\tag{3.11}
$$

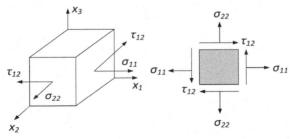

(a) 3D Infinitesimal Element b) 2D infinitesimal Element (plan view)

FIGURE 3.9 State of plane stress of infinitesimal element (positive stress): (a) 3D infinitesimal element and (b) 2D infinitesimal element (plan view).

The strain and stress relation in Eq. (3.11) can be separated into in-plane and out-of-plane values as in Eqs. (3.12.1 and 3.12.2). In general, the [3×3] compliance matrix $[S_{ij}]$ in Eq. (3.12.1) is called the reduced compliance matrix. The elements of the reduced compliance matrix are the same as the elements of the [6×6] full compliance matrix.

For in-plane response:

$$\begin{bmatrix} \varepsilon_{11} \\ \varepsilon_{22} \\ \gamma_{12} \end{bmatrix} = \begin{bmatrix} S_{11} & S_{12} & S_{16} \\ S_{21} & S_{22} & S_{26} \\ S_{61} & S_{62} & S_{66} \end{bmatrix} \begin{bmatrix} \sigma_{11} \\ \sigma_{22} \\ \tau_{12} \end{bmatrix} \qquad (3.12.1)$$

For out-of-plane response:

$$\begin{bmatrix} \varepsilon_{33} \\ \gamma_{23} \\ \gamma_{13} \end{bmatrix} = \begin{bmatrix} S_{31} & S_{32} & S_{36} \\ S_{41} & S_{42} & S_{46} \\ S_{51} & S_{52} & S_{56} \end{bmatrix} \begin{bmatrix} \sigma_{11} \\ \sigma_{22} \\ \tau_{12} \end{bmatrix} \qquad (3.12.2)$$

From Eq. (3.12.2), it is clear that the normal strain (ε_{33}) in the perpendicular direction to (X_1–X_2) plane is not equal to zero under the plane stress condition. Inversion of in-plane strain–stress relation in Eq. (3.12.1) is given as:

$$\begin{bmatrix} \sigma_{11} \\ \sigma_{22} \\ \tau_{12} \end{bmatrix} = \begin{bmatrix} Q_{11} & Q_{12} & Q_{16} \\ Q_{21} & Q_{22} & Q_{26} \\ Q_{61} & Q_{62} & Q_{66} \end{bmatrix} \begin{bmatrix} \varepsilon_{11} \\ \varepsilon_{22} \\ \gamma_{12} \end{bmatrix} = \begin{bmatrix} S_{11} & S_{12} & S_{16} \\ S_{21} & S_{22} & S_{26} \\ S_{61} & S_{62} & S_{66} \end{bmatrix}^{-1} \begin{bmatrix} \varepsilon_{11} \\ \varepsilon_{22} \\ \gamma_{12} \end{bmatrix} \qquad (3.13)$$

The [3×3] stiffness matrix $[Q_{ij}]$ in Eq. (3.13) is called the reduced stiffness matrix. It should be noted that all elements of the reduced stiffness matrix $[Q_{ij}]$ are different from the elements of the [6×6] stiffness matrix $[C_{ij}]$ in Eq. (3.6.2).

$$\begin{bmatrix} Q_{11} & Q_{12} & Q_{16} \\ Q_{21} & Q_{22} & Q_{26} \\ Q_{61} & Q_{62} & Q_{66} \end{bmatrix} \neq \begin{bmatrix} C_{11} & C_{12} & C_{16} \\ C_{21} & C_{22} & C_{26} \\ C_{61} & C_{62} & C_{66} \end{bmatrix} \qquad (3.14)$$

For monoclinic materials, the reduced compliance matrix is obtained by simplifying the [6×6] compliance matrix $[Q_{ij}]$ in Eq. (3.7.1). The reduced compliance matrix of monoclinic materials is the same as the one for generally anisotropic materials, except that the independent constants are reduced to 13. If plane (X_1–X_2) is defined as the in-plane of a problem, then the monoclinic strain–stress relation in Eq. (3.7.1) under the plane stress condition can be simplified as: $\sigma_{33} = \tau_{23} = \tau_{13} = 0$

$$\begin{bmatrix} \varepsilon_{11} \\ \varepsilon_{22} \\ \varepsilon_{33} \\ \gamma_{23} \\ \gamma_{13} \\ \gamma_{12} \end{bmatrix} = \begin{bmatrix} S_{11} & S_{12} & S_{13} & 0 & 0 & S_{16} \\ S_{21} & S_{22} & S_{23} & 0 & 0 & S_{26} \\ S_{31} & S_{32} & S_{33} & 0 & 0 & S_{36} \\ 0 & 0 & 0 & S_{44} & S_{45} & 0 \\ 0 & 0 & 0 & S_{54} & S_{55} & 0 \\ S_{61} & S_{62} & S_{63} & 0 & 0 & S_{66} \end{bmatrix} \begin{bmatrix} \sigma_{11} \\ \sigma_{22} \\ 0 \\ 0 \\ 0 \\ \tau_{12} \end{bmatrix} \qquad (3.15.1)$$

Thus, the reduced compliance matrix of monoclinic materials is given as:

$$
\begin{bmatrix} \varepsilon_{11} \\ \varepsilon_{22} \\ \gamma_{12} \end{bmatrix} =
\begin{bmatrix} S_{11} & S_{12} & S_{16} \\ S_{21} & S_{22} & S_{26} \\ S_{61} & S_{62} & S_{66} \end{bmatrix}
\begin{bmatrix} \sigma_{11} \\ \sigma_{22} \\ \tau_{12} \end{bmatrix}
\tag{3.15.2}
$$

The reduced stiffness matrix of monoclinic materials is the inversion of the monoclinic reduced compliance matrix in Eq. (3.15.2) as:

$$
\begin{bmatrix} \sigma_{11} \\ \sigma_{22} \\ \tau_{12} \end{bmatrix} =
\begin{bmatrix} Q_{11} & Q_{12} & Q_{16} \\ Q_{21} & Q_{22} & Q_{26} \\ Q_{61} & Q_{62} & Q_{66} \end{bmatrix}
\begin{bmatrix} \varepsilon_{11} \\ \varepsilon_{22} \\ \gamma_{12} \end{bmatrix} =
\begin{bmatrix} S_{11} & S_{12} & S_{16} \\ S_{21} & S_{22} & S_{26} \\ S_{61} & S_{62} & S_{66} \end{bmatrix}^{-1}
\begin{bmatrix} \varepsilon_{11} \\ \varepsilon_{22} \\ \gamma_{12} \end{bmatrix}
\tag{3.15.3}
$$

For orthotropic materials, the reduced compliance and stiffness matrices are also obtained in the same manner as the monoclinic materials. The [6×6] orthotropic compliance matrix $[Q_{ij}]$ in Eq. (3.8.1) is simplified for the plane stress condition as:

$$
\begin{bmatrix} \varepsilon_{11} \\ \varepsilon_{22} \\ \gamma_{12} \end{bmatrix} =
\begin{bmatrix} S_{11} & S_{12} & 0 \\ S_{21} & S_{22} & 0 \\ 0 & 0 & S_{66} \end{bmatrix}
\begin{bmatrix} \sigma_{11} \\ \sigma_{22} \\ \tau_{12} \end{bmatrix} =
\begin{bmatrix} \dfrac{1}{E_{11}} & -\dfrac{v_{21}}{E_{22}} & 0 \\ -\dfrac{v_{12}}{E_{11}} & \dfrac{1}{E_{22}} & 0 \\ 0 & 0 & \dfrac{1}{G_{12}} \end{bmatrix}
\begin{bmatrix} \sigma_{11} \\ \sigma_{22} \\ \tau_{12} \end{bmatrix}
\tag{3.16.1}
$$

By inversion of the [3×3] reduced compliance matrix $[S_{ij}]$ of the orthotropic materials, the [3×3] reduced stiffness matrix $[Q_{ij}]$ of orthotropic materials under plane stress condition is:

$$
\begin{bmatrix} \sigma_{11} \\ \sigma_{22} \\ \tau_{12} \end{bmatrix} =
\begin{bmatrix} \dfrac{1}{E_{11}} & -\dfrac{v_{21}}{E_{22}} & 0 \\ -\dfrac{v_{12}}{E_{11}} & \dfrac{1}{E_{22}} & 0 \\ 0 & 0 & \dfrac{1}{G_{12}} \end{bmatrix}^{-1}
\begin{bmatrix} \varepsilon_{11} \\ \varepsilon_{22} \\ \gamma_{12} \end{bmatrix} =
\begin{bmatrix} S_{11} & S_{12} & 0 \\ S_{21} & S_{22} & 0 \\ 0 & 0 & S_{66} \end{bmatrix}^{-1}
\begin{bmatrix} \varepsilon_{11} \\ \varepsilon_{22} \\ \gamma_{12} \end{bmatrix}
\tag{3.16.2}
$$

$$
\begin{bmatrix} \sigma_{11} \\ \sigma_{22} \\ \tau_{12} \end{bmatrix} =
\begin{bmatrix} Q_{11} & Q_{12} & 0 \\ Q_{21} & Q_{22} & 0 \\ 0 & 0 & Q_{66} \end{bmatrix}
\begin{bmatrix} \varepsilon_{11} \\ \varepsilon_{22} \\ \gamma_{12} \end{bmatrix} =
\begin{bmatrix} \dfrac{E_{11}}{1-v_{12}v_{21}} & \dfrac{v_{12}E_{22}}{1-v_{12}v_{21}} & 0 \\ \dfrac{v_{21}E_{11}}{1-v_{12}v_{21}} & \dfrac{E_{22}}{1-v_{12}v_{21}} & 0 \\ 0 & 0 & G_{12} \end{bmatrix}
\begin{bmatrix} \varepsilon_{11} \\ \varepsilon_{22} \\ \gamma_{12} \end{bmatrix}
$$

$$
\tag{3.16.3}
$$

For transversely isotropic materials, the [3×3] reduced compliance matrix $[S_{ij}]$ and stiffness matrix $[Q_{ij}]$ are also simplified from the orthotropic case as mentioned before. The [3×3] reduced compliance matrix $[S_{ij}]$ and stiffness matrix $[Q_{ij}]$ of transversely isotropic are the same as those of the orthotropic materials. For isotropic materials, only two elastic constants (E and ν) are independent. The $[3\times3]$ reduced compliance matrix $[S_{ij}]$ of isotropic materials in terms of elastic constants (E and ν) is given as:

$$\begin{bmatrix} \varepsilon_{11} \\ \varepsilon_{22} \\ \gamma_{12} \end{bmatrix} = \begin{bmatrix} S_{11} & S_{12} & 0 \\ S_{12} & S_{11} & 0 \\ 0 & 0 & 2(S_{11}-S_{12}) \end{bmatrix} \begin{bmatrix} \sigma_{11} \\ \sigma_{22} \\ \tau_{12} \end{bmatrix} = \begin{bmatrix} \dfrac{1}{E} & -\dfrac{\nu}{E} & 0 \\ -\dfrac{\nu}{E} & \dfrac{1}{E} & 0 \\ 0 & 0 & 2\left(\dfrac{1+\nu}{E}\right) \end{bmatrix} \begin{bmatrix} \sigma_{11} \\ \sigma_{22} \\ \tau_{12} \end{bmatrix}$$

(3.17.1)

The $[3\times3]$ reduced stiffness matrix $[Q_{ij}]$ of isotropic materials in terms of elastic constants (E and ν) is given as:

$$\begin{bmatrix} \sigma_{11} \\ \sigma_{22} \\ \tau_{12} \end{bmatrix} = \begin{bmatrix} Q_{11} & Q_{12} & 0 \\ Q_{12} & Q_{22} & 0 \\ 0 & 0 & Q_{66} \end{bmatrix} \begin{bmatrix} \varepsilon_{11} \\ \varepsilon_{22} \\ \gamma_{12} \end{bmatrix} = \begin{bmatrix} \dfrac{E}{1-\nu^2} & \dfrac{\nu E}{1-\nu^2} & 0 \\ \dfrac{\nu E}{1-\nu^2} & \dfrac{E}{1-\nu^2} & 0 \\ 0 & 0 & \dfrac{E}{2(1+\nu)} \end{bmatrix} \begin{bmatrix} \varepsilon_{11} \\ \varepsilon_{22} \\ \gamma_{12} \end{bmatrix}$$

(3.17.2)

Example 3.2

The in-plane properties of orthotropic lamina (carbon fiber) are: $E_{11} = 150\,\text{GPa}$, $E_{22} = 20\,\text{GPa}$, $G_{12} = 15\,\text{GPa}$, and $\nu_{12} = 0.25$.

a. Determine reduced compliance and stiffness matrices of orthotropic lamina in the coordinate system (X_1–X_2)
b. Determine the state of strains in the coordinate system (X_1–X_2) under the state of stresses as: $\sigma_{11} = 250\,\text{MPa}$, $\sigma_{22} = 20\,\text{MPa}$, and $\tau_{12} = 10\,\text{MPa}$
c. Determine the state of stresses in the coordinate system (X_1–X_2) under the state of strains as: $\varepsilon_{11} = -1{,}000\,\mu\text{m/m}$, $\varepsilon_{22} = 1{,}500\,\mu\text{m/m}$ and $\gamma_{12} = 1{,}000\,\mu\,\text{rad/rad}$

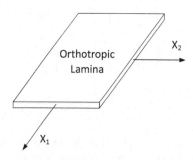

EXAMPLE FIGURE 3.2.1   Orthotropic lamina.

Solution

a. Reduced compliance and stiffness matrices of orthotropic lamina:
   The [3×3] reduced compliance $[Q_{ij}]$ of orthotropic materials can be determined from Eq. (3.16.1) as:

$$-\frac{v_{12}}{E_{11}} = -\frac{v_{21}}{E_{22}} = -\frac{0.25}{150}$$

$$
\begin{bmatrix} S_{11} & S_{12} & 0 \\ S_{12} & S_{22} & 0 \\ 0 & 0 & S_{66} \end{bmatrix} = \begin{bmatrix} \dfrac{1}{E_{11}} & -\dfrac{v_{21}}{E_{22}} & 0 \\ -\dfrac{v_{12}}{E_{11}} & \dfrac{1}{E_{22}} & 0 \\ 0 & 0 & \dfrac{1}{G_{12}} \end{bmatrix} = \begin{bmatrix} \dfrac{1}{150} & -\dfrac{0.25}{150} & 0 \\ -\dfrac{0.25}{150} & \dfrac{1}{20} & 0 \\ 0 & 0 & \dfrac{1}{15} \end{bmatrix} \dfrac{1}{\text{GPa}}
$$

As explained in Section 3.3, the elements of [3×3] reduce compliance matrix $[S_{ij}]$ are equal to those of the three-dimensional compliance matrix as shown in Example 3.1. Since the elements of [3×3] reduced compliance matrix $[S_{ij}]$ are directly taken from the elements of [6×6] full compliance matrix $[S_{ij}]$.
   The [3×3] reduced stiffness $[Q_{ij}]$ can be determined from either Eq. (3.16.2) or inversion of the [3×3] reduced compliance $[Q_{ij}]$.

$$
\begin{bmatrix} Q_{11} & Q_{12} & 0 \\ Q_{21} & Q_{22} & 0 \\ 0 & 0 & Q_{66} \end{bmatrix} = \begin{bmatrix} \dfrac{1}{150} & -\dfrac{0.25}{150} & 0 \\ -\dfrac{0.25}{150} & \dfrac{1}{20} & 0 \\ 0 & 0 & \dfrac{1}{15} \end{bmatrix}^{-1} = \begin{bmatrix} 151.3 & 5.04 & 0 \\ 5.04 & 20.2 & 0 \\ 0 & 0 & 15 \end{bmatrix} \text{GPa}
$$

Elements of [3×3] reduced stiffness matrix $[Q_{ij}]$ are not equal to those of [6×6] 3-D full stiffness matrix $[C_{ij}]$ of order 6 as given in Eq. (3.6.2). The inversion of [3×3] reduced compliance matrix $[S_{ij}]$ is not equal to the inversion of [6×6] full compliance matrix $[S_{ij}]$ as below:

$$
\begin{bmatrix} Q_{11} & Q_{12} & 0 \\ Q_{21} & Q_{22} & 0 \\ 0 & 0 & Q_{66} \end{bmatrix} \neq \begin{bmatrix} C_{11} & C_{12} & 0 \\ C_{21} & C_{22} & 0 \\ 0 & 0 & C_{66} \end{bmatrix} \quad \text{or} \quad \begin{bmatrix} 151.3 & 5.04 & 0 \\ 5.04 & 20.2 & 0 \\ 0 & 0 & 15 \end{bmatrix} \text{GPa}
$$

$$
\neq \begin{bmatrix} 157.1 & 14.35 & 0 \\ 14.35 & 34.9 & 0 \\ 0 & 0 & 15 \end{bmatrix} \text{GPa}
$$

b. The state of strains in the coordinate system $(X_1–X_2)$ under the state of stresses:
   From Eq. (3.16.1):

$$
\begin{bmatrix} \varepsilon_{11} \\ \varepsilon_{22} \\ \gamma_{12} \end{bmatrix} = \begin{bmatrix} \dfrac{1}{150} & -\dfrac{0.25}{150} & 0 \\ -\dfrac{0.25}{150} & \dfrac{1}{20} & 0 \\ 0 & 0 & \dfrac{1}{15} \end{bmatrix} \dfrac{1}{\text{GPa}} \begin{bmatrix} 250 \\ 20 \\ 10 \end{bmatrix} \text{MPa} = \begin{bmatrix} 1,633 \\ 583 \\ 667 \end{bmatrix} \mu
$$

Note: Unit: normal strain (ε_{ii}) m/m and shear strain (γ_{ij}) rad/rad

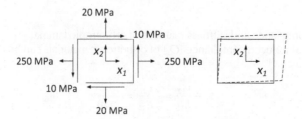

EXAMPLE FIGURE 3.2.2   State of stress and strain for Example 3.2b.

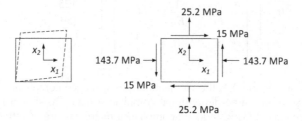

EXAMPLE FIGURE 3.2.3   State of stress and strain for Example 3.2c.

    c. The state of stress in the coordinate system $(X_1–X_2)$ under the state of strains:
From Eq. (3.16.3):

$$
\begin{bmatrix} \sigma_{11} \\ \sigma_{22} \\ \tau_{12} \end{bmatrix} = \begin{bmatrix} Q_{11} & Q_{12} & 0 \\ Q_{21} & Q_{22} & 0 \\ 0 & 0 & Q_{66} \end{bmatrix} \begin{bmatrix} \varepsilon_{11} \\ \varepsilon_{22} \\ \gamma_{12} \end{bmatrix}
$$

$$
= \begin{bmatrix} 151.3 & 5.04 & 0 \\ 5.04 & 20.2 & 0 \\ 0 & 0 & 15 \end{bmatrix} \text{GPa} \begin{bmatrix} -1{,}000 \\ 1{,}500 \\ 1{,}000 \end{bmatrix} \mu = \begin{bmatrix} -143.7 \\ 25.2 \\ 15 \end{bmatrix} \text{MPa}
$$

Example 3.3

The in-plane properties of orthotropic lamina (carbon fiber) are $E_{11} = 150\,\text{GPa}$, $E_{22} = 20\,\text{GPa}$, $G_{12} = 15\,\text{GPa}$, and $\nu_{12} = 0.25$ (use the reduced stiffness matrix from Example 3.2):

    a. Determine the state of strains in the coordinate system $(X_1–X_2)$ under the applied stress σ_{11} of 250 MPa and fixed boundary condition as shown in Example Figure 3.3.1.
    b. Determine the state of strains in the coordinate system $(X_1–X_2)$ under applied normal stress σ_{22} of 20 MPa and fixed boundary condition as shown in Example Figure 3.3.2.

Fixed

250 MPa ← ▢ → 250 MPa

Fixed

EXAMPLE FIGURE 3.3.1   Applied stress σ_{11} and boundary conditions.

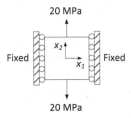

EXAMPLE FIGURE 3.3.2 Applied stress σ_{22} and boundary conditions.

Solution

a. The state of strains under the applied stress (σ_{11}):
From Eq. (3.16.1):

$$\begin{bmatrix} \varepsilon_{11} \\ 0 \\ \gamma_{12} \end{bmatrix} = \begin{bmatrix} \dfrac{1}{E_{11}} & -\dfrac{v_{21}}{E_{22}} & 0 \\ -\dfrac{v_{12}}{E_{11}} & \dfrac{1}{E_{22}} & 0 \\ 0 & 0 & \dfrac{1}{G_{12}} \end{bmatrix} \begin{bmatrix} \sigma_{11} \\ \sigma_{22} \\ \tau_{12} \end{bmatrix} = \begin{bmatrix} \dfrac{1}{150} & -\dfrac{0.25}{150} & 0 \\ -\dfrac{0.25}{150} & \dfrac{1}{20} & 0 \\ 0 & 0 & \dfrac{1}{15} \end{bmatrix} \dfrac{1}{GPa} \begin{bmatrix} 250 \\ \sigma_{22} \\ 0 \end{bmatrix} MPa$$

The second equation of the above simultaneous equations yields (σ_{22}):

$$-\frac{0.25 \times 250}{150}\frac{MPa}{GPa} + \frac{\sigma_{22}}{20}\frac{1}{GPa} = 0$$

$$\sigma_{22} = 8.33\,MPa$$

From the first and third equations of the simultaneous equations, normal strain (ε_{11}) and shear strain (γ_{12}) are:

$$\varepsilon_{11} = \frac{250\,MPa}{150\,GPa} - \frac{8.33 \times 0.25\,MPa}{150\,GPa} = 1{,}653\,\mu m/m$$

$$\gamma_{12} = 0\,\mu\,rad$$

When the applied normal stress is aligned along the orthotropic axis, then the shear strain is not induced. However, if the applied normal stress is aligned off-axis of the orthotropic coordinate system, then normal and shear strains are induced.

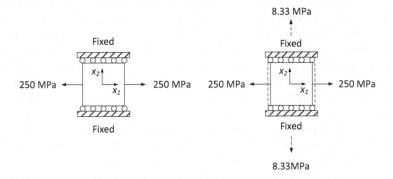

EXAMPLE FIGURE 3.3.3 State of stresses and strain under applied stress σ_{22} and boundary conditions.

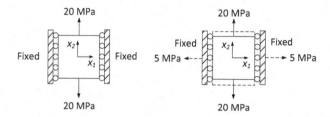

EXAMPLE FIGURE 3.3.4　State of stresses and strain under applied stress σ_{11} and boundary conditions.

b. The state of strains under the applied stress (σ_{22}):
From Eq. (3.16.1):

$$
\begin{bmatrix} 0 \\ \varepsilon_{22} \\ \gamma_{12} \end{bmatrix} =
\begin{bmatrix}
\dfrac{1}{E_{11}} & -\dfrac{v_{21}}{E_{22}} & 0 \\[2mm]
-\dfrac{v_{12}}{E_{11}} & \dfrac{1}{E_{22}} & 0 \\[2mm]
0 & 0 & \dfrac{1}{G_{12}}
\end{bmatrix}
\begin{bmatrix} \sigma_{11} \\ \sigma_{22} \\ \tau_{12} \end{bmatrix} =
\begin{bmatrix}
\dfrac{1}{150} & -\dfrac{0.25}{150} & 0 \\[2mm]
-\dfrac{0.25}{150} & \dfrac{1}{20} & 0 \\[2mm]
0 & 0 & \dfrac{1}{15}
\end{bmatrix}
\dfrac{1}{\text{GPa}}
\begin{bmatrix} \sigma_{11} \\ 20 \\ 0 \end{bmatrix} \text{MPa}
$$

From the first equation of the above simultaneous equations yield (σ_{11}) as:

$$
\frac{\sigma_{11}}{150}\frac{1}{\text{GPa}} - \frac{20 \times 0.25}{150}\frac{\text{MPa}}{\text{GPa}} = 0
$$

$$
\sigma_{11} = 5\,\text{MPa}
$$

From the first and third equations of the simultaneous equations, normal strain (ε_{22}) and shear strain (γ_{12}) are:

$$
\varepsilon_{22} = -\frac{0.25 \times 5}{150}\frac{\text{MPa}}{\text{GPa}} + \frac{20}{20}\frac{\text{MPa}}{\text{GPa}} = 992\,\mu\text{m/m}
$$

$$
\gamma = 0\,\mu\,\text{rad}
$$

3.4　HYGROTHERMAL EFFECTS

In this section, the effects of temperature and moisture on the stress–strain of composite materials are discussed. Under the unrestrained structural condition, materials expand or contract under varying moisture and temperature levels. This phenomenon is independent of the applied load. However, when the applied loads are induced simultaneously with temperature and moisture variations, then deformations under both applied load and hygrothermal effects are influenced for constrained boundary conditions. Unlike isotropic materials, generally anisotropic materials have different coefficients of expansion/contraction in different directions for thermal and moisture variations.

An infinitesimal rectangular parallelepiped is considered under free stress of all faces with no boundary restraints at reference (ambient) temperature. At this temperature, the element has dimensions Δ_1, Δ_2, and Δ_3 and at right angles to the element faces, i.e., θ_{23}, θ_{13}, and θ_{12}. When the element is heated or cooled, then the element sizes have small deformations (dilatation) $\delta\Delta_1$, $\delta\Delta_2$, and $\delta\Delta_3$ and changes of right angles are $\delta\theta_{23}$, $\delta\theta_{13}$, and $\delta\theta_{12}$. Thus, free thermal strain under a linear elastic system can be defined, using strain definition as given in Eqs. (3.18.1 and 3.18.2).

As with free thermal strain case (commonly referred to as induced strain under temperature change only), free moisture strain (referred to as induced strain due to moisture sorption only) can also be generated in the same manner as given in Eqs. (3.18.3 and 3.18.4).

$$\varepsilon_{11}^T = \frac{\delta\Delta_1}{\Delta_1} = \alpha_{11}\Delta T \qquad \varepsilon_{22}^T = \frac{\delta\Delta_2}{\Delta_2} = \alpha_{22}\Delta T \qquad \varepsilon_{33}^T = \frac{\delta\Delta_3}{\Delta_3} = \alpha_{33}\Delta T \qquad (3.18.1)$$

$$\gamma_{23}^T = \delta\theta_{23} = \alpha_{23}\Delta T \qquad \gamma_{13}^T = \delta\theta_{13} = \alpha_{13}\Delta T \qquad \gamma_{12}^T = \delta\theta_{12} = \alpha_{12}\Delta T \qquad (3.18.2)$$

$$\varepsilon_{11}^M = \frac{\delta\Delta_1}{\Delta_1} = \beta_{11}\Delta M \qquad \varepsilon_{22}^M = \frac{\delta\Delta_2}{\Delta_2} = \beta_{22}\Delta M \qquad \varepsilon_{33}^M = \frac{\delta\Delta_3}{\Delta_3} = \beta_{33}\Delta M \qquad (3.18.3)$$

$$\gamma_{23}^M = \delta\theta_{23} = \beta_{23}\Delta M \qquad \gamma_{13}^M = \delta\theta_{13} = \beta_{13}\Delta M \qquad \gamma_{12}^M = \delta\theta_{12} = \beta_{12}\Delta M \qquad (3.18.4)$$

where ΔT and ΔM = temperature difference and moisture difference with respect to the reference temperature and the reference moisture uptake, respectively

ε_{ii}^T and γ_{ij}^T = free thermal normal and shear strain, respectively
α_{ij} = coefficients of thermal expansion (CTEs)
ε_{ii}^M and γ_{ij}^M = free moisture normal and shear strain
β_{ij} = linear moisture expansion

Therefore, both free thermal and moisture strains of generally anisotropic materials are written in matrix form under the Cartesian coordinate system as:
From Eqs. (3.18.1 and 3.18.4):

$$\begin{bmatrix} \varepsilon_{11}^T \\ \varepsilon_{22}^T \\ \varepsilon_{33}^T \\ \gamma_{23}^T \\ \gamma_{13}^T \\ \gamma_{12}^T \end{bmatrix} = \begin{bmatrix} \alpha_{11} \\ \alpha_{22} \\ \alpha_{33} \\ \alpha_{23} \\ \alpha_{13} \\ \alpha_{12} \end{bmatrix}\Delta T \quad \text{and} \quad \begin{bmatrix} \varepsilon_{11}^M \\ \varepsilon_{22}^M \\ \varepsilon_{33}^M \\ \gamma_{23}^M \\ \gamma_{13}^M \\ \gamma_{12}^M \end{bmatrix} = \begin{bmatrix} \beta_{11} \\ \beta_{22} \\ \beta_{33} \\ \beta_{23} \\ \beta_{13} \\ \beta_{12} \end{bmatrix}\Delta M \qquad (3.18.5)$$

It should be noted that Young's modulus of materials decreases in polymer composites with an increase in temperature and moisture with reference to ambient levels. It should be noted, however, that such modulus reductions are negligible up to about 30°F below the glass transition temperature. In addition, moisture increase may lead to an increase in thermal coefficients. The couplings between mechanical and hygrothermal properties are neglected in all the analyses presented herein. Thus, the induced unrestrained (free) strains (ε_{ii}^T, γ_{ij}^T, ε_{ii}^M and γ_{ij}^M) under hygrothermal effects are independent of the load-induced mechanical strains.

If fibers are aligned parallel to the symmetrical plane (X_1–X_2) in monoclinic materials, then thermal and moisture coefficients (α_{23}, α_{13}, β_{23}, and β_{13}) are taken to be zero because of the plane of symmetry as discussed in Section 3.2.2. The unrestrained (free of boundary restraints) thermal and moisture strains of monoclinic materials can be represented by modifying Eq. (3.18.5) as:

$$\begin{bmatrix} \varepsilon_{11}^T \\ \varepsilon_{22}^T \\ \varepsilon_{33}^T \\ \gamma_{23}^T \\ \gamma_{13}^T \\ \gamma_{12}^T \end{bmatrix} = \begin{bmatrix} \alpha_{11} \\ \alpha_{22} \\ \alpha_{33} \\ 0 \\ 0 \\ \alpha_{12} \end{bmatrix}\Delta T \quad \text{and} \quad \begin{bmatrix} \varepsilon_{11}^M \\ \varepsilon_{22}^M \\ \varepsilon_{33}^M \\ \gamma_{23}^M \\ \gamma_{13}^M \\ \gamma_{12}^M \end{bmatrix} = \begin{bmatrix} \beta_{11} \\ \beta_{22} \\ \beta_{33} \\ 0 \\ 0 \\ \beta_{12} \end{bmatrix}\Delta M \qquad (3.18.6)$$

For orthotropic materials, dimensional changes (dilatation) are induced when those materials are under hygrothermal influences with unrestrained (free) boundary conditions. Any deformations corresponding to shear strain are equal to zero under three mutually perpendicular symmetrical planes $(X_1–X_2)$, $(X_2–X_3)$, and $(X_1–X_3)$ for orthotropic materials. The free thermal and moisture strains of orthotropic materials are given in Eq. (3.18.7).

$$
\begin{bmatrix} \varepsilon_{11}^T \\ \varepsilon_{22}^T \\ \varepsilon_{33}^T \\ \gamma_{23}^T \\ \gamma_{13}^T \\ \gamma_{12}^T \end{bmatrix} = \begin{bmatrix} \alpha_{11} \\ \alpha_{22} \\ \alpha_{33} \\ 0 \\ 0 \\ 0 \end{bmatrix} \Delta T \quad \text{and} \quad \begin{bmatrix} \varepsilon_{11}^M \\ \varepsilon_{22}^M \\ \varepsilon_{33}^M \\ \gamma_{23}^M \\ \gamma_{13}^M \\ \gamma_{12}^M \end{bmatrix} = \begin{bmatrix} \beta_{11} \\ \beta_{22} \\ \beta_{33} \\ 0 \\ 0 \\ 0 \end{bmatrix} \Delta M \tag{3.18.7}
$$

In the case of transversely isotropic materials, if fibers are aligned parallel to the direction X_1 and the plane $(X_2–X_3)$ of materials is taken to be isotropic, then coefficients of thermal and moisture expansions on the isotropic plane $(X_2–X_3)$ are assumed to be independent of direction. Thus, coefficients of thermal and moisture expansions are modified of Eq. (3.18.5) and given in Eq. (3.18.8) as:

$$
\begin{bmatrix} \varepsilon_{11}^T \\ \varepsilon_{22}^T \\ \varepsilon_{33}^T \\ \gamma_{23}^T \\ \gamma_{13}^T \\ \gamma_{12}^T \end{bmatrix} = \begin{bmatrix} \alpha_{11} \\ \alpha_{22} \\ \alpha_{22} \\ 0 \\ 0 \\ 0 \end{bmatrix} \Delta T \quad \text{and} \quad \begin{bmatrix} \varepsilon_{11}^M \\ \varepsilon_{22}^M \\ \varepsilon_{33}^M \\ \gamma_{23}^M \\ \gamma_{13}^M \\ \gamma_{12}^M \end{bmatrix} = \begin{bmatrix} \beta_{11} \\ \beta_{22} \\ \beta_{22} \\ 0 \\ 0 \\ 0 \end{bmatrix} \Delta M \tag{3.18.8}
$$

where

$\alpha_{33} = \alpha_{22}$
$\beta_{33} = \beta_{22}.$

For isotropic materials, the properties of materials are directional independent. Only one thermal or moisture coefficient is necessary to define the free hygrothermal strain. For orthogonal and transversely isotropic materials, the hygrothermal effects will induce a change of dilatation (normal strains).

$$
\begin{bmatrix} \varepsilon_{11}^T \\ \varepsilon_{22}^T \\ \varepsilon_{33}^T \\ \gamma_{23}^T \\ \gamma_{13}^T \\ \gamma_{12}^T \end{bmatrix} = \begin{bmatrix} \alpha \\ \alpha \\ \alpha \\ 0 \\ 0 \\ 0 \end{bmatrix} \Delta T \quad \text{and} \quad \begin{bmatrix} \varepsilon_{11}^M \\ \varepsilon_{22}^M \\ \varepsilon_{33}^M \\ \gamma_{23}^M \\ \gamma_{13}^M \\ \gamma_{12}^M \end{bmatrix} = \begin{bmatrix} \beta \\ \beta \\ \beta \\ 0 \\ 0 \\ 0 \end{bmatrix} \Delta M \tag{3.18.9}
$$

where $\alpha = \alpha_{11} = \alpha_{22} = \alpha_{33}$

$$\beta = \beta_{11} = \beta_{22} = \beta_{33}.$$

3.5 IN-PLANE STRESS AND STRAIN RELATIONSHIP WITH HYGROTHERMAL EFFECTS

The infinitesimal element under free thermal and moisture strain (no constraint) does not lead to any induced stress. It means that hygrothermal strains will be induced under hygrothermal load (moisture and temperature variation), but no stresses are induced under unrestrained boundary conditions (zero constraint). Thus, strains under applied mechanical loads resulting in internal stresses that are coupled with hygrothermal effects are the additional resultants of induced mechanical and free hygrothermal strains, which are:

$$\varepsilon_{Total} = \varepsilon_{ii}^{mech} + \varepsilon_{ii}^{T} + \varepsilon_{ii}^{M} \tag{3.19.1}$$

$$\gamma_{Total} = \gamma_{ij}^{mech} + \gamma_{ij}^{T} + \gamma_{ij}^{M} \tag{3.19.2}$$

where

ε_{ii}^{mech} and γ_{ii}^{mech} = induced mechanical strains
ε_{ii}^{T} and γ_{ii}^{T} = free thermal strains
ε_{ii}^{M} and γ_{ii}^{M} = free moisture strains

For generally anisotropic materials, total strains under in-plane stress conditions are written by using superposition principles corresponding to applied loads and hygrothermally induced strains.

From Eqs. (3.19.1 and 3.19.2):

Total strain:

$$\begin{bmatrix} \varepsilon_{11} \\ \varepsilon_{22} \\ \gamma_{12} \end{bmatrix} = \begin{bmatrix} S_{11} & S_{12} & S_{16} \\ S_{21} & S_{22} & S_{26} \\ S_{61} & S_{62} & S_{66} \end{bmatrix} \begin{bmatrix} \sigma_{11} \\ \sigma_{22} \\ \tau_{12} \end{bmatrix} + \begin{bmatrix} \varepsilon_{11}^{T} \\ \varepsilon_{22}^{T} \\ \gamma_{12}^{T} \end{bmatrix} + \begin{bmatrix} \varepsilon_{11}^{M} \\ \varepsilon_{22}^{M} \\ \gamma_{12}^{M} \end{bmatrix} \tag{3.19.3}$$

$$\begin{bmatrix} \varepsilon_{11} \\ \varepsilon_{22} \\ \gamma_{12} \end{bmatrix} = \begin{bmatrix} S_{11} & S_{12} & S_{16} \\ S_{21} & S_{22} & S_{26} \\ S_{61} & S_{62} & S_{66} \end{bmatrix} \begin{bmatrix} \sigma_{11} \\ \sigma_{22} \\ \tau_{12} \end{bmatrix} + \begin{bmatrix} \alpha_{11} \\ \alpha_{22} \\ \alpha_{12} \end{bmatrix} \Delta T + \begin{bmatrix} \beta_{11} \\ \beta_{22} \\ \beta_{12} \end{bmatrix} \Delta M \tag{3.19.4}$$

The induced mechanical strains are subtracted from total strains with free hygrothermal strains. From Eqs. (3.19.1 and 3.19.2):

$$\varepsilon_{ii}^{mech} = \varepsilon_{Total} - \left(\varepsilon_{ii}^{T} + \varepsilon_{ii}^{M} \right) \tag{3.19.5}$$

$$\gamma_{ij}^{mech} = \gamma_{Total} - \left(\gamma_{ij}^{T} + \gamma_{ij}^{M} \right) \tag{3.19.6}$$

From Eqs. (3.19.5 and 3.19.6), it is clear that strains from the stress–strain relation in Sections (3.2 and 3.3) are the mechanical strains (not the total strains). However, total strains are equal to the mechanical strains under applied loads when the effects of hygrothermal strains are neglected. The induced mechanical strain of generally anisotropic materials under the plane stress condition is presented below:

From Eqs. (3.19.5 and 3.19.6):

$$\begin{bmatrix} \varepsilon_{11}^{mech} \\ \varepsilon_{22}^{mech} \\ \gamma_{12}^{mech} \end{bmatrix} = \begin{bmatrix} \varepsilon_{11} \\ \varepsilon_{22} \\ \gamma_{12} \end{bmatrix} - \begin{bmatrix} \alpha_{11} \\ \alpha_{22} \\ \alpha_{12} \end{bmatrix} \Delta T - \begin{bmatrix} \beta_{11} \\ \beta_{22} \\ \beta_{12} \end{bmatrix} \Delta M \tag{3.20.1}$$

$$
\begin{bmatrix} \varepsilon_{11}^{mech} \\ \varepsilon_{22}^{mech} \\ \gamma_{12}^{mech} \end{bmatrix} = \begin{bmatrix} S_{11} & S_{12} & S_{16} \\ S_{21} & S_{22} & S_{26} \\ S_{61} & S_{62} & S_{66} \end{bmatrix} \begin{bmatrix} \sigma_{11} \\ \sigma_{22} \\ \tau_{12} \end{bmatrix} - \begin{bmatrix} \alpha_{11} \\ \alpha_{22} \\ \alpha_{12} \end{bmatrix} \Delta T - \begin{bmatrix} \beta_{11} \\ \beta_{22} \\ \beta_{12} \end{bmatrix} \Delta M \quad (3.20.2)
$$

Inversion of Eq. (3.20.2) relates the stress–strain relation of the generally anisotropic material with hygrothermal effects under the in-plane stress condition and such relation is given as:

$$
\begin{bmatrix} \sigma_{11} \\ \sigma_{22} \\ \tau_{12} \end{bmatrix} = \begin{bmatrix} Q_{11} & Q_{12} & Q_{16} \\ Q_{21} & Q_{22} & Q_{26} \\ Q_{61} & Q_{62} & Q_{66} \end{bmatrix} \begin{bmatrix} \varepsilon_{11}^{mech} \\ \varepsilon_{22}^{mech} \\ \gamma_{12}^{mech} \end{bmatrix} \quad (3.20.3)
$$

$$
\begin{bmatrix} \sigma_{11} \\ \sigma_{22} \\ \tau_{12} \end{bmatrix} = \begin{bmatrix} Q_{11} & Q_{12} & Q_{16} \\ Q_{21} & Q_{22} & Q_{26} \\ Q_{61} & Q_{62} & Q_{66} \end{bmatrix} \begin{bmatrix} \varepsilon_{11} - \varepsilon_{11}^{T} - \varepsilon_{11}^{M} \\ \varepsilon_{22} - \varepsilon_{22}^{T} - \varepsilon_{22}^{M} \\ \gamma_{12} - \gamma_{12}^{T} - \gamma_{12}^{M} \end{bmatrix} \quad (3.20.4)
$$

$$
\begin{bmatrix} \sigma_{11} \\ \sigma_{22} \\ \tau_{12} \end{bmatrix} = \begin{bmatrix} Q_{11} & Q_{12} & Q_{16} \\ Q_{21} & Q_{22} & Q_{26} \\ Q_{61} & Q_{62} & Q_{66} \end{bmatrix} \begin{bmatrix} \varepsilon_{11} - \alpha_{11}\Delta T - \beta_{11}\Delta M \\ \varepsilon_{22} - \alpha_{22}\Delta T - \beta_{22}\Delta M \\ \gamma_{12} - \alpha_{12}\Delta T - \beta_{12}\Delta M \end{bmatrix} \quad (3.20.5)
$$

Example: Let us consider a case with free hygrothermal expansion under unrestrained (free) boundaries as shown in Figure 3.10, with no stresses induced under applied loads. Hence, total strains are equal to hygrothermal strains of unrestrained (free) boundaries only. Thus, hygrothermal strains are presented by considering Eq. (3.18.5) as:

$$
\begin{bmatrix} \varepsilon_{11} \\ \varepsilon_{22} \\ \gamma_{12} \end{bmatrix} = \begin{bmatrix} \alpha_{11} \\ \alpha_{22} \\ \alpha_{12} \end{bmatrix} \Delta T + \begin{bmatrix} \beta_{11} \\ \beta_{22} \\ \beta_{12} \end{bmatrix} \Delta M \quad (3.21.1)
$$

To evaluate induced mechanical strains, the total strains in Eq. (3.21.1) are substituted into Eq. (3.20.1). Thus, it is clear that mechanical strains under free hygrothermal expansion are equal to zero.

$$
\begin{bmatrix} \varepsilon_{11}^{mech} \\ \varepsilon_{22}^{mech} \\ \gamma_{12}^{mech} \end{bmatrix} = \begin{bmatrix} \begin{bmatrix} \alpha_{11} \\ \alpha_{22} \\ \alpha_{12} \end{bmatrix} \Delta T + \begin{bmatrix} \beta_{11} \\ \beta_{22} \\ \beta_{12} \end{bmatrix} \Delta M - \begin{bmatrix} \alpha_{11} \\ \alpha_{22} \\ \alpha_{12} \end{bmatrix} \Delta T - \begin{bmatrix} \beta_{11} \\ \beta_{22} \\ \beta_{12} \end{bmatrix} \Delta M \end{bmatrix} = \begin{bmatrix} 0 \\ 0 \\ 0 \end{bmatrix} \quad (3.21.2)
$$

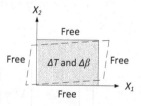

FIGURE 3.10   Free hygrothermal expansion under plane stress condition.

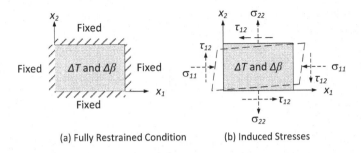

(a) Fully Restrained Condition   (b) Induced Stresses

FIGURE 3.11 Fully restrained element under hygrothermal loads: (a) fully restrained condition and (b) induced stresses.

Example: If the boundary conditions of an infinitesimal element are fully restrained against any deformation as shown in Figure 3.11, and it is under hygrothermal loads only, then the total strain on this element is taken as zero. To satisfy the boundary conditions, mechanical strains in the element are induced with the same magnitude of unrestrained hygrothermal strains, but in the opposite direction of the unrestrained (free) hygrothermal strains. Thus, the mechanical stresses are also induced due to boundary effects. These concepts are illustrated below:

From Eq. (3.20.1), total strains are taken to be zero.

$$\begin{bmatrix} \varepsilon_{11}^{mech} \\ \varepsilon_{22}^{mech} \\ \gamma_{12}^{mech} \end{bmatrix} = \begin{bmatrix} \varepsilon_{11} - \varepsilon_{11}^{T} - \varepsilon_{11}^{M} \\ \varepsilon_{22} - \varepsilon_{22}^{T} - \varepsilon_{22}^{M} \\ \gamma_{12} - \gamma_{12}^{T} - \gamma_{12}^{M} \end{bmatrix} = \begin{bmatrix} 0 - \alpha_{11}\Delta T - \beta_{11}\Delta M \\ 0 - \alpha_{22}\Delta T - \beta_{22}\Delta M \\ 0 - \alpha_{12}\Delta T - \beta_{12}\Delta M \end{bmatrix} = -\begin{bmatrix} \alpha_{11} \\ \alpha_{22} \\ \alpha_{12} \end{bmatrix}\Delta T - \begin{bmatrix} \beta_{11} \\ \beta_{22} \\ \beta_{12} \end{bmatrix}\Delta M$$

(3.21.3)

To evaluate the induced mechanical stresses, the induced mechanical strains in Eq. (3.21.3) are substituted into the stress–strain relation in Eq. (3.20.3), and the resulting stresses are:

$$\begin{bmatrix} \sigma_{11} \\ \sigma_{22} \\ \tau_{12} \end{bmatrix} = -\begin{bmatrix} Q_{11} & Q_{12} & Q_{16} \\ Q_{21} & Q_{22} & Q_{26} \\ Q_{61} & Q_{62} & Q_{66} \end{bmatrix}\begin{bmatrix} \alpha_{11}\Delta T + \beta_{11}\Delta M \\ \alpha_{22}\Delta T + \beta_{22}\Delta M \\ \alpha_{12}\Delta T + \beta_{12}\Delta M \end{bmatrix}$$

(3.21.4)

The negative sign of the induced strains (3.21.3) and stresses (3.21.4) represents the elemental response in the opposite direction with unrestrained (free) hygrothermal strains. The above hygrothermal effects on the generally anisotropic element represent a minimum response and a maximum response depending on the type of boundary conditions. In general, practical applications often include both mechanical and hygrothermal loadings with various boundary conditions. Hence, to determine the responses of practical applications, the above procedure is adopted with extreme values.

To obtain the stress–strain relation with reference to hygrothermal effects of other material types (monoclinic, orthotropic, etc.), the compliance and stiffness matrices of the generally anisotropic materials are simplified by neglecting such terms as stiffness or compliance elements, thermal and moisture coefficients corresponding to other material properties. The reduced compliance and stiffness matrices with hygrothermal effects of monoclinic materials are identical to the generally anisotropic materials. For orthotropic materials, the reduced compliance and stiffness matrices with hygrothermal effect are obtained in the same manner as in monoclinic materials.

The [3×3] reduced stiffness matrix $[Q_{ij}]$ of orthotropic materials is obtained by inverting the [3×3] reduced compliance matrix $[S_{ij}]$ of orthotropic materials in Eq. (3.16.1). From

Eqs. (3.19.3 and 3.20.4), the $[3 \times 3]$ reduced stiffness matrix with hygrothermal effects of orthotropic materials is given as:

$$
\begin{bmatrix} \varepsilon_{11} - \varepsilon_{11}^{T} - \varepsilon_{11}^{M} \\ \varepsilon_{22} - \varepsilon_{22}^{T} - \varepsilon_{22}^{M} \\ \gamma_{12} - \gamma_{12}^{T} - \gamma_{12}^{M} \end{bmatrix} = \begin{bmatrix} \varepsilon_{11}^{mech} \\ \varepsilon_{22}^{mech} \\ \gamma_{12}^{mech} \end{bmatrix} = \begin{bmatrix} S_{11} & S_{12} & 0 \\ S_{21} & S_{22} & 0 \\ 0 & 0 & S_{66} \end{bmatrix} \begin{bmatrix} \sigma_{11} \\ \sigma_{22} \\ \tau_{12} \end{bmatrix}
$$

$$
= \begin{bmatrix} \dfrac{1}{E_{11}} & -\dfrac{v_{21}}{E_{22}} & 0 \\ -\dfrac{v_{12}}{E_{11}} & \dfrac{1}{E_{22}} & 0 \\ 0 & 0 & \dfrac{1}{G_{12}} \end{bmatrix} \begin{bmatrix} \sigma_{11} \\ \sigma_{22} \\ \tau_{12} \end{bmatrix} \tag{3.22.1}
$$

$$
\begin{bmatrix} \sigma_{11} \\ \sigma_{22} \\ \tau_{12} \end{bmatrix} = \begin{bmatrix} Q_{11} & Q_{12} & 0 \\ Q_{21} & Q_{22} & 0 \\ 0 & 0 & Q_{66} \end{bmatrix} \begin{bmatrix} \varepsilon_{11} - \varepsilon_{11}^{T} - \varepsilon_{11}^{M} \\ \varepsilon_{22} - \varepsilon_{22}^{T} - \varepsilon_{22}^{M} \\ \gamma_{12} - \gamma_{12}^{T} - \gamma_{12}^{M} \end{bmatrix} \tag{3.22.2}
$$

$$
\begin{bmatrix} \sigma_{11} \\ \sigma_{22} \\ \tau_{12} \end{bmatrix} = \begin{bmatrix} \dfrac{E_{11}}{1 - v_{12}v_{21}} & \dfrac{v_{12}E_{22}}{1 - v_{12}v_{21}} & 0 \\ \dfrac{v_{21}E_{11}}{1 - v_{12}v_{21}} & \dfrac{E_{22}}{1 - v_{12}v_{21}} & 0 \\ 0 & 0 & G_{12} \end{bmatrix} \begin{bmatrix} \varepsilon_{11} - \varepsilon_{11}^{T} - \varepsilon_{11}^{M} \\ \varepsilon_{22} - \varepsilon_{22}^{T} - \varepsilon_{22}^{M} \\ \gamma_{12} - \gamma_{12}^{T} - \gamma_{12}^{M} \end{bmatrix} \tag{3.22.3}
$$

For transversely isotropic materials, the reduced compliance and stiffness matrices are also simplified from the orthotropic compliance and stiffness matrices as given in Eqs. (3.21.1–3.21.3). The reduced compliance and stiffness matrices of transversely isotropic materials are given as those of the orthotropic materials.

For isotropic materials, only two elastic constants (E and v) are independent. The reduced compliance and stiffness matrices with hygrothermal effect in terms of both the elastic constants are given as:

$$
\begin{bmatrix} \varepsilon_{11} - \varepsilon_{11}^{T} - \varepsilon_{11}^{M} \\ \varepsilon_{22} - \varepsilon_{22}^{T} - \varepsilon_{22}^{M} \\ \gamma_{12} - \gamma_{12}^{T} - \gamma_{12}^{M} \end{bmatrix} = \begin{bmatrix} \dfrac{1}{E} & -\dfrac{v}{E} & 0 \\ -\dfrac{v}{E} & \dfrac{1}{E} & 0 \\ 0 & 0 & 2\left(\dfrac{1+v}{E}\right) \end{bmatrix} \begin{bmatrix} \sigma_{11} \\ \sigma_{22} \\ \tau_{12} \end{bmatrix} \tag{3.23.1}
$$

$$
\begin{bmatrix} \sigma_{11} \\ \sigma_{22} \\ \tau_{12} \end{bmatrix} = \begin{bmatrix} \dfrac{E}{1 - v^2} & \dfrac{vE}{1 - v^2} & 0 \\ \dfrac{vE}{1 - v^2} & \dfrac{E}{1 - v^2} & 0 \\ 0 & 0 & \dfrac{E}{2(1+v)} \end{bmatrix} \begin{bmatrix} \varepsilon_{11} - \varepsilon_{11}^{T} - \varepsilon_{11}^{M} \\ \varepsilon_{22} - \varepsilon_{22}^{T} - \varepsilon_{22}^{M} \\ \gamma_{12} - \gamma_{12}^{T} - \gamma_{12}^{M} \end{bmatrix} \tag{3.23.2}
$$

Example 3.4

The in-plane properties of orthotropic lamina (carbon fiber) are presented as below: $E_{11} = 150\,\text{GPa}$, $E_{22} = 20\,\text{GPa}$, $G_{12} = 15\,\text{GPa}$, $\nu_{12} = 0.25$, $\alpha_{11} = 6\,\mu\text{m/m°C}$, $\alpha_{22} = 20\,\mu\text{m/m°C}$, $\beta_{11} = 100\,\mu\text{m/m \%M}$ and $\beta_{22} = 1{,}200\,\mu\text{m/m \%M}$:

a. Determine the state of strains (total strain) in the coordinate system $(X_1–X_2)$ under the state of stresses and hygrothermal loads $(\Delta T = 80°C)$ and $(\Delta M = 2\%)$
b. Determine the state of stresses in the coordinate system $(X_1–X_2)$ under the state of strains and hygrothermal loads $(\Delta T = 80°C)$ and $(\Delta M = 2\%)$ as: $\varepsilon_{11} = -1{,}000\,\mu\text{m/m}$, $\varepsilon_{22} = 1{,}500\,\mu\text{m/m}$ and $\gamma_{12} = 1{,}000\,\mu\text{rad}$
c. Determine the state of stress in the coordinate system $(X_1–X_2)$ under fixed boundary conditions and hygrothermal loads $(\Delta T = 80°C)$ and $(\Delta M = 2\%)$

Solution

a. State of strains (total strain) under the state of stresses and hygrothermal loads:
From Eqs. (3.16.1 and 3.16.2):

$$
\begin{bmatrix} S_{11} & S_{12} & 0 \\ S_{12} & S_{22} & 0 \\ 0 & 0 & S_{66} \end{bmatrix} = \begin{bmatrix} \dfrac{1}{150} & -\dfrac{0.25}{150} & 0 \\ -\dfrac{0.25}{150} & \dfrac{1}{20} & 0 \\ 0 & 0 & \dfrac{1}{15} \end{bmatrix} \dfrac{1}{\text{GPa}}
$$

$$
\begin{bmatrix} Q_{11} & Q_{12} & 0 \\ Q_{21} & Q_{22} & 0 \\ 0 & 0 & Q_{66} \end{bmatrix} = \begin{bmatrix} 151.3 & 5.04 & 0 \\ 5.04 & 20.2 & 0 \\ 0 & 0 & 15 \end{bmatrix} \text{GPa}
$$

From Eq. (3.21.1), unrestrained (free) hygrothermal strains are:

$$
\begin{bmatrix} \varepsilon_{11}^T \\ \varepsilon_{22}^T \\ \gamma_{12}^T \end{bmatrix} + \begin{bmatrix} \varepsilon_{11}^M \\ \varepsilon_{22}^M \\ \gamma_{12}^M \end{bmatrix} = \begin{bmatrix} 6 \\ 20 \\ 0 \end{bmatrix}(80) + \begin{bmatrix} 100 \\ 1{,}200 \\ 0 \end{bmatrix}(2) = \begin{bmatrix} 480 \\ 1{,}600 \\ 0 \end{bmatrix} + \begin{bmatrix} 200 \\ 2{,}400 \\ 0 \end{bmatrix} = \begin{bmatrix} 680 \\ 4{,}000 \\ 0 \end{bmatrix} \mu
$$

Note: Unit: normal strain ε_{ii} (m/m) and shear strain (γ_{ij}) rad/rad

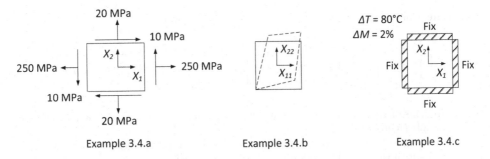

| Example 3.4.a | Example 3.4.b | Example 3.4.c |

EXAMPLE FIGURE 3.4.1 State of stresses and strains and boundary conditions in Example 3.4.

From Eq. (3.22.1), the total strain in an orthotropic lamina is:

$$
\begin{bmatrix} \varepsilon_{11} \\ \varepsilon_{22} \\ \gamma_{12} \end{bmatrix} = \begin{bmatrix} \dfrac{1}{150} & -\dfrac{0.25}{150} & 0 \\ -\dfrac{0.25}{150} & \dfrac{1}{20} & 0 \\ 0 & 0 & \dfrac{1}{15} \end{bmatrix} \dfrac{1}{GPa} \begin{bmatrix} 250 \\ 20 \\ 10 \end{bmatrix} MPa + \begin{bmatrix} 480 \\ 1,600 \\ 0 \end{bmatrix} \mu + \begin{bmatrix} 200 \\ 2,400 \\ 0 \end{bmatrix} \mu
$$

$$
= \begin{bmatrix} 2,313 \\ 4,583 \\ 667 \end{bmatrix} \mu
$$

Note: Unit: normal strain ε_{ii} (m/m) and shear strain (γ_{ij}) rad/rad

The orthotropic lamina is free to deform; hence, the total strain in the lamina is the sum of the induced strains from both mechanical and hygrothermal loads. However, induced stresses of orthotropic lamina under hygrothermal load with unrestrained (free) boundary conditions do not exist. Thus, the induced stresses of lamina are generated due to mechanical strains only.

From Eq. (3.22.1):

$$
\begin{bmatrix} \varepsilon_{11}^{mech} \\ \varepsilon_{22}^{mech} \\ \gamma_{12}^{mech} \end{bmatrix} = \begin{bmatrix} 2,313 - 480 - 200 \\ 4,583 - 1,600 - 2,400 \\ 667 \end{bmatrix} \mu = \begin{bmatrix} \dfrac{1}{150} & -\dfrac{0.25}{150} & 0 \\ -\dfrac{0.25}{150} & \dfrac{1}{20} & 0 \\ 0 & 0 & \dfrac{1}{15} \end{bmatrix} \dfrac{1}{GPa} \begin{bmatrix} 250 \\ 20 \\ 10 \end{bmatrix} MPa
$$

$$
= \begin{bmatrix} 1,633 \\ 583 \\ 667 \end{bmatrix} \mu
$$

Note: Unit: normal strain ε_{ii} (m/m) and shear strain (γ_{ij}) rad/rad

b. State of stresses under the state of strains and hygrothermal loads:

From Eq. (3.22.1):

$$
\begin{bmatrix} \varepsilon_{11}^{mech} \\ \varepsilon_{22}^{mech} \\ \gamma_{12}^{mech} \end{bmatrix} = \begin{bmatrix} \varepsilon_{11} - \varepsilon_{11}^{T} - \varepsilon_{11}^{M} \\ \varepsilon_{22} - \varepsilon_{22}^{T} - \varepsilon_{22}^{M} \\ \gamma_{12} - \gamma_{12}^{T} - \gamma_{12}^{M} \end{bmatrix} = \begin{bmatrix} -1,000 - 480 - 200 \\ 1,500 - 1,600 - 2,400 \\ 1,000 \end{bmatrix} = \begin{bmatrix} -1,680 \\ -2,500 \\ 1,000 \end{bmatrix} \mu
$$

Note: Unit: normal strain ε_{ii} (m/m) and shear strain (γ_{ij}) rad/rad

$$
\begin{bmatrix} \sigma_{11} \\ \sigma_{22} \\ \tau_{12} \end{bmatrix} = \begin{bmatrix} 151.3 & 5.04 & 0 \\ 5.04 & 20.2 & 0 \\ 0 & 0 & 15 \end{bmatrix} GPa \begin{bmatrix} -1,680 \\ -2,500 \\ 1,000 \end{bmatrix} \mu = \begin{bmatrix} -267 \\ 59.0 \\ 15 \end{bmatrix} MPa
$$

The state of stresses under the state of strains and hygrothermal loads is shown in Example Figure 3.4.2

c. State of stress under fixed boundary condition and hygrothermal loads:

The total strains of the orthotropic lamina are equal to zero due to fixed boundary condition. However, mechanical and hygrothermal strains are not equal to zero.

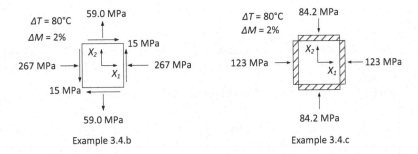

Example 3.4.b Example 3.4.c

EXAMPLE FIGURE 3.4.2 State of stresses in Example 3.4b and 3.4C.

From Eq. (3.22.1):

$$
\begin{bmatrix} \varepsilon_{11}^{mech} \\ \varepsilon_{22}^{mech} \\ \gamma_{12}^{mech} \end{bmatrix} = \begin{bmatrix} \varepsilon_{11} - \varepsilon_{11}^{T} - \varepsilon_{11}^{M} \\ \varepsilon_{22} - \varepsilon_{22}^{T} - \varepsilon_{22}^{M} \\ \gamma_{12} - \gamma_{12}^{T} - \gamma_{12}^{M} \end{bmatrix} = \begin{bmatrix} 0 - 480 - 200 \\ 0 - 1{,}600 - 2{,}400 \\ 0 \end{bmatrix} = \begin{bmatrix} -680 \\ -4{,}000 \\ 0 \end{bmatrix} \mu
$$

Note: Unit: normal strain ε_{ii} (m/m) and shear strain (γ_{ij}) rad/rad

$$
\begin{bmatrix} \sigma_{11} \\ \sigma_{22} \\ \tau_{12} \end{bmatrix} = \begin{bmatrix} 151.3 & 5.04 & 0 \\ 5.04 & 20.2 & 0 \\ 0 & 0 & 15 \end{bmatrix} GPa \begin{bmatrix} -680 \\ -4{,}000 \\ 0 \end{bmatrix} \mu = \begin{bmatrix} -123 \\ -84.2 \\ 0 \end{bmatrix} MPa
$$

The state of stresses under the boundary condition and hygrothermal loads is shown in Example Figure 3.4.2

Example 3.5

The in-plane properties of orthotropic lamina (carbon fiber) are presented as below: $E_{11} = 150\,GPa$, $E_{22} = 20\,GPa$, $G_{12} = 15\,GPa$, $\nu_{12} = 0.25$, $\alpha_{11} = 6\,\mu m/m\,°C$, $\alpha_{22} = 20\,\mu m/m\,°C$, $\beta_{11} = 100\,\mu m/m\,\%M$ and $\beta_{22} = 1{,}200\,\mu m/m\,\%M$:

a. Determine the state of strains in the coordinate system under applied stress (σ_{11}) of 250 MPa, hygrothermal loads ($\Delta T = 80°C$ and $\Delta M = 2\%$). The boundaries are fixed along the X_1 direction.
b. Determine the state of strains in the coordinate system under applied stress (σ_{22}) of 20 MPa and hygrothermal loads ($\Delta T = 80°C$ and $\Delta M = 2\%$). The boundaries are fixed along the X_2 direction.

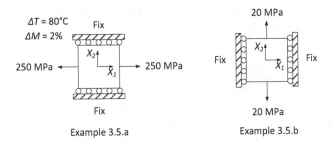

Example 3.5.a Example 3.5.b

EXAMPLE FIGURE 3.5.1 In-plane properties of orthotropic lamina under stresses and boundary conditions.

Solution

a. State of strains under applied stress (σ_{11}) and hygrothermal loads:

Using the reduced compliance matrix of Example 3.4, normal stress (σ_{22}) is induced due to fixed boundary condition along the X_1 direction. The total normal strain (ε_{22}) is equal to zero.

From Eq. (3.21.1), unrestrained (free) hygrothermal strains are:

$$
\begin{bmatrix} \varepsilon_{11}^{T} \\ \varepsilon_{22}^{T} \\ \gamma_{12}^{T} \end{bmatrix} + \begin{bmatrix} \varepsilon_{11}^{M} \\ \varepsilon_{22}^{M} \\ \gamma_{12}^{M} \end{bmatrix} = \begin{bmatrix} 6 \\ 20 \\ 0 \end{bmatrix}(80) + \begin{bmatrix} 100 \\ 1,200 \\ 0 \end{bmatrix}(2) = \begin{bmatrix} 480 \\ 1,600 \\ 0 \end{bmatrix} + \begin{bmatrix} 200 \\ 2,400 \\ 0 \end{bmatrix} = \begin{bmatrix} 680 \\ 4,000 \\ 0 \end{bmatrix}\mu
$$

From Eq. (3.22.1), the total strain in an orthotropic lamina is:

$$
\begin{bmatrix} \varepsilon_{11} \\ 0 \\ \gamma_{12} \end{bmatrix} = \begin{bmatrix} \dfrac{1}{150} & -\dfrac{0.25}{150} & 0 \\ -\dfrac{0.25}{150} & \dfrac{1}{20} & 0 \\ 0 & 0 & \dfrac{1}{15} \end{bmatrix} \dfrac{1}{\text{GPa}} \begin{bmatrix} 250 \\ \sigma_{22} \\ 10 \end{bmatrix}\text{MPa} + \begin{bmatrix} 480 \\ 1,600 \\ 0 \end{bmatrix}\mu + \begin{bmatrix} 200 \\ 2,400 \\ 0 \end{bmatrix}\mu
$$

Note: Unit: normal strain ε_{ii} (m/m) and shear strain (γ_{ij}) rad/rad

The second equation of the above simultaneous equations yields (σ_{22}) as:

$$
-\frac{0.25 \times 250}{150}\frac{\text{MPa}}{\text{GPa}} + \frac{\sigma_{22}}{20}\frac{1}{\text{GPa}} + (1,600 + 2,400)\mu = 0
$$

$$
\sigma_{22} = -71.7\,\text{MPa}
$$

From the first and third equations of the simultaneous equations, normal strain (ε_{11}) and shear strain (γ_{12}) are:

$$
\varepsilon_{11} = \frac{250}{150}\frac{\text{MPa}}{\text{GPa}} + \frac{71.7 \times 0.25}{150}\frac{\text{MPa}}{\text{GPa}} + (480 + 200)\mu = 2,466\ \mu
$$

$$
\gamma_{12} = 0\ \mu
$$

For orthotropic materials, when applied normal stress is aligned along the orthotropic axis under hygrothermal loads, then the shear strain is not induced. However, if applied normal stress is aligned off-axis of orthotropic coordinate, then normal and shear strains are induced, while hygrothermal responses are induced under normal stresses.

b. State of strains under applied stress (σ_{22}) of 20 MPa and hygrothermal loads:

Normal stress (σ_{11}) is induced due to fixed boundary condition along the X_2 direction. The total normal strain (ε_{11}) is equal to zero.

From Eq. (3.22.1):

$$
\begin{bmatrix} 0 \\ \varepsilon_{22} \\ \gamma_{12} \end{bmatrix} = \begin{bmatrix} \dfrac{1}{150} & -\dfrac{0.25}{150} & 0 \\ -\dfrac{0.25}{150} & \dfrac{1}{20} & 0 \\ 0 & 0 & \dfrac{1}{15} \end{bmatrix} \dfrac{1}{\text{GPa}} \begin{bmatrix} \sigma_{11} \\ 20 \\ 0 \end{bmatrix}\text{MPa} + \begin{bmatrix} 480 \\ 1,600 \\ 0 \end{bmatrix}\mu + \begin{bmatrix} 200 \\ 2,400 \\ 0 \end{bmatrix}\mu
$$

The first equation of the above simultaneous equations yields (σ_{11}) as:

$$\frac{\sigma_{11}}{150}\frac{1}{GPa} - \frac{20 \times 0.25}{150}\frac{MPa}{GPa} + (480 + 200)\mu = 0$$

$$\sigma_{11} = -97 \text{ MPa}$$

Thus, normal (ε_{22}) and shear (γ_{12}) strains are:

$$\varepsilon_{22} = \frac{0.25 \times 97}{150}\frac{MPa}{GPa} + \frac{20}{20}\frac{MPa}{GPa} + (1,600 + 2,400)\mu = 5,648\ \mu$$

$$\gamma = 0\ \mu$$

Note: Unit: normal strain ε_{ii} (m/m) and shear strain (γ_{ij}) rad/rad

3.6 STRESS AND STRAIN RELATIONSHIPS IN GLOBAL COORDINATE SYSTEM

In the previous section, the behavior of composite materials has been discussed in their own (local) orthogonal coordinate system (X_1–X_2–X_3). In general, FRP laminated structural members are formed by combining or stacking numerous lamina (in relation to global axes) that are positioned in different orientations whose coordinate orientation may not coincide with the global (laminate) orientation; as shown in Figure 3.12.

To study the thermomechanical responses of structural members in a structural laminate, another coordinate system is needed to find responses of each laminate with respect to this new coordinate system that relates to the local (lamina) coordinate system (X_1–X_2–X_3). This arbitrary coordinate system of structural members is known as the global (X–Y–Z) coordinate system whereas, the orthogonal coordinate system of each FRP laminate is called the local coordinate system or the principal material coordinate system (X_1–X_2–X_3).

To accomplish the interrelation between global and local coordinate systems, transformations between these two coordinates in terms of stresses and strains are presented under the Cartesian coordinate system. Therefore, the global stress–strain relations of each FRP composite lamina are obtained by using a transformation relation between the local stress–strain relationship of composite laminates.

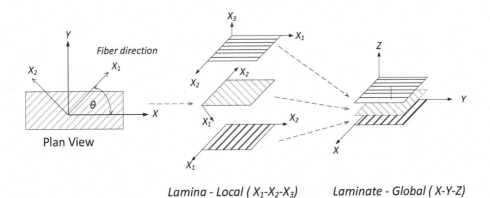

Lamina - Local (X_1-X_2-X_3)   Laminate - Global (X-Y-Z)

FIGURE 3.12 Composite lamina and laminate orientations of fibers in the local and global coordinate system.

3.6.1 Stress Transformation

Transformations of the Cartesian coordinates from local $(X_1–X_2)$ to global $(X–Y)$ coordinate system are highlighted throughout this textbook. We are silent on a polar or curvilinear coordinate system to minimize confusion. The two-dimensional transformation of stresses and strains under the in-plane conditions is formulated later to establish the reduced stiffness in the global coordinate system.

In terms of the plane stress condition, axes of X_3 and Z remain identical as given in Figure 3.13 hence, only two planes $(X_1–X_2)$ and $(X–Y)$ are considered in this case. Thus, stresses in $(X_1–X_2)$ coordinate system are determined from stresses in $(X–Y)$ coordinate by rotating the $(X–Y)$ coordinate about X_3 or Z.

Let us consider $(X–Y)$ element under the plane stress condition as shown in Figure 3.14. The in-plane stresses of the local coordinate system $(X_1–X_2)$ can be written in terms of in-plane stresses of the global coordinate system $(X–Y)$ as follows:

$$\sigma_{11} = \text{Cos}^2\theta\sigma_{xx} + \text{Sin}^2\theta\sigma_{yy} + 2\text{Cos}\theta\text{Sin}\theta\tau_{xy} \tag{3.24.1}$$

$$\sigma_{22} = \text{Sin}^2\theta\sigma_{xx} + \text{Cos}^2\theta\sigma_{yy} - 2\text{Cos}\theta\text{Sin}\theta\tau_{xy} \tag{3.24.2}$$

$$\tau_{12} = -\text{Cos}\theta\text{Sin}\theta\sigma_{xx} + \text{Cos}\theta\text{Sin}\theta\sigma_{yy} + \left(\text{Cos}^2\theta - \text{Sin}^2\theta\right)\tau_{xy} \tag{3.24.3}$$

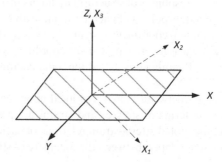

FIGURE 3.13 Coordinate $(X–Y–Z)$ and $(X_1–X_2–X_3)$ systems under plane stress.

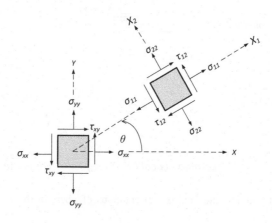

FIGURE 3.14 Transformation of state of plane stress.

where θ is defined as the angle between axis X_1 and X (counterclockwise as shown in Figure 3.14). Therefore, the stress transformation relation in Eqs. (3.24.1–3.24.3) can be written in a matrix form as given in Eq. (3.25), for plane stress condition which is denoted by $\left[T_\sigma^R \right]$.

$$\begin{bmatrix} \sigma_{11} \\ \sigma_{22} \\ \tau_{12} \end{bmatrix} = \left[T_\sigma^R \right] \begin{bmatrix} \sigma_{xx} \\ \sigma_{yy} \\ \tau_{xy} \end{bmatrix} = \begin{bmatrix} \mathrm{Cos}^2\theta & \mathrm{Sin}^2\theta & 2\mathrm{Cos}\theta\mathrm{Sin}\theta \\ \mathrm{Sin}^2\theta & \mathrm{Cos}^2\theta & -2\mathrm{Cos}\theta\mathrm{Sin}\theta \\ -\mathrm{Cos}\theta\mathrm{Sin}\theta & \mathrm{Cos}\theta\mathrm{Sin}\theta & \mathrm{Cos}^2\theta - \mathrm{Sin}^2\theta \end{bmatrix} \begin{bmatrix} \sigma_{xx} \\ \sigma_{yy} \\ \tau_{xy} \end{bmatrix} \quad (3.25.1)$$

In addition, inversion of Eq. (3.25.1) results in stresses in the (X–Y) coordinate system as a function of stresses in the (X_1–X_2) coordinate system as:

$$\begin{bmatrix} \sigma_{xx} \\ \sigma_{yy} \\ \tau_{xy} \end{bmatrix} = \left[T_\sigma^R \right]^{-1} \begin{bmatrix} \sigma_{11} \\ \sigma_{22} \\ \tau_{12} \end{bmatrix} = \begin{bmatrix} \mathrm{Cos}^2\theta & \mathrm{Sin}^2\theta & -2\mathrm{Cos}\theta\mathrm{Sin}\theta \\ \mathrm{Sin}^2\theta & \mathrm{Cos}^2\theta & 2\mathrm{Cos}\theta\mathrm{Sin}\theta \\ \mathrm{Cos}\theta\mathrm{Sin}\theta & -\mathrm{Cos}\theta\mathrm{Sin}\theta & \mathrm{Cos}^2\theta - \mathrm{Sin}^2\theta \end{bmatrix} \begin{bmatrix} \sigma_{11} \\ \sigma_{22} \\ \tau_{12} \end{bmatrix} \quad (3.25.2)$$

3.6.2 Strain Transformation

The strain transformation relation can be formulated in the same manner as the stress transformation relation. The plane strain formulation in the local coordinate (X_1–X_2) system can be written in terms of plane strain in the global coordinate system (X–Y) as shown in Figure 3.15.

$$\varepsilon_{11} = \mathrm{Cos}^2\theta\varepsilon_{xx} + \mathrm{Sin}^2\theta\varepsilon_{yy} + \mathrm{Cos}\theta\mathrm{Sin}\theta\gamma_{xy} \quad (3.26.1)$$

$$\varepsilon_{22} = \mathrm{Sin}^2\theta\varepsilon_{xx} + \mathrm{Cos}^2\theta\varepsilon_{yy} - \mathrm{Cos}\theta\mathrm{Sin}\theta\gamma_{xy} \quad (3.26.2)$$

$$\gamma_{12} = -2\mathrm{Cos}\theta\mathrm{Sin}\theta\varepsilon_{xx} + 2\mathrm{Cos}\theta\mathrm{Sin}\theta\varepsilon_{yy} + \left(\mathrm{Cos}^2\theta - \mathrm{Sin}^2\theta\right)\gamma_{xy} \quad (3.26.3)$$

where θ is defined as the angle between axis X_1 and X (counterclockwise as shown in Figure 3.15). Therefore, the strain transformation relation in Eqs. (3.26.1–3.26.3) can be written in a matrix form as in Eq. (3.27.1) for plane strain condition which is denoted by $\left[T_\varepsilon^R \right]$.

$$\begin{bmatrix} \varepsilon_{11} \\ \varepsilon_{22} \\ \gamma_{12} \end{bmatrix} = \left[T_\varepsilon^R \right] \begin{bmatrix} \varepsilon_{xx} \\ \varepsilon_{yy} \\ \gamma_{xy} \end{bmatrix} = \begin{bmatrix} \mathrm{Cos}^2\theta & \mathrm{Sin}^2\theta & \mathrm{Cos}\theta\mathrm{Sin}\theta \\ \mathrm{Sin}^2\theta & \mathrm{Cos}^2\theta & -\mathrm{Cos}\theta\mathrm{Sin}\theta \\ -2\mathrm{Cos}\theta\mathrm{Sin}\theta & 2\mathrm{Cos}\theta\mathrm{Sin}\theta & \mathrm{Cos}^2\theta - \mathrm{Sin}^2\theta \end{bmatrix} \begin{bmatrix} \varepsilon_{xx} \\ \varepsilon_{yy} \\ \gamma_{xy} \end{bmatrix} \quad (3.27.1)$$

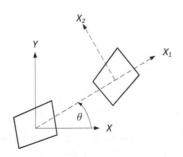

FIGURE 3.15 Strain transformation of the state of plane stress.

In addition, inversion of Eq. (3.27.1) results in strains in the global coordinate (X–Y) system in terms of strains in the local coordinates (X_1–X_2) system, as given in Eq. (3.27.2).

$$
\begin{bmatrix} \varepsilon_{xx} \\ \varepsilon_{yy} \\ \gamma_{xy} \end{bmatrix} = \left[T_\varepsilon^R \right]^{-1} \begin{bmatrix} \varepsilon_{11} \\ \varepsilon_{22} \\ \gamma_{12} \end{bmatrix} = \begin{bmatrix} \cos^2\theta & \sin^2\theta & -\cos\theta\sin\theta \\ \sin^2\theta & \cos^2\theta & \cos\theta\sin\theta \\ 2\cos\theta\sin\theta & -2\cos\theta\sin\theta & \cos^2\theta - \sin^2\theta \end{bmatrix} \begin{bmatrix} \varepsilon_{11} \\ \varepsilon_{22} \\ \gamma_{12} \end{bmatrix}
$$

$$(3.27.2)$$

3.6.3 TRANSFORMATION OF REDUCED COMPLIANCE MATRIX

To obtain the reduced compliance matrix in an arbitrary coordinate system, the strain–stress relations in both global (X–Y) and local (X_1–X_2) coordinates are considered. The stress–strain transformations in Eqs. (3.25.1 and 3.27.1) are substituted into the reduced strain and stress relation in Eq. (3.15.2). The transformation of reduced compliance is presented in the matrix form as:

By substituting Eq. (3.27.1) into Eq. (3.15.2):

$$
\begin{bmatrix} c^2 & s^2 & cs \\ s^2 & c^2 & -cs \\ -2cs & 2cs & c^2 - s^2 \end{bmatrix} \begin{bmatrix} \varepsilon_{xx} \\ \varepsilon_{yy} \\ \gamma_{xy} \end{bmatrix} = \begin{bmatrix} S_{11} & S_{12} & S_{16} \\ S_{21} & S_{22} & S_{26} \\ S_{61} & S_{62} & S_{66} \end{bmatrix} \begin{bmatrix} \sigma_{11} \\ \sigma_{22} \\ \tau_{12} \end{bmatrix}
$$

$$(3.28.1)$$

By substituting Eq. (3.25.1) into Eq. (3.28.1):

$$
\begin{bmatrix} c^2 & s^2 & cs \\ s^2 & c^2 & -cs \\ -2cs & 2cs & c^2 - s^2 \end{bmatrix} \begin{bmatrix} \varepsilon_{xx} \\ \varepsilon_{yy} \\ \gamma_{xy} \end{bmatrix} = \begin{bmatrix} S_{11} & S_{12} & S_{16} \\ S_{21} & S_{22} & S_{26} \\ S_{61} & S_{62} & S_{66} \end{bmatrix} \begin{bmatrix} c^2 & s^2 & 2cs \\ s^2 & c^2 & -2cs \\ -cs & cs & c^2 - s^2 \end{bmatrix} \begin{bmatrix} \sigma_{xx} \\ \sigma_{yy} \\ \tau_{xy} \end{bmatrix}
$$

$$(3.28.2)$$

$$
\begin{bmatrix} \varepsilon_{xx} \\ \varepsilon_{yy} \\ \gamma_{xy} \end{bmatrix} = \begin{bmatrix} c^2 & s^2 & cs \\ s^2 & c^2 & -cs \\ -2cs & 2cs & c^2 - s^2 \end{bmatrix}^{-1} \begin{bmatrix} S_{11} & S_{12} & S_{16} \\ S_{21} & S_{22} & S_{26} \\ S_{61} & S_{62} & S_{66} \end{bmatrix} \begin{bmatrix} c^2 & s^2 & 2cs \\ s^2 & c^2 & -2cs \\ -cs & cs & c^2 - s^2 \end{bmatrix} \begin{bmatrix} \sigma_{xx} \\ \sigma_{yy} \\ \tau_{xy} \end{bmatrix}
$$

$$
= \left[\bar{S}_{ij}^R \right] \begin{bmatrix} \sigma_{xx} \\ \sigma_{yy} \\ \tau_{xy} \end{bmatrix}
$$

$$(3.28.3)$$

where
$c = \cos\theta$
$s = \sin\theta$
θ = defined as the angle between axis X_1 and X.

For most FRP composite materials, the properties of each lamina of a structural laminate are usually classified as orthotropic materials. The transformation of the reduced compliance matrix in Eq. (3.28.3) is simplified by replacing the generally anisotropic reduced compliance with the orthogonal of the reduced compliance matrix. Transformation of the reduced compliance matrix of orthotropic materials is used to establish the laminate compliance matrix, which is shown in Chapter 4. The transformation of the reduced compliance matrix of orthotropic materials is presented as:

$$\left[\bar{S}_{ij}^R\right] = \begin{bmatrix} \bar{S}_{11} & \bar{S}_{12} & \bar{S}_{16} \\ \bar{S}_{12} & \bar{S}_{22} & \bar{S}_{26} \\ \bar{S}_{16} & \bar{S}_{26} & \bar{S}_{66} \end{bmatrix}$$

$$= \begin{bmatrix} c^2 & s^2 & -cs \\ s^2 & c^2 & cs \\ 2cs & -2cs & (c^2-s^2) \end{bmatrix} \begin{bmatrix} \dfrac{1}{E_{11}} & -\dfrac{v_{21}}{E_{22}} & 0 \\ -\dfrac{v_{12}}{E_{11}} & \dfrac{1}{E_{22}} & 0 \\ 0 & 0 & \dfrac{1}{G_{12}} \end{bmatrix} \begin{bmatrix} c^2 & s^2 & 2cs \\ s^2 & c^2 & -2cs \\ -cs & cs & (c^2-s^2) \end{bmatrix} \quad (3.29)$$

where
$c = \text{Cos } \theta$
$s = \text{Sin } \theta$
$\theta = $ angle between axis X_1 and X.

3.6.4 Transformation of Reduced Compliance Matrix with Hygrothermal Effect

The hygrothermal strain in an arbitrary coordinate system can be determined in the same manner as the mechanical strains. The hygrothermal strains in the local $(X_1$–$X_2)$ coordinate system in Eq. (3.21.1) are substituted into Eq. (3.27.2) as:

$$\begin{bmatrix} \varepsilon_{xx} \\ \varepsilon_{yy} \\ \gamma_{xy} \end{bmatrix}^{T \text{ and } M} = \left[T_\varepsilon^R\right]^{-1} \begin{bmatrix} \varepsilon_{11}^T + \varepsilon_{11}^M \\ \varepsilon_{22}^T + \varepsilon_{22}^M \\ \gamma_{12}^T + \gamma_{12}^M \end{bmatrix} = \left[T_\varepsilon^R\right]^{-1} \begin{bmatrix} \alpha_{11}\Delta T + \beta_{11}\Delta M \\ \alpha_{22}\Delta T + \beta_{22}\Delta M \\ \alpha_{12}\Delta T + \beta_{12}\Delta M \end{bmatrix} \quad (3.30.1)$$

$$\begin{bmatrix} \varepsilon_{xx} \\ \varepsilon_{yy} \\ \gamma_{xy} \end{bmatrix}^{T \text{ and } M} = \begin{bmatrix} c^2 & s^2 & -cs \\ s^2 & c^2 & cs \\ 2cs & -2cs & c^2-s^2 \end{bmatrix} \begin{bmatrix} \alpha_{11}\Delta T + \beta_{11}\Delta M \\ \alpha_{22}\Delta T + \beta_{22}\Delta M \\ \alpha_{12}\Delta T + \beta_{12}\Delta M \end{bmatrix} \quad (3.30.2)$$

For orthotropic materials, hygrothermal responses are only induced in the material coordinate directions. Equation (3.30.2) can be simplified by taking shear strain to be zero, as explained before in Section 3.4.

$$\begin{bmatrix} \varepsilon_{xx} \\ \varepsilon_{yy} \\ \gamma_{xy} \end{bmatrix}^{T \text{ and } M} = \Delta T \begin{bmatrix} \alpha_{xx} \\ \alpha_{yy} \\ \alpha_{xy} \end{bmatrix} + \Delta M \begin{bmatrix} \beta_{xx} \\ \beta_{yy} \\ \beta_{xy} \end{bmatrix} = \begin{bmatrix} c^2 & s^2 & -cs \\ s^2 & c^2 & cs \\ 2cs & -2cs & c^2-s^2 \end{bmatrix} \begin{bmatrix} \alpha_{11}\Delta T + \beta_{11}\Delta M \\ \alpha_{22}\Delta T + \beta_{22}\Delta M \\ 0 \end{bmatrix}$$

$$(3.30.3)$$

Thus, the transformation of the reduced strain–stress relation with hygrothermal effects under plane stress condition are obtained from the transformation of the reduced strains in Eqs. (3.28.3 and 3.30.3) as:

$$\begin{bmatrix} \varepsilon_{xx} \\ \varepsilon_{yy} \\ \gamma_{xy} \end{bmatrix} = \begin{bmatrix} \varepsilon_{xx}^{mech} + \varepsilon_{xx}^T + \varepsilon_{xx}^M \\ \varepsilon_{yy}^{mech} + \varepsilon_{yy}^T + \varepsilon_{yy}^M \\ \gamma_{xy}^{mech} + \gamma_{xy}^T + \gamma_{xy}^M \end{bmatrix} = \begin{bmatrix} \bar{S}_{11} & \bar{S}_{12} & \bar{S}_{16} \\ \bar{S}_{12} & \bar{S}_{22} & \bar{S}_{26} \\ \bar{S}_{16} & \bar{S}_{26} & \bar{S}_{66} \end{bmatrix} \begin{bmatrix} \sigma_{xx} \\ \sigma_{yy} \\ \tau_{xy} \end{bmatrix} + \Delta T \begin{bmatrix} \alpha_{xx} \\ \alpha_{yy} \\ \alpha_{xy} \end{bmatrix} + \Delta M \begin{bmatrix} \beta_{xx} \\ \beta_{yy} \\ \beta_{xy} \end{bmatrix}$$

$$(3.30.4)$$

Equation (3.30.4) reveals that the thermomechanical responses of the composite materials are more complex than those of conventional engineering materials such as metals. The thermomechanical response of composite materials is not only presented in parallel and perpendicular to the fiber direction but also exists in both parallel and perpendicular (directions) to fiber alignment.

If normal stress (σ_{xx}) is applied to an orthotropic layer, then not only normal strains (ε_{xx} and ε_{yy}) are induced but also the shear strains (γ_{xy}). The off-axis mechanical response is due to $\bar{S}_{16}$ and $\bar{S}_{26}$ of the transformation of the reduced compliance matrix. In addition, the off-axis hygrothermal responses are due to the coefficients of α_{xy} and β_{xy}.

3.6.5 TRANSFORMATION OF REDUCED STIFFNESS MATRIX

The transformation of the reduced stiffness matrix can be easily obtained by inverting Eq. (3.29). Similarly, the transformation of the reduced compliance matrix is presented herein.

By substituting Eq. (3.25.1) into Eq. (3.13):

$$
\begin{bmatrix} \sigma_{11} \\ \sigma_{22} \\ \tau_{12} \end{bmatrix} = \begin{bmatrix} c^2 & s^2 & 2cs \\ s^2 & c^2 & -2cs \\ -cs & cs & (c^2-s^2) \end{bmatrix} \begin{bmatrix} \sigma_{xx} \\ \sigma_{yy} \\ \tau_{xy} \end{bmatrix} = \begin{bmatrix} Q_{11} & Q_{12} & Q_{16} \\ Q_{21} & Q_{22} & Q_{26} \\ Q_{61} & Q_{62} & Q_{66} \end{bmatrix} \begin{bmatrix} \varepsilon_{11} \\ \varepsilon_{22} \\ \gamma_{12} \end{bmatrix} \tag{3.31.1}
$$

By substituting Eq. (3.27.1) into Eq. (3.31.1):

$$
\begin{bmatrix} c^2 & s^2 & 2cs \\ s^2 & c^2 & -2cs \\ -cs & cs & (c^2-s^2) \end{bmatrix} \begin{bmatrix} \sigma_{xx} \\ \sigma_{yy} \\ \tau_{xy} \end{bmatrix} = \begin{bmatrix} Q_{11} & Q_{12} & Q_{16} \\ Q_{21} & Q_{22} & Q_{26} \\ Q_{61} & Q_{62} & Q_{66} \end{bmatrix} \begin{bmatrix} c^2 & s^2 & cs \\ s^2 & c^2 & -cs \\ -2cs & 2cs & (c^2-s^2) \end{bmatrix} \begin{bmatrix} \varepsilon_{xx} \\ \varepsilon_{yy} \\ \gamma_{xy} \end{bmatrix}
$$

$$\tag{3.31.2}$$

$$
\begin{bmatrix} \sigma_{xx} \\ \sigma_{yy} \\ \tau_{xy} \end{bmatrix} = \begin{bmatrix} c^2 & s^2 & 2cs \\ s^2 & c^2 & -2cs \\ -cs & cs & (c^2-s^2) \end{bmatrix}^{-1} \begin{bmatrix} Q_{11} & Q_{12} & Q_{16} \\ Q_{21} & Q_{22} & Q_{26} \\ Q_{61} & Q_{62} & Q_{66} \end{bmatrix} \begin{bmatrix} c^2 & s^2 & cs \\ s^2 & c^2 & -cs \\ -2cs & 2cs & (c^2-s^2) \end{bmatrix} \begin{bmatrix} \varepsilon_{xx} \\ \varepsilon_{yy} \\ \gamma_{xy} \end{bmatrix}
$$

$$\tag{3.31.3}$$

where
$c = \text{Cos } \theta$
$s = \text{Sin } \theta$
$\theta = $ angle between axis X_1 and X (refer to Figure 3.14).

Transformation of the reduced stiffness matrix of generally anisotropic materials under in-plane stress condition is presented as:

$$
[\bar{Q}_{ij}^R] = \begin{bmatrix} \bar{Q}_{11} & \bar{Q}_{12} & \bar{Q}_{16} \\ \bar{Q}_{12} & \bar{Q}_{22} & \bar{Q}_{26} \\ \bar{Q}_{16} & \bar{Q}_{26} & \bar{Q}_{66} \end{bmatrix}
$$

$$
= \begin{bmatrix} c^2 & s^2 & 2cs \\ s^2 & c^2 & -2cs \\ -cs & cs & (c^2-s^2) \end{bmatrix}^{-1} \begin{bmatrix} Q_{11} & Q_{12} & Q_{16} \\ Q_{21} & Q_{22} & Q_{26} \\ Q_{61} & Q_{62} & Q_{66} \end{bmatrix} \begin{bmatrix} c^2 & s^2 & cs \\ s^2 & c^2 & -cs \\ -2cs & 2cs & (c^2-s^2) \end{bmatrix} \tag{3.31.4}
$$

As mentioned before, most FRP composite materials are made of laminae of orthotropic material. The transformation of the reduced stiffness matrix in Eq. (3.31.4) is simplified by replacing the generally anisotropic reduced stiffness with the orthogonally reduced stiffness in Eq. (3.16.3) (or taking Q_{16} and Q_{26} to be zero.). Then, the transformation of the reduced stiffness of orthotropic materials is presented as:

$$
\left[\bar{Q}_{ij}^R \right] =
\begin{bmatrix}
\bar{Q}_{11} & \bar{Q}_{12} & \bar{Q}_{16} \\
\bar{Q}_{12} & \bar{Q}_{22} & \bar{Q}_{26} \\
\bar{Q}_{16} & \bar{Q}_{26} & \bar{Q}_{66}
\end{bmatrix}
$$

$$
=
\begin{bmatrix}
c^2 & s^2 & -2cs \\
s^2 & c^2 & 2cs \\
cs & -cs & c^2 - s^2
\end{bmatrix}
\begin{bmatrix}
\dfrac{E_{11}}{1 - v_{12}v_{21}} & \dfrac{v_{12}E_{22}}{1 - v_{12}v_{21}} & 0 \\
\dfrac{v_{21}E_{11}}{1 - v_{12}v_{21}} & \dfrac{E_{22}}{1 - v_{12}v_{21}} & 0 \\
0 & 0 & G_{12}
\end{bmatrix}
\begin{bmatrix}
c^2 & s^2 & cs \\
s^2 & c^2 & -cs \\
-2cs & 2cs & c^2 - s^2
\end{bmatrix}
$$

$$(3.31.5)$$

3.6.6 TRANSFORMATION OF THE REDUCED STIFFNESS MATRIX WITH HYGROTHERMAL EFFECT

The transformation of the reduced stress–strain relation with hygrothermal effects under plane stress condition is obtained by subtracting transformation of the reduced total strains from the hygrothermal strains as given below:

$$
\begin{bmatrix}
\sigma_{xx} \\
\sigma_{yy} \\
\tau_{xy}
\end{bmatrix}
=
\begin{bmatrix}
\bar{Q}_{11} & \bar{Q}_{12} & \bar{Q}_{16} \\
\bar{Q}_{12} & \bar{Q}_{22} & \bar{Q}_{26} \\
\bar{Q}_{16} & \bar{Q}_{26} & \bar{Q}_{66}
\end{bmatrix}
\begin{bmatrix}
\varepsilon_{xx} - \varepsilon_{xx}^T - \varepsilon_{xx}^M \\
\varepsilon_{yy} - \varepsilon_{yy}^T - \varepsilon_{yy}^M \\
\gamma_{xy} - \gamma_{xy}^T - \gamma_{xy}^M
\end{bmatrix}
$$

$$(3.32.1)$$

$$
=
\begin{bmatrix}
\bar{Q}_{11} & \bar{Q}_{12} & \bar{Q}_{16} \\
\bar{Q}_{12} & \bar{Q}_{22} & \bar{Q}_{26} \\
\bar{Q}_{16} & \bar{Q}_{26} & \bar{Q}_{66}
\end{bmatrix}
\begin{bmatrix}
\varepsilon_{xx} - \Delta T \alpha_{xx} - \Delta M \beta_{xx} \\
\varepsilon_{yy} - \Delta T \alpha_{yy} - \Delta M \beta_{yy} \\
\gamma_{xy} - \Delta T \alpha_{xy} - \Delta M \beta_{xy}
\end{bmatrix}
$$

$$
\begin{bmatrix}
\sigma_{xx} \\
\sigma_{yy} \\
\tau_{xy}
\end{bmatrix}
=
\begin{bmatrix}
\bar{Q}_{11} & \bar{Q}_{12} & \bar{Q}_{16} \\
\bar{Q}_{12} & \bar{Q}_{22} & \bar{Q}_{26} \\
\bar{Q}_{16} & \bar{Q}_{26} & \bar{Q}_{66}
\end{bmatrix}
\begin{bmatrix}
\varepsilon_{xx}^{mech} \\
\varepsilon_{yy}^{mech} \\
\gamma_{xy}^{mech}
\end{bmatrix}
$$

$$(3.32.2)$$

As in compliance $\bar{S}_{16}$ and $\bar{S}_{26}$, not only the axial stress responses are induced under the axial strains but also the shear stresses are induced due to the effects of $\bar{Q}_{16}$ and $\bar{Q}_{26}$. In addition, the coefficients $\bar{Q}_{16}$ and $\bar{Q}_{26}$ of the reduced stiffness matrix will lead to the shear-extension (A_{16} and A_{26}), extension-twist (B_{16} and B_{26}), and bending-twist couplings (D_{16} and D_{26}). The discussion of the coupling effects is elaborated in Section 4.6.

Example 3.6

The properties of unidirectional FRP lamina 45° off-axis are presented as: $E_{11} = 180\,\text{GPa}$, $E_{22} = 40\,\text{GPa}$, $G_{12} = 20\,\text{GPa}$, $\nu_{12} = 0.25$, $\alpha_{11} = 7\,\mu\text{m/m °C}$, $\alpha_{22} = 25\,\mu\text{m/m °C}$, $\beta_{11} = 120\,\mu\text{m/m \%M}$, and $\beta_{22} = 1{,}000\,\mu\text{m/m \%M}$

   a. Determine the state of strains in the global coordinate system (X–Y) under the state of stresses as: $\sigma_{11} = 25\,\text{MPa}$, $\sigma_{22} = 10\,\text{MPa}$, $\tau_{12} = 15\,\text{MPa}$, $\Delta T = 80°\text{C}$ and $\Delta M = 2\%$.

   b. Determine the state of stresses in global coordinate system (X–Y) under the state of strains as: $\varepsilon_{11} = -1{,}000\,\mu\text{m/m}$, $\varepsilon_{22} = 1{,}500\,\mu\text{m/m}$ and $\gamma_{12} = 1{,}000\,\mu\text{rad}$, $\Delta T = 80°\text{C}$ and $\Delta M = 2\%$.

   c. Determine the state of stress in the global (x–y) coordinate system for fixed boundary condition under hygrothermal loads ($\Delta T = 80°\text{C}$) and ($\Delta M = 2\%$)

Solution

a. State of strains under the state of stresses:

For lamina 45° (where $\theta = 45°$, $c = \text{Cos } 45°$ and $s = \text{Sin } 45°$), the transformation of reduced compliance of orthotropic materials can be determined from the relation in Eq. (3.29) as:

$$
\begin{bmatrix} \bar{S}_{11} & \bar{S}_{12} & \bar{S}_{16} \\ \bar{S}_{12} & \bar{S}_{22} & \bar{S}_{26} \\ \bar{S}_{16} & \bar{S}_{26} & \bar{S}_{66} \end{bmatrix} = \begin{bmatrix} 0.0194 & -5.55 \times 10^{-3} & -9.72 \times 10^{-3} \\ -5.55 \times 10^{-3} & 0.0194 & -9.72 \times 10^{-3} \\ -9.72 \times 10^{-3} & -9.72 \times 10^{-3} & 0.033 \end{bmatrix} \frac{1}{\text{GPa}}
$$

From Eq. (3.30.3), unrestrained (free) hygrothermal strains in the global coordinate system (X–Y) are determined as:

$$
\begin{bmatrix} \varepsilon_{xx} \\ \varepsilon_{yy} \\ \gamma_{xy} \end{bmatrix}^{T \text{ and } M} = \Delta T \begin{bmatrix} \alpha_{xx} \\ \alpha_{yy} \\ \alpha_{xy} \end{bmatrix} + \Delta M \begin{bmatrix} \beta_{xx} \\ \beta_{yy} \\ \beta_{xy} \end{bmatrix}
$$

$$
= \begin{bmatrix} 0.5 & 0.5 & -0.5 \\ 0.5 & 0.5 & 0.5 \\ 1 & -1 & 0 \end{bmatrix} \begin{bmatrix} 7(80) + 120(2) \\ 25(80) + 1{,}000(2) \\ 0 \end{bmatrix} = \begin{bmatrix} 2{,}400 \\ 2{,}400 \\ -3{,}200 \end{bmatrix} \mu
$$

Note: Unit: normal strain ε_{ii} (m/m) and shear strain (γ_{ij}) rad/rad

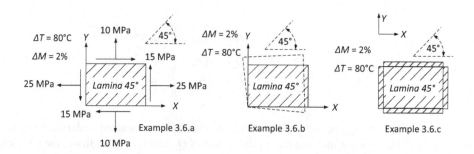

EXAMPLE FIGURE 3.6.1 Unidirectional FRP lamina 45° off-axis under loadings.

From Eq. (3.30.4):

$$
\begin{bmatrix} \varepsilon_{xx} \\ \varepsilon_{yy} \\ \gamma_{xy} \end{bmatrix} = \begin{bmatrix} 0.0194 & -5.55 \times 10^{-3} & -9.72 \times 10^{-3} \\ -5.55 \times 10^{-3} & 0.0194 & -9.72 \times 10^{-3} \\ -9.72 \times 10^{-3} & -9.72 \times 10^{-3} & 0.033 \end{bmatrix} GPa^{-1} \begin{bmatrix} 25 \\ 10 \\ 15 \end{bmatrix} MPa + \begin{bmatrix} 2,400 \\ 2,400 \\ -3,200 \end{bmatrix} \mu
$$

$$
\begin{bmatrix} \varepsilon_{xx} \\ \varepsilon_{yy} \\ \gamma_{xy} \end{bmatrix} = \begin{bmatrix} 2,685 \\ 2,310 \\ -3,040 \end{bmatrix} \mu
$$

For off-axis orthotropic lamina, in-plane shear strain is induced under hygrothermal strains. However, the in-plane shear strain does not exist for orthotropic lamina that has a material coordinate system aligned with the same orientation of the global coordinate system when the off-axis orthotropic lamina is under unrestrained (free) boundary condition. Therefore, the total strain of a lamina is the sum of induced strains from both mechanical and hygrothermal loads. However, induced stresses of off-axis orthotropic lamina due to hygrothermal load (strains) under free boundary condition do not exist, and the induced stresses of lamina are due to the applied load-induced mechanical strains only.

From Eq. (3.30.4):

$$
\begin{bmatrix} \varepsilon_{xx}^{mech} \\ \varepsilon_{yy}^{mech} \\ \gamma_{xy}^{mech} \end{bmatrix} = \begin{bmatrix} \varepsilon_{xx} - \varepsilon_{xx}^{T} - \varepsilon_{xx}^{M} \\ \varepsilon_{yy} - \varepsilon_{yy}^{T} - \varepsilon_{yy}^{M} \\ \gamma_{xy} - \gamma_{xy}^{T} - \gamma_{xy}^{M} \end{bmatrix} = \begin{bmatrix} 2,685 - 2,400 \\ 2,310 - 2,400 \\ -3,040 + 3,200 \end{bmatrix} \mu = \begin{bmatrix} 285 \\ -90 \\ 160 \end{bmatrix} \mu
$$

$$
\begin{bmatrix} \varepsilon_{xx}^{mech} \\ \varepsilon_{yy}^{mech} \\ \gamma_{xy}^{mech} \end{bmatrix} = \begin{bmatrix} \bar{S}_{11} & \bar{S}_{12} & \bar{S}_{16} \\ \bar{S}_{12} & \bar{S}_{22} & \bar{S}_{26} \\ \bar{S}_{16} & \bar{S}_{26} & \bar{S}_{66} \end{bmatrix} \begin{bmatrix} \sigma_{xx} \\ \sigma_{yy} \\ \tau_{xy} \end{bmatrix} = \begin{bmatrix} \dfrac{1}{150} & -\dfrac{0.25}{150} & 0 \\ -\dfrac{0.25}{150} & \dfrac{1}{20} & 0 \\ 0 & 0 & \dfrac{1}{15} \end{bmatrix} \begin{bmatrix} 25 \\ 10 \\ 15 \end{bmatrix} = \begin{bmatrix} 285 \\ -90 \\ 160 \end{bmatrix} \mu
$$

b. **State of stresses under the state of strains:**

From Eqs. (3.30.3 and 3.30.4), mechanical strains in the global coordinate system (X–Y) can be calculated as:

$$
\begin{bmatrix} \varepsilon_{xx}^{mech} \\ \varepsilon_{yy}^{mech} \\ \gamma_{xy}^{mech} \end{bmatrix} = \begin{bmatrix} \varepsilon_{xx} - \varepsilon_{xx}^{T} - \varepsilon_{xx}^{M} \\ \varepsilon_{yy} - \varepsilon_{yy}^{T} - \varepsilon_{yy}^{M} \\ \gamma_{xy} - \gamma_{xy}^{T} - \gamma_{xy}^{M} \end{bmatrix} = \begin{bmatrix} -1,000 - 2,400 \\ 1,500 - 2,400 \\ 1,000 + 3,200 \end{bmatrix} = \begin{bmatrix} -3,400 \\ -900 \\ 4,200 \end{bmatrix} \mu
$$

From Eq. (3.31.5) or an inversion of the reduced compliance matrix in the global coordinate system, the reduced stiffness matrix in the global coordinate system is:

$$
\begin{bmatrix} \bar{Q}_{11} & \bar{Q}_{12} & \bar{Q}_{16} \\ \bar{Q}_{12} & \bar{Q}_{22} & \bar{Q}_{26} \\ \bar{Q}_{16} & \bar{Q}_{26} & \bar{Q}_{66} \end{bmatrix} = \begin{bmatrix} \bar{S}_{11} & \bar{S}_{12} & \bar{S}_{16} \\ \bar{S}_{12} & \bar{S}_{22} & \bar{S}_{26} \\ \bar{S}_{16} & \bar{S}_{26} & \bar{S}_{66} \end{bmatrix}^{-1}
$$

$$
\begin{bmatrix} \bar{Q}_{11} & \bar{Q}_{12} & \bar{Q}_{16} \\ \bar{Q}_{12} & \bar{Q}_{22} & \bar{Q}_{26} \\ \bar{Q}_{16} & \bar{Q}_{26} & \bar{Q}_{66} \end{bmatrix} = \begin{bmatrix} 0.0194 & -5.55\times10^{-3} & -9.72\times10^{-3} \\ -5.55\times10^{-3} & 0.0194 & -9.72\times10^{-3} \\ -9.72\times10^{-3} & -9.72\times10^{-3} & 0.033 \end{bmatrix}^{-1} \text{GPa}
$$

$$
\begin{bmatrix} \bar{Q}_{11} & \bar{Q}_{12} & \bar{Q}_{16} \\ \bar{Q}_{12} & \bar{Q}_{22} & \bar{Q}_{26} \\ \bar{Q}_{16} & \bar{Q}_{26} & \bar{Q}_{66} \end{bmatrix} = \begin{bmatrix} 80.8 & 40.8 & 35.5 \\ 40.8 & 80.8 & 35.5 \\ 35.5 & 35.5 & 50.7 \end{bmatrix} \text{GPa}
$$

From Eq. (3.32.1), the state of stress in the global coordinate system can be determined as:

$$
\begin{bmatrix} \sigma_{xx} \\ \sigma_{yy} \\ \tau_{xy} \end{bmatrix} = \begin{bmatrix} 80.8 & 40.8 & 35.5 \\ 40.8 & 80.8 & 35.5 \\ 35.5 & 35.5 & 50.7 \end{bmatrix} \begin{bmatrix} -3,400 \\ -900 \\ 4,200 \end{bmatrix} = \begin{bmatrix} -163 \\ -62.6 \\ -60.3 \end{bmatrix} \text{MPa}
$$

The state of stresses under the state of strains and hygrothermal loads as shown in Example Figure 3.6.2

c. **State of stress for fixed boundary condition under hygrothermal loads**

Total strains in an orthotropic lamina are equal to zero due to fixed boundary condition. However, mechanical and hygrothermal strains are not equal to zero as illustrated below:

From Eq. (30.4)

$$
\begin{bmatrix} \varepsilon_{xx} \\ \varepsilon_{yy} \\ \gamma_{xy} \end{bmatrix} = \begin{bmatrix} 0 \\ 0 \\ 0 \end{bmatrix} = \begin{bmatrix} \varepsilon_{xx}^{mech} + \varepsilon_{xx}^{T} + \varepsilon_{xx}^{M} \\ \varepsilon_{yy}^{mech} + \varepsilon_{yy}^{T} + \varepsilon_{yy}^{M} \\ \gamma_{xy}^{mech} + \gamma_{xy}^{T} + \gamma_{xy}^{M} \end{bmatrix}
$$

$$
\begin{bmatrix} \varepsilon_{xx}^{mech} \\ \varepsilon_{yy}^{mech} \\ \gamma_{xy}^{mech} \end{bmatrix} = \begin{bmatrix} \varepsilon_{xx} - \alpha_{xx}\Delta T - \beta_{xx}\Delta M \\ \varepsilon_{yy} - \alpha_{yy}\Delta T - \beta_{yy}\Delta M \\ \gamma_{xy} - \alpha_{xy}\Delta T - \beta_{xy}\Delta M \end{bmatrix} = \begin{bmatrix} 0-2,400 \\ 0-2,400 \\ 0+3,200 \end{bmatrix} = \begin{bmatrix} -2,400 \\ -2,400 \\ 3,200 \end{bmatrix} \mu
$$

Note: Unit: normal strain ε_{ij} (m/m) and shear strain (γ_{ij}) rad/rad

By substituting the mechanical strains and the reduced stiffness values into Eq. (3.32.1), we obtain:

$$
\begin{bmatrix} \sigma_{xx} \\ \sigma_{yy} \\ \tau_{xy} \end{bmatrix} = \begin{bmatrix} \bar{Q}_{11} & \bar{Q}_{12} & \bar{Q}_{16} \\ \bar{Q}_{12} & \bar{Q}_{22} & \bar{Q}_{26} \\ \bar{Q}_{16} & \bar{Q}_{26} & \bar{Q}_{66} \end{bmatrix} \begin{bmatrix} \varepsilon_{xx} - \Delta T\alpha_{xx} - \Delta M\beta_{xx} \\ \varepsilon_{yy} - \Delta T\alpha_{yy} - \Delta M\beta_{yy} \\ \gamma_{xy} - \Delta T\alpha_{xy} - \Delta M\beta_{xy} \end{bmatrix}
$$

$$
= \begin{bmatrix} 80.8 & 40.8 & 35.5 \\ 40.8 & 80.8 & 35.5 \\ 35.5 & 35.5 & 50.7 \end{bmatrix} \begin{bmatrix} -2400 \\ -2400 \\ 3200 \end{bmatrix} = \begin{bmatrix} -178 \\ -178 \\ -8.16 \end{bmatrix} \text{MPa}
$$

The state of stresses under hygrothermal loads as shown in Example Figure 3.6.2

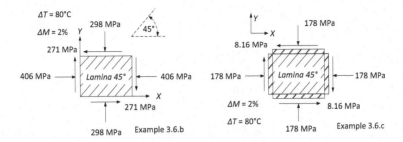

EXAMPLE FIGURE 3.6.2   State of stresses in Example 3.4b and 3.4c.

Example 3.7

The unidirectional FRP lamina 45° off-axis in Example 3.6

    a. Determine the state of strains in the global coordinate system under applied stress (σ_{xx}) of 10 MPa for fixed boundary condition.

    b. Determine the state of strains under applied stress (σ_{xx}) of 10 MPa with hygrothermal loads ($\Delta T = 80°C$ and $\Delta M = 2\%$) for fixed boundary conditions

Solution

a. State of strains under the applied stress and boundary condition:

From the reduced compliance matrix of Example 3.6 and Eq. (3.28.3):

$$\begin{bmatrix} \varepsilon_{xx} \\ \varepsilon_{yy} \\ \gamma_{xy} \end{bmatrix} = \begin{bmatrix} 0.0194 & -5.55\times10^{-3} & -9.72\times10^{-3} \\ -5.55\times10^{-3} & 0.0194 & -9.72\times10^{-3} \\ -9.72\times10^{-3} & -9.72\times10^{-3} & 0.033 \end{bmatrix} \frac{1}{\text{GPa}} \begin{bmatrix} 10 \\ \sigma_{yy} \\ 0 \end{bmatrix} \text{MPa}$$

From the second equation of the above simultaneous equations, induced stress (σ_{yy}) is determined as:

$$-5.55\times10^{-3}(10)\frac{\text{MPa}}{\text{GPa}} + 0.0194\left(\sigma_{yy}\right) = 0$$

$$\sigma_{yy} = 2.86\,\text{MPa}$$

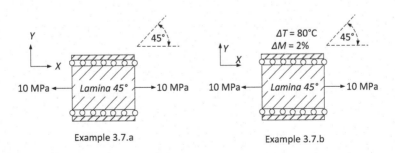

Example 3.7.a

Example 3.7.b

EXAMPLE FIGURE 3.7.1   Unidirectional FRP lamina 45° off-axis under stresses and boundary conditions.

From the first and third equations of the above simultaneous equations, strains (ε_{xx} and γ_{xy}) are determined as:

$$\varepsilon_{xx} = 10(0.0194)\frac{\text{MPa}}{\text{GPa}} - (2.86)5.55\left(10^{-3}\right)\frac{\text{MPa}}{\text{GPa}} = 178.1\ \mu$$

$$\gamma_{xy} = -(10)9.72(10^{-3})\frac{\text{MPa}}{\text{GPa}} - (2.86)9.72\left(10^{-3}\right)\frac{\text{MPa}}{\text{GPa}} = -125\mu$$

When applied normal stress is off-axis of the material coordinate axis, then both in-plane normal and shear strains are induced.

b. **State of strains under applied stress with hygrothermal loads for fixed boundary condition:**

From Eq. (3.30.3), unrestrained (free) hygrothermal strains in the global coordinate system (X–Y) are determined as:

$$\begin{bmatrix} \varepsilon_{xx} \\ \varepsilon_{yy} \\ \gamma_{xy} \end{bmatrix}^{T \text{ and } M} = \Delta T \begin{bmatrix} \alpha_{xx} \\ \alpha_{yy} \\ \alpha_{xy} \end{bmatrix} + \Delta M \begin{bmatrix} \beta_{xx} \\ \beta_{yy} \\ \beta_{xy} \end{bmatrix}$$

$$= \begin{bmatrix} 0.5 & 0.5 & -0.5 \\ 0.5 & 0.5 & 0.5 \\ 1 & -1 & 0 \end{bmatrix} \begin{bmatrix} 7(80) + 120(2) \\ 25(80) + 1,000(2) \\ 0 \end{bmatrix} = \begin{bmatrix} 2,400 \\ 2,400 \\ -3,200 \end{bmatrix} \mu$$

Note: Unit: normal strain ε_{ii} (m/m) and shear strain (γ_{ij}) rad/rad

By substituting Eq. (3.30.4) and the reduced compliance matrix of Example 3.6, the normal stress (σ_{22}) is induced along the fixed boundary condition in the X_1 direction.

$$\begin{bmatrix} \varepsilon_{xx} \\ \varepsilon_{yy} \\ \gamma_{xy} \end{bmatrix} = \begin{bmatrix} \overline{S}_{11} & \overline{S}_{12} & \overline{S}_{16} \\ \overline{S}_{12} & \overline{S}_{22} & \overline{S}_{26} \\ \overline{S}_{16} & \overline{S}_{26} & \overline{S}_{66} \end{bmatrix} \begin{bmatrix} \sigma_{xx} \\ \sigma_{yy} \\ \tau_{xy} \end{bmatrix} + \Delta T \begin{bmatrix} \alpha_{xx} \\ \alpha_{yy} \\ \alpha_{xy} \end{bmatrix} + \Delta M \begin{bmatrix} \beta_{xx} \\ \beta_{yy} \\ \beta_{xy} \end{bmatrix}$$

$$\begin{bmatrix} \varepsilon_{xx} \\ \varepsilon_{yy} \\ \gamma_{xy} \end{bmatrix} = \begin{bmatrix} 0.0194 & -5.55\times10^{-3} & -9.72\times10^{-3} \\ -5.55\times10^{-3} & 0.0194 & -9.72\times10^{-3} \\ -9.72\times10^{-3} & -9.72\times10^{-3} & 0.033 \end{bmatrix} \begin{bmatrix} 10 \\ \sigma_{yy} \\ 0 \end{bmatrix} + \begin{bmatrix} 2,400 \\ 2,400 \\ -3,200 \end{bmatrix} \mu$$

From the second equation of the above simultaneous equations, the induced stress (σ_{yy}) is determined as:

$$-5.55\left(10^{-3}\right)10\frac{\text{MPa}}{\text{GPa}} + 0.0194\sigma_{yy}\frac{1}{\text{GPa}} + 2,400\mu = 0$$

$$\sigma_{yy} = -123.4\ \text{MPa}$$

From the first and third equations of the above simultaneous equations, the induced strains (ε_{xx} and γ_{xy}) are determined as:

$$\varepsilon_{xx} = 10(0.0194)\frac{\text{MPa}}{\text{GPa}} + 123.4(5.55)\left(10^{-3}\right)\frac{\text{MPa}}{\text{GPa}} + 2,400\mu = 3,279\ \mu$$

$$\gamma_{xy} = -10(9.72)\left(10^{-3}\right)\frac{\text{MPa}}{\text{GPa}} + 123.4(9.72)\left(10^{-3}\right)\frac{\text{MPa}}{\text{GPa}} - 3,200\mu = -2,098\ \mu$$

Note: Unit: normal strain ε_{ii} (m/m) and shear strain (γ_{ij}) rad/rad

3.7   ENGINEERING CONSTANTS IN GLOBAL COORDINATE SYSTEM

For orthotropic materials, the engineering constants in the global (X–Y) coordinate system can be written in the principle material (X_1–X_2) coordinate system. In this section, elastic extensional (E) modulus, shear modulus (G), and Poisson's ratio (ν) are presented. Let the global and principle material coordinate systems be defined in the Cartesian coordinates (X–Y) and (X_1–X_2), respectively. Consider an orthotropic lamina that has off-axis fiber orientation under unidirectional stress (σ_{xx}) in Figure 3.16. Then, strain responses of orthotropic lamina include both dilatation and distortion.

From Eq. (3.28.3):

$$
\begin{bmatrix} \varepsilon_{xx} \\ \varepsilon_{yy} \\ \gamma_{xy} \end{bmatrix} =
\begin{bmatrix} \overline{S}_{11} & \overline{S}_{12} & \overline{S}_{16} \\ \overline{S}_{12} & \overline{S}_{22} & \overline{S}_{26} \\ \overline{S}_{16} & \overline{S}_{26} & \overline{S}_{66} \end{bmatrix}
\begin{bmatrix} \sigma_{xx} \\ 0 \\ 0 \end{bmatrix}
\tag{3.33.1}
$$

$$
\varepsilon_{xx} = \overline{S}_{11}\sigma_{xx}
\tag{3.33.2}
$$

$$
E_{xx} = \frac{1}{\overline{S}_{11}}
\tag{3.33.3}
$$

$$
E_{xx} = \frac{1}{S_{11}c^4 + (2S_{12} + S_{66})c^2 s^2 + S_{22}s^4}
\tag{3.33.4}
$$

The compliance elements (S_{11}, S_{12}, and S_{66}) in Eq. (3.16.1) are substituted into Eq. (3.33.4). The elastic modulus (E_{xx}) in the global (X–Y) coordinate system can be presented in terms of engineering constants in the local (X_1–Y_2) coordinate system as:

$$
E_{xx} = \frac{E_{11}}{c^4 + \left(\dfrac{E_{11}}{G_{12}} - 2\nu_{12} \right)c^2 s^2 + \dfrac{E_{11}}{E_{22}}s^4}
\tag{3.33.5}
$$

where $c = \mathrm{Cos}\,\theta$
  $s = \mathrm{Sin}\,\theta$
  θ = angle between axis X_1 and X (refer to Figure 3.14).
Poisson's ratio under applied stress in the global X direction is obtained in the same manner.

$$
\nu_{xy} = -\frac{\varepsilon_{yy}}{\varepsilon_{xx}} = -\frac{\overline{S}_{12}}{\overline{S}_{11}}\frac{\sigma_{xx}}{\sigma_{xx}} = -\frac{(S_{11} + S_{22} - S_{66})c^2 s^2 + S_{12}\left(c^4 + s^4\right)}{S_{11}c^4 + (2S_{12} + S_{66})c^2 s^2 + S_{22}s^4}
\tag{3.34.1}
$$

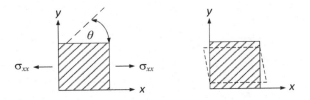

FIGURE 3.16   Orthotropic lamina of off-axis fiber orientations under stress σ_{xx}.

$$v_{xy} = -\frac{v_{12}\left(c^4 + s^4\right) - \left(1 + \dfrac{E_{11}}{E_{22}} - \dfrac{E_{11}}{G_{12}}\right)c^2 s^2}{c^4 + \left(\dfrac{E_{11}}{G_{12}} - 2v_{12}\right)c^2 s^2 + \dfrac{E_{11}}{E_{22}}s^2} \qquad (3.34.2)$$

where
  $c = \cos\theta$
  $s = \sin\theta$
  θ = angle between axis X_1 and X (refer to Figure 3.14).

If the unidirectional stress σ_{xx} on the same lamina of the above case is replaced by the unidirectional stress (σ_{yy}) as shown in Figure 3.17, then the strain responses still include both dilatation and distortion.

The elastic modulus (E_{yy}) in the global Y direction is presented as below:

$$\varepsilon_{yy} = \bar{S}_{22}\sigma_{yy} = \frac{1}{E_{yy}}\sigma_{yy} \qquad (3.35.1)$$

$$E_{yy} = \frac{1}{S_{11}s^4 + \left(2S_{12} + S_{66}\right)c^2 s^2 + S_{22}c^4} \qquad (3.35.2)$$

where
  $c = \cos\theta$
  $s = \sin\theta$
  θ = angle between axis X_1 and X (refer to Figure 3.14).

The compliance elements in Eq. (3.16.1) are substituted into Eq. (3.43.2). Thus, the elastic modulus in the global Y direction is presented as:

$$E_{yy} = \frac{E_{22}}{c^4 + \left(\dfrac{E_{22}}{G_{12}} - 2v_{21}\right)c^2 s^2 + \dfrac{E_{22}}{E_{11}}s^4} \qquad (3.35.3)$$

Poisson's ratio (v_{yx}) under unidirectional stress in the global y-direction is obtained in the same manner.

$$\left(\bar{S}_{21} = \bar{S}_{12}\right)$$

$$v_{yx} = -\frac{\varepsilon_{xx}}{\varepsilon_{yy}} = -\frac{\bar{S}_{21}}{\bar{S}_{22}}\frac{\sigma_{yy}}{\sigma_{yy}} = -\frac{\left(S_{11} + S_{22} - S_{66}\right)c^2 s^2 + S_{12}\left(c^4 + s^4\right)}{S_{11}s^4 + \left(2S_{12} + S_{66}\right)c^2 s^2 + S_{22}c^4} \qquad (3.36.1)$$

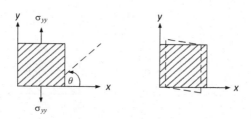

FIGURE 3.17 Orthotropic lamina of off-axis fiber orientations under stress σ_{yy}.

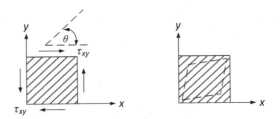

FIGURE 3.18 Orthotropic lamina of off-axis fiber orientations under stress (τ_{xy}).

$$v_{yx} = -\frac{v_{21}\left(c^4 + s^4\right) - \left(1 + \dfrac{E_{22}}{E_{11}} - \dfrac{E_{22}}{G_{12}}\right)c^2 s^2}{c^4 + \left(\dfrac{E_{22}}{G_{12}} - 2v_{12}\right)c^2 s^2 + \dfrac{E_{22}}{E_{11}}s^2} \tag{3.36.2}$$

Only shear stress (τ_{xy}) is applied on the composite lamina as in the previous case (see in Figure 3.18).

The in-plane shear strain in the global (X–Y) coordinate system is given as:

$$\gamma_{xy} = \bar{S}_{66}\tau_{xy} = \frac{1}{G_{xy}}\tau_{xy} \tag{3.37.1}$$

From Eq. (3.28.3):

$$G_{xy} = \frac{1}{2\left(2S_{11} + 2S_{12} - 4S_{12} - S_{66}\right)c^2 s^2 - S_{66}\left(c^4 + s^4\right)} \tag{3.37.2}$$

where

$c = \text{Cos}\,\theta$

$s = \text{Sin}\,\theta$

$\theta =$ angle between axis X_1 and X (refer to Figure 3.14).

The compliance elements in Eq. (3.16.1) are substituted into Eq. (3.37.2). Thus, the elastic shear modulus in the global (XY) plane is presented as:

$$G_{xy} = \frac{G_{12}}{2\left(2\dfrac{G_{12}}{E_{11}}(1 + 2v_{12}) + 2\dfrac{G_{12}}{E_{22}} - 1\right)c^2 s^2 + \left(c^4 + s^4\right)} \tag{3.37.3}$$

In addition, the linear coefficients of the hygrothermal effects in the global (X–Y) coordinate system are also written in terms of the additional resultant of the hygrothermal coefficients in the local (X_1–X_2) coordinate system. The relations of linear thermal and moisture expansion coefficients between the local (X_1–X_2) and global (X–Y) coordinate system are presented in Eq. (3.30.3) as:

For thermal expansion coefficients:

$$\alpha_{xx} = \alpha_{11}c^2 + \alpha_{22}s^2 \tag{3.38.1}$$

$$\alpha_{yy} = \alpha_{11}s^2 + \alpha_{22}c^2 \tag{3.38.2}$$

$$\alpha_{xy} = 2cs\left(\alpha_{11} - \alpha_{22}\right) \tag{3.38.3}$$

For moisture expansion coefficients:

$$\beta_{xx} = \beta_{11}c^2 + \beta_{22}s^2 \tag{3.38.4}$$

$$\beta_{yy} = \beta_{11}s^2 + \beta_{22}c^2 \tag{3.38.5}$$

$$\beta_{xy} = 2cs(\beta_{11} - \beta_{22}) \tag{3.38.6}$$

where $c = \text{Cos }\theta$
  $s = \text{Sin }\theta$
  θ = angle between axis X_1 and X (refer to Figure 3.14).

3.8 SUMMARY

The constitutive relations of composite lamina of orthotropic materials are presented in this chapter. In addition, constitutive relations of composite lamina of plane stress condition are used to establish both laminate stiffness and compliance matrices. Similarly, the hygro-thermomechanical responses of composite lamina in terms of stiffness matrix are also presented. The stress and strain relationships in global coordinates through transformations including the hygro-thermal effects are also highlighted. The knowledge of the mechanics of FRP composite lamina is the basis for understanding the structural response equations for multilayers of composite lamina, which is presented in Chapter 4.

EXERCISES

Problem 3.1: The in-plane properties of orthotropic lamina (carbon fiber) are presented as below: elastic modulus: $E_{11} = 180\,\text{GPa}$ and $E_{22} = 25\,\text{GPa}$, shear modulus: $G_{12} = 20\,\text{GPa}$, $\nu_{12} = 0.25$, thermal expansion: $\alpha_{11} = 7\,\mu\text{m/m °C}$, $\alpha_{22} = 18\,\mu\text{m/m °C}$, moisture expansion: $\beta_{11} = 120\,\mu\text{m/m \%M}$, $\beta_{22} = 1{,}250\,\mu\text{m/m \%M}$.

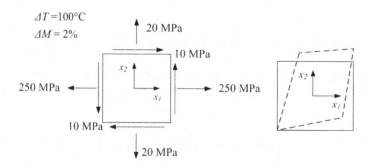

a. Determine the state of strains (total strain) in the coordinate system (x_1–x_2) under the state of stresses and hygrothermal loads ($\Delta T = 100°C$) and ($\Delta M = 2\%$), (b) Determine state of stresses in the coordinate system (x_1–x_2) under state of strains and hygrothermal loads ($\Delta T = 100°C$) and ($\Delta M = 2\%$) as follows: $\begin{bmatrix} \varepsilon_{11} \\ \varepsilon_{22} \\ \gamma_{12} \end{bmatrix} = \begin{bmatrix} -1{,}000\,\mu\text{m/m} \\ 1{,}500\,\mu\text{m/m} \\ 1{,}000\,\mu\text{rad} \end{bmatrix}$.

Problem 3.2: A unidirectional continuous glass fiber composite bonded with epoxy has a 50% fiber volume fraction. Using the constituent material properties given in Chapter 1, find the composites longitudinal tensile strength at failure.

Problem 3.3: A 2 ksi normal stress (σ_x) is applied on a unidirectional angle-ply epoxy composite lamina with glass fibers at 30° to the horizontal (x-axis). Find the stresses in the directions parallel and perpendicular to the fiber direction. Use material properties given in Chapter 1.

Problem 3.4: Find the stiffness matrix for an angle-ply composite lamina containing 70% glass fiber volume fraction with vinyl ester matrix. Assume that the fibers are in +/–45° and $E_f = 10$ msi, $v_f = 0.25$, $E_m = 0.5$ msi, and $v_m = 0.35$.

Problem 3.5: The in-plane properties of orthotropic lamina (carbon fiber) are presented as below: elastic modulus: $E_{11} = 180$ GPa and $E_{22} = 25$ GPa, shear modulus: $G_{12} = 20$ GPa, $v_{12} = 0.25$, thermal expansion: $\alpha_{11} = 7$ μm/m °C, $\alpha_{22} = 18$ μm/m °C, moisture expansion: $\beta_{11} = 120$ μm/m %M, and $\beta_{22} = 1{,}250$ μm/m %M. Determine the state of stress in the coordinate system $(x_1$–$x_2)$ under fixed boundary condition and hygrothermal loads $(\Delta T = 100°C)$ and $(\Delta M = 2\%)$.

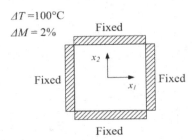

Problem 3.6: The in-plane properties of orthotropic lamina (carbon fiber) are presented as below: Elastic modulus: $E_{11} = 180$ GPa and $E_{22} = 25$ GPa, Shear modulus: $G_{12} = 20$ GPa, $v_{12} = 0.25$, Thermal expansion: $\alpha_{11} = 7$ μm/m °C, $\alpha_{22} = 18$ μm/m °C, Moisture expansion: $\beta_{11} = 120$ μm/m %M, and $\beta_{22} = 1{,}250$ μm/m %M. Determine the state of strains in the coordinate system $(x_1$–$x_2)$ under applied stress (σ_{11}) of 250 MPa, hygrothermal loads $(\Delta T = 100°C$ and $\Delta M = 2\%)$, and fixed boundary condition.

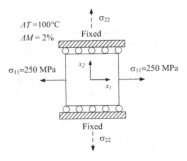

Problem 3.7: The properties of unidirectional FRP lamina 45° are presented as below: elastic modulus: $E_{11} = 200$ GPa and $E_{22} = 30$ GPa; shear modulus: $G_{12} = 15$ GPa and $v_{12} = 0.25$; thermal expansion: $\alpha_{11} = 8$ μm/m °C and $\alpha_{22} = 20$ μm/m °C; and moisture expansion: $\beta_{11} = 150$ μm/m %M and $\beta_{22} = 1{,}050$ μm/m %M.

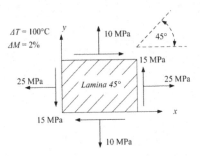

Determine the state of strains (total strain) in the global coordinate system (x–y) under the state

of stresses $\begin{bmatrix} \sigma_{xx} \\ \sigma_{yy} \\ \sigma_{xy} \end{bmatrix} = \begin{bmatrix} 25 \\ 10 \\ 15 \end{bmatrix}$ MPa and hygrothermal condition ($\Delta T = 100°C$) and ($\Delta M = 2\%$).

Problem 3.8: The properties of unidirectional FRP lamina 45° are presented as below: elastic modulus: $E_{11} = 200\,\text{GPa}$ and $E_{22} = 30\,\text{GPa}$; shear modulus: $G_{12} = 15\,\text{GPa}$ and $\nu_{12} = 0.25$; thermal expansion: $\alpha_{11} = 8\,\mu\text{m/m °C}$ and $\alpha_{22} = 20\,\mu\text{m/m °C}$; moisture expansion: $\beta_{11} = 150\,\mu\text{m/m \%M}$ and $\beta_{22} = 1{,}050\,\mu\text{m/m \%M}$.

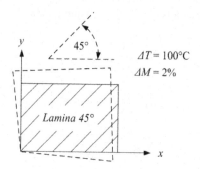

Determine the state of stresses in the global coordinate system (x–y) under the state of strains

$\begin{bmatrix} \varepsilon_{xx} \\ \varepsilon_{yy} \\ \gamma_{xy} \end{bmatrix} = \begin{bmatrix} -1{,}000\,\mu\text{m/m} \\ 1{,}500\,\mu\text{m/m} \\ 1{,}000\,\mu\text{rad} \end{bmatrix}$ and hygrothermal load ($\Delta T = 100°C$) and ($\Delta M = 2\%$)

Problem 3.9: The properties of unidirectional FRP lamina 45° are presented as below: elastic modulus: $E_{11} = 250\,\text{GPa}$ and $E_{22} = 50\,\text{GPa}$; shear modulus: $G_{12} = 25\,\text{GPa}$ and $\nu_{12} = 0.25$; thermal expansion: $\alpha_{11} = 12\,\mu\text{m/m °C}$ and $\alpha_{22} = 28\,\mu\text{m/m °C}$; and moisture expansion: $\beta_{11} = 120\,\mu\text{m/m \%M}$ and $\beta_{22} = 1{,}000\,\mu\text{m/m \%M}$. Determine the state of strains under applied stress (σ_{xx}) of 10 MPa with hygrothermal loads ($\Delta T = 100°C$ and $\Delta M = 2\%$) for fixed boundary conditions.

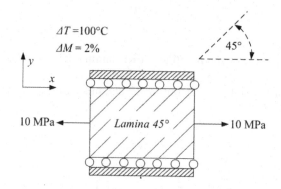

REFERENCES AND SELECTED BIOGRAPHY

Altenbach, H., Altenbach, J., and Kissing, W., *Mechanics of Composite Structural Elements*, Springer-Verlag, Berlin, Germany, 2004.

Barbero, E.J., *Introduction to Composite Materials Design*, Taylor &Francis, Inc., Philadelphia, PA, 1998.

Callister, W.D. Jr, *Fundamentals of Materials Science and Engineering*, 5th edition, John Wiley & Sons, Inc., New York, NY, 2000.

Daniel, I.M., and Ishai, O., *Engineering Mechanics of Composite Materials*, 2nd edition, Oxford University Press, Inc., New York, NY, 2006.

Gibson, R.F., *Principles of Composite Material Mechanics*, McGraw-Hill, Inc, New York, NY 1994.

Hyer, M.W., *Stress Analysis of Fiber-Reinforced Composite Materials*, International edition, WCB/McGraw-Hill, Singapore, 1998.

Jones, R.M., *Mechanics of Composite Materials*, Hemisphere Publishing, New York, NY, 1975.

Kaw, A.K., *Mechanics of Composite Materials*, 2nd edition, CRC Press, Taylor & Francis Group, Boca Raton, FL, 2006.

Kollar, L.P., and Springer, G.S., *Mechanics of Composite Structures*, Cambridge University Press, New York, NY, 2003.

Swanson, S.R., *Introduction to Design and Analysis with Advance Composite Materials*, International edition, Prentice-Hall, Inc., Upper Saddle River, NJ, 1997.

REFERENCES AND SELECTED BIBLIOGRAPHY

4 Mechanics of FRP Composite Laminates

The structural response equations of composite lamina (thin single layer) of Chapter 3 are the basis to establish the behavior of thin composite laminates, which are a combination or stack of several laminae. In general, the FRP structural members are made from multilayers of a single thin composite lamina. Different fiber orientations, materials (fiber/fabric types), and stacking sequences for a laminate formulation are a few of the many parameters necessary to obtain the required properties of structural elements (flanges and webs) or complete cross sections such as wide flange, circular, hollow rectangular, or other shapes. The FRP structural element or member properties vary with several parameters, as stated previously. To study the structural response influence from these parameters, the classical lamination theory (CLT) was developed (Jones, 1975). Technical details of CLT relating to strength and stiffness equations as functions of constituent material properties, and associated examples for different elements and members are presented in this chapter.

4.1 CLASSICAL LAMINATION THEORY

Several assumptions needed for the CLT development are (1) each lamina (layer) of a composite is orthotropic, (2) each layer is assumed to be homogenous (fiber and matrix are smeared together), (3) each layer satisfies the plane stress condition, (4) each layer is perfectly bonded to the next layer (no slip) with no voids, (5) responses under linear elastic and small deformation conditions are valid in the analysis, and (6) the Kirchhoff hypothesis (as given in Section 4.1.1) is valid.

4.1.1 KIRCHHOFF'S HYPOTHESIS

Kirchhoff's hypothesis was introduced in the mid-1800s. This hypothesis was initially applied to metallic structural members such as beams, plates, and shells. The Kirchhoff hypothesis states that a straight line that is perpendicular to the reference plane before deformation remains straight and perpendicular to the plane after deformation. Let us consider a laminated plate with multiple layers, as shown in Figure 4.1. The straight line (*AB*) passes through each layer. The line (*AB*) is

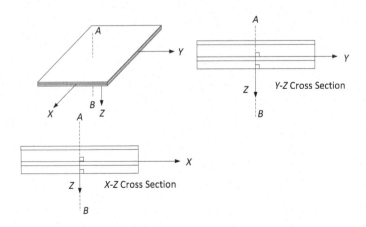

FIGURE 4.1 Undeformed laminated composites.

DOI: 10.1201/9781003196754-4

also perpendicular to the (X–Y) plane of each layer. After deformation, the straight line (AB) does not deform and remains perpendicular to the planes of each layer. It is assumed herein that the derivation is valid for small deformations and also valid for thin laminated plates, i.e., the smallest dimension (thickness) is no greater than around 5% of the next smallest of the two in-plane dimensions of a plate (say, width).

This hypothesis implies that out-of-plane shear strains (γ_{xz} and γ_{yz}) and normal strain (ε_z, perpendicular to X–Z plane) in the thickness direction of the laminated composites are assumed to be zero, as described in Section 3.3. It should be noted that normal strain (ε_z) under plane stress condition still exists. However, the validity of the normal strain (ε_z) assumption is not directly accounted for in applying Kirchhoff's hypothesis, in the strain–displacement relationship.

4.1.2 LAMINATED STRAIN AND DISPLACEMENT RELATIONSHIPS

From Kirchhoff's hypothesis, displacements at any point on a cross section are written in terms of displacements and rotations with respect to the reference surface. From Figures 4.2 and 4.3, if an angle between the reference surface before and after deformation is assumed to remain the same, then the tangent function of this rotated angle can be approximated by its own rotated angle. Thus, displacements through a cross section are obtained from two parts: (1) displacements at the reference surface and (2) displacements due to the rotated angle, away from the reference surface.

The displacements under Kirchhoff's hypothesis are linearly dependent on the vertical distance with respect to the reference surface. In addition, the vertical displacement is assumed to be independent of the thickness. Thus, the abovementioned displacements of both planes XZ and YZ are presented as follows:

$$u(x,y,z) = u^o(x,y) - z\frac{\partial w^o(x,y)}{\partial x} \tag{4.1.1}$$

$$v(x,y,z) = v^o(x,y) - z\frac{\partial w^o(x,y)}{\partial y} \tag{4.1.2}$$

where u and v are the in-plane displacements in X and Y directions, respectively.

w = vertical displacement in direction Z.

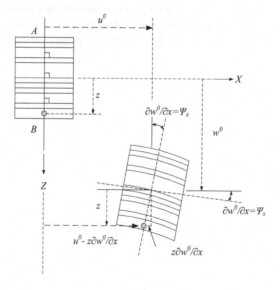

FIGURE 4.2 Deformed laminated plate: plane (X–Z).

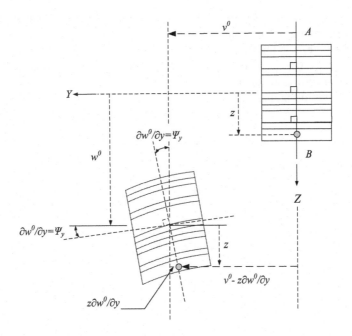

FIGURE 4.3 Deformed laminated plate: plane (Y–Z).

In-plane strains at any position of a laminated plate can be found by substituting the in-plane displacement functions (u and v) of Eqs. (4.1.1 and 4.1.2) into the strain–displacement relations as presented below:

$$
\begin{bmatrix} \varepsilon_{xx} \\ \varepsilon_{yy} \\ \gamma_{xy} \end{bmatrix} = \begin{bmatrix} \dfrac{\partial u(x,y,z)}{\partial x} \\[2mm] \dfrac{\partial v(x,y,z)}{\partial y} \\[2mm] \dfrac{\partial u(x,y,z)}{\partial y} + \dfrac{\partial v(x,y,z)}{\partial x} \end{bmatrix} = \begin{bmatrix} \dfrac{\partial u^{o}(x,y)}{\partial x} \\[2mm] \dfrac{\partial v^{o}(x,y)}{\partial y} \\[2mm] \dfrac{\partial u^{o}(x,y)}{\partial y} + \dfrac{\partial v^{o}(x,y)}{\partial x} \end{bmatrix} - z \begin{bmatrix} \dfrac{\partial^{2} w^{o}(x,y)}{\partial x^{2}} \\[2mm] \dfrac{\partial^{2} w^{o}(x,y)}{\partial y^{2}} \\[2mm] 2\dfrac{\partial^{2} w^{o}(x,y)}{\partial x \partial y} \end{bmatrix}
$$

$$(4.2)$$

From Eq. (4.2), the in-plane strain of laminated plates can be divided into two parts. The first part corresponds to the in-plane strain of the reference surface, while the second part is the curvature of the reference surface multiplied by the vertical distance Z. It should be noted that the negative sign of the second part corresponds to the definition of curvature (corresponding to in-plane displacement in Eqs. (4.1.1 and 4.1.2)). In addition, in-plane strain in Eq. (4.2) can be written as:

$$
\begin{bmatrix} \varepsilon_{xx} \\ \varepsilon_{yy} \\ \gamma_{xy} \end{bmatrix} = \begin{bmatrix} \varepsilon_{xx}^{o} \\ \varepsilon_{yy}^{o} \\ \gamma_{xy}^{o} \end{bmatrix} + z \begin{bmatrix} \kappa_{xx}^{o} \\ \kappa_{yy}^{o} \\ \kappa_{xy}^{o} \end{bmatrix} = \begin{bmatrix} \dfrac{\partial u^{o}(x,y)}{\partial x} \\[2mm] \dfrac{\partial v^{o}(x,y)}{\partial y} \\[2mm] \dfrac{\partial u^{o}(x,y)}{\partial y} + \dfrac{\partial v^{o}(x,y)}{\partial x} \end{bmatrix} + z \begin{bmatrix} -\dfrac{\partial^{2} w^{o}(x,y)}{\partial x^{2}} \\[2mm] -\dfrac{\partial^{2} w^{o}(x,y)}{\partial y^{2}} \\[2mm] -2\dfrac{\partial^{2} w^{o}(x,y)}{\partial x \partial y} \end{bmatrix} \quad (4.3)
$$

where $\varepsilon_{xx}^o, \varepsilon_{yy}^o$ and γ_{xy}^o=normal strains in X and Y directions and in-plane shear strain, respectively $\kappa_{xx}^o, \kappa_{yy}^o$ and κ_{xy}^o curvatures at the reference surface.

As mentioned before, Kirchhoff's hypothesis implies that out-of-plane shear strains (γ_{xz} and γ_{yz}) are assumed to be zero. This is valid for thin plates only. However, out-of-plane shear strains at the reference plane can be presented in a similar way as in Eqs. (4.1 and 4.2) and shown below:

For thin plate (refer to Figures 4.2 and 4.3):

$$
\left[\begin{array}{c} \psi_x \\ \psi_y \end{array} \right] = \left[\begin{array}{c} \dfrac{\partial u(x,y,z)}{\partial z} \\ \dfrac{\partial v(x,y,z)}{\partial z} \end{array} \right] = \left[\begin{array}{c} -\dfrac{\partial w^o(x,y)}{\partial x} \\ -\dfrac{\partial w^o(x,y)}{\partial y} \end{array} \right] \tag{4.4.1}
$$

$$
\left[\begin{array}{c} \gamma_{xz}^o \\ \gamma_{yz}^o \end{array} \right] = \left[\begin{array}{c} \dfrac{\partial u(x,y,z)}{\partial z} + \dfrac{\partial w(x,y,z)}{\partial x} \\ \dfrac{\partial v(x,y,z)}{\partial z} + \dfrac{\partial w(x,y,z)}{\partial y} \end{array} \right]
$$

$$
= \left[\begin{array}{c} \left(\dfrac{\partial u^o(x,y)}{\partial z} + \dfrac{\partial w^o(x,y)}{\partial x} \right) - z\left(\dfrac{\partial^2 w^o(x,y)}{\partial x \partial z} \right) \\ \left(\dfrac{\partial v^o(x,y)}{\partial z} + \dfrac{\partial w^o(x,y)}{\partial y} \right) - z\left(\dfrac{\partial^2 w^o(x,y)}{\partial y \partial z} \right) \end{array} \right] = \left[\begin{array}{c} 0 \\ 0 \end{array} \right] \tag{4.4.2}
$$

4.2 LAMINATE STRESSES AND STRAINS

In this section, stresses through the cross section of a laminated composite are presented using stress and strain relations developed and presented in Chapter 3. In general, stresses are considered to be piecewise linear, which correspond to linear variation of the reduced stiffness and curvature along the vertical distance Z. Thus, stresses through a laminated section are usually discontinuous at the bond line of lamina. In addition, stresses and strains are presented in the principal material coordinates.

4.2.1 LAMINATED STRAIN AND STRESS IN GLOBAL COORDINATE

In global coordinate, the in-plane strains in Eq. (4.3) are substituted into the in-plane total strains in Eq. (3.30.4) as:

$$
\left[\begin{array}{c} \varepsilon_{xx}^o \\ \varepsilon_{yy}^0 \\ \gamma_{xy}^o \end{array} \right] + z \left[\begin{array}{c} \kappa_{xx}^o \\ \kappa_{yy}^o \\ \kappa_{xy}^o \end{array} \right] = \left[\begin{array}{c} \varepsilon_{xx}^{mech} \\ \varepsilon_{yy}^{mech} \\ \gamma_{xy}^{mech} \end{array} \right] + \Delta T \left[\begin{array}{c} \alpha_{xx} \\ \alpha_{yy} \\ \alpha_{xy} \end{array} \right] + \Delta M \left[\begin{array}{c} \beta_{xx} \\ \beta_{yy} \\ \beta_{xy} \end{array} \right] \tag{4.5.1}
$$

$$
\left[\begin{array}{c} \varepsilon_{xx}^{mech} \\ \varepsilon_{yy}^{mech} \\ \gamma_{xy}^{mech} \end{array} \right] = \left[\begin{array}{c} \varepsilon_{xx}^o \\ \varepsilon_{yy}^o \\ \gamma_{xy}^o \end{array} \right] + z \left[\begin{array}{c} \kappa_{xx}^o \\ \kappa_{yy}^o \\ \kappa_{xy}^o \end{array} \right] - \Delta T \left[\begin{array}{c} \alpha_{xx} \\ \alpha_{yy} \\ \alpha_{xy} \end{array} \right] - \Delta M \left[\begin{array}{c} \beta_{xx} \\ \beta_{yy} \\ \beta_{xy} \end{array} \right] \tag{4.5.2}
$$

Then, the strain results in Eq. (4.5.2) are substituted into the transformation of the reduced stress–strain relation with hygrothermal effects in Eq. (3.32.2) resulting in the in-plane stresses through a cross section as:

$$
\begin{bmatrix} \sigma_{xx} \\ \sigma_{yy} \\ \tau_{xy} \end{bmatrix} = \begin{bmatrix} \bar{Q}_{11} & \bar{Q}_{12} & \bar{Q}_{16} \\ \bar{Q}_{12} & \bar{Q}_{22} & \bar{Q}_{26} \\ \bar{Q}_{16} & \bar{Q}_{26} & \bar{Q}_{66} \end{bmatrix} \begin{bmatrix} \varepsilon_{xx}^o + z\kappa_{xx}^o - \Delta T\alpha_{xx} - \Delta M\beta_{xx} \\ \varepsilon_{yy}^o + Z\kappa_{yy}^o - \Delta T\alpha_{yy} - \Delta M\beta_{yy} \\ \gamma_{xy}^o + z\kappa_{xy}^o - \Delta T\alpha_{xy} - \Delta M\beta_{xy} \end{bmatrix} \tag{4.6}
$$

Similarly, the ply stresses can also be calculated using this procedure. By specifying the vertical distance (z) at the top (or bottom) of any ply, the ply in-plane strains at distance k can be determined from Eq. (4.5.2). The ply in-plane strains at a specific z value are substituted into the transformation of the reduced stress–strain relation in Eq. (3.32.2). Thus, the ply in-plane stress is given as follows:

$$
\begin{bmatrix} \sigma_{xx}^k \\ \sigma_{yy}^k \\ \tau_{xy}^k \end{bmatrix} = \begin{bmatrix} \bar{Q}_{11}^k & \bar{Q}_{12}^k & \bar{Q}_{16}^k \\ \bar{Q}_{12}^k & \bar{Q}_{22}^k & \bar{Q}_{26}^k \\ \bar{Q}_{16}^k & \bar{Q}_{26}^k & \bar{Q}_{66}^k \end{bmatrix}
$$
$$
\times \left(\begin{bmatrix} \varepsilon_{xx}^o \\ \varepsilon_{yy}^o \\ \gamma_{xy}^o \end{bmatrix} + k\begin{bmatrix} \kappa_{xx}^o \\ \kappa_{yy}^o \\ \kappa_{xy}^0 \end{bmatrix} - \Delta T \begin{bmatrix} \alpha_{xx}^k \\ \alpha_{yy}^k \\ \alpha_{xy}^k \end{bmatrix} - \Delta M \begin{bmatrix} \beta_{xx}^k \\ \beta_{yy}^k \\ \beta_{xy}^k \end{bmatrix} \right)_{(z=k)} \tag{4.7.1}
$$

$$
\begin{bmatrix} \varepsilon_{xx}^{mech} \\ \varepsilon_{yy}^{mech} \\ \gamma_{xy}^{mech} \end{bmatrix}_{(z=k)} = \left(\begin{bmatrix} \varepsilon_{xx}^o \\ \varepsilon_{yy}^o \\ \gamma_{xy}^o \end{bmatrix} + k\begin{bmatrix} \kappa_{xx}^o \\ \kappa_{yy}^o \\ \kappa_{xy}^0 \end{bmatrix} - \Delta T \begin{bmatrix} \alpha_{xx}^k \\ \alpha_{yy}^k \\ \alpha_{xy}^k \end{bmatrix} - \Delta M \begin{bmatrix} \beta_{xx}^k \\ \beta_{yy}^k \\ \beta_{xy}^k \end{bmatrix} \right)_{(z=k)} \tag{4.7.2}
$$

The transformation of the reduced stiffness matrix $\bar{Q}_{ij}^R$ for orthogonal materials is given in Eq. (3.31.5).

4.2.2 Laminated Strain and Stress in Local Coordinate

Ply strains and stresses in the local (X_1–X_2) coordinate system can be determined using the transformation relations. The global in-plane strains are substituted into Eq. (3.27.1). Thus, the ply in-plane strains in the local coordinate system are:

$$
\begin{bmatrix} \varepsilon_{11}^k \\ \varepsilon_{22}^k \\ \gamma_{12}^k \end{bmatrix} = \begin{bmatrix} \cos^2\theta_k & \sin^2\theta_k & \cos\theta_k\sin\theta_k \\ \sin^2\theta_k & \cos^2\theta_k & -\cos\theta_k\sin\theta_k \\ -2\cos\theta_k\sin\theta_k & 2\cos\theta_k\sin\theta_k & \cos^2\theta_k - \sin^2\theta_k \end{bmatrix} \begin{bmatrix} \varepsilon_{xx}^k \\ \varepsilon_{yy}^k \\ \gamma_{xy}^k \end{bmatrix} \tag{4.8.1}
$$

where θ_k is defined as the angle between local and global axes (X_1 and X as in Figure 3.15) at the vertical distance ($z=k$).

By substituting Eq. (4.7.2) into (4.8.1), the ply in-plane strains in the local coordinate are:

$$\begin{bmatrix} \varepsilon_{11}^k \\ \varepsilon_{22}^k \\ \gamma_{12}^k \end{bmatrix} = \begin{bmatrix} Cos^2\theta_k & Sin^2\theta_k & Cos\theta_k Sin\theta_k \\ Sin^2\theta_k & Cos^2\theta_k & -Cos\theta_k Sin\theta_k \\ -2Cos\theta_k Sin\theta_k & 2Cos\theta_k Sin\theta_k & Cos^2\theta_k - Sin^2\theta_k \end{bmatrix} \begin{bmatrix} \varepsilon_{xx}^k \\ \varepsilon_{yy}^k \\ \gamma_{xy}^k \end{bmatrix} \times$$

$$\left(\begin{bmatrix} \varepsilon_{xx}^o \\ \varepsilon_{yy}^o \\ \gamma_{xy}^0 \end{bmatrix} + k \begin{bmatrix} \kappa_{xx}^o \\ \kappa_{yy}^o \\ \kappa_{xy}^o \end{bmatrix} - \Delta T \begin{bmatrix} \alpha_{xx}^k \\ \alpha_{yy}^k \\ \alpha_{xy}^k \end{bmatrix} - \Delta M \begin{bmatrix} \beta_{xx}^k \\ \beta_{yy}^k \\ \beta_{xy}^k \end{bmatrix} \right)_{(z=k)} \qquad (4.8.2)$$

Similarly, the global in-plane stresses in Eq. (4.7.1) are substituted into Eq. (3.25.1). Thus, the ply in-plane stresses in the local coordinate system are:

$$\begin{bmatrix} \sigma_{11}^k \\ \sigma_{22}^k \\ \tau_{12}^k \end{bmatrix} = \begin{bmatrix} Cos^2\theta_k & Sin^2\theta_k & 2Cos\theta_k Sin\theta_k \\ Sin^2\theta_k & Cos^2\theta_k & -2Cos\theta_k Sin\theta_k \\ -Cos\theta_k Sin\theta_k & Cos\theta_k Sin\theta_k & Cos^2\theta_k - Sin^2\theta_k \end{bmatrix} \begin{bmatrix} \sigma_{xx}^k \\ \sigma_{yy}^k \\ \tau_{xy}^k \end{bmatrix} \quad (same\ as\ 3.25.1)$$

$$(4.8.3)$$

and, further substitution of Eq. (4.6) into Eq. (4.8.3) gives:

$$\begin{bmatrix} \sigma_{11}^k \\ \sigma_{22}^k \\ \tau_{12}^k \end{bmatrix} = \begin{bmatrix} Cos^2\theta_k & Sin^2\theta_k & 2Cos\theta_k Sin\theta_k \\ Sin^2\theta_k & Cos^2\theta_k & -2Cos\theta_k Sin\theta_k \\ -Cos\theta_k Sin\theta_k & Cos\theta_k Sin\theta_k & Cos^2\theta_k - Sin^2\theta_k \end{bmatrix}$$

$$\begin{bmatrix} \bar{Q}_{11}^k & \bar{Q}_{12}^k & \bar{Q}_{16}^k \\ \bar{Q}_{12}^k & \bar{Q}_{22}^k & \bar{Q}_{26}^k \\ \bar{Q}_{16}^k & \bar{Q}_{26}^k & \bar{Q}_{66}^k \end{bmatrix} \begin{bmatrix} \varepsilon_{xx}^{mech} \\ \varepsilon_{yy}^{mech} \\ \gamma_{xy}^{mech} \end{bmatrix}_{(z=k)} \qquad (4.8.4)$$

or

$$\begin{bmatrix} \sigma_{11}^k \\ \sigma_{22}^k \\ \tau_{12}^k \end{bmatrix} = \begin{bmatrix} Cos^2\theta_k & Sin^2\theta_k & 2Cos\theta_k Sin\theta_k \\ Sin^2\theta_k & Cos^2\theta_k & -2Cos\theta_k Sin\theta_k \\ -Cos\theta_k Sin\theta_k & Cos\theta_k Sin\theta_k & Cos^2\theta_k - Sin^2\theta_k \end{bmatrix} \begin{bmatrix} \bar{Q}_{11}^k & \bar{Q}_{12}^k & \bar{Q}_{16}^k \\ \bar{Q}_{12}^k & \bar{Q}_{22}^k & \bar{Q}_{26}^k \\ \bar{Q}_{16}^k & \bar{Q}_{26}^k & \bar{Q}_{66}^k \end{bmatrix} \times$$

$$\left(\begin{bmatrix} \varepsilon_{xx}^o \\ \varepsilon_{yy}^o \\ \gamma_{xy}^o \end{bmatrix} + k \begin{bmatrix} \kappa_{xx}^o \\ \kappa_{yy}^o \\ \kappa_{xy}^o \end{bmatrix} - \Delta T \begin{bmatrix} \alpha_{xx}^k \\ \alpha_{yy}^k \\ \alpha_{xy}^k \end{bmatrix} - \Delta M \begin{bmatrix} \beta_{xx}^k \\ \beta_{yy}^k \\ \beta_{xy}^k \end{bmatrix} \right)_{(z=k)} \qquad (4.8.5)$$

The procedure of ply stress analysis under in-plane strains is summarized and given below:

> **Given:** The reference in-plane strains (ε_{xx}^o, ε_{yy}^o, and γ_{xy}^o) and curvatures (κ_{xx}^o, κ_{yy}^o, and κ_{xy}^o), ΔT and ΔM
>
> **Objective:** find global strains: (ε_{xx}^k, ε_{yy}^k, and γ_{xy}^k) for each layer
> **Solutions:** using the strain relation in Eq. (4.7.2)
> **Objective:** find the in-plane stresses: (σ_{xx}^k, σ_{yy}^k, and τ_{xy}^k) for each layer in the global coordinate
> **Solutions:** using the stress–strain relation in Eq. (4.7.1)
> **Objective:** find the in-plane strains: (ε_{11}^k, ε_{22}^k, and γ_{12}^k) for each layer in the material coordinate
> **Solutions:** using the strain transformation in Eq. (4.8.1)
> **Objective:** find the in-plane stresses: (σ_{11}^k, σ_{22}^k, and τ_{12}^k) for each layer in the material coordinate
> **Solutions:** using the stress–strain relation in Eq. (4.8.4)

Example 4.1

Determine the stress and strain of a symmetrical laminated composite $[0°/45°]_s$ under normal strain and induced curvature as given below. Each lamina is 1 mm thick as shown in Example Figure 4.1.1. Properties of unidirectional FRP lamina 0° and 45° are: $E_{11} = 180\,\text{GPa}$, $E_{22} = 40\,\text{GPa}$, $G_{12} = 20\,\text{GPa}$, $\nu_{12} = 0.25$, $\alpha_{11} = 7\,\mu\text{m/m/°C}$, $\alpha_{22} = 25\,\mu\text{m/m/°C}$, $\beta_{11} = 120\,\mu\text{m/m \%M}$, and $\beta_{22} = 1{,}000\,\mu\text{m/m \%M}$.

   a. Laminated composite is under normal strain ($\varepsilon_{xx}^o = 1{,}000\,\mu\text{m/m}$) in the global direction (X) at the reference surface.
   b. Laminated composite is subjected to induced curvature ($\kappa_{xx}^o = 0.1\,\text{m}^{-1}$) at the reference surface.

Solution

a. Stress and strain under normal strain:

   The strains of each layer are determined from Eq. (4.7.2). Hygrothermal loads are neglected. Total strain without hygrothermal loads (or mechanical strain) is constant through the cross section because curvatures at the reference surface are not applied in the laminated composite.

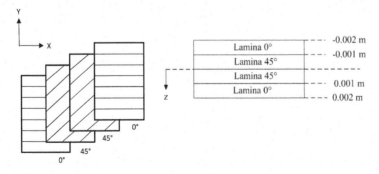

EXAMPLE FIGURE 4.1.1 Laminated composite $[0°/45°]s$.

$$\begin{bmatrix} \varepsilon_{xx}^{mech} \\ \varepsilon_{yy}^{mech} \\ \gamma_{xy}^{mech} \end{bmatrix} = \begin{bmatrix} \varepsilon_{xx}^{o} \\ \varepsilon_{yy}^{0} \\ \gamma_{xy}^{o} \end{bmatrix} + z \begin{bmatrix} \kappa_{xx}^{o} \\ \kappa_{yy}^{o} \\ \kappa_{xy}^{o} \end{bmatrix} = \begin{bmatrix} 1,000 \\ 0 \\ 0 \end{bmatrix} + z \begin{bmatrix} 0 \\ 0 \\ 0 \end{bmatrix} = \begin{bmatrix} 1,000 \\ 0 \\ 0 \end{bmatrix} \mu$$

Normal strain in the global direction (X) is induced in the laminated composite. From Eq. (4.8.1), strain in the local coordinate system can be determined as:
For lamina 0°:

$$\begin{bmatrix} \varepsilon_{11} \\ \varepsilon_{22} \\ \gamma_{12} \end{bmatrix} = \begin{bmatrix} Cos^2 0° & Sin^2 0° & Cos0°Sin0° \\ Sin^2 0° & Cos^2 0° & -2Cos0°Sin0° \\ -2Cos0°Sin0° & Cos0°Sin0° & Cos^2 0° - Sin^2 0° \end{bmatrix} \begin{bmatrix} 1,000\mu \\ 0 \\ 0 \end{bmatrix} = \begin{bmatrix} 1,000 \\ 0 \\ 0 \end{bmatrix} \mu$$

For lamina 45°:

$$\begin{bmatrix} \varepsilon_{11} \\ \varepsilon_{22} \\ \gamma_{12} \end{bmatrix} = \begin{bmatrix} Cos^2 45° & Sin^2 45° & Cos45°Sin45° \\ Sin^2 45° & Cos^2 45° & -2Cos45°Sin45° \\ -2Cos45°Sin45° & 2Cos45°Sin45° & Cos^2 45° - Sin^2 45° \end{bmatrix}$$

$$\begin{bmatrix} 1,000\mu \\ 0 \\ 0 \end{bmatrix} = \begin{bmatrix} 500 \\ 500 \\ -1,000 \end{bmatrix} \mu$$

Units: normal strain ε_{ii} (m/m) and shear strain (γ_{ij}) rad/rad

Strain profiles of a laminated composite $[0°/45°]_s$ in both the global and local coordinate systems are presented in Example Figure 4.1.2.

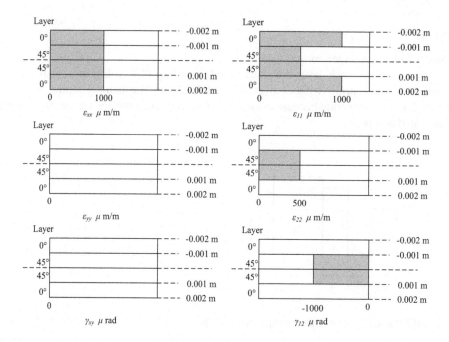

EXAMPLE FIGURE 4.1.2 Strain profiles of laminated composite $[0°/45°]_s$ under normal strain.

The reduced compliance matrix in the material coordinate system (X_1–X_2) of both the 0° and 45° laminae can be determined from Eq. (3.16.1) as:

$$-\frac{v_{12}}{E_{11}} = -\frac{v_{21}}{E_{22}} = -\frac{0.25}{180}$$

$$
\begin{bmatrix} S_{11} & S_{12} & 0 \\ S_{12} & S_{22} & 0 \\ 0 & 0 & S_{66} \end{bmatrix} = \begin{bmatrix} \dfrac{1}{E_{11}} & -\dfrac{v_{21}}{E_{22}} & 0 \\ -\dfrac{v_{12}}{E_{11}} & \dfrac{1}{E_{22}} & 0 \\ 0 & 0 & \dfrac{1}{G_{12}} \end{bmatrix} = \begin{bmatrix} \dfrac{1}{180} & -\dfrac{0.25}{180} & 0 \\ -\dfrac{0.25}{180} & \dfrac{1}{40} & 0 \\ 0 & 0 & \dfrac{1}{20} \end{bmatrix} \dfrac{1}{\text{GPa}}
$$

The reduced stiffness in the material coordinate system (X_1–X_2) is determined from either Eq. (3.16.3) or inversion of the reduced compliance matrix. The reduced stiffness matrix of both the 0° and 45° laminae is identical to each other.

$$
\begin{bmatrix} Q_{11} & Q_{12} & 0 \\ Q_{21} & Q_{22} & 0 \\ 0 & 0 & Q_{66} \end{bmatrix} = \begin{bmatrix} 182.5 & 10.14 & 0 \\ 10.14 & 40.6 & 0 \\ 0 & 0 & 20 \end{bmatrix} \text{GPa}
$$

From the transformation of the reduced stress–strain relation in Eq. (3.32.1), the strain results for each layer are substituted into the reduced stress–strain relation. The reduced stiffness matrix of lamina 0° and 45° is determined as (or using Eq. (4.7.1)):
For lamina of 0°:

$$
\begin{bmatrix} \sigma_{xx} \\ \sigma_{yy} \\ \tau_{xy} \end{bmatrix} = \begin{bmatrix} 182.5 & 10.14 & 0 \\ 10.14 & 40.6 & 0 \\ 0 & 0 & 20 \end{bmatrix} \text{GPa} \begin{bmatrix} 1{,}000 \\ 0 \\ 0 \end{bmatrix} \mu = \begin{bmatrix} 182.5 \\ 10.14 \\ 0 \end{bmatrix} \text{MPa}
$$

For lamina of 45°:

$$
\begin{bmatrix} \sigma_{xx} \\ \sigma_{yy} \\ \tau_{xy} \end{bmatrix} = \begin{bmatrix} 80.8 & 40.8 & 35.5 \\ 40.8 & 80.8 & 35.5 \\ 35.5 & 35.5 & 50.7 \end{bmatrix} \text{GPa} \begin{bmatrix} 1{,}000 \\ 0 \\ 0 \end{bmatrix} \mu = \begin{bmatrix} 80.8 \\ 40.8 \\ 35.5 \end{bmatrix} \text{MPa}
$$

From Eq. (4.8.4), stresses in the local coordinate system are:
For lamina of 0°:

$$
\begin{bmatrix} \sigma_{11} \\ \sigma_{22} \\ \tau_{12} \end{bmatrix} = \begin{bmatrix} \text{Cos}^2 0° & \text{Sin}^2 0° & 2\text{Cos}0°\text{Sin}0° \\ \text{Sin}^2 0° & \text{Cos}^2 0° & -2\text{Cos}0°\text{Sin}0° \\ -\text{Cos}0°\text{Sin}0° & \text{Cos}0°\text{Sin}0° & \text{Cos}^2 0° - \text{Sin}^2 0° \end{bmatrix} \begin{bmatrix} 182.5 \\ 10.14 \\ 0 \end{bmatrix}
$$

$$
= \begin{bmatrix} 182.5 \\ 10.14 \\ 0 \end{bmatrix} \text{MPa}
$$

For lamina of 45°:

$$\begin{bmatrix} \sigma_{11} \\ \sigma_{22} \\ \tau_{12} \end{bmatrix} = \begin{bmatrix} \text{Cos}^2 45° & \text{Sin}^2 45° & 2\text{Cos}\,45°\text{Sin}\,45° \\ \text{Sin}^2 45° & \text{Cos}^2 45° & -2\text{Cos}45°\text{Sin}\,45° \\ -\text{Cos}\,45°\text{Sin}\,45° & \text{Cos}\,45°\text{Sin}\,45° & \text{Cos}^2 45° - \text{Sin}^2 45° \end{bmatrix} \begin{bmatrix} 80.8 \\ 40.8 \\ 35.5 \end{bmatrix}$$

$$= \begin{bmatrix} 96.3 \\ 25.3 \\ -20 \end{bmatrix} \text{MPa}$$

Stress profiles of laminated composite [0°/45°]$_s$ on the cross section in both the global and local coordinate systems are presented in Example Figure 4.1.3.

b. Stress and strain under induced curvature:

From Eq. (4.7.2):

$$\begin{bmatrix} \varepsilon_{xx}^{mech} \\ \varepsilon_{yy}^{mech} \\ \gamma_{xy}^{mech} \end{bmatrix} = \begin{bmatrix} \varepsilon_{xx}^{o} \\ \varepsilon_{yy}^{o} \\ \gamma_{xy}^{o} \end{bmatrix} + z \begin{bmatrix} \kappa_{xx}^{o} \\ \kappa_{yy}^{o} \\ \kappa_{xy}^{0} \end{bmatrix} = \begin{bmatrix} 0 \\ 0 \\ 0 \end{bmatrix} + z \begin{bmatrix} 0.1^{-1} \\ 0 \\ 0 \end{bmatrix} = \begin{bmatrix} 0.1z \\ 0 \\ 0 \end{bmatrix} \times 10^6 \mu$$

Normal strain in the global direction (X) is induced in the laminated composite. From Eq. (4.8.1), the strain in the local coordinate system can be determined as:

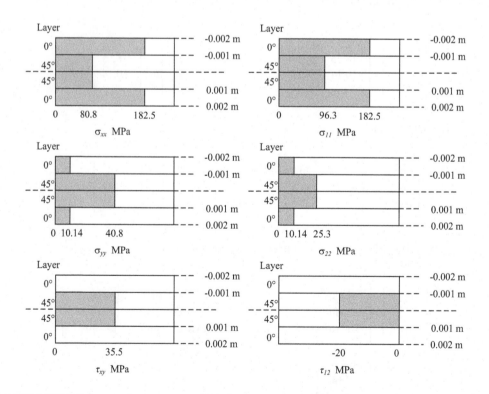

EXAMPLE FIGURE 4.1.3   Stress profiles of laminated composite [0°/45°]$_s$ under normal strain.

For lamina of 0°:

$$
\begin{bmatrix} \varepsilon_{11} \\ \varepsilon_{22} \\ \gamma_{12} \end{bmatrix} = \begin{bmatrix} \text{Cos}^2 0° & \text{Sin}^2 0° & \text{Cos}0°\text{Sin}0° \\ \text{Sin}^2 0° & \text{Cos}^2 0° & -\text{Cos}0°\text{Sin}0° \\ -2\text{Cos}0°\text{Sin}0° & 2\text{Cos}0°\text{Sin}0° & \text{Cos}^2 0° - \text{Sin}^2 0° \end{bmatrix}
$$

$$
\begin{bmatrix} 0.1z \\ 0 \\ 0 \end{bmatrix} = \begin{bmatrix} 0.1z \\ 0 \\ 0 \end{bmatrix} \times 10^6 \, \mu
$$

For lamina of 45°:

$$
\begin{bmatrix} \varepsilon_{11} \\ \varepsilon_{22} \\ \gamma_{12} \end{bmatrix} = \begin{bmatrix} \text{Cos}^2 45° & \text{Sin}^2 45° & \text{Cos}45°\text{Sin}45° \\ \text{Sin}^2 45° & \text{Cos}^2 45° & -\text{Cos}45°\text{Sin}45° \\ -2\text{Cos}45°\text{Sin}45° & 2\text{Cos}45°\text{Sin}45° & \text{Cos}^2 45° - \text{Sin}^2 45° \end{bmatrix} \begin{bmatrix} 0.1z \\ 0 \\ 0 \end{bmatrix}
$$

$$
= \begin{bmatrix} 0.05z \\ 0.05z \\ -0.1z \end{bmatrix} \times 10^6 \, \mu
$$

Units: normal strain ε_{ii} (m/m) and shear strain (γ_{ij}) rad/rad

Strain profiles of laminated composite $[0°/45°]_s$ in both the global and local coordinate systems are presented in Example Figure 4.1.4.

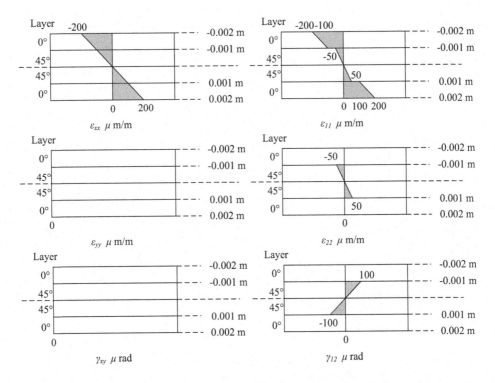

EXAMPLE FIGURE 4.1.4 Strain profiles of laminated composite $[0°/45°]_s$ under induced curvature.

From the transformation of the reduced stress–strain relation in Eq. (3.32.1), strain results of each layer are substituted into the reduced stress–strain relation. The reduced stiffness matrix of lamina 0° and 45° is determined as (or using Eq. (4.7.1)):

For lamina of 0°:

$$
\begin{bmatrix} \sigma_{xx} \\ \sigma_{yy} \\ \tau_{xy} \end{bmatrix} = \begin{bmatrix} 182.5 & 10.14 & 0 \\ 10.14 & 40.6 & 0 \\ 0 & 0 & 20 \end{bmatrix} \text{GPa} \begin{bmatrix} 0.1z \\ 0 \\ 0 \end{bmatrix} \times 10^6 \mu = \begin{bmatrix} 18.25z \\ 1.014z \\ 0 \end{bmatrix} \times 10^3 \text{MPa}
$$

For lamina of 45°:

$$
\begin{bmatrix} \sigma_{xx} \\ \sigma_{yy} \\ \tau_{xy} \end{bmatrix} = \begin{bmatrix} 80.8 & 40.8 & 35.5 \\ 40.8 & 80.8 & 35.5 \\ 35.5 & 35.5 & 50.7 \end{bmatrix} \text{GPa} \begin{bmatrix} 0.1z \\ 0 \\ 0 \end{bmatrix}
$$

$$
\times 10^6 \mu = \begin{bmatrix} 8.08z \\ 4.08z \\ 3.55z \end{bmatrix} \times 10^3 \text{MPa}
$$

From Eq. (4.8.4), stresses in the local coordinate system are:
For lamina of 0°:

$$
\begin{bmatrix} \sigma_{11} \\ \sigma_{22} \\ \tau_{12} \end{bmatrix} = \begin{bmatrix} \text{Cos}^2 0° & \text{Sin}^2 0° & 2\text{Cos}0°\text{Sin}0° \\ \text{Sin}^2 0° & \text{Cos}^2 0° & -2\text{Cos}0°\text{Sin}0° \\ -\text{Cos}0°\text{Sin}0° & \text{Cos}0°\text{Sin}0° & \text{Cos}^2 0° - \text{Sin}^2 0° \end{bmatrix} \begin{bmatrix} 18.25z \\ 1.014z \\ 0 \end{bmatrix} \times 10^3
$$

$$
= \begin{bmatrix} 18.25z \\ 1.014z \\ 0 \end{bmatrix} \times 10^3 \text{MPa}
$$

For lamina of 45°:

$$
\begin{bmatrix} \sigma_{11} \\ \sigma_{22} \\ \tau_{12} \end{bmatrix} = \begin{bmatrix} \text{Cos}^2 45° & \text{Sin}^2 45° & 2\text{Cos}45°\text{Sin}45° \\ \text{Sin}^2 45° & \text{Cos}^2 45° & -2\text{Cos}45°\text{Sin}45° \\ -\text{Cos}45°\text{Sin}45° & \text{Cos}45°\text{Sin}45° & \text{Cos}^2 45° - \text{Sin}^2 45° \end{bmatrix} \begin{bmatrix} 8.08z \\ 4.08z \\ 3.55z \end{bmatrix} \times 10^3
$$

$$
= \begin{bmatrix} 9.63z \\ 2.53z \\ -2.0z \end{bmatrix} \times 10^3 \text{MPa}
$$

Stress profiles of laminated composite $[0°/45°]_s$ in both the global and local coordinate systems are presented in Example Figure 4.1.5.

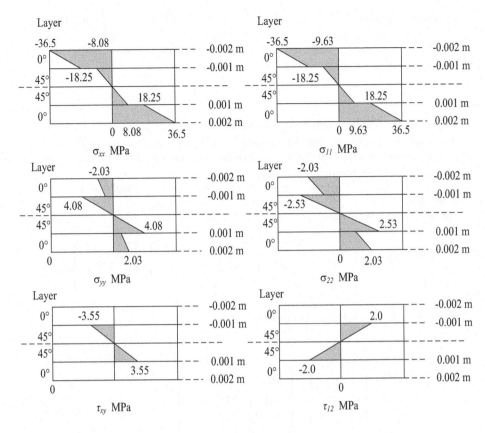

EXAMPLE FIGURE 4.1.5   Stress profiles of laminated composite $[0°/45°]_s$ under induced curvature.

Example 4.2

Determine the stress and strain of a laminated composite $[0°/45°]_s$ under the loads given in (a) and (b) below. Each lamina is 1 mm thick as shown in Example Figure 4.2.1. Properties of unidirectional FRP lamina 0° and 45° are: $E_{11} = 180\,GPa$, $E_{22} = 40\,GPa$, $G_{12} = 20\,GPa$, $\nu_{12} = 0.25$, $\alpha_{11} = 7\,\mu m/m/°C$, $\alpha_{22} = 25\,\mu m/m/°C$, $\beta_{11} = 120\,\mu m/m\ \%M$, and $\beta_{22} = 1{,}000\,\mu m/m\ \%M$.

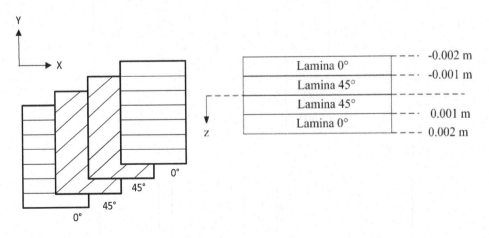

EXAMPLE FIGURE 4.2.1   Laminated composite $[0°/45°]s$.

a. Laminated composite is under normal strain (ε^o_{xx}=1,000 µm/m) and thermal load (ΔT=100°C) in the global direction (x) at the reference surface

b. Laminated composite is subjected to induced curvature (κ^o_{xx}=0.1 m^{-1}) and thermal load (ΔT=100°C) in the global direction (x) at the reference surface.

Solution

a. Stress and strain under normal strain and thermal load:

The strains of each layer are determined from Eq. (4.7.2). Total strain (or mechanical strain) is constant through lamina thickness because curvatures at the reference surface are not applied in the laminated composite.

From Eq. (3.30.2), free hygrothermal strains in the global coordinate system (x–y) are determined as:

For lamina of 0° (where θ=0°, c=Cos 0° and s=Sin 0°):

$$
\begin{bmatrix} \varepsilon_{xx} \\ \varepsilon_{yy} \\ \gamma_{xy} \end{bmatrix}^T = \Delta T \begin{bmatrix} \alpha_{xx} \\ \alpha_{yy} \\ \alpha_{xy} \end{bmatrix} = \begin{bmatrix} 1 & 0 & 0 \\ 0 & 1 & 0 \\ 0 & 0 & 1 \end{bmatrix} \begin{bmatrix} 7\times100 \\ 25\times100 \\ 0 \end{bmatrix}\mu = \begin{bmatrix} 700 \\ 2,500 \\ 0 \end{bmatrix}\mu
$$

Units: normal strain ε_{ii} (m/m) and shear strain (γ_{ij}) rad/rad

From Eq. (4.7.2):

$$
\begin{bmatrix} \varepsilon^{mech}_{xx} \\ \varepsilon^{mech}_{yy} \\ \gamma^{mech}_{xy} \end{bmatrix} = \begin{bmatrix} \varepsilon^o_{xx} \\ \varepsilon^o_{yy} \\ \gamma^o_{xy} \end{bmatrix} + z\begin{bmatrix} \kappa^o_{xx} \\ \kappa^o_{yy} \\ \kappa^o_{xy} \end{bmatrix} - \Delta T\begin{bmatrix} \alpha_{xx} \\ \alpha_{yy} \\ \alpha_{xy} \end{bmatrix}
$$

$$
= \begin{bmatrix} 1,000\mu \\ 0 \\ 0 \end{bmatrix} + z\begin{bmatrix} 0 \\ 0 \\ 0 \end{bmatrix} - \begin{bmatrix} 700\mu \\ 2,500\mu \\ 0 \end{bmatrix} = \begin{bmatrix} 300 \\ -2,500 \\ 0 \end{bmatrix}\mu
$$

For lamina of 45° (where θ=45°, c=Cos 45° and s=Sin 45°)

From Eq. (3.30.2):

$$
\begin{bmatrix} \varepsilon_{xx} \\ \varepsilon_{yy} \\ \gamma_{xy} \end{bmatrix}^T = \Delta T\begin{bmatrix} \alpha_{xx} \\ \alpha_{yy} \\ \alpha_{xy} \end{bmatrix} = \begin{bmatrix} 0.5 & 0.5 & -0.5 \\ 0.5 & 0.5 & 0.5 \\ 1 & -1 & 0 \end{bmatrix}\begin{bmatrix} 7\times100 \\ 25\times100 \\ 0 \end{bmatrix}
$$

$$
= \begin{bmatrix} 1,600 \\ 1,600 \\ -1,800 \end{bmatrix}\mu
$$

From Eq. (4.7.2):

$$
\begin{bmatrix} \varepsilon^{mech}_{xx} \\ \varepsilon^{mech}_{yy} \\ \gamma^{mech}_{xy} \end{bmatrix} = \begin{bmatrix} 1,000 \\ 0 \\ 0 \end{bmatrix} + z\begin{bmatrix} 0 \\ 0 \\ 0 \end{bmatrix} - \begin{bmatrix} 1,600 \\ 1,600 \\ -1,800 \end{bmatrix} = \begin{bmatrix} -600 \\ -1,600 \\ 1,800 \end{bmatrix}\mu
$$

Units: normal strain ε_{ii} (m/m) and shear strain (γ_{ij}) rad/rad

From Eqs. (4.8.1 and 4.8.2), in-plane strains in the local coordinate system can be determined as:
For both lamina of 0°:

$$\begin{bmatrix} \varepsilon_{11} \\ \varepsilon_{22} \\ \gamma_{12} \end{bmatrix} = \begin{bmatrix} \text{Cos}^2 0° & \text{Sin}^2 0° & \text{Cos} 0° \text{Sin} 0° \\ \text{Sin}^2 0° & \text{Cos}^2 0° & -\text{Cos} 0° \text{Sin} 0° \\ -2\text{Cos} 0° \text{Sin} 0° & 2\text{Cos} 0° \text{Sin} 0° & \text{Cos}^2 0° - \text{Sin}^2 0° \end{bmatrix}$$

$$\begin{bmatrix} 300 \\ -2,500 \\ 0 \end{bmatrix} = \begin{bmatrix} 300 \\ -2,500 \\ 0 \end{bmatrix} \mu$$

For both lamina of 45°:

$$\begin{bmatrix} \varepsilon_{11} \\ \varepsilon_{22} \\ \gamma_{12} \end{bmatrix} = \begin{bmatrix} \text{Cos}^2 45° & \text{Sin}^2 45° & \text{Cos} 45° \text{Sin} 45° \\ \text{Sin}^2 45° & \text{Cos}^2 45° & -\text{Cos} 45° \text{Sin} 45° \\ -2\text{Cos} 45° \text{Sin} 45° & 2\text{Cos} 45° \text{Sin} 45° & \text{Cos}^2 45° - \text{Sin}^2 45° \end{bmatrix}$$

$$\begin{bmatrix} -600 \\ -1,600 \\ 1,800 \end{bmatrix} = \begin{bmatrix} -200 \\ -2,000 \\ -1,000 \end{bmatrix} \mu$$

Units: normal strain ε_{ii} (m/m) and shear strain (γ_{ij}) rad/rad

Strain profiles of laminated composite $[0°/45°]_s$ in both the global and local coordinate systems are presented in Example Figure 4.2.2.

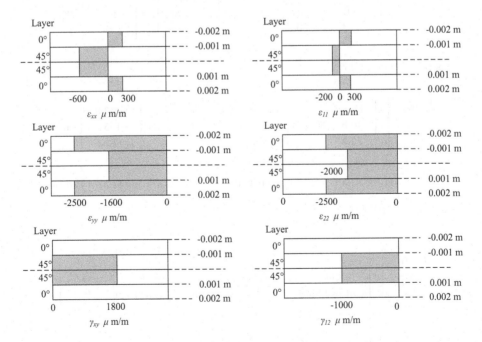

EXAMPLE FIGURE 4.2.2 Strain profiles of laminated composite $[0°/45°]_s$ under normal strain and thermal load.

The reduced compliance matrix in the material coordinate system $(X_1–X_2)$ of both the 0° and 45° laminae can be determined from Eq. (3.16.1) as:

$$-\frac{\nu_{12}}{E_{11}} = -\frac{\nu_{21}}{E_{22}} = -\frac{0.25}{180}$$

$$
\begin{bmatrix} S_{11} & S_{12} & 0 \\ S_{12} & S_{22} & 0 \\ 0 & 0 & S_{66} \end{bmatrix} =
\begin{bmatrix} \dfrac{1}{E_{11}} & -\dfrac{\nu_{21}}{E_{22}} & 0 \\ -\dfrac{\nu_{12}}{E_{11}} & \dfrac{1}{E_{22}} & 0 \\ 0 & 0 & \dfrac{1}{G_{12}} \end{bmatrix} =
\begin{bmatrix} \dfrac{1}{180} & -\dfrac{0.25}{180} & 0 \\ -\dfrac{0.25}{180} & \dfrac{1}{40} & 0 \\ 0 & 0 & \dfrac{1}{20} \end{bmatrix} \dfrac{1}{GPa}
$$

The reduced stiffness in the material coordinate system $(X_1–X_2)$ is determined from either Eq. (3.16.3) or inversion of the reduced compliance matrix. The reduced stiffness matrix of both the 0° and 45° laminae are identical to each other.

$$
\begin{bmatrix} Q_{11} & Q_{12} & 0 \\ Q_{21} & Q_{22} & 0 \\ 0 & 0 & Q_{66} \end{bmatrix} =
\begin{bmatrix} 182.5 & 10.14 & 0 \\ 10.14 & 40.6 & 0 \\ 0 & 0 & 20 \end{bmatrix} GPa
$$

From Eq. (4.7.1), strain results of each layer are substituted into the reduced stress–strain relation as:
For both lamina of 0°:

$$
\begin{bmatrix} \sigma_{xx} \\ \sigma_{yy} \\ \tau_{xy} \end{bmatrix} =
\begin{bmatrix} 182.5 & 10.14 & 0 \\ 10.14 & 40.6 & 0 \\ 0 & 0 & 20 \end{bmatrix} GPa
\begin{bmatrix} 300 \\ -2,500 \\ 0 \end{bmatrix} \mu =
\begin{bmatrix} 29.4 \\ -98.5 \\ 0 \end{bmatrix} MPa
$$

For both lamina of 45°:

$$
\begin{bmatrix} \sigma_{xx} \\ \sigma_{yy} \\ \tau_{xy} \end{bmatrix} =
\begin{bmatrix} 80.8 & 40.8 & 35.5 \\ 40.8 & 80.8 & 35.5 \\ 35.5 & 35.5 & 50.7 \end{bmatrix} GPa
\begin{bmatrix} -600 \\ -1,600 \\ 0 \end{bmatrix} \mu =
\begin{bmatrix} -113.8 \\ -153.8 \\ -85.2 \end{bmatrix} MPa
$$

From Eqs. (4.8.4 and 4.8.5), the stress in the local coordinate system can be determined as:
For both lamina of 0°:

$$
\begin{bmatrix} \sigma_{11} \\ \sigma_{22} \\ \tau_{12} \end{bmatrix} =
\begin{bmatrix} Cos^2 0° & Sin^2 0° & 2Cos0°Sin0° \\ Sin^2 0° & Cos^2 0° & -2Cos0°Sin0° \\ -Cos0°Sin0° & Cos0°Sin0° & Cos^2 0° - Sin^2 0° \end{bmatrix}
\begin{bmatrix} 29.4 \\ -98.5 \\ 0 \end{bmatrix}
$$

$$
= \begin{bmatrix} 29.4 \\ -98.5 \\ 0 \end{bmatrix} MPa
$$

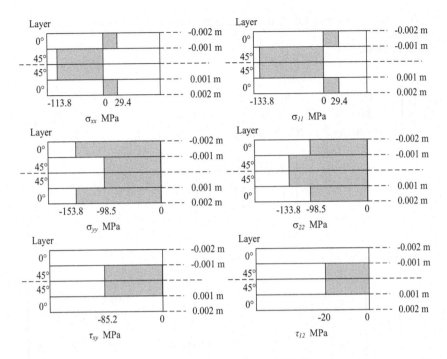

EXAMPLE FIGURE 4.2.3 Stress profiles of laminated composite $[0°/45°]_s$ under normal strain and thermal load.

For both lamina of 45°:

$$
\begin{bmatrix} \sigma_{11} \\ \sigma_{22} \\ \tau_{12} \end{bmatrix} = \begin{bmatrix} \text{Cos}^2 45° & \text{Sin}^2 45° & 2\text{Cos}\,45°\text{Sin}\,45° \\ \text{Sin}^2 45° & \text{Cos}^2 45° & -2\text{Cos}\,45°\text{Sin}\,45° \\ -\text{Cos}\,45°\text{Sin}\,45° & \text{Cos}\,45°\text{Sin}\,45° & \text{Cos}^2 45° - \text{Sin}^2 45° \end{bmatrix} \begin{bmatrix} -113.8 \\ -153.8 \\ -85.2 \end{bmatrix}
$$

$$
= \begin{bmatrix} -133.8 \\ -133.8 \\ -20 \end{bmatrix} \text{MPa}
$$

Stress profiles of laminated composite $[0°/45°]_s$ in both the global and local coordinate systems are presented in Example Figure 4.2.3.

b. Stress and strain under induced curvature and thermal load:

The strains of each layer are determined from Eq. (4.7.2). Curvatures at the reference surface are applied to the laminated composite. Thus, strains linearly vary through laminate thickness according to Eq. (4.3).

From Eq. (4.7.2) and free thermal strains in part (a):

$$
\begin{bmatrix} \varepsilon_{xx}^{mech} \\ \varepsilon_{yy}^{mech} \\ \gamma_{xy}^{mech} \end{bmatrix} = \begin{bmatrix} \varepsilon_{xx}^{o} \\ \varepsilon_{yy}^{o} \\ \gamma_{xy}^{o} \end{bmatrix} + z \begin{bmatrix} \kappa_{xx}^{o} \\ \kappa_{yy}^{o} \\ \kappa_{xy}^{o} \end{bmatrix} - \Delta T \begin{bmatrix} \alpha_{xx} \\ \alpha_{yy} \\ \alpha_{xy} \end{bmatrix} = \begin{bmatrix} 0 \\ 0 \\ 0 \end{bmatrix}
$$

$$
+ z \begin{bmatrix} 0.1 \\ 0 \\ 0 \end{bmatrix} \frac{1}{m} - \begin{bmatrix} 700 \\ 2,500 \\ 0 \end{bmatrix} \mu = \begin{bmatrix} 0.1z\left(10^6\right) - 700 \\ -2,500 \\ 0 \end{bmatrix} \mu
$$

From Eqs. (4.8.1 and 4.8.2), the strain in the local coordinate system can be determined as:

For both lamina of 0°:

$$
\begin{bmatrix} \varepsilon_{11} \\ \varepsilon_{22} \\ \gamma_{12} \end{bmatrix} = \begin{bmatrix} \cos^2 0° & \sin^2 0° & \cos 0° \sin 0° \\ \sin^2 0° & \cos^2 0° & -\cos 0° \sin 0° \\ -2\cos 0° \sin 0° & 2\cos 0° \sin 0° & \cos^2 0° - \sin^2 0° \end{bmatrix} \begin{bmatrix} 0.1z\left(10^6\right) - 700 \\ -2,500 \\ 0 \end{bmatrix}
$$

$$
= \begin{bmatrix} 0.1z\left(10^6\right) - 700 \\ -2,500 \\ 0 \end{bmatrix} \mu
$$

For both lamina of 45°:

$$
\begin{bmatrix} \varepsilon_{11} \\ \varepsilon_{22} \\ \gamma_{12} \end{bmatrix} = \begin{bmatrix} \cos^2 45° & \sin^2 45° & \cos 45° \sin 45° \\ \sin^2 45° & \cos^2 45° & -\cos 45° \sin 45° \\ -2\cos 45° \sin 45° & 2\cos 45° \sin 45° & \cos^2 45° - \sin^2 45° \end{bmatrix}
$$

$$
\begin{bmatrix} 0.1z\left(10^6\right) - 700 \\ -2,500 \\ 0 \end{bmatrix}
$$

$$
\begin{bmatrix} \varepsilon_{11} \\ \varepsilon_{22} \\ \gamma_{12} \end{bmatrix} = \begin{bmatrix} 0.05z\left(10^6\right) - 1,600 \\ 0.05z\left(10^6\right) - 1,600 \\ -0.1z\left(10^6\right) - 1,800 \end{bmatrix} \mu
$$

Units: normal strain ε_{ii} (m/m) and shear strain (γ_{ij}) rad/rad

Strain profiles of laminated composite [0°/45°]$_s$ in both the global and local coordinate systems are presented in Example Figure 4.2.4.

From Eq. (4.7.2) and the reduced stiffness matrix in part (a), the strain results of each layer are substituted into the reduced stress and strain relation in Eq. (4.8.4). For both lamina of 0°:

$$
\begin{bmatrix} \sigma_{xx} \\ \sigma_{yy} \\ \tau_{xy} \end{bmatrix} = \begin{bmatrix} 182.5 & 10.14 & 0 \\ 10.14 & 40.6 & 0 \\ 0 & 0 & 20 \end{bmatrix} \text{GPa} \begin{bmatrix} 0.1z\left(10^6\right) - 700 \\ -2,500 \\ 0 \end{bmatrix} \mu
$$

$$
= \begin{bmatrix} 18.25z\left(10^3\right) - 153.1 \\ 1.014z\left(10^3\right) - 108.6 \\ 0 \end{bmatrix} \text{MPa}
$$

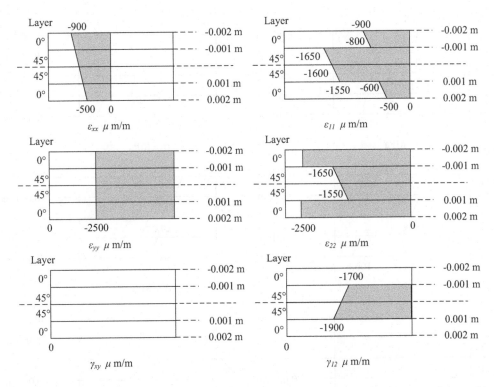

EXAMPLE FIGURE 4.2.4 Strain profiles of composite $[0°/45°]_s$ under induced curvature and thermal load.

For both lamina of 45°:

$$
\begin{bmatrix} \sigma_{xx} \\ \sigma_{yy} \\ \tau_{xy} \end{bmatrix} = \begin{bmatrix} 80.8 & 40.8 & 35.5 \\ 40.8 & 80.8 & 35.5 \\ 35.5 & 35.5 & 50.7 \end{bmatrix} \text{GPa} \begin{bmatrix} 0.1z\left(10^6\right) - 700 \\ -2{,}500 \\ 0 \end{bmatrix} \mu
$$

$$
= \begin{bmatrix} 8.08z\left(10^3\right) - 158.6 \\ 4.08z\left(10^3\right) - 231 \\ 3.55z\left(10^3\right) - 113.6 \end{bmatrix} \text{MPa}
$$

From Eq. (4.8.4), stresses in the local coordinate system can be determined as:
For both lamina 0°:

$$
\begin{bmatrix} \sigma_{11} \\ \sigma_{22} \\ \tau_{12} \end{bmatrix} = \begin{bmatrix} \cos^2 0° & \sin^2 0° & 2\cos0°\sin0° \\ \sin^2 0° & \cos^2 0° & -2\cos0°\sin0° \\ -\cos0°\sin0° & \cos0°\sin0° & \cos^2 0° - \sin^2 0° \end{bmatrix} \begin{bmatrix} 18.25z\left(10^3\right) - 153.1 \\ 1.014z\left(10^3\right) - 108.6 \\ 0 \end{bmatrix}
$$

$$
\begin{bmatrix} \sigma_{11} \\ \sigma_{22} \\ \tau_{12} \end{bmatrix} = \begin{bmatrix} 18.25z\left(10^3\right) - 153.1 \\ 1.014z\left(10^3\right) - 108.6 \\ 0 \end{bmatrix} \text{MPa}
$$

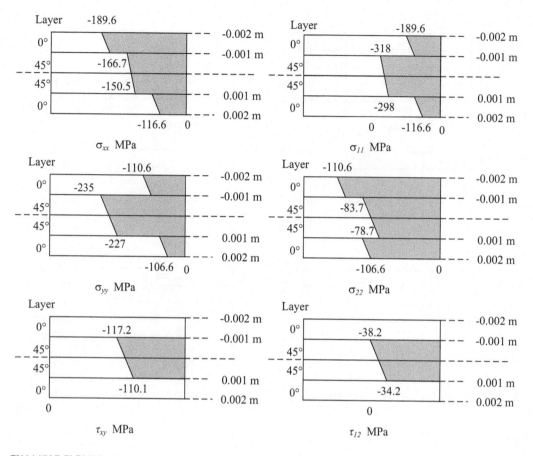

EXAMPLE FIGURE 4.2.5   Stress profiles of composite $[0°/45°]_s$ under induced curvature and thermal load.

For both lamina 45°:

$$
\begin{bmatrix} \sigma_{11} \\ \sigma_{22} \\ \tau_{12} \end{bmatrix} =
\begin{bmatrix}
Cos^2 45° & Sin^2 45° & 2Cos\,45°Sin\,45° \\
Sin^2 45° & Cos^2 45° & -2Cos\,45°Sin\,45° \\
-Cos\,45°Sin\,45° & Cos\,45°Sin\,45° & Cos^2 45° - Sin^2 45°
\end{bmatrix}
$$

$$
\begin{bmatrix}
8.08z\left(10^3\right) - 158.6 \\
4.08k\left(10^3\right) - 231 \\
3.55z\left(10^3\right) - 113.6
\end{bmatrix}
$$

$$
\begin{bmatrix} \sigma_{11} \\ \sigma_{22} \\ \tau_{12} \end{bmatrix} =
\begin{bmatrix}
9.63z\left(10^3\right) - 308 \\
2.53z\left(10^3\right) - 81.2 \\
-2z\left(10^3\right) - 36.2
\end{bmatrix} MPa
$$

Stress profiles of laminated composite $[0°/45°]_s$ in both the global and local coordinate systems are presented in Example Figure 4.2.5.

4.3 FORCE AND MOMENT RESULTANTS

In this section, force and moment resultants of thin composite plates under plane stress conditions are derived using the following assumptions: constant and uniform loads (forces and moments) are applied along the edges of a thin-plate, in-plane forces (N_{xx}, N_{yy}, and N_{xy}) and moments (M_{xx}, M_{yy}, and M_{xy}) are considered under in-plane stresses.

In general, forces (both normal and shear) applied on a specific plane act in the positive direction while the direction of the plane where loads act on that plane is in the positive direction with respect to the coordinate system. Thus, forces on that specific plane are taken to be positive. The positive normal force and shear resultants are shown in Figure 4.4.

Note: The positive and negative normal forces are usually in tension and compression, respectively, as shown in Figure 4.4. The positive sign conventions for bending and torsional moments are shown in Figure 4.5. This sign convention is consistently used throughout this textbook. These plate bending moments (M_{xx} and M_{yy} as in Figure 4.5) are distributed and identified in the positive direction, i.e., the moment that causes tension at the bottom, in the positive face of the plate where positive Z-direction is downward. The distributed in-plane moments (M_{xy} and M_{yx}) along the plate edges as in Figure 4.5, are related to in-plane shear stresses (τ_{xy} and τ_{yx}) and their absolute values are equal in thin plates. A positive torque causes positive shear stress in the bottom or lower face of the plate, as per the X–Y–Z convention shown in Figure 4.5, i.e., shear stress direction at the bottom face in the positive X and Y directions.

In-plane force resultants of a laminated composite plate can be determined by adding in-plane force resultants of each layer through the thickness of a laminated plate as shown in Figure 4.6. The force resultants of laminated plates are assumed to act on the reference plane at ($z=0$).

From Figure 4.6, the force resultant is differentiated with the area. Let the area (A) be represented by unit width and thickness as z, then the force resultant can be rewritten as:

$$\text{where, width is unity} \quad dN = \sigma dA = \sigma(dz \times 1) = \sigma dz \tag{4.9.1}$$

where N, σ, and A are defined as force resultant, applied stress and area corresponding to unit, respectively.

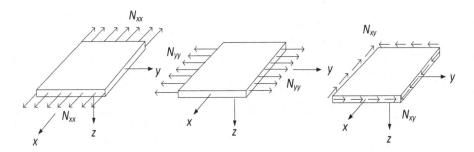

FIGURE 4.4 Positive normal forces and shear resultants.

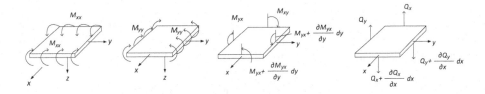

FIGURE 4.5 Positive bending and twisting moment resultants.

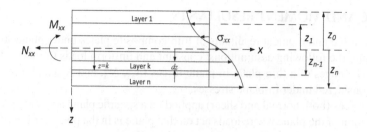

FIGURE 4.6 Force and moment resultants with normal stress in plane (X–Z).

The force resultants of a laminated composite can be obtained by integrating Eq. (4.9.1) through the thickness ($-h/2$ to $h/2$) of a laminated plate.

$$N = \int_{-\frac{h}{2}}^{\frac{h}{2}} dN = \int_{-\frac{h}{2}}^{\frac{h}{2}} \sigma \, dz \tag{4.9.2}$$

where h is denoted as the laminate thickness.

From Eq. (4.9.2), in-plane force resultants for all lamina (1–n) are defined as follows:

$$N_{xx} = \int_{-\frac{h}{2}}^{\frac{h}{2}} \sigma_{xx}^k \, dz = \int_{z0}^{z1} \sigma_{xx}^1 \, dz + \int_{z1}^{z2} \sigma_{xx}^2 \, dz + \cdots + \int_{zn-1}^{zn} \sigma_{xx}^n \, dz = \sum_{k=1}^{n} \int_{zk-1}^{zk} \sigma_{xx}^k \, dz \tag{4.10.1}$$

$$N_{yy} = \int_{-\frac{h}{2}}^{\frac{h}{2}} \sigma_{yy}^k \, dz = \int_{z0}^{z1} \sigma_{yy}^1 dz + \int_{z1}^{z2} \sigma_{yy}^2 \, dz + \cdots + \int_{zn-1}^{zn} \sigma_{yy}^n \, dz = \sum_{k=1}^{n} \int_{zk-1}^{zk} \sigma_{yy}^k \, dz \tag{4.10.2}$$

$$N_{xy} = \int_{-\frac{h}{2}}^{\frac{h}{2}} \tau_{xy}^k \, dz = \int_{z0}^{z1} \tau_{xy}^1 \, dz + \int_{z1}^{z2} \tau_{xy}^2 \, dz + \cdots + \int_{zn-1}^{zn} \tau_{xy}^n \, dz = \sum_{k=1}^{n} \int_{zk-1}^{zk} \tau_{xy}^k \, dz \tag{4.10.3}$$

where n is the number of layers through the thickness of a composite plate.

The in-plane force resultants in Eqs. (4.10.1–4.10.3) are uniformly distributed over the unit width. Therefore, in-plane force resultants are represented in the units of force per (plate) unit length. Similarly, the moment resultants of laminated composite plates can be calculated by adding the moment resultants of each layer through its thickness. The moment resultants of laminated plates are still assumed to act on the reference plane at ($z=0$) as in the case of in-plane force resultants.

By taking the moment of in-plane force in a specific layer at distance z about the edge of the laminated reference plane, the moment resultant in Eq. (4.11.1) is represented in (dA). Using Eq. (4.9.2), the moment resultant, with unit width and thickness of Z, can be rewritten as:

$$dM = dN = z\sigma dA = z\sigma d(z \times 1) = z\sigma dz \tag{4.11.1}$$

The moment resultant of a laminated composite can be obtained by integrating Eq. (4.11.1) through the thickness of the laminated plate ($-h/2$ to $h/2$).

$$M = \int_{-\frac{h}{2}}^{\frac{h}{2}} dM = \int_{-\frac{h}{2}}^{\frac{h}{2}} z\sigma \, dz \tag{4.11.2}$$

where h is denoted as the total plate thickness.

From Eq. (4.11.2), moment resultants are derived as:

$$M_{xx} = \int_{-\frac{h}{2}}^{\frac{h}{2}} z_k \sigma_{xx}^k \, dz = \int_{z_0}^{z_1} z_1 \sigma_{xx}^1 dz + \int_{z_1}^{z_2} z_2 \sigma_{xx}^2 \, dz + \cdots + \int_{z_{n-1}}^{z_n} z_n \sigma_{xx}^n \, dz$$

$$= \sum_{k=1}^{n} \int_{z_{k-1}}^{z_k} z_k \sigma_{xx}^k \, dz \qquad\qquad (4.12.1)$$

$$M_{yy} = \int_{-\frac{h}{2}}^{\frac{h}{2}} z_k \sigma_{xx}^k \, dz = \int_{z_0}^{z_1} z_1 \sigma_{yy}^1 \, dz + \int_{z_1}^{z_2} z_2 \sigma_{yy}^2 \, dz + \cdots + \int_{z_{n-1}}^{z_n} z_n \sigma_{yy}^n \, dz$$

$$= \sum_{k=1}^{n} \int_{z_{k-1}}^{z_k} z_k \sigma_{yy}^k \, dz \qquad\qquad (4.12.2)$$

$$M_{xy} = \int_{-\frac{h}{2}}^{\frac{h}{2}} z_k \tau_{xy}^k \, dz = \int_{z_0}^{z_1} z_1 \tau_{xy}^1 \, dz + \int_{z_1}^{z_2} z_2 \tau_{xy}^2 \, dz + \cdots + \int_{z_{n-1}}^{z_n} z_n \tau_{xy}^n \, dz$$

$$= \sum_{k=1}^{n} \int_{z_{k-1}}^{z_k} z_k \tau_{xy}^k \, dz \qquad\qquad (4.12.3)$$

where n is the number of layers through the thickness of a composite plate.

The moment resultants in Eqs. (4.12.1–4.12.3) are uniformly distributed over the unit width. Therefore, unit of these moment resultants is the force-length per plate length (or force unit).

4.4 LAMINATE STIFFNESS (*ABD*) AND COMPLIANCE MATRIX

Let us consider the laminated composite of n layers as shown in Figure 4.7. The characteristics of laminated composites are presented through the laminate properties. The laminate stiffness (or compliance) of laminated composite materials is developed to study the response under applied loading and deformations. The transformation of reduced stiffness matrices in Eq. (3.31.5) (for orthotropic materials) of each composite layer is combined using relations of force and moment resultants.

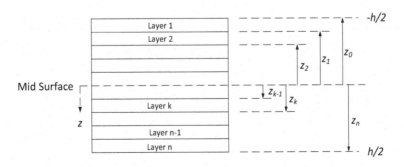

FIGURE 4.7 Laminated composite of n layers.

4.4.1 In-Plane Force Resultant Relation

To present relations between in-plane force resultants and deformations of laminated composites, Eq. (4.7.1) with zero hygrothermal effects are integrated, with respect to variable z, through ($-h/2$ to $h/2$) with reference to the mid-surface (Figure 4.7). The in-plane force resultant is given as:

$$
\begin{bmatrix} N_{xx} \\ N_{yy} \\ N_{xy} \end{bmatrix} = \int_{-\frac{h}{2}}^{\frac{h}{2}} \begin{bmatrix} \sigma_{xx}^k \\ \sigma_{yy}^k \\ \tau_{xy}^k \end{bmatrix} dz = \left(\int_{-\frac{h}{2}}^{\frac{h}{2}} \begin{bmatrix} \bar{Q}_{11}^k & \bar{Q}_{12}^k & \bar{Q}_{16}^k \\ \bar{Q}_{12}^k & \bar{Q}_{22}^k & \bar{Q}_{26}^k \\ \bar{Q}_{16}^k & \bar{Q}_{26}^k & \bar{Q}_{66}^k \end{bmatrix} dz \right) \begin{bmatrix} \varepsilon_{xx}^o \\ \varepsilon_{yy}^o \\ \gamma_{xy}^o \end{bmatrix}
$$
$$
+ \left(\int_{-\frac{h}{2}}^{\frac{h}{2}} \begin{bmatrix} \bar{Q}_{11}^k & \bar{Q}_{12}^k & \bar{Q}_{16}^k \\ \bar{Q}_{12}^k & \bar{Q}_{22}^k & \bar{Q}_{26}^k \\ \bar{Q}_{16}^k & \bar{Q}_{26}^k & \bar{Q}_{66}^k \end{bmatrix} z\, dz \right) \begin{bmatrix} \kappa_{xx}^o \\ \kappa_{yy}^o \\ \kappa_{xy}^o \end{bmatrix} \tag{4.13.1}
$$

The deformation at the reference surface is independent of z, then in-plane strains and curvatures at the reference surface are taken out from integrands of Eq. (4.13.1). Equations (4.13.4 and 4.13.5) are written after designating terms of integration as:

$$
\int_{-\frac{h}{2}}^{\frac{h}{2}} \bar{Q}_{ij}^k \, dz \tag{4.13.2}
$$

$$
\int_{-\frac{h}{2}}^{\frac{h}{2}} \bar{Q}_{ij}^k z \, dz \tag{4.13.3}
$$

where i and $j = 1, 2, 6$.

The elements $\bar{Q}_{ij}^k$ of the reduced stiffness matrix are constant only in their own layers but the elements $\bar{Q}_{ij}^k$ may be different from one layer to another depending upon property variations in the materials and their fiber orientations. The elements $\bar{Q}_{ij}^k$ of each layer can be found by carrying out the integration, and Eqs. (4.13.2 and 4.13.3) are simplified as:

$$
\int_{-\frac{h}{2}}^{\frac{h}{2}} \bar{Q}_{ij}^k \, dz = \sum_{k=1}^{n} \bar{Q}_{ij}^k (z_k - z_{k-1}) = \sum_{k=1}^{n} \bar{Q}_{ij}^k h_k \tag{4.13.4}
$$

$$
\int_{-\frac{h}{2}}^{\frac{h}{2}} \bar{Q}_{ij}^k \, dz = \sum_{k=1}^{n} \bar{Q}_{ij}^k \frac{\left(z_k^2 - z_{k-1}^2 \right)}{2} \tag{4.13.5}
$$

Note: Terms $(z_k - z_{k-1})$ in a summation in Eq. (4.13.4) is the thickness of the kth layer. Equation (4.13.4) is denoted by elements A_{ij} of the laminate in-plane stiffness $[A_{ij}]$ related to the in-plane deformation at the reference surface while the results of Eq. (4.13.5) are denoted by elements B_{ij} of the laminate stiffness $[B_{ij}]$ related to the in-plane force resultants (N_x, N_y, N_{xy}) and curvatures $\left(\kappa_{xx}^o, \kappa_{yy}^o, \kappa_{xy}^o \right)$.

$$
A_{ij} = \int_{-\frac{h}{2}}^{\frac{h}{2}} \bar{Q}_{ij}^k \, dz = \sum_{k=1}^{n} \bar{Q}_{ij}^k (z_k - z_{k-1}) = \sum_{k=1}^{n} \bar{Q}_{ij}^k h_k \tag{4.14.1}
$$

$$B_{ij} = \int_{-\frac{h}{2}}^{\frac{h}{2}} \bar{Q}_{ij}^k z \, dz = \sum_{k=1}^{n} \bar{Q}_{ij}^k \frac{\left(z_k^2 - z_{k-1}^2\right)}{2} \qquad (4.14.2)$$

where i and $j = 1, 2, 6$.

Elements A_{ij} and B_{ij} in Eqs. (4.14.1 and 4.14.2) are substituted into the in-plane force resultants and deformation relation in Eq. (4.13.1). Thus, the in-plane force resultants and deformation relations in terms of laminate stiffness matrix (A and B) are presented as:

$$\begin{bmatrix} N_{xx} \\ N_{yy} \\ N_{xy} \end{bmatrix} = \begin{bmatrix} A_{11} & A_{12} & A_{16} \\ A_{12} & A_{22} & A_{26} \\ A_{16} & A_{26} & A_{66} \end{bmatrix} \begin{bmatrix} \varepsilon_{xx}^o \\ \varepsilon_{yy}^0 \\ \gamma_{xy}^o \end{bmatrix} + \begin{bmatrix} B_{11} & B_{12} & B_{16} \\ B_{12} & B_{22} & B_{26} \\ B_{16} & B_{26} & B_{66} \end{bmatrix} \begin{bmatrix} \kappa_{xx}^o \\ \kappa_{yy}^o \\ \kappa_{xy}^o \end{bmatrix} \qquad (4.15)$$

Equation (4.15) is presented in a matrix form by separating the effects of in-plane deformation and curvature at the reference surface.

4.4.2 MOMENT RESULTANT RELATION

The following relations between moment resultants and deformations of laminated composites are presented herein. Equation (4.13.1) is multiplied with variable z on both sides of equation. Using definitions of the in-plane force moments in Eqs. (4.12.1–4.12.3):

$$\begin{bmatrix} M_{xx} \\ M_{yy} \\ M_{xy} \end{bmatrix} = \int_{-\frac{h}{2}}^{\frac{h}{2}} \begin{bmatrix} \sigma_{xx}^k \\ \sigma_{yy}^k \\ \tau_{xy}^k \end{bmatrix} z \, dz = \left(\int_{-\frac{h}{2}}^{\frac{h}{2}} \begin{bmatrix} \bar{Q}_{11}^k & \bar{Q}_{12}^k & \bar{Q}_{16}^k \\ \bar{Q}_{12}^k & \bar{Q}_{22}^k & \bar{Q}_{26}^k \\ \bar{Q}_{16}^k & \bar{Q}_{26}^k & \bar{Q}_{66}^k \end{bmatrix} z \, dz \right) \begin{bmatrix} \varepsilon_{xx}^o \\ \varepsilon_{yy}^o \\ \gamma_{xy}^o \end{bmatrix}$$

$$+ \left(\int_{-\frac{h}{2}}^{\frac{h}{2}} \begin{bmatrix} \bar{Q}_{11}^k & \bar{Q}_{12}^k & \bar{Q}_{16}^k \\ \bar{Q}_{12}^k & \bar{Q}_{22}^k & \bar{Q}_{26}^k \\ \bar{Q}_{16}^k & \bar{Q}_{26}^k & \bar{Q}_{66}^k \end{bmatrix} z^2 \, dz \right) \begin{bmatrix} \kappa_{xx}^o \\ \kappa_{yy}^o \\ \kappa_{xy}^o \end{bmatrix} \qquad (4.16.1)$$

Considering terms of integration as:

$$\int_{-\frac{h}{2}}^{\frac{h}{2}} \bar{Q}_{ij}^k Z \, dz \qquad (4.16.2)$$

$$\int_{-\frac{h}{2}}^{\frac{h}{2}} \bar{Q}_{ij}^k Z^2 \, dz \qquad (4.16.3)$$

where i and $j = 1, 2, 6$.

The elements $\bar{Q}_{ij}^k$ of each layer can be found by carrying out the integration. Equations (4.16.2 and 4.16.3) are simplified as:

$$\int_{-\frac{h}{2}}^{\frac{h}{2}} \bar{Q}_{ij}^k z \, dz = \sum_{k=1}^{n} \bar{Q}_{ij}^k \frac{\left(z_k^2 - z_{k-1}^2\right)}{2} \quad \left(\text{Same as Eq. 4.14.2}\right) \qquad (4.16.4)$$

$$\int_{-\frac{h}{2}}^{\frac{h}{2}} \overline{Q}_{ij}^k z^2 \, dz = \sum_{k=1}^{n} \overline{Q}_{ij}^k \frac{\left(z_k^3 - z_{k-1}^3\right)}{3} \tag{4.16.5}$$

Equation (4.16.4) is similar to Eq. (4.13.4). Integration of Eq. (4.13.4) is giving the stiffness of a laminate in-plane, and out-of-plane coupling stiffness $[B_{ij}]$ between the in-plane force resultants (N_{xx}, N_{yy}, and N_{xy}) and curvatures (κ_{xx}^o, κ_{yy}^o, κ_{xy}^o); also providing coupling stiffness $[B_{ij}]$ between moment resultants (M_{xx}, M_{yy}, and M_{xy}) and in-plane deformations (ε_{xx}^o, ε_{yy}^o, γ_{xy}^o), as seen in Eq. (4.16.4).

Integration of Eq. (4.16.5) is denoted by elements D_{ij} of the laminate's out-of-plane stiffness $[D_{ij}]$ which is related to moment resultants (M_{xx}, M_{yy}, and M_{xy}) and the curvature at reference surface (κ_{xx}^o, κ_{yy}^o, κ_{xy}^o).

$$B_{ij} = \int_{-\frac{h}{2}}^{\frac{h}{2}} \overline{Q}_{ij}^k z \, dz = \frac{1}{2} \sum_{k=1}^{n} \overline{Q}_{ij}^k \left(z_k^2 - z_{k-1}^2\right) \tag{4.17.1}$$

$$D_{ij} = \int_{-\frac{h}{2}}^{\frac{h}{2}} \overline{Q}_{ij}^k z^2 \, dz = \frac{1}{3} \sum_{k=1}^{n} \overline{Q}_{ij}^k \left(z_k^3 - z_{k-1}^3\right) \tag{4.17.2}$$

where i and $j = 1$, 2, and 6.

Elements of B and D in Eqs. (4.17.1 and 4.17.2) are substituted into the moment resultants and deformation relations in Eq. (4.16.1). The moment resultants and deformation relations in terms of laminated stiffness matrix (B and D) are presented as:

$$\begin{bmatrix} M_{xx} \\ M_{yy} \\ M_{xy} \end{bmatrix} = \begin{bmatrix} B_{11} & B_{12} & B_{16} \\ B_{12} & B_{22} & B_{26} \\ B_{16} & B_{26} & B_{66} \end{bmatrix} \begin{bmatrix} \varepsilon_{xx}^o \\ \varepsilon_{yy}^0 \\ \gamma_{xy}^o \end{bmatrix} + \begin{bmatrix} D_{11} & D_{12} & D_{16} \\ D_{12} & D_{22} & D_{26} \\ D_{16} & D_{26} & D_{66} \end{bmatrix} \begin{bmatrix} \kappa_{xx}^o \\ \kappa_{yy}^o \\ \kappa_{xy}^o \end{bmatrix} \tag{4.18}$$

Equation (4.18) is presented in a matrix form by separating the effects of in-plane deformation and curvature at the reference surface.

4.4.3 IN-PLANE FORCE AND MOMENT RESULTANT RELATION

To formulate the laminate stiffness under in-plane force and moment resultants, Eqs. (4.15 and 4.18) are combined in a matrix form. This stiffness matrix is known as laminate stiffness (ABD) relating to force-moment resultants and in-plane strain-curvature relation. In addition, the laminate stiffness (ABD) can be written in an abbreviated form as:

$$\begin{bmatrix} N_{xx} \\ N_{yy} \\ N_{xy} \\ M_{xx} \\ M_{yy} \\ M_{xy} \end{bmatrix} = \begin{bmatrix} A_{11} & A_{12} & A_{16} & B_{11} & B_{12} & B_{16} \\ A_{12} & A_{22} & A_{26} & B_{12} & B_{22} & B_{26} \\ A_{16} & A_{26} & A_{66} & B_{16} & B_{26} & B_{66} \\ B_{11} & B_{12} & B_{16} & D_{11} & D_{12} & D_{16} \\ B_{12} & B_{22} & B_{26} & D_{12} & D_{22} & D_{26} \\ B_{16} & B_{26} & B_{66} & D_{16} & D_{26} & D_{66} \end{bmatrix} \begin{bmatrix} \varepsilon_{xx}^o \\ \varepsilon_{yy}^0 \\ \gamma_{xy}^o \\ \kappa_{xx}^o \\ \kappa_{yy}^o \\ \kappa_{xy}^o \end{bmatrix}$$

$$\text{or} \begin{bmatrix} N \\ M \end{bmatrix} = \begin{bmatrix} A & B \\ B & D \end{bmatrix} \begin{bmatrix} \varepsilon^o \\ \kappa^o \end{bmatrix} \tag{4.19.1}$$

In general, laminate stiffness (ABD) in Eq. (4.19.1) is symmetric. Inversion of Eq. (4.19.1) gives the compliance matrix of laminate composites.

$$
\begin{bmatrix} \varepsilon_{xx}^{o} \\ \varepsilon_{yy}^{0} \\ \gamma_{xy}^{o} \\ \kappa_{xx}^{o} \\ \kappa_{yy}^{o} \\ \kappa_{xy}^{o} \end{bmatrix} = \begin{bmatrix} a_{11} & a_{12} & a_{16} & b_{11} & b_{12} & b_{16} \\ a_{12} & a_{22} & a_{26} & b_{12} & b_{22} & b_{26} \\ a_{16} & a_{26} & a_{66} & b_{16} & b_{26} & b_{66} \\ b_{11} & b_{12} & b_{16} & d_{11} & d_{12} & d_{16} \\ b_{12} & b_{22} & b_{26} & d_{12} & d_{22} & d_{26} \\ b_{16} & b_{26} & b_{66} & d_{16} & d_{26} & d_{66} \end{bmatrix} \begin{bmatrix} N_{xx} \\ N_{yy} \\ N_{xy} \\ M_{xx} \\ M_{yy} \\ M_{xy} \end{bmatrix}
$$

$$
\text{or} \begin{bmatrix} \varepsilon^{o} \\ \kappa^{o} \end{bmatrix} = \begin{bmatrix} a & b \\ b & d \end{bmatrix} \begin{bmatrix} N \\ M \end{bmatrix} \tag{4.19.2}
$$

4.5 LAMINATED HYGROTHERMAL IN-PLANE FORCE AND MOMENT RESULTANTS

The hygrothermal effects are accounted for in the transformation of reduced in-plane stress and strain relation. In this section, hygrothermal effects on laminated composites are illustrated. The in-plane strain in Eq. (4.7.1) includes total strain and hygrothermal strain. Equation (4.7.1) can be separated into two parts using strain resultants as total strain and hygrothermal strain.

$$
\begin{bmatrix} \sigma_{xx}^{k} \\ \sigma_{yy}^{k} \\ \tau_{xy}^{k} \end{bmatrix} = \begin{bmatrix} \bar{Q}_{11}^{k} & \bar{Q}_{12}^{k} & \bar{Q}_{16}^{k} \\ \bar{Q}_{12}^{k} & \bar{Q}_{22}^{k} & \bar{Q}_{26}^{k} \\ \bar{Q}_{16}^{k} & \bar{Q}_{26}^{k} & \bar{Q}_{66}^{k} \end{bmatrix} \left(\begin{bmatrix} \varepsilon_{xx}^{o} \\ \varepsilon_{yy}^{0} \\ \gamma_{xy}^{o} \end{bmatrix} + z \begin{bmatrix} \kappa_{xx}^{o} \\ \kappa_{yy}^{o} \\ \kappa_{xy}^{o} \end{bmatrix} \right)_{(z=k)}
$$
$$
+ \begin{bmatrix} \bar{Q}_{11}^{k} & \bar{Q}_{12}^{k} & \bar{Q}_{16}^{k} \\ \bar{Q}_{12}^{k} & \bar{Q}_{22}^{k} & \bar{Q}_{26}^{k} \\ \bar{Q}_{16}^{k} & \bar{Q}_{26}^{k} & \bar{Q}_{66}^{k} \end{bmatrix} \left(-\Delta T \begin{bmatrix} \alpha_{xx}^{k} \\ \alpha_{yy}^{k} \\ \alpha_{xy}^{k} \end{bmatrix} - \Delta M \begin{bmatrix} \beta_{xx}^{k} \\ \beta_{yy}^{k} \\ \beta_{xy}^{k} \end{bmatrix} \right)_{(z=k)} \tag{4.20.1}
$$

If hygrothermal strains are neglected, then the in-plane strain part of Eq. (4.20.1) corresponds to total strain and is equal to the mechanical or load-induced strain. In the case of hygrothermal effects, the first part of the in-plane strain means the total strain and the second part is the hygrothermal strain. Thus, the strain resultant of this case is the mechanical strain (total strains minus hygrothermal strains).

4.5.1 IN-PLANE FORCE RESULTANT RELATION

To determine the in-plane force resultant, Eq. (4.20.1) is integrated with respect to z, ranging from $-h/2$ to $h/2$.

$$
\begin{bmatrix} N_{xx} \\ N_{yy} \\ N_{xy} \end{bmatrix} = \int_{-\frac{h}{2}}^{\frac{h}{2}} \begin{bmatrix} \bar{Q}_{11}^{k} & \bar{Q}_{12}^{k} & \bar{Q}_{16}^{k} \\ \bar{Q}_{12}^{k} & \bar{Q}_{22}^{k} & \bar{Q}_{26}^{k} \\ \bar{Q}_{16}^{k} & \bar{Q}_{26}^{k} & \bar{Q}_{66}^{k} \end{bmatrix} dz \begin{bmatrix} \varepsilon_{xx}^{o} \\ \varepsilon_{yy}^{0} \\ \gamma_{xy}^{o} \end{bmatrix} + \int_{-\frac{h}{2}}^{\frac{h}{2}} \begin{bmatrix} \bar{Q}_{11}^{k} & \bar{Q}_{12}^{k} & \bar{Q}_{16}^{k} \\ \bar{Q}_{12}^{k} & \bar{Q}_{22}^{k} & \bar{Q}_{26}^{k} \\ \bar{Q}_{16}^{k} & \bar{Q}_{26}^{k} & \bar{Q}_{66}^{k} \end{bmatrix} z\, dz \begin{bmatrix} \kappa_{xx}^{o} \\ \kappa_{yy}^{o} \\ \kappa_{xy}^{o} \end{bmatrix}
$$
$$
- \int_{-\frac{h}{2}}^{\frac{h}{2}} \begin{bmatrix} \bar{Q}_{11}^{k} & \bar{Q}_{12}^{k} & \bar{Q}_{16}^{k} \\ \bar{Q}_{12}^{k} & \bar{Q}_{22}^{k} & \bar{Q}_{26}^{k} \\ \bar{Q}_{16}^{k} & \bar{Q}_{26}^{k} & \bar{Q}_{66}^{k} \end{bmatrix} \begin{bmatrix} \alpha_{xx}^{k} \\ \alpha_{yy}^{k} \\ \alpha_{xy}^{k} \end{bmatrix} dz \Delta T - \int_{-\frac{h}{2}}^{\frac{h}{2}} \begin{bmatrix} \bar{Q}_{11}^{k} & \bar{Q}_{12}^{k} & \bar{Q}_{16}^{k} \\ \bar{Q}_{12}^{k} & \bar{Q}_{22}^{k} & \bar{Q}_{26}^{k} \\ \bar{Q}_{16}^{k} & \bar{Q}_{26}^{k} & \bar{Q}_{66}^{k} \end{bmatrix} \begin{bmatrix} \beta_{xx}^{k} \\ \beta_{yy}^{k} \\ \beta_{xy}^{k} \end{bmatrix} dz \Delta M
$$

$$
\tag{4.20.2}
$$

If thermal and moisture differences (ΔT and ΔM) are treated as independent of z, then elements $\bar{Q}_{ij}^k$ and thermal and moisture expansion coefficients (α_{ij}^k and β_{ij}^k) are constant only in their own layers and are different from layer to layer. Integration of Eq. (4.20.2) results in:

$$
\begin{bmatrix} N_{xx} \\ N_{yy} \\ N_{xy} \end{bmatrix} = \begin{bmatrix} A_{11} & A_{12} & A_{16} \\ A_{12} & A_{22} & A_{26} \\ A_{16} & A_{26} & A_{66} \end{bmatrix} \begin{bmatrix} \varepsilon_{xx}^o \\ \varepsilon_{yy}^0 \\ \gamma_{xy}^0 \end{bmatrix} + \begin{bmatrix} B_{11} & B_{12} & B_{16} \\ B_{12} & B_{22} & B_{26} \\ B_{16} & B_{26} & B_{66} \end{bmatrix}
$$

$$
\times \begin{bmatrix} \kappa_{xx}^o \\ \kappa_{yy}^o \\ \kappa_{xy}^o \end{bmatrix} - \begin{bmatrix} N_{xx}^T \\ N_{yy}^T \\ N_{xy}^T \end{bmatrix} - \begin{bmatrix} N_{xx}^M \\ N_{yy}^M \\ N_{xx}^M \end{bmatrix} \tag{4.21.1}
$$

where

$$
\begin{bmatrix} N_{xx}^T \\ N_{yy}^T \\ N_{xy}^T \end{bmatrix} = \begin{bmatrix} \sum_{k=1}^{n} \left(\bar{Q}_{11}^k \alpha_{xx}^k + \bar{Q}_{12}^k \alpha_{yy}^k + \bar{Q}_{16}^k \alpha_{xy}^k \right)(z_k - z_{k-1}) \\ \sum_{k=1}^{n} \left(\bar{Q}_{12}^k \alpha_{xx}^k + \bar{Q}_{22}^k \alpha_{yy}^k + \bar{Q}_{26}^k \alpha_{xy}^k \right)(z_k - z_{k-1}) \\ \sum_{k=1}^{n} \left(\bar{Q}_{11}^k \beta_{xx}^k + \bar{Q}_{12}^k \beta_{yy}^k + \bar{Q}_{16}^k \beta_{xy}^k \right)(z_k - z_{k-1}) \end{bmatrix} \Delta T \tag{4.21.2}
$$

$$
\begin{bmatrix} N_{xx}^M \\ N_{yy}^M \\ N_{xx}^M \end{bmatrix} = \begin{bmatrix} \sum_{k=1}^{n} \left(\bar{Q}_{11}^k \beta_{xx}^k + \bar{Q}_{12}^k \beta_{yy}^k + \bar{Q}_{16}^k \beta_{xy}^k \right)(z_k - z_{k-1}) \\ \sum_{k=1}^{n} \left(\bar{Q}_{12}^k \beta_{xx}^k + \bar{Q}_{22}^k \beta_{yy}^k + \bar{Q}_{26}^k \beta_{xy}^k \right)(z_k - z_{k-1}) \\ \sum_{k=1}^{n} \left(\bar{Q}_{16}^k \beta_{xx}^k + \bar{Q}_{26}^k \beta_{yy}^k + \bar{Q}_{66}^k \beta_{xy}^k \right)(z_k - z_{k-1}) \end{bmatrix} \Delta M \tag{4.21.3}
$$

and in-plane thermal and moisture-induced force resultants (N^T and N^M) are functions of z, ΔT, and ΔM.

4.5.2 MOMENT RESULTANT RELATION

To determine the moment resultants, Eq. (4.20.1) is multiplied with z and integrated with respect to z, ranging from $-h/2$ to $h/2$.

$$
\begin{bmatrix} M_{xx} \\ M_{yy} \\ M_{xy} \end{bmatrix} = \int_{-\frac{h}{2}}^{\frac{h}{2}} \begin{bmatrix} \bar{Q}_{11}^k & \bar{Q}_{12}^k & \bar{Q}_{16}^k \\ \bar{Q}_{12}^k & \bar{Q}_{22}^k & \bar{Q}_{26}^k \\ \bar{Q}_{16}^k & \bar{Q}_{26}^k & \bar{Q}_{66}^k \end{bmatrix} \left(\begin{bmatrix} \varepsilon_{xx}^o \\ \varepsilon_{yy}^0 \\ \gamma_{xy}^0 \end{bmatrix} + z^2 \begin{bmatrix} \kappa_{xx}^o \\ \kappa_{yy}^o \\ \kappa_{xy}^0 \end{bmatrix} \right) dz
$$

$$
- \int_{-\frac{h}{2}}^{\frac{h}{2}} \begin{bmatrix} \bar{Q}_{11}^k & \bar{Q}_{12}^k & \bar{Q}_{16}^k \\ \bar{Q}_{12}^k & \bar{Q}_{22}^k & \bar{Q}_{26}^k \\ \bar{Q}_{16}^k & \bar{Q}_{26}^k & \bar{Q}_{66}^k \end{bmatrix} \left(z\Delta T \begin{bmatrix} \alpha_{xx}^k \\ \alpha_{yy}^k \\ \alpha_{xy}^k \end{bmatrix} + z\Delta M \begin{bmatrix} \beta_{xx}^k \\ \beta_{yy}^k \\ \beta_{xy}^k \end{bmatrix} \right) dz \tag{4.22.1}
$$

$$
\begin{bmatrix} M_{xx} \\ M_{yy} \\ M_{xy} \end{bmatrix} = \begin{bmatrix} B_{11} & B_{12} & B_{16} \\ B_{12} & B_{22} & B_{26} \\ B_{16} & B_{26} & B_{66} \end{bmatrix} \begin{bmatrix} \varepsilon_{xx}^o \\ \varepsilon_{yy}^0 \\ \gamma_{xy}^o \end{bmatrix} + \begin{bmatrix} D_{11} & D_{12} & D_{16} \\ D_{12} & D_{22} & D_{26} \\ D_{16} & D_{26} & D_{66} \end{bmatrix}
$$

$$
\times \begin{bmatrix} \kappa_{xx}^o \\ \kappa_{yy}^o \\ \kappa_{xy}^o \end{bmatrix} - \begin{bmatrix} M_{xx}^T \\ M_{yy}^T \\ M_{xy}^T \end{bmatrix} - \begin{bmatrix} M_{xx}^M \\ M_{yy}^M \\ M_{xx}^M \end{bmatrix} \tag{4.22.2}
$$

where

$$
\begin{bmatrix} M_{xx}^T \\ M_{yy}^T \\ M_{xy}^T \end{bmatrix} = \begin{bmatrix} \frac{1}{2}\sum_{k=1}^{n}\left(\bar{Q}_{11}^k\alpha_{xx}^k + \bar{Q}_{12}^k\alpha_{yy}^k + \bar{Q}_{16}^k\alpha_{xy}^k\right)\left(z_k^2 - z_{k-1}^2\right) \\ \frac{1}{2}\sum_{k=1}^{n}\left(\bar{Q}_{12}^k\alpha_{xx}^k + \bar{Q}_{22}^k\alpha_{yy}^k + \bar{Q}_{26}^k\alpha_{xy}^k\right)\left(z_k^2 - z_{k-1}^2\right) \\ \frac{1}{2}\sum_{k=1}^{n}\left(\bar{Q}_{16}^k\alpha_{xx}^k + \bar{Q}_{26}^k\alpha_{yy}^k + \bar{Q}_{66}^k\alpha_{xy}^k\right)\left(z_k^2 - z_{k-1}^2\right) \end{bmatrix}\Delta T \tag{4.22.3}
$$

$$
\begin{bmatrix} M_{xx}^M \\ M_{yy}^M \\ M_{xy}^M \end{bmatrix} = \begin{bmatrix} \frac{1}{2}\sum_{k=1}^{n}\left(\bar{Q}_{11}^k\beta_{xx}^k + \bar{Q}_{12}^k\beta_{yy}^k + \bar{Q}_{16}^k\beta_{xy}^k\right)\left(z_k^2 - z_{k-1}^2\right) \\ \frac{1}{2}\sum_{k=1}^{n}\left(\bar{Q}_{12}^k\beta_{xx}^k + \bar{Q}_{22}^k\beta_{yy}^k + \bar{Q}_{26}^k\beta_{xy}^k\right)\left(z_k^2 - z_{k-1}^2\right) \\ \frac{1}{2}\sum_{k=1}^{n}\left(\bar{Q}_{16}^k\beta_{xx}^k + \bar{Q}_{26}^k\beta_{yy}^k + \bar{Q}_{66}^k\beta_{xy}^k\right)\left(z_k^2 - z_{k-1}^2\right) \end{bmatrix}\Delta M \tag{4.22.4}
$$

and in-plane thermal and moisture-induced force resultants (N^T and N^M) are functions of z, ΔT and ΔM.

4.5.3 In-Plane and Moment Resultant Relation

To formulate the laminate stiffness under the in-plane force and moment resultants, Eqs. (4.21.1 and 4.22.2) are combined and written in a matrix form. This stiffness matrix is known as the laminate stiffness (*ABD*) matrix relating to force-moment resultants and in-plane strain-curvature relations with hygrothermal effects.

$$
\begin{bmatrix} N_{xx} \\ N_{yy} \\ N_{xy} \\ M_{xx} \\ M_{yy} \\ M_{xy} \end{bmatrix} = \begin{bmatrix} A_{11} & A_{12} & A_{16} & B_{11} & B_{12} & B_{16} \\ A_{12} & A_{22} & A_{26} & B_{12} & B_{22} & B_{26} \\ A_{16} & A_{26} & A_{66} & B_{16} & B_{26} & B_{66} \\ B_{11} & B_{12} & B_{16} & D_{11} & D_{12} & D_{16} \\ B_{12} & B_{22} & B_{26} & D_{12} & D_{22} & D_{26} \\ B_{16} & B_{26} & B_{66} & D_{16} & D_{26} & D_{66} \end{bmatrix} \begin{bmatrix} \varepsilon_{xx}^o \\ \varepsilon_{yy}^0 \\ \gamma_{xy}^o \\ \kappa_{xx}^o \\ \kappa_{yy}^o \\ \kappa_{xy}^o \end{bmatrix} - \begin{bmatrix} N_{xx}^T \\ N_{yy}^T \\ N_{xy}^T \\ M_{xx}^T \\ M_{yy}^T \\ M_{xy}^T \end{bmatrix} - \begin{bmatrix} N_{xx}^M \\ N_{yy}^M \\ N_{xy}^M \\ M_{xx}^M \\ M_{yy}^M \\ M_{xy}^M \end{bmatrix} \tag{4.23.1}
$$

or,

$$
\begin{bmatrix} N \\ M \end{bmatrix} = \begin{bmatrix} A & B \\ B & D \end{bmatrix} \begin{bmatrix} \varepsilon^o \\ \kappa^o \end{bmatrix} - \begin{bmatrix} N^T \\ M^T \end{bmatrix} - \begin{bmatrix} N^M \\ M^M \end{bmatrix} \tag{4.23.2}
$$

Laminated stiffness (ABD) matrix in Eq. (4.23.1) is symmetric. Inversion of Eq. (4.23.2) represents the laminate compliance (abd) matrix of a laminated composite, which is also symmetric.

$$
\begin{bmatrix}
\varepsilon_{xx}^o \\
\varepsilon_{yy}^o \\
\varepsilon_{xy}^o \\
\kappa_{xx}^o \\
\kappa_{yy}^o \\
\kappa_{xy}^o
\end{bmatrix}
=
\begin{bmatrix}
a_{11} & a_{12} & a_{16} & b_{11} & b_{12} & b_{16} \\
a_{12} & a_{22} & a_{26} & b_{12} & b_{22} & b_{26} \\
a_{16} & a_{26} & a_{66} & b_{16} & b_{26} & b_{66} \\
b_{11} & b_{12} & b_{16} & d_{11} & d_{12} & d_{16} \\
b_{12} & b_{22} & b_{26} & d_{12} & d_{22} & d_{26} \\
b_{16} & b_{26} & b_{66} & d_{16} & d_{26} & d_{66}
\end{bmatrix}
\begin{bmatrix}
N_{xx} + N_{xx}^T + N_{xx}^M \\
N_{yy} + N_{yy}^T + N_{yy}^M \\
N_{xy} + N_{xy}^T + N_{xy}^M \\
M_{xx} + M_{xx}^T + M_{xx}^M \\
M_{yy} + M_{yy}^T + M_{yy}^M \\
M_{xy} + M_{xy}^T + M_{xy}^M
\end{bmatrix}
$$

$$
\text{or} \quad
\begin{bmatrix}
\varepsilon^o \\
\kappa^o
\end{bmatrix}
=
\begin{bmatrix}
a & b \\
b & d
\end{bmatrix}
\begin{bmatrix}
N + N^T + N^M \\
M + M^T + M^M
\end{bmatrix}
\tag{4.23.3}
$$

At this point, the laminate stiffness (ABD) and compliance (abd) matrix of FRP laminated composites are developed using the stress–strain relations, transformation, in-plane stress condition, definition of the force-moment resultants, etc. The procedure of laminated stiffness formulation is quite long. However, the procedure can easily be presented as given below:

> **Given:** The properties of FRP lamina, fiber orientation, and fiber architecture.
> **Objective:** find the reduced stiffness in the global coordinated system.
> **Solutions:** using the transformed reduced stiffness Eq. (3.31.5).
> **Objective:** find the laminate stiffness (ABD) matrix elements.
> **Solutions:** using the laminated stiffness Eqs. (4.14.1, 4.14.2, and 4.17.2).

Example 4.3

Determine the stiffness (ABD) of the composite lay-ups given below:

a. Laminated composite of $[0°_2/45°_2]$ with each lamina thickness of 2.5 mm

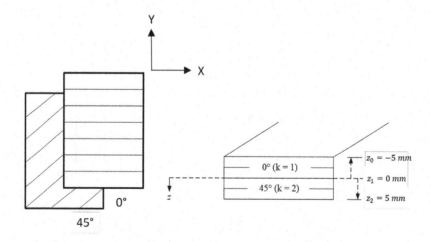

EXAMPLE FIGURE 4.3.1   Laminated composite of $[0°_2/45°_2]$.

b. Laminated composite of [0°/45°] with 6 and 2 mm of lamina for 0° and 45°, respectively

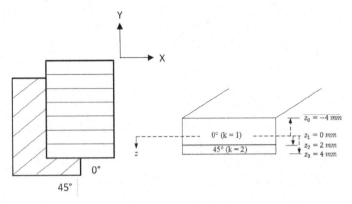

EXAMPLE FIGURE 4.3.2   Laminated composite of [0°/45°].

c. Laminated composite of [0°/45°/–30°]$_s$ with lamina thickness of 1 mm

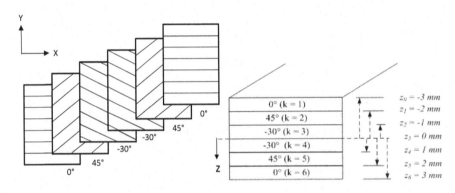

EXAMPLE FIGURE 4.3.3   Laminated composite of [0°/45°/–30°]$_s$.

The properties of unidirectional FRP lamina are $E_{11}=180$ GPa, $E_{22}=40$ GPa, $G_{12}=20$ GPa, $\nu_{12}=0.25$, $\alpha_{11}=7$ μm/m°C, $\alpha_{22}=25$ μm/m°C, $\beta_{11}=120$ μm/m %M, and $\beta_{22}=1{,}000$ μm/m %M.

Solution

a. *ABD* matrix of the laminated composite [0°$_2$/45°$_2$]:

The laminated composite is made from composite laminae of the same materials with different fiber orientations. From the reduced stiffness matrix Eq. (3.31.5) of the ortho-tropic lamina, the reduced stiffness matrix of lamina 0°, 30°, and 45° is determined as:

$$\left[\bar{Q}_{ij}\right]_0 = \begin{bmatrix} 182.5 & 10.14 & 0 \\ 10.14 & 40.6 & 0 \\ 0 & 0 & 20 \end{bmatrix} \text{GPa}$$

$$\left[\bar{Q}_{ij}\right]_{45} = \begin{bmatrix} 80.8 & 40.8 & 35.5 \\ 40.8 & 80.8 & 35.5 \\ 35.5 & 35.5 & 50.7 \end{bmatrix} \text{GPa}$$

$$\left[\bar{Q}_{ij}\right]_{-30} = \begin{bmatrix} 73.8 & -221 & 83.1 \\ -221 & 43 & 208 \\ 83.1 & 208 & 26.7 \end{bmatrix} \text{GPa}$$

The laminated composite is made of composite laminae of the same materials with different fiber orientation. However, the two composite layers of each fiber orientation can be simplified by considering it as a single ply. Thus, the laminated composite includes only two plies of 0° and 45° lamina with 5 mm thickness.

From Eq. (4.14.1):

$$A_{ij} = \bar{Q}_{ij}^1 (z_1 - z_0) + \bar{Q}_{ij}^2 (z_2 - z_1)$$

$$A_{ij} = \begin{bmatrix} 182.5 & 10.14 & 0 \\ 10.14 & 40.6 & 0 \\ 0 & 0 & 20 \end{bmatrix} 0.005 + \begin{bmatrix} 80.8 & 40.8 & 35.5 \\ 40.8 & 80.8 & 35.5 \\ 35.5 & 35.5 & 50.7 \end{bmatrix} 0.005$$

$$A_{ij} = \begin{bmatrix} 1.317 & 0.255 & 0.1775 \\ 0.255 & 0.607 & 0.1775 \\ 0.1775 & 0.1775 & 0.354 \end{bmatrix} \frac{GN}{m}$$

From Eq. (4.14.2):

$$B_{ij} = \frac{1}{2} \bar{Q}_{ij}^1 (z_1^2 - z_0^2) + \frac{1}{2} \bar{Q}_{ij}^2 (z_2^2 - z_1^2)$$

$$B_{ij} = \frac{1}{2} \begin{bmatrix} 182.5 & 10.14 & 0 \\ 10.14 & 40.6 & 0 \\ 0 & 0 & 20 \end{bmatrix} (0^2 - (-0.005)^2) + \frac{1}{2} \begin{bmatrix} 80.8 & 40.8 & 35.5 \\ 40.8 & 80.8 & 35.5 \\ 35.5 & 35.5 & 50.7 \end{bmatrix} (0.005^2 - 0^2)$$

$$B_{ij} = \begin{bmatrix} -12.71 & 3.84 & 4.44 \\ 3.84 & 5.04 & 4.44 \\ 4.44 & 4.44 & 3.84 \end{bmatrix} 10^{-4} GN$$

From Eq. (4.17.2):

$$D_{ij} = \frac{1}{3} \bar{Q}_{ij}^1 (z_1^3 - z_0^3) + \frac{1}{3} \bar{Q}_{ij}^2 (z_2^3 - z_1^3)$$

$$D_{ij} = \frac{1}{3} \begin{bmatrix} 182.5 & 10.14 & 0 \\ 10.14 & 40.6 & 0 \\ 0 & 0 & 20 \end{bmatrix} (0^3 - (-0.005)^3) + \frac{1}{3} \begin{bmatrix} 80.8 & 40.8 & 35.5 \\ 40.8 & 80.8 & 35.5 \\ 35.5 & 35.5 & 50.7 \end{bmatrix} (0.005^3 - 0^3)$$

$$D_{ij} = \begin{bmatrix} 10.97 & 2.12 & 1.479 \\ 2.12 & 5.06 & 1.479 \\ 1.479 & 1.479 & 2.95 \end{bmatrix} 10^{-6} GN.m$$

b. *ABD* matrix of laminated composite [0°/45°]:

From Eq. (4.14.1):

$$A_{ij} = \bar{Q}_{ij}^1 (z_2 - z_0) + \bar{Q}_{ij}^2 (z_3 - z_2)$$

$$A_{ij} = \begin{bmatrix} 182.5 & 10.14 & 0 \\ 10.14 & 40.6 & 0 \\ 0 & 0 & 20 \end{bmatrix} 0.006 + \begin{bmatrix} 80.8 & 40.8 & 35.5 \\ 40.8 & 80.8 & 35.5 \\ 35.5 & 35.5 & 50.7 \end{bmatrix} 0.002$$

$$A_{ij} = \begin{bmatrix} 1.257 & 0.1425 & 0.0071 \\ 0.1425 & 0.405 & 0.0071 \\ 0.0071 & 0.0071 & 0.221 \end{bmatrix} \frac{GN}{m}$$

From Eq. (4.14.2):

$$B_{ij} = \frac{1}{2}\bar{Q}_{ij}^1 (z_2^2 - z_0^2) + \frac{1}{2}\bar{Q}_{ij}^2 (z_3^2 - z_2^2)$$

$$B_{ij} = \frac{1}{2} \begin{bmatrix} 182.5 & 10.14 & 0 \\ 10.14 & 40.6 & 0 \\ 0 & 0 & 20 \end{bmatrix} (0.002^2 - (0.004)^2)$$

$$+ \frac{1}{2} \begin{bmatrix} 80.8 & 40.8 & 35.5 \\ 40.8 & 80.8 & 35.5 \\ 35.5 & 35.5 & 50.7 \end{bmatrix} (0.004^2 - 0.002^2)$$

$$B_{ij} = \begin{bmatrix} -6.10 & 1.842 & 2.13 \\ 1.842 & 2.42 & 2.13 \\ 2.13 & 2.13 & 1.842 \end{bmatrix} 10^{-4} GN$$

From Eq. (4.17.2):

$$D_{ij} = \frac{1}{3}\bar{Q}_{ij}^1 (z_2^3 - z_0^3) + \frac{1}{3}\bar{Q}_{ij}^2 (z_3^3 - z_2^3)$$

$$D_{ij} = \frac{1}{3} \begin{bmatrix} 182.5 & 10.14 & 0 \\ 10.14 & 40.6 & 0 \\ 0 & 0 & 20 \end{bmatrix} (0.002^3 - (-0.004)^3)$$

$$+ \frac{1}{3} \begin{bmatrix} 80.8 & 40.8 & 35.5 \\ 40.8 & 80.8 & 35.5 \\ 35.5 & 35.5 & 50.7 \end{bmatrix} (0.004^3 - 0.002^3)$$

$$D_{ij} = \begin{bmatrix} 5.89 & 1.006 & 0.663 \\ 1.006 & 2.48 & 0.663 \\ 0.663 & 0.663 & 1.426 \end{bmatrix} 10^{-6} GN.m$$

c. *ABD* matrix of laminated composite [0°/45°/−30°]$_s$:
   From Eq. (4.14.1):

$$A_{ij} = \bar{Q}_{ij}^1 (z_1 - z_0) + \bar{Q}_{ij}^2 (z_2 - z_1) + \bar{Q}_{ij}^3 (z_3 - z_2) \ldots + \bar{Q}_{ij}^6 (z_6 - z_5)$$

$$A_{ij} = \begin{bmatrix} 6.742 & -3.40 & 2.37 \\ -3.40 & 3.29 & 4.87 \\ 2.37 & 4.87 & 1.95 \end{bmatrix} 10^{-1} \frac{GN}{m}$$

From Eq. (4.14.2):

$$B_{ij} = \frac{1}{2}\bar{Q}_{ij}^1 \left(z_1^2 - z_0^2\right) + \frac{1}{2}\bar{Q}_{ij}^2 \left(z_2^2 - z_1^2\right) + \frac{1}{2}\bar{Q}_{ij}^3 \left(z_3^2 - z_2^2\right)\ldots + \frac{1}{2}\bar{Q}_{ij}^6 \left(z_6^2 - z_5^2\right)$$

$$B_{ij} = \begin{bmatrix} 0 & 0 & 0 \\ 0 & 0 & 0 \\ 0 & 0 & 0 \end{bmatrix}$$

From Eq. (4.17.2):

$$D_{ij} = \frac{1}{3}\bar{Q}_{ij}^1 \left(z_1^3 - z_0^3\right) + \frac{1}{3}\bar{Q}_{ij}^2 \left(z_2^3 - z_1^3\right) + \frac{1}{3}\bar{Q}_{ij}^3 \left(z_3^3 - z_2^3\right)\ldots + \frac{1}{3}\bar{Q}_{ij}^6 \left(z_6^3 - z_5^3\right)$$

$$D_{ij} = \begin{bmatrix} 27.4 & 1.71 & 2.21 \\ 1.71 & 9.20 & 3.04 \\ 2.21 & 3.04 & 5.08 \end{bmatrix} 10^{-7} \, GN.m$$

Note: when lay-ups of laminated composites are symmetric, then the laminated stiffness $[B_{ij}]$ is equal to zero. The laminated compliance (abd) matrix can be determined from the inversion of the laminated stiffness (ABD) matrix.

Example 4.4

Determine force and moment resultants of laminated composite lay-ups (Example 4.3a) under applied in-plane strain: normal strain $\left(\varepsilon_{xx}^o = 1{,}000\,\mu m/m\right)$, curvature $\left(\kappa_{xx}^o = 0.1\,m^{-1}\right)$, and thermal load $(\Delta T = 100°C)$
　　Lamina of 0° $(\alpha_{xx}=7\,\mu m/m\,°C, \alpha_{yy}=25\,\mu m/m\,°C, \alpha_{xy}=0\,\mu\,rad\,°C)$
　　Lamina of 45° $(\alpha_{xx}=16\,\mu m/m\,°C, \alpha_{yy}=16\,\mu m/m\,°C, \alpha_{xy}=32\,\mu\,rad\,°C)$

Solution

From Example 4.3:

$$\left[\bar{Q}_{ij}\right]_0 = \begin{bmatrix} 182.5 & 10.14 & 0 \\ 10.14 & 40.6 & 0 \\ 0 & 0 & 20 \end{bmatrix} GPa \quad and \quad \left[\bar{Q}_{ij}\right]_{45} = \begin{bmatrix} 80.8 & 40.8 & 35.5 \\ 40.8 & 80.8 & 35.5 \\ 35.5 & 35.5 & 50.7 \end{bmatrix} GPa$$

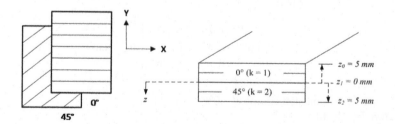

EXAMPLE FIGURE 4.4.1　Laminate composite $[0°_2/45°_2]$.

From Eq. (4.21.2):

$$
\begin{bmatrix} N_{xx}^T \\ N_{yy}^T \\ N_{xy}^T \end{bmatrix} = \begin{bmatrix} \sum_{k=1}^{2}\left(\overline{Q}_{11}^{k}\alpha_{xx}^{k}+\overline{Q}_{12}^{k}\alpha_{yy}^{k}+\overline{Q}_{16}^{k}\alpha_{xy}^{k}\right)(z_k-z_{k-1}) \\ \sum_{k=1}^{2}\left(\overline{Q}_{12}^{k}\alpha_{xx}^{k}+\overline{Q}_{22}^{k}\alpha_{yy}^{k}+\overline{Q}_{26}^{k}\alpha_{xy}^{k}\right)(z_k-z_{k-1}) \\ \sum_{-k=1}^{2}\left(\overline{Q}_{16}^{k}\alpha_{xx}^{k}+\overline{Q}_{26}^{k}\alpha_{yy}^{k}+\overline{Q}_{66}^{k}\alpha_{xy}^{k}\right)(z_k-z_{k-1}) \end{bmatrix} 100 = \begin{bmatrix} 1.420 \\ 1.197 \\ 0.1115 \end{bmatrix} 10^{6}\,\frac{N}{m}
$$

From Eq. (4.22.3):

$$
\begin{bmatrix} M_{xx}^T \\ M_{yy}^T \\ M_{xy}^T \end{bmatrix} = \begin{bmatrix} \frac{1}{2}\sum_{k=1}^{2}\left(\overline{Q}_{11}^{k}\alpha_{xx}^{k}+\overline{Q}_{12}^{k}\alpha_{yy}^{k}+\overline{Q}_{16}^{k}\alpha_{xy}^{k}\right)(z_k^2-z_{k-1}^2) \\ \frac{1}{2}\sum_{k=1}^{2}\left(\overline{Q}_{12}^{k}\alpha_{xx}^{k}+\overline{Q}_{22}^{k}\alpha_{yy}^{k}+\overline{Q}_{26}^{k}\alpha_{xy}^{k}\right)(z_k^2-z_{k-1}^2) \\ \frac{1}{2}\sum_{k=1}^{2}\left(\overline{Q}_{16}^{k}\alpha_{xx}^{k}+\overline{Q}_{26}^{k}\alpha_{yy}^{k}+\overline{Q}_{66}^{k}\alpha_{xy}^{k}\right)(z_k^2-z_{k-1}^2) \end{bmatrix} 100 = \begin{bmatrix} -279 \\ 279 \\ 279 \end{bmatrix} N
$$

From the stiffness (ABD) of laminated composites [0°$_2$/45°$_2$] in Example 4.3a:

$$
A_{ij} = \begin{bmatrix} 1.317 & 0.255 & 0.1775 \\ 0.255 & 0.607 & 0.1775 \\ 0.1775 & 0.1775 & 0.354 \end{bmatrix} \frac{GN}{m} \qquad B_{ij} = \begin{bmatrix} -12.71 & 3.84 & 4.44 \\ 3.84 & 5.04 & 4.44 \\ 4.44 & 4.44 & 3.84 \end{bmatrix} 10^{-4}\,GN
$$

$$
D_{ij} = \begin{bmatrix} 10.97 & 2.12 & 1.479 \\ 2.12 & 5.06 & 1.479 \\ 1.479 & 1.479 & 2.95 \end{bmatrix} 10^{-6}\,GN.m
$$

From Eq. (4.23.1):

$$
\begin{bmatrix} N_{xx} \\ N_{yy} \\ N_{xy} \\ M_{xx} \\ M_{yy} \\ M_{xy} \end{bmatrix} = \begin{bmatrix} A_{11} & A_{12} & A_{16} & B_{11} & B_{12} & B_{16} \\ A_{12} & A_{22} & A_{26} & B_{12} & B_{22} & B_{26} \\ A_{16} & A_{26} & A_{66} & B_{16} & B_{26} & B_{66} \\ B_{11} & B_{12} & B_{16} & D_{11} & D_{12} & D_{16} \\ B_{12} & B_{22} & B_{26} & D_{12} & D_{22} & D_{26} \\ B_{16} & B_{26} & B_{66} & D_{16} & D_{26} & D_{66} \end{bmatrix}
$$

$$
\begin{bmatrix} \varepsilon_{xx}^{o} \\ \varepsilon_{yy}^{0} \\ \gamma_{xy}^{o} \\ \kappa_{xx}^{o} \\ \kappa_{yy}^{o} \\ \kappa_{xy}^{o} \end{bmatrix} - \begin{bmatrix} N_{xx}^T \\ N_{yy}^T \\ N_{xy}^T \\ M_{xx}^T \\ M_{yy}^T \\ M_{xy}^T \end{bmatrix} - \begin{bmatrix} N_{xx}^M \\ N_{yy}^M \\ N_{xy}^M \\ M_{xx}^M \\ M_{yy}^M \\ M_{xy}^M \end{bmatrix}
$$

$$
= \begin{bmatrix}
1.317 & 0.255 & 0.1775 & -12.71 \times 10^{-4} & 3.84 \times 10^{-4} & 4.44 \times 10^{-4} \\
0.255 & 0.607 & 0.1775 & 3.84 \times 10^{-4} & 5.04 \times 10^{-4} & 4.44 \times 10^{-4} \\
0.1775 & 0.1775 & 0.354 & 4.44 \times 10^{-4} & 4.44 \times 10^{-4} & 3.84 \times 10^{-4} \\
-12.71 \times 10^{-4} & 3.84 \times 10^{-4} & 4.44 \times 10^{-4} & 10.97 \times 10^{-6} & 2.12 \times 10^{-6} & 1.479 \times 10^{-6} \\
3.84 \times 10^{-4} & 5.04 \times 10^{-4} & 4.44 \times 10^{-4} & 2.12 \times 10^{-6} & 5.06 \times 10^{-6} & 1.479 \times 10^{-6} \\
4.44 \times 10^{-4} & 4.44 \times 10^{-4} & 3.84 \times 10^{-4} & 1.479 \times 10^{-6} & 1.479 \times 10^{-6} & 2.95 \times 10^{-6}
\end{bmatrix}
$$

$$
\times \left(\begin{bmatrix}
1,000 \times 10^{-6} \\
0 \\
0 \\
0.1 \\
0 \\
0
\end{bmatrix} - \begin{bmatrix}
1.420 \times 10^{3} \\
1.197 \times 10^{3} \\
0.1115 \times 10^{3} \\
-0.279 \\
0.279 \\
0.279
\end{bmatrix} \right)
$$

$$
\begin{bmatrix}
N_{xx} \\
N_{yy} \\
N_{xy} \\
M_{xx} \\
M_{yy} \\
M_{xy}
\end{bmatrix} = \begin{bmatrix}
1.194 \, \text{MN/m} \\
0.292 \, \text{MN/m} \\
0.222 \, \text{MN/m} \\
-173.7 \, \text{N} \\
596 \, \text{N} \\
592 \, \text{N}
\end{bmatrix}
$$

Example 4.5

Determine the stresses and strains of each layer for the following laminated composites:

    a. Laminated composite $[0°/45°/-30°]_s$ under $N_{xx} = 1$ MN/m
    b. Laminated composite $[0°_2/45°_2]$ under $N_{xx} = 1$ MN/m, $M_{xx} = 500$ N and thermal load ($\Delta T = 100°C$)

Solution

a. Stresses and strains in each layer of laminated composite $[0°/45°/-30°]_s$:

    From Eq. (4.19.1):

    From the (ABD) stiffness of laminated composites $[0°/45°/-30°]_s$ in Example 4.3c:

$$
\begin{bmatrix}
a & b \\
b & d
\end{bmatrix} = \begin{bmatrix}
A & B \\
B & D
\end{bmatrix}^{-1}
$$

$$
\begin{bmatrix}
a & b \\
b & d
\end{bmatrix} = \begin{bmatrix}
0.733 & -0.769 & 1.031 & 0 & 0 & 0 \\
-0.769 & -0.319 & 1.732 & 0 & 0 & 0 \\
1.031 & 1.732 & -0.450 & 0 & 0 & 0 \\
0 & 0 & 0 & 378,527 & -19,872.1 & -152,782 \\
0 & 0 & 0 & -19,872.1 & 1.356 \times 10^{6} & -802,767 \\
0 & 0 & 0 & -152,782 & -802,767 & 2.515 \times 10^{6}
\end{bmatrix}
$$

From Eq. (4.19.2):

$$
\begin{bmatrix} \varepsilon_{xx}^o \\ \varepsilon_{yy}^0 \\ \varepsilon_{xy}^o \\ \kappa_{xx}^o \\ \kappa_{yy}^o \\ \kappa_{xy}^0 \end{bmatrix} =
\begin{bmatrix}
0.733 & -0.769 & 1.031 & 0 & 0 & 0 \\
-0.769 & -0.319 & 1.732 & 0 & 0 & 0 \\
1.031 & 1.732 & -0.450 & 0 & 0 & 0 \\
0 & 0 & 0 & 378,527 & -19,872.1 & -152,782 \\
0 & 0 & 0 & -19,872.1 & 1.356\times10^6 & -802,767 \\
0 & 0 & 0 & -152,782 & -802,767 & 2.515\times10^6
\end{bmatrix}
$$

$$
\times \begin{bmatrix} 1\times10^6\ \text{N/m} \\ 0 \\ 0 \\ 0 \\ 0 \\ 0 \end{bmatrix}
$$

$$
\begin{bmatrix} \varepsilon_{xx}^o \\ \varepsilon_{yy}^0 \\ \varepsilon_{xy}^0 \\ \kappa_{xx}^o \\ \kappa_{yy}^o \\ \kappa_{xy}^o \end{bmatrix} =
\begin{bmatrix} 733 \\ -769 \\ 1,031 \\ 0 \\ 0 \\ 0 \end{bmatrix} \mu
$$

Units: normal strain ε_{ii} (m/m) and shear strain (γ_{ij}) rad/rad

In-plane strains of each layer are determined from Eq. (4.7.2). After neglecting terms corresponding to hygrothermal loads. The in-plane strains are constant through the cross section without curvatures.

$$
\begin{bmatrix} \varepsilon_{xx}^{mech} \\ \varepsilon_{yy}^{mech} \\ \gamma_{xy}^{mech} \end{bmatrix} =
\begin{bmatrix} \varepsilon_{xx}^o \\ \varepsilon_{yy}^o \\ \gamma_{xy}^o \end{bmatrix} + z
\begin{bmatrix} \kappa_{xx}^o \\ \kappa_{yy}^o \\ \kappa_{xy}^o \end{bmatrix} =
\begin{bmatrix} 733 \\ -769 \\ 1,031 \end{bmatrix} + z
\begin{bmatrix} 0 \\ 0 \\ 0 \end{bmatrix} =
\begin{bmatrix} 733 \\ -769 \\ 1,031 \end{bmatrix} \mu
$$

From Eq. (4.8.1), strains in the local coordinate system can be determined as:
For both lamina of 0° (k=1 and 6):

$$
\begin{bmatrix} \varepsilon_{11} \\ \varepsilon_{22} \\ \gamma_{12} \end{bmatrix} =
\begin{bmatrix}
\text{Cos}^2 0° & \text{Sin}^2 0° & \text{Cos}0°\text{Sin}0° \\
\text{Sin}^2 0° & \text{Cos}^2 0° & -\text{Cos}0°\text{Sin}0° \\
-2\text{Cos}0°\text{Sin}0° & 2\text{Cos}0°\text{Sin}0° & \text{Cos}^2 0° - \text{Sin}^2 0°
\end{bmatrix}
\begin{bmatrix} 733 \\ -769 \\ 1,031 \end{bmatrix} =
\begin{bmatrix} 733 \\ -769 \\ 1,031 \end{bmatrix} \mu
$$

Units: normal strain ε_{ii} (m/m) and shear strain (γ_{ij}) rad/rad
For both lamina of 45° ($k=2$ and 5):

$$\begin{bmatrix} \varepsilon_{11} \\ \varepsilon_{22} \\ \gamma_{12} \end{bmatrix} = \begin{bmatrix} \text{Cos}^2 45° & \text{Sin}^2 45° & \text{Cos}\,45°\text{Sin}\,45° \\ \text{Sin}^2 45° & \text{Cos}^2 45° & -\text{Cos}\,45°\text{Sin}\,45° \\ -2\text{Cos}\,45°\text{Sin}\,45° & 2\text{Cos}\,45°\text{Sin}\,45° & \text{Cos}^2 45° - \text{Sin}^2 45° \end{bmatrix}$$

$$\begin{bmatrix} 733 \\ -769 \\ 1{,}031 \end{bmatrix} = \begin{bmatrix} 498 \\ -534 \\ -36 \end{bmatrix}\mu$$

For both lamina of –30° ($k=3$ and 4):

$$\begin{bmatrix} \varepsilon_{11} \\ \varepsilon_{22} \\ \gamma_{12} \end{bmatrix} = \begin{bmatrix} \text{Cos}^2 30° & \text{Sin}^2 30° & -\text{Cos}\,30°\text{Sin}\,30° \\ \text{Sin}^2 30° & \text{Cos}^2 30° & -\text{Cos}\,30°\text{Sin}\,30° \\ 2\text{Cos}\,30°\text{Sin}\,30° & -2\text{Cos}\,30°\text{Sin}\,30° & \text{Cos}^2 30° - \text{Sin}^2 30° \end{bmatrix}$$

$$\begin{bmatrix} 733 \\ -769 \\ 1{,}031 \end{bmatrix} = \begin{bmatrix} -88.9 \\ 52.9 \\ 1{,}816 \end{bmatrix}\mu$$

Units: normal strain ε_{ii} (m/m) and shear strain (γ_{ij}) rad/rad
Strain profiles of laminated composite $[0°/45°/-30°]_s$ in both the global and local coordinate systems are presented in Example Figure 4.5.1.
From Eq. (4.7.1), strains in each layer are substituted into the reduced stress and strain relation. The reduced stiffness matrices of lamina 0°, 45°, and –30° are determined and given in Example 4.3c.
For lamina of 0° ($k=1$ and 6):
From Eq. (4.7.1):

$$\begin{bmatrix} \sigma_{xx} \\ \sigma_{yy} \\ \tau_{xy} \end{bmatrix} = \begin{bmatrix} 182.5 & 10.14 & 0 \\ 10.14 & 40.6 & 0 \\ 0 & 0 & 20 \end{bmatrix}\begin{bmatrix} 733 \\ -769 \\ 1{,}031 \end{bmatrix} = \begin{bmatrix} 126 \\ -23.8 \\ 20.6 \end{bmatrix}\text{MPa}$$

From Eq. (4.8.3)

$$\begin{bmatrix} \sigma_{11} \\ \sigma_{22} \\ \tau_{12} \end{bmatrix} = \begin{bmatrix} 1 & 0 & 0 \\ 0 & 1 & 0 \\ 0 & 0° & 1 \end{bmatrix}\begin{bmatrix} 126 \\ -23.8 \\ 20.6 \end{bmatrix} = \begin{bmatrix} 126 \\ -23.8 \\ 20.6 \end{bmatrix}\text{MPa}$$

For lamina of 45° ($k=2$ and 5):
From Eq. (4.7.1):

$$\begin{bmatrix} \sigma_{xx} \\ \sigma_{yy} \\ \tau_{xy} \end{bmatrix} = \begin{bmatrix} 80.8 & 40.8 & 35.5 \\ 40.8 & 80.8 & 35.5 \\ 35.5 & 35.5 & 50.7 \end{bmatrix}\begin{bmatrix} 733 \\ -769 \\ 1{,}031 \end{bmatrix} = \begin{bmatrix} 64.5 \\ 4.37 \\ 51.0 \end{bmatrix}\text{MPa}$$

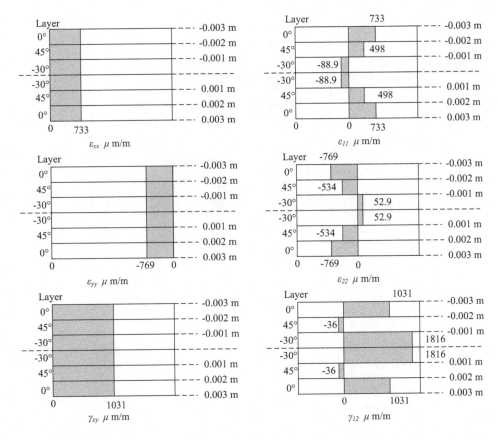

EXAMPLE FIGURE 4.5.1 Strain profiles of laminated composite [0°/45°/–30°]$_s$ under normal force resultant.

From Eq. (4.8.3)

$$\begin{bmatrix} \sigma_{11} \\ \sigma_{22} \\ \tau_{12} \end{bmatrix} = \begin{bmatrix} 1/2 & 1/2 & 1 \\ 1/2 & 1/2 & -1 \\ -1/2 & 1/2 & 0 \end{bmatrix} \begin{bmatrix} 733 \\ -769 \\ 1,031 \end{bmatrix} = \begin{bmatrix} 1,013 \\ -1,049 \\ -751 \end{bmatrix} \text{MPa}$$

For lamina of –30° ($k = 3$ and 4):
From Eq. (4.7.1):

$$\begin{bmatrix} \sigma_{xx} \\ \sigma_{yy} \\ \tau_{xy} \end{bmatrix} = \begin{bmatrix} 73.8 & -221 & 83.1 \\ -221 & 43 & 208 \\ 83.1 & 208 & 26.7 \end{bmatrix} \begin{bmatrix} 733 \\ -769 \\ 1,031 \end{bmatrix} = \begin{bmatrix} 309.7 \\ 19.39 \\ -71.5 \end{bmatrix} \text{MPa}$$

From Eq. (4.8.3):

$$\begin{bmatrix} \sigma_{11} \\ \sigma_{22} \\ \tau_{12} \end{bmatrix} = \begin{bmatrix} 3/4 & 1/4 & -\sqrt{3}/2 \\ 1/4 & 3/4 & \sqrt{3}/2 \\ \sqrt{3}/4 & -\sqrt{3}/4 & 1/2 \end{bmatrix} \begin{bmatrix} 309.7 \\ 19.39 \\ -71.5 \end{bmatrix} = \begin{bmatrix} 299 \\ 30.0 \\ 90.0 \end{bmatrix} \text{MPa}$$

Stress profiles of laminated composite [0°/45°/–30°]$_s$ in both the global and local coordinate systems are presented in Example Figure 4.5.2.

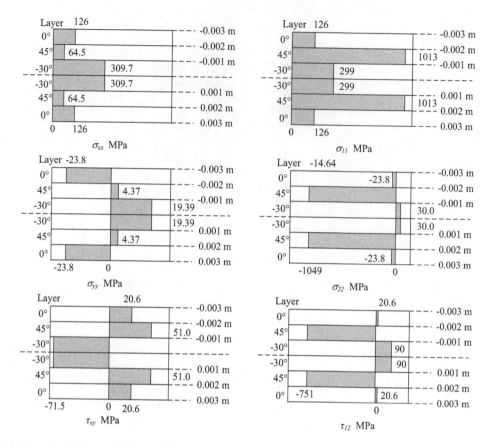

EXAMPLE FIGURE 4.5.2 Stress profiles of laminated composite $[0°/45°/-30°]_s$ under normal force resultant.

b. Stresses and strains in each layer of the laminated composite $[0°_2/45°_2]$:
   From Eq. (4.19.1):
   From the (ABD) stiffness of laminated composites $[0°_2/45°_2]$ in Example 4.3a:

$$
\begin{bmatrix} a & b \\ b & d \end{bmatrix} = \begin{bmatrix} A & B \\ B & D \end{bmatrix}^{-1}
$$

$$
= \begin{bmatrix}
1.119 & -0.323 & -0.421 & 186.3 & -55.7 & -130.4 \\
-0.323 & 2.16 & -0.613 & -56.0 & -74.8 & -130.4 \\
-0.421 & -0.613 & 3.91 & -130.5 & -130.5 & -223 \\
186.3 & -56.0 & -130.5 & 134,261 & -38,573 & -50,602 \\
-55.7 & -74.8 & -130.5 & -38,573 & 258,428 & -73,602 \\
-130.4 & -130.4 & -223 & -50,602 & -73,602 & 469,497
\end{bmatrix}
$$

From Example 4.4, in-plane thermal force–moment resultant is:

$$
\begin{bmatrix} N_{xx}^T \\ N_{yy}^T \\ N_{xy}^T \end{bmatrix} = \begin{bmatrix} 1.420 \\ 1.197 \\ 0.1115 \end{bmatrix} 10^6 \frac{N}{m} \qquad \begin{bmatrix} M_{xx}^T \\ M_{yy}^T \\ M_{xy}^T \end{bmatrix} = \begin{bmatrix} -279 \\ 279 \\ 279 \end{bmatrix} N
$$

From Eq. (4.23.3):

$$
\begin{bmatrix} \varepsilon^o \\ \kappa^o \end{bmatrix} = \begin{bmatrix} a & b \\ b & d \end{bmatrix} \begin{bmatrix} N + N^T + N^M \\ M + M^T + M^M \end{bmatrix}
$$

$$
\begin{bmatrix} \varepsilon_{xx}^o \\ \varepsilon_{yy}^o \\ \varepsilon_{xy}^o \\ \kappa_{xx}^o \\ \kappa_{yy}^o \\ \kappa_{xy}^o \end{bmatrix} = \begin{bmatrix}
1.119 & -0.323 & -0.421 & 186.3 & -55.7 & -130.4 \\
-0.323 & 2.16 & -0.613 & -56.0 & -74.8 & -130.4 \\
-0.421 & -0.613 & 3.91 & -130.5 & -130.5 & -223 \\
186.3 & -56.0 & -130.5 & 134,261 & -38,573 & -50,602 \\
-55.7 & -74.8 & -130.5 & -38,573 & 258,428 & -73,602 \\
-130.4 & -130.4 & -223 & -50,602 & -73,602 & 469,497
\end{bmatrix}
$$

$$
\begin{bmatrix}
1 + 1.420 \, \text{MN/m} \\
1.197 \, \text{MN/m} \\
0.1115 \, \text{MN/m} \\
500 - 279 \, \text{N} \\
279 \, \text{N} \\
279 \, \text{N}
\end{bmatrix}
$$

From Eq. (4.7.1):

$$
\begin{bmatrix} \varepsilon_{xx}^{mech} \\ \varepsilon_{yy}^{mech} \\ \gamma_{xy}^{mech} \end{bmatrix} = \begin{bmatrix} \varepsilon_{xx}^o \\ \varepsilon_{yy}^o \\ \gamma_{xy}^0 \end{bmatrix} + z \begin{bmatrix} \kappa_{xx}^o \\ \kappa_{yy}^o \\ \kappa_{xy}^o \end{bmatrix} - \Delta T \begin{bmatrix} \alpha_{xx} \\ \alpha_{yy} \\ \alpha_{xy} \end{bmatrix}
$$

$$
\begin{bmatrix} \varepsilon_{xx}^{mech} \\ \varepsilon_{yy}^{mech} \\ \gamma_{xy}^{mech} \end{bmatrix} = \begin{bmatrix} 2,263 \\ 1,660 \\ -1,446 \end{bmatrix} \mu + z \begin{bmatrix} 3.74 \times 10^{-1} \\ -1.960 \times 10^{-1} \\ -3.97 \times 10^{-1} \end{bmatrix} 10^6 - 100 \begin{bmatrix} \alpha_{xx} \\ \alpha_{yy} \\ \alpha_{xy} \end{bmatrix}
$$

$$
= \begin{bmatrix}
2,263 + 3.74z\left(10^5\right) - 100\alpha_{xx} \\
1,660 - 1.962z\left(10^5\right) - 100\alpha_{yy} \\
-1,446 - 3.97z\left(10^5\right) - 100\alpha_{xy}
\end{bmatrix} \mu
$$

For lamina of 0° (k=1):
@ Bottom surface (z=−0.005 m)   @ Top surface (z=−0.0025 m)

$$
\begin{bmatrix} \varepsilon_{xx}^{mech} \\ \varepsilon_{yy}^{mech} \\ \gamma_{xy}^{mech} \end{bmatrix} = \begin{bmatrix} -307 \\ 140 \\ 539 \end{bmatrix} \mu
\qquad
\begin{bmatrix} \varepsilon_{xx}^{mech} \\ \varepsilon_{yy}^{mech} \\ \gamma_{xy}^{mech} \end{bmatrix} = \begin{bmatrix} 628 \\ -350 \\ -453 \end{bmatrix} \mu
$$

For lamina of 0° ($k=2$):

@ Bottom surface ($z=-0.0025$ m)   @ Top surface ($z=0$ m)

$$\begin{bmatrix} \varepsilon_{xx}^{mech} \\ \varepsilon_{yy}^{mech} \\ \gamma_{xy}^{mech} \end{bmatrix} = \begin{bmatrix} 628 \\ -350 \\ -453 \end{bmatrix} \mu \qquad \begin{bmatrix} \varepsilon_{xx}^{mech} \\ \varepsilon_{yy}^{mech} \\ \gamma_{xy}^{mech} \end{bmatrix} = \begin{bmatrix} 1,563 \\ -840 \\ -1,446 \end{bmatrix} \mu$$

Units: normal strain ε_{ii} (m/m) and shear strain (γ_{ij}) rad/rad

For lamina of 45° ($k=3$):

@ Bottom surface ($z=0$ m)   @ Top surface ($z=0.0025$ m)

$$\begin{bmatrix} \varepsilon_{xx}^{mech} \\ \varepsilon_{yy}^{mech} \\ \gamma_{xy}^{mech} \end{bmatrix} = \begin{bmatrix} 663 \\ 60 \\ 354 \end{bmatrix} \mu \qquad \begin{bmatrix} \varepsilon_{xx}^{mech} \\ \varepsilon_{yy}^{mech} \\ \gamma_{xy}^{mech} \end{bmatrix} = \begin{bmatrix} 1,598 \\ -430 \\ -639 \end{bmatrix} \mu$$

For lamina of 45° ($k=4$):

@ Bottom surface ($z=0.0025$ m)   @ Top surface ($z=0.005$ m)

$$\begin{bmatrix} \varepsilon_{xx}^{mech} \\ \varepsilon_{yy}^{mech} \\ \gamma_{xy}^{mech} \end{bmatrix} = \begin{bmatrix} 1,598 \\ -430 \\ -639 \end{bmatrix} \mu \qquad \begin{bmatrix} \varepsilon_{xx}^{mech} \\ \varepsilon_{yy}^{mech} \\ \gamma_{xy}^{mech} \end{bmatrix} = \begin{bmatrix} 2,534 \\ -920 \\ -1,632 \end{bmatrix} \mu$$

Units: normal strain ε_{ii} (m/m) and shear strain (γ_{ij}) rad/rad

From Eq. (3.27.1), strains in the local coordinate system can be determined as:

For lamina of 0° ($k=1$) @ bottom surface ($z=-0.005$ m):

$$\begin{bmatrix} \varepsilon_{11} \\ \varepsilon_{22} \\ \gamma_{12} \end{bmatrix} = \begin{bmatrix} Cos^2 0° & Sin^2 0° & Cos0°Sin0° \\ Sin^2 0° & Cos^2 0° & -Cos0°Sin0° \\ -2Cos0°Sin0° & 2Cos0°Sin0° & Cos^2 0° - Sin^2 0° \end{bmatrix} \begin{bmatrix} -307 \\ 140 \\ 539 \end{bmatrix} = \begin{bmatrix} -307 \\ 140 \\ 539 \end{bmatrix} \mu$$

For lamina of 0° ($k=2$) @ top surface ($z=0$ m):

$$\begin{bmatrix} \varepsilon_{11} \\ \varepsilon_{22} \\ \gamma_{12} \end{bmatrix} = \begin{bmatrix} Cos^2 0° & Sin^2 0° & Cos0°Sin0° \\ Sin^2 0° & Cos^2 0° & -Cos0°Sin0° \\ -2Cos0°Sin0° & 2Cos0°Sin0° & Cos^2 0° - Sin^2 0° \end{bmatrix} \begin{bmatrix} 1,563 \\ -840 \\ -1,446 \end{bmatrix} = \begin{bmatrix} 1,563 \\ -840 \\ -1,446 \end{bmatrix} \mu$$

For lamina of 45° ($k=3$) @ bottom surface ($z=0$ m):

$$\begin{bmatrix} \varepsilon_{11} \\ \varepsilon_{22} \\ \gamma_{12} \end{bmatrix} = \begin{bmatrix} Cos^2 45° & Sin^2 45° & Cos45°Sin45° \\ Sin^2 45° & Cos^2 45° & -Cos45°Sin45° \\ -2Cos45°Sin45° & 2Cos45°Sin45° & Cos^2 45° - Sin^2 45° \end{bmatrix} \begin{bmatrix} 663 \\ 599 \\ 354 \end{bmatrix} = \begin{bmatrix} 808 \\ 454 \\ -64 \end{bmatrix} \mu$$

Units: normal strain ε_{ii} (m/m) and shear strain (γ_{ij}) rad/rad

For lamina of 45° ($k=4$) @ top surface ($z=0.005$ m):

$$
\begin{bmatrix} \varepsilon_{11} \\ \varepsilon_{22} \\ \gamma_{12} \end{bmatrix} =
\begin{bmatrix}
\cos^2 45° & \sin^2 45° & \cos 45° \sin 45° \\
\sin^2 45° & \cos^2 45° & -\cos 45° \sin 45° \\
-2\cos 45° \sin 45° & 2\cos 45° \sin 45° & \cos^2 45° - \sin^2 45°
\end{bmatrix}
$$

$$
\times \begin{bmatrix} 2{,}534 \\ -920 \\ -1{,}632 \end{bmatrix} =
\begin{bmatrix} -9 \\ 1{,}623 \\ -3{,}454 \end{bmatrix} \mu
$$

Units: normal strain ε_{ii} (m/m) and shear strain (γ_{ij}) rad/rad

Strain profiles of laminated composite $[0°_2/45°_2]$ in both the global and local coordinate systems are presented in Example Figure 4.5.3

From Eq. (4.7.1), strains in each layer are substituted into the reduced stress and strain relation. The reduced stiffness matrices of lamina 0° and 45° are determined and given in Example 4.3c.

For lamina of 0° ($k=1$) @ bottom surface ($z=-0.005$ m):

From Eq. (4.7.1):

$$
\begin{bmatrix} \sigma_{xx} \\ \sigma_{yy} \\ \tau_{xy} \end{bmatrix} =
\begin{bmatrix}
182.5 & 10.14 & 0 \\
10.14 & 40.6 & 0 \\
0 & 0 & 20
\end{bmatrix} \text{GPa}
\begin{bmatrix} -307 \\ 140 \\ 539 \end{bmatrix} \mu =
\begin{bmatrix} -54.6 \\ 2.57 \\ 10.78 \end{bmatrix} \text{MPa}
$$

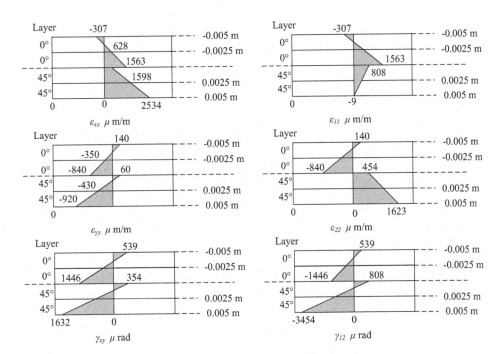

EXAMPLE FIGURE 4.5.3 Strain profiles of laminated composite $[0°_2/45°_2]$ under applied forces and thermal load.

For lamina of 0° ($k=2$) @ top surface ($z=0$ m):
From Eq. (4.7.1):

$$\begin{bmatrix} \sigma_{xx} \\ \sigma_{yy} \\ \tau_{xy} \end{bmatrix} = \begin{bmatrix} 182.5 & 10.14 & 0 \\ 10.14 & 40.6 & 0 \\ 0 & 0 & 20 \end{bmatrix} \text{GPa} \begin{bmatrix} 1{,}563 \\ -840 \\ -1{,}446 \end{bmatrix} \mu = \begin{bmatrix} 277 \\ -18.26 \\ -28.9 \end{bmatrix} \text{MPa}$$

For lamina of 45° ($k=3$) @ bottom surface ($z=0$ m):
From Eq. (4.7.1):

$$\begin{bmatrix} \sigma_{xx} \\ \sigma_{yy} \\ \tau_{xy} \end{bmatrix} = \begin{bmatrix} 80.8 & 40.8 & 35.5 \\ 40.8 & 80.8 & 35.5 \\ 35.5 & 35.5 & 50.7 \end{bmatrix} \text{GPa} \begin{bmatrix} 663 \\ 599 \\ 354 \end{bmatrix} \mu = \begin{bmatrix} 90.6 \\ 88.0 \\ 62.7 \end{bmatrix} \text{MPa}$$

For lamina of 45° ($k=4$) @ top surface ($z=0.005$ m):
From Eq. (4.7.1):

$$\begin{bmatrix} \sigma_{xx} \\ \sigma_{yy} \\ \tau_{xy} \end{bmatrix} = \begin{bmatrix} 80.8 & 40.8 & 35.5 \\ 40.8 & 80.8 & 35.5 \\ 35.5 & 35.5 & 50.7 \end{bmatrix} \text{GPa} \begin{bmatrix} 2{,}534 \\ -920 \\ -1{,}632 \end{bmatrix} \mu = \begin{bmatrix} 109.3 \\ -28.9 \\ -25.4 \end{bmatrix} \text{MPa}$$

From Eq. (4.8.3), the stress in the local coordinate system can be determined as:
For lamina of 0° ($k=1$) @ bottom surface ($z=-0.005$ m):
From Eq. (4.8.3):

$$\begin{bmatrix} \sigma_{11} \\ \sigma_{22} \\ \tau_{12} \end{bmatrix} = \begin{bmatrix} \text{Cos}^2 0° & \text{Sin}^2 0° & \text{Cos}0°\text{Sin}0° \\ \text{Sin}^2 0° & \text{Cos}^2 0° & -2\text{Cos}0°\text{Sin}0° \\ -\text{Cos}0°\text{Sin}0° & \text{Cos}0°\text{Sin}0° & \text{Cos}^2 0° - \text{Sin}^2 0° \end{bmatrix} \begin{bmatrix} -54.6 \\ 2.57 \\ 10.78 \end{bmatrix} = \begin{bmatrix} -54.6 \\ 2.57 \\ 10.78 \end{bmatrix} \text{MPa}$$

For lamina of 0° ($k=2$) @ top surface ($z=0$ m):
From Eq. (4.8.3):

$$\begin{bmatrix} \sigma_{11} \\ \sigma_{22} \\ \tau_{12} \end{bmatrix} = \begin{bmatrix} \text{Cos}^2 0° & \text{Sin}^2 0° & 2\text{Cos}0°\text{Sin}0° \\ \text{Sin}^2 0° & \text{Cos}^2 0° & -2\text{Cos}0°\text{Sin}0° \\ -2\text{Cos}0°\text{Sin}0° & \text{Cos}0°\text{Sin}0° & \text{Cos}^2 0° - \text{Sin}^2 0° \end{bmatrix} \begin{bmatrix} 277 \\ -18.26 \\ -28.9 \end{bmatrix} = \begin{bmatrix} 277 \\ -18.26 \\ -28.9 \end{bmatrix} \text{MPa}$$

For lamina of 45° ($k=3$) @ bottom surface ($z=0$ m):
From Eq. (4.8.3):

$$\begin{bmatrix} \sigma_{11} \\ \sigma_{22} \\ \tau_{12} \end{bmatrix} = \begin{bmatrix} \text{Cos}^2 45° & \text{Sin}^2 45° & 2\text{Cos}45°\text{Sin}45° \\ \text{Sin}^2 45° & \text{Cos}^2 45° & -2\text{Cos}45°\text{Sin}45° \\ -\text{Cos}45°\text{Sin}45° & \text{Cos}45°\text{Sin}45° & \text{Cos}^2 45° - \text{Sin}^2 45° \end{bmatrix} \begin{bmatrix} 90.6 \\ 88.0 \\ 62.7 \end{bmatrix}$$

$$= \begin{bmatrix} 152 \\ 25.8 \\ -0.45 \end{bmatrix} \text{MPa}$$

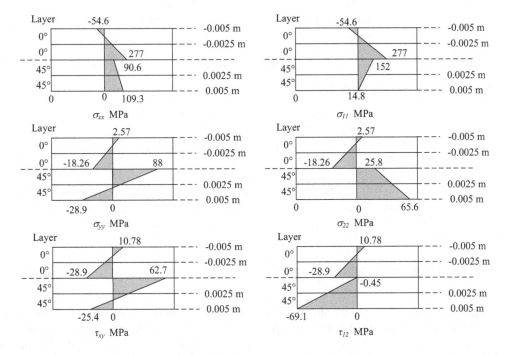

EXAMPLE FIGURE 4.5.4   Stress profiles of laminated composite $[0°_2/45°_2]$ under applied forces and thermal load.

For lamina of 45° (k=4) @ top surface (z=0.005 m):
From Eq. (4.8.3):

$$
\begin{bmatrix} \sigma_{11} \\ \sigma_{22} \\ \tau_{12} \end{bmatrix} =
\begin{bmatrix}
\text{Cos}^2 45° & \text{Sin}^2 45° & 2\text{Cos}\,45°\text{Sin}\,45° \\
\text{Sin}^2 45° & \text{Cos}^2 45° & -2\text{Cos}\,45°\text{Sin}\,45° \\
-\text{Cos}\,45°\text{Sin}\,45° & \text{Cos}\,45°\text{Sin}\,45° & \text{Cos}^2 45° - \text{Sin}^2 45°
\end{bmatrix}
\begin{bmatrix} 109.3 \\ -28.9 \\ -25.4 \end{bmatrix}
$$

$$
= \begin{bmatrix} 14.8 \\ 65.6 \\ -69.1 \end{bmatrix} \text{MPa}
$$

Stress profiles of laminated composite $[0°_2/45°_2]$ on the cross section in the global and local coordinate systems are presented in Example Figure 4.5.4.

4.6   SIGNIFICANCE OF ELASTIC COUPLINGS

In this section, coupling characteristics of laminated composites are discussed. When laminated structural members are under applied unidirectional loadings, then not only the response of those members is due to the extension or contraction in both parallel and perpendicular direction to loading, but also distortions (shear, bending, and twisting coupling), which often exist depending on the types of composite laminates. Similarly, when laminated members are under shear or bending, responses of structural members are not only shear or bending responses corresponding to the direction of loadings but also from deformations (extension or contraction) that do not correspond to the direction of loadings.

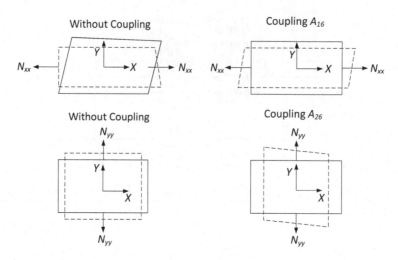

FIGURE 4.8 Extension-shear couplings (A_{16} and A_{26}).

Such behavior is identical to the characteristics of laminated composite materials that are generated because of the elements of laminate stiffness (*ABD*) known as couplings, such as the unsymmetric fiber architecture in a cross section. In general, couplings of all elements of a laminated stiffness (*ABD* matrix) can be separated into seven groups, which are: (1) extension-shear couplings, (2) bending-twisting couplings, (3) extension-twisting couplings, (4) bending–shear couplings, (5) in-plane and out-of-plane couplings, (6) extension-extension couplings, and (7) bending–bending couplings. It should be noted that both the extension–extension and bending–bending couplings usually exist for homogenous and isotropic materials, whereas the other five couplings are not present in isotropic materials. Additional details on classifications of symmetric, balanced, antisymmetric, etc. laminates are discussed in Appendix A of this textbook, entitled "Classification of Laminated Composite Stacking Sequence."

4.6.1 Extension-Shear Couplings

The extension-shear couplings result in nonzero elements (A_{16} and A_{26}) in the laminated stiffness matrix. When such couplings (A_{16} and A_{26}) of laminated stiffness matrix exist, then in-plane force resultants (N_{xx} and N_{yy}) will induce in-plane shear strains (γ_{xy}^o), while in-plane shear resultants (N_{xy}) will cause deformation (extension and contraction) in global coordinate system as shown in Figure 4.8.

4.6.2 Bending-Twisting Couplings

The bending-twisting couplings are identified through non-zero elements (D_{16} and D_{26}) of the laminated stiffness matrix. When such couplings (D_{16} and D_{26}) exist, then twisting curvature $\left(\kappa_{xy}^o\right)$ is induced by either bending moment (M_{xx} or M_{yy}), whereas the bending curvatures (κ_{xx}^o and κ_{yy}^o) are also induced by the twisting moment (M_{xy}). The behavior of bending-twisting couplings is illustrated in Figure 4.9.

4.6.3 Extension-Twisting Couplings

The extension-twisting couplings of laminated composites are identified through elements (B_{16} and B_{26}). When extension-twisting couplings (B_{16} and B_{26}) are not zero, then in-plane force resultants (N_{xx} and N_{yy}) can induce the twisting curvature $\left(\kappa_{xy}^o\right)$ as shown in Figure 4.10.

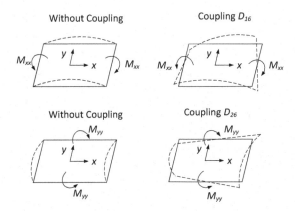

FIGURE 4.9   Bending-twisting couplings (D_{16} and D_{26}).

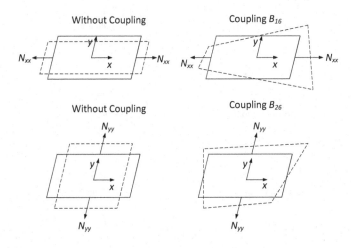

FIGURE 4.10   Extension-twisting couplings (B_{16} and B_{26}).

4.6.4   Bending–Shear Couplings

The bending–shear couplings of laminated composites are also identified through elements (B_{16} and B_{26}) as well as the extension-twisting couplings. When bending–shear couplings (B_{16} and B_{26}) are not zero, bending moments (M_{xx} and M_{yy}) can induce the in-plane shear strain $\left(\gamma_{xy}^{o}\right)$ as shown in Figure 4.11.

4.6.5   In-Plane and Out-of-Plane Couplings

The in-plane and out-of-plane couplings are identified through a non-zero matrix $[B_{ij}]$. In general, the couplings are elements (B_{11}, B_{12}, and B_{66}). When elements of couplings $[B_{ij}]$ exist then, the in-plane force resultants (N_{xx}, N_{yy}, and N_{xy}) and moment resultants (M_{xx}, M_{yy}, and M_{xy}) cause curvatures (κ_{xx}^{o}, κ_{yy}^{o} and κ_{xy}^{o}) and in-plane strain (ε_{xx}^{o}, ε_{yy}^{o} and γ_{xy}^{o}), respectively. The in-plane and out-of-plane couplings are illustrated in Figure 4.12.

4.6.6   Extension–Extension Couplings

The extension–extension coupling corresponds to element A_{12} of the in-plane laminated stiffness $[A_{ij}]$. The coupling A_{12} is also known as the effect of Poisson's ratio for homogenous isotropic

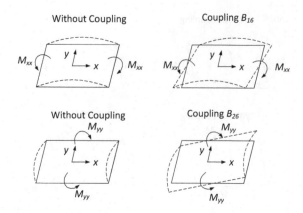

FIGURE 4.11 Bending–shear couplings (B_{16} and B_{26}).

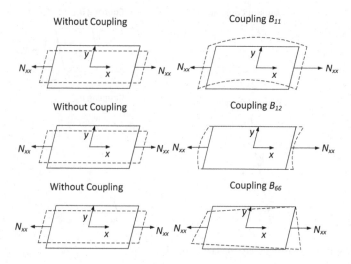

FIGURE 4.12 In-plane and out-of-plane couplings (B_{11}, B_{12}, and B_{66}).

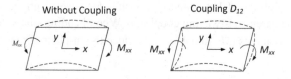

FIGURE 4.13 Extension-extension coupling (A_{12}).

materials. The coupling will lead to the deformation in the perpendicular direction to in-plane force resultants. The in-plane force resultants (N_{xx} and N_{yy}) cause the deformations (ε_{xx}^o and ε_{yy}^o), respectively. The extension--extension coupling is illustrated in Figure 4.13.

4.6.7 BENDING–BENDING COUPLINGS

The bending–bending coupling corresponds to element D_{12} of the bending-twisting laminated stiffness $[D_{ij}]$. The coupling will lead to the curvature perpendicular to the applied moment resultants. The moment resultants (M_{xx} and M_{yy}) cause the curvatures (κ_{xx}^o and κ_{yy}^o), respectively. The bending–bending coupling is illustrated in Figure 4.14.

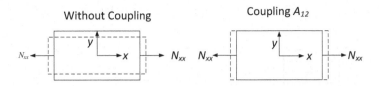

FIGURE 4.14   Bending-bending coupling (D_{12}).

4.7   SUMMARY

This chapter presents the analysis methods based on the CLT to determine the stresses and strains in each layer of a multilayer composite laminate. Technical details and examples in determining the force and moment resultants in the laminates and considering the hygrothermal effects and elastic couplings are also presented. A detailed understanding of the mechanics of composite laminates provides a base knowledge of reliable and safe analysis of different composite structures subject to different individual loading actions and under combined loading actions.

EXERCISES

Problem 4.1: Find A, B, and D matrices for +45/−45/−45/+45 laminate structure where each lamina is 0.25″ thick with 70% fiber volume fraction in a glass fiber composite with vinyl ester.

Problem 4.2: Determine the stress and strain of laminated composite $[0°/30°]_s$ where each lamina is 1 mm thick. The properties of unidirectional FRP lamina 0° and 30° are presented as: elastic modulus: $E_{11}=150\,\text{GPa}$ and $E_{22}=30\,\text{GPa}$, shear modulus: $G_{12}=25\,\text{GPa}$, $\nu_{12}=0.25$, thermal expansion: $\alpha_{11}=8\,\mu\text{m/m}\,°\text{C}$, $\alpha_{22}=25\,\mu\text{m/m}\,°\text{C}$, moisture expansion: $\beta_{11}=120\,\mu\text{m/m}\,\%\text{M}$, $\beta_{22}=1{,}000\,\mu\text{m/m}\,\%\text{M}$. (a) The laminated composite is under normal strain ($\varepsilon_{xx}^o=1{,}000\,\mu\text{m/m}$) in the global direction (x) at the reference surface.

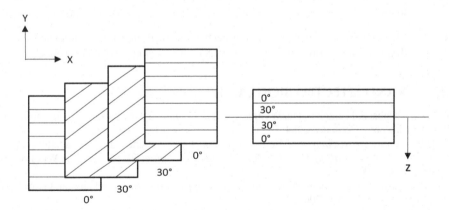

Problem 4.3: Compare the stiffnesses of laminate +45/−45/−45/+45 with that of 0/30/30/0 and comment on the advantages of one laminate over another.

Problem 4.4: The properties of unidirectional FRP lamina are presented as: elastic modulus: $E_{11}=210\,\text{GPa}$ and $E_{22}=30\,\text{GPa}$, shear modulus: $G_{12}=30\,\text{GPa}$, $\nu_{12}=0.25$, thermal expansion: $\alpha_{11}=8\,\mu\text{m/m}\,°\text{C}$, $\alpha_{22}=24\,\mu\text{m/m}\,°\text{C}$, moisture expansion: $\beta_{11}=130\,\mu\text{m/m}\,\%\text{M}$, $\beta_{22}=1{,}100\,\mu\text{m/m}\,\%\text{M}$. Determine the stiffness (ABD) of a composite lay-up which is as below:

a. The laminated composite of [0°/45°] with 6 and 2 mm of lamina 0° and 45°, respectively

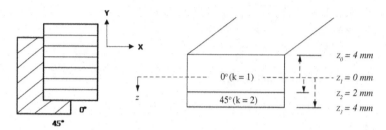

b. The laminated composite of [0°/45°/−30°]$_s$ with lamina thickness of 1 mm.

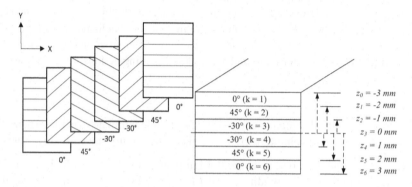

Problem 4.5: Determine force and moment resultants of laminated composite lay-ups in Problem 4.4a, (a) under applied in-plane strain: Laminated composites [0°/45°] under normal strain ($\varepsilon_{xx}^o = 1,000\,\mu\text{m/m}$), curvature ($\kappa_{xx}^o = 0.1\,\text{m}^{-1}$) and thermal load ($\Delta T = 60°\text{C}$).

Problem 4.6: Determine stresses and strains of each layer for laminated composites [0°/45°/−30°]$_s$ in Problem 4.4b under normal force and moment resultant ($N_{xx} = 1.5$ MN/m and $M_{xx} = 500\,\text{N}$).

Problem 4.7: Determine stresses and strains of each layer for Laminated composite [0°/45°/−30°]$_s$ in Problem 4.4b under moment resultant ($M_{yy} = 750\,\text{N}$) with thermal load ($\Delta T = 100°\text{C}$).

REFERENCES AND SELECTED BIOGRAPHY

Altenbach, H., Altenbach, J., and Kissing, W., *Mechanics of Composite Structural Elements*, Springer-Verlag, Berlin, Heidelberg, Germany, 2004.

Barbero, E.J., *Introduction to Composite Materials Design*, Taylor &Francis, Inc., Philadelphia, PA, 1998.

Callister, W.D. Jr, *Fundamentals of Materials Science and Engineering*, 5th edition, John Wiley & Sons, Inc., New York, NY, 2000.

Daniel, I.M., and Ishai, O., *Engineering Mechanics of Composite Materials*, 2nd edition, Oxford University Press, Inc., New York, NY, 2006.

Gibson, R.F., *Principles of Composite Material Mechanics*, McGraw-Hill, Inc., New York, NY, 1994.

Hyer, M.W., *Stress Analysis of Fiber-Reinforced Composite Materials*, International edition, WCB/McGraw-Hill, Singapore, 1998.

Jones, R.M., *Mechanics of Composite Materials*, Hemisphere Publishing, New York, NY, 1975.

Kaw, A.K., *Mechanics of Composite Materials*, 2nd edition, CRC Press, Taylor & Francis Group, Boca Raton, FL, 2006.

Kollar, L.P., and Springer, G.S., *Mechanics of Composite Structures*, Cambridge University Press, New York, NY, 2003.

Swanson, S.R., *Introduction to Design and Analysis with Advance Composite Materials*, International edition, Prentice-Hall, Inc., Upper Saddle River, NJ, 1997.

5 Analysis of FRP Composite Beams

One of the most widely used structural elements in engineering is the beam. Beams are structural members that carry mainly bending loads and have one dimension (length) much larger than the other two dimensions (width and height). To date, most fiber composite beams have taken the form of the accepted structural shapes in steel and aluminum such as I-beams and box beams. With the advent of high-strength composite materials, novel structural shapes with maximum structural efficiency (flexural rigidity/unit cross-sectional weight) are needed. For example, trapezoidal cross sections with extended (winged) top flange projections such as hat sections or hexagonal sections do fit into the category of high sectional efficiency. Some of these sections have been field implemented, albeit, on a demonstration basis only (GangaRao and Siva, 2003; Lopez-Anido et al., 2002).

5.1 GENERAL ASSUMPTIONS OF FRP COMPOSITE BEAM RESPONSE UNDER TRANSVERSE LOADS

In this chapter, FRP composite beams are analyzed systematically and simply, using the following assumptions: (1) beam length is much larger than the cross-sectional dimensions (>10), (2) beams are prismatic, i.e., cross sections are constant throughout the span length, and (3) beams satisfy Bernoulli–Navier hypothesis, i.e., plane sections remain plane and perpendicular to the longitudinal axis of a beam before and after loading. This assumption implies that the shear deformation effect is neglected. However, the shear effect can also be accounted for in our derivation using the first-order Timoshenko's shear deformation theory. The basic governing equation of isotropic beams as shown in Figure 5.1 under transverse (bending) loads and coupling loads is reviewed before developing differential equations for a composite beam under bending.

The governing equation of an isotropic beam under axial loads is:

$$\text{Axial deformation} = u^o = \frac{1}{AE} \int_0^l P(x)\,dx \tag{5.1}$$

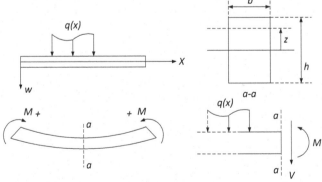

FIGURE 5.1 Beam under bending.

DOI: 10.1201/9781003196754-5

where

$P(x)$ = axial load varying along x
A = cross-sectional area
E = elastic modulus

The differential governing equation for an isotropic beam deflection under transverse bending is:

$$EI\frac{d^2w}{dx^2} = -M(x)$$ (5.2.1)

where

$M(x)$ = bending moment
I = moment of inertia or second moment of area
E = elastic modulus

Note: It should be noted that a positive moment provides a negative curvature with respect to $(X–W)$ coordinate shown in Figure 5.1.

Curvature (κ_x) under beam bending and the beam bending stress is also given as:

$$-\frac{d^2w}{dx^2} = \kappa_x$$ (5.2.2)

$$\sigma = -\frac{M(x)z}{I}$$ (5.2.3)

where z = vertical direction as shown in Figure 5.1.

Note: A positive bending moment $M(x)$ causes negative (compressive) stress above the neutral axis and positive (tensile) stress below the neutral axis. Positive directions of X and W are shown in Figure 5.1.

The relations between bending moment, shear force, and applied transverse load ($q(x)$) are given below:

$$\frac{dM}{dx} = V$$ (5.2.4)

$$\frac{dV}{dx} = -q(x)$$ (5.2.5)

By substituting Eqs. (5.2.4 and 5.2.5) into (5.2.1), the classic differential equation of isotropic prismatic (constant I) beam deflection under transverse load is presented as:

$$EI\frac{d^4w}{dx^4} = q(x)$$ (5.2.6)

The classic differential governing equation of an isotropic circular bar under torsion (Timoshenko, 1970; Boresi and Schmidt, 2003) as shown in Figure 5.2 is:

$$\frac{dT}{dx} = GC\frac{d\phi}{dx}$$ (5.3)

where

T = torque
G = shear modulus
C = torsional constant (polar moment of inertia for a circular section)
ϕ = angle of twist/unit length.

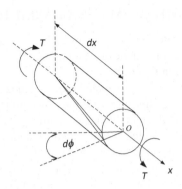

FIGURE 5.2 An isotropic circular bar under torsion.

From the classical lamination theory (CLT), the in-plane force and moment resultants are defined to be force or moment resultants per unit width of a plate. To obtain the in-plane force and moment resultants of beams, the original in-plane force and moment resultant are modified by multiplying with beam width (b). The in-plane force and moment of a beam are defined as (refer to Eq. (4.19.1)):

$$
\begin{bmatrix}
N_{xx}^b \\
N_{yy}^b \\
N_{xy}^b \\
M_{xx}^b \\
M_{yy}^b \\
M_{xy}^b
\end{bmatrix}
= b
\begin{bmatrix}
N_{xx} \\
N_{yy} \\
N_{xy} \\
M_{xx} \\
M_{yy} \\
M_{xy}
\end{bmatrix}
= b
\begin{bmatrix}
A_{11} & A_{12} & A_{16} & B_{11} & B_{12} & B_{16} \\
A_{12} & A_{22} & A_{26} & B_{12} & B_{22} & B_{26} \\
A_{16} & A_{26} & A_{66} & B_{16} & B_{26} & B_{66} \\
B_{11} & B_{12} & B_{16} & D_{11} & D_{12} & D_{16} \\
B_{12} & B_{22} & B_{26} & D_{12} & D_{22} & D_{26} \\
B_{16} & B_{26} & B_{66} & D_{16} & D_{26} & D_{66}
\end{bmatrix}
\begin{bmatrix}
\varepsilon_{xx}^o \\
\varepsilon_{yy}^o \\
\gamma_{xy}^o \\
\kappa_{xx}^o \\
\kappa_{yy}^o \\
\kappa_{xy}^o
\end{bmatrix}
\qquad (5.4.1)
$$

The above definition of in-plane force and moment resultants in Eq. (5.4.1) is used throughout this chapter. The in-plane force and moment resultants in the laminated constitutive relations (compliance (*abd*) matrix) are written in terms of the beam force and moment resultants as (refer to Eq. (4.19.2)):

$$
\begin{bmatrix}
\varepsilon_{xx}^o \\
\varepsilon_{yy}^o \\
\gamma_{xy}^o \\
\kappa_{xx}^o \\
\kappa_{yy}^o \\
\kappa_{xy}^o
\end{bmatrix}
=
\begin{bmatrix}
a_{11} & a_{12} & a_{16} & b_{11} & b_{12} & b_{16} \\
a_{12} & a_{22} & a_{26} & b_{12} & b_{22} & b_{26} \\
a_{16} & a_{26} & a_{66} & b_{16} & b_{26} & b_{66} \\
b_{11} & b_{12} & b_{16} & d_{11} & d_{12} & d_{16} \\
b_{12} & b_{22} & b_{26} & d_{12} & d_{22} & d_{26} \\
b_{16} & b_{26} & b_{66} & d_{16} & d_{26} & d_{66}
\end{bmatrix}
\frac{1}{b}
\begin{bmatrix}
N_{xx}^b \\
N_{yy}^b \\
N_{xy}^b \\
M_{xx}^b \\
M_{yy}^b \\
M_{xy}^b
\end{bmatrix}
\qquad (5.4.2)
$$

The governing equation of composite beams can be directly established by using the CLT. The behavior of laminated composite beams under applied axial loads, independent of bending and/ or torsional loads (not combined loads) is separately discussed under Sections 5.2, 5.3, and 5.5. In addition, the composite beam behavior under various load combinations is also presented at the end of this chapter.

5.2   LAMINATED COMPOSITE BEAM UNDER AXIAL LOAD

In this section, laminated composite beam responses under applied axial load are presented. The laminated lay-ups of composite beams can be oriented in both the perpendicular and parallel directions to the (X–Z) plane as shown in Figure 5.3.

The applied axial load can be either distributed axially along the length or concentrated at a specific location. To satisfy force equilibrium conditions, the applied axial loads and in-plane force resultant must be of the same magnitude in the opposite direction as shown in Figure 5.4.

The in-plane force resultant is an induced internal stress under an external (applied) load. The applied axial load is taken to be positive when the load direction corresponds to the normal line outward from the plane of the cross section at the applied load location. Such load direction results in the applied axial load inducing tension or compression in a portion along the axial length of a member (from the load location to supports), depending on the left or right side of the load location, along the member length.

A negative applied load causes negative (compressive) stress on a cross section. For additional clarifications on sign conventions, the reader should refer to Section 4.3.

From Figure 5.4, the applied axial load and in-plane force resultant are related as:

$$P = N_{xx}^{b} \qquad\qquad (5.5.1)$$

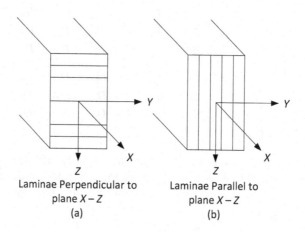

FIGURE 5.3   Laminated lay-up orientation of a composite beam.

$$P = N_{xx}^{b}$$

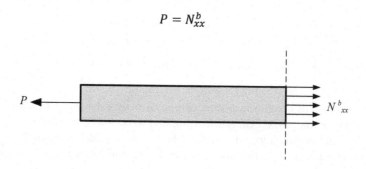

FIGURE 5.4   Applied axial load and in-plane force resultant.

5.2.1 Case of Laminated Layers Perpendicular to (X–Z) Plane (Figure 5.3)

For symmetric lay-ups, when the composite beams are under uniform applied load effects (unidirectional normal loads $P(x)$), then curvatures are not induced from coupling effects, i.e., $[B_{ij}]$ is zero. As discussed before, both the in-plane and out-of-plane couplings $[B_{ij}]$ mainly cause nonuniform in-plane strains that linearly vary through a cross section, resulting in nonuniform deformation, but they are zero because of symmetric lay-ups.

By substituting Eq. (5.5.1) into Eq. (5.4.2), the in-plane force-moment resultants ($N_{yy}^b, N_{xy}^b, M_{yy}^b$ and M_{xy}^b) and compliance $[b_{ij}]$ are taken as zero.

$$
\begin{bmatrix}
\varepsilon_{xx}^o \\
\varepsilon_{yy}^o \\
\gamma_{xy}^o \\
\kappa_{xx}^o \\
\kappa_{yy}^o \\
\kappa_{xy}^o
\end{bmatrix}
=
\begin{bmatrix}
a_{11} & a_{12} & a_{16} & 0 & 0 & 0 \\
a_{12} & a_{22} & a_{26} & 0 & 0 & 0 \\
a_{16} & a_{26} & a_{66} & 0 & 0 & 0 \\
0 & 0 & 0 & d_{11} & d_{12} & d_{16} \\
0 & 0 & 0 & d_{12} & d_{22} & d_{26} \\
0 & 0 & 0 & d_{16} & d_{26} & d_{66}
\end{bmatrix}
\frac{1}{b}
\begin{bmatrix}
P \\
0 \\
0 \\
0 \\
0 \\
0
\end{bmatrix}
\qquad (5.5.2)
$$

or

$$
\begin{bmatrix}
\varepsilon_{xx}^o \\
\varepsilon_{yy}^0 \\
\gamma_{xy}^o \\
\kappa_{xx}^o \\
\kappa_{yy}^o \\
\kappa_{xy}^o
\end{bmatrix}
=
\begin{bmatrix}
a_{11} \\
a_{12} \\
a_{16} \\
0 \\
0 \\
0
\end{bmatrix}
\frac{P}{b}
\qquad (5.5.3)
$$

From Eq. (4.3), Eq. (5.5.3) is derived as:

$$
\begin{bmatrix}
\varepsilon_{xx}^o \\
\varepsilon_{yy}^o \\
\gamma_{xy}^o \\
\kappa_{xx}^o \\
\kappa_{yy}^o \\
\kappa_{xy}^o
\end{bmatrix}
=
\begin{bmatrix}
\dfrac{\partial u^o}{\partial x} \\[2mm]
\dfrac{\partial v^o}{\partial y} \\[2mm]
\dfrac{\partial u^o}{\partial y}+\dfrac{\partial v^o}{\partial x} \\[2mm]
-\dfrac{\partial^2 w^o}{\partial x^2} \\[2mm]
-\dfrac{\partial^2 w^o}{\partial y^2} \\[2mm]
-2\dfrac{\partial^2 w^o}{\partial x \partial y}
\end{bmatrix}
=
\begin{bmatrix}
a_{11} \\
a_{12} \\
a_{16} \\
0 \\
0 \\
0
\end{bmatrix}
\frac{P}{b}
\qquad (5.6)
$$

Equation (5.6) reveals that in-plane deformations are induced whereas curvatures are not induced under the applied axial load over the cross section in direction (x) for symmetric lay-ups of composite beams. The axial strain $\left(\varepsilon_{xx}^b\right)$ of a composite beam is uniform and equals to the in-plane strain $\left(\varepsilon_{xx}^o\right)$ through the cross section since curvature $\left(\kappa_{xx}^o\right)$ is zero.

The beam deformation in direction (X) can be written as (refer to Eqs. (4.3 and 5.6)):

$$\varepsilon_{xx} = \varepsilon_{xx}^0 + z\kappa_{xx}^0 \quad \text{or} \quad \varepsilon_{xx}^b = \varepsilon_{xx}^o = \frac{a_{11}}{b}P \tag{5.7.1}$$

where superscript "b" stands for beam

$$\text{or,} \quad \frac{b}{a_{11}}\frac{du^b}{dx} = \frac{b}{a_{11}}\frac{du^0}{dx} = P \tag{5.7.2}$$

where $\varepsilon_{xx}^b = \dfrac{du^b}{dx}$

where u^b is axial displacement of a laminated composite beam in direction (x)

u^0 is the axial displacement of a laminated composite beam at the reference surface.

Thus, the axial displacement of composite beams with symmetric lay-ups can be calculated by solving the above differential equation. By comparing with the axial displacement Eq. (5.1) of isotropic materials, the equivalent axial stiffness can be presented as:

$$u^o = \frac{a_{11}}{b}\int_0^l P\,dx = \frac{1}{(AE)_{eq}}\int_0^l P\,dx \tag{5.7.3}$$

The equality between $\dfrac{a_{11}}{b} = \dfrac{1}{(AE)_{eq}}$ in Eq. (5.7.3) is intuitively understood from basic theories of mechanics of materials.

For uniform load distribution (p_x) along the beam length, an infinitesimal element of a composite beam is considered as shown in Figure 5.5.

$$(P + dP) + p_x dx - P = 0 \tag{5.8.1}$$

$$\frac{dP}{dx} = -p_x \tag{5.8.2}$$

where p_x is the intensity of the uniform axial load (load per unit length):

By substituting Eq. (5.7.2) into Eq. (5.8.2), the governing equation of a composite beam under uniformly distributed axial load, after integration with respect to x, is given:

$$\frac{b}{a_{11}}\frac{du^o}{dx} = -p_x x + C_1 \tag{5.8.3}$$

Equation (5.8.3) means that the uniformly distributed load of a beam portion is equal to the axially concentrated force on the cut section, but in the opposite directions as shown in Figure 5.6.

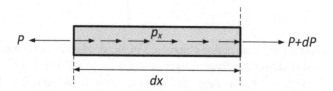

FIGURE 5.5 Uniformly distributed load (p_x) along the composite beam length.

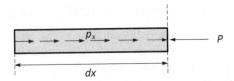

FIGURE 5.6   Uniformly distributed axial load and axial concentrated force on cut section.

$$\frac{b}{a_{11}}\frac{du^o}{dx} = -p_x x + C_1 = P \qquad (5.8.4)$$

Note: constant (C_1) can be determined from the boundary conditions of a composite beam.

5.2.2   CASE OF LAMINATED LAYERS PARALLEL TO $(X–Z)$ PLANE

For symmetric lay-ups, the principle material plane of laminated layers is parallel to the structural plane $(X–Z)$ as shown in Figure 5.3 and is under applied external axial load $P(x)$ in the longitudinal (X) direction. Only the in-plane strains are induced for each layer whereas other curvatures are equal to zero in the absence of couplings $[B_{ij}]$. Then, in-plane strains at any distance Y (in the structural coordinate system) are uniform across the beam width and through the beam cross section, still validating Eq. (5.7.3) for the axial displacement of a composite beam under axial loads.

However, the orientation of the coordinated system in the derivation of Eq. (5.7.1) is different from the structural coordinated system of the composite beam. Fortunately, the direction of axial strain $\left(\varepsilon_{xx}^o\right)$ in Eq. (5.7.1) corresponds to the (X) direction of the structural coordinate system. In addition, components of other strains are not of interest in this case, which are insignificant, hence neglected. Then, it is not necessary to modify any equations of composite beams with laminated layers perpendicular to the structural plane $(X–Z)$. Thus, composite beams with laminated layers perpendicular and parallel to the structural plane $(X–Z)$ provide the same governing equation under axial loads and symmetric lay-ups.

5.3   LAMINATED COMPOSITE RECTANGULAR BEAM UNDER BENDING

For beams under bending, applied bending moments can be directly applied onto a composite beam or through arbitrary transverse loads. In this section, two different laminated forms, orthogonal and parallel to the $(X–Z)$ plane, are shown in Figure 5.7. Beam stiffness of various cross sections that can be built from a rectangular shape are formulated in Eqs. (5.9.1–5.9.8).

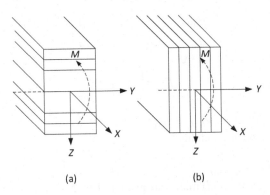

(a)   (b)

FIGURE 5.7   Laminated forms of composite beam under bending: (a) laminae perpendicular to plane $X–Z$ and (b) laminae parallel to plane $X–Z$.

5.3.1 CASE OF LAMINATED LAYERS PERPENDICULAR TO (X–Z) PLANE

For laminated layers aligned on a plane perpendicular to the structural (X–Z) plane, composite beams with symmetric lay-ups are subjected to induced moment resultants due to both the applied bending moments and transverse loads on the (X–Z) plane. For symmetric lay-ups, when the composite beams are under applied bending moment, then axial deformations and curvatures are not induced from coupling effects, i.e., $[B_{ij}]$ is zero. The in-plane force-moment resultants and deformation-curvature relations (inversion of ABD matrix) in Eq. (5.4.2) can be presented as:

From Eq. (5.4.2), the in-plane force-moment resultants ($N_{xx}^b, N_{yy}^b, N_{xy}^b, M_{yy}^b$ and M_{xy}^b) and compliance $[b_{ij}]$ are taken as zero.

$$
\begin{bmatrix} \varepsilon_{xx}^o \\ \varepsilon_{yy}^o \\ \gamma_{xy}^o \\ \kappa_{xx}^o \\ \kappa_{yy}^o \\ \kappa_{xy}^o \end{bmatrix} = \begin{bmatrix} a_{11} & a_{12} & a_{16} & 0 & 0 & 0 \\ a_{12} & a_{22} & a_{26} & 0 & 0 & 0 \\ a_{16} & a_{26} & a_{66} & 0 & 0 & 0 \\ 0 & 0 & 0 & d_{11} & d_{12} & d_{16} \\ 0 & 0 & 0 & d_{12} & d_{22} & d_{26} \\ 0 & 0 & 0 & d_{16} & d_{26} & d_{66} \end{bmatrix} \frac{1}{b} \begin{bmatrix} 0 \\ 0 \\ 0 \\ M_{xx}^b \\ 0 \\ 0 \end{bmatrix} \tag{5.9.1}
$$

From Eq. (5.9.1), only curvature terms exist for symmetric lay-ups. The elements of laminated compliance are reduced due to nonzero term $\left(M_{xx}^b\right)$ of the moment resultants. If the composite beams are narrow, then the effect of curvatures (κ_{yy}^o and κ_{xy}^o) is very small. Thus, the curvatures (κ_{yy}^o and κ_{xy}^o) can be neglected without much loss of accuracy. Only the major bending curvature $\left(\kappa_{xx}^o\right)$ is considered. Thus, the governing equation of composite beams with symmetric lay-up under bending is simplified as:

$$
\kappa_{xx}^o = d_{11} \frac{M_{xx}^b}{b} \tag{5.9.2}
$$

From Eq. (5.2.2):

$$
\kappa_{xx}^o = -\frac{d^2 w^0}{dx^2} \tag{5.9.3}
$$

$$
\kappa_{xx}^o = -\frac{d^2 w^0}{dx^2} = d_{11} \frac{M_{xx}^b}{b} \tag{5.9.4}
$$

Comparing Eq. (5.9.4) with isotropic beam case in Eq. (5.2.1):

$$
\frac{d^2 w^0}{dx^2} = -\frac{d_{11}}{b} M_{xx}^b = -\frac{M_{xx}^b}{(EI)_{eq}} \tag{5.9.5}
$$

$$
E_{eq} = \left(\frac{b}{d_{11}}\right) \Big/ I \tag{5.9.6}
$$

where equivalent flexural rigidity $(EI)_{eq} = (b/d_{11})$,

E_{eq} = equivalent elastic beam modulus,

b = beam width,

I = beam moment of inertia, and

d_{11} = compliance element of (abd) matrix as given in Eq. (5.4.2); however, $[b] = 0$ for symmetric composite lay-up.

Again, Eq. (5.9.6) is obvious from the mechanics of materials knowledge base.

In addition, Eq. (5.9.5) can be written in terms of transverse load distribution $q(x)$ in a manner similar to isotropic beams. The equations corresponding to shear force-bending moment relation and shear force-transverse load distribution for isotropic materials are still valid for any type of symmetric lay-up of composite since no material property term appears in those relations.

$$\frac{d^4 w^0}{dx^4} = -\left(\frac{d_{11}}{b}\right)\frac{d^2 M_{xx}^b}{d^2 x} = -\left(\frac{d_{11}}{b}\right)\frac{dV_{xx}^b}{dx} \tag{5.9.7}$$

$$\frac{d^4 w^0}{dx^4} = \frac{d_{11}}{b} q(x) = \frac{q(x)}{(EI)_{eq}} \tag{5.9.8}$$

Note: Beam sections under applied transverse loading are generally subjected to both internal bending and transverse shear on a cross section. In the beam deflection analysis, if the span length of the beam is long in relation to depth (approximately 12), then the effect of shear is less significant (<4% of bending effect) and can be neglected. Thus, Eq. (5.9.8) is typically used for the analysis of long composite beams with symmetric lay-ups.

To determine induced stresses of each layer at any cross section, the bending moment as a function of "x" is substituted into Eq. (5.9.1). An identical procedure can be followed with the laminated analysis as given in Sections 4.4 and 4.5.

5.3.2 Case of Laminated Layers Parallel to (X–Z) Plane

In this section, composite beams with laminated layers parallel to the structural (X–Z) plane are discussed. When those composite beams are under applied bending moment or arbitrary transverse loads on the structural plane (X–Z), the internal moment resultants will not be induced since the structural and laminated planes are different. The external bending moment is applied parallel to the plane (X–Y) of the global laminated plane corresponding to the structural plane (X–Z) as shown in Figure 5.8 (or the right side of Figure 5.7).

The assumptions of stiffness formulation of laminated beams do not account for bending moment in the (X–Y) plane of laminated layers as per the theory of plates. Since the external bending is under equilibrium with the internal bending moment and it induces internal in-plane normal force, which varies linearly along the depth of a cross section as shown in Figure 5.8; then, Eq. (5.4.2) can be written as:

$$\begin{bmatrix} \varepsilon_{xx}^o \\ \varepsilon_{yy}^o \\ \gamma_{xy}^o \\ \kappa_{xx}^o \\ \kappa_{yy}^o \\ \kappa_{xy}^0 \end{bmatrix} = \begin{bmatrix} a_{11} & a_{12} & a_{16} & 0 & 0 & 0 \\ a_{12} & a_{22} & a_{26} & 0 & 0 & 0 \\ a_{16} & a_{26} & a_{66} & 0 & 0 & 0 \\ 0 & 0 & 0 & d_{11} & d_{12} & d_{16} \\ 0 & 0 & 0 & d_{12} & d_{22} & d_{26} \\ 0 & 0 & 0 & d_{16} & d_{26} & d_{66} \end{bmatrix} \begin{bmatrix} N_{xx} \\ 0 \\ 0 \\ 0 \\ 0 \\ 0 \end{bmatrix} \tag{5.10.1}$$

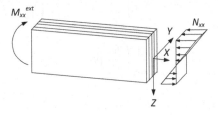

FIGURE 5.8 Internal in-plane normal force.

From Eq. (5.10.1):

$$
\begin{bmatrix} \varepsilon_{xx}^o \\ \varepsilon_{yy}^o \\ \gamma_{xy}^o \end{bmatrix} = \begin{bmatrix} a_{11} \\ a_{12} \\ a_{16} \end{bmatrix} N_{xx}
\tag{5.10.2}
$$

To satisfy the equilibrium condition, the external bending moment must be equal to the internal bending moment from the in-plane normal force in each layer of the beam cross section.

$$
M_{xx}^{ext}(x) = \int_{-\frac{h}{2}}^{\frac{h}{2}} N_{xx} z^b \, dz
\tag{5.10.3}
$$

To establish the governing equation under bending, the strain-curvature relation of a composite beam is utilized. In the case of isotropic beams, the cross sections are assumed to remain on the plane before and after loadings. In addition, the geometric centroid of a cross section relies on the neutral plane. Thus, it precisely presents that the normal strain of a cross section at a distance (z^b) is the product between curvature and vertical distance from the neutral plane.

$$
\varepsilon_{xx}^b = z^b \kappa_{xx}^b
\tag{5.10.4}
$$

In the case of symmetric lay-ups, the in-plane and out-of-plane coupling [B_{ij} or b_{ij}] is zero. From Eqs. (4.3 and 5.10.1), the total normal strain in the global coordinate system is determined for each layer as below ($\kappa_{xx}^o = 0$):

$$
\varepsilon_{xx} = \varepsilon_{xx}^o + z\kappa_{xx}^o = \varepsilon_{xx}^b = a_{11} N_{xx}
\tag{5.10.5}
$$

Equation (5.10.5) presents that the induced strain (ε_{xx}) is uniform across the beamwidth. In addition, the induced strain (ε_{xx}) linearly varies through the depth (z^b) of a cross section. In the case of symmetric laminated lay-ups, the assumption (3) in Section 5.1 can be accepted for composite beams.

From Eqs. (5.10.4 and 5.10.5), (N_{xx}) can be determined as:

$$
N_{xx} = \frac{Z^b \kappa_{xx}^b}{a_{11}} = \frac{\varepsilon_{xx}^b}{a_{11}}
\tag{5.10.6}
$$

By substituting (N_{xx}) of Eqs. (5.10.6) into Eq. (5.10.3), the governing equation of the composite beam with symmetric laminated lay-ups parallel to the structural (X–Z) plane can be presented in Eq. (5.11.1). To evaluate the equivalent bending rigidity, Eq. (5.11.2) is compared to the governing equation of isotropic beams.

$$
M_{xx}^{ext}(x) = \frac{1}{a_{11}} \kappa_{xx}^b \int_{-\frac{b}{2}}^{\frac{b}{2}} (z^b)^2 \, dz = \frac{1}{a_{11}} \left(\frac{b^3}{12} \right) \kappa_{xx}^b
\tag{5.11.1}
$$

$$
M_{xx}^{ext}(x) = \frac{1}{a_{11}} \left(\frac{b^3}{12} \right) \kappa_{xx}^b = (EI)_{eq} \kappa_{xx}^b
\tag{5.11.2}
$$

where equivalent flexural rigidity $(EI)_{eq} = (b^3/12a_{11})$
  E_{eq} = equivalent elastic beam modulus
  b = beam width

I = moment of inertia

a_{11} = compliance element of (abd) matrix.

To determine deflections of composite beams, Eq. (5.11.2) can be written in terms of the displacement (w^0) at the neutral plane by substituting the relation between curvature $\left(\kappa_{xx}^b\right)$ and displacement (w^0) as:

$$\frac{1}{a_{11}}\left(\frac{b^3}{12}\right)\frac{d^2 w^0}{dx^2} = -M_{xx}^{ext}(x) \tag{5.11.3}$$

In addition, Eq. (5.11.3) can also be presented in terms of transverse distribution of loads on the structural (X–Z) plane as:

$$\frac{1}{a_{11}}\left(\frac{b^3}{12}\right)\frac{d^4 w^0}{dx^4} = -\frac{d^2 M_{xx}^{ext}(x)}{dx^2} = -\frac{d^3 V_{xx}^{ext}(x)}{dx^3} = q(x) \tag{5.11.4}$$

where a_{11} for complex laminates (i.e., lamina of varying fabrics and fabric orientations) can be obtained along the lines given in Examples 4.1 and 4.2.

5.3.3 LAMINATED COMPOSITE RECTANGULAR BEAM UNDER BENDING WITH SHEAR DEFORMATION

In general, shear deformations are significant in composite beams due to low shear stiffness of laminates (about 20 times lower than in steel and 7–10 times lower than the longitudinal stiffness of the composite beam itself). However, the significance of shear deformation on the beam deflection will be reduced with the increase in the span-to-depth ratio.

In this section, the shear deformation effect of composite beams with laminated lay-up forms, orthogonal to the (X–Z) plane under bending, is presented. The total deflection (w^{total}) of a composite beam is the sum of deflections due to pure bending (w^b) and shear deformation ($w^{total} = w^b + w^s$).

For beam deflection due to pure bending (w^b), the governing equation is presented in Eq. (5.12) with no shear effect.

$$\frac{d^2 w^{total}}{dx^2} = -\frac{M_{xx}^b}{(EI)_{eq}} \tag{5.12}$$

where $(EI)_{eq}$ = equivalent flexural rigidity of a composite beam

Note: Defection due to bending (w^b) is positive (downward) according to the positive direction of (Z) axis. However, deflection shape under a positive moment (in Figure 5.1) provides a negative sign in terms of curvature. To obtain a downward deflection as positive, the negative sign is taken with the governing equation of beam under bending for this coordinate system. For deflection due to shear (w^s), positive shear deflection (Figure 5.1) corresponds with positive bending deflection.

From Timoshenko beam theory, the effect of transverse shear deformation needs to be corrected using shear coefficient (k), which varies with the shape of a cross section including the width to depth ratio where applicable. The shear coefficient is a result of the nonuniform distribution of shear stresses and strains over a cross section. In the case of flanged beams, such as I-, T-, Box-, neglecting out-of-plane shear rigidity of flanges does not result in a major loss of accuracy. The general governing equations of a composite beam with shear deformation can be generated using CLT. Considering composite beam lay-ups with symmetrical laminates and no in-plane force effects, the constitutive relation, for out-of-plane shear stresses (interlaminar shear) with first-order shear deformation can be presented using Eqs. (4.4 and 3.8.3) by changing (axes 2–3 and 1–3) to (yz and xz), respectively, as (Whitney, 1987):

$$
\begin{bmatrix} \tau_{yz}^k \\ \tau_{xz}^k \end{bmatrix} = \begin{bmatrix} \bar{Q}_{44}^k & \bar{Q}_{45}^k \\ \bar{Q}_{45}^k & \bar{Q}_{55}^k \end{bmatrix} \begin{bmatrix} \dfrac{\partial w^o}{\partial y} + \psi_y \\ \dfrac{\partial w^o}{\partial x} + \psi_x \end{bmatrix} = \begin{bmatrix} \bar{Q}_{44}^k & \bar{Q}_{45}^k \\ \bar{Q}_{45}^k & \bar{Q}_{55}^k \end{bmatrix} \begin{bmatrix} \dfrac{\partial w^o}{\partial y} + \dfrac{\partial u}{\partial z} \\ \dfrac{\partial w^o}{\partial x} + \dfrac{\partial v}{\partial z} \end{bmatrix}
\tag{5.13.1}
$$

where $\bar{Q}_{ij}^k$ can be obtained from the reduced stiffness matrix Eq. (3.31.5) of an orthotropic lamina [special case of Eq. (3.6.2)].

To present relations between out-of-plane shear resultants and deformations of laminated composites, Eq. (5.13.1) is integrated, with respect to variable z, through ($-h/2$ to $h/2$). The relation of out-of-plane shear resultant and deformation is:

$$
\begin{bmatrix} Q_y \\ Q_x \end{bmatrix} = \begin{bmatrix} kA_{44} & kA_{45} \\ kA_{54} & kA_{55} \end{bmatrix} \begin{bmatrix} \dfrac{\partial w^o}{\partial y} + \psi_y \\ \dfrac{\partial w^o}{\partial x} + \psi_x \end{bmatrix}
\tag{5.13.2}
$$

$$
\text{where,} \quad A_{ij} = \int_{-\frac{h}{2}}^{\frac{h}{2}} \bar{Q}_{ij}^k \, dz = \sum_{n}^{k=1} \bar{Q}_{ij}^k (z_k - z_{k-1}) = \sum_{n}^{k=1} \bar{Q}_{ij}^k h_k
\tag{5.13.3}
$$

Noting [B] is zero for symmetric laminates, bending moment and shear per unit width are substituted into Eqs. (4.18 and 5.13.2) as:

$$
\frac{1}{b} \begin{bmatrix} M_{xx}^b \\ M_{yy}^b \\ M_{xy}^b \end{bmatrix} = \begin{bmatrix} D_{11} & D_{12} & D_{16} \\ D_{12} & D_{22} & D_{26} \\ D_{16} & D_{26} & D_{66} \end{bmatrix} \begin{bmatrix} \dfrac{\partial \psi_x}{\partial x} \\ \dfrac{\partial \psi_y}{\partial y} \\ \dfrac{\partial \psi_x}{\partial y} + \dfrac{\partial \psi_y}{\partial x} \end{bmatrix}
\tag{5.13.4}
$$

$$
\frac{1}{b} \begin{bmatrix} Q_y^b \\ Q_x^b \end{bmatrix} = \begin{bmatrix} kA_{44} & kA_{45} \\ kA_{45} & kA_{55} \end{bmatrix} \begin{bmatrix} \dfrac{\partial w^o}{\partial y} + \psi_y \\ \dfrac{\partial w^o}{\partial x} + \psi_x \end{bmatrix}
\tag{5.13.5}
$$

For beam bending problems under bending moment (M_{xx}), the only curvature in the beam direction is considered while curvature in the other direction is less significant and can be neglected, i.e., Q_x^b relation with $kA_{45}\left(\dfrac{\partial w^o}{\partial y} + \psi_y\right)$ is neglected in comparison with $kA_{55}\left(\dfrac{\partial w^o}{\partial x} + \psi_x\right)$. Then, the constitutive relation of composite beam with shear deformation can be simplified from Eqs. (5.13.4 and 5.13.5) as:

$$
\frac{\partial \psi_x}{\partial x} = \frac{d_{11}}{b} M_{xx}^b = d_{11} M_{xx}
\tag{5.14.1}
$$

$$\left(\frac{\partial w^o}{\partial x} + \psi_x \right) = \frac{a_{55}}{bk} Q_x^b = \frac{a_{55}}{k} Q_x \tag{5.14.2}$$

where M_{xx}^b and Q_x^b are beam bending moment and shear

M_{xx} and Q_x are bending moment and shear per unit width of the beam.

From the theory of elasticity, the equilibrium equation of the elastic body and applied force-stress relations are used to formulate the equilibrium equation of a laminate plate. Thus, the equilibrium equation of a beam with shear deformation can be simplified from the equilibrium equation of a laminate plate as:

$$\frac{\partial Q_x}{\partial x} + \frac{\partial Q_y}{\partial y} + p_z = 0 \tag{5.14.3}$$

$$Q_x = \frac{\partial M_{xx}}{\partial x} + \frac{\partial M_{xy}}{\partial y} \tag{5.14.4}$$

Contributions of (Q_y and M_{xy}) are neglected; then, equilibrium equations can be presented as:

$$\frac{\partial Q_x^b}{\partial x} + p_z = 0 \tag{5.14.5}$$

$$Q_x = \frac{\partial M_{xx}^b}{\partial x} \tag{5.14.6}$$

Substituting Eqs. (5.14.1 and 5.14.2) into Eqs. (5.14.5 and 5.14.6):

$$\frac{\partial Q_x^b}{\partial x} + p_z = \frac{bk}{a_{55}} \left(\frac{\partial^2 w^o}{\partial x^2} + \frac{\partial \psi_x}{\partial x} \right) + p_z = k(GA)_{eq} \left(\frac{\partial^2 w^o}{\partial x^2} + \frac{\partial \psi_x}{\partial x} \right) + p_z = 0$$

$$\tag{5.14.7}$$

$$Q_x^b = \frac{bk}{a_{55}} \left(\frac{\partial w^o}{\partial x} + \psi_x \right) = k(GA)_{eq} \left(\frac{\partial w^o}{\partial x} + \psi_x \right)$$

and

$$\frac{b}{d_{11}} \frac{\partial^2 \psi_x}{\partial x^2} = (EI)_{eq} \frac{\partial^2 \psi_x}{\partial x^2} \tag{5.14.8}$$

where $(EI)_{eq}$ and $(GA)_{eq}$ are equivalent bending and shear rigidity of a composite beam.

Equations (5.14.7 and 5.14.8) are integrated with respect to variable x. The transverse beam deflection ($w^0 = w^b + w^s$) can be separated into two parts: (1) deflection due to shear deformation (w^s) and (2) deflection from bending (w^b), which are given below.

$$w^s = -\frac{1}{k(GA)_{eq}} \left(\iint p_z \, dx + C_0 x \right) \tag{5.15.1}$$

$$w^b = \frac{1}{(EI)_{eq}} \left(\iiint p_z \, dx + C_1 \frac{x^3}{6} + C_2 \frac{x^2}{2} + C_3 x + C_4 \right) \tag{5.15.2}$$

In Eqs. (5.15.1 and 5.15.2), constants (C_0, C_1, C_2, C_3, and C_4) can be determined from the boundary conditions of an orthotropic beam under bending loads. The transverse beam deflection with shear deformation for various loads and boundary conditions are summarized in Table 5.1.

TABLE 5.1

Transverse Deflection with Shear Deformation (Kollar, 2003; others)

Load and Boundary Case	Maximum Deflection Due to		Ratio
	Pure Bending (w^b)	Shear Deformation (w^s)	$(w^s)/(w^b)$
	$\dfrac{PL^3}{48(EI)_{eq}}$	$\dfrac{PL}{4k(GA)_{eq}}$	$\dfrac{12(EI)_{eq}}{L^2 k(GA)_{eq}}$
	$\dfrac{PL^3}{192(EI)_{eq}}$	$\dfrac{PL}{4k(GA)_{eq}}$	$\dfrac{48(EI)_{eq}}{L^2 k(GA)_{eq}}$
	$\dfrac{19PL^3}{1,536(EI)_{eq}}$	$\dfrac{PL}{8k(GA)_{eq}}$	$\dfrac{192(EI)_{eq}}{19L^2 k(GA)_{eq}}$
			(Continued)

TABLE 5.1 (*Continued*)
Transverse Deflection with Shear Deformation (Kollar, 2003; others)

Load and Boundary Case	Maximum Deflection Due to		Ratio
	Pure Bending (w^b)	Shear Deformation (w^s)	$(w^s)/(w^b)$
	$\dfrac{PL^3}{3(EI)_{eq}}$	$\dfrac{PL}{k(GA)_{eq}}$	$\dfrac{3(EI)_{eq}}{L^2 k(GA)_{eq}}$
	$\dfrac{qL^4}{8(EI)_{eq}}$	$\dfrac{qL^2}{2k(GA)_{eq}}$	$\dfrac{4(EI)_{eq}}{L^2 k(GA)_{eq}}$
	$\dfrac{5qL^4}{384(EI)_{eq}}$	$\dfrac{qL^2}{8k(GA)_{eq}}$	$\dfrac{48(EI)_{eq}}{5L^2 k(GA)_{eq}}$
	$\dfrac{qL^4}{384(EI)_{eq}}$	$\dfrac{qL^2}{8k(GA)_{eq}}$	$\dfrac{48(EI)_{eq}}{L^2 k(GA)_{eq}}$

In the following section, composite beams under combined loads will be discussed. The concept of the neutral plane is also presented to establish the beam stiffness of unsymmetric laminates. A composite beam under axial and bending load combination is elaborated. The results will be used to formulate beam stiffness of general cross sections that can be formed by using rectangular solid segments such as T, L, hollow box, etc.

5.4 LAMINATED COMPOSITE BEAM UNDER BENDING AND AXIAL LOADS

Responses of composite beams with unsymmetric lay-ups under bending and axial loads are discussed herein. The force and moment resultants (N_{xx} and M_{xx}) induce in-plane strains and curvatures in a composite beam. However, the composite beams are assumed to be of long span beams, i.e., $l/d > 10$; hence, normal strain $\left(\varepsilon_{xx}^o\right)$ and curvature $\left(\kappa_{xx}^o\right)$ are considered as significant, whereas other in-plane strains (ε_{yy}^o and ε_{xy}^o) and curvatures (κ_{yy}^o and κ_{xy}^o) are neglected. The laminated beam compliance Eq. (5.4.2) is presented as:

$$
\begin{bmatrix} \varepsilon_{xx}^O \\ \varepsilon_{yy}^O \\ \varepsilon_{xy}^O \\ \kappa_{xx}^O \\ \kappa_{yy}^O \\ \kappa_{xy}^O \end{bmatrix} =
\begin{bmatrix}
a_{11} & a_{12} & a_{16} & b_{11} & b_{12} & b_{16} \\
a_{12} & a_{22} & a_{26} & b_{12} & b_{22} & b_{26} \\
a_{16} & a_{26} & a_{66} & b_{16} & b_{26} & b_{66} \\
b_{11} & b_{12} & b_{16} & d_{11} & d_{12} & d_{16} \\
b_{12} & b_{22} & b_{26} & d_{12} & d_{22} & d_{26} \\
b_{16} & b_{26} & b_{66} & d_{16} & d_{26} & d_{66}
\end{bmatrix}
\frac{1}{b}
\begin{bmatrix} N_{xx}^b \\ 0 \\ 0 \\ M_{xx}^b \\ 0 \\ 0 \end{bmatrix}
$$

$$
=
\begin{bmatrix}
a_{11} & b_{11} \\
a_{12} & b_{12} \\
a_{16} & b_{16} \\
b_{11} & d_{11} \\
b_{12} & d_{12} \\
b_{16} & d_{16}
\end{bmatrix}
\begin{bmatrix} \dfrac{N_{xx}^b}{b} \\ \dfrac{M_{xx}^b}{b} \end{bmatrix}
\tag{5.16.1}
$$

As mentioned above, all in-plane strains and curvatures still exist and could be significant due to interrelationship between normal and in-plane strains and curvatures of the generally anisotropic materials. Considering only in-plane strain $\left(\varepsilon_{xx}^o\right)$ and curvature $\left(\kappa_{xx}^o\right)$, Eq. (5.16.1) is simplified as:

$$
\varepsilon_{xx}^o = a_{11}\frac{N_{xx}^b}{b} + b_{11}\frac{M_{xx}^b}{b}
\tag{5.16.2}
$$

$$
\kappa_{xx}^o = b_{11}\frac{N_{xx}^b}{b} + d_{11}\frac{M_{xx}^b}{b}
\tag{5.16.3}
$$

To derive the governing equation, in-plane and out-of-plane components must be decoupled. For decoupling, the neutral plane of cross section will be utilized. In general, the neutral plane of a composite beam is not located at the same plane as the geometric centroid of a cross section. However, the neutral plane of composite beams $\varepsilon_{xx}^o = 0$ with symmetric lay-ups corresponds to the geometric centroid of cross section. In addition, the curvature $\left(\kappa_{xx}^o\right)$ will also be induced by axial force, due to anisotropy.

At the neutral plane, Eqs. (5.16.2 and 5.16.3) can be simplified as given in Eqs. (5.16.4 and 5.16.5). Assuming only normal strain $\left(\varepsilon_{xx}^o\right)$ is acting, then the neutral plane can be determined with respect to the mid-plane of a cross section as:

$$b_{11}\frac{N_{xx}^b}{b} + d_{11}\frac{M_{xx}^b}{b} = 0 \tag{5.16.4}$$

$$z^n = \frac{M_{xx}^b}{N_{xx}^b} = -\frac{b_{11}}{d_{11}} \tag{5.16.5}$$

The laminated compliance (*abd*) and stiffness (*ABD*) matrices are recalculated using the neutral plane to be the reference plane. The normal and bending resultants are located on the neutral plane.

$$\begin{bmatrix} \varepsilon_{xx}^n \\ \kappa_{xx}^n \end{bmatrix} = \begin{bmatrix} a_{11}^n & 0 \\ 0 & d_{11}^n \end{bmatrix} \begin{bmatrix} \dfrac{N_{xx}^b}{b} \\ \dfrac{M_{xx}^b}{b} \end{bmatrix} \tag{5.16.6}$$

where superscript "*n*" stands for the neutral plane

In addition, relations between laminated stiffness and compliance at the neutral plane are presented as:

$$\begin{bmatrix} A_{11}^n & A_{12}^n & A_{16}^n & 0 & B_{12}^n & B_{16}^n \\ A_{12}^n & A_{22}^n & A_{26}^n & B_{12}^n & B_{22}^n & B_{26}^n \\ A_{16}^n & A_{26}^n & A_{66}^n & B_{16}^n & B_{26}^n & B_{66}^n \\ 0 & B_{12}^n & B_{16}^n & D_{11}^n & D_{12}^n & D_{16}^n \\ B_{12}^n & B_{22}^n & B_{26}^n & D_{12}^n & D_{22}^n & D_{26}^n \\ B_{16}^n & B_{26}^n & B_{66}^n & D_{16}^n & D_{26}^n & D_{66}^n \end{bmatrix}^{-1} = \begin{bmatrix} a_{11}^n & a_{12}^n & a_{16}^n & 0 & b_{12}^n & b_{16}^n \\ a_{12}^n & a_{22}^n & a_{26}^n & b_{12}^n & b_{22}^n & b_{26}^n \\ a_{16}^n & a_{26}^n & a_{66}^n & b_{16}^n & b_{26}^n & b_{66}^n \\ 0 & b_{12}^n & b_{16}^n & d_{11}^n & d_{12}^n & d_{16}^n \\ b_{12}^n & b_{22}^n & b_{26}^n & d_{12}^n & d_{22}^n & d_{26}^n \\ b_{16}^n & b_{26}^n & b_{66}^n & d_{16}^n & d_{26}^n & d_{66}^n \end{bmatrix} \tag{5.17}$$

From Eq. (5.17), the axial displacement and deflection of composite beams with unsymmetric lay-ups can be calculated in a manner similar to symmetric lay-up case as:

$$\varepsilon_{xx}^n = a_{11}^n \frac{N_{xx}^b}{b} \tag{5.18.1}$$

$$\kappa_{xx}^n = d_{11}^n \frac{M_{xx}^b}{b} \tag{5.18.2}$$

The governing equations of composite beams are given as:

$$\frac{du^n}{dx} = \frac{a_{11}^n N_{xx}^n}{b} = \frac{N_{xx}^n}{(AE)_{eq}} \tag{5.19.1}$$

$$\frac{d^4 w^n}{dx^4} = \frac{d^2(M_{xx}^n)}{dx^2} = \frac{q(x)}{\left(\dfrac{12}{d_{11}^n h^3}\right)\left(\dfrac{bh^3}{12}\right)} = \frac{q(x)}{(EI)_{eq}} \tag{5.19.2}$$

$$(AE)_{eq} = \frac{b}{a_{11}^n}, \quad E_{eq} = \frac{12}{d_{11}^n h^3}, \quad I = \frac{bh^3}{12} \tag{5.19.3}$$

The bending rigidity can be written in terms of laminated compliances at the mid-plane of a cross section. Assuming $\left(\varepsilon_{xx}^{o}\right)$ as zero, Eq. (5.16.2) can be simplified as:

$$N_{xx}^{b} = -\frac{b_{11}}{a_{11}}M_{xx}^{b} \tag{5.20.1}$$

Substituting Eq. (5.20.1) into Eq. (5.16.3):

$$\frac{M_{xx}^{b}}{b} = -\left(\frac{a_{11}}{d_{11}a_{11} - b_{11}^{2}}\right)\kappa_{xx} \tag{5.21.1}$$

From Eq. (5.2.2):

$$\frac{d^{2}w^{n}}{dx^{2}} = -\left(\frac{d_{11}a_{11} - b_{11}^{2}}{a_{11}}\right)\frac{M_{xx}^{b}}{b} \tag{5.21.2}$$

$$\frac{d^{4}w^{n}}{dx^{4}} = -\left(\frac{d_{11}a_{11} - b_{11}^{2}}{a_{11}}\right)\frac{q(x)}{b} \tag{5.21.3}$$

Equations (5.21.2 and 5.21.3) are used to determine deflection solutions at the neutral plane using the laminated compliance at the mid-plane of a cross section. Similarly, axial stiffness is also presented in terms of compliance at the mid-plane.

By substituting Eq. (5.16.4) into Eq. (5.16.2), the governing equation for beam displacement at the neutral plane is given as:

$$\frac{N_{xx}^{b}}{b} = \left(\frac{d_{11}}{a_{11}d_{11} - b_{11}^{2}}\right)\varepsilon_{xx}^{o} \tag{5.21.4}$$

$$\frac{du^{n}}{dx} = \left(\frac{a_{11}d_{11} - b_{11}^{2}}{d_{11}}\right)\frac{N_{xx}^{b}}{b} = \frac{N_{xx}^{n}}{(AE)_{eq}} \tag{5.21.5}$$

If a composite beam is made from symmetric lay-ups, then Eqs. (5.21.3 and 5.21.5) are reduced to the same as Eqs. (5.9.8 and 5.7.2), respectively, i.e., the structural responses under axial force and bending moment act independently of each other.

5.5   LAMINATED COMPOSITE BEAM UNDER TORSION

In this section, composite beams with solid rectangular cross section are considered under torsion. Only primary torsion, commonly known as "Saint–Venant torsion", is studied. The cross sections of composite beams are assumed to be free to deform out-of-plane under torsion, which is referred to as free warping. On the other hand, if composite beams are supported in a manner to prevent free warping of a cross section under torsion, then nonuniform warping will be induced.

Nonuniform warping induces additional shearing stresses, which are known as secondary torsional effects (Gjelsvik, 1981; Murray, 1984). The shear stress induced under applied torque is presented in terms of shear flow (q) (force/length). The shear flow is related to torsional moment through equilibrium as shown in Figure 5.9.

The total applied torque (T) is the sum of torques due to shear flow (q) on a cross section. The couple induced by horizontal shear flow (per unit length) is equal to the twisting moment (M_{xy} per unit length) on the laminated composite beam. The couple corresponding to horizontal shear flow is assumed to be equal to that of vertical shear flow. Thus, the total applied torque can be written in terms of the twisting moment as:

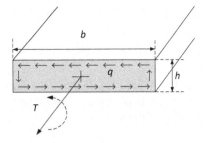

FIGURE 5.9   Shear flow under torsional moment.

$$T = 2qbh = -2bM_{xy} \tag{5.22.1}$$

Compatibility between the twisting curvature and rotational angle of a cross section at the reference plane has been used to establish the applied torque versus twist.

From Eq. (4.3) and Figure 5.10:

$$\kappa_{xy}^o = -2\frac{\partial^2 w^o(x,y)}{\partial x \partial y} = -\frac{2\partial}{\partial x}\left(\frac{\partial w^o}{\partial y}\right) = -2\theta \tag{5.22.2}$$

where rate of twist (θ) is defined as the change in the rotational angle in the plane of a cross section along the beam length.

The twisting curvature $\left(\kappa_{xy}^o\right)$ is related only to the twisting moment (M_{xy}) since our focus is only on primary torsion. The lay-ups of a composite beam can be both symmetric and unsymmetric. Equation (5.4.2) is developed by equating all forces and moments equal to zero except twisting moment (M_{xy}).

$$
\begin{bmatrix}
\varepsilon_{xx}^o \\
\varepsilon_{yy}^o \\
\gamma_{xy}^o \\
\kappa_{xx}^o \\
\kappa_{yy}^o \\
\kappa_{xy}^o
\end{bmatrix}
=
\begin{bmatrix}
a_{11} & a_{12} & a_{16} & b_{11} & b_{12} & b_{16} \\
a_{12} & a_{22} & a_{26} & b_{12} & b_{22} & b_{26} \\
a_{16} & a_{26} & a_{66} & b_{16} & b_{26} & b_{66} \\
b_{11} & b_{12} & b_{16} & d_{11} & d_{12} & d_{16} \\
b_{12} & b_{22} & b_{26} & d_{12} & d_{22} & d_{26} \\
b_{16} & b_{26} & b_{66} & d_{16} & d_{26} & d_{66}
\end{bmatrix}
\begin{bmatrix}
0 \\
0 \\
0 \\
0 \\
0 \\
M_{xy}
\end{bmatrix}
\tag{5.22.3}
$$

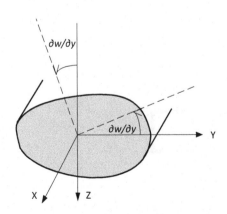

FIGURE 5.10   Rotational angle (twist) of a cross section under torsion.

The constitutive relation of laminated composite beams under primary torsion can be simplified as:

$$\kappa_{xy}^{o} = d_{66}M_{xy} \tag{5.22.4}$$

By substituting Eqs. (5.22.2 and 5.22.4) into Eq. (5.22.1), the governing equation of a laminated composite beam under primary torsion is presented. To evaluate the equivalent torsional rigidity (GC), Eq. (5.22.4) is compared to the equation of isotropic beam as shown below, after substitution for κ_{xy}^{o} from Eq. (5.22.2):

$$T = -2bM_{xy} = -2b\frac{\kappa_{xy}^{o}}{d_{66}} = 2b\left(\frac{2\theta}{d_{66}}\right) \tag{5.22.5}$$

$$T = \frac{4b}{d_{66}}\theta = (GC)_{eq}\theta \tag{5.22.6}$$

$$(GC)_{eq} = (G)_{eq}\left(\frac{bh^{3}}{3}\right) = \frac{4b}{d_{66}} \tag{5.22.7}$$

Equation (5.22.5) is valid in laminated layers perpendicular and parallel to the structural plane $(X–Z)$.

5.5.1 LAMINATED HOLLOW COMPOSITE BEAM UNDER TORSION

A composite beam with a hollow cross section as shown in Figure 5.11 is considered under primary torsion.

The limitations and assumptions considered in evaluating the rigidity of a hollow composite beam are: (1) torque is uniform along the length of a structural member and applied along its longitudinal (beam) axis, (2) cross sections of a composite beam are uniform and have no warping restraint along the beam length (free to warp), (3) cross section warps in the beam direction and does not distort from its original plane, and (4) for closed hollow beams, both shear and twisting moment resultants are generally induced on the cross section as shown in Figure 5.12. However, the induced twisting force resultants under applied torque are small compared to induced shear force resultant; only then the induced shear force resultant (N_{xy}) is significant.

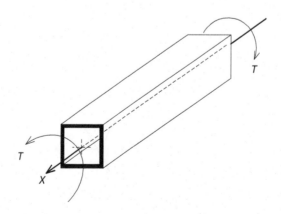

FIGURE 5.11   Hollow beam cross section under torque.

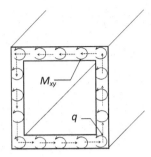

FIGURE 5.12   Shear flow and twisting moment on hollow beam cross section under torque.

From Eq. (5.4.2):

$$
\begin{bmatrix} \varepsilon_{xx}^{o} \\ \varepsilon_{yy}^{o} \\ \gamma_{xy}^{o} \\ \kappa_{xx}^{o} \\ \kappa_{yy}^{o} \\ \kappa_{xy}^{o} \end{bmatrix} = \begin{bmatrix} a_{11} & a_{12} & a_{16} & b_{11} & b_{12} & b_{16} \\ a_{12} & a_{22} & a_{26} & b_{12} & b_{22} & b_{26} \\ a_{16} & a_{26} & a_{66} & b_{16} & b_{26} & b_{66} \\ b_{11} & b_{12} & b_{16} & d_{11} & d_{12} & d_{16} \\ b_{12} & b_{22} & b_{26} & d_{12} & d_{22} & d_{26} \\ b_{16} & b_{26} & b_{66} & d_{16} & d_{26} & d_{66} \end{bmatrix} \begin{bmatrix} 0 \\ 0 \\ N_{xy} \\ 0 \\ 0 \\ M_{xy} \end{bmatrix} \tag{5.23.1}
$$

All strains and curvatures are nonzero. For structural beam, only in-plane shear strain and twisting curvature are considered under primary torsion while other strains and curvatures are insignificant. The above statement is only approximate for generally anisotropic lay-ups, but it provides accurate solutions for orthotropic ($a_{16} = a_{26} = b_{16} = b_{26} = d_{16} = d_{26} = 0$) lay-ups.

From Eq. (5.23.1):

$$
\gamma_{xy}^{o} = a_{66}N_{xy} + b_{66}M_{xy} \tag{5.23.2}
$$

$$
\kappa_{xy}^{o} = b_{66}N_{xy} + d_{66}M_{xy} \tag{5.23.3}
$$

The neutral plane of a cross section under torsion is used to decouple (N_{xy} and M_{xy}) for establishing the governing equation. At the neutral plane, distortion will not be induced under in-plane shear force. Only in-plane shear strain $\left(\gamma_{xy}^{o}\right)$ is valid. Thus, the neutral plane under torsion can be determined with respect to the mid-plane of cross section as:

From Eq. (5.23.3):

$$
\kappa_{xy}^{o} = b_{66}N_{xy} + d_{66}M_{xy} = 0 \tag{5.23.4}
$$

$$
z^{nt} = \frac{M_{xy}}{N_{xy}} = -\frac{b_{66}}{d_{66}} \tag{5.23.5}
$$

By substituting Eq. (5.23.5) into Eq. (5.23.2), the relation of in-plane shear strain and force resultant at the neutral plane is written in terms of the compliance at the mid-plane.

$$
\gamma_{xy}^{o} = \left(a_{66} - \frac{b_{66}^{2}}{d_{66}} \right) N_{xy} \tag{5.24.1}
$$

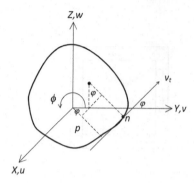

FIGURE 5.13 Coordinate on thin-walled cross section.

Considering Figure 5.13, the compatibility relation between warping displacement and in-plane shear strain at mid-plane of a cross section can be presented in terms of segmental displacements. In general, in-plane shear strain on segments (or walls) of thin-walled structural members is given as:

$$\gamma = \left(\frac{\partial u}{\partial s} + \frac{\partial v}{\partial x} \right) \tag{5.24.2}$$

where

u = displacement along the longitudinal axis (X-direction)
v = displacement along transverse direction (Y-direction)
w = displacement along transverse direction (Z-direction)
v_t = tangential displacement of any point n.

Assume that the cross section of a beam is maintained by closely spaced diaphragms, which are rigid in their own plane. Thus, there is no resistance to axial displacement and the cross section moves as a rigid body in its own plane. Then, the displacement at any point is completely specified by translations (v, w) and rotation (ϕ).

$$v = p\phi + v\text{Cos}\varphi + w\text{Sin}\varphi \tag{5.24.3}$$

By substituting Eq. (5.24.3) into Eq. (5.24.2), in-plane shear strain is:

$$\gamma = \left(\frac{\partial u}{\partial s} + p\frac{\partial \phi}{\partial x} + \frac{\partial v}{\partial x}\text{Cos}\varphi + \frac{\partial w}{\partial x}\text{Sin}\varphi \right) \tag{5.24.4}$$

Integrating Eq. (5.24.4) with respect to s from the chosen origin for s, compatibility relation between warping displacement and in-plane shear strain at mid-plane of the cross section is:

$$\int_0^s \gamma_{xy}^0 \, ds = (u_s - u_0) + 2(A_{0s})\frac{d\phi}{dx} + z_R\frac{dv}{dx}(y_s - y_0) + y_R\frac{dw}{dx}(z_s - z_0) \tag{5.24.5}$$

where
A_{os} = area swept out by a generator, with its center at the origin (R) of the axes, from the origin to any points (S) around the cross section
ϕ = angle of twist as illustrated in Figure 5.14.

It should be noted that the last two terms on the right-hand side of Eq. (5.24.5) represent the effects of warping displacement and an arbitrary origin resulting in axial displacement due to warping. If the origin coincides with the center of twist (R) of the cross section, Eq. (5.24.5) can be simplified by taking the last two terms to be equal to zero. Integration around a closed cross section

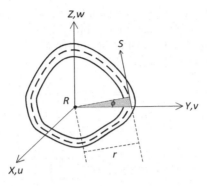

FIGURE 5.14 A_{os} area swept from o to s.

is performed. Due to starting and ending points being identical, difference of axial displacement (u) between both locations is zero ($u_s = u_0$). Therefore, Eq. (5.24.5) is simplified as:

$$\oint \gamma_{xy}^o \, ds = 2\bar{A}\frac{d\phi}{dx} = 2\bar{A}\theta \tag{5.24.6}$$

where $\bar{A}$ = area enclosed by the midline of the cross section as shown in Figure 5.15.

Rate of twist (θ) is defined as the rotational angle about the longitudinal axis per unit length in the longitudinal direction.

This area is not the cross-sectional area of the component under torque (T). An equilibrium equation of a closed thin-walled composite beam is obtained from Bredt-Batho's theory (Magson, 1999). Bredt-Batho's theory presents a relation between shear flow (q) on a cross section and its enclosed area as well as for anisotropic and isotropic thin-walled beams as:

$$T = 2\bar{A}q \tag{5.25.1}$$

where T = applied torque

q = shear flow on cross section, $\bar{A}$, as shown in Figure 5.15.

Only in-plane shear force is significant in structural walls of thin walls, then shear flow on a cross section is defined as:

$$q = \int \tau_{xy} \, dz = N_{xy} \tag{5.25.2}$$

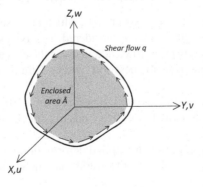

FIGURE 5.15 An enclosed area by the mid-line of a cross section.

To arrive at the torsional rigidity of a closed section, three basic relations are solved in Eqs. (5.24.1, 5.24.6 and 5.25.2) as constitutive relation, compatibility, and equilibrium of forces, respectively. Thus, the relation between torque and rate of twist is given as:

Substituting Eq. (5.25.2) into Eq. (5.25.1):

$$T = 2\bar{A}N_{xy} \tag{5.25.3}$$

Substituting Eq. (5.24.1) into Eq. (5.25.3):

$$\gamma_{xy}^{o} = \frac{T}{2\bar{A}}\left(a_{66} - \frac{b_{66}^2}{d_{66}} \right) \tag{5.25.4}$$

Substituting Eq. (5.25.4) into (5.24.6):

$$\theta = \frac{T}{4\bar{A}^2}\oint\left(a_{66} - \frac{b_{66}^2}{d_{66}} \right)ds \tag{5.25.5}$$

The equivalent torsional rigidity (GC) of composite beams is obtained by comparing with Eq. (5.3) corresponding to an isotropic beam as:

$$T = \frac{4\bar{A}^2}{\oint\left(a_{66} - \dfrac{b_{66}^2}{d_{66}} \right)ds}\theta = (GC)_{eq}\,\ell \tag{5.26.1}$$

For FRP composite circular tubes, the integration around the cross section of a circular tube yields the relation between applied torque and rate of twist (θ) as:

$$T = (GC)_{eq}\,\theta = \frac{2\bar{A}^2}{\pi r\left(a_{66} - \dfrac{b_{66}^2}{d_{66}} \right)}\theta \tag{5.26.2}$$

where r is the radial coordinate measured from the origin to the mid-line of the cross section.

$\bar{A}$ = area enclosed by the midline of the cross section as shown in Figure 5.15

a_{66}, b_{66}, and d_{66} are compliance elements of (abd) matrix.

5.6 LAMINATED COMPOSITE BEAM WITH OPEN CROSS SECTION OF SOLID RECTANGULAR SEGMENTS

In this section, composite beams with cross sections, which are assembled from several solid rectangular segments as shown in Figure 5.16, are considered. The axial stiffness (AE), bending rigidity (EI) and torsional rigidity (GC) are established herein using the concept of stiffness definition (force per unit displacement). In addition, all stiffness and rigidity definitions from Sections 5.2 to 5.5 will be used in this section.

Briefly, the concept of stiffness is presented to evaluate the stiffness of assembled cross section as shown in Figure 5.16. The curvature that corresponds to a specific stiffness will be applied on the neutral plane of a beam.

The force and moment resultants due to applied deformation (or curvature) of each beam segment are determined. By using the solutions of several cases in Sections 5.2–5.5, force equilibrium

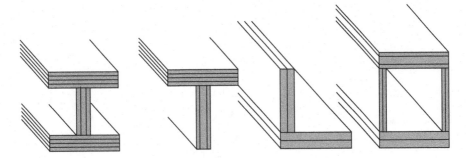

FIGURE 5.16 Composite beam open cross sections of solid rectangular segments.

conditions are invoked, i.e., summation of all force and moment resultants in each segment of a beam cross section are equal to the applied forces (or moments) at the neutral plane of a cross section.

Then, the sum of all force and moment resultants is presented in terms of applied deformations (or curvatures). By comparing the stiffness of isotropic beams, equivalent stiffness of a composite beam has been obtained. The above procedure is elaborated below:

BEAM STIFFNESS (OR RIGIDITY) FORMULATION

Objective: evaluation of equivalent beam stiffness (or rigidity): $(EA)_{eq}$, $(EI)_{eq}$, and $(GC)_{eq}$

Curvature κ^b (or strain ε^b) that corresponds to the beam rigidity (stiffness) is applied on the neutral plane of a beam.

Determine: induced deformation (strain) and curvature of each cross section segment.

Determine: induced in-plane force N and moment M resultants of each cross section segment.

Equilibrium condition is applied: sum of induced in-plane force (or moment) resultants is equal to the external applied load.

$$N^b = \sum_i^n N^i \quad M^b = \sum_i^n M^i \quad T^b = \sum_i^n T^i$$

Equivalent stiffness (or rigidity) is obtained by comparing the governing equation of isotropic case

$$N^b = \sum_i^n N^i = (EA)_{eq}\,\varepsilon^b \quad M^b = \sum_i^n M^i = (EI)_{eq}\,\kappa^b \quad T^b = \sum_i^n T^i = (GC)_{eq}\,\theta$$

The symmetric cross sections of composite beams are illustrated through the examples presented in this section. It should be noted that the structural coordinate is referred to the coordinate system of the cross section, while the global coordinate is referred to that of the laminated segment. The global coordinate is different for each segment as shown in Figure 5.17. In addition, the material principal coordinate is referred to as the coordinate system of each layer for a specific laminated segment.

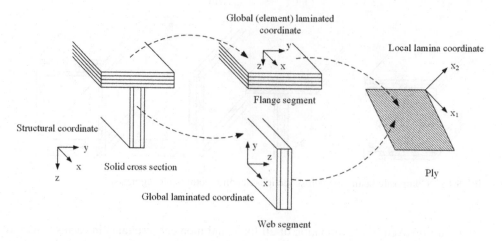

FIGURE 5.17   Structural, global (element) and local coordinates.

Example 5.1

Determine: (a) axial stiffness of a laminated composite I beam and (b) axial stiffness of laminated composite T beam.

A symmetric I section with symmetric lay-ups of each rectangular segment is considered herein. The cross section is symmetric with respect to the structural (X–Z) plane.

When the neutral plane is extended then the curvature will not be induced, and the axial strains are equal throughout cross section. In addition, the composite beam is only bent about the structural coordinate axis y under transverse loading. The induced normal force and moment are also independent due to symmetric lay-ups.

The structural coordinate system is located at the neutral plane of the cross section while the global coordinate systems of each segment (element) are positioned on the mid-planes of each segment as shown in Example Figure 5.1.2.

Solution

a. The axial stiffness of laminated composite I beam:

The axial deformation $\left(\varepsilon_{xx}^n\right)$ of the beam axis at the neutral plane is applied. The curvature of a beam is not induced due to the symmetric nature of lay-up and that of the

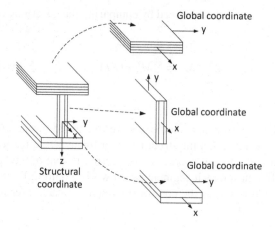

EXAMPLE FIGURE 5.1.1   Laminate composite I beam.

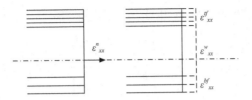

EXAMPLE FIGURE 5.1.2   Applied axial deformation at the neutral plane.

cross section. The axial strain (ε_{xx}) of each segment is same, all through the cross section as shown in Example Figure 5.1.2.

$$\varepsilon_{xx}^{n} = \varepsilon_{xx}^{tf} = \varepsilon_{xx}^{w} = \varepsilon_{xx}^{bf}$$

where ε_{xx}^{n} = axial strain on the neutral plane

$\varepsilon_{xx}^{tf}, \varepsilon_{xx}^{w}, \varepsilon_{xx}^{bf}$ = axial strains of top flange, web, and bottom flange segments at the mid-plane of each segment, respectively.

Only the force resultants (N_{xx}^{tf}, N_{xx}^{w}, and N_{xx}^{bf}) at the mid-plane of each segment are induced due to the applied axial strains $\left(\varepsilon_{xx}^{tf}, \varepsilon_{xx}^{w}, \varepsilon_{xx}^{bf}\right)$ and the symmetric lay-up and cross section. The force resultants of each segment are obtained from Section 5.2.

From Eq. (5.7.1):

$$N_{xx}^{tf} = \frac{b^{tf}}{a_{11}^{tf}} \varepsilon_{xx}^{tf} \quad N_{xx}^{w} = \frac{b^{w}}{a_{11}^{w}} \varepsilon_{xx}^{w} \quad N_{xx}^{bf} = \frac{b^{bf}}{a_{11}^{bf}} \varepsilon_{xx}^{bf}$$

The total force resultant on a cross section is obtained by adding force resultants of each segment as shown in Example Figure 5.1.3:

$$N_{xx}^{b} = N_{xx}^{tf} + N_{xx}^{w} + N_{xx}^{bf} = \frac{b^{tf}}{a_{11}^{tf}} \varepsilon_{xx}^{tf} + \frac{b^{w}}{a_{11}^{w}} \varepsilon_{xx}^{w} + \frac{b^{bf}}{a_{11}^{bf}} \varepsilon_{xx}^{bf} = \left(\frac{b^{tf}}{a_{11}^{tf}} + \frac{b^{w}}{a_{11}^{w}} + \frac{b^{bf}}{a_{11}^{bf}}\right) \varepsilon_{xx}^{n} \qquad (5.27.1)$$

From beam axial stiffness formulation given in Section 5.2.1, the equivalent axial stiffness of an I-shaped beam with symmetric lay-ups is given as:

$$(EA)_{eq}^{b} \varepsilon_{xx}^{n} = \left(\frac{b^{tf}}{a_{11}^{tf}} + \frac{b^{w}}{a_{11}^{w}} + \frac{b^{bf}}{a_{11}^{bf}}\right) \varepsilon_{xx}^{n} \qquad (5.27.2)$$

To determine a centroidal plane, the balancing moment of the bottom fiber of a cross section is considered as shown in Example Figure 5.1.3.

$$z_{c}^{b} N_{xx}^{b} = z^{tf} N_{xx}^{tf} + z^{w} N_{xx}^{w} + z^{bf} N_{xx}^{bf} \qquad (5.27.3)$$

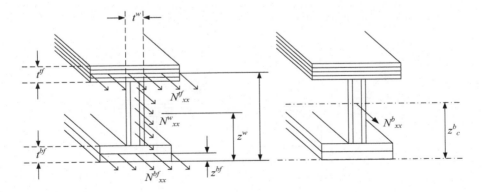

EXAMPLE FIGURE 5.1.3   Force resultant on I cross section.

By substituting force resultant in Eq. (5.27.1) into Eq. (5.27.3):

$$z_c^b = \frac{z^{tf}\dfrac{b^{tf}}{a_{11}^{tf}} + z^w\dfrac{b^w}{a_{11}^w} + z^{bf}\dfrac{b^{bf}}{a_{11}^{bf}}}{\left(\dfrac{b^{tf}}{a_{11}^{tf}} + \dfrac{b^w}{a_{11}^w} + \dfrac{b^{bf}}{a_{11}^{bf}}\right)} \qquad (5.27.4)$$

Axial stress can be determined using the axial stress definition as studied in mechanics of materials as:

From Eq. (5.27.1):

$$\sigma_{xx}^b = \frac{N^b}{A^b} = \frac{\left(\dfrac{b^{tf}}{a_{11}^{tf}} + \dfrac{b^w}{a_{11}^w} + \dfrac{b^{bf}}{a_{11}^{bf}}\right)}{b^{tf}t^{tf} + b^w t^w + b^{bf}t^{bf}}\varepsilon_{xx}^b \qquad (5.27.5)$$

If the top and bottom flanges are identical in width (b), thickness (t), and laminate lay-ups, then the axial stiffness, centroid, and axial stress of a composite beam can be simplified as:

$$\frac{b^{tf}}{a_{11}^{tf}} = \frac{b^{bf}}{a_{11}^{bf}} \qquad (5.27.6)$$

By substituting Eq. (5.27.6) into Eqs. (5.27.2, 5.27.4 and 5.27.5):

$$(EA)_{eq}^b = \left(\frac{2b^{tf}}{a_{11}^{tf}} + \frac{b^w}{a_{11}^w}\right) \qquad (5.27.7)$$

Equations (5.27.4 and 5.27.5) are rewritten below by invoking Eq. (5.27.6) as:

$$z_c^b = \frac{\left(z^{tf} + z^{bf}\right)\dfrac{b^{tf}}{a_{11}^{tf}} + z^w\dfrac{b^w}{a_{11}^w}}{\left(\dfrac{2b^{tf}}{a_{11}^{tf}} + \dfrac{b^w}{a_{11}^w}\right)} \qquad (5.27.8)$$

$$\sigma_{xx}^b = \frac{\left(\dfrac{2b^{tf}}{a_{11}^{tf}} + \dfrac{b^w}{a_{11}^w}\right)}{2b^{tf}t^{tf} + b^w t^w}\varepsilon_{xx}^b \qquad (5.27.9)$$

b. The axial stiffness of laminated composite T beam:

For the composite beams with T cross section in Example Figure 5.1.4, Eq. (5.27.1) can be simplified by neglecting the terms of the bottom flange.

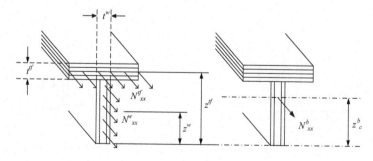

EXAMPLE FIGURE 5.1.4 Force resultant on T cross section.

From Eqs. (5.27.2, 5.27.4, and 5.27.5), the axial stiffness, centroid, and axial stress of composite beams are presented as:

$$\frac{b^{bf}}{a_{11}^{bf}} = 0 \tag{5.28.1}$$

$$\left(EA\right)_{eq}^{b} = \left(\frac{b^{tf}}{a_{11}^{tf}} + \frac{b^{w}}{a_{11}^{w}}\right) \tag{5.28.2}$$

$$z_{c}^{b} = \frac{z^{tf}\dfrac{b^{tf}}{a_{11}^{tf}} + z^{w}\dfrac{b^{w}}{a_{11}^{w}}}{\left(\dfrac{b^{tf}}{a_{11}^{tf}} + \dfrac{b^{w}}{a_{11}^{w}}\right)} \tag{5.28.3}$$

$$\sigma_{xx}^{b} = \frac{\left(\dfrac{b^{tf}}{a_{11}^{tf}} + \dfrac{b^{w}}{a_{11}^{w}}\right)}{b^{tf}t^{tf} + b^{w}t^{w}}\varepsilon_{xx}^{b} \tag{5.28.4}$$

Example 5.2

Determine: (a) flexural rigidity of laminated composite WF beam (as given in Example 5.1) and (b) flexural rigidity of laminated composite T beam.

A symmetric I-section with symmetric lay-ups of each rectangular segment is considered herein. The cross section is symmetric with respect to the structural (X–Z) plane. The composite beam is only bent about the structural coordinate axis y under transverse loading.

The induced normal force and moment are also independent due to symmetric lay-ups. The structural coordinate system is located at the neutral plane of the cross section while the global coordinate systems of each segment are positioned on the mid-planes of each segment as shown in Example 5.1.

Solution

a. Flexural rigidity of laminated composite I beam:

When a composite beam is bent in (X–Z) plane, then the equivalent bending rigidity $(EI)_{eq}$ about the Y-axis is determined. The curvature of the beam axis at the neutral plane (centroid) of the cross section is applied, then normal strains and curvatures of each segment of the cross section are induced under applied curvature as shown in Example Figure 5.2.1.

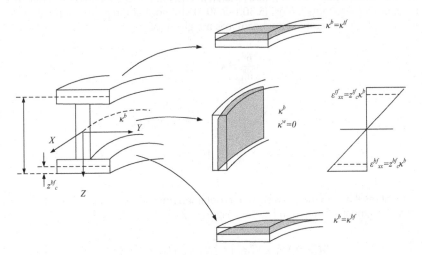

EXAMPLE FIGURE 5.2.1   Curvature and strain distribution of WF cross section.

By applying the curvature $\left(\kappa^b\right)$, the total strain of the cross section is given as:

$$\varepsilon_{xx}^Z = \varepsilon_{xx}^o + z\kappa^b = z\kappa^b \tag{5.29.1}$$

where $\varepsilon_{xx}^o = 0$ because of neutral plane strain under bending

For the top flange portion, curvature and normal strain at the mid-plane of the top flange are induced resulting in force and moment resultants on the top portion as well as the bottom flange portion. The curvatures of the top and bottom flanges (elements of a cross section) at the mid-plane are also equal to the applied curvature at the beam axis.

$$\varepsilon_{xx}^{tf} = z_c^{tf}\kappa^b \quad \kappa_{xx}^{tf} = \kappa^b \tag{5.29.2}$$

$$\varepsilon_{xx}^{bf} = z_c^{bf}\kappa^b \quad \kappa_{xx}^{bf} = \kappa^b \tag{5.29.3}$$

The normal strain is induced on the web because the laminate layers of the web portion are oriented parallel to the structural (X–Z) plane. This issue was discussed in Section 5.3.

$$\varepsilon_{xx}^w = Z\kappa^b \quad \kappa_{xx}^w = 0 \tag{5.29.4}$$

For top and bottom flange portions using Eqs. (5.7.1 and 5.9.2), the induced force and moment resultants of each segment of cross section are presented as:

$$N_{xx}^{tf} = \frac{b^{tf}}{a_{11}^{tf}}\varepsilon_{xx}^{tf} \quad M_{xx}^{tf} = \frac{b^{tf}}{d_{11}^{tf}}\kappa^b \tag{5.30.1}$$

$$N_{xx}^{bf} = \frac{b^{bf}}{a_{11}^{bf}}\varepsilon_{xx}^{bf} \quad M_{xx}^{bf} = \frac{b^{bf}}{d_{11}^{bf}}\kappa^b \tag{5.30.2}$$

For web portion using Eq. (5.7):

$$N_{xx}^w = \frac{1}{a_{11}^w}\varepsilon_{xx}^w \tag{5.30.3}$$

By substituting strain and curvature relation in Eqs. (5.29.2–5.29.4) into the induced force and moment resultants in Eqs. (5.30.1–5.30.3), the normal force and moment resultants in terms of the curvature at the beam axis are given as:

$$N_{xx}^{tf} = \frac{b^{tf}}{a_{11}^{tf}}z_c^{tf}\kappa^b \quad M_{xx}^{tf} = \frac{b^{tf}}{d_{11}^{tf}}\kappa^b \tag{5.31.1}$$

$$N_{xx}^{bf} = \frac{b^{bf}}{a_{11}^{bf}}z_c^{bf}\kappa^b \quad M_{xx}^{bf} = \frac{b^{bf}}{d_{11}^{bf}}\kappa^b \tag{5.31.2}$$

$$N_{xx}^w = \frac{1}{a_{11}^w}z\kappa^b \tag{5.31.3}$$

Moment resultant on the cross section is obtained from the additional moment resultants of each segment as shown in Example Figure 5.2.2.

$$M_{xx}^b = \left(M_{xx}^{tf} + N_{xx}^{tf}z_c^{tf}\right) + \int_{b^w} N_{xx}^w z\,dz + \left(M_{xx}^{bf} + N_{xx}^{bf}z_c^{bf}\right) \tag{5.31.4}$$

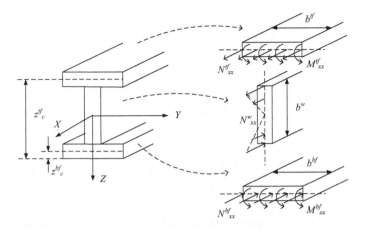

EXAMPLE FIGURE 5.2.2   Force and moment resultants on WF cross section.

By substituting the normal force and moment resultants in Eqs. (5.31.1–5.31.3) into Eq. (5.31.4):

$$M_{xx}^b = \left(\frac{b^{tf}}{d_{11}^{tf}} \kappa^b + \frac{b^{tf}}{a_{11}^{tf}} \left(z_c^{tf} \right)^2 \kappa^b \right) + \int_{b^w} \frac{1}{a_{11}^w} \kappa^b z^2 dz + \left(\frac{b^{bf}}{d_{11}^{bf}} \kappa^b + \frac{b^{bf}}{a_{11}^{bf}} \left(z_c^{bf} \right)^2 \kappa^b \right)$$ (5.31.5)

After integration and simplification, Eq. (5.31.5) yields:

$$M_{xx}^b = \left[\left(\frac{b^{tf}}{d_{11}^{tf}} + \frac{b^{tf}}{a_{11}^{tf}} \left(z_c^{tf} \right)^2 \right) + \frac{1}{a_{11}^w} \left(\frac{\left(b^w \right)^3}{12} + b^w \left(z_c^w \right)^2 \right) + \left(\frac{b^{bf}}{d_{11}^{bf}} + \frac{b^{bf}}{a_{11}^{bf}} \left(z_c^{bf} \right)^2 \right) \right] \kappa^b$$ (5.31.6)

From beam stiffness formulation given earlier, the equivalent bending stiffness of the composite beam with symmetric lay-ups for I section is given below:

$$\frac{M_{xx}^b}{\kappa^b} = (EI)_{eq}^b = \left(\frac{b^{tf}}{d_{11}^{tf}} + \frac{b^{tf}}{a_{11}^{tf}} \left(z_c^{tf} \right)^2 \right) + \frac{1}{a_{11}^w} \left(\frac{\left(b^w \right)^3}{12} + b^w \left(z_c^w \right)^2 \right) + \left(\frac{b^{bf}}{d_{11}^{bf}} + \frac{b^{bf}}{a_{11}^{bf}} \left(z_c^{bf} \right)^2 \right)$$ (5.31.7)

If the top and bottom flanges are identical in width (b), thickness (t), and laminated lay-ups, then the bending and axial rigidities can be simplified as:

$$\frac{b^{tf}}{d_{11}^{tf}} = \frac{b^{bf}}{d_{11}^{bf}} \qquad \frac{b^{tf}}{a_{11}^{tf}} = \frac{b^{bf}}{a_{11}^{bf}}$$ (5.32.1)

By substituting Eq. (5.32.1) into Eq. (5.31.7), the equivalent bending rigidity is:

$$(EI)_{eq}^b = 2 \left(\frac{b^{bf}}{d_{11}^{tf}} + \frac{b^{tf}}{a_{11}^{tf}} \left(z_c^{tf} \right)^2 \right) + \frac{\left(b^w \right)^3}{12 a_{11}^w}$$ (5.32.2)

b. Flexural rigidity of laminated composite T beam:
   For a composite beam with T section in Example Figure 5.2.3, Eq. (5.31.7) can be simplified by neglecting (b^{bf}) to zero. The equivalent bending rigidity of a composite beam is presented as:

$$(EI)_{eq}^b = \left(\frac{b^{tf}}{d_{11}^{tf}} + \frac{b^{tf}}{a_{11}^{tf}} \left(z_c^{tf} \right)^2 \right) + \frac{1}{a_{11}^w} \left(\frac{\left(b^w \right)^3}{12} + b^w \left(z_c^w \right)^2 \right)$$ (5.32.3)

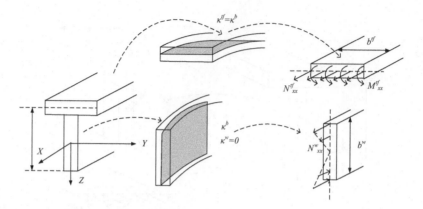

EXAMPLE FIGURE 5.2.3   Curvature, force and moment resultants of T cross section.

Example 5.3

Determine: (a) torsional rigidity of laminated composite WF beam (as given in Example 5.1) and (b) torsional rigidity of laminated composite T beam.

A symmetric I section with symmetric lay-ups of each rectangular element (flange, web) is considered herein. The cross section is symmetric with respect to the structural (*X–Z*) plane.

The induced normal force and moment are also independent due to symmetric lay-ups. The structural coordinate system is located at the neutral plane of the cross section, while the global coordinate systems of each element are positioned on the mid-planes of each segment as shown in Example 5.1.

Solution

a. Torsional rigidity of laminated composite WF beam:

The equivalent torsional rigidity (*GC*) of an open cross section about the beam axis is obtained from the summation of torsional rigidities of each segment. The twisting curvature is applied to the beam axis and all the beam segments are also twisted at the same twist angle. Only twisting curvatures (κ_{xy}) of each beam element (flange, web) are considered for this analysis.

$$\kappa_{xy}^{b} = \kappa_{xy}^{tf} = \kappa_{xy}^{w} = \kappa_{xy}^{bf}$$

Thus, twisting moments of each beam element are shown in Example Figure 5.3.1.

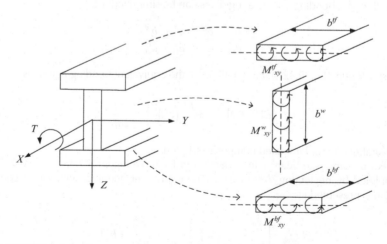

EXAMPLE FIGURE 5.3.1   Torque and twisting moment resultants on WF cross section.

From Eq. (5.22.4):

$$M_{xy}^{tf} = \frac{\kappa_{xy}^{tf}}{d_{66}^{tf}} \quad M_{xy}^{w} = \frac{\kappa_{xy}^{w}}{d_{66}^{w}} \quad M_{xy}^{bf} = \frac{\kappa_{xy}^{bf}}{d_{66}^{bf}}$$

(5.33.1)

From Eqs. (5.22.4 and 5.22.5), torque on each element can be obtained as:

$$T_{xx}^{tf} = -2b^{tf}\frac{\kappa_{xy}^{tf}}{d_{66}^{tf}} \quad T_{xx}^{w} = -2b^{w}\frac{\kappa_{xy}^{w}}{d_{66}^{w}} \quad T_{xx}^{bf} = -2b^{bf}\frac{\kappa_{xy}^{bf}}{d_{66}^{bf}}$$

(5.33.2)

Total applied torque on a cross section is obtained by adding torque resultants of each segment as:

$$T_{xx}^{b} = -2\left(\frac{b^{tf}}{d_{66}^{tf}} + \frac{b^{w}}{d_{66}^{w}} + \frac{b^{bf}}{d_{66}^{bf}}\right)\kappa_{xy}^{b}$$

(5.33.3)

The compatibility relation between the twisting curvature and rotational angle of the cross section at the reference plane is given in Eq. (5.22.2).

$$\kappa_{xy}^{b} = -2\theta$$

(5.33.4)

By substituting Eq. (5.33.4) into Eq. (5.33.3), the governing equation of a laminated composite beam under primary torsion is given below:

$$T_{xx}^{b} = (GC)_{eq}^{b}\,\theta = 4\left(\frac{b^{tf}}{d_{66}^{tf}} + \frac{b^{w}}{d_{66}^{w}} + \frac{b^{bf}}{d_{66}^{bf}}\right)\theta$$

(5.33.5)

Equivalent twisting rigidity of a composite beam with symmetric lay-ups for a WF section is:

$$(GC)_{eq}^{b} = 4\left(\frac{b^{tf}}{d_{66}^{tf}} + \frac{b^{w}}{d_{66}^{w}} + \frac{b^{bf}}{d_{66}^{bf}}\right)$$

(5.33.6)

If the top and bottom flanges are identical in width (b), thickness (t), and laminated lay-ups, then the torsional rigidity can be simplified as:

$$\frac{b^{tf}}{d_{66}^{tf}} = \frac{b^{bf}}{d_{66}^{bf}}$$

(5.33.7)

By substituting Eq. (5.33.7) into Eq. (5.33.6), the equivalent torsional rigidity is:

$$(GC)_{eq}^{b} = 4\left(\frac{2b^{tf}}{d_{66}^{tf}} + \frac{b^{w}}{d_{66}^{w}}\right)$$

(5.33.8)

b. Torsional rigidity of laminated composite T beam:
   For a composite beam with T section, Eq. (5.33.8) can be simplified by neglecting (b^{bf}) to zero. The torsional rigidity of composite beams is presented as:

$$(GC)_{eq}^{b} = 4\left(\frac{b^{tf}}{d_{66}^{tf}} + \frac{b^{w}}{d_{66}^{w}}\right)$$

(5.33.9)

5.7   LAMINATED COMPOSITE RECTANGULAR BOX BEAM

In this section, the box beam with different thicknesses at the top and bottom flanges is shown in Figure 5.18. The procedure to find stiffness is similar to those given in Section 5.6. The curvature or deformation that corresponds to specific stiffness will be applied on the neutral plane of a beam. The force and moment resultants under applied deformation (or curvature) of each beam segment are determined.

The sum of all force and moment resultants in each segment of a beam is equal to the applied forces (or moments) at the neutral plane of a cross section. Then, the sum of all force and moment resultants is presented in terms of applied deformations (or curvatures). By comparing the stiffness of isotropic beams (or using stiffness definition), the equivalent stiffness of a composite beam has been obtained.

Example 5.4

Determine: (a) the axial stiffness of laminated composite box beam with orthotropic symmetric lay-ups (as shown in Figure 5.18)

A composite beam of a rectangular box section is made from orthotropic symmetric lay-ups. The equivalent axial stiffness $(EA)_{eq}$ is determined herein. When the beam axis is elongated, the curvatures will not be induced, and the axial strains are equal through the cross section. The induced normal force, both the bending and twisting moments are also independent due to symmetric lay-ups.

Solution

a. The axial stiffness of laminated composite box beam

The axial deformation $\left(\varepsilon_{xx}^{n}\right)$ is induced at the neutral plane. The beam curvature is not induced due to the symmetric nature of lay-up and that of the cross section. The axial strain (ε_{xx}) of each segment is the same through the cross section as shown in Example Figure 5.4.1.

$$\varepsilon_{xx}^{n} = \varepsilon_{xx}^{tf} = \varepsilon_{xx}^{rw} = \varepsilon_{xx}^{lw} = \varepsilon_{xx}^{bf}$$

Only the force resultants at the mid-plane of each segment are induced due to the applied axial strains and the symmetric lay-ups.

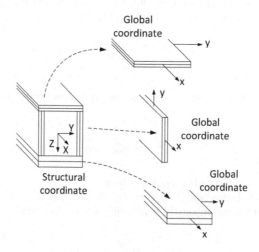

FIGURE 5.18   Rectangular laminate composite box beam.

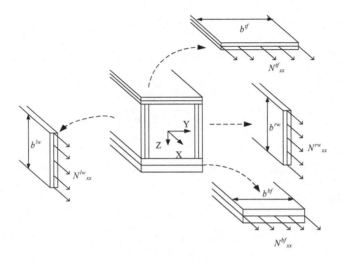

EXAMPLE FIGURE 5.4.1 Axial force on box beam segments.

From Eq. (5.4.2):

$$N_{xx}^{tf} = \frac{b^{tf}}{a_{11}^{tf}} \varepsilon_{xx}^{tf} \quad N_{xx}^{rw} = \frac{b^{rw}}{a_{11}^{rw}} \varepsilon_{xx}^{w} \quad N_{xx}^{lw} = \frac{b^{lw}}{a_{11}^{lw}} \varepsilon_{xx}^{w} \quad N_{xx}^{bf} = \frac{b^{bf}}{a_{11}^{bf}} \varepsilon_{xx}^{bf} \tag{5.34.1}$$

Total force resultant on a cross section is obtained by adding force resultants of each segment in a manner similar to the I-Section shown in Example Figure 5.1.3:

$$N_{xx}^{b} = \frac{b^{tf}}{a_{11}^{tf}} \varepsilon_{xx}^{tf} + \frac{b^{rw}}{a_{11}^{rw}} \varepsilon_{xx}^{rw} + \frac{b^{lw}}{a_{11}^{lw}} \varepsilon_{xx}^{lw} + \frac{b^{bf}}{a_{11}^{bf}} \varepsilon_{xx}^{bf} \tag{5.34.2}$$

If the right and left web portions are identical, then the total force resultant on the cross section is simplified as:

$$\frac{b^{w}}{a_{11}^{w}} = \frac{b^{rw}}{a_{11}^{rw}} = \frac{b^{lw}}{a_{11}^{lw}} \tag{5.34.3}$$

$$N_{xx}^{b} = (EA)_{eq}^{b} \varepsilon_{xx}^{n} = \left(\frac{b^{tf}}{a_{11}^{tf}} + \frac{2b^{w}}{a_{11}^{w}} + \frac{b^{bf}}{a_{11}^{bf}} \right) \varepsilon_{xx}^{n} \tag{5.34.4}$$

$$(EA)_{eq}^{b} = \left(\frac{b^{tf}}{a_{11}^{tf}} + \frac{2b^{w}}{a_{11}^{w}} + \frac{b^{bf}}{a_{11}^{bf}} \right) \tag{5.34.5}$$

To determine the location of the centroid plane of a cross section, the balancing moment of the bottom fiber of the cross section is considered as shown in Example Figure 5.4.2.

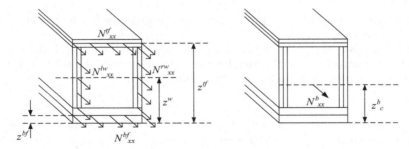

EXAMPLE FIGURE 5.4.2 Force resultants on box beam.

$$z_c^b N_{xx}^b = z^{tf} N_{xx}^{tf} + 2z^w N_{xx}^w + z^{bf} N_{xx}^{bf} \tag{5.34.6}$$

By substituting Eq. (5.34.1) into Eq. (5.34.6):

$$z_c^b = \frac{z^{tf}\dfrac{b^{tf}}{a_{11}^{tf}} + 2z^w\dfrac{b^w}{a_{11}^w} + z^{bf}\dfrac{b^{bf}}{a_{11}^{bf}}}{\left(\dfrac{b^{tf}}{a_{11}^{tf}} + \dfrac{2b^w}{a_{11}^w} + \dfrac{b^{bf}}{a_{11}^{bf}}\right)} \tag{5.34.7}$$

Axial stress can be determined using the axial stress definition as studied in the mechanics of materials class as:

From Eq. (5.34.4):

$$\sigma_{xx}^b = \frac{N^b}{A^b} = \frac{\left(\dfrac{b^{tf}}{a_{11}^{tf}} + \dfrac{2b^w}{a_{11}^w} + \dfrac{b^{bf}}{a_{11}^{bf}}\right)}{b^{tf}t^{tf} + 2b^w t^w + b^{bf}t^{bf}}\,\varepsilon_{xx}^b \tag{5.34.8}$$

If the top and bottom flanges are identical in width (b), thickness (t), and laminated lay-ups, then the axial stiffness, centroid, and axial stress of a composite beam can be simplified as:

$$\frac{b^{tf}}{a_{11}^{tf}} = \frac{b^{bf}}{a_{11}^{bf}} \tag{5.35.1}$$

By substituting Eq. (5.35.1) into Eqs. (5.34.5, 5.34.7, and 5.34.8):

$$(EA)_{eq}^b = \left(\frac{2b^f}{a_{11}^f} + \frac{2b^w}{a_{11}^w}\right) \tag{5.35.2}$$

$$z_c^b = \frac{z^f\dfrac{b^{tf}}{a_{11}^{tf}} + z^w\dfrac{b^w}{a_{11}^w}}{\left(\dfrac{b^f}{a_{11}^f} + \dfrac{b^w}{a_{11}^w}\right)} \tag{5.35.3}$$

$$\sigma_{xx}^b = \frac{\left(\dfrac{b^f}{a_{11}^f} + \dfrac{b^w}{a_{11}^w}\right)}{b^{tf}t^{tf} + b^w t^w}\,\varepsilon_{xx}^b \tag{5.35.4}$$

Example 5.5

Determine: Equivalent flexural rigidity $(EI)_{eq}$ of laminated composite box beam with symmetric lay-ups (as shown in Figure 5.18)

A composite box beam is made from symmetric lay-up, which is bending about the structural coordinate axis Y under transverse loading. The induced normal forces, due to bending and twisting moments are independent due to symmetric lay-ups.

Solution

This composite beam is bent only on the structural (X–Z) plane. The equivalent bending rigidity $(EI)_{eq}^{yy}$ about the structural axis Y is determined. The curvature (κ^b) is applied at the beam axis. Then, both the normal strain (ε_{xx}) and curvature (κ_{yy}) for each segment are induced due to applied curvature as shown in Example Figure 5.5.1.

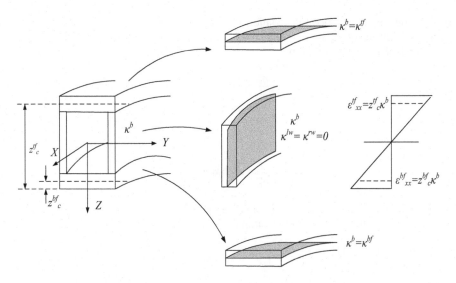

EXAMPLE FIGURE 5.5.1 Curvature and strain distribution for box beam.

By adding curvature (κ^b) effects to axial effects, the total strain is given as:

$$\varepsilon_{xx}^z = \varepsilon_{xx}^o + z\kappa^b = z\kappa^b \qquad (5.36.1)$$

where $\varepsilon_{xx}^o = 0$, then the bending effects are:

For the top flange portion, curvature and normal strain at the mid-plane of the top flange are induced, resulting in force and moment resultants on both the top and bottom flanges of the box beam. The curvatures of the top and bottom flanges at the mid-plane are also equal to the applied curvature at the beam axis.

$$\varepsilon_{xx}^{tf} = z_c^{tf}\kappa^b \qquad \kappa_{xx}^{tf} = \kappa^b \qquad (5.36.2)$$

$$\varepsilon_{xx}^{bf} = z_c^{bf}\kappa^b \qquad \kappa_{xx}^{bf} = \kappa^b \qquad (5.36.3)$$

The normal strain is induced on the web because the laminate layers of the web portion are oriented parallel to the structural (X–Z) plane. This issue is discussed in Section 5.3.

$$\varepsilon_{xx}^{rw} = z\kappa^b \qquad \kappa_{xx}^{rw} = 0 \qquad (5.36.4)$$

$$\varepsilon_{xx}^{lw} = z\kappa^b \qquad \kappa_{xx}^{lw} = 0 \qquad (5.36.5)$$

For both the top and bottom flange portions using Eqs. (5.18.1 and 5.9.2), the induced force and moment resultants of each segment of the cross section are presented as:

$$N_{xx}^{tf} = \frac{b^{tf}}{a_{11}^{tf}}\varepsilon_{xx}^{tf} \qquad M_{xx}^{tf} = \frac{b^{tf}}{d_{11}^{tf}}\kappa^b \qquad (5.36.6)$$

$$N_{xx}^{bf} = \frac{b^{bf}}{a_{11}^{bf}}\varepsilon_{xx}^{bf} \qquad M_{xx}^{bf} = \frac{b^{bf}}{d_{11}^{bf}}\kappa^b \qquad (5.36.7)$$

For web portion using Eq. (5.10.6):

$$N_{xx}^{rw} = \frac{1}{a_{11}^{rw}}\varepsilon_{xx}^{rw} \qquad N_{xx}^{lw} = \frac{1}{a_{11}^{lw}}\varepsilon_{xx}^{lw} \qquad (5.36.8)$$

By substituting strain and curvature relation in Eqs. (5.36.2–5.36.5) into the induced force and moment resultants in Eqs. (5.36.6–5.36.8), the normal force and moment resultants in terms of the curvature at the beam axis are given as:

$$N_{xx}^{tf} = \frac{b^{tf}}{a_{11}^{tf}} z_c^{tf} \kappa^b \quad M_{xx}^{tf} = \frac{b^{tf}}{d_{11}^{tf}} \kappa^b \tag{5.37.1}$$

$$N_{xx}^{bf} = \frac{b^{bf}}{a_{11}^{bf}} z_c^{bf} \kappa^b \quad M_{xx}^{bf} = \frac{b^{bf}}{d_{11}^{bf}} \kappa^b \tag{5.37.2}$$

$$N_{xx}^{rw} = \frac{1}{a_{11}^{rw}} z \kappa^b \quad N_{xx}^{lw} = \frac{1}{a_{11}^{lw}} z \kappa^b \tag{5.37.3}$$

Total moment resultant on a cross section is obtained by adding moment resultants of each segment is similar to the one shown in Example Figure 5.2.2.

$$M_{xx}^b = \left(M_{xx}^{tf} + N_{xx}^{tf} z_c^{tf} \right) + \int_{b^{rw}} N_{xx}^{rw} z\, dz + \int_{b^{rw}} N_{xx}^{lw} z\, dz + \left(M_{xx}^{bf} + N_{xx}^{bf} z_c^{bf} \right) \tag{5.37.4}$$

By substituting Eqs. (5.37.1–5.37.3) into Eq. (5.37.4):

$$M_{xx}^b = \left(\frac{b^{tf}}{d_{11}^{tf}} \kappa^b + \frac{b^{tf}}{a_{11}^{tf}} \left(z_c^{tf} \right)^2 \kappa^b \right) + \int_{b^{rw}} \frac{1}{a_{11}^{rw}} \kappa^b z^2\, dz + \int_{b^{rw}} \frac{1}{a_{11}^{lw}} \kappa^b z^2\, dz + \left(\frac{b^{bf}}{d_{11}^{bf}} \kappa^b + \frac{b^{bf}}{a_{11}^{bf}} \left(z_c^{bf} \right)^2 \kappa^b \right) \tag{5.37.5}$$

If the right and left web portions are identical in both properties and geometries, then the equation is simplified as:

$$\frac{1}{a_{11}^w} = \frac{1}{a_{11}^{rw}} = \frac{1}{a_{11}^{lw}} \quad t^w = t^{rw} = t^{lw} \quad b^w = b^{rw} = b^{lw} \tag{5.37.6}$$

By substituting Eq. (5.37.6) into Eq. (5.37.5):

$$M_{xx}^b = \left(EI \right)_{eq}^b \kappa^b = \left[\begin{array}{c} \left(\dfrac{b^{tf}}{d_{11}^{tf}} + \dfrac{b^{tf}}{a_{11}^{tf}} \left(z_c^{tf} \right)^2 \right) + \dfrac{2}{a_{11}^w} \left(\dfrac{\left(b^w \right)^3}{12} + b^w \left(z_c^w \right)^2 \right) \\[4mm] + \left(\dfrac{b^{bf}}{d_{11}^{bf}} + \dfrac{b^{bf}}{a_{11}^{bf}} \left(z_c^{bf} \right)^2 \right) \end{array} \right] \kappa^b \tag{5.37.7}$$

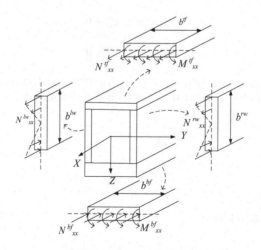

EXAMPLE FIGURE 5.5.2 Force and moment resultants of box beam.

The equivalent bending stiffness of a composite box beam with symmetric lay is given below:

$$(EI)^b_{eq} = \left(\frac{b^{tf}}{d^{tf}_{11}} + \frac{b^{tf}}{a^{tf}_{11}} \left(z^{tf}_c \right)^2 \right) + \frac{2}{a^w_{11}} \left(\frac{\left(b^w \right)^3}{12} + b^w \left(z^w_c \right)^2 \right) + \left(\frac{b^{bf}}{d^{bf}_{11}} + \frac{b^{bf}}{a^{bf}_{11}} \left(z^{bf}_c \right)^2 \right) \tag{5.37.8}$$

If the top and bottom flanges are identical in width (b), thickness (t), and laminated lay-ups, then the bending rigidity can be simplified as:

$$(EI)^b_{eq} = 2 \left(\frac{b^f}{d^f_{11}} + \frac{b^f}{a^f_{11}} \left(z^f_c \right)^2 \right) + \frac{\left(b^w \right)^3}{6a^w_{11}} \tag{5.37.9}$$

Example 5.6

Determine the torsional rigidity of a laminated composite box beam with orthotropic symmetric lay-ups. A rectangular composite beam is made from orthotropic symmetric lay-ups of laminates. The induced normal force, bending, and twisting moment are also independent due to symmetric lay-ups.

Solution

A rectangular composite box beam assumes that it is made from thin plates joined along their edges. The thickness of a plate is small compared with the cross-sectional dimensions, i.e., <10%. In addition, the assumptions used for composite hollow beams in Section 5.5.1, still hold true for thin-walled rectangular cross sections, except warping terms become predominant.

The rectangular box composite beam under uniform torque is considered herein. When the composite beams are twisted about the neutral plane, the equivalent torsional rigidity $(GC)_{eq}$ is obtained by integrating Eq. (5.25.5) at the mid-plane of a segment about the box cross section.

From symmetric lay-ups $[B_{ij}] = 0$, Eq. (5.25.5) is simplified as:

$$\theta = \frac{T}{4\bar{A}^2} \oint a_{66} \, ds = \frac{T}{4\bar{A}^2} \left[\int_{tf} a_{66} \, ds + \int_{rw} a_{66} \, ds + \int_{bf} a_{66} \, ds + \int_{lw} a_{66} \, ds \right] \tag{5.38.1}$$

$$T = (GC)_{eq} \theta = \frac{4\bar{A}^2}{a^{tf}_{66} s^{tf} + a^{rw}_{66} s^{rw} + a^{bf}_{66} s^{bf} + a^{lw}_{66} s^{lw}|} \theta \tag{5.38.2}$$

$$(GC)_{eq} = \frac{4\bar{A}^2}{a^{tf}_{66} s^{tf} + a^{rw}_{66} s^{rw} + a^{bf}_{66} s^{bf} + a^{lw}_{66} s^{lw}} \tag{5.38.3}$$

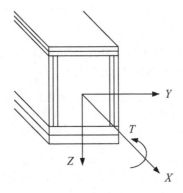

EXAMPLE FIGURE 5.6.1 Laminated composite box beam under torsion.

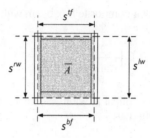

EXAMPLE FIGURE 5.6.2   Enclosed area surrounded by the mid-line of segments.

where s = mid-line length of the segment from the center to the center of the contiguous segments;
$\bar{A}$ = enclosed area that is surrounded by the mid-line of the segment of rectangular cross
section as shown in Example Figure 5.6.2.

If the right and left web portions are identical in properties and geometries, then equivalent
torsional rigidity can be simplified as:

$$\frac{1}{a_{11}^w} = \frac{1}{a_{11}^{rw}} = \frac{1}{a_{11}^{lw}} \tag{5.38.4}$$

$$s^w = s^{rw} = s^{lw} \tag{5.38.5}$$

By substituting Eqs. (5.38.4 and 5.38.5) into Eq. (5.38.3):

$$(GC)_{eq} = \frac{4\bar{A}^2}{a_{66}^{tf}s^{tf} + 2a_{66}^w s^w + a_{66}^{bf}s^{bf}} \tag{5.38.6}$$

If the top and bottom flanges are identical in width and thickness and laminated lay-ups, then the
bending rigidity can be simplified as:

$$\frac{1}{a_{66}^f s^f} = \frac{1}{a_{66}^{tf} s^{tf}} = \frac{1}{a_{66}^{bf} s^{bf}} \tag{5.38.7}$$

$$\frac{1}{a_{11}^w s^w} = \frac{1}{a_{11}^{rw} s^{rw}} = \frac{1}{a_{11}^{lw} s^{lw}} \tag{5.38.8}$$

By substituting Eqs. (5.38.7 and 5.38.8) into Eq. (5.38.6):

$$(GC)_{eq} = \frac{2\bar{A}^2}{a_{66}^f s^f + a_{66}^w s^w} \tag{5.38.9}$$

5.8   LAMINATED COMPOSITE RECTANGULAR BOX BEAM WITH UNSYMMETRIC LAY-UPS

As with the symmetrical lay-up responses in Section 5.7, the box section bending and torsion
responses with unsymmetrical lay-ups are considered in this section. The composite beam has the
top flange thickness different from the bottom flange thickness. The composite beam is made from
orthotropic laminates.

Under the axial deformation, not only the normal force resultant is induced but also the moment
resultant is induced in each segment. For arbitrary transverse loadings in the structural (X–Z) plane,

the bending moment of the structural axes Y and Z are induced due to unsymmetric cross section. Three different equivalent bending rigidities of this composite beam include:

1. bending rigidity $(EI)_{eq}^{yy}$ about axis Y,
2. bending rigidity $(EI)_{eq}^{zz}$ about Z, and
3. bending rigidity coupling $(EI)_{eq}^{yz}$ between axis Y and Z

Example 5.7

Determine: the axial stiffness of a laminated composite box beam with orthotropic unsymmetrical lay-ups of laminates.

A composite beam of rectangular box section is made from the orthotropic unsymmetric lay-ups of laminates. When the beam axis is free to elongate, then the curvatures will not be induced, and the axial strains are equal throughout the cross section. Both normal force and bending moment are induced due to unsymmetric lay-ups.

Solution

The axial deformation $\left(\varepsilon_{xx}^{b}\right)$ is applied. All curvatures are equal to zero. The axial strain (ε_{xx}) of each segment is identical to the cross section.

$$\varepsilon_{xx}^{b} = \varepsilon_{xx}^{tf} = \varepsilon_{xx}^{rw} = \varepsilon_{xx}^{lw} = \varepsilon_{xx}^{bf} \tag{5.39.1}$$

The force and moment resultants at the mid-plane of each segment are determined under the applied axial strains in Eq. (5.39.1) using inversion of Eqs. (5.16.2 and 5.16.3) as:

$$\begin{bmatrix} \dfrac{N_{xx}^{b}}{b} \\[2mm] \dfrac{M_{xx}^{b}}{b} \end{bmatrix} = \frac{1}{a_{11}d_{11} - b_{11}^{2}} \begin{bmatrix} d_{11} & -b_{11} \\ -b_{11} & a_{11} \end{bmatrix} \begin{bmatrix} \varepsilon_{xx}^{b} \\ 0 \end{bmatrix} \tag{5.39.2}$$

From Eq. (5.39.2):

$$N_{xx}^{tf} = \left(\frac{d_{11}}{a_{11}d_{11} - b_{11}^{2}} \right)_{tf} b^{tf}\varepsilon_{xx}^{tf} \quad M_{xx}^{tf} = -\left(\frac{b_{11}}{a_{11}d_{11} - b_{11}^{2}} \right)_{tf} b^{tf}\varepsilon_{xx}^{tf} \tag{5.39.3}$$

$$N_{xx}^{rw} = \left(\frac{d_{11}}{a_{11}d_{11} - b_{11}^{2}} \right)_{rw} b^{rw}\varepsilon_{xx}^{rw} \quad M_{xx}^{rw} = -\left(\frac{b_{11}}{a_{11}d_{11} - b_{11}^{2}} \right)_{rw} b^{rw}\varepsilon_{xx}^{rw} \tag{5.39.4}$$

$$N_{xx}^{lw} = \left(\frac{d_{11}}{a_{11}d_{11} - b_{11}^{2}} \right)_{lw} b^{lw}\varepsilon_{xx}^{lw} \quad M_{xx}^{lw} = -\left(\frac{b_{11}}{a_{11}d_{11} - b_{11}^{2}} \right)_{lw} b^{lw}\varepsilon_{xx}^{lw} \tag{5.39.5}$$

$$N_{xx}^{bf} = \left(\frac{d_{11}}{a_{11}d_{11} - b_{11}^{2}} \right)_{bf} b^{bf}\varepsilon_{xx}^{bf} \quad M_{xx}^{bf} = -\left(\frac{b_{11}}{a_{11}d_{11} - b_{11}^{2}} \right)_{tf} b^{bf}\varepsilon_{xx}^{bf} \tag{5.39.6}$$

The total force resultant on the box section is obtained by adding force resultants of each segment as (see Example Figure 5.7.1):

$$N_{xx}^{b} = N_{xx}^{tf} + N_{xx}^{rw} + N_{xx}^{lw} + N_{xx}^{bf} \tag{5.40.1}$$

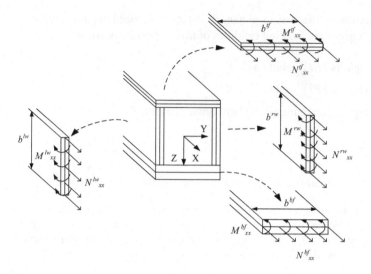

EXAMPLE FIGURE 5.7.1 Axial force on box beam segments (unsymmetric lay-ups).

$$N_{xx}^b = (EA)_{eq}^b \varepsilon_{xx}^b = \left[\begin{array}{l} \left(\dfrac{d_{11}}{a_{11}d_{11} - b_{11}^2} \right)_{tf} b^{tf} + \left(\dfrac{d_{11}}{a_{11}d_{11} - b_{11}^2} \right)_{rw} b^{rw} \\[3mm] + \left(\dfrac{d_{11}}{a_{11}d_{11} - b_{11}^2} \right)_{lw} b^{lw} + \left(\dfrac{d_{11}}{a_{11}d_{11} - b_{11}^2} \right)_{bf} b^{bf} \end{array} \right] \varepsilon_{xx}^b \qquad (5.40.2)$$

$$(EA)_{eq}^b = \left(\frac{d_{11}}{a_{11}d_{11} - b_{11}^2} \right)_{tf} b^{tf} + \left(\frac{d_{11}}{a_{11}d_{11} - b_{11}^2} \right)_{rw} b^{rw} + \left(\frac{d_{11}}{a_{11}d_{11} - b_{11}^2} \right)_{lw} b^{lw} + \left(\frac{d_{11}}{a_{11}d_{11} - b_{11}^2} \right)_{bf} b^{bf} \qquad (5.40.3)$$

The distance of the centroid plane of a cross section is evaluated by balancing moments of the bottom fiber (Example Figure 5.7.2) as below:

$$z_c^b N_{xx}^b = z^{tf} N_{xx}^{tf} + z^{rw} N_{xx}^{rw} + z^{lw} N_{xx}^{lw} + z^{bf} N_{xx}^{bf} + M_{xx}^{tf} + M_{xx}^{bf} \qquad (5.40.4)$$

$$y_c^b N_{xx}^b = y^{tf} N_{xx}^{tf} + y^{rw} N_{xx}^{rw} + y^{lw} N_{xx}^{lw} + y^{bf} N_{xx}^{bf} + M_{xx}^{rw} + M_{xx}^{lw} \qquad (5.40.5)$$

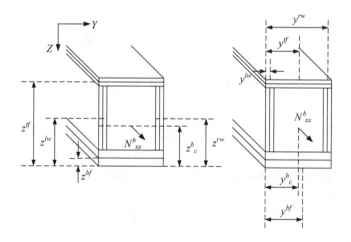

EXAMPLE FIGURE 5.7.2 Centroid of composite box beam with unsymmetric lay-ups.

By substituting Eqs. (5.39.3–5.39.6) into Eqs. (5.40.4 and 5.40.5) results in:

$$
z_c^b = \frac{\left[\left(\dfrac{d_{11}z^{tf}-b_{11}}{a_{11}d_{11}-b_{11}^2}\right)_{tf}b^{tf} + \left(\dfrac{d_{11}}{a_{11}d_{11}-b_{11}^2}\right)_{rw}z^{rw}b^{rw} + \left(\dfrac{d_{11}}{a_{11}d_{11}-b_{11}^2}\right)_{w}z^{lw}b^{lw} + \left(\dfrac{d_{11}z^{bf}-b_{11}}{a_{11}d_{11}-b_{11}^2}\right)_{bf}b^{bf}\right]}{\left[\left(\dfrac{d_{11}}{a_{11}d_{11}-b_{11}^2}\right)_{tf}b^{tf} + \left(\dfrac{d_{11}}{a_{11}d_{11}-b_{11}^2}\right)_{rw}b^{rw} + \left(\dfrac{d_{11}}{a_{11}d_{11}-b_{11}^2}\right)_{lw}b^{lw} + \left(\dfrac{d_{11}}{a_{11}d_{11}-b_{11}^2}\right)_{bf}b^{bf}\right]}
$$

(5.40.6)

$$
y_c^b = \frac{\left[\left(\dfrac{d_{11}}{a_{11}d_{11}-b_{11}^2}\right)_{tf}y^{tf}b^{tf} + \left(\dfrac{d_{11}y^{w}-b_{11}}{a_{11}d_{11}-b_{11}^2}\right)_{rw}y^{rw}b^{rw} + \left(\dfrac{d_{11}y^{w}-b_{11}}{a_{11}d_{11}-b_{11}^2}\right)_{lw}y^{lw}b^{lw} + \left(\dfrac{d_{11}}{a_{11}d_{11}-b_{11}^2}\right)_{bf}y^{bf}b^{bf}\right]}{\left[\left(\dfrac{d_{11}}{a_{11}d_{11}-b_{11}^2}\right)_{tf}b^{tf} + \left(\dfrac{d_{11}}{a_{11}d_{11}-b_{11}^2}\right)_{rw}b^{rw} + \left(\dfrac{d_{11}}{a_{11}d_{11}-b_{11}^2}\right)_{lw}b^{lw} + \left(\dfrac{d_{11}}{a_{11}d_{11}-b_{11}^2}\right)_{bf}b^{bf}\right]}
$$

(5.40.7)

The axial stress of the composite box beam can be determined using the axial stress definition as:

$$
\sigma_{xx}^b = \frac{N_{xx}^b}{A} = \frac{\left[\left(\dfrac{d_{11}}{a_{11}d_{11}-b_{11}^2}\right)_{tf}b^{tf} + \left(\dfrac{d_{11}}{a_{11}d_{11}-b_{11}^2}\right)_{rw}b^{rw} + \left(\dfrac{d_{11}}{a_{11}d_{11}-b_{11}^2}\right)_{lw}b^{lw} + \left(\dfrac{d_{11}}{a_{11}d_{11}-b_{11}^2}\right)_{bf}b^{bf}\right]}{\left[\begin{array}{c} b^{tf}t^{tf} + b^{rw}t^{rw} \\ + b^{lw}t^{lw} + b^{bf}t^{bf} \end{array}\right]}\varepsilon_{xx}^b
$$

(5.40.8)

If the top and bottom flange portions are identical in width, thickness, and laminated lay-ups as well as the right and left web portions, then the axial stiffness, centroid, and axial stress of composite beams can be simplified as:

$$
(EA)_{eq}^b = 2\left(\frac{d_{11}}{a_{11}d_{11}-b_{11}^2}\right)_f b^f + 2\left(\frac{d_{11}}{a_{11}d_{11}-b_{11}^2}\right)_w b^w
$$

(5.41.1)

$$
z_c^b = \frac{\left(\dfrac{d_{11}z^{tf}-b_{11}}{a_{11}d_{11}-b_{11}^2}\right)_f b^f + \left(\dfrac{d_{11}}{a_{11}d_{11}-b_{11}^2}\right)_w z^w b^w}{\left(\dfrac{d_{11}}{a_{11}d_{11}-b_{11}^2}\right)_f b^f + \left(\dfrac{d_{11}}{a_{11}d_{11}-b_{11}^2}\right)_w b^w}
$$

(5.41.2)

$$
y_c^b = \frac{\left(\dfrac{d_{11}}{a_{11}d_{11}-b_{11}^2}\right)_f y^f b^f + \left(\dfrac{d_{11}y^{w}-b_{11}}{a_{11}d_{11}-b_{11}^2}\right)_w y^w b^w}{\left(\dfrac{d_{11}}{a_{11}d_{11}-b_{11}^2}\right)_f b^f + \left(\dfrac{d_{11}}{a_{11}d_{11}-b_{11}^2}\right)_w b^w}
$$

(5.41.3)

$$
\sigma_{xx}^b = \frac{\left(\dfrac{d_{11}}{a_{11}d_{11}-b_{11}^2}\right)_f b^f + \left(\dfrac{d_{11}}{a_{11}d_{11}-b_{11}^2}\right)_w b^w}{b^f t^f + b^w t^w}\varepsilon_{xx}^b
$$

(5.41.4)

Example 5.8

Determine the equivalent bending stiffness of a laminated composite box beam of Example 5.7, where orthotropic laminates have unsymmetric lay-ups.

The composite beam is bent about both of the structural axes (Y and Z). The equivalent bending rigidities $(EI)_{eq}^{yy}$, $(EI)_{eq}^{zz}$ and $(EI)_{eq}^{yz}$ are determined for unsymmetric cross section.

Solution

The applied curvature (κ_{yy}) of the beam axis about the structural axis Y causes both the normal strain $\left(\varepsilon_{xx}^o\right)$ and curvature (κ_{yy}), while the applied curvature about the structural axis Z also induces normal strain $\left(\varepsilon_{xx}^o\right)$ and curvature (κ_{zz}) as shown in Example Figure 5.8.1.

The curvature (κ_{yy}) of the beam axis (about the structural axis Y), the curvatures at mid-plane for both the top and bottom flanges are induced, and are equal to applied curvature (κ_{yy}) at the beam axis. Since, both the right and left web portions are oriented parallel to the structural (X–Z) plane, only normal strain is induced on the web portions. Thus, the curvature and strain of all portions of the laminated box beam are presented as:

$$\varepsilon_{xx}^{tf} = z_c^{tf}\kappa_{yy} \qquad \kappa_{xx}^{tf} = \kappa_{yy} \tag{5.42.1}$$

$$\varepsilon_{xx}^{bf} = z_c^{bf}\kappa_{yy} \qquad \kappa_{xx}^{bf} = \kappa_{yy} \tag{5.42.2}$$

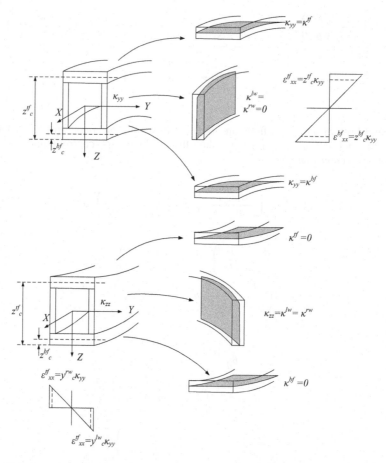

EXAMPLE FIGURE 5.8.1 Curvatures of composite beam with unsymmetric lay-ups.

$$\varepsilon_{xx}^{rw} = z\kappa_{yy} \quad \kappa_{xx}^{w} = 0 \tag{5.42.3}$$

$$\varepsilon_{xx}^{fw} = z\kappa_{yy} \quad \kappa_{xx}^{w} = 0 \tag{5.42.4}$$

The curvature (κ_{zz}) at the beam axis is applied. The curvatures at mid-plane for both the top and bottom flanges are not induced because the flange portions are oriented parallel to the structural (Y–Z) plane. While both the right and left web portions are oriented perpendicular to the structural (Y–Z) plane, both the normal strain and curvature are induced on the web portion. The curvature and strain of all portions are presented as:

$$\varepsilon_{xx}^{tf} = y\kappa_{zz} \quad \kappa_{xx}^{tf} = 0 \tag{5.42.5}$$

$$\varepsilon_{xx}^{bf} = y\kappa_{zz} \quad \kappa_{xx}^{bf} = 0 \tag{5.42.6}$$

$$\varepsilon_{xx}^{rw} = y_c^{rw}\kappa_{zz} \quad \kappa_{xx}^{rw} = \kappa_{zz} \tag{5.42.7}$$

$$\varepsilon_{xx}^{lw} = y_c^{lw}\kappa_{zz} \quad \kappa_{xx}^{lw} = \kappa_{zz} \tag{5.42.8}$$

The force and moment resultants at the mid-plane of each segment are determined under the applied axial strains using inversion of Eqs. (5.16.2 and 5.16.3) as:

$$\begin{bmatrix} \dfrac{N_{xx}^{b}}{b} \\[2mm] \dfrac{M_{xx}^{b}}{b} \end{bmatrix} = \frac{1}{a_{11}d_{11} - b_{11}^2} \begin{bmatrix} d_{11} & -b_{11} \\ -b_{11} & a_{11} \end{bmatrix} \begin{bmatrix} \varepsilon_{xx}^{o} \\ \kappa_{xx}^{o} \end{bmatrix} \tag{5.43}$$

In the case of applied curvature (κ_{yy}), the strain and curvature are substituted into the induced force and moment resultants for each segment. Thus, the induced force and moment resultants under the applied curvature (κ_{yy}) are presented as:
By substituting Eqs. (5.42.1–5.42.4) into Eq. (5.43):

For the top flange portion:

$$\frac{N_{xx}^{tf}}{b^{tf}} = \left(\frac{d_{11}}{a_{11}d_{11} - b_{11}^2}\right)_{tf} z_c^{tf}\kappa_{yy} - \left(\frac{b_{11}}{a_{11}d_{11} - b_{11}^2}\right)_{tf}\kappa_{yy}$$

$$\frac{M_{xx}^{tf}}{b^{tf}} = -\left(\frac{b_{11}}{a_{11}d_{11} - b_{11}^2}\right)_{tf} z_c^{tf}\kappa_{yy} + \left(\frac{a_{11}}{a_{11}d_{11} - b_{11}^2}\right)_{tf}\kappa_{yy} \tag{5.44.1}$$

For the bottom flange portion:

$$\frac{N_{xx}^{bf}}{b^{bf}} = \left(\frac{d_{11}}{a_{11}d_{11} - b_{11}^2}\right)_{bf} z_c^{bf}\kappa_{yy} - \left(\frac{b_{11}}{a_{11}d_{11} - b_{11}^2}\right)_{bf}\kappa_{yy}$$

$$\frac{M_{xx}^{bf}}{b^{bf}} = -\left(\frac{b_{11}}{a_{11}d_{11} - b_{11}^2}\right)_{bf} z_c^{bf}\kappa_{yy} + \left(\frac{a_{11}}{a_{11}d_{11} - b_{11}^2}\right)_{bf}\kappa_{yy} \tag{5.44.2}$$

For the right web portion:

$$N_{xx}^{rw} = \left(\frac{d_{11}}{a_{11}d_{11} - b_{11}^2}\right)_{rw} z\kappa_{yy} \quad \text{and} \quad M_{xx}^{rw} = -\left(\frac{b_{11}}{a_{11}d_{11} - b_{11}^2}\right)_{rw} z\kappa_{yy} \tag{5.44.3}$$

For the left web portion:

$$N_{xx}^{lw} = \left(\frac{d_{11}}{a_{11}d_{11} - b_{11}^2} \right)_{lw} z\kappa_{yy} \quad \text{and} \quad M_{xx}^{lw} = -\left(\frac{b_{11}}{a_{11}d_{11} - b_{11}^2} \right)_{lrw} z\kappa_{yy} \tag{5.44.4}$$

In the case of applied curvature (κ_{zz}), the strain and curvature equations are substituted into the induced force and moment resultants for each segment. Thus, the induced force and moment resultants under the applied curvature (κ_{zz}) are presented as:

By substituting Eqs. (5.42.5–5.42.8) into Eq. (5.43):

For the top flange portion:

$$N_{xx}^{tf} = \left(\frac{d_{11}}{a_{11}d_{11} - b_{11}^2} \right)_{tf} y\kappa_{zz} \quad \text{and} \quad M_{xx}^{tf} = -\left(\frac{b_{11}}{a_{11}d_{11} - b_{11}^2} \right)_{tf} y\kappa_{zz} \tag{5.44.5}$$

For the bottom flange portion:

$$N_{xx}^{bf} = \left(\frac{d_{11}}{a_{11}d_{11} - b_{11}^2} \right)_{bf} y\kappa_{zz} \quad \text{and} \quad M_{xx}^{bf} = -\left(\frac{b_{11}}{a_{11}d_{11} - b_{11}^2} \right)_{bf} y\kappa_{zz} \tag{5.44.6}$$

For the right web portion:

$$\frac{N_{xx}^{rw}}{b^{rw}} = \left(\frac{d_{11}}{a_{11}d_{11} - b_{11}^2} \right)_{rw} y_c^{rw} \kappa_{zz} - \left(\frac{b_{11}}{a_{11}d_{11} - b_{11}^2} \right)_{rw} \kappa_{zz}$$

$$\frac{M_{xx}^{rw}}{b^{rw}} = -\left(\frac{b_{11}}{a_{11}d_{11} - b_{11}^2} \right)_{rw} y_c^{rw} \kappa_{zz} + \left(\frac{a_{11}}{a_{11}d_{11} - b_{11}^2} \right)_{rw} \kappa_{zz} \tag{5.44.7}$$

For the left web portion:

$$\frac{N_{xx}^{lw}}{b^{lw}} = \left(\frac{d_{11}}{a_{11}d_{11} - b_{11}^2} \right)_{lw} y_c^{lw} \kappa_{zz} - \left(\frac{b_{11}}{a_{11}d_{11} - b_{11}^2} \right)_{lw} \kappa_{zz}$$

$$\frac{M_{xx}^{lw}}{b^{lw}} = -\left(\frac{b_{11}}{a_{11}d_{11} - b_{11}^2} \right)_{lw} y_c^{lw} \kappa_{zz} + \left(\frac{a_{11}}{a_{11}d_{11} - b_{11}^2} \right)_{lw} \kappa_{zz} \tag{5.44.8}$$

Equivalent bending stiffness $(EI)_{eq}^{yy}$

To evaluate the equivalent bending stiffness $(EI)_{eq}^{yy}$, the total moment resultant of the Y-axis under the curvature (κ_{yy}) is considered and obtained by adding moment resultants on the structural (X–Z) plane of each segment as shown in Example Figure 5.8.2.

$$\bar{M}_y = \left(M_{xx}^{tf} + N_{xx}^{tf} z_c^{tf} \right) + \int_{b^{rw}} N_{xx}^{rw} z \, dz + \int_{b^{lw}} N_{xx}^{lw} z \, dz + \left(M_{xx}^{bf} + N_{xx}^{bf} z_c^{bf} \right) \tag{5.45.1}$$

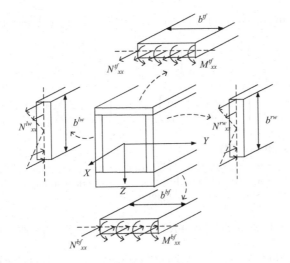

EXAMPLE FIGURE 5.8.2 Force and moment resultants due to curvature (κ_{yy}).

By substituting Eqs. (5.44.1–5.44.4) into Eq. (5.45.1):

$$\bar{M}_y = (EI)_{eq}^{yy} \kappa_{yy} = \left[\begin{array}{l} b^{tf}\left(\dfrac{a_{11} - 2b_{11}z_c^{tf} + d_{11}\left(z_c^{tf}\right)^2}{a_{11}d_{11} - b_{11}^2} \right)_{tf} + \left(\dfrac{d_{11}}{a_{11}d_{11} - b_{11}^2} \right)_{rw} \left(\dfrac{\left(b^{rw}\right)^3}{12} + b^{rw}\left(z_c^w\right)^2 \right) \\[4mm] + \left(\dfrac{d_{11}}{a_{11}d_{11} - b_{11}^2} \right)_{lw} \left(\dfrac{\left(b^{lw}\right)^3}{12} + b^{lw}\left(z_c^w\right)^2 \right) + b^{bf}\left(\dfrac{a_{11} - 2b_{11}z_c^{bf} + d_{11}\left(z_c^{bf}\right)^2}{a_{11}d_{11} - b_{11}^2} \right)_{bf} \end{array} \right] \kappa_{yy}$$

(5.45.2)

Equivalent bending stiffness $(EI)_{eq}^{yz}$

To evaluate equivalent bending stiffness $(EI)_{eq}^{yz}$, total moment resultant $\bar{M}_z$ on the structural (X–Y) plane is determined under induced curvature (κ_{yy}). Then, total moment $\bar{M}_z$ is obtained from additional resultants of the total moment resultants on the structural (X–Y) plane corresponding to each segment as shown in Example Figure 5.8.2.

$$\bar{M}_z = N_{xx}^{tf} y_c^{tf} + \int_{b^{rw}} \left(N_{xx}^{rw} y_c^{rw} + M_{xx}^{rw} \right) dz + \int_{b^{lw}} \left(N_{xx}^{lw} y_c^{lw} + M_{xx}^{lw} \right) dz + N_{xx}^{bf} y_c^{bf}$$

(5.45.3)

By substituting Eqs. (5.44.5–5.44.8) into Eq. (5.45.3):

$$\bar{M}_z = (EI)_{eq}^{yz} \kappa_{yy} = \left[\begin{array}{l} \left(\dfrac{d_{11}z^{tf} - b_{11}}{a_{11}d_{11} - b_{11}^2} \right)_{tf} b^{tf} y_c^{tf} + \left(\dfrac{d_{11}y_c^{rw} - b_{11}}{a_{11}d_{11} - b_{11}^2} \right)_{rw} b^{rw} z_c^{rw} \\[4mm] + \left(\dfrac{d_{11}y_c^{lw} - b_{11}}{a_{11}d_{11} - b_{11}^2} \right)_{lw} b^{lw} z_c^{lw} + \left(\dfrac{d_{11}z^{bf} - b_{11}}{a_{11}d_{11} - b_{11}^2} \right)_{bf} b^{bf} y_c^{bf} \end{array} \right] \kappa_{yy}$$

(4.45.4)

The equivalent bending rigidity $(EI)_{eq}^{yz}$ of the composite beam with unsymmetric lay-ups is:

$$(EI)_{eq}^{yz} = \begin{bmatrix} \left(\dfrac{d_{11}z^{tf}-b_{11}}{a_{11}d_{11}-b_{11}^2}\right)_{tf} b^{tf}y_c^{tf} + \left(\dfrac{d_{11}y_c^{rw}-b_{11}}{a_{11}d_{11}-b_{11}^2}\right)_{rw} b^{rw}z_c^{rw} \\[3mm] + \left(\dfrac{d_{11}y_c^{lw}-b_{11}}{a_{11}d_{11}-b_{11}^2}\right)_{lw} b^{lw}z_c^{lw} + \left(\dfrac{d_{11}z^{bf}-b_{11}}{a_{11}d_{11}-b_{11}^2}\right)_{bf} b^{bf}y_c^{bf} \end{bmatrix} \tag{5.45.5}$$

Equivalent bending stiffness $(EI)_{eq}^{zz}$

To evaluate equivalent bending stiffness $(EI)_{eq}^{zz}$, total moment resultant of Z-axis under induced curvature (κ_{zz}) is considered and obtained by adding moment resultants on the structural $(X–Y)$ plane of each segment as shown in Example Figure 5.8.3.

$$\bar{M}_z = \int_{b^{tf}} N_{xx}^{tf} y\,dy + \left(M_{xx}^{rw} + N_{xx}^{rw}y_c^{rw}\right) + \left(M_{xx}^{lw} + N_{xx}^{lw}y_c^{lw}\right) + \int_{b^{bf}} N_{xx}^{bf} y\,dy \tag{5.45.6}$$

Substitution of Eqs. (5.44.5–5.44.8) in Eq. (5.45.6) gives the equivalent bending rigidity $(EI)_{eq}^{zz}$ of the composite beam with unsymmetric lay-ups is:

$$\bar{M}_z = (EI)_{eq}^{zz}\kappa_{zz} = \begin{bmatrix} \left(\dfrac{d_{11}}{a_{11}d_{11}-b_{11}^2}\right)_{tf} \left(\dfrac{(b^{tf})^3}{12} + b^{tf}\left(z_c^{tf}\right)^2\right) + \left(\dfrac{a_{11}-2b_{11}y_c^{rw}+d_{11}\left(y_c^{rw}\right)^2}{a_{11}d_{11}-b_{11}^2}\right)_{rw} b^{rw} \\[3mm] + \left(\dfrac{a_{11}-2b_{11}y_c^{lw}+d_{11}\left(y_c^{lw}\right)^2}{a_{11}d_{11}-b_{11}^2}\right)_{lw} b^{lw} + \left(\dfrac{d_{11}}{a_{11}d_{11}-b_{11}^2}\right)_{bf} \left(\dfrac{(b^{bf})^3}{12} + b^{bf}\left(z_c^{bf}\right)^2\right) \end{bmatrix}\kappa_{zz} \tag{5.45.7}$$

Equivalent bending stiffness $(EI)_{eq}^{zy}$

To evaluate equivalent bending stiffness $(EI)_{eq}^{zy}$, total moment resultant $\bar{M}_y$ on the structural $(X–Z)$ plane is determined under induced curvature (κ_{zz}). Then, total moment $\bar{M}_y$ is obtained from the additional resultants of the moment resultants on the structural $(X–Z)$ plane corresponding to each segment as shown in Example Figure 5.8.3.

$$\bar{M}_y = \int_{b^{tf}} \left(N_{xx}^{tf}z_c^{tf} + M_{xx}^{tf}\right)dy + N_{xx}^{rw}z_c^{rw} + N_{xx}^{lw}z_c^{lw} + \int_{bb} \left(N_{xx}^{bf}z_c^{bf} + M_{xx}^{bf}\right)dy \tag{5.45.8}$$

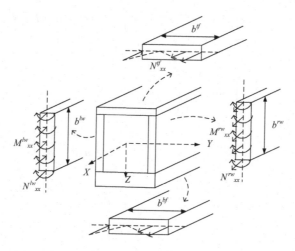

EXAMPLE FIGURE 5.8.3 Force and moment resultants due to curvature (κ_{zz}).

By substituting Eqs. (5.44.1–5.44.4) into Eq. (5.45.8):

$$\bar{M}_y = (EI)^{zy}_{eq} \kappa_{zz} = \left[\begin{array}{l} \left(\dfrac{d_{11}z^{tf}_c - b_{11}}{a_{11}d_{11} - b^2_{11}} \right)_{tf} b^{tf} y^{tf}_c \kappa_{zz} + \left(\dfrac{d_{11}y^{rw} - b_{11}}{a_{11}d_{11} - b^2_{11}} \right)_{rw} b^{rw} z^{rw}_c \\[4mm] + \left(\dfrac{d_{11}y^{lw} - b_{11}}{a_{11}d_{11} - b^2_{11}} \right)_{lw} b^{lw} z^{lw}_c + \left(\dfrac{d_{11}z^{bf}_c - b_{11}}{a_{11}d_{11} - b^2_{11}} \right)_{bf} b^{bf} y^{bf}_c \end{array} \right] \kappa_{zz} \qquad (5.45.9)$$

The equivalent bending rigidities of both $(EI)^{yz}_{eq}$ and $(EI)^{zy}_{eq}$ are equal (symmetry of section). When the top and bottom flange portions are identical in the width, thickness, and laminated lay-ups, then the bending rigidities $(EI)^{yy}_{eq}, (EI)^{yz}_{eq}$, and $(EI)^{zz}_{eq}$ of a composite beam can be simplified as:

$$(EI)^{yy}_{eq} = \left[\begin{array}{l} 2b^f \left(\dfrac{a_{11} - 2b_{11}z^f_c + d_{11}\left(z^f_c\right)^2}{a_{11}d_{11} - b^2_{11}} \right)_f + \left(\dfrac{d_{11}}{a_{11}d_{11} - b^2_{11}} \right)_{rw} \left(\dfrac{\left(b^{rw}\right)^3}{12} + b^{rw}\left(z^{rw}_c\right)^2 \right) \\[5mm] + \left(\dfrac{d_{11}}{a_{11}d_{11} - b^2_{11}} \right)_{lw} \left(\dfrac{\left(b^{lw}\right)^3}{12} + b^{lw}\left(z^{lw}_c\right)^2 \right) \end{array} \right] \qquad (5.46.1)$$

$$(EI)^{zz}_{eq} = \left[\begin{array}{l} 2\left(\dfrac{d_{11}}{a_{11}d_{11} - b^2_{11}} \right)_f \left(\dfrac{\left(b^f\right)^3}{12} + b^f\left(z^f_c\right)^2 \right) + \left(\dfrac{a_{11} - 2b_{11}y^{rw}_c + d_{11}\left(y^{rw}_c\right)^2}{a_{11}d_{11} - b^2_{11}} \right)_{rw} b^{rw} \\[5mm] + \left(\dfrac{a_{11} - 2b_{11}y^{lw}_c + d_{11}\left(y^{lw}_c\right)^2}{a_{11}d_{11} - b^2_{11}} \right)_{lw} b^{lw} \end{array} \right] \qquad (5.46.2)$$

$$(EI)^{yz}_{eq} = (EI)^{zy}_{eq} = \left[2\left(\dfrac{d_{11}z^f_c - b_{11}}{a_{11}d_{11} - b^2_{11}} \right)_f b^f y^f_c + \left(\dfrac{d_{11}y^{rw}_c - b_{11}}{a_{11}d_{11} - b^2_{11}} \right)_{rw} b^{rw} z^{rw}_c + \left(\dfrac{d_{11}y^{lw}_c - b_{11}}{a_{11}d_{11} - b^2_{11}} \right)_{lw} b^{lw} z^{lw}_c \right] \qquad (5.46.3)$$

Example 5.9

Determine the equivalent torsional stiffness of a laminated composite box beam with the ortho-tropic unsymmetric lay-ups (in Example 5.7). The rectangular box composite beam assumes that it is made of thin plates joined along their edges. The thickness of a plate is small (L/t and $s/t > 10$) compared to the cross-sectional dimensions, which are small compared to the overall member length.

Solution

As in the symmetric lay-up case, the equivalent torsional rigidity of a composite beam with orthotropic unsymmetric lay-ups is obtained using Eq. (5.25.5).

Equation (5.25.5) is integrated at the mid-plane of segment around the box cross section as:

$$\theta = \frac{T}{4\bar{A}^2} \oint \left(a_{66} - \frac{b^2_{66}}{d_{66}} \right) ds$$

$$\theta = \frac{T}{4\bar{A}^2} \left[\int_{tf} \left(a_{66} - \frac{b^2_{66}}{d_{66}} \right)_{tf} ds + \int_{rw} \left(a_{66} - \frac{b^2_{66}}{d_{66}} \right)_{rw} ds + \int_{bf} \left(a_{66} - \frac{b^2_{66}}{d_{66}} \right)_{bf} ds + \int_{lw} \left(a_{66} - \frac{b^2_{66}}{d_{66}} \right)_{lw} ds \right] \qquad (5.47.1)$$

$$T = (GC)_{eq} \theta$$

$$T = \frac{4\bar{A}^2}{\left(a_{66} - \dfrac{b_{66}^2}{d_{66}}\right)_{tf} s^{tf} + \left(a_{66} - \dfrac{b_{66}^2}{d_{66}}\right)_{rw} s^{rw} + \left(a_{66} - \dfrac{b_{66}^2}{d_{66}}\right)_{bf} s^{bf} + \left(a_{66} - \dfrac{b_{66}^2}{d_{66}}\right)_{lw} s^{lw}} \theta \qquad (5.47.2)$$

where

s = mid-line length of the segment from the center to the center of the contiguous segments

$\bar{A}$ = enclosed area that is surrounded by the mid-line of each portion of the rectangular cross section as shown in Example Figure 5.6.2.

If the right and left web portions are identical in both properties and geometries, then the equivalent torsional rigidity can be simplified as:

$$(GC)_{eq} = \frac{4\bar{A}^2}{\left(a_{66} - \dfrac{b_{66}^2}{d_{66}}\right)_{tf} s^{tf} + 2\left(a_{66} - \dfrac{b_{66}^2}{d_{66}}\right)_{w} s^{w} + \left(a_{66} - \dfrac{b_{66}^2}{d_{66}}\right)_{bf} s^{bf}} \qquad (5.47.3)$$

If the top and bottom flanges are identical in width, thickness, and laminated lay-ups, the right and left web portions are also identical in width, thickness, and laminated lay-ups. Then, the torsional rigidity can be simplified as:

$$(GC)_{eq} = \frac{2\bar{A}^2}{\left(a_{66} - \dfrac{b_{66}^2}{d_{66}}\right)_{f} s^{f} + \left(a_{66} - \dfrac{b_{66}^2}{d_{66}}\right)_{w} s^{w}} \qquad (5.47.4)$$

5.9   GENERAL GOVERNING EQUATION OF COMPOSITE BEAMS

The axial, bending, and torsional stiffnesses are established in the previous section. The stiffness and rigidity components of composite beams are developed for symmetric cross sections while laminated lay-ups are also devoted to symmetric and unsymmetric lay-ups.

However, the cases of composite beams with unsymmetric lay-ups are restricted to orthotropic materials and not to anisotropic materials. The couplings (a_{16}, a_{26}, b_{16}, b_{26}, d_{16}, and d_{26}) are neglected from the laminated compliance (or stiffness) relation of unsymmetric lay-ups because the axial and bending effects are decoupled to formulate separate axial, bending, and torsional stiffnesses.

Under the linear elastic system, the behavior of a composite beam under combined loadings (axial force, bending, and torsion) can be presented using the superposition of each loading case as shown in Figure 5.19. The general governing equation of a composite beam with orthotropic materials can be written in the matrix form as follows:

$$\begin{bmatrix} N_{xx}^b \\ \bar{M}_y \\ \bar{M}_z \\ T_{xx}^b \end{bmatrix} = \begin{bmatrix} (EA)_{eq} & 0 & 0 & 0 \\ 0 & (EI)_{eq}^{yy} & (EI)_{eq}^{yz} & 0 \\ 0 & (EI)_{eq}^{zy} & (EI)_{eq}^{zz} & 0 \\ 0 & 0 & 0 & (GC)_{eq} \end{bmatrix} \begin{bmatrix} \varepsilon_{xx}^b \\ \kappa_{yy}^b \\ \kappa_{zz}^b \\ \theta^b \end{bmatrix} \qquad (5.48.1)$$

where equivalent axial stiffness $(EA)_{eq}$, bending rigidity $(EI)_{eq}$, and torsional rigidity $(GC)_{eq}$ are determined using the same procedure as given in Sections 5.2–5.8. The inversion of Eq. (5.48.1) gives the general governing equation in terms of compliance (flexibility) as:

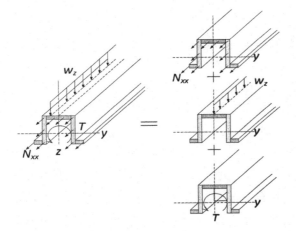

FIGURE 5.19 Composite beams under loading combinations.

$$
\begin{bmatrix}
\varepsilon_{xx}^{b} \\[4pt]
\kappa_{yy}^{b} \\[4pt]
\kappa_{zz}^{b} \\[4pt]
\theta^{b}
\end{bmatrix}
=
\begin{bmatrix}
\dfrac{1}{(EA)_{eq}} & 0 & 0 & 0 \\[12pt]
0 & \dfrac{(EI)_{eq}^{zz}}{(EI)_{eq}^{yy}(EI)_{eq}^{zz}-\left((EI)_{eq}^{yz}\right)^{2}} & \dfrac{(EI)_{eq}^{yz}}{(EI)_{eq}^{yy}(EI)_{eq}^{zz}-\left((EI)_{eq}^{yz}\right)^{2}} & 0 \\[14pt]
0 & \dfrac{(EI)_{eq}^{yz}}{(EI)_{eq}^{yy}(EI)_{eq}^{zz}-\left((EI)_{eq}^{yz}\right)^{2}} & \dfrac{(EI)_{eq}^{yy}}{(EI)_{eq}^{yy}(EI)_{eq}^{zz}-\left((EI)_{eq}^{yz}\right)^{2}} & 0 \\[14pt]
0 & 0 & 0 & (GC)_{eq}
\end{bmatrix}
\begin{bmatrix}
N_{xx}^{b} \\[4pt]
\bar{M}_{y} \\[4pt]
\bar{M}_{z} \\[4pt]
T_{xx}^{b}
\end{bmatrix}
\quad (5.48.2)
$$

Example 5.10

Determine: (a) axial deformation, (b) maximum deflection, and (c) twist of a composite rectangular box beam with symmetric lay-ups under uniformly distributed transverse load, axial load distributed uniformly across the cross section, and uniform torque (eccentricity = e) as shown in Example Figure 5.10.1 (neglect shear deformation).

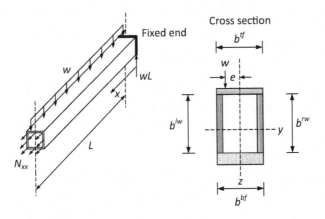

EXAMPLE FIGURE 5.10.1 Cantilever composite beam under applied loads.

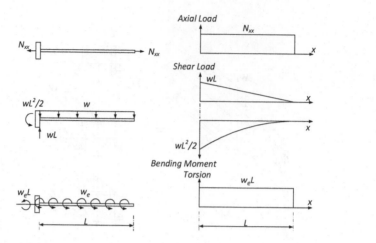

EXAMPLE FIGURE 5.10.2 Axial, shear, bending moment, and torsional moment diagram.

Solution

The combination of applied load on the cantilever beam can be separated for each load type as (Example Figure 5.10.2):

Maximum normal force $\left(N_{xx}^b\right) = N_{xx}$
Maximum shear force $(V^b) = wL$
Maximum bending moment $\left(M_{xx}^b\right) = wL^2/2$
Maximum torque $(T^b) = w_eL$.

From symmetric cross section and laminate lay-ups, the induced curvature of beam occurs about Y axis, Eq. (5.48.1) can be simplified as:

$$
\begin{bmatrix} N_{xx}^b \\ \overline{M}_y \\ T_{xx}^b \end{bmatrix} = \begin{bmatrix} (EA)_{eq} & 0 & 0 \\ 0 & (EI)_{eq}^{yy} & 0 \\ 0 & 0 & (GC)_{eq} \end{bmatrix} \begin{bmatrix} \varepsilon_{xx}^b \\ \kappa_{yy}^b \\ \theta^b \end{bmatrix} \tag{5.48.3}
$$

The equivalent axial, bending, and torsional rigidities (Examples 5.4–5.6) are substituted into the above governing equation as: by substituting Eqs. (5.34.5, 5.37.8, and 5.38.6) into Eq. (5.48.3):

$$
\begin{bmatrix} N_{xx}^b \\ \overline{M}_y \\ T_{xx}^b \end{bmatrix} = \begin{bmatrix} \left(\dfrac{b^{tf}}{a_{11}^{tf}} + \dfrac{2b^w}{a_{11}^w} + \dfrac{b^{bf}}{a_{11}^{bf}} \right) & 0 & 0 \\[3ex] 0 & \left(\dfrac{b^{tf}}{d_{11}^{tf}} + \dfrac{b^{tf}}{a_{11}^{tf}} z_c^{tf2} \right) + \dfrac{2}{a_{11}^w}\left(\dfrac{(b^w)^3}{12} + b^w\left(z_c^w\right)^2 \right) + \left(\dfrac{b^{bf}}{d_{11}^{bf}} + \dfrac{b^{bf}}{a_{11}^{bf}} z_c^{bf2} \right) & 0 \\[3ex] 0 & 0 & \dfrac{4\overline{A}^2}{a_{66}^{tf} s^{tf} + 2a_{66}^w s^w + a_{66}^{bf} s^{bf}} \end{bmatrix}
$$

$$
\times \begin{bmatrix} \varepsilon_{xx}^b \\ \kappa_{yy}^b \\ \theta^b \end{bmatrix} \tag{5.48.4}
$$

where the centroid of beam cross section can be found from Eq. (5.34.7) as:

$$z_c^b = \frac{z^{tf}\dfrac{b^{tf}}{a_{11}^{tf}} + 2z^w\dfrac{b^w}{a_{11}^w} + z^{bf}\dfrac{b^{bf}}{a_{11}^{bf}}}{\left(\dfrac{b^{tf}}{a_{11}^{tf}} + \dfrac{2b^w}{a_{11}^w} + \dfrac{b^{bf}}{a_{11}^{bf}}\right)}$$

From Eq. (5.48.4), the axial deformation of the composite beam is:

$$\varepsilon_{xx}^b = \left(\frac{b^{tf}}{a_{11}^{tf}} + \frac{2b^w}{a_{11}^w} + \frac{b^{bf}}{a_{11}^{bf}}\right)^{-1} N_{xx}$$

By substituting maximum bending moment into Eq. (5.48.4), the curvature of the composite beam is:

$$\kappa_{yy}^b = \left(\left(\frac{b^{bf}}{d_{11}^{tf}} + \frac{b^{tf}}{a_{11}^{tf}}\left(z_c^{tf}\right)^2\right) + \frac{2}{a_{11}^w}\left(\frac{\left(b^w\right)^3}{12} + b^w\left(z_c^w\right)^2\right) + \left(\frac{b^{bf}}{d_{11}^{bf}} + \frac{b^{bf}}{a_{11}^{bf}}\left(z_c^{bf}\right)^2\right)\right)^{-1}\left[\bar{M}_y = \frac{wL^2}{2}\right]$$

By substituting maximum torsion into Eq. (5.48.4), the angle of twist of a composite beam is:

$$\theta^b = \left(\frac{4\bar{A}^2}{a_{66}^{tf}s^{tf} + 2a_{66}^w s^w + a_{66}^{bf}s^{bf}}\right)^{-1}\left[T = w_e L\right]$$

where $\bar{A}$ is the enclosed area that is surrounded by the mid-line of the segment about the rectangular cross section as shown in Example Figure 5.6.2.

Maximum deflection of a composite beam can be determined from Eq. (5.9.5) or results in Table 5.1. From Eq. (5.9.5), the equivalent bending rigidity of a solid beam section is replaced by that of rectangular box beam as:

$$\frac{d^2 w^0}{dx^2} = -\frac{M_{xx}^b}{(EI)_{eq}} = \frac{1}{(EI)_{eq}}\left(\frac{wL^2}{2} + \frac{wx^2}{2} - wLx\right)$$

$$\frac{dw^0}{dx} = \frac{1}{(EI)_{eq}}\left(\frac{wL^2 x}{2} + \frac{wx^3}{6} - \frac{wLx^2}{2}\right) + C_1$$

$$w^0(x) = \frac{1}{(EI)_{eq}}\left(\frac{wL^2 x^2}{4} + \frac{wx^4}{24} - \frac{wLx^3}{6}\right) + C_1 x + C_2$$

The boundary conditions are applied as:

@ $x = 0$ beam deflection $w = 0$ then, $C_2 = 0$
@ $x = 0$ beam slope $dw/dx = 0$ then, $C_1 = 0$
@ $x = L$ maximum deflection.

5.10 SUMMARY

This chapter provides a detailed analysis of laminated composite beams. The governing bending equations for composite beams were presented considering the effect of shear deformation. Furthermore, analysis of the behavior of laminated beams subjected to loads causing bending, axial and torsional effects, and their load combination(s) is separately discussed. Examples were provided to help better understand the laminated composite beam behavior. Similar analysis techniques for laminated plates are presented in the next chapter.

EXERCISES

Problem 5.1: Determine: axial stiffness of the symmetrical laminated composite beam. **Hint:** The axial deformation ε_{xx}^n of the beam axis at the neutral plane is applied. The curvature of the beam is not induced due to the symmetric nature of both lay-ups and cross section.

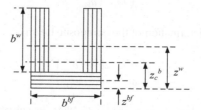

Problem 5.2: Determine: bending rigidity of laminated composite U beam in Problem 5.1. **Hint:** A composite beam is bent only on the structural plane (x–z) then, the equivalent bending rigidity about the structural axis y is determined. The curvature κ^b is applied at the beam axis then, both normal strain ε_{xx} and curvature κ_{yy} of each segment of a cross section are induced due to applied curvature.

Problem 5.3: A composite box beam with segment laminates $[0°_2/45°_2/0°_2]$ and each lamina thickness of 2.5 mm. The beam is 1.5 m long and is under a uniformly distributed load (2.5 kN/m) with a simply supported condition at both ends. The effect of axial restraint is neglected. The properties of unidirectional FRP lamina are presented as: elastic modulus: $E_{11} = 180$ GPa and $E_{22} = 40$ GPa, shear modulus: $G_{12} = 20$ GPa, $\nu_{12} = 0.25$. Determine: (a) bending rigidity, and (b) maximum deflection.

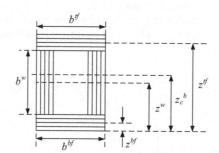

Problem 5.4: A composite box beam with segment laminates $[0°_2/45°_2/0°_2]$ and each lamina thickness of 2.5 mm in Problem 5.3. The beam is 1.5 m long and is under a uniformly distributed load (2.5 kN/m) and an axial load (15 kN) with a simply supported condition at both ends. Determine: (a) bending rigidity, (b) axial rigidity, and (c) maximum deflection.

Problem 5.5: A composite box beam with segment laminates $[0°_2/45°_2/0°_2]$ and each lamina thickness of 2.5 mm in Problem 5.3. The beam is 1.5 m long and is under a uniform torque (1.5 kN.m) with built-in at each end. Determine: (a) torsional rigidity and (b) maximum twist.

REFERENCES AND SELECTED BIOGRAPHY

Altenbach, H., Altenbach, J., and Kissing, W., *Mechanics of Composite Structural Elements*, Springer-Verlag, Berlin, Germany, 2004.

Barbero, E.J., *Introduction to Composite Materials Design*, Taylor & Francis, Inc., Philadelphia, PA, 1998.

Bauld, N.R. Jr., and Tzeng, L.S., A Vlasov theory for fiber-reinforced beams with thin-walled open cross sections, *International Journal of Solids Structures*, 20(3), 1984, pp. 277–297.

Boresi, A.P., and Schmidt, R.J., *Advanced Mechanics of Materials*, 6th edition, John Wiley & Sons, New York, NY, 2003.

GangaRao, H.V., and Siva, H., Light Weight FRP Composite Modular Panel, US Patent No. 6,591,156 B2, issued July 15, 2003.

Gjelsvik, A., *The Theory of Thin Walled Bars*, John Wiley & Sons, Inc., New York, NY, 1981.

Kirchhoff, G., Uber das Gleichgwich und die Bewegung einer Elastichen Scheibe, *Journal of angewandte Mathematik*, 40, 1850, pp. 51–88.

Kollar, L.P., and Pluzsik, A., Analysis of thin walled composite beams with arbitrary layup, *Journal of Reinforced Plastics and Composites*, 21(16), 2002, pp. 1423–1465.

Kollar, L.P., and Springer, G.S., *Mechanics of Composite Structures*, Cambridge University Press, New York, NY, 2003.

Lopez-Anido, R., GangaRao, H.V., and Barbero, E., Modular FRP Composite Deck System, US Patent No. 6,455,131, issued Sept. 24, 2002.

Massa, J.C., and Barbero, E.J., A strength of materials formulation for thin walled composite beams with torsion, *Journal of Composite Materials*, 32(17), 1998, pp. 1560–1594.

Megson, T.H.G., *Aircraft Structures for Engineering Students*, 3rd edition, Elsevier, (Singapore) Pte Ltd, Singapore, 1999.

Murray, N.W., *Introduction to the Theory of Thin-Walled Structures*, Oxford University Press, New York, NY, 1984.

Pilkey, W.D., *Analysis and Design of Elastic Beams, Computational Methods*, John Wiley & Sons, Inc., New York, NY, 2002.

Reddy, J.N., *Mechanics of Laminated Composite Plates and Shells Theory and Analysis*, 2nd edition, John Wiley, New York, NY, 2003.

Skudra, A.M., Bulavs, F.Ya., Gurvich, M.R., and Kruklinsh, A.A., *Structural Analysis of Composite Beam Systems*, Technomic Publishing Company, Inc., Lancaster, PA, 1991.

Timoshenko, S., and Goodier, J., *Theory of Elasticity*, 3rd edition, McGraw Hill, New York, NY, 1970.

Vasiliev, V.V., and Morozov, E.V., *Mechanics and Analysis of Composite Materials*, Elsevier Science Ltd., Oxford, UK, 2001.

Whitney, J.M., *Structural Analysis of Laminated Anisotropic Plates*, Technomic Publishing Company, Inc., Lancaster, PA, 1987.

6 Analysis of FRP Composite Plates

FRP composite plates have two of the three dimensions (length and width) significantly larger than the third dimension (thickness), while the beams have only one of the three dimensions, i.e., length, much larger than width and thickness. A thin plate is defined as the one having a width-to-thickness ratio higher than 10. The length-to-width ratio of a composite plate depends upon the plate stiffness, but a typical plate length-to-thickness ratio should be under 40, preferably arond 20 to 25, to avoid geometric nonlinear responses.

FRP composite plates are used in a wide range of engineering applications such as sheathing in building walls, bridge decks, floor systems, and other structural components for aerospace, military, and automotive industries. An understanding of shear influence in relation to the bending influence of composite plates' structural behavior is essential, but it is beyond the scope of this textbook. Therefore, readers are encouraged to review the shear deformation theory of laminated anisotropic plates from the literature (Whitney, 1969).

6.1 INTRODUCTION

The governing differential equations of a thin FRP laminate composite plate (Figure 6.1) under bending are developed using Kirchhoff's hypotheses and a few assumptions, which are as follows: (1) the transverse section of a thin plate remains in the same plane before and after deformation; (2) no changes in plate dimensions (inextensible); (3) the smaller of the two spans of a plate to its thickness is around 20 (no less than 10) and no more than 40; and (4) FRP plate thickness to deflection is greater than 10 for small defection theory to hold good. The above assumptions are similar to those assumed while developing the classical plate theory (CPT) (Timoshenko, 1959; Whitney, 1987). The reader should be alerted that assumption (3) is the one employed in CPT; however, the real limits of span-to-thickness ratio are less revealing in terms of orthotropic plate behavior than the one corresponding to span to bending stiffness of a plate in different directions. Such in-depth discussion is beyond the scope of this textbook.

The hypothesis dealing with no dimensional changes implies that normal strain (ε_z) and vertical shear strains (γ_{xz}) and (γ_{yz}) are negligible and the transverse deflection $w(x, y)$ is independent of the thickness direction z (thin plate). The classical lamination theory (CLT) inaccurately predicts the results of moderate and thick plates because shear deformation influence has been neglected in the development of the CLT-based governing differential equation for thin plates.

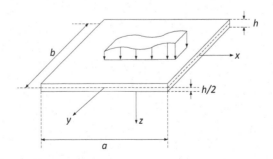

FIGURE 6.1 Orthogonal coordinate system (x–y–z) of plate.

DOI: 10.1201/9781003196754-6

The absence of shear deformation influence leads to under-prediction of deflections and over-prediction on buckling loads in relation to the results from the CPT. The CPT including shear deformation is necessary for moderate and thick plate problems and not so important for thin plate problems from the viewpoint of the accuracy of results.

Two approaches, including both the theory of elasticity and energy methods, are adopted herein to derive the governing differential equations. Initially, the theory of the elasticity approach is presented. Stresses must satisfy the equilibrium conditions at every point. Similarly, displacement compatibility and force-displacement (stress–strain) relations along with the boundary conditions must be satisfied. The energy approach, a sequel to the elasticity approach, is considered using classical variational principles, wherein the internal strain energy (U) and external potential energy (W) are used to formulate the governing equations under bending. The abovementioned approaches result in identical governing equations and boundary conditions.

6.2   THEORY OF ELASTICITY APPROACH

The equilibrium equations of an elastic body without body forces (Figure 6.2) are considered, by summing forces in three different directions.

$$\frac{\partial \sigma_{xx}}{\partial x} + \frac{\partial \tau_{xy}}{\partial y} + \frac{\partial \tau_{xz}}{\partial z} = 0 \tag{6.1.1}$$

$$\frac{\partial \tau_{xy}}{\partial x} + \frac{\partial \sigma_{yy}}{\partial y} + \frac{\partial \tau_{yz}}{\partial z} = 0 \tag{6.1.2}$$

$$\frac{\partial \tau_{xz}}{\partial x} + \frac{\partial \tau_{yz}}{\partial y} + \frac{\partial \sigma_{zz}}{\partial z} = 0 \tag{6.1.3}$$

The in-plane and out-of-plane stress and moment resultants as shown in Figure 6.3 are defined as:

$$\begin{bmatrix} N_{xx} \\ N_{yy} \\ N_{xy} \end{bmatrix} = \int_{-\frac{h}{2}}^{\frac{h}{2}} \begin{bmatrix} \sigma_{xx}^{k} \\ \sigma_{yy}^{k} \\ \tau_{xy}^{k} \end{bmatrix} dz \tag{6.2.1}$$

$$\begin{bmatrix} M_{xx} \\ M_{yy} \\ M_{xy} \end{bmatrix} = \int_{-\frac{h}{2}}^{\frac{h}{2}} \begin{bmatrix} \sigma_{xx}^{k} \\ \sigma_{yy}^{k} \\ \tau_{xy}^{k} \end{bmatrix} z \, dz \tag{6.2.2}$$

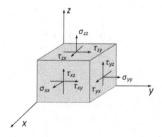

FIGURE 6.2   An elastic body under the state of stresses without body forces.

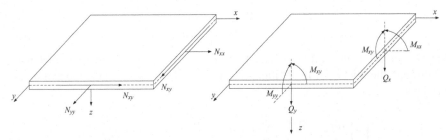

a) In - plane force resultants   b) Moment and transverse shear resultants

FIGURE 6.3 In-plane and out-of-plane stress and moment resultants: (a) in-plane force resultants and (b) moment and transverse shear resultants.

$$\begin{bmatrix} Q_x \\ Q_y \end{bmatrix} = \int_{-\frac{h}{2}}^{\frac{h}{2}} \begin{bmatrix} \tau_{xz}^k \\ \tau_{yz}^k \end{bmatrix} dz \tag{6.2.3}$$

Integrating the equilibrium Eqs. (6.1.1–6.1.3) with respect to z through the thickness of a constant cross section ($-h/2$ to $+h/2$), the force resultants are:

$$\int_{-\frac{h}{2}}^{\frac{h}{2}} \left(\frac{\partial \sigma_{xx}^k}{\partial x} + \frac{\partial \tau_{xy}^k}{\partial y} + \frac{\partial \tau_{xz}^k}{\partial z} \right) dz = \frac{\partial N_{xx}}{\partial x} + \frac{\partial N_{xy}}{\partial y} + \frac{\partial Q_x}{\partial z} = 0 \tag{6.3.1}$$

$$\int_{-\frac{h}{2}}^{\frac{h}{2}} \left(\frac{\partial \tau_{xy}^k}{\partial x} + \frac{\partial \sigma_{yy}^k}{\partial y} + \frac{\partial \tau_{yz}^k}{\partial z} \right) dz = \frac{\partial N_{xy}}{\partial x} + \frac{\partial N_{yy}}{\partial y} + \frac{\partial Q_y}{\partial z} = 0 \tag{6.3.2}$$

$$\int_{-\frac{h}{2}}^{\frac{h}{2}} \left(\frac{\partial \tau_{xz}^k}{\partial x} + \frac{\partial \tau_{yz}^k}{\partial y} + \frac{\partial \sigma_{zz}^k}{\partial z} \right) dz = \frac{\partial Q_x}{\partial x} + \frac{\partial Q_y}{\partial y} + p_z = 0 \tag{6.3.3}$$

where $\int_{-\frac{h}{2}}^{-\frac{h}{2}} \left(\frac{\partial \sigma_z^k}{\partial z} \right) dz = p_z$, N_{ij} and M_{ij}=in-plane force resultants and moment resultants in Eqs. (6.2.1 and 6.2.2), Q_i=transverse shear resultants in Eq. (6.2.3). For additional clarification on sign convention, readers need to refer to Section 4.3.

From Eqs. (6.3.1–6.3.3), $\partial Q_x/\partial z$ and $\partial Q_y/\partial z$ are the in-plane shear resultants on a plate surface corresponding to the out-of-plane shear through the plate thickness. $\partial Q_x/\partial z$ and $\partial Q_y/\partial z$ are the differences of out-of-plane shear (τ_{xz} and τ_{yz}) between the top and bottom surfaces, respectively. In general, in thin-plate bending problems, transverse shear stresses (τ_{xz}) and (τ_{yz}) are zero on the surface ($h/2$ and $-h/2$). Thus, $\partial Q_x/\partial z$ and $\partial Q_y/\partial z$ are neglected. Hence, Eqs. (6.3.1–6.3.3) can be simplified as:

$$\frac{\partial N_{xx}}{\partial x} + \frac{\partial N_{xy}}{\partial y} = 0 \tag{6.4.1}$$

$$\frac{\partial N_{xy}}{\partial x} + \frac{\partial N_{yy}}{\partial y} = 0 \tag{6.4.2}$$

$$\frac{\partial Q_x}{\partial x} + \frac{\partial Q_y}{\partial y} + p_z = 0 \qquad (6.4.3)$$

Considering the moment equilibrium equations, Eqs. (6.1.1 and 6.1.2) are multiplied with z and integrated through the thickness. In addition, (τ_{xz}) and (τ_{yz}), being zero on the top and bottom surfaces, are neglected.

$$\int_{-\frac{h}{2}}^{\frac{h}{2}} \left(\frac{\partial \sigma_{xx}^k}{\partial x} + \frac{\partial \tau_{xy}^k}{\partial y} + \frac{\partial \tau_{xz}^k}{\partial z} \right) z\,dz = \frac{\partial M_{xx}}{\partial x} + \frac{\partial M_{xy}}{\partial y} - Q_x + \frac{h}{2} p_x = 0 \qquad (6.5.1)$$

$$Q_x = \frac{\partial M_{xx}}{\partial x} + \frac{\partial M_{xy}}{\partial y} \qquad (6.5.2)$$

$$\int_{-\frac{h}{2}}^{\frac{h}{2}} \left(\frac{\partial \tau_{xy}^k}{\partial x} + \frac{\partial \sigma_{yy}^k}{\partial y} + \frac{\partial \tau_{yz}^k}{\partial z} \right) z\,dz = \frac{\partial M_{xy}}{\partial x} + \frac{\partial M_{yy}}{\partial y} - Q_y + \frac{h}{2} p_y = 0 \qquad (6.5.3)$$

$$Q_y = \frac{\partial M_{xy}}{\partial x} + \frac{\partial M_y}{\partial y} \qquad (6.5.4)$$

(Q_x) and (Q_y) of Eqs. (6.5.2 and 6.5.4) substituted into Eq. (6.4.3) yield the differential governing equations of a thin FRP laminate composite plate as:

$$\frac{\partial^2 M_{xx}}{\partial x^2} + 2\frac{\partial^2 M_{xy}}{\partial x \partial y} + \frac{\partial^2 M_{yy}}{\partial y^2} + p_z = 0 \qquad (6.6.1)$$

$$\frac{\partial N_{xx}}{\partial x} + \frac{\partial N_{xy}}{\partial y} = 0 \qquad (6.6.2)$$

$$\frac{\partial N_{xy}}{\partial x} + \frac{\partial N_{yy}}{\partial y} = 0 \qquad (6.6.3)$$

The above governing equations are identical to the governing equation of a thin homogenous plate and do not include any terms corresponding to the material properties. Hence, these governing equations are valid for thin rectangular plates with material properties being arbitrary in different directions in the plane of a plate.

6.3 ENERGY METHOD

The total potential energy under static condition is given as $\Pi = U + W$. By applying the first order of the variational principle with the total potential energy as:

$$\delta \Pi = \delta U + \delta W = 0 \qquad (6.7)$$

The differential governing equations and boundary conditions can be obtained by substituting internal strain energy (U) of the elastic body and potential energy under external transverse loads on the surface of a plate into Eq. (6.7). In addition, the previous assumption of constitutive relations assumes that transverse normal (ε_z) and out-of-plane shear $(\gamma_{xz}$ and $\gamma_{yz})$ strains are neglected. Therefore,

$$\delta(U+W)=\int\left(\left[\begin{array}{c} \sigma_{xx}^{k} \\ \sigma_{yy}^{k} \\ \tau_{xy}^{k} \end{array}\right]^{T}\delta\left[\begin{array}{c} \varepsilon_{xx} \\ \varepsilon_{yy} \\ \gamma_{xy} \end{array}\right]\right)dV-\int p_{z}\delta w^{o}dA=0 \tag{6.8}$$

The in-plane shear stresses in Eq. (4.7.2) (by neglecting hygrothermal effect) are substituted into the variational Eq. (6.8).

$$\delta(U+W)=\int\left(\left[\begin{array}{ccc} \bar{Q}_{11} & \bar{Q}_{12} & \bar{Q}_{16} \\ \bar{Q}_{12} & \bar{Q}_{22} & \bar{Q}_{26} \\ \bar{Q}_{16} & \bar{Q}_{26} & \bar{Q}_{66} \end{array}\right]\left[\begin{array}{c} \varepsilon_{xx}^{o}+z\kappa_{xx}^{o} \\ \varepsilon_{yy}^{o}+z\kappa_{yy}^{o} \\ \gamma_{xy}^{o}+z\kappa_{xy}^{o} \end{array}\right]\right)^{T}\delta\left[\begin{array}{c} \varepsilon_{xx}^{o}+z\kappa_{xx}^{o} \\ \varepsilon_{yy}^{o}+z\kappa_{yy}^{o} \\ \gamma_{xy}^{o}+z\kappa_{xy}^{o} \end{array}\right]dV$$

$$-\int p_{z}\delta w^{o}dA=0 \tag{6.9}$$

By integrating the first term of the right side of Eq. (6.9) and substituting force-moment resultants in Eqs. (4.13.1 and 4.16.1) into Eq. (6.9):

$$\delta(U+W)=\int_{A}\left[\begin{array}{c} N_{x} \\ N_{y} \\ N_{xy} \\ M_{x} \\ M_{y} \\ M_{xy} \end{array}\right]^{T}\delta\left[\begin{array}{c} \varepsilon_{xx}^{o} \\ \varepsilon_{yy}^{o} \\ \varepsilon_{xy}^{o} \\ \kappa_{xx} \\ \kappa_{yy} \\ \kappa_{xy} \end{array}\right]dA-\int p_{z}\delta wdA \tag{6.10.1}$$

By substituting the strain-displacement relation in Eq. (4.3), the variational equations and force-moment resultants are simplified as:

$$\delta(U+W)=\int_{A}\left[\begin{array}{c} N_{xx} \\ N_{yy} \\ N_{xy} \\ M_{xx} \\ M_{yy} \\ M_{xy} \end{array}\right]^{T}\delta\left[\begin{array}{c} \dfrac{\partial u^{o}}{\partial x} \\[2mm] \dfrac{\partial v^{o}}{\partial y} \\[2mm] \dfrac{\partial u^{o}}{\partial y}+\dfrac{\partial v^{o}}{\partial x} \\[2mm] -\dfrac{\partial^{2}w^{o}}{\partial x^{2}} \\[2mm] -\dfrac{\partial^{2}w^{o}}{\partial y^{2}} \\[2mm] -2\dfrac{\partial^{2}w^{o}}{\partial x\partial y} \end{array}\right]dA-\int p_{z}\delta w^{o}dA=0 \tag{6.10.2}$$

Equation (6.10.2) can be rewritten in matrix form as:

$$\delta(U+W)=\int_A \begin{bmatrix} \dfrac{\partial N_{xx}}{\partial x}+\dfrac{\partial N_{xy}}{\partial y} \\[2mm] \dfrac{\partial N_{yy}}{\partial y}+\dfrac{\partial N_{xy}}{\partial x} \\[2mm] \dfrac{\partial^2 M_{xx}}{\partial x^2}+2\dfrac{\partial^2 M_{xy}}{\partial x\partial y}+\dfrac{\partial^2 M_{yy}}{\partial y^2}+p_z \end{bmatrix}^T \begin{bmatrix} \delta u^o \\ \delta v^o \\ \delta w^o \end{bmatrix} dA=0 \qquad (6.10.3)$$

The differential governing equations using the energy method are identical to Eqs. (6.6.1–6.6.3).

6.4 GOVERNING EQUATIONS IN TERMS OF DISPLACEMENTS

It is convenient to present the governing equations in terms of displacements and can be related to the physical response of composite structural behavior. To accomplish this objective, the stress resultant in Eq. (4.19.1) and strain-displacement in Eq. (4.3) relations are substituted into the governing differential Eqs. (6.6.1–6.6.3). Thus, the governing differential equations of a thin FRP laminate plate can be written in terms of displacements (u^o, v^o, and w^o) at the reference surface (Figures 4.2 and 4.3) and the global stiffness (ABD) matrix as:

From Eqs. (4.19.1 and 4.3):

$$\begin{bmatrix} N_x \\ N_y \\ N_{xy} \\ M_x \\ M_y \\ M_{xy} \end{bmatrix} = \begin{bmatrix} A_{11} & A_{12} & A_{16} & B_{11} & B_{12} & B_{16} \\ A_{12} & A_{22} & A_{26} & B_{12} & B_{22} & B_{26} \\ A_{16} & A_{26} & A_{66} & B_{16} & B_{26} & B_{66} \\ B_{11} & B_{12} & B_{16} & D_{11} & D_{12} & D_{16} \\ B_{12} & B_{22} & B_{26} & D_{12} & D_{22} & D_{26} \\ B_{16} & B_{26} & B_{66} & D_{16} & D_{26} & D_{66} \end{bmatrix} \begin{bmatrix} \dfrac{\partial u^o}{\partial x} \\[2mm] \dfrac{\partial v^o}{\partial y} \\[2mm] \dfrac{\partial u^o}{\partial y}+\dfrac{\partial v^o}{\partial x} \\[2mm] -\dfrac{\partial^2 w^o}{\partial x^2} \\[2mm] -\dfrac{\partial^2 w^o}{\partial y^2} \\[2mm] -2\dfrac{\partial^2 w^o}{\partial x\partial y} \end{bmatrix} \qquad (6.11)$$

Substituting Eq. (6.11) into Eq. (6.6) and differentiating with respect to x or y results in:

$$A_{11}\frac{\partial^2 u^o}{\partial x^2}+2A_{16}\frac{\partial^2 u^o}{\partial x\partial y}+A_{66}\frac{\partial^2 u^o}{\partial y^2}+A_{16}\frac{\partial^2 v^o}{\partial x^2}+(A_{12}+A_{66})\frac{\partial^2 v^o}{\partial x\partial y}+A_{26}\frac{\partial^2 v^o}{\partial y^2}$$

$$-B_{11}\frac{\partial^3 w^o}{\partial x^3}-3B_{16}\frac{\partial^3 w^o}{\partial x^2\partial y}-(B_{12}+2B_{66})\frac{\partial^3 w^o}{\partial x\partial y^2}-B_{26}\frac{\partial^3 w^o}{\partial y^3}=0 \qquad (6.12.1)$$

$$A_{16}\frac{\partial^2 u^o}{\partial x^2}+(A_{12}+A_{16})\frac{\partial^2 u^o}{\partial x\partial y}+A_{26}\frac{\partial^2 u^o}{\partial y^2}+A_{66}\frac{\partial^2 v^o}{\partial x^2}+2A_{26}\frac{\partial^2 v^o}{\partial x\partial y}+A_{22}\frac{\partial^2 v^o}{\partial y^2}$$

$$-B_{16}\frac{\partial^3 w^o}{\partial x^3}-(B_{12}+2B_{66})\frac{\partial^3 w^o}{\partial x^2\partial y}-3B_{26}\frac{\partial^3 w^o}{\partial x\partial y^2}-B_{22}\frac{\partial^3 w^o}{\partial y^3}=0 \qquad (6.12.2)$$

$$D_{11}\frac{\partial^4 w^o}{\partial x^4} + 4D_{16}\frac{\partial^4 w^o}{\partial x^3 \partial y} + 2(D_{12} + 2D_{66})\frac{\partial^4 w^o}{\partial x^2 \partial y^2} + 4D_{26}\frac{\partial^4 w^o}{\partial x \partial y^3} + D_{22}\frac{\partial^4 w^o}{\partial y^4}$$

$$-B_{11}\frac{\partial^3 u^o}{\partial x^3} - 3B_{16}\frac{\partial^3 u^o}{\partial x^2 \partial y} - (B_{12} + 2B_{66})\frac{\partial^3 u^o}{\partial x \partial y^2} - B_{26}\frac{\partial^3 u^o}{\partial y^3}$$

$$-B_{16}\frac{\partial^3 v^o}{\partial x^3} - (B_{12} + 2B_{66})\frac{\partial^3 v^o}{\partial x^2 \partial y} - 3B_{26}\frac{\partial^3 v^o}{\partial x \partial y^2} - B_{22}\frac{\partial^3 v^o}{\partial y^3} = p_z \qquad (6.12.3)$$

6.5   BOUNDARY CONDITIONS

To obtain unique solutions from the differential Eq. (6.6 or 6.12), appropriate boundary conditions are invoked, which correspond to a physical system. For rectangular plate problems, four pairs of boundary conditions are available. However, only one of the four pairs of boundaries for each edge is satisfied through natural conditions (displacements and rotations), while the others are satisfied by essential boundary conditions (forces and moments). The four pairs of boundary conditions for each edge are presented below:

i. For edges parallel to y-direction (perpendicular to x direction):
    (i.1.1) u^o=specific value (i.1.2) N_{xx}=total applied in-plane normal x force on the edge
    (i.2.1) v^0=specific value (i.2.2) N_{xy}=total applied in-plane shear force on the edge

(i.3.1) w^o=specific value (i.3.2) $\dfrac{\partial M_{xx}}{\partial x} + 2\dfrac{\partial M_{xy}}{\partial y} = Q_x + \dfrac{\partial M_{xy}}{\partial y}$ (Kirchhoff-free-edge)

(i.4.1) $\dfrac{\partial w^o}{\partial x}$=specific value (i.4.2) M_x=total applied moment x on the edge

ii. For edges parallel to x direction (perpendicular to y-direction):
    (ii.1.1) v^0= specific value (ii.1.2) N_{yy}=total applied in-plane normal y force on the edge
    (ii.2.1) u^o=specific value (ii.2.2) N_{xy}=total applied in-plane shear force on the edge

(ii.3.1) w^o= specific value (ii.3.2) $\dfrac{\partial M_{yy}}{\partial y} + 2\dfrac{\partial M_{xy}}{\partial x} = Q_y + \dfrac{\partial M_{yx}}{\partial x}$ (Kirchhoff-free-edge)

(ii.4.1) $\dfrac{\partial w^o}{\partial y}$=specific value (ii.4.2) M_{yy}=total applied moment y on the edge

In the next section, one- and two-dimensional plate problems are discussed. Initially, one-dimensional plate behavior, cylindrical bending, is presented. Deflection solutions with various boundary conditions are determined and used to evaluate force and moment resultants later. Two-dimensional plate problems with various types of fabric lay-ups, loading, and boundary conditions are considered. Fourier series of two-dimensional plate solutions are developed, which have increasing complexity when compared to solving problems of one independent variable. This complexity can be further compounded because of increasing generality in lay-up scenarios. However, the emphasis herein is on fast converging series solutions for deflection functions by accounting for only the first term of the infinite series solution. The primary idea is to arrive at one-term solutions for computational ease without losing accuracy.

6.6   LONG LAMINATED FRP PLATES (CYLINDRICAL BENDING)

Long rectangular laminated composite plates resulting in a cylindrical deformation (Figure 6.4) under transverse loading are designated when the ratio of longer to shorter edge exceeds 8. Derivations of the differential governing equations are independent of plate properties along the longer sides, thus resulting in composite plate displacements being only functions of plate properties

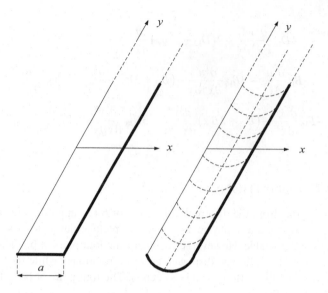

FIGURE 6.4 Long plate geometry and deflection shape.

on the shorter side. Plates with longer edges and uniformly supported through the length ($x=0$ and a) in the y-direction are considered in this section.

Herein, general solutions of a long anisotropic laminated thin plate are presented. When a general anisotropic plate solution is established, it can be reduced to various solutions of different types of laminated plates by considering appropriate bending stiffnesses and boundary conditions. Additional discussion is given at the end of this section. Terms y and v^o in the differential governing Eqs. (6.12.1–6.12.3) are equated to zero. Thus, all partial derivatives with respect to x are changed to ordinary derivatives with respect to x (length direction) by neglecting response variations in the y-direction.

$$A_{11}\frac{d^2u^o}{dx^2} + A_{16}\frac{d^2v^o}{dx^2} - B_{11}\frac{d^3w^o}{dx^3} = 0 \tag{6.13.1}$$

$$A_{16}\frac{d^2u^o}{dx^2} + A_{66}\frac{d^2v^o}{dx^2} - B_{16}\frac{d^3w^o}{dx^3} = 0 \tag{6.13.2}$$

$$D_{11}\frac{d^4w^o}{dx^4} - B_{11}\frac{d^3u^o}{dx^3} - B_{16}\frac{d^3v^o}{dx^3} = p_z \tag{6.13.3}$$

Equations (6.13.1) and (6.13.2) are differentiated with respect to x and simultaneously solved in terms of the transverse displacement w^o as:

$$\frac{d^3u^o}{dx^3} = \left(\frac{A_{66}B_{11} - A_{16}B_{16}}{A_{11}A_{66} - A_{16}A_{16}}\right)\frac{d^4w^o}{dx^4} = \frac{B}{A}\frac{d^4w^o}{dx^4} \tag{6.14.1}$$

$$\frac{d^3v^o}{dx^3} = \left(\frac{A_{11}B_{16} - A_{16}B_{11}}{A_{11}A_{66} - A_{16}A_{16}}\right)\frac{d^4w^o}{dx^4} = \frac{C}{A}\frac{d^4w^o}{dx^4} \tag{6.14.2}$$

where $A = A_{11}A_{66} - A_{16}A_{16}$, $B = A_{66}B_{11} - A_{16}B_{16}$ and $C = A_{11}B_{16} - A_{16}B_{11}$

By substituting Eqs. (6.14.1 and 6.14.2) into Eq. (6.13.3):

$$D_{11}\frac{d^4w^o}{dx^4} - B_{11}\left(\frac{A_{66}B_{11} - A_{16}B_{16}}{A_{11}A_{66} - A_{16}A_{16}}\right)\frac{d^4w^o}{dx^4} - B_{16}\left(\frac{A_{11}B_{16} - A_{16}B_{11}}{A_{11}A_{66} - A_{16}A_{16}}\right)\frac{d^4w^o}{dx^4} = p_z \qquad (6.15.1)$$

Rewriting Eq. (6.15.1) results in:

$$\left(\frac{D_{11}A - B_{11}B - B_{16}C}{A}\right)\frac{d^4w^o}{dx^4} = \left(\frac{D}{A}\right)\frac{d^4w^o}{dx^4} = p_z \qquad (6.15.2)$$

where $D = D_{11}A - B_{11}B - B_{16}C$; $A = A_{11}A_{66} - A_{16}A_{16}$; $B = A_{66}B_{11} - A_{16}B_{16}$; $C = A_{11}B_{16} - A_{16}B_{11}$.

For a special case where (p_z)=constant, the transverse displacement $w^o(x)$ is given as:

$$w^o(x) = \frac{A}{D}\left(p_z\frac{x^4}{24} + C_1\frac{x^3}{6} + C_2\frac{x^2}{2} + C_3x + C_4\right) \qquad (6.16)$$

where constants $(C_1 - C_4)$ are obtained from the boundary conditions.

Example 6.1

Find the deflection function w^0 of a long laminated composite plate with fixed end boundary conditions, under triangular loading function $p_z = p_0 x/a$ as shown in Example Figure 6.1.

Solution

From Eq. (6.15.2), deflection of a long laminate composite plate under $p_z = p_0 x/a$ is determined as:

$$\left(\frac{D_{11}A - B_{11}B - B_{16}C}{A}\right)\frac{d^4w^o}{dx^4} = \left(\frac{D}{A}\right)\frac{d^4w^o}{dx^4} = p_0\frac{x}{a}$$

where $A = A_{11}A_{66} - A_{16}A_{16}$ $B = A_{66}B_{11} - A_{16}B_{16}$ $C = A_{11}B_{16} - A_{16}B_{11}$.

By integrating the above equation:

$$\left(\frac{D}{A}\right)w^o(x) = p_0\frac{x^5}{120a} + C_1\frac{x^3}{6} + C_2\frac{x^2}{2} + C_3x + C_4$$

For fixed boundaries, deformation conditions are:

   i. $w^o(0) = 0$: $C_4 = 0$ and
  ii. $dw^o(0)/dx = 0$: $C_3 = 0$

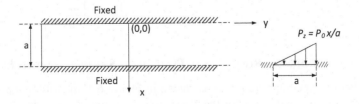

EXAMPLE FIGURE 6.1   Long laminated composite plate with fixed end condition.

iii. $w^o (a) = 0$: $p_0 \dfrac{a^5}{120a} + C_1 \dfrac{a^3}{6} + C_2 \dfrac{a^2}{2} = 0$ and

iv. $dw^o(a)/dx = 0$: $p_0 \dfrac{a^4}{24a} + C_1 \dfrac{a^2}{2} + C_2 a = 0$

From boundary conditions (iii) and (iv), constants (C_1 and C_2) are found to be: $C_1 = -3\, p_0\, a/20$ and $C_2 = 2\, p_0\, a^2/30$. Thus, the transverse displacement $w^o (x)$ is given as:

$$w^o(x) = p_0 x^2 \left(\frac{A}{D} \right) \left(\frac{x^3}{120a} - \frac{ax}{40} + \frac{a^2}{60} \right)$$

where $A = A_{11}A_{66} - A_{16}A_{16}$ $\qquad B = A_{66}B_{11} - A_{16}B_{16}$ $\qquad C = A_{11}B_{16} - A_{16}B_{11}.$

6.7 SPECIALLY ORTHOTROPIC RECTANGULAR PLATES

In general, the two-dimensional problems of the laminated FRP composite rectangular plates have complex stress and strain relationships in terms of constitutive relations (*ABD* matrix). In this section, analytical solutions of simple laminated plates, i.e., specially orthotropic lay-ups, are presented. The specially orthotropic lay-ups are defined herein as orthotropic where A_{16}, A_{26}, B_{16}, B_{26}, D_{16} and D_{26} are equal to zero with symmetrical [$B_{ij}=0$] lay-up. Thus, the in-plane shear stiffness [A_{ij}] is decoupled from bending stiffness [D_{ij}].

The specially orthotropic laminated rectangular plates with all four simple supports are analyzed (solved for deflection) using Navier's method. Levy's method has been employed commonly to develop analytical solutions for plate problems with two opposite edges simply supported. In addition, the energy approach based on Ritz's method is presented to derive approximate solutions of plates with clamped supports (Whitney, 1987).

6.7.1 THE GOVERNING DIFFERENTIAL EQUATIONS

The general governing differential equation of orthotropic plates can be reduced from Eq. (6.12.3) as (Szilard, 2004):

$$D_{11} \frac{\partial^4 w^o}{\partial x^4} + 2(D_{12} + 2D_{66}) \frac{\partial^4 w^o}{\partial x^2 \partial y^2} + D_{22} \frac{\partial^4 w^o}{\partial y^4} = p_z \qquad (6.17.1)$$

The bending moment resultants in Eq. (6.11) are simplified as:

$$\begin{bmatrix} M_{xx} \\ M_{yy} \\ M_{xy} \end{bmatrix} = \begin{bmatrix} D_{11} & D_{12} & 0 \\ D_{12} & D_{22} & 0 \\ 0 & 0 & D_{66} \end{bmatrix} \begin{bmatrix} -\dfrac{\partial^2 w^o}{\partial x^2} \\ -\dfrac{\partial^2 w^o}{\partial y^2} \\ -2\dfrac{\partial^2 w^o}{\partial x \partial y} \end{bmatrix} \qquad (6.17.2)$$

6.7.2 SPECIALLY ORTHOTROPIC RECTANGULAR PLATES WITH SIMPLY SUPPORTED EDGES

To solve these laminated composite plates as shown in Figure 6.5, the differential equation solution presented by Navier is still valid for the classical homogenous rectangular plates with simply supported edges. In Navier's method, Fourier series solutions are developed. Under transverse load

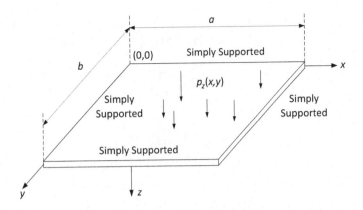

FIGURE 6.5 Specially orthotropic rectangular plates with simply supported edges.

$p_z(x, y)$, the displacement function $w(x, y)$ is expanded into double trigonometric series. To ensure convergence, the displacement series solutions are selected to satisfy the boundary conditions. For additional understanding of the application of trigonometric series to solve differential equations including orthogonality conditions, readers should refer to Timoshenko and Woinowsky (1959) or Szilard (2004).

The boundary conditions of rectangular plates with all edges simply supported are given as:

$$w^o(0,y) = w^o(a,y) = w^o(x,0) = w^o(x,b) = 0 \tag{6.18.1}$$

$$M_{xx}(0,y) = M_{xx}(a,y) = M_{yy}(x,0) = M_{yy}(x,a) = 0 \tag{6.18.2}$$

The applied transverse load $p_z(x, y)$ is often expanded into double sine series as:

$$p_z(x,y) = \sum_{m=1}^{\infty}\sum_{n=1}^{\infty} P_{mn} \mathrm{Sin}\left[\frac{m\pi x}{a}\right]\mathrm{Sin}\left[\frac{n\pi y}{b}\right] \tag{6.19}$$

The coefficient (P_{mn}) of load expansion is determined by multiplying both sides of the above equation with double sine series. Integration ranging from $(0, a)$ and $(0, b)$ is performed on variables x and y, respectively. The orthogonal condition is also applied to obtain (P_{mn}).

$$P_{mn} = \frac{4}{ab}\int_0^b\int_0^a p_z(x,y)\mathrm{Sin}\left[\frac{m\pi x}{a}\right]\mathrm{Sin}\left[\frac{n\pi y}{b}\right]dx\,dy \tag{6.20}$$

The assumed deflection solution in terms of double sine series satisfies all boundaries as given in Eq. (6.21).

$$w^o(x,y) = \sum_{m=1}^{\infty}\sum_{n=1}^{\infty} W_{mn} \mathrm{Sin}\left[\frac{m\pi x}{a}\right]\mathrm{Sin}\left[\frac{n\pi y}{b}\right] \tag{6.21}$$

Substituting the assumed defection solution (Eq. 6.21) and applied transverse load function (Eq. 6.19) into the governing differential Eq. (6.17.1) and then involving orthogonality conditions can be written as:

$$\sum_{m=1}^{\infty}\sum_{n=1}^{\infty}\left[W_{mn}\left(\begin{array}{c}D_{11}\left(\dfrac{m\pi}{a}\right)^4+2(D_{12}+2D_{66})\left(\dfrac{m\pi}{a}\right)^2\left(\dfrac{n\pi}{b}\right)^2\\[2mm]+D_{22}\left(\dfrac{n\pi}{b}\right)^4\end{array}\right)-P_{mn}\right]$$
$$\times\mathrm{Sin}\left[\dfrac{m\pi x}{a}\right]\mathrm{Sin}\left[\dfrac{n\pi y}{b}\right]=0 \qquad (6.22.1)$$

$$W_{mn}\left(D_{11}\left(\dfrac{m\pi}{a}\right)^4+2(D_{12}+2D_{66})\left(\dfrac{m\pi}{a}\right)^2\left(\dfrac{n\pi}{b}\right)^2+D_{22}\left(\dfrac{n\pi}{b}\right)^4\right)-P_{mn}=0 \qquad (6.22.2)$$

$$W_{mn}=\left(\dfrac{a}{\pi}\right)^4\dfrac{P_{mn}}{D_{mn}} \qquad (6.23)$$

where $D_{mn}=\left(D_{11}m^4+2(D_{12}+2D_{66})(mnR)^2+D_{22}(nR)^4\right)$ and $R=a/b$.

Thus, the general deflection (w^o) solution of thin FRP specially orthotropic laminated plates with all edges simply supported for various lateral loadings is obtained by substituting Eq. (6.23) into Eq. (6.21). In addition, the general in-plane displacement solutions can be determined by substituting the general deflection (w^o) solution into the in-plane displacement function as:

$$w^o(x,y)=\sum_{m=1}^{\infty}\sum_{n=1}^{\infty}\left(\dfrac{a}{\pi}\right)^4\dfrac{P_{mn}}{D_{mn}}\mathrm{Sin}\left[\dfrac{m\pi x}{a}\right]\mathrm{Sin}\left[\dfrac{n\pi y}{b}\right] \qquad (6.24.1)$$

$$u=u^o-z\dfrac{\partial w^o}{\partial x}=u^o-z\sum_{m=1}^{\infty}\sum_{n=1}^{\infty}\left(\dfrac{a}{\pi}\right)^3 m\dfrac{P_{mn}}{D_{mn}}\mathrm{Cos}\left[\dfrac{m\pi x}{a}\right]\mathrm{Sin}\left[\dfrac{n\pi y}{b}\right] \qquad (6.24.2)$$

$$v=v^o-z\dfrac{\partial w^o}{\partial y}=v^o-z\sum_{m=1}^{\infty}\sum_{n=1}^{\infty}\left(\dfrac{a}{\pi}\right)^4\left(\dfrac{\pi}{b}\right)n\dfrac{P_{mn}}{D_{mn}}\mathrm{Sin}\left[\dfrac{m\pi x}{a}\right]\mathrm{Cos}\left[\dfrac{n\pi y}{b}\right] \qquad (6.24.3)$$

Bending and twisting moments can be determined by substituting the deflection solution (Eqs. 6.24.1–6.24.3) into the bending moment resultants as given in Eq. (6.17.2).

$$M_{xx}=\sum_{m=1}^{\infty}\sum_{n=1}^{\infty}\left(\dfrac{a}{\pi}\right)^2\dfrac{P_{mn}}{D_{mn}}\left(m^2D_{11}+n^2R^2D_{12}\right)\mathrm{Sin}\left[\dfrac{m\pi x}{a}\right]\mathrm{Sin}\left[\dfrac{n\pi y}{b}\right] \qquad (6.25.1)$$

$$M_{yy}=\sum_{m=1}^{\infty}\sum_{n=1}^{\infty}\left(\dfrac{a}{\pi}\right)^2\dfrac{P_{mn}}{D_{mn}}\left(m^2D_{12}+n^2R^2D_{22}\right)\mathrm{Sin}\left[\dfrac{m\pi x}{a}\right]\mathrm{Sin}\left[\dfrac{n\pi y}{b}\right] \qquad (6.25.2)$$

$$M_{xy}=-2RD_{66}\sum_{m=1}^{\infty}\sum_{n=1}^{\infty}\left(\dfrac{a}{\pi}\right)^2\left(\dfrac{mn}{D_{mn}}\right)P_{mn}\mathrm{Cos}\left[\dfrac{m\pi x}{a}\right]\mathrm{Cos}\left[\dfrac{n\pi y}{b}\right] \qquad (6.25.3)$$

To evaluate in-plane stresses for each ply, in-plane stress-strain relation in Eq. (4.3) is substituted into Eq. (4.6) to arrive at:

$$
\begin{bmatrix} \sigma_{xx}^k \\ \sigma_{yy}^k \\ \tau_{xy}^k \end{bmatrix} = \begin{bmatrix} \bar{Q}_{11}^k & \bar{Q}_{12}^k & 0 \\ \bar{Q}_{12}^k & \bar{Q}_{22}^k & 0 \\ 0 & 0 & \bar{Q}_{66}^k \end{bmatrix} \begin{bmatrix} -z\dfrac{\partial^2 w^o}{\partial x^2} \\ -z\dfrac{\partial^2 w^o}{\partial y^2} \\ -2z\dfrac{\partial^2 w^o}{\partial x \partial y} \end{bmatrix}
\tag{6.26.1}
$$

By substituting Eqs. (6.24.1–6.24.3) into Eq. (6.26.1):

$$
\sigma_x^k = z \sum_{m=1}^{\infty} \sum_{n=1}^{\infty} \left(\frac{a}{\pi}\right)^2 \frac{P_{mn}}{D_{mn}} \left(m^2 Q_{11}^k + n^2 R^2 Q_{12}^k\right) \mathrm{Sin}\left[\frac{m\pi x}{a}\right] \mathrm{Sin}\left[\frac{n\pi y}{b}\right]
\tag{6.26.2}
$$

$$
\sigma_y^k = z \sum_{m=1}^{\infty} \sum_{n=1}^{\infty} \left(\frac{a}{\pi}\right)^2 \frac{P_{mn}}{D_{mn}} \left(m^2 Q_{12}^k + n^2 R^2 Q_{22}^k\right) \mathrm{Sin}\left[\frac{m\pi x}{a}\right] \mathrm{Sin}\left[\frac{n\pi y}{b}\right]
\tag{6.26.3}
$$

$$
\tau_{xy}^k = -2zRQ_{66}^k \sum_{m=1}^{\infty} \sum_{n=1}^{\infty} \left(\frac{a}{\pi}\right)^2 \left(\frac{mn}{D_{mn}}\right) P_{mn} \mathrm{Cos}\left[\frac{m\pi x}{a}\right] \mathrm{Cos}\left[\frac{n\pi y}{b}\right]
\tag{6.26.4}
$$

6.7.3   Specially Orthotropic Plates with Two Opposite Edges Simply Supported

Specially orthotropic rectangular plates with two opposite edges simply supported at $y=0$ and b under $p_z(x, y)$ are analyzed (refer to Figure 6.6 for general boundaries). Navier method does not lend itself to obtaining a closed-form deflection function satisfying all the general boundary conditions. In addition, the bending-twisting moment and stresses are of slow convergence, unless special correction factors such as those derived by Lancoz are applied to the final Fourier series solutions (GangaRao and Prachasaree, 2007).

To overcome the above disadvantages, another approach called Levy's method is more appropriate, which was initially introduced by Levy (1899) for homogenous materials. Levy's method can be extended to solve the specially orthotropic laminated plate problems, and it is elaborated below:

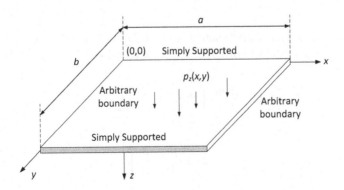

FIGURE 6.6   Specially orthotropic rectangular plate for Levy's method.

Fundamental concepts and assumptions of Levy's method are given as follows:

- The assumed deflection function and applied loading are represented in a single trigonometric series form. The two opposite edges ($y=0$ and b) are required to be simply supported as shown in Figure 6.6. In addition, the applied loading functions are identical for all sections parallel to y-direction.
- The solutions of the governing differential equation can be separated into two parts: one as a homogenous (w_H) solution and another as a particular (w_P) solution. It is found that the boundary conditions along edges $y=0$ and b (required two opposite edges simply supported) and the other two boundaries ($x=0$ and a) are arbitrary, corresponding to the particular and homogenous solutions, respectively. Thus, the particular solutions must be specific for arbitrary applied loading.

To satisfy the simply supported boundary condition on two opposite edges ($y=0$ and b), the other two opposite edges are assumed to be separated by a very long distance, i.e., plate aspect ratio (length to width) of 10 or more. In such a case, the particular solution can be easily determined as in the previous case, which is similar to the cylindrical bending phenomenon. The differential governing equation (6.17.1) is simplified to be a function of variable y only and it is given below:

$$D_{22}\frac{\partial^4 w_p^o}{\partial y^4} = p_z \tag{6.27.1}$$

The assumed particular solution in Eq. (6.27.2) satisfies the simply supported condition and it is written in terms of sine series as:

$$w_p^o = \sum_{n=1}^{\infty}\phi_n^p(x)\mathrm{Sin}\left[\frac{n\pi y}{b}\right] \tag{6.27.2}$$

The arbitrary applied transverse loads can also be represented in terms of single sine series as follows:

$$p_z = \sum_{n=1}^{\infty}P_n\mathrm{Sin}\left[\frac{n\pi y}{b}\right] \tag{6.28.1}$$

$$P_n = \frac{2}{b}\int_0^b p_z\mathrm{Sin}\left[\frac{n\pi y}{b}\right]dy \tag{6.28.2}$$

By substituting the assumed particular solution in Eq. (6.27.2) and arbitrary transverse loads in Eqs. (6.28.1 and 6.28.2) into the modified differential governing Eq. (6.27.1):

$$D_{22}\sum_{n=1}^{\infty}\phi_n^p(x)\left(\frac{n\pi}{b}\right)^4\mathrm{Sin}\left[\frac{n\pi y}{b}\right] = \sum_{n=1}^{\infty}P_n\mathrm{Sin}\left[\frac{n\pi y}{b}\right] \tag{6.29.1}$$

$$\phi_n^p(x) = \frac{P_n}{D_{22}}\left(\frac{b}{n\pi}\right)^4 \tag{6.29.2}$$

The particular solutions are obtained by substituting Eq. (6.29) into Eq. (6.27.1), which results in:

$$w_p^o = \sum_{n=1}^{\infty} \frac{P_n}{D_{22}} \left(\frac{b}{n\pi}\right)^4 \mathrm{Sin}\left[\frac{n\pi y}{b}\right]$$

(6.30)

Note: Equation (6.30) is identical to the deflection function of beam under transverse loading, where

$$D_{22} = EI|_{beam}$$

The particular solutions under arbitrary transverse applied loads can be determined by substituting coefficients P_n of those applied loads into Eq. (6.30). To obtain the homogeneous solution, the assumed deflection function in series is made identical with the particular solution as given in Eq. (6.27.2).

$$w_H^o = \sum_{n=1}^{\infty} \phi_n^H(x) \mathrm{Sin}\left[\frac{n\pi y}{b}\right]$$

(6.31)

This assumed solution is substituted into the governing differential Eq. (6.17.1) and the applied transverse loading term (p_z) is taken to be zero (homogenous solution for ∇^4 bi-harmonic equation). By substituting Eq. (6.31) into Eq. (6.17.1):

$$D_{11}\frac{d^4\phi_n^H}{dx^4} - 2(D_{12}+2D_{66})\left(\frac{n\pi}{b}\right)^2 \frac{d^2\phi_n^H}{dx^2} + D_{22}\left(\frac{n\pi}{b}\right)^4 \phi_n^H = 0$$

(6.32.1)

The homogenous differential Eq. (6.32.1) solution is presented as an exponential function.

$$\phi_n^H = e^{\left(\frac{n\pi\lambda x}{b}\right)}$$

(6.32.2)

By substituting the exponential solution in Eq. (6.32.2) into the homogenous differential Eq. (6.32.1), the algebraic equation with a fourth degree of polynomials is obtained as:

$$\left(D_{11}\lambda^4 - 2(D_{12}+2D_{66})\lambda^2 + D_{22}\right) = 0$$

(6.33)

The characteristic Eq. (6.33) is a combination of four roots (λ), and constants $(A_n, B_n, C_n,$ and $D_n)$ can be presented in a general form of quadratic solution as:

$$\lambda^2 = \frac{1}{D_{11}}\left(D_{12}+2D_{66} \pm \sqrt{(D_{12}+2D_{66})^2 - D_{11}D_{22}}\right)$$

(6.34)

Roots, λ, can be either real or complex depending on plate stiffnesses $[D_{ij}]$. However, three different cases are considered herein: (1) real number and unequal, (2) real number and equal, and (3) complex number.

Case I: Real number and unequal
This case is valid when the term in the square root of Eq. (6.34) is positive. Thus, the roots are given by $\pm\lambda_1$ and $\pm\lambda_2$ $(\lambda_1, \lambda_2 > 0)$. The solution of homogenous Eq. (6.34) can be presented as:

$$\lambda_1^2 = \frac{1}{D_{11}}\left(D_{12}+2D_{66} - \sqrt{(D_{12}+2D_{66})^2 - D_{11}D_{22}}\right)$$

(6.35.1)

$$\lambda_2^2 = \frac{1}{D_{11}}\left(D_{12}+2D_{66} + \sqrt{(D_{12}+2D_{66})^2 - D_{11}D_{22}}\right)$$

(6.35.2)

$$\phi_n^H = A_n \text{Cosh} \frac{n\pi\lambda_1 x}{b} + B_n \text{Sinh} \frac{n\pi\lambda_1 x}{b} + C_n \text{Cosh} \frac{n\pi\lambda_2 x}{b} + D_n \text{Sinh} \frac{n\pi\lambda_2 x}{b} \qquad (6.35.3)$$

By substituting Eq. (6.35.2) into the assumed homogenous deflection function Eq. (6.31):

$$w_H^o = \sum_{n=1}^{\infty} \left(A_n \text{Cosh} \frac{n\pi\lambda_1 x}{b} + B_n \text{Sinh} \frac{n\pi\lambda_1 x}{b} + C_n \text{Cosh} \frac{n\pi\lambda_2 x}{b} + D_n \text{Sinh} \frac{n\pi\lambda_2 x}{b} \right) \text{Sin} \left[\frac{n\pi y}{b} \right] \qquad (6.35.4)$$

The long plate solution corresponding to cylindrical bending can be obtained by summing: (1) homogenous solution (w_H) in Eq. (6.35.4) and (2) particular solution (w_P) in Eq. (6.30):

$$w^o = \sum_{n=1}^{\infty} \left(\begin{array}{c} A_n \text{Cosh} \dfrac{n\pi\lambda_1 x}{b} + B_n \text{Sinh} \dfrac{n\pi\lambda_1 x}{b} + \\[2mm] C_n \text{Cosh} \dfrac{n\pi\lambda_2 x}{b} + D_n \text{Sinh} \dfrac{n\pi\lambda_2 x}{b} + \dfrac{P_n}{D_{22}} \left(\dfrac{b}{n\pi} \right)^4 \end{array} \right) \text{Sin} \left[\frac{n\pi y}{b} \right] \qquad (6.35.5)$$

Case II: Real number and equal

This case is valid when the term in the square root of Eq. (6.34) is zero. Thus, the roots are given by $\pm\lambda$ ($\lambda > 0$). The solution of homogenous Eq. (6.31) can be presented as follows:

$$\lambda^2 = \frac{1}{D_{11}} (D_{12} + 2D_{66}) \qquad (6.36.1)$$

$$\phi_n^H = (A_n + B_n x) \text{Cosh} \frac{n\pi\lambda x}{b} + (C_n + D_n x)_n \text{Sinh} \frac{n\pi\lambda x}{b} \qquad (6.36.2)$$

The plate solution can be obtained by summing both the homogenous (w_H) and particular (w_P) solutions.

$$w^o = \sum_{n=1}^{\infty} \left((A_n + B_n x) \text{Cosh} \frac{n\pi\lambda x}{b} + (C_n + D_n x)_n \text{Sinh} \frac{n\pi\lambda x}{b} + \frac{P_n}{D_{22}} \left(\frac{b}{n\pi} \right)^4 \right) \text{Sin} \left[\frac{n\pi y}{b} \right] \qquad (3.36.3)$$

Case III: Complex number

This case is valid when the term in the square root of Eq. (6.34) is negative. Thus, the roots are given by $\lambda_1 \pm i\lambda_2$ and $-\lambda_1 \pm i\lambda_2$ ($\lambda_1, \lambda_2 > 0$). The solution of homogenous Eq. (6.31) can be presented as:

$$\lambda_1^2 = \frac{1}{2D_{11}} \left(D_{12} + 2D_{66} + \sqrt{D_{11}D_{22}} \right) \qquad (3.37.1)$$

$$\lambda_2^2 = \frac{1}{2D_{11}} \left(-(D_{12} + 2D_{66}) + \sqrt{D_{11}D_{22}} \right) \qquad (3.37.2)$$

$$\phi_n^H = \left(A_n \text{Cos} \frac{n\pi\lambda_2 x}{b} + B_n \text{Sin} \frac{n\pi\lambda_2 x}{b} \right) \text{Cosh} \frac{n\pi\lambda_1 x}{b}$$

$$+ \left(C_n \text{Cos} \frac{n\pi\lambda_2 x}{b} + D_n \text{Sin} \frac{n\pi\lambda_2 x}{b} \right) \text{Sinh} \frac{n\pi\lambda_1 x}{b} \qquad (3.37.3)$$

The plate solution can be obtained by summing both the homogenous (w_H) and particular (w_P) solutions.

$$w^o = \sum_{n=1}^{\infty} \left\{ \begin{array}{l} \left(A_n \text{Cos} \dfrac{n\pi\lambda_2 x}{b} + B_n \text{Sin} \dfrac{n\pi\lambda_2 x}{b} \right) \text{Cosh} \dfrac{n\pi\lambda_1 x}{b} \\[3mm] + \left(C_n \text{Cos} \dfrac{n\pi\lambda_2 x}{b} + D_n \text{Sin} \dfrac{n\pi\lambda_2 x}{b} \right)_n \text{Sinh} \dfrac{n\pi\lambda_1 x}{b} \\[3mm] + \dfrac{P_n}{D_{22}} \left(\dfrac{b}{n\pi} \right)^4 \end{array} \right\} \text{Sin} \left[\dfrac{n\pi y}{b} \right] \qquad (6.37.4)$$

where A_n, B_n, C_n, and D_n will be determined from the non-simply supported boundary conditions on two opposite sides.

For specially orthotropic plates with various boundary conditions, an approximate solution based on the energy method (Ritz's method) can be used. In this approach, the assumed solutions containing several arbitrary constants must satisfy the kinematic boundary conditions of the problem. In general, the assumed solutions are often written in terms of polynomial or trigonometric functions. Then, the stationary potential energy theorem is used to determine the optimum values of the arbitrary constants. It should be noted that the accuracy of the solution depends on how closely the assumed shape function represents the exact solution, thus minimizing errors (Whitney, 1987; Reddy, 2003).

The full treatment of other laminate plate problems (i.e., antisymmetric angle plates, mid-plane symmetric plates, laminate composite plates with thermal load, etc.) is beyond the scope of this textbook. More information and solutions to various laminate composite plate problems are provided in textbooks by Lekhnitskii (1968), Whitney (1987), Reddy (2003), and Vinson (2005).

Example 6.2

Determine the deflection function of a specially orthotropic laminated composite plate under uniform load p_o: (a) all simply supported edges and (b) clamped and simply supported edges as shown in Example Figure 6.2. (Assume: $(D_{12}+2D_{66})^2-D_{11}D_{22}>0$.)

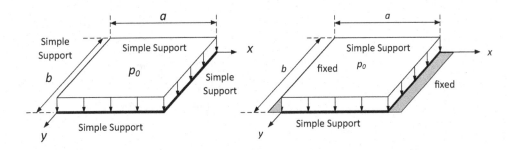

a) plate with all simply supported edges opposite

b) plate with fixed and simply supported on edges

EXAMPLE FIGURE 6.2 Specially orthotropic laminated composite plate: (a) plate with all simply supported edges and (b) plate with fixed and simply supported on opposite edges.

Solution

a. Plate with all simply supported edges:
   From Eq. (6.35.5):

$$w^o = \sum_{n=1}^{\infty} \left(\begin{array}{c} A_n \text{Cosh} \dfrac{n\pi\lambda_1 x}{b} + B_n \text{Sinh} \dfrac{n\pi\lambda_1 x}{b} + \\[2mm] C_n \text{Cosh} \dfrac{n\pi\lambda_2 x}{b} + D_n \text{Sinh} \dfrac{n\pi\lambda_2 x}{b} + \dfrac{P_n}{D_{22}} \left(\dfrac{b}{n\pi} \right)^4 \end{array} \right) \text{Sin} \left[\dfrac{n\pi y}{b} \right]$$

From Eq. (6.35.5) with a boundary condition at $x=0$: $w^o=0$

$$A_n + C_n + \frac{P_n}{D_{22}} \left(\frac{b}{n\pi} \right)^4 = 0 \tag{6.38.1}$$

From Eq. (6.35.5) with boundary condition at $x=a$: $w^o=0$

$$A_n \text{Cosh} \frac{n\pi\lambda_1 a}{b} + B_n \text{Sinh} \frac{n\pi\lambda_1 a}{b} + C_n \text{Cosh} \frac{n\pi\lambda_2 a}{b} + D_n \text{Sinh} \frac{n\pi\lambda_2 a}{b} + \frac{P_n}{D_{22}} \left(\frac{b}{n\pi} \right)^4 = 0 \quad (6.38.2)$$

From Eq. (6.35.5) with a boundary condition at $x=0$: $M_x=0$

$$M_x = -\left(D_{11} \frac{\partial^2 w^o}{\partial x^2} + D_{12} \frac{\partial^2 w^o}{\partial y^2} \right)$$

$$= D_{11} \left(A_n \left(\frac{n\pi\lambda_1}{b} \right)^2 + C_n \left(\frac{n\pi\lambda_2}{b} \right)^2 \right)$$

$$- D_{12} \left(A_n + C_n + \frac{P_n}{D_{22}} \left(\frac{b}{n\pi} \right)^4 \right) = 0 \tag{6.38.3}$$

From Eq. (6.35.5) with boundary condition at $x=0$: $M_x=0$

$$M_x = -\left(D_{11} \frac{\partial^2 w^o}{\partial x^2} + D_{12} \frac{\partial^2 w^o}{\partial y^2} \right)$$

$$= D_{11} \left(\begin{array}{c} A_n \left(\dfrac{n\pi\lambda_1}{b} \right)^2 \text{Cosh} \dfrac{n\pi\lambda_1 a}{b} + B_n \left(\dfrac{n\pi\lambda_1}{b} \right)^2 \text{Sinh} \dfrac{n\pi\lambda_1 a}{b} \\[3mm] + C_n \left(\dfrac{n\pi\lambda_2}{b} \right)^2 \text{Cosh} \dfrac{n\pi\lambda_2 a}{b} + D_n \left(\dfrac{n\pi\lambda_2}{b} \right)^2 \text{Sinh} \dfrac{n\pi\lambda_2 a}{b} \end{array} \right)$$

$$- D_{12} \left(\begin{array}{c} A_n \text{Cosh} \dfrac{n\pi\lambda_1 a}{b} + B_n \text{Sinh} \dfrac{n\pi\lambda_1 a}{b} \\[3mm] + C_n \text{Cosh} \dfrac{n\pi\lambda_2 a}{b} + D_n \text{Sinh} \dfrac{n\pi\lambda_2 a}{b} + \dfrac{P_n}{D_{22}} \left(\dfrac{b}{n\pi} \right)^4 \end{array} \right) = 0 \tag{6.38.4}$$

By solving Eqs. (6.38.1–6.38.4) simultaneously, constants A_n, B_n, C_n, and D_n are given as:

$$A_n = -\frac{4b^4 P_n}{D_{22}(n\pi)^5}\left(\frac{\lambda_2^2}{\lambda_2^2 - \lambda_1^2}\right)$$

$$B_n = -\frac{4b^4 P_n}{D_{22}(n\pi)^5}\left(\frac{\lambda_2^2}{\lambda_2^2 - \lambda_1^2}\right)\left(\frac{1 - \text{Cosh}\left(\frac{n\pi\lambda_1 a}{b}\right)}{\text{Sinh}\left(\frac{n\pi\lambda_1 a}{b}\right)}\right)$$

$$C_n = \frac{4b^4 P_n}{D_{22}(n\pi)^5}\left(\frac{\lambda_1^2}{\lambda_2^2 - \lambda_1^2}\right)$$

$$D_n = \frac{4b^4 P_n}{D_{22}(n\pi)^5}\left(\frac{\lambda_2^2}{\lambda_2^2 - \lambda_1^2}\right)\left(\frac{1 - \text{Cosh}\left(\frac{n\pi\lambda_2 a}{b}\right)}{\text{Sinh}\left(\frac{n\pi\lambda_2 a}{b}\right)}\right)$$

where $P_n = \dfrac{4p_0}{\pi^2 n}$; $n = 1, 3, 5\ldots$

b. Plate with clamped and simply supported edges:
   From Eq. (6.35.5):

$$w^o = \sum_{n=1}^{\infty}\left(\begin{array}{c} A_n \text{Cosh}\dfrac{n\pi\lambda_1 x}{b} + B_n \text{Sinh}\dfrac{n\pi\lambda_1 x}{b} + C_n \text{Cosh}\dfrac{n\pi\lambda_2 x}{b} \\[2mm] + D_n \text{Sinh}\dfrac{n\pi\lambda_2 x}{b} + \dfrac{P_n}{D_{22}}\left(\dfrac{b}{n\pi}\right)^4 \end{array}\right) \text{Sin}\left[\frac{n\pi y}{b}\right]$$

From Eq. (6.36.5) with a boundary condition at $x=0$: $w^o = 0$

$$A_n + C_n + \frac{P_n}{D_{22}}\left(\frac{b}{n\pi}\right)^4 = 0 \qquad (6.38.5)$$

From Eq. (6.35.5) with a boundary condition at $x=a$: $w^o = 0$

$$A_n \text{Cosh}\frac{n\pi\lambda_1 a}{b} + B_n \text{Sinh}\frac{n\pi\lambda_1 a}{b} + C_n \text{Cosh}\frac{n\pi\lambda_2 a}{b} + D_n \text{Sinh}\frac{n\pi\lambda_2 a}{b} + \frac{P_n}{D_{22}}\left(\frac{b}{n\pi}\right)^4 = 0 \quad (6.38.6)$$

From Eq. (6.35.5) with a boundary condition at $x=0$: $\partial w^o / \partial x = 0$

$$A_n\left(\frac{n\pi\lambda_1}{b}\right) + C_n\left(\frac{n\pi\lambda_2}{b}\right) = 0 \qquad (6.38.7)$$

From Eq. (6.35.5) with a boundary condition at $x=a$: $\partial w^o / \partial x = 0$

$$A_n \left(\frac{n\pi\lambda_1}{b} \right) \text{Cosh} \frac{n\pi\lambda_1 a}{b} + B_n \left(\frac{n\pi\lambda_1}{b} \right) \text{Sinh} \frac{n\pi\lambda_1 a}{b}$$

$$+ C_n \left(\frac{n\pi\lambda_2}{b} \right) \text{Cosh} \frac{n\pi\lambda_2 a}{b} + D_n \left(\frac{n\pi\lambda_2}{b} \right) \text{Sinh} \frac{n\pi\lambda_2 a}{b} = 0 \qquad (6.38.8)$$

By solving Eqs. (6.38.5–6.38.8) simultaneously, constants A_n, B_n, C_n, and D_n are given as:

$$A_n = \frac{4b^4 P_n \lambda_2^2}{H_n D_{22} (n\pi)^5} \left(\begin{array}{c} \lambda_1 \left(\text{Cosh} \dfrac{n\pi\lambda_1 a}{b} - \text{Cosh} \dfrac{n\pi\lambda_2 a}{b} \right) \left(\text{Cosh} \dfrac{n\pi\lambda_2 a}{b} - 1 \right) \\[2ex] - \left(\lambda_2 \text{Sinh} \dfrac{n\pi\lambda_1 a}{b} - \lambda_1 \text{Sinh} \dfrac{n\pi\lambda_2 a}{b} \right) \left(\text{Sinh} \dfrac{n\pi\lambda_2 a}{b} \right) \end{array} \right)$$

$$B_n = \frac{4b^4 P_n \lambda_2^2}{H_n D_{22} (n\pi)^5} \left(\begin{array}{c} \lambda_2 \left(\text{Cosh} \dfrac{n\pi\lambda_1 a}{b} - \text{Cosh} \dfrac{n\pi\lambda_2 a}{b} \right) \left(\text{Sinh} \dfrac{n\pi\lambda_2 a}{b} \right) \\[2ex] - \left(\lambda_1 \text{Sinh} \dfrac{n\pi\lambda_1 a}{b} - \lambda_2 \text{Sinh} \dfrac{n\pi\lambda_2 a}{b} \right) \left(\text{Cosh} \dfrac{n\pi\lambda_2 a}{b} - 1 \right) \end{array} \right)$$

$$C_n = - \left(\frac{4b^4 P_n}{H_n D_{22} (n\pi)^5} + A_n \right)$$

$$D_n = - \frac{\lambda_1}{\lambda_2} B_n$$

where $P_n = \dfrac{4 p_0}{\pi^2 n}$; $n = 1, 3, 5 \ldots$

Note: The focus is on developing Fourier series solutions that converge fast. The idea is that a designer can consider only the first term of the infinite series solution without loss of much accuracy to arrive at deflections and rotations. Typically, first-term accuracies of the infinite series corresponding to deformations are adequate from a designer viewpoint.

6.8 SUMMARY

The governing differential equations to accurately analyze the behavior of an FRP laminate composite plate are presented in this chapter. Two approaches, namely, the theory of elasticity and energy methods, are adopted to formulate the governing equations for FRP plates under bending and considering shear deformation. Several cases including long laminated FRP plates and specially orthotropic rectangular plates with different boundary conditions were presented together with examples to provide an accurate approach to the understanding of the structural behavior of this FRP component. The examples presented in this chapter are rudimentary and solution forms will get extremely complex for higher order boundary conditions, for varying material properties in different directions, and for inclusion of shear deformations.

EXERCISES

Problem 6.1: Determine the deflection function w^0 of a long laminated composite plate $[0°_2/45°_2]$ with fixed end condition, under loading function $p_z = p_0 \, x/a$.

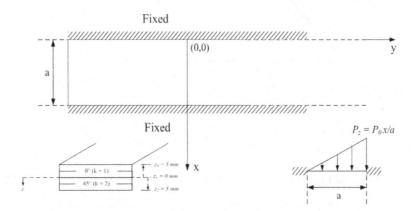

Problem 6.2: Determine bending and twisting moments of laminated composite plate $[0°_2/45°_2]$ in Problem 6.1. The properties of unidirectional FRP lamina are presented as: elastic modulus: $E_{11} = 180\,\text{GPa}$ and $E_{22} = 40\,\text{GPa}$, shear modulus: $G_{12} = 20\,\text{GPa}$, $\nu_{12} = 0.25$. ($[0°_2/45°_2]$ with each lamina thickness of 2.5 mm)

Problem 6.3: Determine in-plane stresses (for each ply) of laminated composite plate $[0°_2/45°_2]$ in Problem 6.1. The properties of unidirectional FRP lamina are presented as: elastic modulus: $E_{11} = 180\,\text{GPa}$ and $E_{22} = 40\,\text{GPa}$, shear modulus: $G_{12} = 20\,\text{GPa}$, $\nu_{12} = 0.25$. ($[0°_2/45°_2]$ with each lamina thickness of 2.5 mm)

Problem 6.4: Determine deflection function of a specially orthotropic laminated composite plate under sinusoidal distributed load: $p_z(x) = p_o \, \text{Sin} \, [\pi y/b]$. **Hint:** To evaluate constants (A_n, B_n, C_n and D_n), boundary conditions ($x=0$ and a) of the homogenous solution are used for this purpose. $\left(D_{12} + 2D_{66}\right)^2 - D_{11}D_{22} > 0$

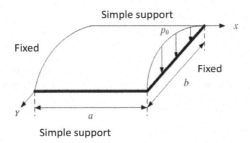

Problem 6.5: Determine the first-term approximation of displacements for antisymmetric cross-ply thin square plate $[0°/90°/0°/90°]$ under sinusoidal transverse load $p_o \text{Sin}\left[\dfrac{\pi x}{a}\right] \text{Sin}\left[\dfrac{\pi y}{b}\right]$ with hinged support for all edges. The properties of unidirectional FRP lamina are presented as: elastic modulus: $E_{11} = 185\,\text{GPa}$ and $E_{22} = 30\,\text{GPa}$, shear modulus: $G_{12} = 25\,\text{GPa}$, $\nu_{12} = 0.25$. ($[0°/90°/0°/90°]$ with each lamina thickness of 2.0 mm).

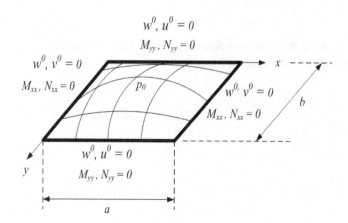

REFERENCES AND SELECTED BIOGRAPHY

Ambartsumyan, S.A., *Theory of Anisotropic Plates*, translated from Russian by T. Cheron, Technomic, Stamford, CT, 1969.

Ashton, J.E., Approximate solutions for unsymmetrically laminated plates, *Journal of Composite Materials*, 3(1), 1969, pp. 189–191.

GangaRao, H.V., and Prachasaree, W., Correction factors in series solutions for one- and two-dimensional boundary value problems, *Journal of Engineering Mechanics, ASCE*, 133(11), 2007, pp. 1151–1161.

Jones, R.M., *Mechanics of Composite Materials*, Hemisphere Publishing, New York, NY, 1975.

Kirchhoff, G., Uber das Gleichgwich und die Bewegung einer Elastichen Scheibe, *Journal of angewandte Mathematik*, 40, 1850, pp. 51–88.

Langhaar, H.L., *Energy Methods in Applied Mechanics*, Krieger, Melbourne, FL, 1989.

Lekhnitskii, S.G., *Anisotropic Plates*, translated from Russian by S.W. Tsai and T. Cheron, Gordon and Breach, Newark, NJ, 1968.

Levy, M., Sur L' E'quibre E'lastique D'une Plaque Rectangulaire, *Comptes Rendus De l'Académie Des Sciences*, 129, 1899, pp. 535–539.

Pagano, N.J., Exact solutions for composite laminates in cylindrical bending, *Journal of Composite Materials*, 3(3), 1969, pp. 398–411.

Reddy, J.N., *Mechanics of Laminated Composite Plates and Shells Theory and Analysis*, 2nd Edition, John Wiley, New York, NY, 2003.

Szilard, R., *Theories and Applications of Plate Analysis: Classical Numerical and Engineering Methods*, John Wiley & Sons, Inc., Hoboken, NJ, 2004.

Timoshenko, S., and Woinowsky, K.S., *Theory of Plates and Shells*, 2nd edition, McGraw-Hill, New York, NY, 1959.

Vinson, J.R., *Plate and Panel Structures of Isotropic, Composite and Piezoelectric Materials, Including Sandwich Construction (Solid Mechanics and Its Applications)*, Springer, New York, NY, 2005.

Vlasov, V.Z., and Leontev, N.N., *Beams, Plates and Shells on Elastic Foundation (translation of Balki, Plity, I Oblochki Na Uprugom Osnavanii), NASA TT F-357*, National Aeronautics and Space Administration, Washington, D.C., 1966.

Whitney, J.M., Cylindrical bending of unsymmetrically laminated plates, *Journal of Composite Materials*, 3(4), 1969, pp. 715–719.

Whitney, J.M., *Structural Analysis of Laminated Anisotropic Plates*, Technomic Publishing Company, Inc., Lancaster, PA, 1987.

7 Design Philosophy and Basis of FRP Composite Structural Members

7.1 DESIGN OF FRP COMPOSITE STRUCTURAL MEMBERS

Most structural members are subjected to individual loadings such as axial load (tension or compression), transverse loads, torsional loads, or even combined loading such as axial load along with the transverse load. In general, structural members can be classified according to the characteristics of externally applied and environmentally induced loads. A structural member under axial tension or compression loads acting typically at its geometric centroid of the cross section, with little induced bending moment or torsional moment, is referred to as a tie or column, respectively. Beams are members carrying loads transverse to their longitudinal axes, whereas shafts are members subjected to torsional loads along with other loads. In addition, structural members under axial and transverse loads and/or other bending moments are referred to as beam-columns. Schematic diagrams of different loads acting on members are shown in Figure 7.1.

The structural design of pultruded FRP members includes: (1) axial tension members or ties under forces acting at the center of gravity (CG) of members; if forces are not acting at CG or shear center, additional moments are induced, (2) columns under axial compression forces acting at the CG or shear center of the FRP members, (3) beams under transverse loads, (4) beam columns, and (5) bolted and bonded adhesive connections are presented and discussed through various design examples. The design of pultruded FRP structural members is developed based on both the allowable stress design (ASD) and load and resistance factor design (LRFD) methods. However, there is still no universally accepted design standard available yet for pultruded FRP structural members. However, CUR-CROW Recommendations (96: 2019) on "FRPs in Buildings and Civil Engineering Structures" is the most welcoming and timely document on hand, to date. The design procedure of FRP structural members can be followed in a manner similar to the design of conventional

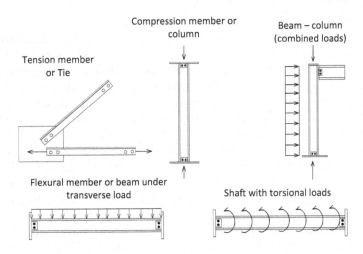

FIGURE 7.1 Classification of structural members.

DOI: 10.1201/9781003196754-7

structural members (steel or timber), using equivalent FRP member stiffness properties provided by the manufacturer. However, one must recognize the difference in material responses of FRPs versus conventional materials, before initiating any structural designs with FRPs.

7.2 DESIGN PHILOSOPHY AND BASIS

The objective of the design philosophy of FRP composite structural members (one-dimensional structural elements) is to maintain the same design approach as in isotropic materials except for using composite material properties (strengths and stiffnesses) in different directions.

At present, there are two different design philosophies of FRP structural members. These are (1) Allowable Stress Design (ASD – Structural Plastics Design Manual, ASCE, 1984) and (2) Load & Resistance Factor Design (LRFD). For LRFD, the draft EUROCOMP design code and handbook were released in 1996, which is yet to be adopted formally. In 2021, the American Composites Manufacturing Association (ACMA) and the American Society of Civil Engineers (ASCE) have released jointly the pre-standard for LRFD of pultruded fiber-reinforced polymer (FRP) structures. However, LRFD design equations based on the pre-standard were developed for FRP profiles or flat plates whose fibers are arranged perpendicular to each other (orientation in 0° and 90°). The design equations are mainly developed based on pultruded glass FRP composite profiles with fiber architectures of symmetric and balanced laminated lay-ups. Many other countries are in the process of developing standardized specifications for designing structures with FRP shapes and joints, but final design specs are yet to be approved, and released as standards.

The fiber orientation in each portion or element (flange and web) of pultruded FRP profiles is laid at least in two directions; however varying in many fiber orientations. To avoid catastrophic failures and to provide quasi (or even semi) isotropic material properties, the minimum fiber volume fraction of each pultruded FRP structural portion or element (web or flange) is suggested to be not less than 30%. In addition, appropriate resistance factors have been determined for pultruded FRP members based on available experimental data which may have to be refined as the availability of data becomes wider than the current level of availability of thermomechanical properties of FRP components. In general, the ASD method has been widely used for FRP structural members because of simple design methodology and because of linear stress–strain responses till about 90% of the material failure strength. In this section, both design philosophies (ASD and LRFD) are presented, recognizing the limitations associated with current manufacturing procedures and design anomalies.

7.2.1 ALLOWABLE STRESS DESIGN (ASD)

In the ASD philosophy, safety factors for rupture are introduced to arrive at working or allowable stress limits. In addition, those allowable stresses must be higher than the induced stresses obtained under external design loads. Typically, safety factors for working stresses will result in design stresses that are of the order (approximately) of 20% of the minimum guaranteed (or nominal or characteristic) long-term laminate stress limits. In addition, the stresses in laminated composites are limited based on an appropriate failure criterion incorporating an appropriate safety margin. Many design codes represent the material characteristic (resistance) strength as the lower fifth percentile of the test data under a given loading condition or corresponding to a specific failure mode. For example, ASTM D30 on composite materials defines the characteristic value as a material property representing 80% lower confidence bound interval on the fifth percentile of specified population and coefficient of variation (COV). A nominal value is computed at five percentile using scale parameters. The COV is found from the Gamma function, and the characteristic value is found by further reducing the nominal value by the confidence factor (Bharil, 2020).

Frequently, the knockdown (environmental) factors are incorporated to account for the loss of strength or stiffness during the service life of laminated composites due to physical or chemical aging effects, moisture effects, size effects, and sustained stresses in a manner similar to those by timber member design codes (NDS, National Design Specification for Wood Construction, 2018). An allowable strength (R_a) is obtained by dividing the nominal strength (R_n) by a safety factor (Ω), and R_a must be greater than or equal to basic load combination (Q):

$$R_a = R_n/\Omega \geq Q \tag{7.1}$$

where

R_a = allowable strength or required strength ($= R_n/\Omega$)

R_n = nominal strength (computed based on member's nominal or characteristic properties)

Ω = factor of safety (based on prior experience, see Table 7.1)

Q = basic load combination defined per applicable design codes.

The allowable or required strength (R_a) must be greater than the sum of stresses induced by service loads (Q). The basic load combination of the ASD method is provided by ASCE 7 as follows:

Load combination 1: D

Load combination 2: $D + L$

Load combination 3: $D + (L_r$ or S or $R)$

Load combination 4: $D + 0.75L + 0.75$ $(L_r$ or S or $R)$

Load combination 5: $D \pm (1.0W$ or $0.7E)$

Load combination 6: $D + (1.0W$ or $0.7E) + 0.75L + 0.75$ $(L_r$ or S or $R)$

Load combination 7: $0.6D \pm (1.0W$ or $0.7E)$

where

D = dead load

L = live load due to occupancy

L_r = roof live load

S = snow load

R = load due to initial rainwater or ice exclusive of the ponding condition

W = wind load

E = earthquake load.

For the ASD approach, the safety factors, Ω, recommended by FRP manufacturers (e.g., Bedford Reinforced Plastics, Creative Pultrusions Inc, Strongwell Corporation, Fiberline, and others) are applied in the design procedures as given in Table 7.1.

For both ASD and LRFD approaches, the nominal strength (R_a) is generally provided by manufacturers or can be computed using laboratory test data. For example, ASTM Committee D30 on Composite Material (2017) describes the nominal value of FRP composites as a statistically based material property representing the 80% lower confidence bound interval (or 20% significance level) on the fifth percentile (Zureick et al., 2006) value of a specified population (n) and COVs. The "3σ-Rule" (mean-3(standard deviation)), applicable to normal distribution discards 99.9%

TABLE 7.1

Safety Factors for Allowable Stress Design (ASD)

Members	Strength	Safety Factor
Pultruded beam	Flexural strength	2.5–3.0
	Shear strength	2.5–3.0
	Buckling strength	2.5
Pultruded column	Compressive strength	3.0
Axial tension member	Tension strength	2.0
Joint and connection	Bearing strength	4.0

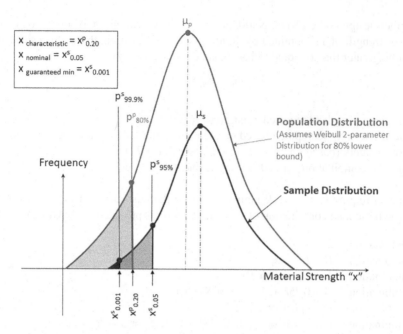

FIGURE 7.2 FRP nominal and characteristics values per ASTM D30 (Bharil, 2020).

values that are unsafe from a design point of view, is used most often. The nominal z-value (number of standard deviations (σ) from mean) is located at 2.33σ for normal distribution functions with a confidence interval of 99% (fifth percentile of lower 20%). The "3σ-Rule" guarantees a 99.9% confidence interval and can serve as a characteristic value in reliability-based design. For additional details, refer to Chapters 2 and 3 of Bharil (2020). The minimum sample size (n) is generally stated as 10, but 50 or more is preferred. This distinction between the nominal strength versus average ultimate strength of structural composites is important for proper application of safety factors (ASD approach) and strength reduction factors (LRFD approach). In lieu of published nominal or characteristic strength, minimum guaranteed strength (sample mean strength–3 × standard deviations) is often used. Figure 7.2 graphically represents the statistical significance of nominal strength. Variable "x" represents the tested material strength, "p" is the probability, and "μ" is the mean value. Sample distribution deals with the material strength test "data-spread" of samples tested from the same manufactured batch. Population distribution is the "data-spread" of samples tested by different manufacturers, at different time periods, and not necessarily from the same manufactured batch.

The ASD procedure is as follows:

1. Load combination:
   - Design loads acting on structural members must be determined. Those design loads include live load, dead load (beam weight + sustained load), impact load, and other loads depending on the function of a structural element under design.
   - To obtain the dead load of FRP members, the size and shape of a beam will be initially assumed based on the designer's previous experience. After assuming the size and shape of FRP elements, equivalent bending and shear rigidity (or stiffness) of the assumed cross section will be determined. In general, equivalent structural properties for design purposes are provided in the manufacturer's design guide.
2. Structural FRP element analysis:
   - After evaluating design loads, maximum load response induced by external load combinations is determined to compute required strength (R_a)

3. Computation of allowable stress and serviceability:
   - Maximum deflection of pultruded FRP members is determined to satisfy deflection and/or rotational limit states. If the maximum deflection or rotation is higher than the prescribed deflection or rotation limit, then the structural members must be redesigned with larger cross-sectional moments of inertia.
   - Strength of pultruded FRP members based on manufacturer's data is defined as the structural strength (ultimate strength). Using the ASD methodology, the allowable stress of pultruded FRP members is determined, using design equations recommended in the specifications.
   - For long-term serviceability (e.g., sustained loads), only 20% of the failure strain is considered as allowable design strain (which includes reductions corresponding to knock-down factors). The knockdown factors are used to account for aging, moisture effects, size effects, sustained stress, etc. in a design. In addition, stress concentration factors have to be incorporated in a design for cross sections with re-entrant angles or concentrated loads or any other structural aspects with holes, notches, etc.
   - If induced stress under external loads is higher than the allowable stress, then structural members must be redesigned.
   - Global and local failures such as lateral torsional buckling, flange buckling, and web crippling must be checked to overcome premature failure under major and multiple loading responses.

7.2.2 LOAD AND RESISTANCE FACTOR DESIGN (LRFD)

The LRFD philosophy accounts for the uncertainty of an event's occurrence in nature. In general, there are two main limit states that need to be considered in any design. These are: (1) serviceability limit state that includes the instantaneous and long-term serviceability of pultruded FRP members controlled by material and structural degradation, vibration and deflection, rotation, human response, etc. and (2) ultimate limit state relates to the structural life safety controlled by the strength of the composite, its stability, failures, and collapse of both the structural component and system. By the acceptable probability of failure approach, FRP composite members are designed to sustain single or multiple loadings (bending, shear, and combinations) that may occur during service life.

To ensure the induced or applied load being less than the failure load, load factors are introduced and applied to service loads. In general, load factors are higher than unity. The sum of all service loads including their corresponding load factors is called the factored load (Q_u). The factored load is a failure load greater than the total actual service load. In addition, nominal strength (R_n) of structural members is reduced using a resistance factor (φ), as in Table 7.3, and time effect factor (λ), as in Table 7.4. These strength reduction factors are typically less than unity.

$$\lambda \varphi R_n \geq Q_u \qquad (7.2)$$

where
  R_n = nominal strength based on the reference (nominal or characteristic) strength of the structural member (adjusted as per end use conditions)
  Q_u = factored load (minimum required member resistance)
  φ = resistance factor
  λ = time effect factor.

It should be noted that the LRFD design equations and load factors of the Pre-Standard (ACMA, 2021) are based primarily on glass fibers, fabrics, and mats only, and not enough test data is available yet to base them on carbon or aramid. The adjustment factors are based on end use conditions which are moisture condition, temperature variation, pH variations including times of exposure, and others.

As mentioned above, the central theme of the LRFD approach is to make sure that the product of the nominal resistance of a material and statistically derived (characteristic value) resistance factor i.e., $(\lambda\phi R_n)$ is greater than the sum of the products of different load factors multiplied with the corresponding load types (factored load effect or Q_u). Typical resistance factors correspond to uncertainties in constituent material properties and the manufacturing processes in addition to the accuracy of equations predicting different stresses induced in a structure. In addition, resistance factors account for the consequences of failure.

Similarly, uncertainties in load factors, in any design, consider type and magnitude of external loads acting on a structure and their combinations. For example, proper accounting of statistical design data for static, wind, and earthquake forces acting simultaneously at peak design magnitudes is a design challenge. Additionally, the design limit states in LRFD account for service limit states (limits on deformations and vibrations), fatigue and fracture limit states, and strength and extreme event limit states.

A complete treatment of LRFD probability theory is beyond the scope of this textbook. Only basic principles of LRFD are briefly explained below. Typically, the resistance part $(\lambda\phi R_n)$ and the load part (Q_u) are statically independent random variables with typical probability distribution shown in Figure 7.3.

By considering the probability of resistance and load combination, the distribution of combined functions can be represented as the margin of safety factor (M) as:

$$M = (\lambda\varphi R_n) - Q_u \qquad (7.3.1)$$

If M is positive, then a margin of safety exists. Thus, structures or members can survive under such load effects. On the opposite side, if M is negative, it means that structures or members cannot sustain the maximum induced loads, and failure occurs under such load scenarios as shown in Figure 7.4.

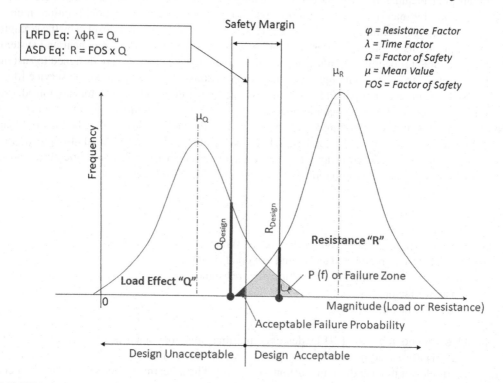

FIGURE 7.3   Probability of the failure defined by relative magnitudes of resistance (R) or load effect (Q) (Bharil, 2020).

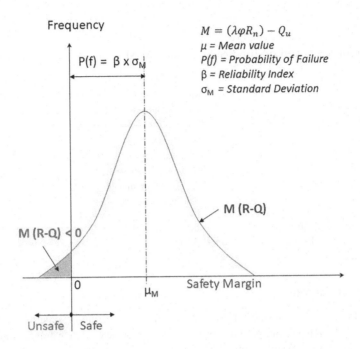

FIGURE 7.4   Probability of combined effect of the resistance and load (Bharil, 2020).

The probability of failure can be presented in the predetermined quantity by specifying that the mean value of safety margin (M_m) is the product of reliability index (β) and the standard deviations (σ_M) as:

$$M_m = \beta\sigma_M \qquad (7.3.2)$$

Thus, the resistance and time factors (φ and λ) depend on the reliability index (β). The selection of reliability index determines the value of resistance and time factor for each limit state. Normally, the probability of failure will be decreased corresponding with an increase in the reliability index (β), resulting in lower resistance and time factor (φ and λ) (Table 7.2).

TABLE 7.2

Probability of Failures Associated with Various Reliability Indices (β) (Bharil, 2020)

Reliability Index (β)	Assessment	Probability of Failure (Pf)
0	Unacceptable	0.5 (1 in 2)
1	Too low	0.159 (1 in 6)
1.5		0.0668 (1 in 15)
2	Low	0.0228 (1 in 44)
2.5		0.00621 (1 in 161)
3	Acceptable	0.00135 (1 in 741)
3.5 (common target) value)		0.000233 (1 in 4,292)
4		0.0000317 (1 in 31,546)
4.5	High	0.0000034 (1 in 294,118)
5		0.000000287 (1 in 3.5 million)
5.5	Too high	0.0000000190 (1 in 53 million)
6		0.000000000987 (1 in 1 billon)

TABLE 7.3
Resistance Factor (φ) Based on the Pre-Standard (ACMA, 2021)

Members	Strength	Resistance Factor φ
Pultruded member (flexural and shear)	Rupture strength	0.65
	Local buckling strength	0.80
	Global buckling strength	0.55
	Local crippling	0.70
	Torsional strength	0.70
Axial compression member	Flexural buckling strength	0.70
	Local buckling strength	0.80
Axial tension member	Tension rupture strength	0.65

Note: For additional details, readers need to consult, "Time-Dependent Reliability Framework for Durability Design of FRP Composites," by R.K. Bharil, Ph.D. Dissertation, WVU, 2020.

7.2.3 RESISTANCE FACTOR

The resistance factor (φ) is considered for uncertainties in any one or several of structural resistances of pultruded FRP members. The uncertainties in structural resistances can be attributed to the influence of variability inherent in the material and geometrical properties, manufacturing, environments, or others. The resistance factor (φ) is used to reduce the predicted nominal strength of structural members. The appropriate resistance factors (φ) provided in the Pre-Standard for LRFD (ACMA, 2021) are presented in Table 7.3.

7.2.4 LOAD COMBINATIONS

The nominal load (or service load) represents the mean maximum lifetime value of each load type. The nominal loads stipulated in ASCE Standard 7 are the Minimum Design Loads for Buildings and Other Structures are accepted herein. For design purposes, the design loads are the product of the nominal loads and the appropriate load factors. Typically, the load factors account for the variability of the design loads due to: (1) uncertainties of load magnitude and then positions on a structure, (2) lack of accuracy in structural analysis, (3) load variations with time, (4) design idealization, and (5) associated assumptions. The load factors provided in ASCE Standard 7 were developed for a 50-year service life of a structure. Pultruded FRP members shall be designed so that their design strength equals or exceeds the required strength (the effects of the factored loads) in the basic load combinations as:

Load combination 1: $1.4D$
Load combination 2: $1.2D + 1.6L + 0.5$ (L_r or S or R)
Load combination 3: $1.2D + 1.6$ (L_r or S or R) + ($0.5L$ or $0.8W$)
Load combination 4: $1.2D + 1.0W + 0.5L + 0.5$ (L_r or S or R)
Load combination 5: $1.2D + 1.0E + 0.5L + 0.2S$
Load combination 6: $0.9D + 1.0W$
Load combination 7: $0.9D + 1.0E$

where
   D = dead load,
   L = live load due to occupancy,
   L_r = roof live load,
   S = snow load,

R = load due to initial rainwater or ice exclusive of the ponding condition,
W = wind load, and
E = earthquake load.

7.2.5 TIME EFFECT FACTOR

The fiber in an FRP composite is generally considered to be elastic to failure, but the viscoelastic properties responsible for creep behavior are typically found in pultruded FRP structural members due to the influence of matrix (cured resin) that helps transfer loads from one fiber to another. The strengths and moduli of pultruded FRP structural members will decrease over time under sustained loads. Bharil (2020) concluded (Figure 7.5) that as an FRP member ages, not only its material strength decreases (shifting of the curve) but also standard deviation increases (flattening of the curve). Both effects can significantly increase the probability of failure and lead to a member's premature failure (by exceeding its severability or strength limit states) if not accounted for in design. The time effect factor is commonly used to compensate for the stiffness reduction during the entire service life of pultruded FRP structural members. The time effect factors for each basic load combination based on the Pre-Standard (ACMA, 2021) are presented in Table 7.4.

7.2.6 OTHER RESISTANCE AND LOAD FACTORS – EUROCOMP (1996)

The resistance factors provided by EUROCOMP (1996) may be used to analyze and design FRP structural elements and systems. The EUROCOMP has a slightly different design philosophy when compared with the pre-standard ACMA – ASCE approach. The material (partial) factor is derived from: (1) material properties, (2) material part production, and (3) environmental effects and duration of loading. The material (partial) factor (γ_m) is defined as (EUROCOMP, 1996):

$$\gamma_m = \gamma_{m,1}\gamma_{m,2}\gamma_{m,3} > 1.5 \tag{7.4.1}$$

where
γ_m = material (partial) factor
$\gamma_{m,1}$ = factor for the evaluation method in which the material properties were obtained
$\gamma_{m,2}$ = the manufacturing process factor
$\gamma_{m,3}$ = the environmental and duration of loading effect on the material property factor.

More information for the FRP resistance factors from the EUROCOMP code is shown in Table 7.5.

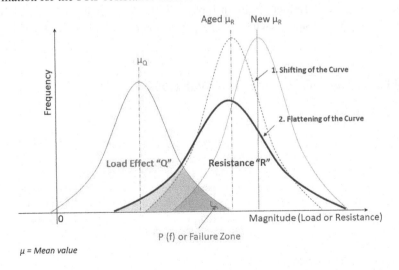

FIGURE 7.5   Effect of time on the combined effect of the resistance and load (Bharil, 2020).

TABLE 7.4

Time Effect Factor λ Based on the Pre-Standard (ACMA, 2021)

Basic Load Combination		Time Effect λ
Load combination 1: $1.4D$		0.4
Load combination 2: $1.2D + 1.6L + 0.5$ (L_r or S or R)	L is from occupancy	0.8
	L is from storage	0.6
	L is from impact	1.0
Load combination 3: $1.2D + 1.6$ (L_r or S or R) + ($0.5L$ or $0.8W$)		0.75
Load combination 4: $1.2D + 1.6W + 0.5L + 0.5$ (L_r or S or R)		1.0
Load combination 5: $1.2D + 1.0E + 0.5L + 0.2S$		1.0
Load combination 6: $0.9D + 1.6W$		1.0
Load combination 7: $0.9D + 1.0E$		1.0
The full design load acts during the entire service life equal to or exceeding 50 years		0.4

Note: For additional information, readers are referred to, "FRP in Buildings and Civil Engineering Structures," CUR-CROW Recommendations, 96: 2019.

To ensure structural safety, partial safety factors for various actions (γ_f, γ_G, γ_Q, and γ_A) are introduced and applied to service loads. The partial safety factors for actions are accounted for the possibility of unfavorable deviations of the actions, uncertainties in the assessment of the effect of actions and the limit states considered, herein, and the possibility of inaccurate modeling of deviations of the actions. In general, the partial safety factors for actions are higher than unity. The sum of all design actions with their own partial safety factors is referred to as the total design scenario. The strength of structural members is reduced using partial safety factors for materials.

$$\left(\gamma_f, \gamma_G, \gamma_Q, \text{and } \gamma_A\right) S < R/\gamma_m \tag{7.4.2}$$

where

S = design value of composite section under actions

R = resistance value corresponding to design actions (loads).

Note that the material (partial) safety factor in Table 7.5 is the denominator for Eq. (7.4.2). The product of the material (partial) safety factors ($\gamma_{m,1}\gamma_{m,2}\gamma_{m,3}$) is the reciprocal of the reduction factor (φ).

TABLE 7.5

γ_m **Material Partial Safety Factor by EUROCOMP (1996)**

Factor	Type		Value
$\gamma_{m,1}$ Material properties	Experimental test		1.15
	Laminate theory		1.5
$\gamma_{m,2}$ Manufacturing process	Fully cured		1.1
	Non-fully cured		1.7
$\gamma_{m,3}$ Environment and loading duration	Short-term-loading	$0 < T < 25°C$ and $T_g > 80°C$	1.0
		$25 < T < 50°C$ and $T_g > 90°C$	
	Long-term-loading	$0 < T < 25°C$ and $T_g > 80°C$	2.5
		$25 < T < 50°C$ and $T_g > 90°C$	

Note: T_g = glass transition temperature of pultruded FRP materials.

(More information on the material (partial) safety factor is provided in Chapter 2, entitled "Basis of Design" (EUROCOMP, 1996).

For infrastructural applications as highway bridges, the AASHTO bridge design specifications (AASHTO LRFD, 2007) provide different limit states to control failures or serviceability needs. There are four different limit states in AASHTO that need to be satisfied. These limit states are (1) strength, (2) extreme event, (3) service, and (4) fatigue and fracture.

In each type of four different limit states, sub-limit states of load combinations are given to cover different design situations such as (AASHTO LRFD, 2007):

1) **Strength limit state:** the strength or stability of structures are to be satisfied to resist the combinations of specific statistically significant loads under design life.
2) **Extreme-event limit state:** the structures must survive a major natural disaster (earthquake or flood) or an accidental event (i.e., when collided by a vessel, vehicle, etc.).
3) **Service limit state:** stress, deformation, and crack width induced under regular service conditions are restricted to certain values to satisfy human comfort or to minimize crack growth rates or other features.
4) **Fatigue and fracture limit state:** restrictions on the stress range due to a design truck occur at the number of stress range excursions during the design life.

The total design load effect for applications of highway bridges is:

$$Q_n = \sum \eta_i \gamma_f Q_i \tag{7.5}$$

- For loads with a maximum value of load factors: $\eta_i = \eta_D \eta_R \eta_i \geq 0.95$
- For loads with a minimum value of load factors: $\eta_i = 1/(\eta_D \eta_R \eta_i) \leq 1.00$.

where
Q_i = load effect
η_i = load modifier.

The load modifiers according to the AASHTO specification are given in Table 7.6. It should be noted that the load modifier (η_i) for the strength limit state should be considered to be 1.05 ($\eta_D = 1.05$ for nonductile components, $\eta_R = 1.00$ and $\eta_I = 1.00$). In addition, a summary of load combinations and factors for bridge applications are given in Tables 7.6–7.8.

For example, load combinations of highway bridges with FRP composite decks, the flexural capacity requirements are determined as follows: For more information on each sub-limit states, interested readers should refer to available bridge design handbooks, textbooks (Fu, 2013; Tonias and Zhao, 2007; Taly, 1997; etc. and AASHTO LRFD, 2007).

Strength I limit state
Design load Combination: $\eta [1.25DL + 1.5DW + 1.75 (LL + IM)]$
- The load modifier (η_i) for the strength limit state is considered to be 1.05.

Service I limit state
Design load Combination: $\eta [1.3 (LL + IM)]$ for deflection
- The load modifier (η_i) for the service limit state is considered to be 1.0.

Fatigue II limit state
Design load Combination: $\eta [0.75 (LL + IM)]$ for deflection
- The load modifier (η_i) for the service limit state is considered to be 1.0.

TABLE 7.6
Load Modifier According to Ductility (η_D), Redundancy (η_R), and Operational Classification (η_I)

Load Modifier	Limit State	Conditions	Value
Ductility (η_D)	Strength	For nonductile components and connections	≥1.05
		For conventional designs and details complying with these specifications	=1.00
		For components and connections for which additional ductility enhancing measures have been specified beyond those required by the specifications	≥0.95
	All others		=1.00
Redundancy (η_R)	Strength	For non-redundant members	≥1.05
		For conventional levels of redundancy, foundation elements where the resistance factors already account for redundancy as specified	=1.00
		For exceptional levels of redundancy beyond girder continuity and a torsionally closed cross section	≥0.95
	All others		=1.00
Operational classification (η_I)	Strength	For critical or essential bridges	≥1.05
		For typical bridges	=1.00
		For relatively less important bridges	≥0.95
	All others		=1.00

Note: Table 7.6 is modified from AASHTO LRFD Bridge Design Specifications (2007) by the American Association of State Highway and Transportation Officials, Washington, DC.

TABLE 7.7
Load Factors for Load Combinations @ Different Limit States

State Limit	DC DD DW EH EV ES	LL IM CE BR PL LS	WA	WS	WL	FR	TU CR SH	TG	SE	EQ	BL	IC	CT	CV
Strength I	γ_p	1.75	1	–	–	1	0.5/1.2	γ_{TG}	γ_{SE}	–	–	–	–	–
Strength II	γ_p	1.35	1	–	–	1	0.5/1.2	γ_{TG}	γ_{SE}	–	–	–	–	–
Strength III	γ_p	–	1	1.4	–	1	0.5/1.2	γ_{TG}	γ_{SE}	–	–	–	–	–
Strength IV	γ_p	–	1	–	–	1	0.5/1.2	–	–	–	–	–	–	–
Strength V	γ_p	1.35	1	0.4	1	1	0.5/1.2	γ_{TG}	γ_{SE}	–	–	–	–	–
Extreme I	γ_p	γ_{EQ}	1	–	–	1	–	–	–	1*	–	–	–	–
Extreme II	γ_p	0.5	1	–	–	1	–	–	–	–	1	1*	1*	1*
*Use one at a time.														
Service I	1	1	1	0.3	1	1	1/1.2	γ_{TG}	γ_{SE}	–	–	–	–	–
Service II	1	1.3	1	–	–	1	1/1.2	–	–	–	–	–	–	–
Service III	1	0.8	1	–	–	1	1/1.2	γ_{TG}	γ_{SE}	–	–	–	–	–
Service IV	1	–	1	0.7	–	1	1/1.2	–	1	–	–	–	–	–
Fatigue I	–	1.5♦	–	–	–	–	–	–	–	–	–	–	–	–
Fatigue II	–	0.75♦	–	–	–	–	–	–	–	–	–	–	–	–

♦ Use for LL, IM, and CE only.

Note: Table 7.7 is modified from AASHTO LRFD Bridge Design Specifications (2007) by the American Association of State Highway and Transportation Officials, Washington, DC.

TABLE 7.8

Load Factors γ_p for Permanent Load Effects

Load Type	Load Factors γ_p	
	Maximum	Minimum
DC		
Component and attachments	1.25	0.90
Strength IV only	1.50	0.90
DW		
Wearing surfaces and utilities	1.50	0.65
EH (Horizontal Earth Pressure)		
Active	1.50	0.90
At rest	1.35	0.90
EL		
Locked in construction stresses	1.00	1.00
EV (Vertical Earth Pressure)		
Overall stability	1.00	–
Retaining walls and abutments	1.35	1.00
Rigid buried structure	1.30	0.90
Rigid frames	1.35	0.90
Flexible buried structures other than metal box culverts	1.95	0.90
Flexible metal box culverts	1.50	0.90
ES		
Earth surcharge	1.50	0.75

Note: Table 7.8 is modified from AASHTO LRFD Bridge Design Specifications (2007) by the American Association of State Highway and Transportation Officials, Washington, DC.

The general design procedure using the LRFD method is as follows:

1. Design load combinations:
   - Design loads with load factors that will be carried by a structural member must be determined using conventional analysis procedure or other numerical approaches. These design loads include live load, dead load (member weight + sustained load), impact load, and any other load as appropriate. Design load combinations for different limit states have to be calculated to obtain load-induced internal forces and deformations of a structure.
   - Size, shape, and fiber architecture will be initially approximated for self-weight computations. After assuming a certain size, shape, and fiber architecture of pultruded FRP members, equivalent rigidity (or stiffness) will be determined. Typically, equivalent structural properties for design purposes are provided in the manufacture's design guide. Both the flexural and shear strengths are provided through the manufacturer's design guide. It should be noted that the theoretical flexural and shear rigidities determined from the fiber architecture may be different from those values obtained through laboratory testing. Rely on Manufacturer data; if not, rely on the computations as per the steps given in Chapter 5 of this book.
2. Structural analysis:
   - After computing design loads for each limit state, load-induced structural responses will be calculated. Member forces induced by the design loads are used to check for the sizes and shapes of members that are assumed in Step 1 above.
3. Computation of strength and serviceability limit state:

- Deflection of FRP members under design loads is determined. If the design load-induced deflection is higher than the deflection limit state, then the structural member has to be redesigned with a larger cross section having a higher moment of inertia than the cross section used in the first trial.
- Based on the manufacturer's data, the ultimate strength that will be resisted by structural members will be determined, using the design load combinations, as per the standard codes (The pre-standard, ACMA – ASCE, 2021; EUROCOMP, 1996 and others).
- The load-induced stresses must be less than the structural strength of an FRP member. If the load-induced stresses are higher than the strength of a structural member, then the structural member will be redesigned with a higher moment of inertia.
4. Check for local stability and deflection limit in service state:
- Global and local failures such as lateral torsional buckling, web crippling, flange buckling, etc. must be checked to overcome failure under varying design load combinations.

7.3   BASIC ASSUMPTION

Many sectional profiles of pultruded FRP structural members are available in the market today. Typically, these profiles are formed with multidirectional fabric-based FRP laminated thin plates (e.g., perpendicular orientation between the flange and web portions). In addition, the structural properties of each lamina rely on fiber architecture, material (fiber and matrix) properties, fiber orientation, etc. However, the most common commercial FRP members can be classified into nearly orthotropic materials. Using such assumptions, pultruded FRP members can be designed in a manner similar to isotropic materials to arrive at equivalent beam rigidities (or stiffnesses) by neglecting the influence of "extensional-shear" and "extensional–flexural" coupling effects for structural responses. The advantage of this assumption is quite simple and very handy to a practical designer. The equivalent stiffness of an FRP member can be presented in terms of constitutive relations as given in Eq. (3.16.3), which is repeated herein as Eq. (7.6).

$$
\begin{bmatrix} \sigma_L \\ \sigma_T \\ \tau_{LT} \end{bmatrix} = \begin{bmatrix} Q_L & Q_{LT} & 0 \\ Q_{LT} & Q_T & 0 \\ 0 & 0 & Q_S \end{bmatrix} \begin{bmatrix} \varepsilon_L \\ \varepsilon_T \\ \gamma_{LT} \end{bmatrix} = \begin{bmatrix} \dfrac{E_L}{1-v_{LT}v_{TL}} & \dfrac{v_{LT}E_T}{1-v_{LT}v_{TL}} & 0 \\ \dfrac{v_{TL}E_L}{1-v_{LT}v_{TL}} & \dfrac{E_T}{1-v_{LT}v_{TL}} & 0 \\ 0 & 0 & G_{LT} \end{bmatrix} \begin{bmatrix} \varepsilon_L \\ \varepsilon_T \\ \gamma_{LT} \end{bmatrix}
$$

(7.6)

For structural profiles, E_L = the longitudinal modulus

E_T = the transverse modulus
v_{LT} = major Poisson ratio
v_{TL} = minor Poisson ratio
G_{LT} = the shear modulus.

The longitudinal (L) and transverse (T) structural coordinates are defined as the axial direction and perpendicular direction to the beam length, respectively, as shown in Figure 7.6.

Stiffness of pultruded FRP profiles can be analytically obtained using classical lamination theory as described in Chapters 4 and 5. Otherwise, experimental tests of pultruded FRP profiles at different levels (i.e., coupon and component level) are performed according to standard test methods to obtain stiffness values. It should be noted that most of the structural stiffnesses provided by the manufacturers are based on experimental results. For design purposes, the web-flange junctions of pultruded

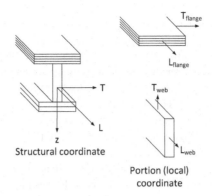

FIGURE 7.6 Structural coordinate in protruded profiles.

FRP profiles are assumed to be the same structural properties as the connecting portions of a structural profile. The load-deformation relations of pultruded FRP members are assumed to be linear in response in the service stage, i.e., up to about 20%–25% of the ultimate response (Prachasaree et.al., 2006).

The minimum requirements of the characteristic values for mechanical properties of pultruded FRP composite structural members are recommended for structural profiles and plates using the Pre-Standard information (ACMA, 2021), as shown in Table 7.9. These properties (Table 7.9) typically are lower than the experimental values for strength and stiffness of different structural FRP shapes.

TABLE 7.9

Minimum Required Characteristic Properties for FRP Composite Profiles (ACMA, 2021)

Mechanical Properties	Minimum Requirement	
	Profiles	**Plates**
Longitudinal tensile strength	30,000	20,000
Longitudinal compressive strength	30,000	24,000
Longitudinal flexural strength	–	30,000
Longitudinal pin-bearing strength	21,000	21,000
Longitudinal tensile modulus	3,000,000	1,800,000
Longitudinal compressive modulus	3,000,000	1,800,000
Longitudinal flexural modulus	–	1,600,000
Transverse tensile strength	7.000	7,000
Transverse compressive strength	–	15,500
Transverse flexural strength	–	13,000
Transverse pin-bearing strength	18,000	13,000
Transverse tensile modulus	800,000	700,000
Transverse compressive modulus	1,000,000	1,000,000
Transverse flexural modulus	–	900,000
In-plane shear strength	8,000	6,000
In-plane shear modulus	400,000	400,000
Interlaminar shear strength	3,500	3,500
Pull through Strength Per Fastener		
$t = 3/8$ in	650 lbs	650 lbs
$t = 1/2$ in	900 lbs	900 lbs
$t = 3/4$ in	1,250 lbs	1,250 lbs

Note: The minimum requirement may vary from one manufacturer to another, and manufacturer-suggested requirements need to be incorporated in designs by designers.

7.4 SUMMARY

FRP components were designed, manufactured, fabricated, and repaired utilizing general codes issued by standard specifying bodies (ASCE, EUROCOMP, CEN, etc.) with design manuals published by individual manufacturers based upon typical products they manufacture following extensive testing and evaluation. In Europe, EUROCOMP Design Code and Handbook (Clarke, 1996) is extensively used to design FRP composite structures, while others are available to a lesser extent; for example, Reinforced Plastic Composites: Specifications for Pultruded Profiles by the European Committee for Standardization (CEN, 2000), and Guide for the Design and Construction of Structures Made of Thin FRP Pultruded Elements by the National Research Council of Italy (CNR, 2008). In the United States, ASCE has released the Structural Plastics Design Manual for the standard for pultruded profiles in 1984. However, it is no longer in print. In 2021, the Pre-Standard for LRFD of pultruded FRP structures was published jointly by ACMA and ASCE. As FRP codes and standards are rapidly evolving, any codes, standards, or design guides should be verified for their current applicability before use. Typically, the design of an FRP structural component based on any codes and specifications are broadly summarized for guidance as follows (GangaRao, 2014):

- The structural and performance requirements (such as size and shape of a component subjected to service and other loads) are identified under service environments (i.e., corrosion, fire, UV degradation, moisture uptake, abrasion resistance, etc.)
- The production and cost targets are established for volume and production rates, fabrication and installation anomalies if any, strength-to-weight ratios including cost per unit structural performance, etc.
- An FRP system configuration and preliminary design are carried out including the selection of constituent materials, manufacturing process, structural response computations targeting design criteria (limit states) before proportioning structural components and developing details for final analyses and design. Detailed evaluations have to be carried out, including full-scale prototype structural testing if needed, to make sure that all design limit states are satisfied. In other words, check for any research, development, and/or implementation work performed through past activities, under a given set of service environments, and prorate for the design at hand.

REFERENCES AND SELECTED BIOGRAPHY

AASHTO LRFD Bridge Design Specifications SI units, 3rd, American Association of State Highway and Transportation Officials (AASHTO), Washington, D.C., 2007.

AASHTO Standard Bridge Design Specifications, American Association of State Highway and Transportation Officials, Washington, D.C., 2004.

ACMA, Pre-Standard for Load & Resistance Factor Design (LRFD) of Pultruded Fiber Reinforced Polymer (FRP) Structures, Arlington, VA, 2021.

ANSI/AISC 360–05, An American National Standard, Specification for Structural Steel Buildings, AISC, Chicago, IL, 2005.

ASCE, Structural Plastics Design Manual, Alexander Bell Drive, Reston, VA, 1984.

ASCE/SEI7–10, Minimum Design Loads for Buildings and Other Structures, Alexander Bell Drive, Reston, VA, 2010.

Bharil, R., Time-Dependent Reliability Framework for Durability Design of FRP Composites – Doctoral Dissertation, West Virginia University, WA, 2020.

Clarke, J.L., *Structural Design of Polymeric Composites – EUROCOMP Design Code and Handbook*, E & FN Spon, London, UK, 1996.

CEN, *Reinforced Plastic Composites: Specifications for Pultruded Profiles Parts 1–3 EN 13706*, European Committee for Standardization (CEN), Brussels, Belgium, 2000.CNR, *Guide for the Design and Construction of Structures made of FRP Pultruded Elements,* National Research Council of Italy, Rome, Italy, 2008.

CUR-CROW, FRP in Buildings and Civil Engineering Structures, 96, 2019.

Fu, G., *Bridge Design and Evaluation: LRFD and LRFR*, John Wiley & Sons, Inc., New York, NY, 2013.

GangaRao, H.V.S., *Chapter 1: The International Handbook of FRP Composites in Civil Engineering* (Zoghi, M., Editor), CRC Press, Taylor & Francis Group, Boca Raton, FL, 2014, pp. 3–12.

Gutowski, T.G., *Advanced Composites Manufacturing*, John Wiley & Sons, Inc., New York, NY, 1997.

Jones, R.M., *Mechanics of Composite Materials*, Hemisphere Publishing, New York, NY, 1975.

Le Riche, R., Saouab, A., and Bréard, J., Coupled compression RTM and composite layup optimization, *Composites Science and Technology*, 63(15), 2003, pp. 2277–2287.

NDS, *National Design Specification (NDS) for Wood Construction*, American Wood Council, Leesburg, VA, 2018.

Nuplex Composites, http://www.nuplex.com/composites/processes/hot-moulding-processes, Australia, 2014.

Prachasaree, W., GangaRao, H.V.S., and Shekar, V., Performance evaluation of FRP bridge deck component under torsion, *Journal of Bridge Engineering, ASCE*, 11(4), 2006, pp. 430–442.

Taly, N., *Design of Modern Highway Bridges*, McGraw-Hill Companies, New York, NY, 1997.

Tonias, D., and Zhao, J.J., *Bridge Engineering: Design, Rehabilitation and Maintenance of Modern Highway Bridges*, 2nd edition, McGraw-Hill Professional, New York, NY, 2007.

Zureick, A.H., Bennett, R.M., and Ellingwood, B.R., Statistical Characterization of Fiber-Reinforced Polymer Composite Material Properties for Structural Design, *The Journal of Structural Engineering*, 132(8), ASCE, 2006, pp. 1320–1327.

8 Design of Pultruded FRP Axial Tension Members

8.1 AXIAL TENSION MEMBERS

Tension member is defined as a structural member subjected to axial loading, which induces only uniform tensile stresses across the cross section of an FRP member. The section is considered to have a symmetric shape about the geometric centroid and is reinforced with a symmetric fabric architecture, i.e. inducing only uniform axial tensile stresses under uniform tensile loads across the cross section. The tension members to be designed herein are assumed to be under uniform tensile stress through the cross section and along the entire length of a member, i.e., any moment induced due to out-of-plane misalignment from manufacturing errors or residual stress build-up is neglected.

Unlike the isotropic members, FRP composite members can be easily subjected to bending and torsion under uniform axial loads if the fabric architecture of a cross section is not symmetric while the cross sectional geometry may be symmetric with reference to its geometric centroid. Therefore, the center of gravity of a cross section is a function of not only the geometric quantities such as flange and web widths and thicknesses but also a function of their axial stiffnesses. The axial stiffness is dependent on the fabric architecture in the web and flange elements. For additional details to compute axial stiffnesses of FRP elements, readers are referred to Parambil (2010) or Chapters 3 and 4 of this textbook.

The pultruded FRP structural members can be field-implemented to resist axial tension in various shapes such as I-shape, L or angle, WF or H-shape, rods, plates, built-up sections, etc. H- and I-shapes are commonly used as columns to carry tension loads. Channels or angles are widely employed as purlins for roofing systems. Rods are often used as hangers in walkways and balconies as tension members, sag rods for supporting purlin in a roof system, or as roof bolts in underground mining operations. Other tension members, over and above the ones identified above, are also found in practice.

For axial tension members, the design procedure is simple and straightforward, assuming that the center of gravity of an FRP section coincides with its geometric center of gravity. Herein, the focus is on sections where the geometric center of gravity coincides with the axial stiffness-based center of gravity (Parambil, 2010) of axial FRP tension members and those members that do not have the stability problems as in compression or flexure. The capacity of pultruded FRP tension members can be obtained from the axial strength limit state. However, great care in the design and detailing of connections of tension members is required to avoid catastrophic failures. In general, the end connections have to be designed to be stronger than the member for the successful design of FRP members under tension.

A tensile load applied along the longitudinal axis of a pultruded FRP member at its center of gravity results in, P_t as:

$$P_t = \frac{P}{A} \tag{8.1}$$

where
P = applied axial tensile load.
A = gross area of a pultruded FRP member.

DOI: 10.1201/9781003196754-8

8.2   NET AREA (A_N)

Typically, a tension member is drilled for bolt holes at its ends for connections. Then, a fraction of the cross-sectional area (gross area) of a tension member is not available to transfer tensile forces. Therefore, a portion of the axial tensile load of a member is carried by the remaining (or net) area. The remaining area of the tension member is defined as the net area (A_n) of the tension member, as shown in Figure 8.1.

It should be noted that any minor shift in the center of gravity of a section due to drilling of holes is neglected in terms of design computations.

Therefore, the axial tensile stress is determined, based on the net area (A_n), which results in simple computation of tensile stress without much loss of accuracy. Typically, the net area is the remaining gross area after removing the material to form holes or bolt connections. The simplest fastening mechanism for end connections contains the same number of holes and arrangement of holes in each transverse section (holes drilled perpendicular to the longitudinal member axis; through the thickness) as shown in Figure 8.1. The net area for such a pattern often refers to the bolt hole arrangements without stagger.

The net cross-sectional area without stagger of pultruded FRP members, shown in Figure 8.2 (Section AA), is determined as follows:

$$A_n = A - \left(\text{area of holes along width}\right)$$
$$A_n = A - \sum_i d_i t_i \tag{8.2}$$

where
d_i = hole diameter (the diameter of a bolt hole is taken as 1/16″ greater than the nominal dimension of the hole and 1/8″ greater than the fastener diameter)
t_i = member thickness at the location of the hole
i = number of holes along the member width.

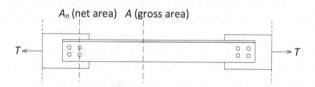

FIGURE 8.1   Gross and net area of the tension member.

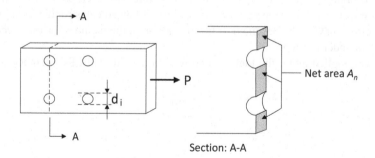

FIGURE 8.2   Net area (A_n) of tension member.

8.3 NET AREA (A_N) WITH STAGGERED BOLT HOLES

For the convenience of fabrication and for adequate load transfer at a joint, several rows of fastener holes are needed in a member. For example, to obtain an increase in the net area of a FRP composite section, the staggered arrangement is possible, as in steel members. However, when a staggered pattern on a member is opted, several possible failure paths must be considered. As shown in Figure 8.3, the distance between the holes perpendicular and parallel to the longitudinal direction of a member is defined as the gage distance (g) and pitch (s), respectively. In general, the net area is determined by subtracting the area of holes from the gross area. The net area is adjusted by adding the expression ($s^2/4g$)t for each diagonal segment in the failure path. The adjustment factor ($s^2/4g$) (t) has been derived and found in the AISC design manual, where "s" is the bolt pitch and "g" is the gage length (transverse spacing), as shown in Figure 8.3.

The net cross-sectional area with stagger of pultruded FRP members is determined as follows:

$$A_n = A - \left(\text{area of holes}\right) + \left(\text{adjusted quantity}\right)$$

$$A_n = A - \sum_i d_i t_i + \sum_j \frac{s_j^2}{4g_j} t_j \tag{8.3}$$

where

d_i = hole diameter (typically 1/8″ to 1/16″ larger than the bolt diameter)

t_i = member thickness at the location of the hole

s = longitudinal center-to-center spacing (pitch) of any two consecutive holes

g = transverse center-to-center spacing (gage) between fastener lines with j number of lines along the width

i = number of holes along the direction perpendicular to the load direction.

The transverse center-to-center spacings (gages) for holes in the adjacent legs of angles shall be the sum of the spacings (gages) from the back of the angles minus the thickness of the angle (ACMA, 2021).

For bolted connections, several failure paths are possible. Thus, it is necessary to consider all possible failure paths to find the critical failure path (i.e., failure path that resists the least amount of induced stress). The critical failure path is typically defined as the failure path providing the minimum net area of the cross section. For several possible failure paths, the net cross-sectional area with stagger in a member is:

$$A_n = \text{Min}\left[A_{n1}, A_{n2}, A_{n3}..A_{nn}\right] \tag{8.4}$$

where

A_{nk} = net area along the possible failure paths ($k = 1, 2, 3... n$).

For example, angles (L-shape) are connected by bolts in both legs for two different bolt arrangements as shown in Figure 8.4. The cross sections of both angles are transformed into equivalent flat

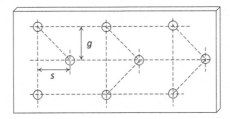

FIGURE 8.3 Staggered holes of tension member.

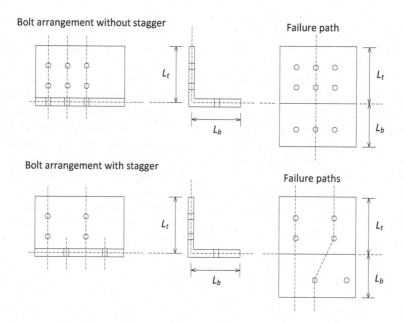

FIGURE 8.4   Examples of failure paths.

plates by using the centerlines of each leg. The intersection of the centerlines is the fixed point for legs of angles unfolded into a plate as shown in Figure 8.4. For bolt arrangement without stagger, the net area at a critical section is the gross area minus the projected area of all bolt holes (or the cross-sectional area lost due to bolt holes). For bolt arrangements with stagger, the net area is obtained from the minimum area of different failure paths. The adjusted quantity ($s^2/4g$) will be added for each gage space with a diagonal segment in the failure path (ACMA, 2021).

8.4   SHEAR LAG

When a structural member is subjected to axial tension, there is only a part of the member cross section that is connecting to the supports. Then the nonconnected parts are free to distort. Through the induced shear loads along the edges under distortion may lead to nonconnected part to warp, albeit very slightly. The stress in the cross section is said to have shear lag due to the shear deformation, as shown in Figure 8.5 (Nagaraj and Gangarao, 1997). Shear lag tends to reduce the structural capacity of the nonconnected part; thus, the effect of shear lag must be considered to compute the nominal design strength of a tension member.

In some situations, the full cross section of a tension member is not connected to the supporting members. When a member is loaded, nonuniform load transfer and deformation are also induced across a section as shown in Figure 8.5. The phenomenon of nonuniform straining in the

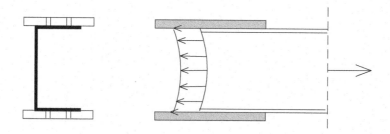

FIGURE 8.5   Shear lag at the end connection of C-channel.

unconnected portion of the web (Fig 8.5) with shear stress concentration near the connection is referred to as shear lag. To account for the effect of shear lag (nonuniform load transfer), the effective net area is to be considered in design.

8.5 EFFECTIVE NET AREA (A_e)

The effective net area (A_e) of a tension member from a design viewpoint is:

$$A_e = UA_n \tag{8.5}$$

where U is shear lag factor as given in Table 8.1 and illustrated in Figure 8.6.

8.6 STRESS CONCENTRATION FACTOR

When notches, openings, copes, and other concentrations exist in a structural member under tension, as shown in Figure 8.7, the stress concentration factor has to be applied to account for strength reduction effects and stress riser effects. Occasionally, experimental tests show that higher strength can be attained (ACMA, 2021) than the strengths including conventional stress concentration factors, in which case(s) appropriate strength adjustments are made. In general, the theoretical stress concentration factor is defined to be the ratio of the maximum stress at the discontinuity to the nominal stress. The nominal stress is calculated based on zero discontinuity of the base materials. In addition, the discontinuity of materials is usually in the form of holes and notches. However, the materials can be damaged near the discontinuous locations under the applied loading. Any damage to the material due to lower stiffness leads to a reduction in the induced maximum

TABLE 8.1
Shear Lag Factor U (ACMA, 2021)

Category	Shear Lag Factor U
Tension transmitted through all of the elements (web as well as flanges) of a cross section	1.0
For members having only two fasteners per line in the direction of tensile stress	0.70
For members having three or more fasteners per line in the direction of tensile stress	0.80

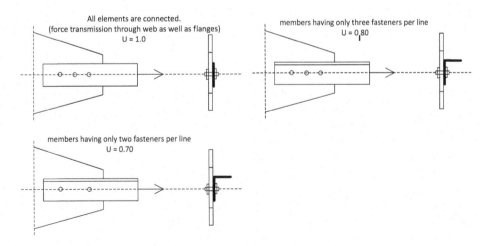

FIGURE 8.6 Detail examples for shear lag factors U.

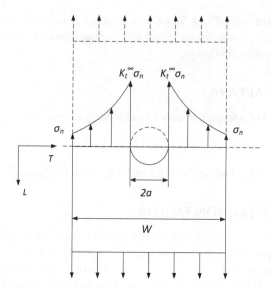

FIGURE 8.7   Stress concentration on a hole.

stress near discontinuous locations. Thus, actual stress concentration factors are lower than the idealized stress concentration factors as presented by Tan (1988). Based on the experimental data, the stress concentration factor for circular hole (K_{SCF}) is recommended to be about 1.6–2.0, for FRP members.

8.7   AXIAL TENSILE STRENGTH

The nominal tensile strength due to rupture of the material is determined as:
   For tension rupture in the gross section:

$$P_n = F_n A \tag{8.6}$$

$$\varphi P_n = \varphi F_n A = 0.65 F_n A \tag{8.7}$$

For tension rupture in the effective net section:

$$P_n = F_n A_e \tag{8.8}$$

$$\varphi P_n = \varphi F_n (U A_n) = 0.65 F_n (U A_n) \tag{8.9}$$

Resistance factor (φ) for failure (rupture) of a section under tension is taken as 0.65 based on existing experimental data.
   For design purpose, the nominal strength with the time effect factor (λ) must be higher than the required strength computed from the factored load, which is given below:
   For tension rupture in the gross section:

$$\lambda \varphi P_n = \lambda \varphi F_n A \tag{8.10}$$

For tension rupture in the effective net section, from Eq. (8.9):

$$\lambda \varphi P_n = \lambda \varphi F_n (U A_n) \tag{8.11}$$

In practice, to account for the strength effect due to bolt holes or openings, the nominal strength (F_n) of a member is reduced using appropriate stress centration factor (K_{SCF}). For members with bolt holes, the nominal strength with the time effect factor and stress concentration factor must be higher than the required strength P_u computed from the factored load, which is:

For tension rupture in the effective net section:

$$\lambda\varphi P_n = \lambda\varphi\left(\frac{F_n}{K_{SCF}}\right)(A_e) = \lambda\varphi\left(\frac{F_n}{K_{SCF}}\right)(UA_n) \geq P_u \qquad (8.12)$$

where
 λ = time factor depending on the load combination (see in Table 7.3)
 F_n = nominal longitudinal tensile strength of pultruded FRP members
 P_n = nominal longitudinal axial capacity (load)
 U = shear lag factor, and
 K_{SCF} = stress concentration factor (1.6–2.0).

Note: the resistance factor φ for the failure of a section under tension rupture is taken as 0.65 based on existing experimental data, which is limited. This number (0.65) may need modification with the availability of additional experimental data of FRP members under tension.

Example 8.1

A single angle pultruded FRP tension member L 4×4×1/2″ is connected to a gusset plate with 7/8″ dia. bolts as shown in Example Figure 8.1. Determine: (a) design strength (LRFD) of the member and (b) the allowable strength (ASD) with 20% of the limiting failure strain for long-term service-ability. Section L 4×4×1/2″: Gross area (A) = 1.94 in² and pitch (s) = 2 in, Stress concentration factor (K_{SCF}) for bolt holes = 1.6, and F_n = 24 ksi (recommended by manufacturer).

Solution

The diameter of a bolt hole is taken as 1/8″ higher than the fastener diameter. Thus, $d_i = (7/8 + 1/8 = 1″)$
 From Eq. (8.2):

$$A_n = A - \sum_i d_i t_i = 1.94 - (1 \times 0.5) = 1.44\,in^2$$

For all members having three or more fasteners per line in the direction of tensile stress, shear lag factor $U = 0.80$. The effective net area (A_e) of a tension member is determined as:

$$A_e = 0.8A_n = 0.8(1.44) = 1.152\,in^2$$

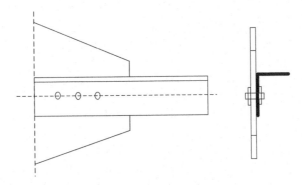

EXAMPLE FIGURE 8.1   Pultruded FRP tension member L4×4×1/2″.

a. Design strength (LRFD):

The time effect factor λ for the design load combination $(1.2D + 1.6L)$ is 0.8. The tensile strength of the pultruded FRP L $4\times4\times1/2''$ is determined as:
From Eq. (8.12):

$$\lambda\varphi P_n = \lambda\varphi\left(F_n/K_{SCF}\right)A_e = 0.8(0.65)\left(24/1.6\,\text{ksi}\right)\left(1.152\,\text{in}^2\right) = 8.98\,\text{kips}$$

b. Determine the allowable strength (ASD):

From Eq. (8.8), the nominal strength effect due to bolt holes is accounted for using K_{SCF}. Thus, the allowable strength (P_a) of the pultruded FRP L $4\times4\times1/2''$ is determined as:

$$P_n = \left(F_n/K_{SCF}\right)A_e = \left(24/1.6\,\text{ksi}\right)\left(1.152\,\text{in}^2\right) = 17.28\,\text{kips thus,}$$

$$P_a = P_n/(\text{F.S.}) = \left(17.28/2\right) = 8.64\,\text{kips}$$

For long-term serviceability, 20% of the limiting failure strain is considered to be corresponding to the allowable design stress (safety factor including the knockdown factor F.S. = 5) (ACMA, 2021). The allowable strength (P_{aL}) including long-term effects is:

$$P_{al} = \left(P_n\right)/\text{F.S.} = \left(17.28/5.0\right) = 3.46\,\text{kips}$$

Note: The allowable strength (P_{aL}) for long-term serviceability is the allowable strength (P_a) with an additional factor of safety (F.S.) (knockdown factor) accounting for long-term serviceability.

Example 8.2

Determine the tensile strength of pultruded FRP WF $8\times8\times1/2''$ beam. The member end connection has two lines of 7/8'' dia. bolts in each flange, three in each line. Self-weight = 9.23 lbs/ft; Area (A) = 11.51 in²; Web area (A_w) = 3.50 in²; flange thickness = 0.5'', and longitudinal tensile strength (F_t) = 30,000 psi. Use both the load resistance factor design (LRFD) and allowable stress design (ASD) methods with 20% of the limiting failure strain for long-term serviceability. Stress concentration factor (K_{SCF}) for bolt holes = 2.0.

Solution

Determine the net area of pultruded FRP member WF $8\times8\times1/2''$:

The diameter of a bolt hole is taken as 1/8'' greater than the fastener diameter. Thus, $d_i = (7/8 + 1/8 = 1'')$. The critical section passes through four bolts (two in each flange).
From Eq. (8.2):

$$A_n = A - \sum_i d_i t_i = 11.51 - \left(4\times1\times0.5\right) = 9.53\,\text{in}^2$$

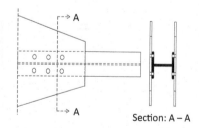

EXAMPLE FIGURE 8.2   Pultruded FRP tension member WF $8\times8\times1/2''$.

For all members having three or more fasteners per line in the direction of tensile stress, shear lag factor $U = 0.80$. The effective net area (A_e) of tension members in Eq. (8.5) is determined as:

$$A_e = 0.8A_n = 0.8(9.53) = 7.62\,\text{in}^2$$

a. Design strength (LRFD):
   For λ (for $1.2D + 1.6L$) = 0.8, the tensile strength of the pultruded FRP WF $8 \times 8 \times 1/2''$ beam is:
   From Eq. (8.12):

$$\lambda\varphi P_n = \lambda\varphi\left(F_n/K_{SCF}\right)A_e = 0.8(0.65)(30/2\,\text{ksi})(7.62\,\text{in}^2) = 59.5\,\text{kips}$$

Note: Load factors have to be applied for comparison with ASD results.

b. Determine the allowable strength (ASD):
   For long-term serviceability, 20% of the limiting failure strain is considered as allowable design strain (safety factor including the knockdown factor F.S. =5.0 and $K_{SCF} = 2.0$). From Eq. (8.8) including K_{SCF}, the long-term allowable strength (P_{aL}) of the pultruded FRP WF $8 \times 8 \times 1/2''$ is determined as:

$$P_n = \left(F_n/K_{SCF}\right)A_e = (30/2\,\text{ksi})(7.62\,\text{in}^2) = 114\,\text{kips}$$

$P_a = (P_n/\text{F.S.}) = (114/2) = 57.2$ kips and P_{aL} (with 20% limiting strain) $= (P_n/\text{F.S.}) = (114/5.0) = 22.9$ kips

Example 8.3

A pultruded FRP tension member $L\ 6 \times 4 \times 1/2''$ with staggered fasteners as shown in Example Figure 8.3.1. Diameter holes are for $1/2''$ dia. fasteners. Determine: (a) design strength (LRFD) and (b) the allowable (ASD) method with 20% of the limiting failure strain for long-term serviceability. Pultruded FRP angle properties (manufacturers):

Longitudinal tensile strength (F_t) = 24,000 psi
Longitudinal modulus (E_L) = 2 (10^6) psi
Stress concentration factor (K_{SCF}) for bolt holes = 1.6.

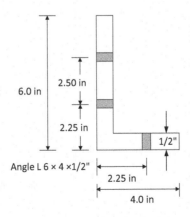

EXAMPLE FIGURE 8.3.1   Angle pultruded FRP member.

Solution

Each angle is flattened out into a single plate then the least net area may be determined. It should be noted that the gross width is taken as the sum of leg widths minus the angle thickness (2.25+2.25−0.5 = 4″). The net cross-sectional area with stagger of pultruded FRP members (see Example Figure 8.3.2) is determined as follows:

Section L 6×4×1/2″: Gross area (A) = 4.75 in², pitch spacing (s) = 1.5″, gage spacing (g) = 2.5″ and 4.0″.

From Eq. (8.3):

$$A_n = A - \sum_i d_i t_i + \sum_j \frac{s_j^2}{4g_j} t_j$$

The diameter of a bolt hole is taken as 1/8″ greater than the fastener diameter. Thus, d_i = 5/8″

Line A–B–C–D–E: A_n = 4.75−3 (0.5×5/8)+[0.5×1.5²/(4×2.5)] + [0.5×1.5²/(4×4.0)] = 4.0 in²
Line F–G–H–I: A_n = 4.75−2 (0.5×5/8) = 4.125 in²
Line J–K–L–M: A_n = 4.75−2 (0.5×5/8)+[0.5×1.5²/(4×2.5)] = 4.24 in²
Line N–O–P–Q: A_n = 4.75−2 (0.5×5/8)+[0.5×1.5²/(4×4.0)] = 4.20 in².

The net area of a pultruded FRP tension member L 6×4×1/2″ with staggered fasteners is determined by following the failure path of Line A–B–C–D–E because it resulted in the smallest value of A_n of the above four cases. For all members having three or more fasteners per line in the direction of tensile stress, shear lag factor U = 0.80 (Table 8.1). The effective net area (A_e) of tension members in Eq. (8.5) is determined as:

$$A_e = 0.8 A_n = 0.8(4.0) = 3.20\,\text{in}^2$$

a. Design strength (LRFD):

The time effect factor λ for the designed load combination (1.2D + 1.6L) is 0.8. The tensile strength of the pultruded FRP L 6×4×1/2″ with staggered fasteners is determined as:

From Eq. (8.12):

$$\varphi \lambda P_n = \varphi \lambda \left(F_n / K_{SCF} \right) A_e = 0.65(0.8)(24/1.6\,\text{ksi})(3.2\,\text{in}^2) = 24.96\,\text{kips}$$

Note: Load factors are applied for comparison with ASD results.

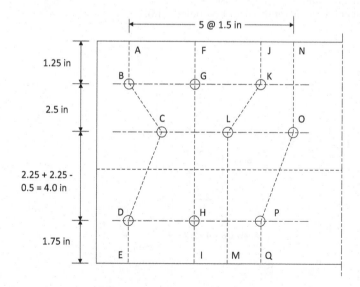

EXAMPLE FIGURE 8.3.2   Failure paths for bolt arrangement with stagger.

b. Allowable strength (ASD):

From Eq. (8.8), the allowable strength (P_a) of the pultruded FRP L 6×4×1/2″ with staggered fasteners and stress concentration factor is determined as:

$$P_n = \left(F_n/K_{SCF}\right)A_e = \left(24/1.6\,\text{ksi}\right)\left(3.20\,\text{in}^2\right) = 48.0\,\text{kips}$$

$$P_a = \left(P_n/\text{F.S.}\right) = \left(48.0/2\right) = 24.0\,\text{kips}$$

For long-term serviceability, 20% of the limiting failure strain is considered as allowable design stress (safety factor including the knockdown factor F.S. = 5.0). The allowable strength (P_{aL}) for long-term serviceability is:

$$P_{aL} = \left(P_n/\text{F.S.}\right) = \left(48.0/5.0\right) = 9.60\,\text{kips}$$

Note: The allowable strength (P_{aL}) for long-term serviceability is the allowable strength (P_{aL}) with an additional F.S. (knockdown factor) accounting for long-term serviceability conditions, including creep and durability.

8.8   SLENDERNESS AND DEFORMATION LIMITATION

The ratio of unbraced length to the least radius of gyration for a cross section of a member is defined as the slenderness ratio of a member. The recommended limit for pultruded FRP axial tension members is provided in the pre-standard (ACMA, 2021) as:

$$L/r \leq 300 \tag{8.13}$$

where
   L = unbraced length of a pultruded FRP member; r = radius of gyration about the weak axis of a cross section.

8.9   BLOCK SHEAR

Block shear failure occurs when one or more segments (or blocks) of plate material tears out from the end of a tension member or a gusset plate. Typically, failure segments are rectangular in shape due to both the shear along the longitudinal axis and tension in the perpendicular direction to the longitudinal axis. Block shear segments on the tension member with single and double bolt rows are illustrated as shown in Figure 8.8. Thus, block shear capacity may be defined as the summation of both parts: (1)

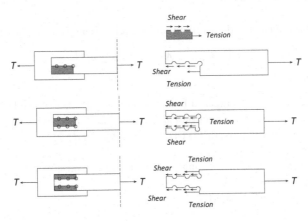

FIGURE 8.8   Block shear failure of the tension member.

tension rupture strength provided by the side of the block perpendicular to the load causing tension and (2) shear rupture strength provided by the side of the block parallel to the load causing tension. The nominal block shear strength based on the pre-standard (ACMA, 2021) is given as:

The connection force is concentric to a group of bolts, tensile and parallel to the direction of FRP material so that lower-bound resistance is computed; thus resulting in the most conservative design values. The block shear resistance of the member is:

$$R_{bs} = \frac{1}{2}\left(A_{ns}F_{sh} + A_{nt}F_L^t\right) \tag{8.14}$$

where the quantities within the parenthesis represent forces resisted by net areas under shear and tension, respectively, which are multiplied by corresponding in-plane shear and tensile strengths; additionally multiplied by (1/2) to correspond to the average force distributions along the planes of failure under shear and tension.

The block shear resistance with stress concentration factor (K_{SCF}) of the member is:

$$R_{bs} = \frac{1}{2}\left(A_{ns}\frac{F_{sh}}{K_{SCF}} + A_{nt}\frac{F_L^t}{K_{SCF}}\right) \tag{8.15}$$

For a bolt group subjected to eccentric in-plane loading, the block shear resistance is:

$$R_{bs,e} = \frac{1}{2}\left(A_{ns}F_{sh} + \frac{A_{nt}F_L^t}{2}\right) \tag{8.16}$$

The block shear resistance with factor (K_{SCF}) for a bolt group subjected to eccentric in-plane loading is:

$$R_{bs,e} = \frac{1}{2}\left(A_{ns}\frac{F_{sh}}{K_{SCF}} + \frac{A_{nt}F_L^t}{2K_{SCF}}\right) \tag{8.17}$$

where

F_L^t = tensile strength in the longitudinal (or pultrusion) direction of FRP material (primary force resistance of FRP section)

F_{sh} = in-plane shear strength of FRP material appropriate to the shear-out failure

A_{nt} = net area subjected to tension

A_{ns} = net area subjected to shear

K_{SCF} = stress concentration factor.

The net area (A_{nt}) under tension is determined for staggered bolted connection scenario as:

$$A_{nt} = \text{Max}\left[A \text{ or } t\left(nd_n - \Sigma b_s\right)\right] \tag{8.18}$$

where

A_{nt} = net area in tension with staggered bolts which is the greater of (1) the maximum of the sectional area (A) in any cross-section (subtract bolt area in a given row at a section) perpendicular to the member axis, or (2) (A) deducted by t (n d_n–Σ b_s)

d_n = nominal diameter of hole

n = number of holes extending in any diagonal or zig-zag line progressively across the number or part of the number ($n_{max} = 3$)

b_s = lesser of ($s^2/4g_s$) or $0.65g_s$ (ACMA, 2021).

For additional explanation, refer to Eq. (8.3).

Example 8.4

A pultruded FRP tension member L 6×6×1/2" with fasteners as shown in Example Figure 8.4. Hole is for 7/8" dia. fasteners. The nominal hole diameter is taken as 1/8" larger than the nominal bolt diameter (7/8"). Determine the block shear resistance by (a) strength design (LRFD) (b) (ASD) methods with 20% of the limiting failure strain for long-term serviceability.

Pultruded FRP angle properties as per (manufacturers):

Longitudinal tensile strength (F_t) = 24,000 psi, longitudinal modulus (E_L) = 2 (10⁶) psi, in-plane shear strength (F_s) = 6,000 psi, stress concentration factor (K_{SCF}) for bolt holes = 1.6.

Solution

The nominal hole diameter is taken as 1/8" larger than the nominal bolt diameter (7/8"). The shear (A_{ns}) and tension (A_{nt}) area of the block shear segment in Example Figure 8.4 are determined as:

Number of 7/8" bolt holes are 2.5 holes for shear. L-shape thickness is 1/2".
Net area subjected to shear A_{ns} = (2.5+3.0+3.0)(1/2)−[2.5×(7/8+1/8)](1/2) = 4.25−1.25 = 3.0 in²
Number of 7/8" bolt holes are 0.5 holes for tension. L-shape thickness is 1/2".
Net area subjected to tension A_{nt} = (2.5 ×1/2)−[0.5×(7/8+1/8)](1/2) = 1.0 in²

The above computations are valid because the full cross section of a tension member is not connected to the supporting members. Thus, the effect of shear lag is considered using the shear lag factor U. The effective net area (A_e) of the tension member is determined as:

For members having three or more fasteners per line in the direction of tensile stress U = 0.80 (Table 8.1).
From Eq. (8.5):

Effective area subjected to shear A_{ns} = 0.80(3.0) = 2.4 in²
Effective area subjected to tension A_{nt} = 0.80(1.0) = 0.8 in².

The connection force is concentric to the group of bolts, tensile and parallel to the direction of FRP material. The block shear resistance with stress concentration factor (K_{SCF}) of the member is determined as:
From Eq. (8.15):

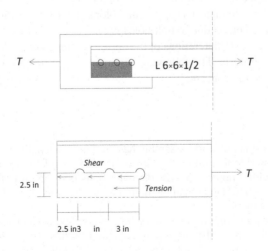

EXAMPLE FIGURE 8.4   Block shear failure of pultruded FRP tension member L 6×6×1/2".

$$R_{bs} = \frac{1}{2}\left(A_{ns} \frac{F_{sh}}{K_{SCF}} + A_{nt} \frac{F_L^t}{K_{SCF}} \right) = \frac{1}{2}\left(2.4 \times \frac{6{,}000}{1.6} + 0.8 \times \frac{24{,}000}{1.6} \right) = 10{,}500\,\text{lbs} = 10.5\,\text{kips}$$

a. LRFD method:

From Table 7.3, λ (for $1.2D + 1.6L$) = 0.8. The block shear strength of the pultruded FRP L $6\times6\times1/2''$ is determined as:

$$\varphi\lambda R_{bs} = 0.65(0.8)(10.5) = 5.46\,\text{kips}$$

Note: Herein, block shear (5.46 kips) has to be divided by load factors for proper comparison with ASD results (as given below) for long-term servicability.

b. ASD method:

The allowable block shear strength of the pultruded FRP L $6\times6\times1/2''$ is determined as (with F.S. = 2.0):

$$P_a = \left(R_{bs}/F.S. \right) = \left(10.5/2 \right) = 5.25\,\text{kips}$$

For long-term serviceability, 20% of the limiting failure strain is considered as allowable design stress (safety factor including the knockdown factor F.S. = 5.0).

For long-term serviceability:

$$P_{al} = \left(P_n/F.S. \right) = \left(10.5/5.0 \right) = 2.1\,\text{kips}$$

Note: The allowable strength (P_{al}) for long-term serviceability is the allowable strength (P_a) with an additional F.S. (knockdown factor) to be incorporated in design for long-term serviceability.

8.10 DESIGN OF PULTRUDED FRP TENSION MEMBER

The design of pultruded FRP tension members involves the inclusion of serviceability and strength limits of tension members. First, the deformation and slenderness limit should be considered for a member to be selected to satisfy the design loads. Then, the trial connection of the selected member must be estimated. Since the actual number of bolts is not known yet at this step, the nominal strengths of the members due to material rupture and block shear failure are to be determined separately. The nominal strength effect due to bolt holes is recommended to account for an appropriate stress concentration factor. Several iterations of the abovementioned procedure may be necessary in order to satisfy all design criteria. The following examples of pultruded FRP tension members are illustrated to lay out the design concept.

Example 8.5

A tension member of a small platform assembly has an effective length of 4 ft (between bolts on either end) and is stressed in tension by 0.7 kips of dead load and 1.4 kips of live load. Select a pultruded FRP angle with 2-number of 1/2″ diameter bolts per line on both ends. The nominal hole diameter is taken as 1/16″ larger than the nominal bolt diameter (1/2″). Use the LRFD method to design the tension member.

Pultruded FRP angle properties (from manufacturers): longitudinal tensile strength (F_t) = 24,000 psi, longitudinal modulus (E_L) = 2 (10^6) psi, shear strength (F_s) = 8,000 psi, stress concentration factor (K_{SCF}) for bolt holes = 1.6 (locations of bolts are assumed to satisfy the minimum design criteria as shown in Example Figure 8.5.1).

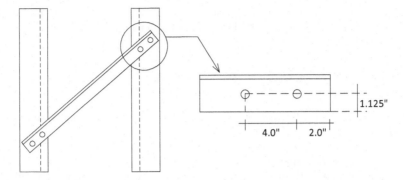

EXAMPLE FIGURE 8.5.1   Tension member.

Solution

a. Determine factored and working loads:
   For the ultimate load state, the load factors for dead and live loads are 1.2 and 1.6, respectively. Then, the total ultimate load (P_u) including load factors is:

$$P_u = 1.2(0.7) + 1.6(1.4) = 3.08 \, \text{kips}$$

The required working design load (P) of an axial tension member is:

$$P_s = 1.0(0.7) + 1.0(1.4) = 2.10 \, \text{kips}$$

b. Trial section with the deformation and slenderness limit:
   The recommended limit for pultruded FRP axial tension members is:
   From Eq. (8.13):

$$\frac{L}{r} = \frac{48}{r} \le 300 \quad \text{and} \quad r = \sqrt{\frac{I_{weak}}{A}} \ge 0.16$$

A pultruded FRP angle of equal legs (L shape) of $2 \times 2 \times 1/4''$ is selected for this design.

$$\frac{L}{r} = \frac{48}{0.61} = 78.7 \le 300$$

The gross area of $2 \times 2 \times 1/4''$ angle is 0.94 in². The radius of gyration (r) is 0.61″. Therefore, the selected pultrude FRP angle (L shape) of $2 \times 2 \times 1/4''$ satisfies the recommended limits of slenderness ratio.

c. Check for tensile strength:
   The pultruded FRP equal leg angle will be connected to other members using two bolts per line in each end. For 1/2″ dia. bolt, the net cross-sectional area of pultruded FRP members is determined as follows (the diameter of a bolt hole is taken as 1/16″ greater than the fastener diameter):
   From Eq. (8.2):

$$A_n = A - dt = 0.94 \, \text{in}^2 - 0.25 \left(\frac{1}{2} + \frac{1}{16} \right) \text{in}^2 = 0.80 \, \text{in}^2$$

The resistance factor φ for axial tension members in the LRFD method is 0.65 (tension rupture). The selected member has only two fasteners per line in the direction of tensile stress. Therefore, the factor accounted for shear lag U is 0.70 (Table 8.1).

Including the time effect factor λ (for $1.2D + 1.6L$) = 0.8 and K_{SCF} = 1.6 in the design, the tensile strength of the pultruded FRP $2 \times 2 \times 1/4''$ angle is determined as:
From Eq. (8.12):

$$\varphi \lambda P_n = \varphi \lambda \frac{F_n}{K_{SCF}} A_e = \varphi \lambda \frac{F_n}{K_{SCF}} (UA_n) = 0.65(0.8)\left(\frac{24}{1.6}\right)(0.7 \times 0.80) = 4.36 \, \text{kips}$$

The tensile strength of the member (4.36 kips) is higher than the required strength (3.08 kips).

d. Check for block shear strength:

The nominal hole diameter is taken as $1/16''$, which is larger than the nominal bolt diameter ($1/2''$). The shear (A_{ns}) and tension (A_{nt}) areas of the block shear segment as shown in Example Figure 8.5.2 are determined as:

Number of $1/2''$ bolt holes are 1.5 holes for shear. The selected L-shape member has a thickness of $1/4''$.

Net area subjected to shear $A_{ns} = (4.0 + 2.0)(1/4) - [1.5 \times (1/2 + 1/16)](1/4) = 1.5 - 0.21 = 1.29$ in^2

Number of $1/2''$ bolt holes are 0.5 holes for tension. The selected L-shape member has a thickness of $1/4''$.

Net area subjected to tension $A_{nt} = (1.125 \times 1/4) - [0.5 \times (1/2 + 1/16)](1/4) = 0.28 - 0.070 = 0.21$ in^2.

Since the full section of the tension member is not connected to the supporting members, the shear lag effect is considered. For all members having only two fasteners per line in the direction of tensile stress $U = 0.70$ (Table 8.1), the effective net area (A_e) of tension members is determined as:

Effective area under shear $A_{ns} = 0.70(1.29) = 0.90$ in^2
Effective area under tension $A_{nt} = 0.70(0.21) = 0.15$ in^2

The connection force under tension is concentric to the group of bolts and parallel to the direction of FRP material. The time effect factor λ for the design load combination ($1.2D + 1.6L$) is 0.8. The block shear strength with a stress concentration factor of the pultruded FRP L $2 \times 2 \times 1/4''$ is determined as:
From Eq. (8.15):

$$R_{bs} = \frac{1}{2}\left(A_{ns}\frac{F_{sh}}{K_{SCF}} + A_{nt}\frac{F_L^t}{K_{SCF}}\right) = \frac{1}{2}\left(0.90 \times \frac{8,000}{1.6} + 0.15 \times \frac{24,000}{1.6}\right) = 3,375 \, \text{lbs}$$
$$= 3.38 \, \text{kips}$$

$$\lambda \varphi R_{bs} = 0.65(0.8)(3.38) = 1.75 \, \text{kips}$$

The nominal resistance of the pultruded FRP member (1.75 kips from the block shear capacity) is lower than the ultimate design load (3.08 kips). Therefore, redesign is needed with a larger section or using two angles $2 \times 2 \times 1/4''$ (back to back).

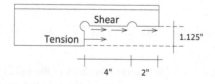

EXAMPLE FIGURE 8.5.2 Block shear segment of $2 \times 2 \times 1/4''$.

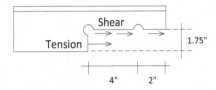

EXAMPLE FIGURE 8.5.3 Block shear segment $3 \times 3 \times 3/8''$.

e. Re-design with a larger section:

A pultruded FRP angle of equal legs (L shape) of $3 \times 3 \times 3/8''$ is selected for re-design. Gross area $(A) = 2.11$ in². Radius of gyration $(r) = 0.91''$. The selected pultruded FRP angle (L shape) of $3 \times 3 \times 1/4''$ satisfies the recommended limits of slenderness ratio due to a larger section. Block shear resistance must only be checked in this redesign (locations of bolts are assumed to satisfy the minimum design criteria as shown in Example Figure 8.5.2).

The nominal hole diameter is taken as 1/16″ larger than the nominal bolt diameter (1/2″). The shear (A_{ns}) and tension (A_{nt}) areas of the block shear segment as shown in Example Figure 8.5.3 are determined as:

Number of 1/2″ dia bolt holes are 1.5 holes for shear. L-shape thickness = 3/8″.
Net area subjected to shear A_{ns}=(4.0+2.0)(3/8)−[1.5×(1/2+1/16)](3/8)=2.25−0.32=1.93 in²
Number of 1/2″ dia bolt holes are 0.5 holes for tension. L-shape thickness = 3/8″.
Net area subjected to tension A_{nt}=(1.75×3/8)−[0.5×(1/2+1/16)](3/8)=0.66−0.11=0.55 in²
Since the full section of the tension member is not connected to the supporting members, the shear lag effect is considered. Using the shear lag factor U (0.70), for all members having only two fasteners per line in the direction of tensile stress (Table 8.1). The effective net area (A_e) of tension members in Eq. (8.5) is determined as:

Effective area under shear A_{ns}=0.70(1.93)=1.35 in²
Effective area under tension A_{nt}=0.70(0.55)=0.39 in²

The connection force under tension is concentric to the group of bolts and parallel to the direction of FRP material. The time effect factor λ for the design load combination (1.2D + 1.6L) is 0.8. The block shear strength with a stress concentration factor of the pultruded FRP L $2 \times 2 \times 1/4''$ is determined as:

From Eq. (8.5):

$$R_{bs} = \frac{1}{2}\left(A_{ns}F_{sh} + A_{nt}F_L^t\right) = \frac{1}{2}\left(1.35 \times \frac{8,000}{1.6} + 0.39 \times \frac{24,000}{1.6}\right) = 6,300\,\text{lbs} = 6.30\,\text{kips}$$

$$\lambda\varphi R_{bs} = 0.65(0.8)(6.30) = 3.28\,\text{kips}$$

The nominal resistance of the selected pultruded FRP member, $3 \times 3 \times 1/4''$ (3.28 kips from the block shear capacity), is higher than the ultimate design load (3.08 kips). The slenderness ratio is still under the suggested limit of pultrude FRP tension profiles.

Example 8.6

Using the LRFD method, design FRP tension rods to support the purlins of a roof having a span of 84 ft and a rise of 21 ft. FRP tension rods are placed at quarter points of each purlin. The span of roof trusses is 24 ft long as shown in Example Figure 8.6.1. The total dead loads due to self-weight of roofing and purlins are estimated to be 12 psf. In addition, the snow load is designed to be 24 psf.

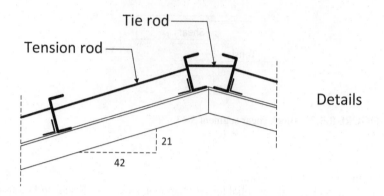

EXAMPLE FIGURE 8.6.1   Roof plan with details.

Solution

   a. Determine design load for tension rod:
        From roof plan in Example Figure 8.6.2, spacing of tension rods = the span of roof trusses/4 = 24/4 = 6 ft. Inclination of roof = $\tan^{-1}(21/42)$ = 26.6°.
   Tributary area A of the roof surface per line of tension rods is determined as:

   $A = (84/2 \times 6)/\cos(26.6°) = 282\,\text{ft}^2$
   Dead loads $(D) = 12\text{ psf} \times 282\,\text{ft}^2 = 3,384\text{ lbs} = 3.38\text{ kips}$
   For snow load, the horizontal projection of the roof area = $42 \times 6 = 252\,\text{ft}^2$
   Snow load, $S = 24\text{ psf} \times 252\,\text{ft}^2 = 6,048\text{ lbs} = 6.05\text{ kips}$
   Service load, $W_s = 3.38 + 6.05 = 9.43\text{ kips}$
   Factored load, $W_u = 1.2(3.38) + 1.6(6.05) = 13.74\text{ kips}$.

   b. Design the tension rod parallel to the roof truss:
        From Example Figure 8.6.3, the tension rods parallel to the roof truss are subjected only to the vertical component of the roof load. Then, the tensile force per rod (a span between purlins) is determined as:

   $$T_s = 9.43\text{ kips} \times \sin(26.6°) = 4.22\text{ kips} \quad \text{and} \quad T_u = 13.74\text{ kips} \times \sin(26.6°) = 6.15\text{ kips}$$

   b.1) Trial section with the slenderness limit:
        The recommended limit for pultruded FRP axial tension members is:
            From Eq. (8.13):

   $$\frac{L}{r} \le 300 \text{ then; } \frac{8 \times 12}{300} = 0.320 \left(\text{should be less than the radius of gyration, } r, \text{of the rod} \right)$$

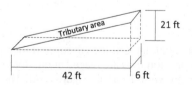

EXAMPLE FIGURE 8.6.2   Tributary area of the roof surface.

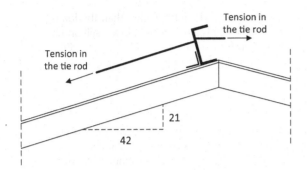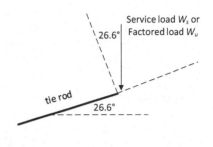

EXAMPLE FIGURE 8.6.3 Tension in the tension and tie rods.

A pultruded FRP round rod of 1.5″ dia. is selected for this application. The gross area (A) of the FRP round rod is 1.768 in². The radius of gyration r is 0.375″. F_n is 50 ksi (as suggested by manufacturers).
b.2) Check for tensile strength:
For $\varphi = 0.65$ (tension rupture) and λ (for $1.2D + 1.6S$) = 0.75, the tensile strength of a 1.5″ dia. pultruded rod is: ($A_e = 0.75A_g$ which accounts for the threaded part). From Eq. (8.10):

$$\phi\lambda P_n = \phi\lambda F_n A_e = 0.65(0.75)(50\,\text{ksi})(0.75\times1.768) = 32.3\,\text{kips}$$

The resistance of the pultruded FRP rod (32.3 kips) is much higher than the ultimate design load (6.15 kips). In addition, the axial deformation and slenderness ratio are still under the suggested limit states of pultruded FRP tension profiles. It should be noted that the proposed cross section is controlled by the serviceability limit state.
c. Design a horizontal tie rod between ridge purlins:
The tie rod between ridge purlins is subjected to the horizontal component of the roof load only and is as shown in Example Figure 8.6.4. The tension load per rod is:
T_s = (4.22 kips)/cos 26.6° = 4.72 kips and T_u = (6.15 kips)/cos 26.6° = 6.88 kips
c.1) Trial section with the slenderness limits:
The length of the tie rod is assumed to be 2.5 ft. The recommended limit for pultruded FRP axial tension members in Eq. (8.13) is:

$$\frac{L}{r} \le 300 \quad \text{then;} \quad \frac{2.5\times12}{300} = 0.10 \le r$$

A pultruded FRP round rod of 1.5″ dia is recommended for design checks. This rod is identical to the tension rod parallel to the roof truss. Gross area (A) = 1.768 in². Radius of gyration (r) = 0.375″.
c.2) Check the tensile strength:
For $\varphi = 0.65$ (tension rupture) and λ (for $1.2D + 1.6S$) = 0.75, the tensile strength of the pultruded FRP rod of 1.5″ dia. is: ($A_e = 0.75A_g$; accounted for threaded parts) From Eq. (8.10):

$$\phi\lambda P_n = \phi\lambda F_n A_e = 0.65(0.75)(50\,\text{ksi})(0.75\times1.768) = 32.3\,\text{kips}$$

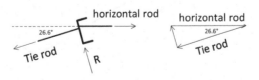

EXAMPLE FIGURE 8.6.4 Free body diagram of a ridge purlin.

The resistance of the pultruded FRP rod (32.3 kips) is much higher than the load-induced value (6.88 kips). The axial deformation and slenderness ratio are still under the suggested limit of pultrude FRP tension profiles.

Note: cross section is controlled by the serviceability limit.

Example 8.7

Through LRFD and ASD methods, design threaded rods to carry a service walkway load through a roof truss. A 24 ft-long and 5 ft-wide service walkway suspender as shown in Example Figure 8.7 is designed for dead load of 20 lbs/ft² and live load of 80 lbs/ft², with 8' spacing. (This example presents the design procedure of the rod under tension, only.)

Solution

a. LRFD method:

   a.1) Determine design load for a threaded rod, in Example Figure 8.7:

   Spacing of threaded rods = 8 ft, length of walkway = 24 ft, and walkway width = 5 ft

   Dead load = $(24 \times 5 \, \text{ft}^2)$ (20 lbs/ft²) = 2.40 kips and live load = $(24 \times 5 \, \text{ft}^2)$ (80 lbs/ft²) = 9.60 kips

   Factored dead load = 1.2 (2.40) = 2.88 kips and factored live load = 1.6 (9.60) = 15.36 kips

   Impact factor =1.3 (as given by standard codes)

   Service load W_s = 1.3 (2.40+9.60) = 15.60 kips and factored load W_u = 1.3 (2.88+15.36) = 23.7 kips

   Number of threaded rods = 8

   Thus, service load W_s per rod = (15.60/8) = 1.95 kips and factored load W_u per rod = (23.7/8) = 2.96 kips

   a.2) Trial section with the slenderness limit:

   The recommended limit for pultruded FRP axial tension members is:

   From Eq. (8.13):

$$\frac{L}{r} \le 300 \quad \text{then;} \quad \frac{8 \times 12}{300} = 0.320 \le r$$

A threaded FRP rod of 1.5" dia. was selected for this design. Gross area (A) = 1.776 in². Radius of gyration (r) = 0.375".

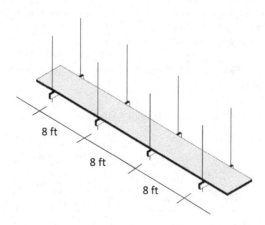

EXAMPLE FIGURE 8.7 Service walkway suspender.

a.3) Check for tensile strength of design section:

For $\varphi = 0.65$ (tension rupture) and λ (for $1.2D + 1.6L$) = 0.8, the tensile strength of the pultruded FRP rod of 1.5″ dia. is:

($A_e = 0.75A_g$; accounted for area reduction due to threads)

From Eq. (8.10):

$$\phi\lambda P_n = \phi\lambda F_n A_e = 0.65(0.75)(50\,\text{ksi})(0.75 \times 1.776) = 32.5\,\text{kips}$$

The resistance of pultruded FRP rod (32.5 kips) is much higher than the required factored load (2.96 kips for each). The axial deformation and slenderness ratio are still well within the suggested limit of pultruded FRP tension profiles.

Note: designed cross section is controlled by the serviceability limit.

b. ASD method:

b.1) Determine design load for each threaded rod:

Spacing of threaded rods = 8 ft, length of walkway = 24 ft, and walkway width = 5 ft

Dead load = $(24 \times 5\,\text{ft}^2)$ (20 lbs/ft^2) = 2.40 kips and live load = $(24 \times 5\,\text{ft}^2)$ (80 lbs/ft^2) = 9.60 kips

Impact factor = 1.3

Service load $W_s = 1.3\,(2.40 + 9.60) = 15.60$ kips

Number of threaded rods = 8

Thus, service load W_s per rod = (15.60/8) = 1.95 kips.

b.2) Trial section with the slenderness limit:

The recommended limit for pultruded FRP axial tension members is:

From Eq. (8.13):

$$\frac{L}{r} \le 300 \quad \text{then;} \quad \frac{8 \times 12}{300} = 0.320 \le r$$

FRP threaded rod of 1.5″ dia. is selected for this design. Gross area (A) = 1.776 in^2. Radius of gyration (r) = 0.375″.

b.3) Check for tensile strength:

The allowable tensile strength (P_a) of the pultruded FRP rod of 1.5″ dia. is:

($A_e = 0.75A_g$; accounted for threaded parts and F.S. = 2.0)

$$P_a = \frac{F_n A_e}{F.S.} = \frac{(50\,\text{ksi})(0.75 \times 1.776)}{2.0} = 33.2\,\text{kips}$$

For long-term serviceability, 20% of the limiting failure strain is considered as allowable design stress (safety factor for long-term including the knockdown factor = 5.0). The allowable tensile strength for long-term serviceability is:

$$P_{aL} = \frac{P_n}{F.S.} = \frac{(50\,\text{ksi})(0.75 \times 1.776)}{5.0} = 13.3\,\text{kips}$$

The allowable resistance of the pultruded FRP rod (13.3 kips) is much higher than the design load (1.95 kips). The axial deformation and slenderness ratio are still under the suggested limits.

Note: The selection of cross section is controlled by the serviceability limit and not by the strength limit state.

EXERCISES:

Problem 8.1: Determine the tensile strength of pultruded FRP WF $8 \times 8 \times 3/8''$ beam. The member end connection has two lines of $7/8''$ dia. bolts in each flange, three in each line. Self-weight = 6.49 lbs/ft; Area (A) = 8.73 in²; Web area (A_w) = 2.72 in²; flange thickness = $3/8''$; longitudinal tensile strength (F_t) = 30,000 psi. Use both the LRFD and ASD methods with 20% of the limiting failure strain for long-term serviceability. Stress concentration factor (K_{SCF}) for bolt holes = 2.0.

Problem 8.2: A tension member of a small platform assembly has an effective length of 6 ft (between bolts on either end) and is stressed in tension by 0.5 kips of dead load and 1.2 kips of live load. Select a pultruded FRP angle with two- number of $1/2''$ diameter bolts per line on both ends. The nominal hole diameter is taken as $1/16''$ larger than the nominal bolt diameter ($1/2''$). Use the LRFD method to design the tension member. Pultruded FRP angle properties (from manufacturers): longitudinal tensile strength (F_t) = 24,000 psi, longitudinal modulus (E_L) = 2 (10⁶) psi, shear strength (F_s) = 8,000 psi, stress concentration factor (K_{SCF}) for bolt holes = 1.6 (locations of bolts are assumed to satisfy the minimum design criteria as shown in Problem Figure 8.2).

Problem 8.3: Using the LRFD method, design FRP tension rods to support the purlins of a roof having a span of 84 ft and a rise of 21 ft. FRP tension rods are placed at quarter points of each purlin. The span of roof trusses is 24 ft long as shown in Problem Figure 8.3. The total dead loads due to self-weight of roofing and purlins are estimated to be 14 psf. In addition, the snow load is designed to be 28 psf.

Problem 8.4: Determine the tensile load capacity of an FRP pultruded angle ($6'' \times 6'' \times 3/8''$ with 50% fvf) made of polyester and E-glass, which is attached as a tension brace in a portal frame. The angle is attached to a gusset plate using $\frac{3}{4}''$ diameter steel bolts through one leg of the angle and the strength reduction factor of 0.6 is recommended rear the hole, which is $1/8''$ larger than the bolt diameter ($3/4''$).

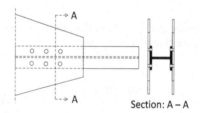

Section: A – A

PROBLEM FIGURE 8.1   Pultruded FRP tension member WF $8 \times 8 \times 3/8''$.

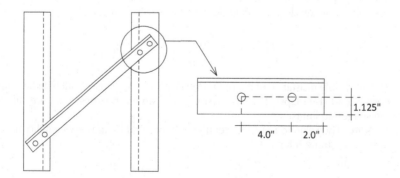

PROBLEM FIGURE 8.2   Tension member.

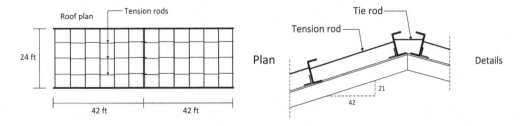

PROBLEM FIGURE 8.3   Roof plan with details.

Problem 8.5: Employing the LRFD approach, design a polyester/E-glass composite tension member to carry 10 kip load (including factored dead and live loads). The member is connected to gussets with ¾″ diameter bolts with 1/8″ oversized holes. The forces are flowing through the center of gravity of the member. The member length is 10′. Assume the maximum out-of-plane crookedness of ¼″, and the Young's Modulus of 3.2 msi.

REFERENCES AND SELECTED BIOGRAPHY

ACMA, Pre-Standard for Load & Resistance Factor Design (LRFD) of Pultruded Fiber Reinforced Polymer (FRP) Structures, Arlington, VA, 2021.

ANSI/AISC 360–05, An American National Standard, Specification for Structural Steel Buildings, AISC, Chicago, IL, 2005.

ASCE/SEI7–10, Minimum Design Loads for Buildings and Other Structures, Alexander Bell Drive, Reston, VA, 2010.

Bank, L.C., *Composites for Construction: Structural Design with FRP Materials*, Hoboken, NJ, John Wiley & Sons, Inc., 2006.

Bedford Reinforced Plastics, *Design Guide*, Bedford Reinforced Plastics, Inc., Bedford, PA, 2012.

Creative Pultrutions, *The Pultex Pultrusion Design Manual of Standard and Custom Fiber Reinforced Structural Profiles*, 5, revision (2), Creative Pultrutions Inc, Alum Bank, PA, 2004.

Fiberline, *Design Manual*, Fiberline, Kolding, Denmark, 2003.

Nagaraj, V., and GangaRao, H.V.S., Static behavior of pultruded GFRP beams, *Journal of Composites for Construction*, 1(3), 1997, pp. 120–129.

Parambil, J.C., Static Analysis of Laminated Composite Beams with I Section, MS Thesis, Submitted to UT-Arlington, 2010.

Strongwell Corporation, Design Manual: EXTERN and Other Proprietary Pultruded Products, Bristol, VA, 2010.

Structural Design of Polymeric Composites – EUROCOMP Design Code and Handbook, E & FN Spon, London, UK, 1996, (Ed. Clarke, J.L.).

Tan, S.C., Effective stress fracture models for unnotched and notched multidirectional laminates, *Journal of Composite Materials*, 22(4), 1988, pp. 322–340.

Tan, S.C., *Stress Concentrations in Laminated Composites*, Technomic Publishing Company, Inc., Lancaster, PA, 1994.

9 Flexural Member Design

FRP composite components subjected to transverse loading (Figure 9.1) induce flexure and shear force, typically. The transverse loading does induce geometric nonlinearities if the span-to-depth ratio is very large, say 50 or above, especially regarding glass fiber-reinforced polymer composites. From the flexural behavior viewpoint, a component (e.g., flange of a beam) resists tension on one face and compression on the opposite face. If the compression face is adequately restrained from buckling which is typical in practice, then the component stability need not be evaluated, and the design checks would focus on flexure-induced stresses, deflections, and rotations. Unlike in the design of steel sections, design checks for deformations must include the shear-related influences due to very low shear modulus in relation to bending (flexure) modulus, which is of the order of 10%–15% of bending modulus. Shear deformation can be as high as 18%–20% of flexural deformations in FRP composite components while this value is no more than 3% in steel under normal design criteria. In addition, stress concentration influences from the corner effects are incorporated in the stress computations to establish realistic strength limit states from a design viewpoint.

The failure modes of FRP components under bending depend on the span-to-depth ratio. For example, if the span-to-depth ratio is in a range of 12–40, bending stress equations derived in this chapter can be used, with appropriate material properties. On the other hand, if the span-to-depth ratio is below 12, then shear dominant failure mode may control the design and design equations representing such a failure mode are derived herein. FRP components with a span-to-depth ratio higher than 40 will result in geometrically nonlinear bending responses, and such analysis is outside the scope of this textbook. In addition, other stability checks incorporating mechanics of materials principles are developed for flange buckling, web crippling, and lateral torsional buckling (LTB) in this chapter, and their applicability is illustrated through many design examples presented in this chapter.

9.1 FLEXURAL MEMBERS

Flexural members (beams under bending loads) carry transverse loads, i.e., loads perpendicular to the plane of bending of a member. Beams are primarily designed to resist transverse load-induced moments as shown in Figure 9.1. The most common beam cross sections are doubly symmetric sections such as I-shape, H-shape, or WF-wide flange sections. In addition, closed sections such as hollow rectangular box cross sections are often used as beam members. Rectangular hollow box sections are used typically to resist large torsional and bending moments where lateral supports are essential to prevent premature failure under the LTB phenomenon. This is discussed in Section 9.5.

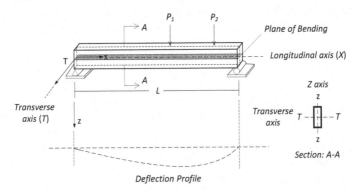

FIGURE 9.1 Beam and deflected shape under transverse loading.

DOI: 10.1201/9781003196754-9

In this chapter, the focus is primarily on uniaxial bending, while multiaxial bending coupled with axial and torsional effects are presented in Chapter 12.

Flexural designs of pultruded FRP profiles are introduced initially. Flexure of FRP members is the most common form of action that a member undergoes under transverse loads. The pultruded FRP beam profiles carry applied loads at right angles to the longitudinal plane (XT of Figure 9.1) of a member. The loading may act either about its major or about its minor axis or both axes. The pultruded FRP beam profiles are symmetric with respect to the loading plane and carry no axial force, i.e., bending forces are decoupled from axial forces. The pultruded FRP beam profiles considered herein are straight and long (10 < span/depth < 40) members with the constant cross-sectional area along the beam length and fiber volume fraction no less than 0.3. If the span-to-depth ratio exceeds 40, the geometrically nonlinear bending response would dominate while that ratio below 10 results in shear dominance response and may even lead to local compression failure at a ratio below 5.

In the design of pultruded FRP beams, the first consideration may be serviceability limit state checks followed by strength limit state checks for both the moment and shear resistances. The serviceability checks are done first because of the low structural stiffness of a section compared to conventional steel structural members. Thus, the design of pultruded FRP structural members is often controlled by deflection and/or rotational limit states. Second, global and local structural stability of pultruded FRP beams must be checked to satisfy the design criteria as mentioned in Chapter 7. It should be noted that the moment-resisting capacity of sections may be limited by the global stability (e.g., LTB as in Section 9.5) particularly when the beam members have inadequate lateral supports. Based on the above, the following equations are presented as illustrations.

From the elastic theory of beam bending, transverse loads are applied perpendicular to the longitudinal direction (X-axis in Figure 9.1) of a structural member and bending about the transverse axis (T-axis, perpendicular to X–T plane) of a member. Induced flexural stress at distance Z from the neutral axis of a cross section can be written as (Gere and Goodno, 2009, also refer to Chapter 5):

$$f_b = \frac{Mz}{I}$$
(9.1)

where

f_b = bending (flexural) stress

M = bending moment. The positive internal moment causes compression in the top fiber of a section, whereas the negative internal moment causes tension at the top

Z = distance from the neural axis to a specific fiber on the cross section

I = moment of inertia with respect to the neutral axis.

When a beam is subjected to transverse loads, both the shear and bending stresses are induced. The induced shear stress is developed parallel to the longitudinal axis of a beam. To remain in equilibrium, an equal magnitude of the shear stresses must be generated in the beam section perpendicular to the longitudinal direction (X-axis). The induced shear stress, f_v (should be the same as in Eq. 9.2), on the beam cross section is (Gere and Goodno, 2009):

$$f_v = \frac{VQ}{It}$$
(9.2)

where

V = shear forced induced by transverse loading

t = thickness of the section at shear stress location

Q = first area moment about the neutral axis of the sectional area above (or below) the depth of shear stress location

I = moment of inertia with respect to the bending neutral axis.

For design purposes, the flanges of a structural profile resist a very small amount of shear stresses, and hence, neglected. Then, it is reasonable to note that the shear stresses are predominant in the web. The average induced web (parabolic, but nearly a rectangular shape) shear stress may be used assuming shear resistance contribution from flange(s) is negligible:

$$f_{v,max} \cong \frac{1.5V}{A_{web}} \text{ at mid} - \text{depth of the section; and } f_{v,avg} \cong \frac{V}{A_{web}} \qquad (9.3)$$

where

f_v = shear stress on a horizontal plane located with reference to the neutral axis
V = vertical shear force
t = thickness of the section at the plane of interest along the section depth
A_{web} = area of the web only.

9.2 NOMINAL STRENGTH DUE TO MATERIAL RUPTURE IN FLEXURE (ACMA, 2021)

To determine the moment capacity of a cross section, the classical area transformation method of isotropic components in a beam with different material layers may be used (ACMA, 2021).

Thus, the nominal flexural moment of members due to material rupture is determined as:

$$M_n = F_t S_t \quad \text{or} \quad F_c S_c \qquad (9.4)$$

where

F_t, F_c = characteristic values of longitudinal tensile and compressive strengths, respectively
S_t, S_c = "transformed section moduli associated with the member extreme fiber in tension and compression, respectively, computed with respect to the axis of bending, taking into account conditions of force equilibrium and strain compatibility" (as stated in ACMA, 2021).

9.3 NOMINAL STRENGTH DUE TO MATERIAL RUPTURE IN SHEAR

Typically, beams under transverse loading are subjected to both the shear and bending forces. In particular, short-span beams tend to fail in shear mode. The nominal shear strength of pultruded FRP structural members due to material rupture in shear is (ACMA, 2021):

$$V_n \cong F_{LT} A_s \qquad (9.5)$$

where

V_n = nominal shear strength under material rupture
F_{LT} = in-plane shear strength of the profile
A_s = shear area.

9.4 DEFLECTION

For the most commercially pultruded FRP profiles in the market, it is found that the longitudinal modulus (E_L) is higher than the shear modulus (G_{LT}) by a factor varying from 7 to 10, which is a function of fiber volume in shear direction and resin type. The influence of resin type on G_{LT} is limited. The shear deformation plays a significant role (up to ~18%–20% of bending deflection) in the deflection response of pultruded FRP profiles (Nagaraj and GangaRao, 1997). Therefore, the total deflection of pultruded FRP profiles can be determined as: (1) deflection due to bending and (2) deflection due to shear. The total deflection can be obtained using Timoshenko's shear deformation theory (Nagaraj and GangaRao, 1997), which is an approximate approach and not as rigorous as the one from the classical lamination theory. The deflection solutions corresponding to different loading and boundary conditions are presented below. The deflection solutions can be found in many structural engineering handbooks.

Case 1: Cantilever beam with a point load at its tip

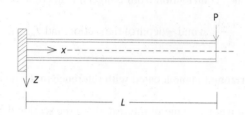

$$w_{max} = \frac{PL^3}{3(E_L I_T)} + \frac{PL}{k(G_{LT}A)} \quad \text{and} \quad w(x) = \frac{P(3Lx^2 - x^3)}{6(E_L I_T)} + \frac{Px}{k(G_{LT}A)} \qquad (9.6.1)$$

Case 2: Cantilever beam with uniformly distributed load throughout its length

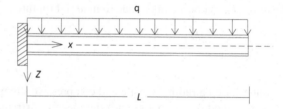

$$w_{max} = \frac{qL^4}{8(E_L I_T)} + \frac{qL^2}{2k(G_{LT}A)} \quad \text{and} \quad w(x) = \frac{q(x^4 - 4Lx^3 + 6L^2x^2)}{24(E_L I_T)} + \frac{q(2Lx - x^2)}{2k(G_{LT}A)} \qquad (9.6.2)$$

Case 3: Simply supported beam with a point load at mid-span

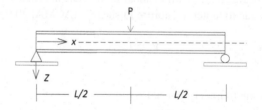

$$w_{max} = \frac{PL^3}{48(E_L I_T)} + \frac{PL}{4k(G_{LT}A)}, \quad \text{and} \quad w(x) = \frac{P(3Lx^2 - 4x^3)}{48(E_L I_T)} + \frac{Px}{2k(G_{LT}A)} \qquad (9.6.3)$$

Case 4: Simply supported beam with uniformly distributed load through its length

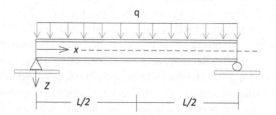

$$w_{max} = \frac{5qL^4}{384(E_L I_T)} + \frac{qL^2}{8k(G_{LT} A)}, \quad \text{and} \quad w(x) = \frac{q(x^4 - 2Lx^3 + L^3 x)}{24(E_L I_T)} + \frac{q(Lx - x^2)}{2k(G_{LT} A)} \quad (9.6.4)$$

where

E_L = longitudinal modulus of beam profiles

G_{LT} = in-plane shear modulus of beam profiles

I = moment of inertia respected to the bending neutral axis

k = Timoshenko's shear coefficient. Timoshenko's shear coefficient varies depending on load distribution, Poisson's ratio, aspect ratio (depth to width) of the cross section, and boundary conditions.

However, a k value of 0.4–0.6 will be adequate for most cases with supports at both ends. It should be noted that $k(G_{LT} A)$ can be approximated with $(G_{LT})_{web} A_{web}$ (Nagaraj and GangaRao, 1997; Bank, 2006). The serviceability limit state in terms of deflection must not exceed $L/240$ for a structural profile attached to brittle nonstructural components (ACMA, 2021; EUROCOMP, 1996 CUR-CRO Recommendations 96: 2019), in building construction. However, the deflection limit state may be taken as $L/400$ for beams under vibration. These limit states do vary for different functional conditions of a structure and vary from one code of practice to another.

9.5 GLOBAL BUCKLING (LTB)

Typically, flexure members are stiffened with lateral bracing at certain intervals on the compressive flange side of a structural member. For members under flexure with adequate lateral bracing, members that are loaded in the transverse plane will only deflect in the horizontal (lateral) direction, when compared with the beams of inadequate lateral bracing. Beams with inadequate lateral bracing tend to vertically deflect and laterally buckle leading the compression flange to laterally move out-of-its plane and the cross section to be twisted from vertical loads that cause torsional moments due to out-of-plane movements (Bank, 2006; Nguyen et al., 2014; Estep et al., 2016). The abovementioned physical instability of flexural members is known as LTB or global lateral buckling as shown in Figures 9.2 and 9.3.

In general, lateral buckling of a beam is common for those that are overly flexible about their transverse axis (perpendicular to the major axis of bending). Therefore, the preferred solution is to

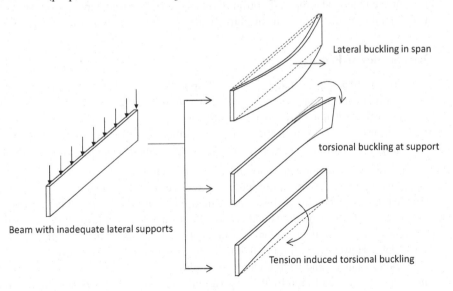

FIGURE 9.2 Lateral displacement and buckling of a beam with inadequate lateral supports.

FIGURE 9.3 Lateral displacement and local flange buckling of a WF beam (Qureshi, 2012).

laterally brace a member to prevent it from LTB under transverse loads. Bracing counteracts the three major buckling forms that can occur in a simple beam under transverse (bending) loads. This structural behavior generally takes place in open sections under bending. It was found that the axial, shear, and flexural strengths rarely dominate the design of structural FRP profiles. Typically, the critical global buckling strength (LTB) controls the design for most pultruded and laterally unrestrained/unbraced open FRP structural profiles of longer spans, carrying significantly lower levels of transverse loads than those failing by the material rupture strength.

The effect of end supports influences the LTB strength of pultruded FRP structural beams. For a pultruded FRP beam supported with torsional restraint at both ends with no intermediate lateral support, the unbraced length of such a beam is equal to the actual length between supports. However, the unbraced length is more than the actual length when torsional restraint is absent.

For example, a beam with different end supports as shown in Figure 9.4 has different LTB strengths. Case (a): both the compression and tension flanges of a beam are laterally restrained at supports and torsional restraint is against twisting of the longitudinal axis. However, both the top and bottom flanges are still free to rotate under bending, as shown in Figure 9.4. Case (b): compression flange of a beam is laterally unrestrained and partial torsional restraint is provided by the bolted connection for the bottom flange at the supports to partially prevent rotation about the longitudinal axis of the beam. The top and bottom flanges are free to rotate away from supports and may result in partial twisting due to out-of-plane bending. This is shown in Figure 9.4, Case (b).

9.5.1 Open Sectional Profiles

For open sectional profiles (e.g., I-shape), a simply supported pultruded FRP doubly symmetrical beam is under applied bending moments at the ends as shown in Figure 9.5. The governing equations under LTB for the pultruded FRP members are derived below. Two fundamental assumptions of small deformations are involved in the derivation, which are: (1) major axis bending rigidity $\left(E_l I_{yy}\right)$ is much larger than minor axis bending rigidity $\left(E_{tf} I_{zz}\right)$ and (2) bending rigidities before and after deformation about the major and minor axes are unaffected.

If the I-sectional member, as shown in Figure 9.5, is under bending moment with both ends free to twist, then the torque induced in the displaced structure is resisted by the member torsional rigidity, which is a product of shear modulus and torsional constant. In an approximate sense, the total torsional constant is the sum of the torsional constants of the two flanges and the web of an I-section (Timoshenko, 1956). However, with one torsionally fixed support, I-section torque is resisted by the sectional rigidity (flanges and web) under torque, plus the warping rigidity offered by the two flanges. Any small resistance offered by the web rigidity is neglected (Boresi and Schmidt, 2002). In addition, moments in the original and displaced axes are assumed to be identical.

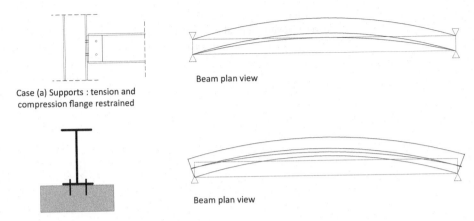

Case (a) Supports : tension and compression flange restrained

Beam plan view

Case (b) Supports : bolts prevent uplift (only)

Beam plan view

FIGURE 9.4   Effect of end supports under lateral–torsional buckling of a beam.

Near the fixed support, a small amount of warping is induced under torque for noncircular sections. In general, the torque has two parts (Figure 9.5C):

a. Torque-induced (partial) lateral shear $= V^{PT}(h)$
b. Torque-induced warping on the section (in the absence of support restraint) $= GC\varphi$

where
   $h =$ depth from the center of the top flange to the center of the bottom flange
   $\varphi = \text{twist}\,(\beta)$ at the geometric center of the cross section divided by the beam span length
   $G =$ shear modulus
   $C =$ torsional constant of a cross section as defined by Timoshenko (1956), Boresi and Schmidt (2002), and others.

Note: V_{PT} is zero at free support with no torsional resistance, and φ is zero at torsionally fixed support.

$$T = V^{PT}h + GC\varphi \tag{9.7.1}$$

where
   $GC =$ torsional rigidity

Using the lateral bending equation of the upper or lower flange of the beam (about the weak axis of the I-section).

$$\left(\frac{E_{T,f}I_z}{2}\right)\frac{d^2y}{dx^2} = -M_z \tag{9.7.2}$$

where
   $M_z =$ lateral bending due to torsion for one flange only.
   $I_z/2 \cong$ weak axis moment of inertia of one of the two flanges about the web of the I-section (neglecting web moment of inertial about the weak axis; however, the stiffening effect from the web at the middle of the flange improves I_z; hence, the apparent I_z is higher resulting in higher LTB moment resistance from the experimental data).

$$y\,(\text{horizontal deflection of flange}) = \frac{h}{2}\beta, \quad \text{or} \quad \frac{dy}{dx} = \frac{h}{2}\frac{d\beta}{dx} \tag{9.7.3}$$

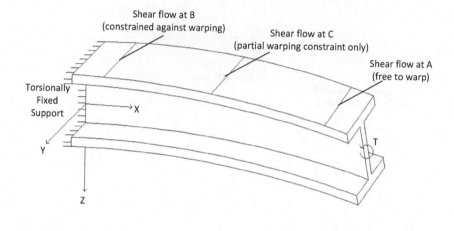

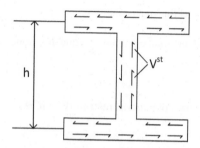

A. Shear from Torsion (location of very low or no torsional resistance) along beam length with zero torsional resistance; free to twist.

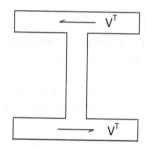

B. Shear due to lateral deformation resistance & zero torsional shear (near torsionally fixed support). This is St. Venant's shear only resulting from torsion and torsional load is transmitted primarily by the shear flanges.

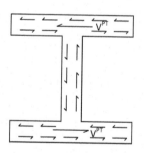

C. Shear from torsion, as well as partial lateral deformation resistance, away from torsionally fixed support. Between cases A and B, i.e. between torsionally fixed and free supports. This is from warping shear plus St. Venant's shear.

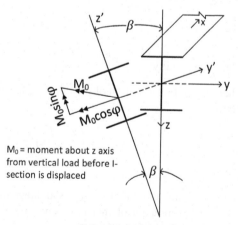

M_0 = moment about z axis from vertical load before I-section is displaced

D. Exaggerated view of displaced I-section

FIGURE 9.5 I-beam under torque.

Note: Lateral bending rigidity of flanges due to lateral bending is $(E_{Tf})\,(I_z)$, where E_{Tf} is the modulus of flange in the direction perpendicular to the beam-span direction. Few researchers show this bending modulus is along the beam-span direction with which the authors disagree.

$$\frac{d^2y}{dx^2} = \left(\frac{h}{2}\right)\frac{d^2\beta}{dx^2} = \left(\frac{h}{2}\right)\frac{d\varphi}{dx} \quad \text{because} \quad \frac{d\beta}{dx} = \varphi \tag{9.7.4}$$

where
β = twist at the center of a section.

Substituting this equation in weak axis bending deflection, Eq. (9.7.2) of the upper or lower flange is:

$$\left(\frac{E_{T,f}I_z}{2}\frac{h}{2}\right)\frac{d\varphi}{dx} = -M_z \tag{9.7.5}$$

or differentiating both sides gives:

$$\frac{d}{dx}\left[\left(\frac{E_{Tf}I_z h}{4}\right)\frac{d\varphi}{dx}\right] = -\frac{dM_z}{dx} = -V^{PT} \tag{9.7.6}$$

Substituting V^{PT} into Eq. (9.7.1) corresponding to "T" and rearranging gives:

$$-\frac{E_{Tf}I_z h^2}{4GC}\frac{d^2\varphi}{dx^2} + \varphi = \frac{T}{GC} \tag{9.7.7}$$

Curvature in the displaced xy plane as shown in Figure (9.5D) is:

$$E_{Tf}I_z\frac{d^2u(x)}{dx^2} = -M_{z'} = -M_o\varphi \tag{9.7.8}$$

where
z' = vertically displaced coordinate of I-section while the undisplaced coordinate is z
$u(x)$ = horizontal deflection or displacement from neutral axis along the beam length, x

$\varphi = \dfrac{d\beta}{dx}$ = variation in beam's angle of twist along its length

= angle of twist per unit length of the member under applied torque,
β = angle of twist at a given cross section of a beam,
$M_{z'}$ = moment about displaced weak axis.
M_0 = moment about z axis as in Figure 9.5.D"

When a beam is displaced under lateral torsional moment, the torsional component of M_o is proportional to the slope of the displaced structure in the xy plane:

$$M_{x'} = -\frac{du(x)}{dx}M_o \tag{9.7.9}$$

Substituting M_x' which is the induced torque from (out-of-plane) displaced section into Eq. (9.7.7) gives:

$$GC\varphi - E_{Tf}I_z\frac{h^2}{4}\left(\frac{d^2\varphi}{dx^2}\right) = -\frac{du}{dx}M_o \tag{9.7.10}$$

Differentiate φ and u with respect to x on both sides of Eq. (9.7.10) to get:

$$GC\frac{d\varphi}{dx} - E_{tf}I_z\frac{h^2}{4}\left(\frac{d^3\varphi}{dx^3}\right) = -\frac{d^2u}{dx^2}M_o \qquad (9.7.11)$$

or

$$GC\frac{d^2\beta}{dx^2} - E_{tf}I_z\frac{h^2}{4}\left(\frac{d^4\beta}{dx^4}\right) = -\frac{d^2u}{dx^2}M_o$$

where

GC = torsional rigidity

$$\varphi = \frac{d\beta}{dx}$$

The governing equation for global buckling for open sectional profiles is:

$$E_{T,f}C_\omega\beta^{iv} - G_{LT}C\beta^{ii} - M_0u^{ii} = 0 \qquad (9.8.1)$$

If the member boundaries are free to warp and cannot deflect or twist, then there is no end moment induced about the z-axis (or the weak axis of the cross section).

At torsionally fixed ends,

$$u(0) = u(L) = \varphi(0) = \varphi(L) = 0$$

or, Eq. (9.7.3) gives:

$$\varphi\left(\begin{array}{c}0\\L\end{array}\right) = \frac{d\beta}{dx}\bigg|_{x=0,L} = \left(\frac{2}{h}\right)\frac{dy}{dx} = 0 \qquad (9.8.2)$$

At torsionally free end,

$$\frac{d\varphi}{dx} = \frac{d^2\beta}{dx^2} = 0 \qquad (9.8.3)$$

At simply supported ends free of torsional resistance

$$u^{ii}(0) = u^{ii}(L) = \varphi^{ii}(0) = \varphi^{ii}(L) = 0 \qquad (9.8.4)$$

where

$E_{T,f}$ = transverse modulus of the flange
G_{LT} = shear modulus of the beam
I_z = beam moment of inertia about z-axis (weak axis)
C_ω = warping constant = $I_z\dfrac{h^2}{4}$
$G_{LT}C$ = torsional rigidity
u = deflection of the shear center in the y-direction
φ = angle of twist.
$C = (1/3)bh_3$ for a rectangular section with h being the smaller dimension

The solutions of governing equation (Eq. 9.7.11) with the boundary conditions (Eq. 9.8.2) provide the critical LTB strength as (Estep et al., 2016):

$$M^{LTB} = \frac{nC_b\pi}{L_b}\sqrt{E_{T,f}I_zG_{LT}C + \frac{\pi^2 E_{T,f}{}^2 I_z C_\omega}{L_b^2}}$$ (9.9)

It should be noted that the first term of the lateral torsional strength expression in Eq. (9.9) represents the St. Venant's torsional strength and the second term represents the lateral bending strength. The shape function of the beam and the lateral deflection u are determined for constant M_o and zero angle of twist at $x = 0, L$, as:

$$\varphi = C_b\mathrm{Sin}\frac{n\pi x}{L}; n = 1, 2, 3,...$$ (9.10.1)

$$u = C_b\frac{ML^2}{(n\pi)^2 E_{T,f}I_z}\mathrm{Sin}\frac{n\pi x}{L}; n = 1, 2, 3,...$$ (9.10.2)

where n is an integer corresponding to first, second, third, and higher harmonics of a complex deformed shape of a beam under bending, coupled with lateral deformation.

Note: From a design viewpoint, for structural systems under static loads, only the first harmonic is of interest, and results in accurate capacity (not exact).

In addition, the cross section rotates about the center of rotation C_{LB} when the lateral buckling of the member has occurred as shown in Figure 9.6. Typically, the center of rotation C_{LB}, positioned below the shear center of the cross section can be approximated by taking C_b as one, as:

$$z_{LB} \approx \frac{u}{\phi} \approx \frac{ML^2}{(n\pi)^2 E_{T,f}I_z}$$ (9.10.3)

For applied bending moments where a lateral moment is less than or equal to the critical LTB moment M^{LTB}, the lateral buckling of the member has not yet occurred. The vertical deflection in the z-direction (Figure 9.6) is determined from the beam bending equation (Eq. 5.9.5).

$$z = \frac{ML^2}{2E_{T,f}I_T}\left[\frac{x}{L} - \left(\frac{x}{L}\right)^2\right]$$ (9.10.4)

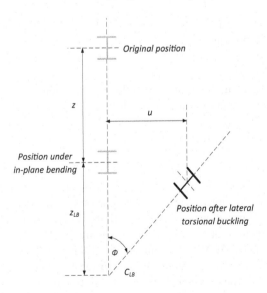

FIGURE 9.6 LTB of the beam cross section.

The minimum critical moment for the LTB can be obtained from the minimum $n = 1$ in Eq. (9.9); thus, the nominal LTB moment of a pultruded FRP member is determined as:

$$M_n^{LTB} = \frac{C_b \pi}{L_b} \sqrt{E_{T,f} I_z G_{LT} C + \frac{\pi^2 E_{T,f}^2 I_z C_\omega}{L_b^2}} \tag{9.11.1}$$

Note: Equation (9.11.1) is identical to Eq. (4.39) of "Structural Design of Polymeric Composites," after grouping the technical contributions of torques induced by lateral shear and that by warping, as given in Euro Code (Eurocomp Design Code, 1996), and Laudiero et al. (2012). However, ACMA Pre-Standard for LRFD of Pultruded FRP Structures (2021) has a similar equation with the elastic modulus along the longitudinal axis (not in the transverse direction) of a member.

$$f_n^{LTB} = \frac{C_b \pi}{S_T L_b} \sqrt{E_{T,f} I_z G_{LT} C + \frac{\pi^2 E_{T,f}^2 I_z C_\omega}{L_b^2}} \tag{9.11.2a}$$

$$f_n^{LTB} = \frac{C_b \pi}{S_T L_b} \sqrt{E_{L,f} I_z G_{LT} C + \frac{\pi^2 E_{L,f}^2 I_z C_\omega}{L_b^2}} \quad \left(\text{AMCA}, 2021 \right) \tag{9.11.2b}$$

Based on experimental data, effective laterally unbraced length ($L_{b,\ eff}$) modification is used as a correction factor to obtain a more accurate prediction of the nominal lateral torsion buckling strength from Eqs. (9.12 and 9.13). The effective laterally unbraced length ($L_{b,\ eff}$) is established as the laterally unbraced length (L_b) multiplied by a constant (0.62 for $d < 6''$ and 0.95 for $d \geq 6''$) (Estep et al., 2016), which is dependent on the web stiffness. Thus, the modified nominal LTB strength of the pultruded FRP members is:

$$M_n^{LTB} = \frac{C_b \pi}{L_{b,eff}} \sqrt{E_{T,f} I_z G_{LT} C + \frac{\pi^2 E_{T,f}^2 I_z C_\omega}{L_{b,eff}^2}} \tag{9.12}$$

$$f_n^{LTB} = \frac{C_b \pi}{S_T L_{b,eff}} \sqrt{E_{T,f} I_z G_{LT} C + \frac{\pi^2 E_{T,f}^2 I_z C_\omega}{L_{b,eff}^2}} \tag{9.13}$$

where
   f_n^{LTB} = nominal LTB strength (stress)
   M_n^{LTB} = nominal LTB moment
   $G_{LT} C$ = torsional rigidity
   C = torsional constant
   I_z = moment of inertia about the weak axis of the cross section (Z-axis)
   S_T = sectional modulus about the strong axis of the cross section (T-axis)
   C_ω = warping constant
   $E_{T,f}$ = transverse modulus of the flange
   G_{LT} = shear modulus of the beam profiles.

Note: For wide-flange or I-section: $J = \Sigma b_i t_i^3/3$, $C_\omega = I_z d^2/4$, or $t_f d^2 b_f^3/24$, t_i = thickness of flange or web portion. d = depth of section. b_i = width of flange or web portion. Since the above equations (9.12 and 9.13) are derived based on certain assumptions as given at the beginning of Section 9.5.1, M_n^{LTB} and f_n^{LTB} result in conservative values when compared with the experimental data, i.e., lower than the experimental values. ACMA Pre-Standard for LRFD of Pultruded FRP Structures (2021) does not consider $L_{b,\ eff}$ or unbraced length reduction due to support constraints. However, it is strongly recommended that the designer should use discretion in establishing the effective length.

The LTB strength depends on the position (varying bending moment magnitudes along the beam length) and regions of different rigidities along the beam span. To account for this effect, the approximate moment modification factors are introduced to minimize computational cumbersomeness (ACMA, 2021):

$$C_b = \frac{12.5M_{max}}{2.5M_{max} + 3M_A + 4M_B + 3M_C} \tag{9.14}$$

where

C_b = moment modification factor for unsupported spans with both ends braced. C_b is permitted to be conservatively taken as 1.0 for all cases. For cantilevers or overhangs where the free end is unbraced, $C_b = 1.0$:

M_{max} = absolute value of maximum moment in the unbraced segment

M_A = absolute value of moment at a quarter of the unbraced segment

M_B = absolute value of moment at the centerline of the unbraced segment

M_C = absolute value of moment at three-quarter point of the unbraced segment.

The moment modification factor, C_b, needs to be further evaluated experimentally to establish an accurate upper limit and its variation as a function of M_{max}, M_A, M_B, and M_C. The moment modification factors can be used for any bending moment diagrams. The moment modification factors for various loading and lateral support cases are presented in Figure 9.7 as design aids (Structural Plastics Manual, ASCE, 1984; Vinnakota, 2005).

Note: C_b is permitted to be conservatively taken as 1.0 for all cases of GFRP sections because of their low bending rigidity (ACMA, 2021).

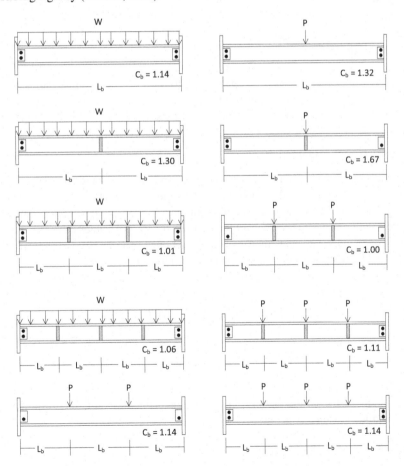

FIGURE 9.7 Moment modification factor C_b for simply supported beam.

9.5.2 Closed Sectional Profiles

For closed sectional profiles, the torsional resistance of these sections is substantially higher than that of open sections. The LTB strength for closed sectional profiles (i.e., rectangular box beam) can be determined by neglecting the second term of the open sectional expression. The LTB strength for closed sectional profiles can be determined as:

$$M_n^{LTB} = \frac{C_b \pi}{L_b} \sqrt{E_{T,f} I_z G_{LT} C} \tag{9.15}$$

$$f_n^{LTB} = \frac{C_b \pi}{S_T L_b} \sqrt{E_{T,f} I_z G_{LT} C} \tag{9.16}$$

$$(\text{Hollow rectangular section}): \quad G_{LT} C = \frac{G_{LT}}{8} \left(\frac{t_f}{b_f} + \frac{t_w}{b_w} \right) 4 \left(b_f - t_w \right)^2 \left(h - t_f \right)^2 \tag{9.17}$$

where

C_b = moment modification factor for unsupported spans with both ends braced
$E_{L,f}$ = longitudinal modulus of the flange
G_{LT} = shear modulus of the beam profiles
L_b = unbraced span length
I_z = moment of inertia about the weak axis of the cross section (Z-axis)
S_T = sectional modulus about the strong axis of the cross section (T-axis)
$G_{LT} C$ = torsional rigidity
$t_f \, or \, t_w$ = thickness of flange or web
h = center to center flange depth of section
$b_f \, or \, b_w$ = width of flange or web.

Note: Eqs. (9.15) and (9.16) are valid for $h/b < 3$, and for $h/t_w < 20$. For higher ratios, the lateral torsional buckling term may contribute significantly; hence closer evaluation is needed. Again, an accurate approach is to relate the above ratios to local bending stiffnesses of both the flange and the web.

9.5.3 Simplified LTB Strength

The LTB expression in Section 9.5.1 is inconvenient for hand computations. The LTB computation can be time-consuming for long hand calculation. This is an important limitation for a designer. To overcome this barrier, a simplified LTB equation (under the assumption that $E_{T,f} = G_{LT}$) developed by the authors is presented herein with acceptable accuracy. From Eq. (9.9), the shear modulus is rewritten in terms of the transverse elastic modulus. The resultant is rearranged for open sections as given in Eq. (9.18).

The simplified LTB strength of the I-shaped or WF pultruded FRP section is:

$$M_n^{LTB} = \frac{\pi E_{T,f} C_b}{2 L_{b,eff}} \sqrt{2 I_z C + \left(\frac{\pi h I_z}{L_{b,eff}} \right)^2} \tag{9.18}$$

$$f_n^{LTB} = \frac{\pi E_{T,f} C_b}{2 S_T L_{b,eff}} \sqrt{2 I_z C + \left(\frac{\pi h I_z}{L_{b,eff}} \right)^2} \tag{9.19}$$

Note: the moment modification factor (C_b) can be incorporated into the simplified expression Eq. (9.18) to obtain a reasonably accurate result.

where

f_n^{LTB} = nominal LTB strength (stress)
M_n^{LTB} = nominal LTB moment

$G_{LT} C$ = torsional rigidity
C = torsional constant
I_z = moment of inertia about the weak axis of the cross section (Z-axis)
S_T = sectional modulus about the strong axis of the cross section (T-axis)
C_ω = warping constant
$E_{T,f}$ = transverse modulus of the flange
G_{LT} = shear modulus of a beam profile
h = full depth of the cross section
C_b = moment magnification factor for unsupported spans with braced ends

Note: For wide-flange or I-section: $J = \Sigma b_i t_i^3/3$, $C_\omega = I_z d^2/4$ or $t_f d^2 b_f^3 /24$, t_i = thickness of flange or web portion. d = depth of section. b_i = width of flange or web portion. The effective laterally unbraced length ($L_{b,\ eff}$) is established as the laterally unbraced (L_b) multiplied by a constant (0.62 for $d < 6''$ and 0.95 for $d \geq 6''$) which is dependent on the web stiffness. These two ratios (0.62 & 0.95) theoretically must be tied to (d/t_i); however, for simplicity, specific ratios are identified as a function of "d" from experimental data (Estep, 2014).

Example 9.1

Determine the lateral torsional strength of a pultruded FRP WF beam 12 × 12 × 1/2″ with a simple span of 24 ft. The pultruded FRP beam is braced with lateral supports at the ends and laterally restrained at the third point of the span.
  Section WF 12 × 12 × 1/2″:
  Self-weight = 13.98 lbs/ft; Area (A) = 17.51 in²; Web area (A_w) = 5.50 in²; Sectional modulus about the strong axis (S_T) = 75.5 in³; Sectional modulus about the weak axis (S_Z) = 24.0 in³; Moment of inertia about strong axis (I_T) = 453 in⁴; moment of inertia about weak axis (I_Z) = 144.1 in⁴; Torsional constant (C) = 1.458 in⁴.
  A pultruded FRP WF beam properties (manufacturers): Longitudinal tensile strength (F_t) = 30,000 psi, shear strength (F_v) = 4,500 psi, Longitudinal modulus (E_L) = 2.6 (10⁶) psi, Transverse modulus (E_T) = 0.8 (10⁶) psi, Shear modulus (G_{LT}) = 0.425 (10⁶) psi.

Solution

a. Determine LTB strength:
  From Eqs. (9.13 and 9.14):
    As shown in Example Figure 9.1.2, moment modification factor C_b for lateral supports at the ends and at the third point of the span is determined as follows: (It should be noted that C_b is permitted to be conservatively taken as 1.0 for all cases.)
    The absolute value of maximum moment in the unbraced segment $M_{max} = WL^2/8$.
    The absolute value of moment at a quarter of the unbraced segment (at 5L/12 from support) $M_A = (5WL^2/24)-(25WL^2/288)$.
    The absolute value of moment at the centerline of the unbraced segment $M_B = WL^2/8$.

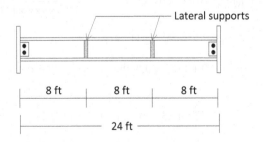

EXAMPLE FIGURE 9.1.1   Pultruded FRP WF beam 12 × 12 × 1/2″.

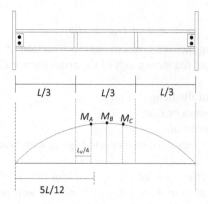

EXAMPLE FIGURE 9.1.2 Moments for C_b.

The absolute value of moment at three-quarter point of the unbraced segment $M_C = (5WL^2/24)-(25WL^2/288)$.

$$C_b = \frac{12.5M_{max}}{2.5M_{max}+3M_A+4M_B+3M_C} = \frac{12.5(1/8)}{2.5(1/8)+3(35/288)+4(1/8)+3(35/288)}$$

$$C_b = 1.014$$

Warping constant (C_ω for wide-flange or I-section):

$$C_\omega = \frac{I_z d^2}{4} = \frac{144.1(12^2)}{4} = 5,188 \text{ in}^6$$

For $d \geq 6''$ then $L_{b, eff} = 0.95L_b$, the LTB strength is determined from Eq. (9.13) as:

$$f_n^{LTB} = \frac{C_b \pi}{S_T L_{b,eff}}\sqrt{E_{T,f}I_z G_{LT}C + \frac{\pi^2 E_{T,f}^2 I_z C_\omega}{L_{b,eff}^2}}$$

$$\text{where,} \quad C = \sum_i \frac{b_i t_i^3}{3}, \quad I_z = 2\left(\frac{1}{12}b_f t_f^3\right)$$

$$f_n^{LTB} = \frac{(1.014)\pi}{75.5(0.95\times 8\times 12)}$$

$$\times\sqrt{(0.8\times 10^6)(144.1)(0.425\times 10^6)(1.458)+\frac{\pi^2(0.8\times 10^6)^2(144.1)(5,188)}{(0.95\times 8\times 12)^2}}$$

$$f_n^{LTB} = 11,697\,\text{psi}$$

For $\varphi = 0.55$ and λ (for $1.2D + 1.6L$) = 0.8, the LTB strength is determined as:

$$\varphi\lambda f_n^{LTB} = (0.55\times 0.8)11,697 = 5,147\,\text{psi}$$

$$M_u = \left(\varphi\lambda f_n^{LTB}\right)S_T = \left(5,147\,\text{psi}\right)\left(75.5\,\text{in}^3\right) = 388,574\,\text{lbs.in} = 32.4\,\text{kips.ft}$$

(**Alternative method**): b) Determine LTB strength using the simplified expression:
The simplified LTB strength is determined from Eq. (9.18) as:

$$M_n^{LTB} = \frac{\pi E_{T,f}}{2L_{b,eff}} \sqrt{2I_z C + \left(\frac{\pi h I_z}{L_{b,eff}}\right)^2}$$

$$M_n^{LTB} = \frac{\pi(0.8 \times 10^6)}{2(0.95 \times 12 \times 8)} \sqrt{(2 \times 144.1 \times 1.458) + \left(\frac{\pi \times 12 \times 144.1}{0.95 \times 12 \times 8}\right)^2} = 867,999 \, \text{lbs.in}$$

$$M_n^{LTB} = 72.3 \, \text{kips.ft}$$

$$\phi \lambda M_n^{LTB} = (0.55 \times 0.8)72.3 = 31.8 \, \text{kips.ft}$$

Note: The difference between the complex and simplified LTB results is about 1.9%.

Note: ACMA Pre-Standard for LRFD of Pultruded FRP Structures (2021) uses a similar equation for LTB strength with the elastic modulus along the longitudinal axis (not in the transverse direction) of a member.

9.6 LOCAL BUCKLING

Another interesting buckling phenomenon in beams under bending is based on plate local buckling (web or flange portion). When portions of a beam profile buckle locally, it may cause premature failure of a structural member. For open sections, such as wide-flange or I-section, the half compression flange is under uniform moment similar to an unstiffened plate subjected to edge compression. Then, the half flange can buckle locally as shown in Figure 9.8.

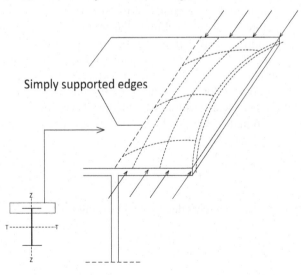

FIGURE 9.8   Local flange buckling.

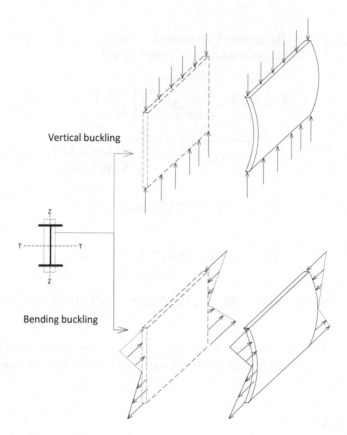

FIGURE 9.9 Web buckling of a member.

In addition, the web portion as a stiffened plate with top and bottom flanges as stiffening elements may buckle locally under flexural compression, potentially leading to two different failure mechanisms (Figure 9.9). The web can buckle as a vertical member under compression or the web can buckle as a plate under horizontal in-plane bending stress. However, local flange buckling tends to occur before local web buckling, depending on the aspect ratios (flange width-to-thickness ratios) of the pultruded FRP cross-sectional profiles. In general, the most pultruded FRP profiles demonstrate the elastic instability behavior (global or local buckling) before the FRP materials reaching their rupture strength. Thus, the term of a compact section as given in standard steel profiles may not be applicable for the pultruded FRP profiles (Kollar, 2003; Bank, 2006).

The nominal strength under local buckling of pultruded FRP profiles is considered from the minimum local buckling strength corresponding to the flange and web portions as (Structural Plastics Manual, ASCE, 1984; ACMA, 2021):

$$f_{cr} = \min\left(f_{cr,f}, f_{cr,w}\right) \tag{9.20}$$

The nominal flexural moment of members due to local bucking is determined as:

$$M_n = S_c f_{cr} \tag{9.21}$$

where
   f_{cr} = critical buckling stress determined for each cross section
   $f_{cr,f}$ and $f_{cr,w}$ = critical buckling stress of the flange and web, respectively

M_n = nominal bending moment

S_c = transformed section modulus, associated with the member extreme fiber in compression, about the axis of bending.

The nominal local buckling strengths of flange and web portions for various FRP pultruded profiles are summarized as presented in Tables 9.1–9.3 (Structural Plastics Manual, ASCE, 1984; Kollar, 2003; Bank, 2006; ACMA, 2021).

where

G_{LT} = characteristic value of in-plane shear modulus

b_f = compression flange width

t_f = compressive flange thickness

$E_{L,f}$ = characteristic value of longitudinal modulus of the flange

$E_{T,f}$ = characteristic value of transverse modulus of the flange

$E_{L,w}$ = characteristic value of longitudinal modulus of the web

$E_{T,w}$ = characteristic value of transverse modulus of the web

d = overall depth of section

t_w = web thickness.

t = thickness of the longest leg (back-to-back angle),

v_{LT} = longitudinal Poisson's ratio.

TABLE 9.1

Nominal Strength under Local Buckling (Bending about Strong Axis of Members) (ACMA, 2021)

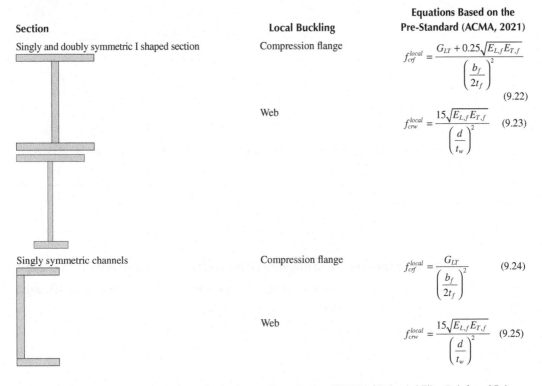

Section	Local Buckling	Equations Based on the Pre-Standard (ACMA, 2021)
Singly and doubly symmetric I shaped section	Compression flange	$f_{crf}^{local} = \dfrac{G_{LT} + 0.25\sqrt{E_{L,f}E_{T,f}}}{\left(\dfrac{b_f}{2t_f}\right)^2}$ (9.22)
	Web	$f_{crw}^{local} = \dfrac{15\sqrt{E_{L,f}E_{T,f}}}{\left(\dfrac{d}{t_w}\right)^2}$ (9.23)
Singly symmetric channels	Compression flange	$f_{crf}^{local} = \dfrac{G_{LT}}{\left(\dfrac{b_f}{2t_f}\right)^2}$ (9.24)
	Web	$f_{crw}^{local} = \dfrac{15\sqrt{E_{L,f}E_{T,f}}}{\left(\dfrac{d}{t_w}\right)^2}$ (9.25)

Note: Based on the Pre-Standard for Load and Resistance Factor Design (LRFD) of Pultruded Fiber Reinforced Polymer (FRP) Structures (ACMA, 2021).

TABLE 9.2

Nominal Strength under Local Buckling (Bending about Strong Axis of Members) (ACMA, 2021)

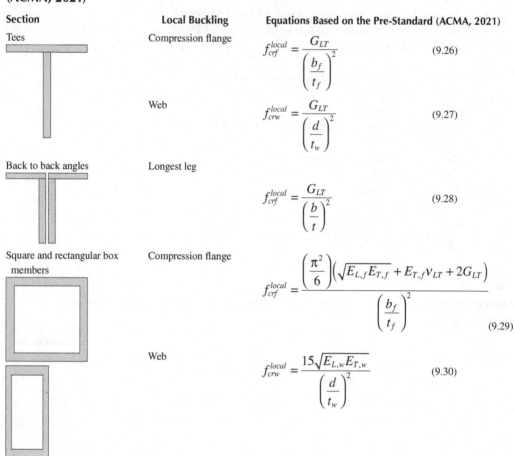

Section	Local Buckling	Equations Based on the Pre-Standard (ACMA, 2021)	
Tees	Compression flange	$$f_{crf}^{local} = \dfrac{G_{LT}}{\left(\dfrac{b_f}{t_f}\right)^2}$$	(9.26)
	Web	$$f_{crw}^{local} = \dfrac{G_{LT}}{\left(\dfrac{d}{t_w}\right)^2}$$	(9.27)
Back to back angles	Longest leg	$$f_{crf}^{local} = \dfrac{G_{LT}}{\left(\dfrac{b}{t}\right)^2}$$	(9.28)
Square and rectangular box members	Compression flange	$$f_{crf}^{local} = \dfrac{\left(\dfrac{\pi^2}{6}\right)\left(\sqrt{E_{L,f}E_{T,f}} + E_{T,f}\nu_{LT} + 2G_{LT}\right)}{\left(\dfrac{b_f}{t_f}\right)^2}$$	(9.29)
	Web	$$f_{crw}^{local} = \dfrac{15\sqrt{E_{L,w}E_{T,w}}}{\left(\dfrac{d}{t_w}\right)^2}$$	(9.30)

Note: Based on the Pre-Standard for Load & Resistance Factor Design (LRFD) of Pultruded Fiber Reinforced Polymer (FRP) Structures (ACMA, 2021).

TABLE 9.3

Nominal Strength under Local Buckling (Bending about Weak Axis of Members) (ACMA, 2021)

Section	Local Buckling	Equations Based on the Pre-Standard (ACMA, 2021)	
Doubly symmetric I-shaped members	Compression flange	$$f_{crf}^{local} = \dfrac{4t_f^2}{b_f^2}G_{LT}$$	(9.31)

(Continued)

TABLE 9.3 (*Continued*)

Nominal Strength under Local Buckling (Bending about Weak Axis of Members) (ACMA, 2021)

Section	Local Buckling	Equations Based on the Pre-Standard (ACMA, 2021)
Singly symmetric channels	Compression flange	$f_{crf}^{local} = \dfrac{t_f^2}{b_f^2} G_{LT}$ (9.32)
	Web	$f_{crw}^{local} = \dfrac{\pi^2 t_w^2}{6d^2}\left(\sqrt{E_{L,w}E_{T,w}} + E_{T,w}\nu_{LT} + 2G_{LT}\right)$ (9.33)

Note: Based on the Pre-Standard for Load & Resistance Factor Design (LRFD) of Pultruded Fiber Reinforced Polymer (FRP) Structures (ACMA, 2021).

9.7   WEB SHEAR BUCKLING

When pultruded FRP beams are under concentrated load or high shear zone then, the web may buckle due to instability under shear. The local web buckling under shear can occur in both the open and closed pultruded FRP beams. For webs of a single and double symmetric I-shaped member, C-channels, back to back channels, square and rectangular box sections with bending about the strong axis, the local shear buckling strength of a web is given as follows (Structural Plastics Manual, ASCE, 1984; Kollar, 2003; Bank, 2006; ACMA, 2021):

$$V_n = F_{cr} A_s \tag{9.34}$$

The critical shear buckling strength for both the stiffened and unstiffened webs of pultruded sections shall be determined as:

$$f_{cr} = k \frac{\sqrt[4]{E_{L,w}\left(E_{T,w}\right)^3}}{\left(\dfrac{d_w}{t_w}\right)^2} \tag{9.35.1}$$

The value of k for web plates without stiffeners is determined as:

$$k = 2.67 + 1.59\beta \tag{9.35.2}$$

$$\beta = \frac{G_{LT}}{\sqrt{E_{L,w}E_{T,w}}} \tag{9.35.3}$$

The value of k for stiffened webs shall be determined as:

$$k = 2.67 + 1.59\beta + 1.47\alpha\beta + 1.49\alpha^3 \tag{9.35.4}$$

$$\alpha = b_s \left(\sqrt[4]{\frac{E_{L,w}}{E_{T,w}}}\right) \tag{9.35.5}$$

where

V_n = nominal shear strength due to local web shear buckling
F_{cr} = critical web shear buckling strength (stress)
A_s = shear area
G_{LT} = characteristic value of in-plane shear modulus
$E_{L,w}$ = characteristic value of longitudinal modulus of the web
$E_{T,w}$ = characteristic value of transverse modulus of the web
d_w = height of web
t_w = thickness of web
k = shear buckling coefficient
b_s = stiffener spacing.

Note: For tees, back-to-back angles, and members bent about their minor axis, the shear buckling strength of the elements perpendicular to the plane of bending shall be determined by analysis (ACMA, 2021).

Example 9.2

A pultruded FRP WF beam 12 × 12 × 1/2″ with a simple span of 24 ft with lateral supports at the ends and at the third point of the span is shown in Example Figure 9.2.
Given: Section WF 12 × 12 × 1/2″: Self-weight = 13.98 lbs/ft, Area (A) = 17.51 in², Web area (A_w) = 5.50 in², Sectional modulus about the strong axis (S_T) = 75.5 in³, Sectional modulus about the weak axis (S_Z) = 24.0 in³, Moment of inertia about strong axis (I_T) = 453 in⁴, Moment of inertia about weak axis (I_Z) = 144 in⁴, Torsional constant (C) = 1.458 in⁴.

For pultruded FRP WF beam properties as per manufacturers: longitudinal tensile strength (F_t) = 30,000 psi, shear strength (F_v) = 4,500 psi, longitudinal modulus of flange (E_{Lf}) = 2.6 (10⁶) psi and transverse modulus of flange (E_{Tf}) = 0.8 (10⁶) psi, longitudinal modulus of web (E_{Lw}) = 2.6 (10⁶) psi and Transverse modulus of web (E_{Tw}) = 0.8 (10⁶) psi, shear modulus (G_{LT}) = 0.425 (10⁶) psi; Poisson's ratio (v_{LT}) = 0.33, web thickness (t_w) = 0.5″, and flange width (b_f) = 12″.

Determine: (a) Local flange buckling strength, (b) local web buckling strength, and (c) local web shear buckling strength.

Solution

a. Local flange buckling:

The local flange buckling strength is determined from Eq. (9.22) as:

$$f_{crf}^{local} = \frac{G_{LT} + 0.25\sqrt{E_{L,f}E_{T,f}}}{\left(\frac{b_f}{2t_f}\right)^2}$$

$$f_{crf}^{local} = \frac{(0.425 \times 10^6) + 0.25\sqrt{(2.6 \times 10^6)(0.8 \times 10^6)}}{\left(\frac{12}{2 \times 0.5}\right)^2} = 5,455\,\text{psi}$$

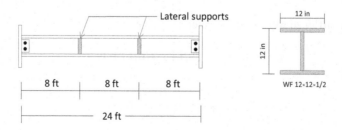

EXAMPLE FIGURE 9.2   Pultruded FRP WF beam 12 × 12 × 1/2″ with lateral supports.

For φ = 0.8 and λ (for 1.2D + 1.6L) = 0.8, the local flange buckling strength is determined as:

$$\varphi\lambda f_{crf}^{local} = (0.8 \times 0.8)5{,}455 = 3{,}491 \text{ psi}$$

$$M_u = \varphi\lambda f_{crf}^{local} S_T = (3{,}491 \text{ psi})(75.5 \text{ in}^3) = 21.97 \text{ kips.ft}$$

b. Local web buckling:
   The local web buckling strength is determined from Eq. (9.23) as:

$$f_{crw}^{local} = \frac{15\sqrt{E_{L,f}E_{T,f}}}{\left(\dfrac{d}{t_w}\right)^2}$$

$$f_{crw}^{local} = \frac{15\sqrt{(2.6 \times 10^6)(0.8 \times 10^6)}}{\left(\dfrac{12}{0.5}\right)^2} = 37{,}558 \text{ psi}$$

For φ = 0.8 and λ (for 1.2D + 1.6L) = 0.8, the local web buckling strength is:

$$\varphi\lambda\left(f_{crw}^{local}\right) = (0.8 \times 0.8)37{,}558 = 24{,}037 \text{ psi}$$

$$M_u = \left(\varphi\lambda\left(f_{crw}^{local}\right)\right)S_T = (24{,}037 \text{ psi})(75.5 \text{ in}^3) = 151 \text{ kips.ft}$$

c. Web shear buckling strength:
   The web shear buckling strength is determined from Eqs. (9.35.1–9.35.3) as:

$$\beta = \frac{G_{LT}}{\sqrt{E_{L,w}E_{T,w}}} = \frac{(0.425 \times 10^6)}{\sqrt{(2.6 \times 10^6)(0.8 \times 10^6)}} = 0.295$$

$$k = 2.67 + 1.59\beta = 2.67 + 1.59(0.295) = 3.14$$

$$f_{cr} = k\frac{\sqrt[4]{E_{L,w}(E_{T,w})^3}}{\left(\dfrac{d_w}{t_w}\right)^2} = 3.14\frac{\sqrt[4]{(2.6 \times 10^6)(0.8 \times 10^6)^3}}{\left(\dfrac{12 - 0.5 - 0.5}{0.5}\right)^2} = 6{,}965 \text{ psi}$$

For φ = 0.8 and λ (for 1.2D + 1.6L) = 0.8, the web shear buckling strength is:

$$\varphi\lambda\left(f_{cr}^{local}\right) = (0.8 \times 0.8)6{,}965 = 4{,}458 \text{ psi}$$

$$V_u = \varphi\lambda\left(f_{cr}^{local}\right)A_w = (4{,}458 \text{ psi})(5.5 \text{ in}^2) = 24.5 \text{ kips}$$

Example 9.3

A pultruded FRP WF beam 12 × 12 × 1/2″ is spanning 24 ft with simply supported boundaries. Determine the uniformly distributed load that the beam can carry if the lateral supports are provided at the ends and at the third point of the span. FRP WF beam properties as per manufacturers are: longitudinal tensile strength (F_t) = 30,000 psi, shear strength (F_v) = 4,500 psi, longitudinal modulus (E_L) = 2.6 (10⁶) psi, transverse modulus (E_T) = 0.8 (10⁶) psi, and shear modulus (G_{LT}) = 0.425 (10⁶) psi.

Given: Section WF 12 × 12 × 1/2″. Self-weight = 13.98 lbs/ft, Area (A) = 17.51 in², Web area (A_w) = 5.50 in², Sectional modulus about the strong axis (S_T) = 75.5 in³, Sectional modulus about the weak axis (S_Z) = 24.0 in³, Moment of inertia about strong axis (I_T) = 453 in⁴, Moment of inertia about weak axis (I_Z) = 144 in⁴, and Torsional constant (C) = 1.458 in⁴.

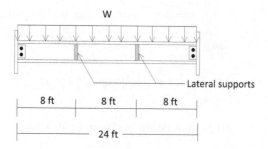

EXAMPLE FIGURE 9.3   Pultruded FRP WF beam 12 × 12 × 1/2″.

Solution

a. Determine nominal material strength:

The resistance of the pultruded FRP WF profiles is determined using $\varphi = 0.65$ for material rupture and time factor λ (for $1.2D + 1.6L) = 0.8$

$$\varphi\lambda F_t = (0.65 \times 0.8)30,000 = 15,600\,\text{psi and } \varphi\lambda F_v = (0.65 \times 0.8)4,500 = 2,340\,\text{psi}$$

$$M_u = (\varphi\lambda F_t)S_T = (15.6\,\text{ksi})(75.5\,\text{in}^3) = 98.2\,\text{kips.ft}$$

$$V_u = (\varphi\lambda F_v)A_w = (2.34\,\text{ksi})(5.5\,\text{in}^2) = 12.87\,\text{kips}$$

b. Determine LTB strength:

From the global buckling result in Example 9.1, the LTB strength of the pultruded FRP beam is: (the resistance factor $\varphi = 0.55$ and time factor λ for $1.2D + 1.6L = 0.8$)

$$\varphi\lambda f_n^{LTB} = (0.55 \times 0.8)11,697 = 5,147\,\text{psi}$$

$$M_u = (\varphi\lambda f_n^{LTB})S_T = (5,147\,\text{psi})(75.5\,\text{in}^3) = 32.4\,\text{kips.ft}$$

c. Determine local buckling strength:

The local buckling strength of a pultruded FRP beam is obtained from the lower of the two values obtained from compressive flange buckling strength and compressive web buckling strength. From Example 9.2, the results of the local buckling strength are presented as follows: (resistance factor $\varphi = 0.8$ and time factor λ for $1.2D + 1.6L = 0.8$)

c.1) Local flange buckling from part a of Example 9.2 determined by Eq. (9.22):

$$\varphi\lambda\left(f_{crf}^{local}\right) = (0.8 \times 0.8)5,455 = 3,491\,\text{psi}$$

$$M_u = \varphi\lambda\left(f_{crf}^{local}\right)S_T = (3,491\,\text{psi})(75.5\,\text{in}^3) = 21.97\,\text{kips.ft}$$

c.2) Local web buckling from part b of Example 9.2 determined by Eq. (9.23):

$$\varphi\lambda\left(f_{crw}^{local}\right) = (0.8 \times 0.8)37,558 = 24,037\,\text{psi}$$

$$M_u = \varphi\lambda\left(f_{crw}^{local}\right)S_T = (24,037\,\text{psi})(75.5\,\text{in}^3) = 151\,\text{kips.ft}$$

The local buckling strength (minimum of C.1 and C.2) is due to compressive local flange buckling failure, which is 3,491 psi.

d. Determine web shear buckling strength from Eqs. (9.35.1–9.35.3):

For $\varphi = 0.8$ and λ (for $1.2D + 1.6L$) = 0.8 (Example 9.2, part c), the web shear buckling strength of the beam is:

$$\varphi\lambda\left(f_{cr}^{local}\right) = (0.8 \times 0.8)6,965 = 4,458\,\text{psi}$$

$$V_u = \varphi\lambda\left(f_{cr}^{local}\right)A_w = (4,458\,\text{psi})(5.5\,\text{in}) = 24.5\,\text{kips}$$

Summary of the nominal strength:

The uniformly distributed load of the pultruded FRP beam is determined from the minimum nominal strength of various failure strengths as:

a) Material strength:
   M_u = 98.2 kips.ft then; W_u = 98.2 × (8/24²) = 1.36 kips/ft
   V_u = 12.87 kips then; W_u = 12.87 × (2/24) = 1.07 kips/ft

b) Lateral torsional strength:
   M_u = 32.4 kips.ft then; W_u = 32.4 × (8/24²) = 0.45 kips/ft

c) Local buckling strength due to compressive flange:
   M_u = 21.97 kips.ft then; W_u = 21.97 × (8/24²) = 0.305 kips/ft

d) Web shear buckling strength:
   V_u = 24.5 kips then; W_u = 24.5 × (2/24) = 2.04 kips/ft

Thus, the factored uniformly distributed load on the pultruded FRP WF beam 12 × 12 × 1/2" is 305 lbs/ft for a simply supported beam of 24' span with lateral supports.

Example 9.4

Design a pultruded FRP beam for a simply supported beam of 20 ft in span length with a dead load of 50 lbs/ft and a live load of 100 lbs/ft. Assume: compression flange is fully supported laterally by the deck by accounting for shear deformation. Using the LRFD method, design an FRP beam with adequate bracing. Refer to Example 9.3 for beam properties as per manufacturer.

Solution

a. Determine factored loads:
   Total service load (W) = (50 lbs/ft) + (100 lbs/ft) = 150 lbs/ft
   Total factored load (W_u) = 1.2 (50 lbs/ft) +1.6 (100 lbs/ft) = 220 lbs/ft

   Try I-beam using the serviceability limit state (deflection limit $L/240$). The shear rigidity $k(G_{LT}A)$ can be approximated with $(G_{LT})_{web}\,A_{web}$. The deflection, including shear deflection of the pultruded FRP beam, is determined from Eq. (9.6.4) as:

At serviceability:

$$w_{max} = \frac{5WL^4}{384(E_L I_T)} + \frac{WL^2}{8k(G_{LT}A)} = \frac{5(150/12)(20\times12)^4}{384(2.6\times10^6)I_T} + \frac{(150/12)(20\times12)^2}{8(1)(0.425\times10^6)A_w}$$

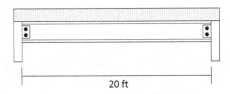

20 ft

EXAMPLE FIGURE 9.4 Pultruded FRP beam with full lateral supports.

$$w_{max} = \frac{208}{I_T} + \frac{0.21}{A_w} \le \frac{L}{240}$$

where $k \cong 1.0$

b. Select $12 \times 6 \times 1/2''$ I - shaped section:

   Given: Longitudinal modulus of flange $(E_{Lf}) = 2.6 \,(10^6)$ psi, Transverse modulus of flange $(E_{Tf}) = 0.8 \,(10^6)$ psi, Longitudinal modulus of web $(E_{Lw}) = 2.6 \,(10^6)$ psi, Transverse modulus of web $(E_{Tw}) = 0.8 \,(10^6)$ psi, Shear modulus $(G_{LT}) = 0.425 \,(10^6)$ psi; Poisson's ratio $(v_{LT}) = 0.33$. Web thickness $(t_w) = 0.5$ inch and flange width $(b_f) = 6$ inch. Self-weight = 9.24 lbs/ft, Area $(A) = 11.73$ in², Web area $(A_w) = 5.50$ in², Sectional modulus about the strong axis $(S_T) = 42.3$ in³, I_T Moment of inertia about strong axis $(I_T) = 254$ in⁴, Moment of inertia about weak axis $(I_Z) = 18.11$ in⁴, and Torsional constant $(C) = 0.958$ in⁴.

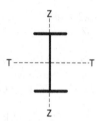

$$w_{max} = \frac{208}{254} + \frac{0.21}{5.50} = 0.857 \le \frac{20(12)}{240} = 1.0 \,\text{in}$$

   The proposed pultruded FRP beam satisfies the deflection limit state.

c. Determine design bending and shear strengths:

   The design bending moment and shear are:

$$M_u = \frac{W_u L^2}{8} = \frac{220(20^2)}{8} = 11,000 \,\text{lbs} - \text{ft} \quad \text{and} \quad V_u = \frac{W_u L}{2} = \frac{220(20)}{2} = 2,200 \,\text{lbs}$$

   Induced flexural and average web shear stresses are determined as:

$$f_{bu} = \frac{M_u}{S_T} = \frac{11,000(12)}{42.3} = 3,121 \,\text{psi} \quad \text{and} \quad f_{vu} = \frac{V}{A_{web}} = \frac{2,200}{5.50} = 400 \,\text{psi}$$

d. Determine nominal material strength:

   For $\varphi = 0.65$ (for material rupture) and λ (for $1.2D + 1.6L$) $= 0.8$, the resistance of the pultruded FRP I-shaped beam is determined as:

$$\varphi \lambda F_t = (0.65 \times 0.8)30,000 = 15,600 \,\text{psi} \quad \text{and} \quad \varphi \lambda F_v = (0.65 \times 0.8)4,500 = 2,340 \,\text{psi}$$

$$\varphi \lambda F_t = 15,600 \,\text{psi} > f_{bu} = 3,121 \,\text{psi} \quad \text{and} \quad \varphi \lambda F_v = 2,340 \,\text{psi} > f_{vu} = 400 \,\text{psi}$$

   It is found that the WF FRP beam satisfies the material strength limit.

e. Check for LTB strength:

   The LTB strength is not considered in this design. The compression flange of the selected FRP I beam is fully supported laterally by the deck.

f. Check for local buckling strength:

The compressive local flange buckling strength is not considered in this design. The compression flange of the selected FRP I beam is fully supported laterally by the deck. Thus, only the local web bucking strength is determined from Eq. (9.23) as:

$$f_{crw}^{local} = \frac{15\sqrt{E_{L,f}E_{T,f}}}{\left(\dfrac{d}{t_w}\right)^2}$$

$$f_{crw}^{local} = \frac{15\sqrt{(2.6\times10^6)(0.8\times10^6)}}{\left(\dfrac{12}{0.5}\right)^2} = 37,558\,\text{psi}$$

$$\varphi\lambda\left(f_{crw}^{local}\right) = (0.8\times0.8)37,558 = 24,037\,\text{psi}$$

The local web buckling strength (24,037 psi) is higher than the design flexural strength (3,121 psi). The cross section of the pultruded FRP I-shaped beam provides enough resistance to the local web bucking.

g. Check for web shear buckling strength:

The web shear buckling strength is determined from Eqs. (9.35.1–9.35.3) as:

$$\beta = \frac{G_{LT}}{\sqrt{E_{L,w}E_{T,w}}} = \frac{(0.425\times10^6)}{\sqrt{(2.6\times10^6)(0.8\times10^6)}} = 0.295$$

$$k = 2.67 + 1.59\beta = 2.67 + 1.59(0.295) = 3.14$$

$$f_{cr} = k\frac{\sqrt[4]{E_{L,w}(E_{T,w})^3}}{\left(\dfrac{d_w}{t_w}\right)^2} = 3.14\frac{\sqrt[4]{(2.6\times10^6)(0.8\times10^6)^3}}{\left(\dfrac{12-0.5-0.5}{0.5}\right)^2} = 6,965\,\text{psi}$$

For $\varphi = 0.8$ and λ (for 1.2D + 1.6L) = 0.8, the web shear buckling strength is:

$$\varphi\lambda\left(f_{cr}^{local}\right) = (0.8\times0.8)6,965 = 4,458\,\text{psi}$$

$$f_{vu} = 400\,\text{psi} < \varphi\lambda\left(f_{cr}^{local}\right) = 4,458\,\text{psi}$$

Note: Web shear buckling strength (4,458 psi) is higher than the design shear strength (400 psi) of the FRP I-beam. The cross section of the FRP I-beam will not reach the web shear buckling value. The resistance of the beam is much higher than the flexural and shear-induced stresses from factored loads. The design of the selected beam is controlled by the serviceability limit state.

Example 9.5

Determine a maximum unbraced length of the selected pultruded FRP I beam in Example 9.4. The pultruded FRP beam has lateral supports at a specific position and at beam supports. (Assuming the beam is controlled by LTB and C_b is permitted to be conservatively taken as 1.0 for all cases.)

Solution

From Example 9.4, the design bending strength (ultimate moment) is:

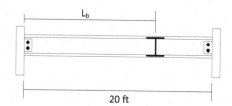

EXAMPLE FIGURE 9.5 Pultruded FRP I beam with a maximum unbraced length.

$$M_u = \frac{W_u L^2}{8} = \frac{220(20^2)}{8} = 11,000\,\text{lbs}-\text{ft}$$

Induced flexural stress is determined as follows:

$$f_{bu} = \frac{M_u}{S_T} = \frac{11,000(12)}{42.3} = 3,121\,\text{psi}$$

The maximum unbraced length of the pultruded FRP I beam is determined from the global LTB strength when the nominal LTB strength is equal to the induced flexural stress under applied loads.

$$\varphi\lambda f_n^{LTB} = f_I$$

The LTB strength is determined as:

C_b (moment modification factor) = 1.0 for a simply supported uniformly beam. Web thickness (t_w) = 0.5 inch and width of flange (b_f) = 6 inch, Sectional modulus about the strong axis (S_T) = 42.3 in³, Moment of inertia about strong axis (I_T) = 254.1 in⁴, Moment of inertia about weak axis (I_Z) = 18.11 in⁴, and Torsional constant (C) = 0.958 in⁴.

From Eq. (9.13):

$$C_\omega = \frac{I_z d^2}{4} = \frac{18.11(12^2)}{4} = 652\,\text{in}^6$$

$$f_n^{LTB} = \frac{C_b \pi}{S_T L_{b,eff}} \sqrt{E_{T,f} I_z G_{LT} C + \frac{\pi^2 E_{T,f}^2 I_z C_\omega}{L_{b,eff}^2}}$$

$$f_n^{LTB} = \frac{(1)\pi}{42.3 L_{b,eff}} \sqrt{\begin{array}{l}(0.8\times10^6)(18.11)(0.425\times10^6)(0.958) \\[2mm] +\dfrac{\pi^2(0.8\times10^6)^2(18.11)(652)}{(L_{b,eff})^2}\end{array}}$$

For φ = 0.55 and λ (for 1.2D + 1.6L) = 0.8, the lateral–torsional buckling strength is determined as:

$$\varphi\lambda f_n^{LTB} = (0.55\times0.8)f_n^{LTB} = f_{bu}$$

$$\frac{0.55\times0.8\times\pi}{42.3 L_{b,eff}} \sqrt{(2.6\times10^6)(18.11)(0.425\times10^6)(0.958) + \frac{\pi^2(0.8\times10^6)^2(18.11)(652)}{(L_{b,eff})^2}}$$

$$= 3,121\,\text{psi}$$

Maximum $L_{b,\,eff}$ = 64.01″ and L_b = 67.34″ (a correction factor of 0.95 is required for d > 6″).

Example 9.6

Using the ASD method, design a pultruded FRP beam for a simply supported beam of 20 ft design span, including shear deformation with a dead load of 50 lbs/ft and a live load of 100 lbs/ft. Assume: compression flange is fully supported laterally by the deck.

Solution

a. Determine service loads and maximum deflection for serviceability limit:
   Total service load $(W) = (50 \text{ lbs/ft}) + (100 \text{ lbs/ft}) = 150 \text{ lbs/ft}$
   Try pultruded I-beam using the serviceability limit (deflection limit $L/240$). The shear rigidity $k(G_{LT}A)$ can be approximated with $(G_{LT})_{web}\, A_{web}$. Deflection of the pultruded FRP beam from Eq. (9.6.4) is:

$$w_{max} = \frac{5WL^4}{384(EI)} + \frac{WL^2}{8k(G_{LT}A)} = \frac{5(150/12)(20\times12)^4}{384(2.6\times10^6)I_T} + \frac{(150/12)(20\times12)^2}{8(0.425\times10^6)A_w} = \frac{208}{I_T}$$

$$w_{max} \le \frac{L}{240}$$

b. Select 12 × 6 × 1/2″ I-shaped section:
   Given: Longitudinal modulus of flange $(E_{Lf}) = 2.6\ (10^6)$ psi, Transverse modulus of flange $(E_{Tf}) = 0.8\ (10^6)$ psi, Longitudinal modulus of web $(E_{Lw}) = 2.6\ (10^6)$ psi, Transverse modulus of web $(E_{Tw}) = 0.8\ (10^6)$ psi, Shear modulus $(G_{LT}) = 0.425\ (10^6)$ psi; Poisson's ratio $(\nu_{LT}) = 0.33$. Web thickness $(t_w) = 0.5$ inch and Flange width $(b_f) = 6″$, Self-weight = 9.24 lbs/ft, Area $(A) = 11.73\ \text{in}^2$, web area $(A_w) = 5.50\ \text{in}^2$, Sectional modulus about the strong axis $(S_T) = 42.3\ \text{in}^3$, Moment of inertia about strong axis $(I_T) = 254\ \text{in}^4$, Moment of inertia about weak axis $(I_Z) = 18.11\ \text{in}^4$, and Torsional constant $(C) = 0.958\ \text{in}^4$.

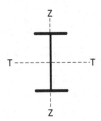

$$w_{max} = \frac{208}{254} + \frac{0.21}{5.50} = 0.857 \le \frac{20(12)}{240} = 1.0\,\text{in}$$

   The proposed pultruded FRP beam satisfies the deflection limit.
c. Determine the design service bending and shear strength:
   The design bending moment and shear are:

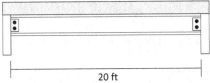

20 ft

EXAMPLE FIGURE 9.6 Pultruded FRP beam with full lateral supports.

Total service load (W) = (50 lbs/ft) + (100 lbs/ft) = 150 lbs/ft

$$M = \frac{WL^2}{8} = \frac{150(20^2)}{8} = 7,500 \text{ lbs} - \text{ft} \quad \text{and} \quad V = \frac{WL}{2} = \frac{150(20)}{2} = 1,500 \text{ lbs}$$

Induced flexural and average web shear stresses are determined as follows:

$$f_b = \frac{M}{S_T} = \frac{7,500(12)}{42.3} = 2,128 \text{ psi} \quad \text{and} \quad f_v = \frac{V}{A_{web}} = \frac{1,500}{5.50} = 273 \text{ psi}$$

d. Determine nominal material strength:

The resistance of the pultruded I–beam is: (F.S. = 2.5 and 3.0 for allowable flexural and shear strength, respectively)

$$(f_b)_{allow} = \frac{F_t}{2.5} = \frac{30,000}{2.5} = 12,000 \text{ psi} \quad \text{and} \quad (f_v)_{allow} = \frac{F_v}{3.0} = \frac{4,500}{3} = 1,500 \text{ psi}$$

$$f_b = 2,128 \text{ psi} < (f_b)_{allow} = 12,000 \text{ psi} \quad \text{and} \quad f_v = 273 \text{ psi} < (f_v)_{allow} = 1,500 \text{ psi}$$

Note: the pultruded FRP I-beam satisfies the allowable material stress limit.

e. Check for LTB strength:

The LTB strength is not considered in this design, because the compression flange of the FRP I-shaped beam is fully supported laterally.

f. Check for local buckling strength:

Using Eq. (9.23):

The compressive local flange buckling strength is not considered in this design. The compression flange of the selected FRP I beam is fully supported laterally by the deck. Thus, only the local web bucking strength is determined as:

$$f_{crw}^{local} = \frac{15\sqrt{E_{L,f}E_{T,f}}}{\left(\dfrac{d}{t_w}\right)^2}$$

$$f_{crw}^{local} = \frac{15\sqrt{(2.6\times10^6)(0.8\times10^6)}}{\left(\dfrac{12}{0.5}\right)^2} = 37,558 \text{ psi}$$

$$\left(f_{crw}^{local}\right)_{allow} = \frac{\left(f_{crw}^{local}\right)}{F.S.} = \frac{37,558}{2.5} = 15,023 \text{ psi}$$

The cross section of the pultruded FRP I-shaped beam satisfies the resistance to the local web bucking.

g. Check for web shear buckling strength:

The web shear buckling strength is determined from Eqs. (9.35.1–9.35.3) as:

$$\beta = \frac{G_{LT}}{\sqrt{E_{L,w}E_{T,w}}} = \frac{(0.425\times10^6)}{\sqrt{(2.6\times10^6)(0.8\times10^6)}} = 0.295$$

$$k = 2.67 + 1.59\beta = 2.67 + 1.59(0.295) = 3.14$$

$$f_{cr} = k\frac{\sqrt[4]{E_{L,w}(E_{T,w})^3}}{\left(\dfrac{d_w}{t_w}\right)^2} = 3.14\frac{\sqrt[4]{(2.6\times10^6)(0.8\times10^6)^3}}{\left(\dfrac{12-0.5-0.5}{0.5}\right)^2} = 6,965 \text{ psi}$$

Allowable web shear buckling stress (with F.S. = 3.0) is determined as:

$$\left(f_{cr}^{local}\right)_{allow} = \frac{f_{cr}^{local}}{F.S.} = \frac{6,965}{3.0} = 2,322\,\text{psi}$$

$$f_v = 273\,\text{psi} < \left(f_{cr}^{local}\right)_{allow} = 2,322\,\text{psi}$$

Note: Allowable web shear buckling strength (2,322 psi) is much higher than the design load-induced shear strength (273 psi) and also higher than the allowable material shear stress (1,500 psi) of the I-shaped beam. Thus, the cross section of the I-beam based on this design will not exceed the web shear buckling value.

Example 9.7

Using the LRFD method, design a simply supported pultruded FRP beam of 16 ft span carrying dead and live loads of 60 and 150 lbs/ft, respectively. The FRP beam is laterally braced at the mid-span and at supports. Refer to Example 9.3 for beam properties as per manufacturer.

Solution

a. Determine factored loads:
   Total service load (W) = (60 lbs/ft) + (150 lbs/ft) = 210 lbs/ft
      Total factored load (W_u) = 1.2 (60 lbs/ft) +1.6 (150 lbs/ft) = 312 lbs/ft
      Try FRP wide flange beam using the serviceability limit (the deflection limit L/240) including shear deformation effects. The shear rigidity $k(G_{LT}A)$ can be approximated with $(G_{LT})_{web}\,A_{web}$.
      The pultruded FRP beam deflection is determined from Eq. (9.6.4) as:

$$w_{max} = \frac{5WL^4}{384(E_L I_T)} + \frac{WL^2}{8k(G_{LT}A)} = \frac{5(210/12)(16\times12)^4}{384(2.6\times10^6)I_T} + \frac{(210/12)(16\times12)^2}{8(0.425\times10^6)A_{web}}$$

$$w_{max} = \frac{119}{I_T} + \frac{0.19}{A_{web}} \le \frac{L}{240}$$

b. Select 10 × 10 × 1/2″ WF section:
   Given: Longitudinal modulus of flange (E_{Lf}) = 2.6 (10⁶) psi, Transverse modulus of flange (E_{Tf}) = 0.8 (10⁶) psi, Longitudinal modulus of web (E_{Lw}) = 2.6 (10⁶) psi, Transverse modulus of web (E_{Tw}) = 0.8 (10⁶) psi, Shear modulus (G_{LT}) = 0.425 (10⁶) psi, Poisson's ratio (v_{LT}) = 0.33. Flange thickness (t_f) = 0.5″ and Web thickness (t_w) = 0.5″ and Flange width (b_f) = 10″, Self-weight = 11.64 lbs/ft, Area (A) = 14.55 in², Web area (A_w) = 4.50 in², Sectional modulus about the strong axis (S_T) = 51.2 in³, Moment of inertia about strong axis (I_T) = 256.2 in⁴, Moment of inertia about weak axis (I_Z) = 83.4 in⁴, and Torsional constant (C) = 1.208 in⁴.

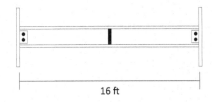

16 ft

EXAMPLE FIGURE 9.7.1 Pultruded FRP beam with partial lateral supports.

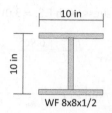

WF 8x8x1/2

$$w_{max} = \frac{119}{256.2} + \frac{0.19}{4.50} = 0.507 \le \frac{16(12)}{240} = 0.8\,\text{in}$$

Note: FRP wide flange beam satisfies the deflection limit.

c. Determine the design bending and shear strengths:

The design bending moment and shear are:

$$M_u = \frac{W_u L^2}{8} = \frac{312(16^2)}{8} = 9,984\,\text{lbs}-\text{ft} \quad \text{and} \quad V_u = \frac{W_u L}{2} = \frac{312(16)}{2} = 2,496\,\text{lbs}$$

Induced flexural and average web shear stresses are:

$$f_{bu} = \frac{M_u}{S_T} = \frac{9,984(12)}{51.2} = 2,340\,\text{psi} \quad \text{and} \quad f_{vu} \cong \frac{V_u}{A_{web}} = \frac{2,496}{4.50} = 555\,\text{psi}$$

d. Determine nominal material strength:

For $\varphi = 0.65$ (for material rupture) and λ (for $1.2D + 1.6L$) = 0.8, the resistance of the FRP wide flange section is:

$$\varphi\lambda F_t = (0.65 \times 0.8)30,000 = 15,600\,\text{psi} \quad \text{and} \quad \varphi\lambda F_v = (0.65 \times 0.8)4,500 = 2,340\,\text{psi}$$

$$\varphi\lambda F_t = 15,600\,\text{psi} > f_{bu} = 2,340\,\text{psi} \quad \text{and} \quad \varphi\lambda F_v = 2,340\,\text{psi} > f_{vu} = 555\,\text{psi}$$

It is found that the FRP wide flange beam satisfies the material strength limit as well.

e. Determine lateral torsional buckling strength:

The FRP wide flange beam is partially supported laterally at the mid-span and at both ends of the span.

As shown in Example Figure 9.7.2, moment modification factor C_b for lateral supports at the ends and at the third point of the span is determined as follows:

The maximum moment in the unbraced segment, $M_{max} = WL^2/8$

Moment at quarter of the unbraced segment, $(L_u/4 = L/8)$, $M_A = (WL^2/16)-(WL^2/128) = (7WL^2/128)$

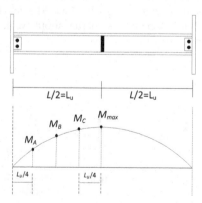

EXAMPLE FIGURE 9.7.2 Moments for C_b.

Moment at the centerline of the unbraced segment, $M_B = (WL^2/8)-(WL^2/32) = (3WL^2/32)$
Moment at three-quarter point of the unbraced segment $M_C = (3WL^2/16)-(9WL^2/128) = (15WL^2/128)$

$$C_b = \frac{12.5M_{max}}{2.5M_{max}+3M_A+4M_B+3M_c}$$

$$C_b = \frac{12.5(1/8)}{2.5(1/8)+3(7/128)+4(3/32)+3(15/128)} = 1.30 \le 3.0$$

e.1) Using the complex expression from Eq. (9.13):
$d \ge 6''$ then $L_{b,\,eff} = 0.95L_b$

$$C_\omega = \frac{I_z d^2}{4} = \frac{83.4(10^2)}{4} = 2,085\,\text{in}^6$$

$$f_n^{LTB} = \frac{C_b \pi}{S_T L_{b,eff}} \sqrt{E_{T,f}I_z G_{LT}C + \frac{\pi^2 E_{T,f}^2 I_z C_\omega}{L_{b,eff}^2}}$$

$$f_n^{LTB} = \frac{1.3 \times \pi}{51.2(0.95 \times 8 \times 12)} \sqrt{\begin{array}{c}(0.8 \times 10^6)(83.4)(0.425 \times 10^6)(1.208) \\ \\ +\frac{\pi^2(0.8 \times 10^6)^2(83.4)(2,085)}{(0.95 \times 8 \times 12)^2}\end{array}}$$

$$f_n^{LTB} = 11,279\,\text{psi}$$

(Alternative method): e.2) Using the simplified expression from Eq. (9.19):

$$f_n^{LTB} = \frac{C_b \pi E_{T,f}}{2S_T L_{b,eff}} \sqrt{2I_z C + \left(\frac{\pi h I_z}{L_{b,eff}}\right)^2}$$

$$f_n^{LTB} = \frac{1.3 \times \pi (0.8 \times 10^6)}{2 \times 51.2(0.95 \times 8 \times 12)} \sqrt{(2 \times 83.4 \times 1.208)+\left(\frac{\pi \times 10 \times 83.4}{0.95 \times 8 \times 12}\right)^2} = 11,211\,\text{psi}$$

Note: The difference in the LTB strength $\left(f_n^{LTB}\right)$ between the complex and simplified expression is less than 1%.

For $\varphi = 0.55$ and λ (for $1.2D + 1.6L) = 0.8$, the LTB strength is:

$$\varphi \lambda f_n^{LTB} = (0.55 \times 0.8)11,279 = 4,962\,\text{psi} \quad \text{and} \quad f_{bu} = 2,340\,\text{psi} < \varphi \lambda f_n^{LTB} = 4,962\,\text{psi}$$

The cross section of the FRP wide flange beam is sufficient to prevent global lateral torsion buckling.

Note: ASCE Pre-Standard for LRFD of Pultruded FRP Structures (2021) uses a similar equation for LTB strength with the elastic modulus along the longitudinal axis (not in the transverse direction) of a member.

f. Check for local buckling strength:

    f.1) The local flange buckling strength is determined from Eq. (9.22) as:

$$f_{crf}^{local} = \frac{G_{LT} + 0.25\sqrt{E_{L,f}E_{T,f}}}{\left(\dfrac{b_f}{2t_f}\right)^2}$$

$$f_{crf}^{local} = \frac{\left(0.425\times10^6\right)+0.25\sqrt{\left(2.6\times10^6\right)\left(0.8\times10^6\right)}}{\left(\dfrac{10}{2\times0.5}\right)^2} = 7,856\,\text{psi}$$

For $\varphi = 0.8$ and λ (for $1.2D + 1.6L$) $= 0.8$, the local flange buckling strength is determined as:

$$\varphi\lambda f_{crf}^{local} = (0.8\times0.8)7,856 = 5,028\,\text{psi}$$

$$f_{bu} = 2,340\,\text{psi} < \phi\lambda\left(f_{crf}^{local}\right) = 5,028\,\text{psi}$$

Hence, the cross section of the FRP wide flange beam is sufficient to prevent compressive local flange buckling.

    f.2) Local web buckling using Eq. (9.23):

$$f_{crw}^{local} = \frac{15\sqrt{E_{L,f}E_{T,f}}}{\left(\dfrac{d}{t_w}\right)^2}$$

$$f_{crw}^{local} = \frac{15\sqrt{\left(2.6\times10^6\right)\left(0.8\times10^6\right)}}{\left(\dfrac{10}{0.5}\right)^2} = 54,083\,\text{psi}$$

For $\varphi = 0.8$ and λ (for $1.2D + 1.6L$) $= 0.8$, the local web buckling strength is:

$$\varphi\lambda\left(f_{crw}^{local}\right) = (0.8\times0.8)54,083 = 34,613\,\text{psi}$$

The local web buckling strength (34,613 psi) is higher than the material tensile (or compressive) strength (30,000 psi). Thus, the cross section of the FRP wide flange beam provides enough buckling resistance to the web.

g. Check for web shear buckling strength:

    The web shear buckling strength is determined from Eqs. (9.35.1–9.35.3) as:

$$\beta = \frac{G_{LT}}{\sqrt{E_{L,w}E_{T,w}}} = \frac{(0.425\times10^6)}{\sqrt{\left(2.6\times10^6\right)\left(0.8\times10^6\right)}} = 0.295$$

$$k = 2.67 + 1.59\beta = 2.67 + 1.59(0.295) = 3.14$$

$$f_{cr} = k\frac{\sqrt[4]{E_{L,w}\left(E_{T,w}\right)^3}}{\left(\dfrac{d_w}{t_w}\right)^2} = 3.14\frac{\sqrt[4]{\left(2.6\times10^6\right)\left(0.8\times10^6\right)^3}}{\left(\dfrac{10-0.5-0.5}{0.5}\right)^2} = 6,969\,\text{psi}$$

For $\varphi = 0.8$ and λ for $(1.2D + 1.6L) = 0.8$, the web shear buckling strength is:

$$\varphi\lambda\left(f_{cr}^{local}\right) = (0.8 \times 0.8)6,969 = 4,460\,\text{psi}$$

The web shear buckling strength (4,460 psi) is higher than the design shear strength (555 psi). Therefore, the cross section based on this design will not reach the web shear buckling value.

Example 9.8

Determine an additional maximum uniformly distributed load for the pultruded FRP beam in Example 9.7. Assume the pultruded FRP beam is braced with lateral supports at the mid-span and also at supports.

Solution

The minimum nominal strength of the pultruded FRP beam in Example 9.7 is obtained from the LTB value by equating to the induced flexural stress.

$$\varphi\lambda f_n^{LTB} = f_{bu}$$

For $\varphi = 0.55$ and λ (for $1.2D + 1.6L) = 0.8$, the compressive local flange buckling strength is determined as:

$$\varphi\lambda f_n^{LTB} = 4,963\,\text{psi} = \frac{M_u}{S_T} = \frac{(9,984 + M_{add})(12)}{51.2}$$

$$\text{then,}\quad M_{add} = 11.2\,\text{kip.ft}\ \left(\text{Factored live Load}\right)$$

$$\left(W_u\right)_{add} = \frac{(11.2)\times 8}{16^2} = 0.350\,\text{kip/ft}\quad\text{and}\quad \left(W_s\right)_{add} = \frac{(11.2)\times 8}{1.6\times 16^2} = 0.219\,\text{kip/ft}$$

Check for deflection limit: ($W_s = 0.210 + 0.219 = 0.429$ kips/ft)
  From Eq. (9.6.4):

$$w_{max} = \frac{5W_sL^4}{384(E_LI_T)} + \frac{W_sL^2}{8(G_{LT}A_{web})} = \frac{5(429/12)(16\times 12)^4}{384(2.6\times 10^6)256.2} + \frac{(429/12)(16\times 12)^2}{8(0.425\times 10^6)4.5}$$

$$w_{max} = 1.04\,\text{in} > \frac{16\times 12}{240} = 0.8\,\text{in}$$

At serviceability: (W_{max} = defection limit @ $L/240$)
  From Eq. (9.6.4):

$$w_{max} = \frac{5W(16\times 12)^4}{384(2.6\times 10^6)256.2} + \frac{W(16\times 12)^2}{8(0.425\times 10^6)4.5} = 0.0266W + 0.0024W = \frac{16\times 12}{240}$$

$$= 0.8\,\text{in}$$

$$W_s = 331\,\text{lbs/ft}\quad\text{and}\quad \left(W_s\right)_{add} = 121\,\text{lbs/ft}$$

Note: Additional uniform load is controlled by the deflection limit under the serviceability conditions.

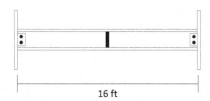

16 ft

EXAMPLE FIGURE 9.8   Pultruded FRP beam with lateral supports at mid-span and supports.

Example 9.9

Using the LRFD method, design a simply supported FRP beam of 11.5 ft span carrying dead and live loads of 50 and 100 lbs/ft, respectively.

Given: Longitudinal tensile strength (F_t) = 30,000 psi, Shear strength (F_v) = 4,500 psi, Longitudinal modulus (E_L) = 2.6 (10^6) psi,
Transverse modulus (E_T) = 0.8 (10^6) psi,
Shear modulus (G_{LT}) = 0.425 (10^6) psi.

Solution

a. Determine factored loads:

Total service load (W) = (50 lbs/ft) + (100 lbs/ft) = 150 lbs/ft.

Total factored load (W_u) = 1.2 (50 lbs/ft) +1.6 (100 lbs/ft) = 220 lbs/ft.

Try wide flange beam using the serviceability limit (the deflection limit L/240) and shear deformation effect. The shear rigidity $k(G_{LT}A)$ can be approximated with $(G_{LT})_{web}$ A_{web}. The FRP wide flange beam deflection is determined from Eq. (9.9.4) as:

$$w_{max} = \frac{5WL^4}{384(E_L I_T)} + \frac{WL^2}{8k(G_{LT}A)} = \frac{5(150/12)(11.5 \times 12)^4}{384(2.6 \times 10^6)I_T} + \frac{(150/12)(11.5 \times 12)^2}{8(0.425 \times 10^6)A_{web}}$$

$$w_{max} = \frac{22.7}{I_T} + \frac{0.070}{A_{web}} \le \frac{L}{240}$$

b. Select 8 × 8 × 1/2″ wide flange section:

Given: Longitudinal modulus of flange (E_{Lf}) = 2.6 (10^6) psi, Transverse modulus of flange (E_{Tf}) = 0.8 (10^6) psi, Longitudinal modulus of web (E_{Lw}) = 2.6 (10^6) psi, Transverse modulus of web (E_{Tw}) = 0.8 (10^6) psi, Shear modulus (G_{LT}) = 0.425 (10^6) psi, Poisson's ratio (ν_{LT}) = 0.33. Flange thickness (t_f) = 0.5″ and Web thickness (t_w) = 0.5″ and Flange width (b_f) = 8″, Self-weight = 9.23 lbs/ft, Area (A) = 11.51 in², Web area (A_w) = 3.50 in², Sectional modulus about the strong axis (S_T) = 31.8 in³, Moment of inertia about strong axis (I_T) = 127.1 in⁴, Moment of inertia about weak axis (I_Z) = 42.7 in⁴, and Torsional constant (C) = 0.958 in⁴.

$$w_{max} = \frac{22.7}{127.1} + \frac{0.070}{3.50} = 0.20 \le \frac{12.5(12)}{240} = 0.625 \, in$$

Note: FRP wide flange beam satisfies the deflection limit.

c. Determine design bending and shear strengths:

The design bending moment and shear are:

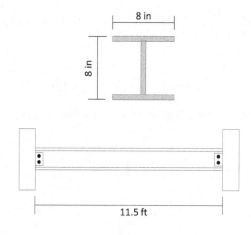

EXAMPLE FIGURE 9.9 Simply supported pultruded FRP beam.

$$M_u = \frac{W_u L^2}{8} = \frac{220(11.5^2)}{8} = 3,637\,\text{lbs}-\text{ft} \quad \text{and} \quad V_u = \frac{W_u L}{2} = \frac{220(11.5)}{2} = 1,265\,\text{lbs}$$

Induced flexural and average web shear stresses are:

$$f_{bu} = \frac{M_u}{S_T} = \frac{3,637(12)}{31.8} = 1,372\,\text{psi} \quad \text{and} \quad f_{vu} = \frac{V}{A_{web}} = \frac{1,265}{3.50} = 361\,\text{psi}$$

d. Determine nominal material strength:

For $\varphi = 0.65$ (for material rupture) and λ (for $1.2D + 1.6L$) = 0.8, the resistance of the FRP wide flange section is determined as:

$$\varphi\lambda F_t = (0.65 \times 0.8)30,000 = 15,600\,\text{psi} \quad \text{and} \quad \varphi\lambda F_v = (0.65 \times 0.8)4,500 = 2,340\,\text{psi}$$

$$\varphi\lambda F_t = 15,600\,\text{psi} > f_{bu} = 1,372\,\text{psi} \quad \text{and} \quad \varphi\lambda F_v = 2,340\,\text{psi} > f_{vu} = 361\,\text{psi}$$

It is found that the FRP wide flange beam satisfies the material strength limit as well.

e. Determine LTB strength:

C_b (moment modification factor) = 1.0 for a simply supported uniformly beam.

e.1) Using the complex expression from Eq. (9.13):

($d \geq 6''$ then $L_{b,\,eff} = 0.95 L_b$).

$$C_\omega = \frac{I_z d^2}{4} = \frac{42.7(8^2)}{4} = 683\,\text{in}^6$$

$$f_n^{LTB} = \frac{C_b \pi}{S_T L_{b,eff}}\sqrt{E_{T,f}I_z G_{LT}C + \frac{\pi^2 E_{T,f}^2 I_z C_\omega}{L_{b,eff}^2}}$$

$$f_n^{LTB} = \frac{\pi}{31.8(0.95 \times 12 \times 12.5)}\sqrt{\begin{array}{c}(0.8\times10^6)(42.7)(0.425\times10^6)(0.958) \\[2mm] + \dfrac{\pi^2(0.8\times10^6)^2(42.7)(683)}{(0.95\times12.5\times12)^2}\end{array}}$$

$$f_n^{LTB} = 3,323\,\text{psi}$$

(**Alternative method**): e.2) Using the simplified expression from Eq. (9.22):

$$f_n^{LTB} = \frac{C_b \pi E_{T,f}}{2 S_T L_{b,eff}}\sqrt{2I_z C + \left(\frac{\pi h I_z}{L_{b,eff}}\right)^2}$$

$$f_n^{LTB} = \frac{\pi(0.8\times10^6)}{2\times31.8(0.95\times12.5\times12)}\sqrt{(2\times42.7\times0.958) + \left(\frac{\pi\times8\times42.7}{0.95\times12.5\times12}\right)^2}$$

$$= 3,264\,\text{psi}$$

Note: The difference in the lateral torsional strength $\left(f_n^{LTB}\right)$ between the complex and simplified expression is less than 1%.

For $\varphi = 0.55$ and λ (for $1.2D + 1.6L$) = 0.8, the LTB strength is determined as:

$$\varphi \lambda f_n^{LTB} = (0.55 \times 0.8)3,323 = 1,462 \, \text{psi} \quad \text{and} \quad f_{bu} = 1,372 \, \text{psi} < \varphi \lambda f_n^{LTB} = 1,462 \, \text{psi}$$

The cross section of the FRP wide flange beam is sufficient to prevent global lateral torsion buckling.

Note: ASCE Pre-Standard for LRFD of Pultruded FRP Structures (2021) uses a similar equation for LTB strength with the elastic modulus along the longitudinal axis (not in the transverse direction) of a member.

   f. Check for local buckling strength:

   f.1) The local flange buckling strength is determined from Eq. (9.22) as:

$$f_{crf}^{local} = \frac{G_{LT} + 0.25\sqrt{E_{L,f}E_{T,f}}}{\left(\dfrac{b_f}{2t_f}\right)^2}$$

$$f_{crf}^{local} = \frac{\left(0.425 \times 10^6\right) + 0.25\sqrt{\left(2.6 \times 10^6\right)\left(0.8 \times 10^6\right)}}{\left(\dfrac{8}{2 \times 0.5}\right)^2} = 12,274 \, \text{psi}$$

For $\varphi = 0.8$ and λ (for $1.2D + 1.6L$) = 0.8, the local flange buckling strength is determined as:

$$\varphi \lambda f_{crf}^{local} = (0.8 \times 0.8)12,274 = 7,856 \, \text{psi}$$

$$f_{bu} = 1,372 \, \text{psi} < \phi\lambda\left(f_{crf}^{local}\right) = 7,856 \, \text{psi}$$

The cross section of the FRP wide flange beam is sufficient to prevent local flange buckling.

   f.2) Local web buckling using Eq. (9.23):

$$f_{crw}^{local} = \frac{15\sqrt{E_{L,f}E_{T,f}}}{\left(\dfrac{d}{t_w}\right)^2}$$

$$f_{crw}^{local} = \frac{15\sqrt{\left(2.6 \times 10^6\right)\left(0.8 \times 10^6\right)}}{\left(\dfrac{8}{0.5}\right)^2} = 84,505 \, \text{psi}$$

For $\varphi = 0.8$ and λ (for $1.2D + 1.6L$) = 0.8, the local web buckling strength is:

$$\varphi \lambda\left(f_{crw}^{local}\right) = (0.8 \times 0.8)84,505 = 54,083 \, \text{psi}$$

The local web buckling strength is higher than the material tensile (or compressive) strength (30,000 psi) of the FRP beam. Thus, the cross section of the FRP wide flange beam provides enough resistance to local web bucking.

   g. Check web shear buckling strength

   The web shear buckling strength is determined from Eqs. (9.35.1–9.35.3) as:

$$\beta = \frac{G_{LT}}{\sqrt{E_{L,w}E_{T,w}}} = \frac{\left(0.425 \times 10^6\right)}{\sqrt{\left(2.6 \times 10^6\right)\left(0.8 \times 10^6\right)}} = 0.295$$

$$k = 2.67 + 1.59\beta = 2.67 + 1.59(0.295) = 3.14$$

$$f_{cr} = k \frac{\sqrt[4]{E_{L,w}\left(E_{T,w}\right)^3}}{\left(\dfrac{d_w}{t_w}\right)^2} = 3.14 \frac{\sqrt[4]{\left(2.6 \times 10^6\right)\left(0.8 \times 10^6\right)^3}}{\left(\dfrac{8 - 0.5 - 0.5}{0.5}\right)^2} = 17,208\,\text{psi}$$

For $\varphi = 0.8$ and λ for $(1.2D + 1.6L) = 0.8$, the web shear buckling strength is:

$$\varphi\lambda\left(f_{cr}^{local}\right) = (0.8 \times 0.8)17,208 = 11,013\,\text{psi}$$

The web shear buckling strength (11,013 psi) is much higher than the design shear strength (393 psi) of the FRP beam. The cross section of the FRP wide flange beam will not exceed the web shear buckling under the present loading scenario.

9.8 PULTRUDED FRP MEMBERS UNDER TORSION

When a cross section of a structural member is subjected to torsional moment, then it will result in twisting and warping for non-circular sections as shown in Figure 9.10. Then, the shear stresses are induced on the cross section. The shear stress variation is often assumed to be linear.

The angle of twist per unit length θ of an arbitrary cross section is governed by:

$$\theta = \frac{T}{GC} \tag{9.36}$$

where
    C = torsional constant of cross section
    G = shear modulus
    T = applied torque.

In addition, displacements in the longitudinal direction on a cross section along the beam length are induced and known as warping displacements for noncircular sections. If uniform displacements

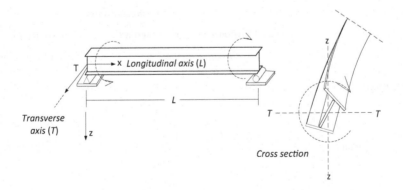

FIGURE 9.10 Beam under torsion.

with no restraint exist throughout the member length, then the uniform stress distribution also takes place on the cross section. It is called *St. Venant's torsion*. If the member is prevented from uniform displacements, then nonuniform warping exists and results in additional shear stresses and torsional stiffness (Prachasaree et al., 2006; Qureshi et al., 2014, 2017). Thus, (nonuniform) warping of an arbitrary cross section is presented in a general form as follows:

$$T = (GC)\frac{d\theta}{dx} + (EC_w)\frac{d^3\theta}{dx^3} \tag{9.37}$$

where

C = torsional constant of a cross section
E = elastic modulus
G = shear modulus
T = applied torque
C_w = warping constant of a cross section
θ = angle of twist per unit length.

The nominal torsional strength of a pultruded FRP structural member under torsional buckling and torsional rupture is determined as follows (ACMA, 2021):

For strength control:

$$T_n = \gamma G_{LT} C = F_n C J \tag{9.38.1}$$

For stiffness (buckling) control:

$$T_n = F_{cr} C \tag{9.38.2}$$

where

C is the St. Venant's torsional moment of inertia as given in Table 9.4
$\bar{C} = C$ divided by the mean radius for tubular section or half the depth for rectangular section
F_n and F_{cr} are critical strengths for strength and stiffness control, respectively, as shown in Table 9.5.

TABLE 9.4
Torsional Constants

Cross Section	Torsional Constant	
	St. Venant Torsional Moment of Inertia C	**Torsional Constant $\bar{C}$**
Wide flange	$C = \left(\dfrac{d_w t_w^3 + 2b_f t_f^3}{3}\right)$	$\bar{C} = \dfrac{1}{t_f}\left(\dfrac{d_w t_w^3 + 2b_f t_f^3}{3}\right) = \dfrac{C}{t_f}$
Rectangular tube	$C \approx \dfrac{2A^2}{\left(\dfrac{d_w}{t_w} + \dfrac{b_f}{t_f}\right)}$	$\bar{C} \approx 2t(b_r - t_f)(h - 2t)$ $\bar{C} \approx 2tb_f(h - 2t)$
Circular tube	$C = \pi\left(\dfrac{R^4 - R_i^4}{2}\right)(\pi/2)tR_m(2R - t)^2$	$\bar{C} \approx \dfrac{\pi t}{2}(2R - t)^2$

where $A =$ and mean of the areas enclosed by the inner-outer boundaries
$b_r =$ outer width of rectangular tube section
$b_f =$ width of the flange between the centers of webs in rectangular tubes, and flange width for others as T, I, L, and C cross section
$d_w =$ clear depth of the web
$h =$ total depth of cross sections
$R =$ outer radius of a circular section
$R_i =$ inner radius of a circular section
$R_m =$ mean radius to the center of wall thickness
$t =$ thickness of an element in the cross section
$t_f =$ thickness of the flange
$t_w =$ thickness of the web.

TABLE 9.5

Critical Strength for the Nominal Torsional Strength (Structural Plastics Manual, ASCE, 1984; ACMA, 2021)

Cross Section	Critical Strength	
	Strength Control F_n	Stiffness Control F_{cr}
Wide flange	$F_n = \gamma G_{LT}$	$F_{cr} = \dfrac{E_L^c}{24 I_o}\left(\dfrac{\pi}{l_b}\right)^2 \left(d_w^2 b_f^3 t_f\right)$ $+ \dfrac{G_{LT}}{3 I_o}\left(d_w t_w^3 + 2 b_f t_f^3\right)$ (9.39.1)
Rectangular tube	$F_n = \gamma G_{LT}$	$F_{cr} = \dfrac{G_{LT} b_f d_w \left(d_w + b_f\right) t}{2 I_o}$ (9.39.2)
Circular tube	$F_n = \gamma G_{LT}$	$\min\left[F_{cr1}, F_{cr2}\right] \le F_{LT}^v$ (9.39.3) $F_{cr1} = \dfrac{0.2\left(E_T^c\right)^{\frac{5}{8}}\left(E_L^c\right)^{\frac{3}{8}}}{\left(\dfrac{R}{t}\right)^{\frac{3}{2}}} \le F_{LT}^v$ (9.39.4) $F_{cr2} = \dfrac{0.7\left(E_T^c\right)^{\frac{5}{8}}\left(E_L^c\right)^{\frac{3}{8}}}{\left(\dfrac{R}{t}\right)^{\frac{5}{4}}\sqrt{\dfrac{L}{R}}} \le F_{LT}^v$ (9.39.5)

Note: The critical torsional buckling strength F_{cr} shall not exceed the in-plane shear strength F_{LT}^v.

where E_L^C and $E_T^C =$ longitudinal and transverse compression modulus
$R =$ circular tube outer radius
$t =$ tube thickness
t_f and $t_w =$ thickness of the flange and web, respectively
$b_f =$ width of the flange between the centers of webs in rectangular tubes, and flange width for others, such as T, I, L, and C cross sections
$d_w =$ clear depth of the web
$I_o =$ sum of moments of inertia about the strong axis of bending and the weak axis of bending

l_b = length between points that are braced against twist of the cross section
γ = shear strain/unit beam length
G_{LT} = in-plane shear modulus
L = unbraced length of a member.

Example 9.10

Determine the torsional strength of a pultruded FRP WF beam 8 × 8 × 3/8″ with 12 ft span. Its lateral supports are provided at the ends and at the mid-span.

Given: Section WF 8 × 8 × 3/8″: self-weight = 6.97 lbs/ft, Area (A) = 8.73 in², Web area (A_w) = 2.72 in², Sectional modulus about the strong axis (S_T) = 24.8 in³, Moment of inertia about strong axis (I_T) = 99.1 in⁴, Moment of inertia about weak axis (I_Z) = 32.0 in⁴, Torsional constant (C) = 0.409 in⁴. Longitudinal tensile strength (F_t) = 30,000 psi, Shear strength (F_v) = 4,500 psi, Longitudinal modulus (E_L) = 2.6 (10⁶) psi, Transverse modulus (E_T) = 0.8 (10⁶) psi, and Shear modulus (G_{LT}) = 0.425 (10⁶) psi.

Solution

The nominal torsional strength of pultruded FRP structural members is influenced by the torsional buckling or torsional rupture strength. The minimum strengths corresponding to torsional buckling (stiffness control) and torsional rupture (material control) are defined as the nominal torsional strength of the cross section.

For strength control:

It should be noted that γ in Eq. (9.38.1) is shear strain per unit beam length, where the shear strength (stress) F_v must be divided by half of the web depth $(d_w/2)$.

From torsional constant (C) in Table 9.4:

$$T_n = F_n C = \left(\gamma G_{LT}\right)\left(\frac{d_w t_w^3 + 2b_f t_f^3}{3}\right) = \left(\frac{4.5\,\text{ksi}}{8/2}\right)\left(\frac{8 \times 0.375^3 + 2 \times 8 \times 0.375^3}{3}\right)$$

$$T_n = 475\,\text{lbs.in} \cong 40\,\text{lbs.ft},$$

Note: γ (shear strain/unit length)

$$\tau = \frac{R}{C}T_n = \left(\gamma_{LT}G_{LT}\right) \quad \text{or} \quad T_n = \left(\gamma_{LT}G_{LT}\right)\frac{C}{R} = \left(\frac{F_v}{R}\right)C$$

where γ is the shear strain per unit beam length is equal to the strain at the rotation center of a cross section constant (C) divided by R.

For stiffness (buckling) control:

From Eq. (9.39.1):

$$F_{cr} = \frac{E_L^c}{24I_o}\left(\frac{\pi}{l_b}\right)^2\left(d_w^2 b_f^3 t_f\right) + \frac{G_{LT}}{3I_o}\left(d_w t_w^3 + 2b_f t_f^3\right)$$

$$F_{cr} = \left(\frac{2.6(10^6)}{24(99.1+32.0)}\left(\frac{\pi}{6\times12}\right)^2\left(8^2 \times 0.375^3 \times 0.375\right)\right)$$

EXAMPLE FIGURE 9.10   Pultruded FRP WF beam 8 × 8 × 3/8″ with lateral restraints at the mid-span and at supports.

$$+\left(\frac{0.425\left(10^6\right)}{3(99.1+32.0)}\left(8\times0.375^3+2\times8\times0.375^3\right)\right)$$

$$F_{cr}=20,700\,\text{psi}$$

$$\bar{C}=\frac{C}{t_f}=\frac{1}{t_f}\left(\frac{d_w t_w^3+2b_f t_f^3}{3}\right)$$

$$\bar{C}=\frac{1}{0.375}\left(\frac{8\times0.375^3+2\times8\times0.375^3}{3}\right)=1.125\,\text{in}^3$$

$$T_n=F_{cr}\bar{C}=20,700\times1.125=23,287\,\text{lbs.in}=1,940\,\text{lbs.ft}$$

The nominal torsional strength is obtained from the torsional strength under material rupture. The nominal torsion strength of a 12 ft pultruded FRP WF $8\times8\times3/8''$ beam with lateral supports at the mid-span and supports is: (resistance factor for torsion $\varphi=0.7$ and time factor for $1.2D+1.6L=0.8$)

$$\varphi\lambda T_n=(0.7\times0.8)1,940=1,087\,\text{lbs.ft}$$

9.9   PULTRUDED FRP MEMBERS UNDER CONCENTRATED LOADS

Beams are subjected to applied concentrated point loads along a beam length, or beams may be supported by connections to other structural members as shown in Figure 9.11. Then, the beam sections at supports or load points are stressed under concentrated forces. Localized failure of a section may be induced and can lead to structural failure of members. Therefore, various modes of failure must be checked to ensure the strength limits of members.

For design purposes, the nominal strength of a pultruded FRP member under concentrated loads is taken as the smallest strength obtained from the following limit states: (1) material rupture of the web (tension or compression), (2) Web crippling (Figure 9.12), (3) Web compression buckling (Figure 9.12), and (4) flange rupture due to bending.

The nominal strength of a pultruded FRP section under concentrated loads is considered from the minimum strength of four different failure modes as follows (ACMA, 2021):

$$R_n=\min\left(R_n^t,R_n^c,R_n^b,R_n^f\right) \tag{9.40}$$

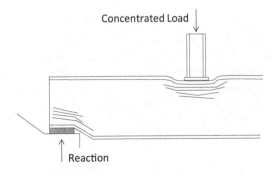

FIGURE 9.11   Member under concentrated loads.

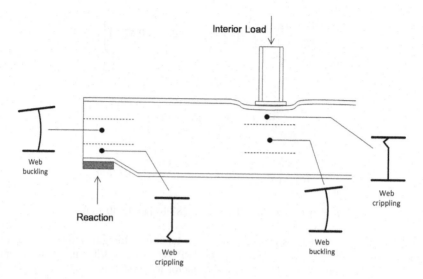

FIGURE 9.12 Localized failures under concentrated loads.

where

R_n^t = nominal strength due to tensile rupture of webs
R_n^c = nominal strength due to web crippling
R_n^b = nominal strength due to web compression buckling
R_n^f = nominal strength due to flange flexural failure.

9.9.1 Tensile Material Rupture

The nominal strength for tensile material rupture in a web due to a concentrated load is (ACMA, 2021):

$$R_n^t = l_e \xi F_{T,w} t_w \tag{9.41}$$

where

l_e = web length over which the applied tensile load is distributed which is the smaller of the depth of a member or the spacing between vertical stiffeners on either side of the tensile load
$F_{T,w}$ = characteristic transverse strength of the web portion
t_w = thickness of the web
ξ = fraction of transverse fibers in the web that continue through the web-to-flange junction.
For additional information, refer to Structural Plastics Manual, ASCE (1984).

9.9.2 Web Crippling

The web crippling or bearing failure can be attributed to high local intensities under concentrated loads. The web crippling behavior of a pultruded FRP member is quite complicated and involves several issues such as elastic stability of web portions, non-uniform stress distribution under the concentrated loads and the adjacent portions of a web, local failure in the immediate zone of load application, and even the eccentric load resulting from bending moment, which is applied on the bearing flange away from the web flange junction. The nominal strength of a member at locations of interior supports and concentrated compressive loads for a member with a depth of $h \leq 12$ inches is determined as (Structural Plastics Manual, ASCE, 1984; ACMA, 2021):

$$R_n^c = 0.7 h t_w F_{sh,int} \left(1 + \frac{2k + 6t_{plate} + b_{plate}}{d_w} \right) \tag{9.42}$$

where

$F_{sh, int}$ = characteristic interlaminar shear strength of the member in ksi

t_w = thickness of the web

k = distance from the top of a member to the bottom of the fillet connect top flange and web portion $(t_f + r)$

t_{plate} = thickness of the bearing plate

b_{plate} = length of the bearing plate along the axis of the section $(b_p \leq 4$ inches$)$

d_w = depth of the web

h = member full depth.

It should be noted that the vertical bearing stiffeners are provided directly under the load at all locations of interior supports and concentrated compressive loads for members with depth $(h > 12$ inches$)$. At the end supports, the length of the bearing plate is at least $(h/2)$.

9.9.3 WEB BUCKLING

The nominal strength of a member, R_n^b, under web compression buckling is (Structural Plastics Manual, ASCE, 1984; Kollar, 2003; Bank, 2006; ACMA, 2021):

$$R_n^b = A_{eff} \left(\frac{\pi^2 t_w^2}{6 l_{eff}^2} \right) \left(\sqrt{E_{L,w} E_{T,w}} + E_{T,w} \nu_{LT} + 2 G_{LT} \right) \tag{9.43}$$

where

A_{eff} = effective area $(l_{eff} t_w)$

l_{eff} = lesser of the web depth (d_w) and the distance between the vertical web stiffeners

t_w = thickness of the web

$E_{L, w}$ = longitudinal modulus of the web

$E_{T, w}$ = transverse modulus of the web

G_{LT} = in-plane shear modulus

ν_{LT} = major (longitudinal) Poisson's ratio (in absence of available data $\nu_{LT} = 0.3$).

9.9.4 FLANGE RUPTURE FROM WEB DUE TO BENDING

The nominal strength of a member, R_n^f, under flexural failure at the web-flange junction of an outstanding flange loaded by an eccentric concentrated load is (Structural Plastics Manual, ASCE, 1984; ACMA, 2021):

$$R_n^f = \frac{F_{T,f} b t_f^2}{6 l_e k_{w,f}} \tag{9.44}$$

where

$F_{T,f}$ = transverse tensile strength of the flange

t_f = thickness of the flange

l_e = distance of concentrated load on the flange from the web

b = projected width of the concentrated load on the web $(2 l_e)$

$k_{w,f}$ = the stress concentration factor at the web-flange junction, which is taken as 1.0, if it is not specified by the manufacturer.

9.10 BEARING STIFFENERS

Typically, a minimum thickness of bearing stiffeners is at least the web thickness of the members. The stiffeners must be extended from the web to the edge of the flange along the entire depth of the web from the inside of the compression flange to the inside of the tension flange (ACMA, 2021).

For design purposes, bearing stiffeners are required when the nominal strength of pultruded FRP members under concentrated load is less than twice the required strength.

$$\lambda \varphi R_n \leq 2R_u \tag{9.45}$$

where
  R_n = minimum nominal strength obtained from the minimum strength of four different failure modes as mentioned above
  R_u = required strength due to concentrated loads
  φ = reduction factor
  λ = time factor

Example 9.11

Determine the pultruded FRP beam WF 8 × 8 × 3/8″ failure capacity under possible reactions or concentrated loads as shown in Example Figure 9.11.

Given: Section WF 8 × 8 × 3/8″: Self-weight = 6.97 lbs/ft, Area (A) = 8.73 in², Web area (A_w) = 2.72 in², Longitudinal tensile strength (F_{tL}) = 30,000 psi, Transverse tensile strength (F_{tT}) = 7,000 psi, Shear strength (F_v) = 4,500 psi, Longitudinal modulus (E_L) = 2.6 (10⁶) psi, Transverse modulus (E_T) = 0.8 (10⁶) psi, and Shear modulus (G_{LT}) = 0.425 (10⁶) psi. Thickness of the bearing plate t_{plate} = 0.5″ and its length along the axis of the section b_{plate} = 4″.

Solution

The nominal strength of a pultruded FRP member under concentrated loads is considered from the minimum strength evaluated under four different failure modes (Section 9.9).
a. Determine strength for tensile material rupture:

The nominal strength for tensile material rupture in the web portions under a concentrated load is determined from Eq. (9.41): (transverse strength $F_{T, W}$ = 7,000 psi)

$$R_n^t = l_e \xi F_{T,w} t_w = 8 \times 1.0 \times 7,000 \times 0.375 = 21 \text{ kips}$$

b. Determine strength due to web crippling:
  The nominal strength of a member at interior supports and under concentrated compressive loads for members with a depth of $h \leq 12$ inches is determined from Eq. (9.42): (interlaminar shear strength $F_{sh, int}$ = 4,500 psi)

$$k = t_f + r = 0.375 + 0.1875 \text{ in}$$

$$R_n^c = 0.7 h t_w F_{sh,int} \left(1 + \frac{2k + 6t_{plate} + b_{plate}}{d_w} \right)$$

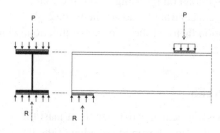

EXAMPLE FIGURE 9.11 FRP beam under concentrated loads.

$$R_n^c = 0.7(8 \times 0.375)(4{,}500)\left(1 + \frac{2(0.375 + 0.1875) + 6(0.5) + 4}{8 - 2(0.375)}\right) = 20.0\,\text{kips}$$

c. Determine web (compression) buckling capacity:

The nominal strength of a member under web compression buckling is determined from Eq. (9.43) as:

$$R_n^b = A_{eff}\left(\frac{\pi^2 t_w^2}{6 l_{eff}^2}\right)\left(\sqrt{E_{L,f}E_{T,f}} + E_{T,w}\nu_{LT} + 2G_{LT}\right)$$

$$R_n^b = (2.72)\left(\frac{\pi^2 0.375^2}{6(8 - 2(0.375))^2}\right)\left(\begin{array}{c}\sqrt{2.6 \times 0.8 \times 10^{12}} + (0.8 \times 0.33 \times 10^6) \\ + (2 \times 0.425 \times 10^6)\end{array}\right)$$

$$R_n^b = 30.6\,\text{kips}$$

d. Determine flange flexural failure strength:

The concentrated load or reaction is applied directly over the web of a pultruded FRP WF beam. The flange is not subjected to an eccentric concentrated load. Therefore, the nominal strength of a member under flexural at the web-flange junction of a flange loaded by an eccentric concentrated load is not considered in this Example.

Answer: Nominal strength under concentrated loads is due to web crippling, which is about 20.0 kips, and the lowest value of items (a), (b), and (c).

EXERCISES:

Problem 9.1: Determine the lateral torsional strength of a pultruded FRP WF beam 12 × 12 × 1/2″ with a simple span of 24 ft. The pultruded FRP beam is braced with lateral supports at the ends and laterally restrained at the third point of the span. Section WF 12 × 12 × 1/2″: Self-weight = 13.98 lbs/ft; Area (A) = 17.51 in²; Web area (A_w) = 5.50 in²; Sectional modulus about the strong axis (S_T) = 75.5 in³; Sectional modulus about the weak axis (S_Z) = 24.0 in³; Moment of inertia about strong axis (I_T) = 453 in⁴; moment of inertia about weak axis (I_Z) = 144 in⁴; Torsional rigidity (J) = 1.458 in⁴. A pultruded FRP WF beam properties (manufacturers): Longitudinal tensile strength (F_t) = 30,000 psi, shear strength (F_v) = 4,500 psi, Longitudinal modulus (E_L) = 2.6 (10⁶) psi, Transverse modulus (E_T) = 0.8 (10⁶) psi, and Shear modulus (G_{LT}) = 0.425 (10⁶) psi.

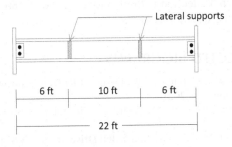

PROBLEM FIGURE 9.1   Pultruded FRP WF beam 12 × 12 × 1/2″.

Problem 9.2: Design a pultruded FRP beam for a simply supported span of 20 ft with a dead load of 75 lbs/ft and a live load of 125 lbs/ft. Assume: compression flange is fully supported laterally by the deck by accounting for shear deformation. Using the LRFD method, design an FRP beam with adequate bracing.

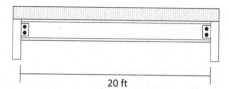

PROBLEM FIGURE 9.2   Pultruded FRP beam with full lateral supports.

Problem 9.3: Using the LRFD method, design a simply supported pultruded FRP beam of 16 ft span carrying dead and live loads of 50 and 200 lbs/ft, respectively. The FRP beam is braced with partial lateral supports at the mid-span and at supports.

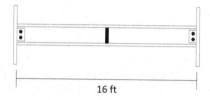

PROBLEM FIGURE 9.3   Pultruded FRP beam with partial lateral supports.

Problem 9.4: Determine a maximum uniformly distributed load for the pultruded FRP beam in Problem 9.3. Assume the pultruded FRP beam is braced with lateral supports at the mid-span and at supports.

Problem 9.5: Find the mid-span deflection including shear deflection of an FRP beam ($8'' \times 8'' \times \frac{1}{2}''$) of 12′ effective span that is subjected to 120 lb/ft of dead load and 160 lb/ft of live load. In addition, find mid-span deflection of 20 years of sustained (dead) load. The beam weight is 3.8 lbs/ft. Also, find the critical torsional buckling stress for the same wide flange section and critical buckling stress under bending loads, and the web in-plane shear buckling stress. Use realistic limit stresses by reviewing the state of practice given by manufacturers of profiles.

Problem 9.6: Design a wide-flange glass fiber-polyester FRP section with 50% fiber volume fraction that is subjected to a uniform load of 100 lb/ft, which includes self-weight and has a span of 20′. The maximum short-term deflection is limited to a value of $L/180$. Find the spacing of lateral bracings to prevent LTB of the wide flange section.

REFERENCES AND SELECTED BIOGRAPHY

ACMA, Pre-Standard for Load & Resistance Factor Design (LRFD) of Pultruded Fiber Reinforced Polymer (FRP) Structures, Arlington, VA, 2021.

ANSI/AISC 360–05, An American National Standard, Specification for Structural Steel Buildings, AISC, Chicago, IL, 2005.

ASCE, Structural Plastics Design Manual, Alexander Bell Drive, Reston, VA, 1984.

ASCE/SEI7–10, Minimum Design Loads for Buildings and Other Structures, Alexander Bell Drive, Reston, VA, 2010.

Bank, L.C., *Composites for Construction – Structural Design with FRP Materials*, John Wiley & Sons, Inc., Hoboken, NJ, 2006.

Bank, L.C., Gentry, T.R., and Nadipelli, M., Local buckling of pultruded FRP beams – Analysis and design, *Journal of Reinforced Plastics and Composites*, 15(3), 1996, pp. 283–294.

Bank, L.C., Nadipelli, M., and Gentry, T.R., Local buckling and failure of pultruded fiber-reinforced plastic beams, *Journal of Engineering Materials and Technology*, 116(2), 1994, pp. 233–237.

Bank, L.C., and Yin, J., Buckling of orthotropic plates with free and rotationally restrained unloaded edges, *Thin-Walled Structures*, 24(1), 1996, pp. 83–96.

Bank, L.C., Yin, J., and Nadipelli, M., Local buckling of pultruded beams – Nonlinearity, anisotropy and inhomogeneity, *Construction and Building Materials*, 9(6), 1995, pp. 325–331.

Barbero, E.J., *Introduction to Composite Materials Design*, 2nd edition, CRC Press, Boca Raton, FL, 2010.

Bedford Reinforced Plastics, Design Guide, Bedford Reinforced Plastics, Inc., Bedford, PA, 2012.

Boresi, A.P., and Schmidt, R.J., *Advanced Mechanics of Materials*, 6th edition, John Wiley & Sons, Inc., New York, NY, 2002.

Creative Pultrutions, *The Pultex Pultrusion Design Manual of Standard and Custom Fiber Reinforced Structural Profiles*, 5, revision (2), Creative Pultrutions Inc., Alum Bank, PA, 2004.

Daniel, I.M., and Ishai, O., *Engineering Mechanics of Composite Materials*, 2nd edition, Oxford University Press, Inc., New York, NY, 2006.

Estep, D.D., Bending and Shear Behavior of Pultruded Glass Fiber Reinforced Polymer Composite Beams with Closed and Open Sections, WVU MSCE Thesis, 2014.

Estep, D.D., GangaRao, H.V.S., Dittenber, D.B., and Qureshi, M.A., Response of pultruded glass composite box beams under bending and shear, *Composite Part B – Engineering*, 88, 2016, pp. 150–161.

Fiberline, *Design Manual*, Fiberline, Kolding, Denmark, 2003.

Gere, J.M., and Goodno, B.J., *Mechanics of Materials,* 7th edition, Cengage Learning, Toronto, ON, 2009.

Kollar, L.P., Buckling of unidirectionally loaded composite plates with one free and one rotationally restrained unloaded edge, *Journal of Structural Engineering, ASCE*, 128(9), 2002, pp. 1202–1211.

Kollar, L.P., Local buckling of fiber reinforced plastic composite structural members with open and closed cross sections, *Journal of Structural Engineering, ASCE*, 129(11), 2003, pp. 1503–1513.

Kollar, L.P., and Pluzsik, A., Analysis of thin walled composite beams with arbitrary layup, *Journal of Reinforced Plastics and Composites*, 21(16), 2002, pp. 1423–1465.

Kollar, L.P., and Springer, G.S., *Mechanics of Composite Structures*, Cambridge University Press, New York, NY, 2003.

Laudiero, F., Minghini, F., Ponara, N., and Tulliui, N., Buckling resistance of pultruded FRP profiles under pure compression or uniform bending – numerical simulation, *Sixth International Conference on FRP Composites in Civil Engineering*, CICE2012, 2012, pp. 1–8.

Nagaraj, V., and GangaRao, H.V.S., Static behavior of pultruded GFRP beams, *Journal of Composites for Construction*, 1(3), 1997, pp. 120–129.

Nguyen, T.T., Chan, T.M., and Mottram, J.T., Lateral torsional buckling resistance by testing for pultruded FRP beams under different loading and displacement boundary conditions, *Composites – Part B*, 2014, pp. 306–318.

Prachasaree, W., GangaRao, H.V.S., and Shekar, V., Performance evaluation of FRP bridge deck component under torsion, *Journal of Bridge Engineering, ASCE*, 11(4), 2006, pp. 430–442.

Prachasaree, W., Limkatanyu, S., Kaewjuea, W., and GangaRao, H.V.S., Simplified buckling-strength determination of pultruded FRP structural beams, *Journal of Practice Periodied on Structural Design and Construction, ASCE*, 24(2), 2019, p. 04018036.

Qureshi, M.A., Failure Behavior of Pultruded GFRP Members under Combined Bending and Torsion, Ph.D. Dissertation, West Virginia University, Morgantown, WV, 26506, 2012.

Qureshi, M.A., and GangaRao, H.V.S., Torsional response of closed FRP composite sections, *Composites – Part B*, 61, 2014, pp. 254–266.

Qureshi, M.A., GangaRao, H.V.S., Hayat, N., and Majjigapu, P., Response of closed glass fiber reinforced polymer sections under combined bending and torsion, *Journal of Composite Materials*, 51(2), 2017, pp. 241–260.

Robert, T.M., and Al-Ubaidi, H., Flexural and torsional properties of pultruded fiber reinforced plastic I – Profiles, *Journal of Composites for Construction*, 6(1), 2002b, pp. 28–34.

Roberts, T.M., Influence of shear deformation on buckling of pultruded fiber reinforced plastic profiles, *Journal of Composites for Construction*, 6(4), 2002, pp. 241–248.

Salmon, C.G., and Johnson, J.E., *Steel Structures, Design and Behavior*, Intext Educational Publishers, New York, NY, 1971.

Strongwell Corporation, Design Manual – EXTREN and Other Proprietary Pultruded Products, Bristol, VA, 2010.

Structural Design of Polymeric Composites – EUROCOMP Design Code and Handbook (Clarke, J.L., Editor), E & FN Spon, London, 1996.

Timoshenko, S., *Strength of Materials – Part II*, D. Van Nostraud Co. Inc., 3rd edition, New York, NY, 1956.

Vinnakota, S., *Steel Structures – Behavior and LRFD*, McGraw-Hill Education, Columbus, OH, 2005.

10 Design of Pultruded FRP Axial Compression Members

This chapter deals with the response of FRP composite components under axial compression forces. If axial forces are induced in the axial direction of a component (e.g., truss member) with a relatively low level of end restraints, and small magnitude of bending, then it can be designed as a concentrically loaded column. It should be noted that very short columns under axial loads fail in crushing, especially at the corners of a cross section. Otherwise, axially loaded columns fail due to local or global buckling, with local buckling being the most likely cause. Likely, FRP columns may even fail catastrophically; such failures can be attributed to instability due to bending induced from out-of-plane deformations multiplied by the axial compression forces. In addition, the bending effects may come from column crookedness or unsymmetrical fabric layup during manufacturing. Even residual stress build-up, especially at the corners of FRP components during the post-curing (after manufacturing) process, can result in uneven stress distribution leading to premature or unwarrented failure. The above premature failures are possible before developing the full crushing strength of the FRP laminates. Therefore, a thorough understanding of FRP column failure modes and stability is essential before designing for axial load responses. Such issues are dealt with in the following sections.

10.1 AXIAL COMPRESSION MEMBERS

An axial compression member is defined as a structural member under compression loading in the longitudinal direction of a member as shown in Figure 10.1. Axial FRP compression members, such as FRP columns or truss members in compression, are mainly used to transmit compression loads to supports or in the longitudinal direction of a member to other members through a joint. Various FRP structural shapes are commonly used as compression (column) members. These shapes are I-, H- or WF, hollow or others.

In general, the compression (column) members may support varying amounts of axial and bending moments simultaneously, either induced by external forces, from residual stress build-up during post-curing, misalignment of fibers, or other manufacturing errors. Herein, the design of a pultruded FRP member under a concentrated axial compression load is introduced in addition to providing design methods of columns under axial and bending loads. A member can be designed as a column without bending when small eccentricities of axial forces are induced longitudinally due to out-of-plane column deformations (crookedness) from nonuniform curing or other out-of-plane distortions after manufacturing. It should be noted that the appropriate safety and resistance factors for ASD and LRFD will compensate for the effect of small bending moments induced from off-center axial loading or column crookedness. The FRP members are more susceptible to instability in terms of excess lateral displacement and/or buckling than steel members due to lower bending and/or shear moduli than steel.

A concentric axial compression load that is applied along the longitudinal axis of a pultruded FRP column profile induces compressive stress, which can be written as:

$$f_z = \frac{P}{A_g} \tag{10.1}$$

where
  f_z = axial compressive stress
  A_g = gross area of the pultruded FRP column.

DOI: 10.1201/9781003196754-10

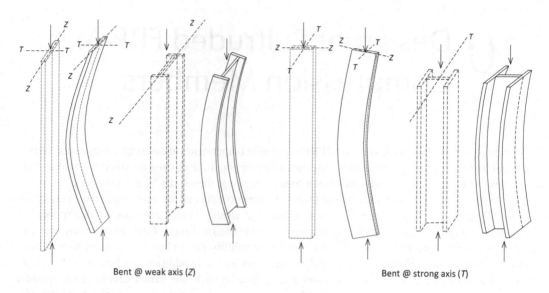

Bent @ weak axis (Z)   Bent @ strong axis (T)

FIGURE 10.1   Axial compression members bent about T- and Z-axis.

Axial deformation (compression) δ under the concentric axial load and no bending is:

$$\delta = \frac{PL}{AE_L}$$

(10.2)

where

L = length of pultruded FRP members

E_L = the longitudinal modulus of structural profile, in compression.

10.2   SLENDERNESS RATIO AND EFFECTIVE LENGTH

As per derivations in many textbooks dealing with the theory of elastic stability of beams, the critical compressive stress before bending of a beam is primarily a function of modulus and slenderness ratio, which is defined as (effective) beam length to the smallest radius of gyration. The radius of gyration (r) is defined as the square root of the ratio of moment of inertia about weak axis divided by the cross-sectional area, refer to Eqs. (10.3 and 10.4). The effective beam length is a function of column boundary conditions and spacing between points of bracings (refer to Eq. (10.3)).

In general, pultruded compression members will fail in different modes (material compression failure, global flexural buckling, torsional buckling, and local flange or web buckling) depending on the slenderness ratio of a section, stress concentrations at reentrant angles or holes in a cross section (Tomblin and Barbero, 1994; Mottram, 2003; Bank, 2006; GangaRao and Blanford, 2014). For smaller slenderness ratios (less than the critical buckling value), the failure mode of a pultruded FRP column is found typically from material failure or local buckling failure. When the slenderness ratio nearly equals the critical slenderness ratio, the pultruded FRP column may fail in coupled modes of failure modes, i.e., local and global buckling, or even bucking and simultaneous material rupture, which is rare.

A pultruded FRP column with a high slenderness ratio (KL_e/r) may fail in global (Euler) flexure buckling. The critical length and slenderness ratio are theoretically determined as (ACMA, 2021):

$$KL_e = \sqrt{\frac{\pi^2 (EI)_{min}}{P}}$$

(10.3)

where
L_e = effective member length of axially loaded compression member which is the center to center distance between lateral supports (ACMA, 2021)
K = member effective length factor.
For the braced system, the effective length factor (K) is equal to unity unless a rational analysis is performed.
P = column buckling load
$(EI)_{min}$ = flexural rigidity about the weak axis of a cross section.

The buckling load of a column is dominated by the end restraints of a member and its cross-sectional shape and dimensions, which is well established through theories on the elastic stability of structures. The end-restraint and member dimensions are denoted as effective slenderness ratio (KL_e/r). For design purpose, the slenderness ratio of a compression member is recommended by the pre-standard (ACMA, 2021) as:

$$\frac{KL_e}{r} \leq 1.4 \sqrt{\frac{E_L A_g}{P_D}} \tag{10.4}$$

or

$$\frac{KL_e}{r} \leq 300 \tag{10.5}$$

where
A_g = gross area of a section
E_L = longitudinal modulus of the flange or web whichever is smaller
P_D = compression load due to unfactored dead load
r = radius of gyration.

For design purposes, the cross section of a compression member can make a difference in effective lengths, between strong axis $(KL)_T$ and weak axis $(KL)_Z$. Typically, the effective length of a member can be considered as follows:
Example: if the buckling strength (P_T) about the strong axis (T) bending is higher than (P_Z) corresponding to bending about the weak axis (Z), then the lower buckling strength (P_Z) is the nominal buckling strength of a column.

$$P_T > P_z \text{ and } \frac{\pi^2 E_L I_T}{(KL)_T^2} > \frac{\pi^2 E_L I_Z}{(KL)_Z^2} \text{ or,}$$

$$\frac{I_T}{I_Z} > \frac{(KL)_T}{(KL)_Z} \text{ where } P_z \left(\text{or } KL_Z \right) \text{controls the buckling failure}$$

Summary:
- $(KL)_Z$ (weak axis) will control when the ratio of r_T/r_z is higher than $(KL)_T/(KL)_Z$
- $(KL)_Z$ is equal to $(KL)_T$, then the cross section has equal strength in both directions (e.g., hollow tubes, square sections).
- $(KL)_T$ (strong axis) will control when the ratio of r_T/r_z is less than $(KL)_T/(KL)_Z$, i.e., the definition of strong versus weak axis will be altered.

10.3   NOMINAL STRENGTH DUE TO MATERIAL RUPTURE IN COMPRESSION

For compression members under axial loading, pultruded FRP columns with a small slenderness ratio tend to fail under material rupture in the case of short columns.

The nominal longitudinal strength of pultruded FRP structural columns due to material rupture in compression is determined as (ACMA, 2021):

$$P_L = 0.3F_L^c A_g \quad \text{serviceability limit} \tag{10.6}$$

$$P_u = 0.7\lambda F_L^c A_g \quad \text{strength limit} \tag{10.7}$$

where
   A_g = gross area of cross section
   F_L^c = minimum longitudinal compression material strength of all elements comprising the cross section
   λ = time factor (ACMA, 2021).

10.4   GLOBAL FLEXURAL (EULER) BUCKLING

Typically, global structural instability of pultruded FRP compression members under axial concentric load is commonly known as the Euler buckling, and the deformed or buckled shape under axial load is shown in Figure 10.2. The entire FRP profile laterally moves out from its vertical plane when a global flexural buckling load is reached. No twisting of the cross-sectional profile is assumed under the flexural buckling. Thus, the global flexural buckling strength of slender pultruded FRP compression members under axial loads can be determined based on the classical Euler column buckling equation.

The governing bending equation under axial loads (Boresi and Schmidt, 2003) is:

$$E_L I \frac{d^2 u}{dx^2} + Pu = 0 \tag{10.8}$$

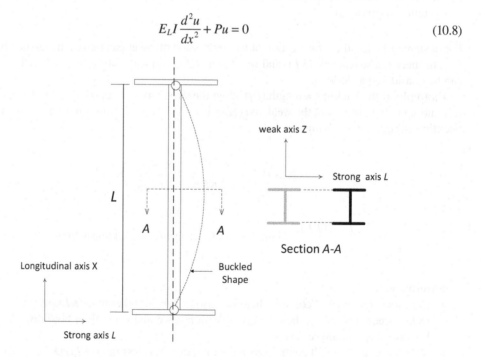

FIGURE 10.2   Global flexural buckling (Euler buckling), under first mode.

The general solution of Eq. (10.8) is:

$$u = A\mathrm{Sin}\left(\frac{n\pi x}{L}\right) + B\mathrm{Cos}\left(\frac{n\pi x}{L}\right) \tag{10.9}$$

where
u = lateral displacement
P = axial compression load
L = unbraced length of the column, $n = 0, 1, 2,...$, corresponding different buckling modes
I = moment of inertia
E_L = longitudinal elastic modulus of a column.
n = mode shape number

From Eqs. (10.8 and 10.9) with boundary conditions ($u = 0$ at the ends $x = 0$ and $x = L$), the solution of Eq. (10.8) provides the global buckling load as follows (where $B = 0$ in Eq. (10.9) for these boundaries):

$$P_{cr}^{GFB} = \frac{\pi^2 E_L I}{(KL)^2} \tag{10.10}$$

where, r = radius of gyration = $\sqrt{I/A}$

Converting the critical global flexural buckling load $\left(P_{cr}^{GFB}\right)$ into critical buckling stress by dividing Eq. (10.10) with the cross-sectional area (A) gives:

$$f_{cr}^{GFB} = \frac{\pi^2 E_L I}{A(KL)^2} = \frac{\pi^2 E_L}{\left(\dfrac{KL}{r}\right)^2} \tag{10.11}$$

For commercially pultruded FRP profiles, the longitudinal modulus is significantly higher than the shear modulus. In addition, the effects of shear deformation and flexural deformations are considered in the evaluation of critical buckling stress. The governing equation based on the energy method of global flexural buckling load with shear effects is (Timoshenko, 1956; Roberts, 2002; Bank, 2006):

$$\int \frac{E_L I}{2}\left(\frac{d^2 u}{dx^2}\right)^2 dx - \int \frac{P}{2}\left(1 - \frac{P}{G_{LT}kA}\right)\left(\frac{du}{dx}\right)^2 dx = 0 \tag{10.12.1}$$

Shear deflection equation is:

$$V = -kAG\left(\frac{dy_1}{dx}\right) \tag{10.12.2}$$

Differentiating both sides of Eq. (10.12.2) with respect to x gives:

$$\frac{d^2 y_1}{dx^2} = -\frac{dV/dx}{kAG} \tag{10.12.3}$$

The solution u_b Sin($\pi x/L$) of Eq. (10.12.1) for simply supported boundaries under 1st mode of buckling provides the global buckling strength with shear effect as:

$$f_{cr,\,shear}^{GFB} = \frac{\pi^2 E_L}{\lambda^2}\left(\frac{1}{1 + \dfrac{1}{kG_{LT}}\left(f_{cr}^{GFB}\right)}\right) = \frac{P_{Euler}/A}{\left(1 + \dfrac{P_{Euler}}{kA_2 G_{LT}}\right)} \tag{10.13}$$

where
   u = displacement due to out-of-plane bending
   K = effective length factor depending on the end conditions of structural members
   L = unbraced length of the column
   k = Timoshenko's shear coefficient
   G = shear modulus
   E = bending modulus

It should be noted that $k(GA)$ can be approximated with $(G_{LT})_{web} A_{web}$ (Nagaraj and GangaRao, 1997). k is 3/4 for circular section, 2/3 for rectangular section and varies from 1/2 to 1/3 for I-section (Timoshenko, 1956).

10.5   EFFECTIVE LENGTH FACTOR

Typically, lateral bracing is provided to axial compression members (columns) for several reasons such as to increase axial strength of members and to resist lateral loads and increase stability. Therefore, the design load of axial compression members can increase with the lateral bracing of members. To account for the restraint against translation and rotation of column ends, effective length factor (K), known as a ratio of the effective column length to unbraced column length, is introduced to determine the maximum axial load under buckling. In practice, the effective length factor for isolated columns, without horizontal displacement, at both the ends of a member is assumed to be one.

Columns in Figure 10.3 are considered isolated, with beam end conditions treated as pin–pin supports for analysis purposes. The effective length factor K is equal to unity for design purposes, even though member ends may be partially restrained in practice, resulting in K being lower than 1.0.

When one or both ends of isolated columns are restrained against rotation and translation, the effective length factor is less than one due to higher fixity of the column ends. Isolated columns are not part of the frame structure. The effective length factor K (Figure 10.4) is normally used to determine the global buckling strength of columns. The effective length factor K for isolated columns may also be used in a preliminary design in frame structures. It should be noted that the effective length factor (K) for columns in an unbraced frame is recommended to be 1.0 unless a rational analysis is performed (ACMA, 2021). Additional discussion on effective length factor can be found in many classical textbooks on mechanics of materials.

Note: The reason for higher K value under practical conditions being higher than the theoretical boundary conditions is due to the impossibility of attaining 100% fixed boundaries in practice.

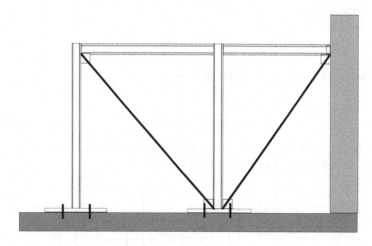

FIGURE 10.3   Typical braced frame structure.

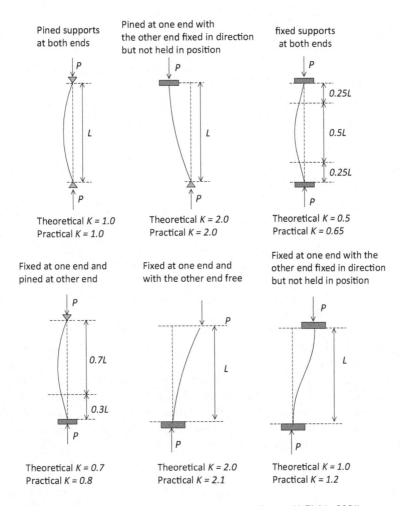

Pined supports at both ends

Theoretical K = 1.0
Practical K = 1.0

Pined at one end with the other end fixed in direction but not held in position

Theoretical K = 2.0
Practical K = 2.0

fixed supports at both ends

Theoretical K = 0.5
Practical K = 0.65

Fixed at one end and pined at other end

Theoretical K = 0.7
Practical K = 0.8

Fixed at one end and with the other end free

Theoretical K = 2.0
Practical K = 2.1

Fixed at one end with the other end fixed in direction but not held in position

Theoretical K = 1.0
Practical K = 1.2

FIGURE 10.4 Effective length factors for sway and non-sway columns (ACMA, 2021).

When pultruded FRP columns are parts of a frame structure, these columns may be restrained with different degrees of stiffness at the top and bottom ends, depending on their connection details to beams and columns. Indeed, the connections in the FRP frame structure are more flexible than in steel frames. The frame structures without bracing systems (Figure 10.5) under lateral and/or unsymmetrical loading are highly sensitive to horizontal displacements resulting in sidesway.

Typically, most of the FRP frame structures are constructed with diagonal bracing. The lateral stability is provided by bracings in diagonal directions. The pre-standard (ACMA, 2021) recommends the effective length factor K for axial compression members to be taken as 1.0 (except for an end with free lateral movement at supports) unless the rational analysis is conducted to justify the use of a smaller value for K.

As mentioned above, most pultruded FRP frame structures must be braced to prevent lateral displacements. Since the rigid frame connection of a pultruded FRP structure is quite difficult to obtain due to inherent structural properties of FRP materials, the unbraced FRP frames with rigid connections are rarely adopted in practice.

However, the pre-standard (ACMA, 2021) suggests that the effective length factor K of axial compression members in unbraced frame structures to be determined using the rational analysis. In addition, the secondary effect, such as the P-Δ effect, due to column self-weight or permanent dead load on a column is transferred to the lateral load resisting system and designed accordingly.

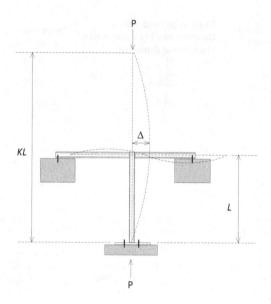

FIGURE 10.5 Buckling of the unbraced frame structure.

To evaluate the effective length factor of an unbraced frame system, the alignment chart method, recommended by the Structural Stability Research Council (SSRC) Guide (Johnston, 1976) is adopted in the absence of any available specific data of pultruded FRP frame structures. The rotational restraint at the ends of columns in a structural frame is provided by the beams and columns connecting at the top and bottom ends. Thus, the rotational restraint is the rotational stiffness of all members connected at the same intersection joint. Theoretically, the effective length factor (K) of unbraced frames is controlled by the ratio (G) of column ($E_{cL}I_c/L_c$) to beam ($E_{bL}I_b/L_b$) stiffness as:

$$G = \frac{\sum \left(\dfrac{E_{cL}I_c}{L_c} \right)}{\sum \left(\dfrac{E_{bL}I_b}{L_b} \right)} \tag{10.14}$$

where
  I_c = moment of inertia of the column section
  L_c = length of a column section
  I_b = moment of inertia of a beam section
  L_b = length of a beam
  I_c and I_b = taken about the axis perpendicular to the plane of buckling (weak axis).

10.5.1 Modified Factor for G

It is important that the modified factor for G, to be considered particularly in pultruded FRP frame structures. For braced frames, the single curvature of the beam is assumed, and the stiffness of beam members is equal to $2EI/L$. For typical FRP frame structures, the far-ends of beam members are treated as pin supports. Their stiffness is considered as $3EI/L$. Therefore, the stiffness of beam members should be multiplied by $1.5 = (3EI/L)/(2EI/L)$.

For an unbraced frame, the beams are considered to bend in double curvature. Thus, the stiffness of beam members is equal to $6EI/L$. However, when the far-ends are assumed to be pin supports, then the stiffness of the beam members is reduced to $3EI/L$. Hence, the stiffness of beam members should be multiplied by $0.5 (= (3EI/L)/(6EI/L))$.

As mentioned above, the modified factor of the beam end condition should be the multiplier modifying the beam stiffness (*EI/L*) with a modifying factor for certain far end conditions (Johnston, 1976).

- Sidesway is prevented: 1.5 for the far end of beam hinged;
- Sidesway is prevented: 2.0 for the far end of beam fixed (rarely found in FRP frames);
- Sidesway is not prevented: 0.5 for the far end of beam hinged;
- Sidesway is not prevented: 0.67 for the far end of beam fixed (rarely found in FRP frames).

10.5.2 CONDITION OF FRAME FOUNDATION

For a rigid column base, factor *G* is theoretically taken as zero (Johnston, 1976). However, it may be taken as 1 for design purpose. For a frictionless hinge column base, factor *G* is about infinity. However, SSRC Guide (Johnston, 1976) suggests a *G* value of 10 for design purposes. It should be noted that the suggested *G* values from the SSRC Guide are adopted for now due to the absence of test data for FRP columns comprising braced and unbraced multi-story frames.

10.5.3 PROCEDURE FOR ALIGNMENT CHART

To determine the effective length factor *K* of a column in frame structures, the *G*-factor with required modifications of a column joint has to be established. By drawing a straight line between G_A and G_B on the alignment chart (Figure 10.6), the effective length factor *K* can be determined as the point of interaction of the vertical line identified as "*K*". This effective length factor varies for cases with and without sidesway as provided in Figure 10.6.

Example: the following three-story unbraced frame structure as shown in Figure 10.7 is used as an example to determine *G*-factor. To determine the effective length factor *K* of column *AB* in

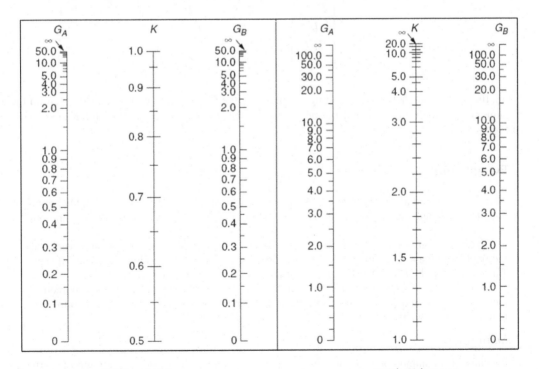

Without Sidesway   With Sidesway

FIGURE 10.6 Alignment charts for an effective length of columns. (SSRC Guide, Courtesy of Jackson and Mooreland Division of United Engineers and Constructors, Inc.)

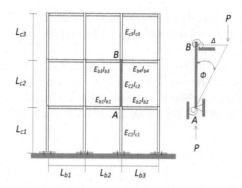

FIGURE 10.7 Typical unbraced frame structures.

an unbraced frame structure, the G_A and G_B factors at joints (A) and (B) are determined after establishing stiffnesses as outlined in Eq. (10.14) as well as shown in Figure 10.6. When the far end of a beam is hinged, the original beam stiffness must be multiplied by a modification factor of 0.5 before computing G-factors. This concept is illustrated below:

$$G_A = \frac{\Sigma\left(\dfrac{E_{cL}I_c}{L_c}\right)}{\Sigma\left(\dfrac{E_{bL}I_b}{L_b}\right)} = \frac{\dfrac{E_{c1}I_{c1}}{L_{c1}} + \dfrac{E_{c2}I_{c2}}{L_{c2}}}{\dfrac{0.5E_{b1}I_{b1}}{L_{b1}} + \dfrac{0.5E_{b2}I_{b2}}{L_{b2}}}$$

$$G_B = \frac{\Sigma\left(\dfrac{E_{cL}I_c}{L_c}\right)}{\Sigma\left(\dfrac{E_{bL}I_b}{L_b}\right)} = \frac{\dfrac{E_{c2}I_{c2}}{L_{c2}} + \dfrac{E_{c3}I_{c3}}{L_{c3}}}{\dfrac{0.5E_{b3}I_{b3}}{L_{b3}} + \dfrac{0.5E_{b4}I_{b4}}{L_{b4}}}$$

Using the alignment chart corresponding to sidesway, the G_A and G_B factors are positioned on the alignment chart to draw a straight line and to find the effective length factor K, on the K factor axis, of Figure 10.6.

Example 10.1

A two-story braced frame of pultruded FRP members is shown in Example Figure 10.1. Determine the effective length factor K of columns using the alignment chart in Figure 10.6. (Assume no out-of-plane buckling)

Given: Column section WF 8 × 8 × 1/2″: Gross area (A) = 11.51 in². Strong and weak axis moment of inertia = 127.1 in⁴ and 42.7 in⁴, respectively. Longitudinal modulus (E_L) = 2.6 (10⁶) psi, Transverse modulus (E_T) = 0.8 (10⁶) psi, Shear modulus (G_{LT}) = 0.425 (10⁶) psi.

Beam section WF 8 × 4 × 1/2″: Gross area (A) = 7.50 in². Strong and weak axis moment of inertia = 70.6 in⁴ and 5.41 in⁴, respectively. Longitudinal modulus (E_L) = 2.6 (10⁶) psi, Transverse modulus (E_T) = 0.8 (10⁶) psi, Shear modulus (G_{LT}) = 0.425 (10⁶) psi.

Solution

Buckling of a column could occur in the plane of the frame. Therefore, the moment of inertia (strong axis) perpendicular to the plane is used. Since sidesway is prevented, the modified factor is equal to 1.5 for the far end, which is hinged.

From Eq. (10.14):

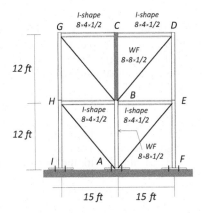

EXAMPLE FIGURE 10.1   Two-story braced frame.

$$G_B = \frac{\Sigma\left(\dfrac{E_{cL}I_c}{L_c}\right)}{\Sigma\left(\dfrac{E_{bL}I_b}{L_b}\right)} = \frac{\left(\dfrac{E_cI_c}{L_c}\right)_{AB} + \left(\dfrac{E_cI_c}{L_c}\right)_{BC}}{\left(\dfrac{1.5E_bI_b}{L_b}\right)_{BH} + \left(\dfrac{1.5E_bI_b}{L_b}\right)_{BE}} = \frac{\left(\dfrac{2E_cI_c}{L_c}\right)}{\left(\dfrac{3E_bI_b}{L_b}\right)} = \frac{2\left(\dfrac{127.1}{12}\right)}{3\left(\dfrac{70.6}{15}\right)} = 1.5$$

$$G_C = \frac{\Sigma\left(\dfrac{E_{cL}I_c}{L_c}\right)}{\Sigma\left(\dfrac{E_{bL}I_b}{L_b}\right)} = \frac{\left(\dfrac{E_cI_c}{L_c}\right)_{BC}}{\left(\dfrac{1.5E_bI_b}{L_b}\right)_{CG} + \left(\dfrac{1.5E_bI_b}{L_b}\right)_{CD}} = \frac{\left(\dfrac{E_cI_c}{L_c}\right)}{\left(\dfrac{3E_bI_b}{L_b}\right)} = \frac{\left(\dfrac{127.1}{12}\right)}{3\left(\dfrac{70.6}{15}\right)} = 0.75$$

Using the alignment chart (Figure 10.6) without sidesway, the effective length factor K is 0.75 for buckling perpendicular to the strong axis.

Example 10.2

A two-story unbraced frame of pultruded FRP members is shown in Example Figure 10.2. Determine the effective length factor K of columns using the alignment chart in Figure 10.6. (Assume no out-of-plane buckling)

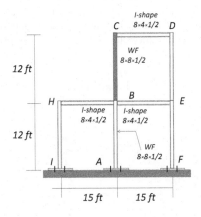

EXAMPLE FIGURE 10.2   Two-story unbraced frame.

Given: Column section WF 8 × 8 × 1/2″: Gross area (A) = 11.51 in². Strong and weak axis moment of inertia = 127.1 in⁴ and 42.7 in⁴, respectively. Longitudinal modulus (E_L) = 2.6 (10⁶) psi, Transverse modulus (E_T) = 0.8 (10⁶) psi, Shear modulus (G_{LT}) = 0.425 (10⁶) psi.

Beam section WF 8 × 4 × 1/2″: Gross area (A) = 7.50 in². Strong and weak axis moment of inertia = 70.6 and 5.41 in⁴, respectively. Longitudinal modulus (E_L) = 2.6 (10⁶) psi, Transverse modulus (E_T) = 0.8 (10⁶) psi, Shear modulus (G_{LT}) = 0.425 (10⁶) psi.

Solution

Buckling of a column occurs in the plane of the frame in Example Figure 10.2. Therefore, the moment of inertia (strong axis) perpendicular to the plane is used. Since sidesway is not prevented, the modified factor for the factor G is equal to 0.5 for the far end, which is hinged.

From Eq. (10.14):

$$G_B = \frac{\Sigma\left(\dfrac{E_{cL}I_c}{L_c}\right)}{\Sigma\left(\dfrac{E_{bL}I_b}{L_b}\right)} = \frac{\left(\dfrac{E_cI_c}{L_c}\right)_{AB} + \left(\dfrac{E_cI_c}{L_c}\right)_{BC}}{\left(\dfrac{0.5E_bI_b}{L_b}\right)_{BH} + \left(\dfrac{0.5E_bI_b}{L_b}\right)_{BE}} = \frac{\left(\dfrac{2E_cI_c}{L_c}\right)}{\left(\dfrac{E_bI_b}{L_b}\right)} = \frac{2\left(\dfrac{127.1}{12}\right)}{\left(\dfrac{70.6}{15}\right)} = 4.5$$

$$G_C = \frac{\Sigma\left(\dfrac{E_{cL}I_c}{L_c}\right)}{\Sigma\left(\dfrac{E_{bL}I_b}{L_b}\right)} = \frac{\left(\dfrac{E_cI_c}{L_c}\right)_{BC}}{\left(\dfrac{0.5E_bI_b}{L_b}\right)_{CD}} = \frac{\left(\dfrac{E_cI_c}{L_c}\right)}{\left(\dfrac{0.5E_bI_b}{L_b}\right)} = \frac{\left(\dfrac{127.1}{12}\right)}{0.5\left(\dfrac{70.6}{15}\right)} = 4.5$$

Using the alignment chart (Figure 10.6) with sidesway, the effective length factor K is 2.1 for buckling perpendicular to the strong axis.

Example 10.3

For FRP column BC in both the braced and unbraced frames of Examples 10.1 and 10.2, determine (a) global buckling strength, (b) nominal material rupture strength, and (c) global buckling strength with the shear deformation effect. Column BC is assumed to buckle about the strong axis of the cross section (Time factor λ is 0.8 for $W = 1.2D + 1.6L$).

Given: For 8 × 8 × 1/2″ WF column, Area (A) of 8 × 8 × 1/2″ = 11.51 in², Web area (A_{web}) = 3.50 in², Major and minor radius of gyration (r) = 3.33″ and 1.93″, respectively. Moment of inertia about the strong (I_T) = 127.1 in⁴, Moment of inertia about the strong (I_T) = 42.7 in⁴, and Torsional rigidity (J) is 0.958 in⁴.

Beam properties (manufacturer): Longitudinal modulus (E_L) = 2.6 (10⁶) psi, Transverse modulus (E_T) = 0.8 (10⁶) psi, Shear modulus (G_{LT}) = 0.425 (10⁶) psi, Longitudinal compressive strength (F_c) = 30,000 psi, Shear strength (F_v) = 4,500 psi, Longitudinal modulus (E_L) = 2.6 (10⁶) psi, Transverse modulus (E_T) = 0.8 (10⁶) psi, and Shear modulus (G_{LT}) = 0.425 (10⁶) psi.

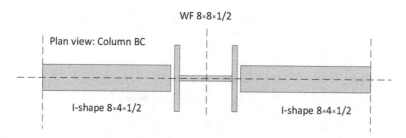

EXAMPLE FIGURE 10.3   FRP column BC in the frame.

Solution

a.1) **For braced frame:**
The effective length factor (K = 0.75) is obtained from the results of Example 10.1. The global buckling strength of pultruded FRP braced column BC is:
From Eq. (10.10):

$$P_{cr}^{GFB} = \frac{\pi^2 E_L I}{(KL)^2} = \frac{\pi^2 (2.6 \times 10^6)(127.1)}{(0.75 \times 12 \times 12)^2} = 279,622\,\text{lbs}$$

a.2) **For unbraced frame:**
The effective length factor (K = 2.1) is obtained from the results of Example 10.2, for buckling perpendicular to the strong axis. The global buckling strength of pultruded FRP unbraced column BC with buckling perpendicular to the strong axis is:
From Eq. (10.10):

$$P_{cr}^{GFB} = \frac{\pi^2 E_L I}{(KL)^2} = \frac{\pi^2 (2.6 \times 10^6)(127.1)}{(2.1 \times 12 \times 12)^2} = 35,666\,\text{lbs}$$

b) **Strength limit state:**
The nominal material rupture strength of column *BC* is:
From Eq. (10.7):

$$P_u = 0.7 \lambda F_L^c A_g = 0.7(0.8)(30,000 \times 11.51) = 193,368\,\text{lbs}$$

c.1) **For braced frame:**
The global flexural buckling load with shear effect is assumed to occur about the strong axis. The radius of gyration is also taken as the major radius of gyration. Slenderness ratio about the strong axis (KL/r) = 0.75(12 × 12/3.33) = 32.4.

$$k(G_{LT}A) = (G_{LT})_{web}\, A_{web} \quad \text{then } kG_{LT} = (G_{LT})_{web}\, A_{web}\,/A$$

$$= 0.425(10^6)(3.50/11.51) = 0.129(10^6)$$

From Eq. (10.13):

$$f_{cr}^{GFB} = \frac{\pi^2 2.6(10^6)}{32.4^2} \left(\cfrac{1}{1 + \cfrac{1}{0.129(10^6)} \left(\cfrac{\pi^2 2.6(10^6)}{32.4^2} \right)} \right) = 24,464(0.841) = 20,564\,\text{psi}$$

$$P_{cr}^{GFB} = f_{cr}^{GFB} A = 20,564 \times 11.51 = 236,696\,\text{lbs}$$

c.2) **For unbraced frame:**
The global flexural buckling load with the shear effect is assumed to occur about the strong axis. The radius of gyration is also taken as the major radius of gyration. Slenderness ratio (KL/r) = 2.1(12 × 12/3.33) = 90.8.

$$k(G_{LT}A) = (G_{LT})_{web}\, A_{web} \quad \text{Thus, } kG_{LT} = (G_{LT})_{web}\, A_{web}\,/A$$

$$= 0.425(10^6)(3.50/11.51) = 0.129(10^6)$$

From Eq. (10.13):

$$f_{cr}^{GFB} = \frac{\pi^2 2.6(10^6)}{90.8^2} \left(\frac{1}{1 + \frac{1}{0.129(10^6)}\left(\frac{\pi^2 2.6(10^6)}{90.8^2}\right)} \right) = 3,115(0.976) = 3,042\,\text{psi}$$

$$P_{cr}^{GFB} = f_{cr}^{GFB} A = 3,042 \times 11.51 = 35,007\,\text{lbs}$$

10.6 TORSIONAL BUCKLING

The global buckling under torsion can be induced in open pultruded FRP column sections. The torsional buckling is due to twisting of the cross section about the longitudinal axis of a column, which will involve warping of a section along with the twisting. In general, the flexural buckling nearly always exists before the buckling failure in torsional mode (refer to Chapter 9). The governing equation of the global torsional buckling about a strong axis is presented as:

$$E_L C_\omega \varphi^{iv} + \left(\frac{P}{A}I_p - CG_{LT}\right)\varphi^{ii} = 0 \tag{10.15.1}$$

The boundary conditions at both ends without warping of the cross section are (Figure 10.8):

$$u(0) = u(L) = \varphi(0) = \varphi(L) = 0 \tag{10.15.2}$$

The solutions of the governing equations with the above boundary conditions (Eq. 10.15.2) provide the global torsional buckling load as follows (Roberts and Al-Ubaidi, 2002):

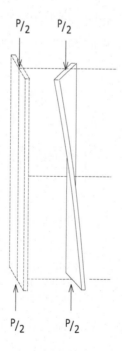

FIGURE 10.8   Torsional buckling of column.

$$f_{cr}^{GTB} = \cfrac{1}{\left(\cfrac{1}{f_{cr}^{TB}} + \cfrac{1}{kG_{LT}} \right)}$$

(10.16.1)

$$f_{cr}^{GTB} = \frac{1}{I_p}\left(CG_{LT} + \frac{\pi^2 E_L C_\omega}{(k_\omega L)^2} \right)$$

(10.16.2)

where

f_{cr}^{TB} = torsion buckling strength (without the effects of shear deformation)

I_p = the polar moment of inertia

C = torsional constant

$C_\omega = \dfrac{I_z d^2}{4}$, warping constant

k_ω = end restraint coefficient for torsional buckling

L = unbraced length of the column

G_{LT} = shear modulus.

The buckling strength about the weak axis of pultruded FRP compression members typically corresponds to a flexural buckling mode, particularly in double symmetric open sections. In practice, the torsional buckling of pultruded FRP columns is often neglected in the structural member design. However, FRP column sections without double symmetry may fail under torsional buckling or a mixed mode of torsional-flexural buckling because these cross sections are significantly lower resistant in local buckling strengths than the global buckling strength (Euler buckling). Therefore, torsional buckling and mixed mode of torsion-flexure buckling mode have to be checked for a column design.

Note: The same resistance factor for global flexural buckling ($\varphi = 0.7$) is used for global torsional buckling unless engineering analysis concludes a different resistance factor is adequate.

Example 10.4

A pultruded FRP 6 × 3 × 3/8″ I-shaped column has an unbraced length of 10 ft as shown in Example Figure 10.4. Determine a nominal torsional strength of the pultruded FRP I-shape column (Assuming both column ends are pin-supports).

Given: Section 6 × 3 × 3/8″ I-shaped column: Area (A) = 4.23 in², Moment of inertia about strong axis (I_T) = 22.3 in⁴, Moment of inertia about weak axis (I_z) = 1.71 in⁴, Torsional constant after neglecting web effect (C) = 0.198 in⁴; Longitudinal modulus of flange (E_{Lf}) = 2.6 (10⁶) psi, Transverse modulus of flange (E_{Tf}) = 0.8 (10⁶) psi, Longitudinal modulus of web (E_{Lw}) = 2.6 (10⁶) psi, Transverse modulus of web (E_{Tw}) = 0.8 (10⁶) psi, and Shear modulus (G_{LT}) = 0.425 (10⁶) psi.

Solution

The global torsional buckling load is determined from Eq. (10.16.2) as:

Polar moment of inertia I_p = 22.3 + 1.71 = 24.0 in⁴, k_ω = 1.0. The time effect factor λ for (W = 1.2D + 1.6L) is 0.8. The resistance factor φ_c for lateral torsional buckling is 0.7.

$$C_\omega = \frac{I_z d^2}{4} = \frac{1.71(6^2)}{4} = 15.39\,\text{in}^6$$

$$f_{cr}^{GTB} = \frac{1}{I_p}\left(CG_{LT} + \frac{\pi^2 E_L C_\omega}{(k_\omega L)^2} \right)$$

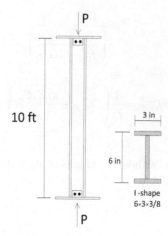

EXAMPLE FIGURE 10.4   FRP I-shape column.

$$f_{cr}^{GTB} = \frac{1}{24.0}\left((0.198\times0.425\times10^6) + \frac{\pi^2\left(2.6\times10^6\times15.39\right)}{\left(12\times10\right)^2}\right) = 4{,}649\,\mathrm{psi}$$

$$\varphi_c\lambda f_{cr}^{GTB} = (0.7\times0.8)4{,}649 = 2{,}603\,\mathrm{psi}$$

$$P_{cr} = \left(\varphi_c\lambda f_{cr}^{GTB}\right)A_g = (2{,}603)(4.23) = 11{,}013\,\mathrm{lbs} = 11.0\,\mathrm{kips}$$

10.7   LOCAL BUCKLING

Pultruded FRP members under compression may fail in local buckling due to relatively low stiffness of column webs and flanges. The local buckling of a compression member is often called

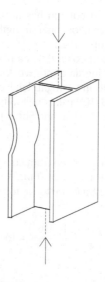

FIGURE 10.9   Local buckling of column flange.

TABLE 10.1

Critical Strength under Buckling of WF, I-Shaped and L-Angle Section (ASCE Str. Plastics Design Manual, 1984; ACMA, 2021)

Section	Buckling	Equations Based on the Pre-Standard (ACMA, 2021)	
Symmetric I shaped sections	Flexural global buckling @ the strong axis	$$f_{crT} = \dfrac{\pi^2 E_L}{\left(\dfrac{K_T L_T}{r_T}\right)^2}$$	(10.17.1)
	Flexural global buckling @ the weak axis	$$f_{crZ} = \dfrac{\pi^2 E_L}{\left(\dfrac{K_Z L_Z}{r_Z}\right)^2}$$	(10.17.2)
	Local flange buckling	$$f_{crf} = \dfrac{G_{LT}}{\left(\dfrac{b_f}{2t_f}\right)^2}$$	(10.17.3)
	Local web buckling	$$f_{crw} = \dfrac{\pi^2 t_w^2}{6h^2}\left(\sqrt{E_{L,w}E_{T,w}} + E_{T,w}\nu_{LT} + 2G_{LT}\right)$$	(10.17.4)
Single angle sections with equal legs	Flexural global buckling @ the strong axis	$$f_{crT} = \dfrac{\pi^2 E_L}{\left(\dfrac{K_T L_T}{r_T}\right)^2}$$	(10.18.1)
	Local flange buckling	$$f_{crf} = \dfrac{G_{LT}}{\left(\dfrac{b}{t}\right)^2}$$	(10.18.2)

crippling, as shown in Figure 10.9. The local buckling of individual element (webs and flanges are considered as individual parts or elements) may occur before an applied load reaches its material strength or the global buckling of a member. Typically, the influence of local buckling on a pultruded FRP column may depend on the cross-sectional profile, slenderness ratio of members, and mechanical properties of elements such as flange or web (Tomblin and Barbero, 1994; Bank 2002; Kollar and Springer, 2003). The local buckling strength of an FRP member under compression for various cross sections is summarized in Table 10.1.

where f_{crT} = elastic flexural buckling strength about the strong axis, f_{crZ} = elastic flexural buckling strength about the weak axis, f_{crf} = local flange buckling strength, f_{crw} = local web buckling strength, K_T = effective length factor corresponding to the strong axis, K_Z = effective length factor corresponding to the weak axis, L = laterally unbraced length of members, r = radius of gyration about the axis of buckling, E_L = longitudinal compression elastic modulus of the flange or web whichever is smaller, $E_{L,w}$ = longitudinal compression elastic modulus of the web, $E_{T,w}$ = transverse compression elastic modulus of the web, G_{LT} = in-plane shear modulus of members, b_f = flange width, b = outside width of a leg in compression, t_f = flange thickness, and t = angle leg thickness.

Note: The factored flexural torsional buckling strength under compression is computed by rational analysis for single angle sections of unequal legs.

TABLE 10.2

Critical Strength under Buckling of T-Section (ASCE Str. Plastics Design Manual, 1984; ACMA, 2021)

Section	Buckling	Equations Based on the Pre-Standard (ACMA, 2021)	
	Flexural buckling @ the strong axis	$f_{crT} = \dfrac{\pi^2 E_L}{\left(\dfrac{K_T L_T}{r_T}\right)^2}$	(10.19.1)
	Local flange buckling	$f_{crf} = \dfrac{G_{LT}}{\left(\dfrac{b_f}{2t_f}\right)^2}$	(10.19.2)
T shaped sections	Local web buckling	$f_{crw} = \dfrac{G_{LT}}{\left(\dfrac{d}{t_w}\right)^2}$	(10.19.3)
(symmetric about vertical axis Z)		$f_{ft} = \dfrac{F_{crz} + F_{cr}^{GTB}}{2H}\left(1 - \sqrt{1 - \dfrac{4HF_{crz}F_{cr}^{GTB}}{\left(F_{crz} + F_{cr}^{GTB}\right)^2}}\right)$	(10.19.4)
		where $f_{crz} = \dfrac{\pi^2 E_L}{\left(\dfrac{K_Z L_Z}{r_Z}\right)^2}$ $\quad F_{cr}^{GTB} = \dfrac{1}{R_P^2}\left(D_J + D_W\left(\dfrac{\pi}{L}\right)^2\right)$	
	Flexural-torsional buckling	$H = 1 - \dfrac{y_P^2}{R_P^2}$	
		$R_P^2 = \dfrac{1}{b_f t_f + h_w t_w}\left(\dfrac{b_f t_f}{12}\left(b_f^2 + t_f^2\right) + h_w t_w\left(\dfrac{h_w^2}{3} + \dfrac{t_w^2}{12}\right)\right)$	
		$y_p = \dfrac{h_w}{2\left(1 + \dfrac{b_f t_f}{h_w t_w}\right)} \quad D_J = \dfrac{G_{LT}}{3}\left(b_f t_f^3 + h_w t_w^3\right)$	
		$D_w = E_L\left(\dfrac{b_f^3 t_f^3}{144} + \dfrac{h_w^3 t_w^3}{36}\right)$	

where f_{crT} = elastic flexural buckling strength about the strong axis, f_{crZ} = elastic flexural buckling strength about the weak axis, f_{cr}^{GTB} = torsional buckling strength, f_{ft} = flexural-torsional buckling strength, f_{crf} = local flange buckling strength, f_{crw} = local web buckling strength, K_T = effective length factor corresponding to the strong axis, K_Z = effective length factor corresponding to the weak axis, L = laterally unbraced length of members, r = radius of gyration about the axis of buckling, E_L = longitudinal compression elastic modulus of the flange or web whichever is smaller, $E_{L,w}$ = longitudinal compression elastic modulus of the web, $E_{T,w}$ = transverse compression elastic modulus of the web, G_{LT} = in-plane shear modulus of the flange, D_J = torsional rigidity of section, D_w = warping rigidity of section, R_p = polar radius of gyration about the center of twisting of the cross section, h_w = distance between the centerline of the flange and the outer face of the stem, b_f = flange width, t_f = flange thickness, and t_w = stem (web) thickness.

TABLE 10.3

Critical Strength under Buckling of Closed Sections (ASCE Str. Plastics Design Manual, 1984; ACMA, 2021)

Section	Buckling	Equations Based on the Pre-Standard (ACMA, 2021)	
Square and rectangular tube sections	Flexural	$f_{cr} = \dfrac{\pi^2 E_L}{\left(\dfrac{KL}{r}\right)^2}$	(10.20.1)
	Web-local	$f_{crw} = \dfrac{\pi^2}{6\beta_w^2}\left(\sqrt{E_{L,w}E_{T,w}} + E_{T,w}\nu_{LT} + 2G_{LT}\right)$	(10.20.2)
Circular tube sections	Flexural	$f_{cr} = \dfrac{\pi^2 E_L}{\left(\dfrac{KL}{r}\right)^2}$	(10.21.1)
	Local	$f_{cr} = \dfrac{2t}{D}\sqrt{\dfrac{2}{3}G_{LT}\sqrt{E_L E_T}} \leq \dfrac{2t}{D}\sqrt{\dfrac{E_L E_T}{3}}$	(10.21.2)
Square, rectangular, and circular solid sections	Flexural	$f_{cr} = \dfrac{\pi^2 E_L}{\left(\dfrac{KL}{r}\right)^2}$	(10.22)

where f_{cr} = elastic flexural buckling strength. f_{crw} = local web buckling strength, K = effective length factor corresponding to the axis of buckling, L = laterally unbraced length of members, r = radius of gyration about the axis of buckling, E_L = longitudinal compression elastic modulus, E_T = transverse compression elastic modulus, $E_{L,w}$ = longitudinal compression elastic modulus of the web, $E_{T,w}$ = transverse compression elastic modulus of the web, G_{LT} = in-plane shear modulus, β_w = maximum width to thickness ratio, whichever is larger, of all elements comprising the tube section, ν_{LT} = Poisson's ratio associated with transverse deformation when compression is applied in the longitudinal direction, D = outer diameter of a tubular section, and t = tube wall thickness.

10.8 DESIGN OF COMPRESSION MEMBERS

For design purpose, the minimum nominal longitudinal strength P_L of all elements comprising a cross section must be higher than the nominal axial strength of the section as a whole under buckling and the required crushing strength under factored loads.

The general provisions for compression members under concentric loading are:

For serviceability limit state (ACMA, 2021):

$$P_s \leq \varphi_0 \frac{\pi^2 E_L}{\left(\dfrac{KL_e}{r}\right)^2} A_g \leq 0.3 F_L^c A_g \tag{10.23}$$

It should be noted that the original factor (φ_0) in Section 4.2 of the pre-standard (ACMA, 2021) may not be appropriately used with the initial out-of-straightness fraction limit (0.05 in/ft). The modified factor (φ_0) including out-of-straightness is presented herein for design purpose.

$$\varphi_0 = 1 - 125\frac{\delta_0}{L} \tag{10.24}$$

where

E_L = longitudinal modulus of the flange or web whichever is smaller

A_g = gross area of cross section

(KL_e/r) = slenderness ratio

F_L^c = minimum longitudinal compression material strength of all elements comprising the cross section

P_s = compression load due to serviceability load combinations

(δ_0/L) = initial out of straightness fraction guaranteed by the manufacturer.

For strength limit state (ACMA, 2021):

$$P_u \leq \lambda \varphi_c F_{cr} A_g \leq 0.7 \lambda F_L^c A_g \tag{10.25}$$

F_{cr} = critical strength under buckling of a member. The critical strength F_{cr} in Eq. (10.25) is taken as the minimum buckling strength under different failure modes as given in Tables 10.1–10.3.

where

P_u = required compression strength due to factored loads

λ = time effect factor as given in Table 7.3

F_L^c = minimum longitudinal compression material strength of all elements comprising the cross section

φ_c = resistance factor.

Example 10.5

Determine the nominal strength of a pultruded FRP 12 × 12 × 1/2″ wide flange column that has an unbraced effective length of 18 and 9 ft with respect to the strong axis and weak axis, respectively as shown in Example Figure 10.5 (Assuming the pultruded FRP wide flange column, pin-supported at both ends and pin-braced at the mid-height).

Given: Section WF 12 × 12 × 1/2″

Self-weight = 13.98 lbs/ft, Area (A) = 17.51 in², Web area (A_w) = 5.50 in²; Sectional modulus about the strong axis (S_T) = 75.5 in³, Sectional modulus about the weak axis (S_Z) = 24.0 in³, Moment of inertia about strong axis (I_T) = 452.7 in⁴, Radius of gyration about the strong axis (r_T) = 5.15 in, Radius of gyration about the weak axis (r_Z) = 2.90 in, Moment of inertia about weak axis (I_Z) = 144.1 in⁴, Torsional constant (C) = 1.458 in⁴, and Poisson's ratio (ν_{LT}) = 0.33.

For beam properties (manufacturer): Longitudinal compressive strength (F_c) = 30,000 psi, Shear strength (F_v) = 4,500 psi, Longitudinal modulus (E_L) = 2.6 (10⁶) psi, Transverse modulus (E_T) = 0.8 (10⁶) psi, and Shear modulus (G_{LT}) = 0.425 (10⁶) psi.

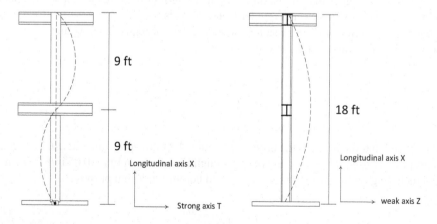

EXAMPLE FIGURE 10.5 Pultruded FRP column with unbraced length.

Solution

a. Determine material strength:
   The material strength of a pultruded FRP WF column is determined from Eq. (10.7) as:

$$P_u = 0.7\lambda F_L^c A_g = 0.7(0.8)(30,000 \times 17.51) = 294 \text{ kips}$$

b. Determine flexural buckling strength about the strong axis:
   The flexural buckling strength about the strong axis with an unbraced effective length of 18 ft, where $K_T = 1.0$, is determined from Eq. (10.17.1) as:

$$f_{crT} = \frac{\pi^2 E_L}{\left(\dfrac{K_T L_T}{r_T}\right)^2} = \frac{\pi^2 (2.6 \times 10^6)}{\left(\dfrac{18 \times 12}{5.15}\right)^2} = 14,587 \text{ psi}$$

c. Determine flexural buckling strength about the weak axis:
   The flexural buckling strength about the weak axis with an unbraced effective length of 9 ft, where $K_Z = 1.0$, is determined from Eq. (10.17.2) as:

$$f_{crz} = \frac{\pi^2 E_L}{\left(\dfrac{K_Z L_z}{r_Z}\right)^2} = \frac{\pi^2 (2.6 \times 10^6)}{\left(\dfrac{9 \times 12}{2.90}\right)^2} = 18,502 \text{ psi}$$

d. Determine local flange buckling:
   The local flange buckling strength is determined from Eq. (10.17.3) as:

$$f_{crf} = \frac{G_{LT}}{\left(\dfrac{b_f}{2t_f}\right)^2} = \frac{(0.425 \times 10^6)}{\left(\dfrac{12}{2 \times 0.5}\right)^2} = 2,951 \text{ psi}$$

e. Determine local web buckling:
   e.1) using the complex expression from Eq. (10.17.4):
   The local flange buckling strength is:

$$f_{crw} = \frac{\pi^2 t_w^2}{6h^2}\left(\sqrt{E_{L,w}E_{T,w}} + E_{T,w}\nu_{LT} + 2G_{LT}\right)$$

$$f_{crw} = \frac{\pi^2 (0.5)^2}{6(12^2)}\left(\sqrt{(2.6 \times 0.8 \times 10^{12})} + (0.33 \times 0.8 \times 10^6) + (2 \times 0.425 \times 10^6)\right)$$

$$f_{crw} = 7,300 \text{ psi}$$

f. Determine global torsional buckling strength:
   For $\varphi_c = 0.7$ and λ (for $1.2D + 1.6L$) =0.8, the global torsional buckling load is:
   Polar moment of inertia $I_p = 452.7 + 144.1 = 596.8 \text{ in}^4$, $k_\omega = 1.0$.

   f.1) Weak axis (an unbraced length $L = 9$ ft) from Eq. (10.16.2):

$$C_\omega = \frac{I_z d^2}{4} = \frac{144.1(12^2)}{4} = 5,188 \text{ in}^6$$

$$f_{cr}^{GTB} = \frac{1}{I_p}\left(CG_{LT} + \frac{\pi^2 E_L C_\omega}{(k_\omega L)^2}\right)$$

$$f_{cr}^{GTB} = \frac{1}{597}\left(\left(1.458 \times 0.425 \times 10^6\right) + \frac{\pi^2\left(2.6 \times 10^6 \times 5,188\right)}{(12 \times 9)^2}\right) = 20,156\,\text{psi}$$

$$\varphi_c \lambda f_{cr}^{GTB} = (0.7 \times 0.8)20,156 = 11,288\,\text{psi}$$

f.2) Strong axis (an unbraced length $L = 18\,\text{ft}$) from Eq. (10.16.2):

$$C_\omega = \frac{I_z d^2}{4} = \frac{144.1\left(12^2\right)}{4} = 5,188\,\text{in}^6$$

$$f_{cr}^{GTB} = \frac{1}{I_p}\left(JG_{LT} + \frac{\pi^2 E_L C_\omega}{\left(k_\omega L\right)^2}\right)$$

$$f_{cr}^{GTB} = \frac{1}{597}\left(\left(1.458 \times 0.425 \times 10^6\right) + \frac{\pi^2\left(2.6 \times 5,188 \times 10^6\right)}{(12 \times 18)^2}\right) = 5,818\,\text{psi}$$

$$\varphi_c \lambda f_{cr}^{GTB} = (0.7 \times 0.8)5,818 = 3,258\,\text{psi}$$

The minimum nominal strength is obtained from the local flange buckling strength (F_{crf}). For φ_c (for local buckling) = 0.8 and λ (for $1.2D + 1.6L$) = 0.8, the nominal strength is:

$$P_{cr} = \lambda \varphi_c f_{crf} A_g = (0.8 \times 0.8)(2,951)(17.51) = 33,074\,\text{lbs} = 33.1\,\text{kips}$$

Example 10.6

Using LRFD method, design a pultruded FRP wide flange 9 ft long column resisting dead load plus column self-weight, which are 6,500 lbs, and live load of 8,500 lbs.

Given: Pultruded FRP column properties (manufacturer): Longitudinal compression strength (F_c) = 30,000 psi, Longitudinal modulus (E_L) = 2.6 (10^6) psi, Transverse modulus (E_T) = 0.8 (10^6) psi, and Shear modulus (G_{LT}) = 0.425 (10^6) psi.

Solution

  a. Determine service and factored load
     Total service load (P_s) = (6.5 kips) + (8.5 kips) = 15 kips
     Total service load (P_u) = 1.2(6.5 kips) + 1.6(8.5 kips) = 21.4 kips
  b. Trial section with the slenderness and deformation limit
     The slenderness ration of an axial compression member is suggested as:
     From Eq. (10.5): (Assume pin-end condition $K = 1.0$)

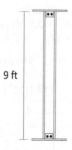

9 ft

EXAMPLE FIGURE 10.6   Simply supported FRP column.

$$\frac{KL_e}{r} = \frac{(9 \times 12)}{\sqrt{\dfrac{I_z}{A}}} \le 300 \quad \text{then} \quad \sqrt{\frac{A}{I_z}} \le 2.78$$

Select WF 8 × 8 × 1/2″ section:

Gross area (A) = 11.51 in², Major and minor radius of gyration (r) = 3.33″ and 1.93″, respectively. Moment of inertia about the strong axis (I_T) = 127.1 in⁴, Moment of inertia about weak axis (I_z) = 42.7 in⁴, and Torsional constant (C) is 0.958 in⁴.

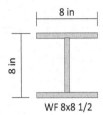

WF 8x8 1/2

At serviceability: P_S = 15 kips

$$\sqrt{\frac{A}{I_z}} = \sqrt{\frac{11.51}{42.7}} = 0.52 \le 2.78$$

The pultrude FRP wide flange column (8 × 8 × 1/2″) is selected for this design.

c. Determine axial compression design strength:

Induced compression stress (f_s) and design strength (f_u) are determined as:

$$P_s = 15,000 \, \text{lbs} \quad \text{and} \quad f_s = \frac{P_s}{A_g} = \frac{15,000}{11.51} = 1,303 \, \text{psi}$$

$$P_u = 21,400 \, \text{lbs} \quad \text{and} \quad f_u = \frac{P_u}{A_g} = \frac{21,400}{11.51} = 1,859 \, \text{psi}$$

d. Determine nominal material strength:

For λ (for 1.2D + 1.6L) = 0.8, the nominal material strength at serviceability and strength limit states are:

$$P_L = 0.3 F_L^c A_g = 0.3(30,000)(11.51) = 103,590 \, \text{lbs} = 103 \, \text{kips} \qquad \text{(Eq.10.6)}$$

$$P_u = 0.7 \lambda F_L^c A_g = 0.7(0.8)(30,000)(11.51) = 193,368 \, \text{lbs} = 193 \, \text{kips} \qquad \text{(Eq.10.7)}$$

e. Determine the allowable and minimum nominal buckling strengths:

e.1) For the allowable buckling strength:

(δ_0/L) initial out of straightness fraction is 0.05 in/ft (manufacturers).

$$(P_{cr})_{allow} = \varphi_0 \frac{\pi^2 E_L}{\left(\dfrac{KL_e}{r}\right)^2} A_g = 0.48 \left[\frac{\pi^2 (2.6 \times 10^6)}{\left(\dfrac{9 \times 12}{1.93}\right)^2} \right] \times (11.51) = 45,275 \, \text{lbs} \qquad \text{(Eq.10.23)}$$

where φ_0, as per Eq. (10.24) is:

$$\varphi_0 = 1 - 125 \frac{\delta_0}{L} = 1 - 125(0.05) = 0.48$$

e.2) For strength limit state:

The critical strength (F_{cr}) is taken from the minimum buckling strength obtained from the different failure modes.

e.2.1) flexural buckling about the strong axis from Eq. (10.17.1):

$$f_{crT} = \frac{\pi^2 E_L}{\left(\frac{K_T L_T}{r_T}\right)^2} = \frac{\pi^2 \left(2.6 \times 10^6\right)}{\left(\frac{9 \times 12}{3.33}\right)^2} = 24,396 \, \text{psi}$$

e.2.2) flexural buckling about the weak axis from Eq. (10.17.2):

$$f_{crz} = \frac{\pi^2 E_L}{\left(\frac{K_z L_z}{r_z}\right)^2} = \frac{\pi^2 \left(2.6 \times 10^6\right)}{\left(\frac{9 \times 12}{1.93}\right)^2} = 8,194 \, \text{psi}$$

e.2.3) local flange buckling from Eq. (10.17.3):

$$f_{crf} = \frac{G_{LT}}{\left(\frac{b_f}{2t_f}\right)^2} = \frac{\left(0.425 \times 10^6\right)}{\left(\frac{8}{2 \times 0.5}\right)^2} = 6,641 \, \text{psi}$$

e.2.4) local web buckling from Eq. (10.17.4):

$$f_{crw} = \frac{\pi^2 t_w^2}{6h^2}\left(\sqrt{E_{L,w} E_{T,w}} + E_{T,w} \nu_{LT} + 2G_{LT}\right)$$

$$f_{crw} = \frac{\pi^2 (0.5)^2}{6(8^2)}\left(\sqrt{\left(2.6 \times 0.8 \times 10^{12}\right)} + \left(0.33 \times 0.8 \times 10^6\right) + \left(2 \times 0.425 \times 10^6\right)\right)$$

$$f_{crw} = 16,425 \, \text{psi}$$

The minimum buckling strength (f_{cr}) is obtained from the local flange buckling strength (f_{crf}). For φ_c (for local buckling) = 0.8 and λ for (1.2D + 1.6L) = 0.8, the nominal buckling strength is:

$$P_{cr} = \lambda \varphi_c f_{cr} A_g = (0.8 \times 0.8)(6,641)(11.51) = 48,918 \, \text{lbs}$$

f. Check for the minimum nominal strength corresponding to both the severability and strength limit states:

The minimum nominal longitudinal strength P_L of all elements comprising the cross section must be higher than the nominal axial strength due to buckling and the required strength due to factored loads.

At serviceability limit state:

From (d) and (e.1):

$$P_s \leq \varphi_0 \frac{\pi^2 E_L}{\left(\frac{KL_e}{r}\right)^2} A_g \leq 0.3 F_L^c A_g$$

$$P_s = 15 \, \text{kips} \leq \left(P_{cr}\right)_{allow} = 45.3 \, \text{kips} \leq 0.3 F_L^c A_g = 103 \, \text{kips}$$

At strength limit state:
From (e.2):

$$P_u \le \lambda \varphi_c F_{cr} A_g \le 0.7 \lambda F_L^c A_g$$

$$P_u = 21.4 \,\text{kips} \,\big(\text{as in (a)}\big) \le \lambda \varphi_c F_{cr} A_g = 48.9 \,\text{kips} \le 0.7 \lambda F_L^c A_g = 193 \,\text{kips}$$

It is found that the FRP wide flange section under axial compression satisfies the allowable material and buckling strength limit states.

g. Check for global torsional buckling strength:
The global torsional buckling load is determined from Eq. (10.16.2) as:
$k_\omega = 1.0$. Polar moment of inertia $I_p = 127.1 + 42.7 = 169.8 \,\text{in}^4$.

$$C_\omega = \frac{I_z d^2}{4} = \frac{42.7\big(8^2\big)}{4} = 683 \,\text{in}^6$$

$$f_{cr}^{GTB} = \frac{1}{I_p}\left(CG_{LT} + \frac{\pi^2 E_L C_\omega}{\big(k_\omega L\big)^2}\right) = \frac{1}{169.8}\left(\big(0.958 \times 0.425 \times 10^6\big) + \frac{\pi^2\big(2.6 \times 683 \times 10^6\big)}{\big(12 \times 9\big)^2}\right)$$

$$f_{cr}^{GTB} = 11,247 \,\text{psi}$$

$$\varphi_c \lambda f_{cr}^{GTB} = \big(0.7 \times 0.8\big)11,247 = 6,298 \,\text{psi}$$

The global torsional buckling strength (6,298 psi) is higher than the design compressive strength (1,859 psi) for the pultruded FRP wide flange column. The pultruded FRP wide flange column section based on the LRFD method satisfies the strength and serviceability limit states.

Example 10.7

Using the ASD method, design a pultruded FRP WF beam 9 ft long column having dead load including column self-weight of 6,500 lbs and live load of 8,500 lbs.

Pultruded FRP column properties (manufacturer): Longitudinal compression strength $(F_c) = 30,000$ psi, Longitudinal modulus $(E_L) = 2.6$ (10⁶) psi, Transverse modulus $(E_T) = 0.8$ (10⁶) psi and Shear modulus $(G_{LT}) = 0.425$ (10⁶) psi.

Solution

a. Determine working design load
Total service load $(P) = (6.5 \,\text{kips}) + (8.5 \,\text{kips}) = 15 \,\text{kips}$

9 ft

EXAMPLE FIGURE 10.7 Simply supported FRP column.

b. Trial section with the slenderness and deformation limit

The slenderness ratio of an axial compression member is suggested as:
From Eq. (10.5): (Assume pin-end condition $K = 1.0$)

$$\text{if } \frac{KL_e}{r} = \frac{(9 \times 12)}{\sqrt{\dfrac{I_z}{A}}} \le 300; \quad \text{then} \quad \sqrt{\frac{A}{I_z}} \le 2.78$$

The total service load including self-weight (P_s) is 15,000 lbs. Unbraced length of a column L is 9 ft. The longitudinal modulus of a column is assumed to be 2.6 (10⁶) psi.

Select WF $8 \times 8 \times 1/2''$ section:

Gross area (A) = 11.51 in², Major and minor radius of gyration (r) = 3.33″ and 1.93″, respectively. Moment of inertia about the strong (I_T) = 127.1 in⁴, Moment of inertia about weak axis = 42.7 in⁴, and Torsional constant (C) is 0.958 in⁴.

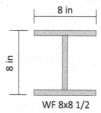

WF 8x8 1/2

At serviceability: $P_S = 15$ kips

$$\sqrt{\frac{A}{I_z}} = \sqrt{\frac{11.51}{42.7}} = 0.52 \le 2.78$$

The pultruded FRP wide flange column ($8 \times 8 \times 1/2''$) is selected for this design.
c. Determine axial compression design strength:

Induced compression stress (f_s) and design strength (f_u) are determined as:

$$P_s = 15,000 \text{ lbs} \quad \text{and} \quad f_s = \frac{P_s}{A_g} = \frac{15,000}{11.51} = 1,303 \text{ psi}$$

d. Determine nominal material strength:

For λ (for $1.2D + 1.6L$) = 0.8, the nominal material strength at serviceability limit state is:
From Eq. (10.6):

$$P_L = 0.3 F_L^c A_g = 0.3(30,000)(11.51) = 103,590 \text{ lbs} = 103 \text{ kips}$$

e. Determine the allowable buckling strength:

(δ_0/L) initial out of straightness fraction is 0.05 in/ft (manufacturers).
From Eqs. (10.23 and 10.24):

$$(P_{cr})_{allow} = \varphi_0 \frac{\pi^2 E_L}{\left(\dfrac{KL_e}{r}\right)^2} A_g = 0.48 \left(\frac{\pi^2 (2.6 \times 10^6)}{\left(\dfrac{9 \times 12}{1.93}\right)^2} \right) \times (11.51) = 45,275 \text{ lbs}$$

where φ_0, as per Eq. (10.24) is:

$$\varphi_0 = 1 - 125 \frac{\delta_0}{L} = 1 - 125(0.05) = 0.48$$

The minimum allowable material strength P_L of all elements comprising the cross section must be higher than the allowable strength under buckling and the design service load. From (d) and (e):

$$P_s \leq \varphi_0 \frac{\pi^2 E_L}{\left(\dfrac{KL_e}{r}\right)^2} A_g \leq 0.3 F_L^c A_g$$

$$P_s = 15\,\text{kips} \leq (P_{cr})_{allow} = 45.3\,\text{kips} \leq 0.3 F_L^c A_g = 103\,\text{kips}$$

The FRP wide flange column under externally applied axial compression load (15 kips, as computed in (a)) satisfies the allowable material and buckling strength limit.

f. Check for the allowable global torsional buckling strength:

$k_\omega = 1.0$. Polar moment of inertia $I_p = 127.1 + 42.7 = 169.8\,\text{in}^4$.

The global torsional buckling load is determined from Eq. (10.16.2) as:

$$C_\omega = \frac{I_z h^2}{4} = \frac{42.7(8^2)}{4} = 683\,\text{in}^6$$

$$f_{cr}^{GTB} = \frac{1}{I_p}\left(CG_{LT} + \frac{\pi^2 E_L C_\omega}{(k_\omega L)^2}\right)$$

$$f_{cr}^{GTB} = \frac{1}{169.8}\left((0.958 \times 0.425 \times 10^6) + \frac{\pi^2(2.6 \times 683 \times 10^6)}{(12 \times 9)^2}\right) = 11{,}247\,\text{psi}$$

$$\left(f_{cr}^{GTB}\right)_{allow} = \frac{f_{cr}^{GTB}}{F.S.} = \frac{11{,}247}{3} = 3{,}749\,\text{psi}$$

The allowable global torsional buckling strength (3,749 psi) under compression is higher than the design compressive strength (1,303 psi).

g. Check for the allowable local buckling strength:

g.1) Allowable local flange buckling strength:

Based on the ASD method, the factor of safety (F.S.) for compression members is taken as 3.

The local flange buckling strength is determined from Eq. (10.17.3) as:

$$f_{crf} = \frac{G_{LT}}{\left(\dfrac{b_f}{2t_f}\right)^2} = \frac{(0.425 \times 10^6)}{\left(\dfrac{8}{2 \times 0.5}\right)^2} = 6{,}641\,\text{psi}$$

$$\left(f_{crf}\right)_{allow} = \frac{F_{crf}}{F.S.} = \frac{6{,}641}{3} = 2{,}214\,\text{psi}$$

g.2) The allowable local web buckling strength from Eq. (10.17.4):

$$f_{crw} = \frac{\pi^2 t_w^2}{6h^2}\left(\sqrt{E_{L,w} E_{T,w}} + E_{T,w} \nu_{LT} + 2G_{LT}\right)$$

$$f_{crw} = \frac{\pi^2(0.5)^2}{6(8^2)}\left(\sqrt{(2.6 \times 0.8 \times 10^{12})} + (0.33 \times 0.8 \times 10^6) + (2 \times 0.425 \times 10^6)\right)$$

$$f_{crw} = 16,425 \, psi$$

$$\left(f_{crw}\right)_{allow} = \frac{f_{crw}}{F.S.} = \frac{16,425}{3} = 5,475 \, psi$$

The minimum allowable local buckling strength (2,214 psi) is higher than the axial compressive strength (1,303 psi), hence the pultruded column based on the ASD method satisfies the strength and serviceability limit states.

Example 10.8

Using the LRFD method, design a pultruded FRP column under 2,500 lbs of dead load including column self-weight, and 10,000 lbs of live load. Assume: effective lengths $(KL)_T$ about the strong axis and $(KL)_Z$ about the weak axis are 10 ft long and 5 ft long, respectively.

Solution

a. Determine service and factored load:
   Total service load (P_s) = (2.5 kips) + (7.5 kips) = 10.0 kips
   Total factored load (P_u) = 1.2(2.5 kips) + 1.6(7.5 kips) = 15.0 kips
b. Trial section with the slenderness and deformation limits:
   The slenderness ratio of an axial compression member is determined from Eq. (10.5):
   (Assuming: pinned condition $K = 1.0$)
   Strong axis:

$$\left(\frac{KL_e}{r}\right)_T = \frac{(10 \times 12)}{\sqrt{\dfrac{I_T}{A}}} \leq 300; \quad or \quad \sqrt{\frac{A}{I_T}} \leq 2.50$$

Weak axis:

$$\left(\frac{KL_e}{r}\right)_Z = \frac{(5 \times 12)}{\sqrt{\dfrac{I_Z}{A}}} \leq 300; \quad or \quad \sqrt{\frac{A}{I_Z}} \leq 5.0$$

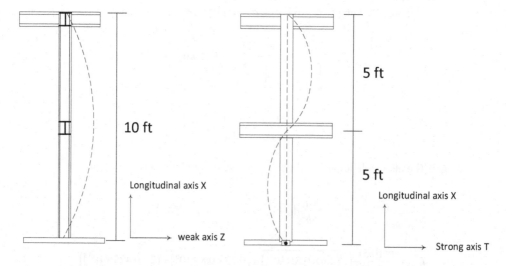

10 ft

5 ft

5 ft

Longitudinal axis X

Longitudinal axis X

weak axis Z

Strong axis T

EXAMPLE FIGURE 10.8   Pultruded FRP column with different effective length in each axis.

Select 10 × 5 × 1/2″ I-shaped section:

Gross area (A) = 9.50 in². Major and minor radius of gyration (r) is 3.88″ and 1.05″, respectively. Moment of inertia about the strong axis (I_T) = 143.3 in⁴ and Moment of inertia about the weak axis (I_z) = 10.51 in⁴. Torsional constant (C) = 0.792 in⁴. Longitudinal modulus (E_L) = 2.6 (10⁶) psi, Transverse modulus (E_T) = 0.8 (10⁶) psi, and Shear modulus (G_{LT}) = 0.425 (10⁶) psi.

$$\frac{(KL)_T}{(KL)_Z} = \frac{10}{5} = 2 \quad < \left(\frac{r_T}{r_Z} = \frac{3.88}{1.05} = 3.70 \right)$$

Thus, $(KL)_Z$ (the weak axis) controls when the ratio of (r_T/r_z) is higher than $(KL)_T/(KL)_Z$.

Weak axis:
At serviceability load: P_S = 10.0 kips

$$\sqrt{\frac{A}{I_z}} = \sqrt{\frac{9.50}{10.51}} = 0.951 \le 5.0$$

I-shaped column (10 × 5 × 1/2″) is selected for this design.

c. Determine axial compression design strength:
   Induced compression stress (f_s) and design strength (f_u) are determined as:

$$P_s = 10,000 \, \text{lbs} \quad \text{and} \quad f_s = \frac{P_s}{A_g} = \frac{10,000}{9.50} = 1,053 \, \text{psi}$$

$$P_u = 15,000 \, \text{lbs} \quad \text{and} \quad f_u = \frac{P_u}{A_g} = \frac{15,000}{9.50} = 1,579 \, \text{psi}$$

d. Determine nominal material strength:
   For λ (for 1.2D + 1.6L) = 0.8, the nominal material strength at serviceability and strength limit state is:
   At serviceability limit state:

$$P_L = 0.3 F_L^c A_g = 0.3(30,000)(9.50) = 85,500 \, \text{lbs} = 85.5 \, \text{kips} \qquad \left(\text{Serviceability Eq.} 10.6\right)$$

At strength limit state:

$$P_E = 0.7 \lambda F_L^c A_g = 0.7(0.8)(30,000)(9.50) = 159,600 \, \text{lbs} = 159.6 \, \text{kips} \qquad \left(\text{Strength Eq.} 10.7\right)$$

e. Determine the allowable and minimum nominal buckling strength:
   e.1) For allowable buckling strength:
   (δ_0/L) initial out of straightness fraction is 0.05 in/ft (manufacturers).
   From Eqs. (10.23 and 10.24):

$$\left(P_{cr}\right)_{T,\,allow} = \varphi_0 \frac{\pi^2 E_L}{\left(\dfrac{KL_{eT}}{r_T}\right)^2} A_g = 0.48 \left(\frac{\pi^2 \left(2.6 \times 10^6\right)}{\left(\dfrac{10 \times 12}{3.88}\right)^2}\right) \times (9.50) = 122{,}331\,\text{lbs}$$

$$\left(P_{cr}\right)_{Z,\,allow} = \varphi_0 \frac{\pi^2 E_L}{\left(\dfrac{KL_{eZ}}{r_Z}\right)^2} A_g = 0.48 \left(\frac{\pi^2 \left(2.6 \times 10^6\right)}{\left(\dfrac{5 \times 12}{1.05}\right)^2}\right) \times (9.50) = 35{,}835\,\text{lbs}$$

where φ_0, as per Eq. (10.24) is:

$$\varphi_0 = 1 - 125 \frac{\delta_0}{L} = 1 - 125(0.05) = 0.48$$

e.2) For strength limit state:
The critical strength (f_{cr}) is taken from the minimum buckling strength obtained from different failure modes.
e.2.1) Flexural buckling @ strong axis from Eq. (10.17.1):

$$f_{crT} = \frac{\pi^2 E_L}{\left(\dfrac{K_T L_T}{r_T}\right)^2} = \frac{\pi^2 \left(2.6 \times 10^6\right)}{\left(\dfrac{10 \times 12}{3.88}\right)^2} = 26{,}827\,\text{psi}$$

e.2.2) Flexural buckling @ weak axis from Eq. (10.17.2):

$$f_{crZ} = \frac{\pi^2 E_L}{\left(\dfrac{K_Z L_Z}{r_Z}\right)^2} = \frac{\pi^2 \left(2.6 \times 10^6\right)}{\left(\dfrac{5 \times 12}{1.05}\right)^2} = 7{,}859\,\text{psi}$$

e.2.3) Local flange buckling from Eq. (10.17.3):

$$f_{crf} = \frac{G_{LT}}{\left(\dfrac{b_f}{2t_f}\right)^2} = \frac{\left(0.425 \times 10^6\right)}{\left(\dfrac{10}{2 \times 0.5}\right)^2} = 4{,}250\,\text{psi}$$

e.2.4) Local web buckling from Eq. (10.17.4):

$$f_{crw} = \frac{\pi^2 t_w^2}{6h^2}\left(\sqrt{E_{L,w} E_{T,w}} + E_{T,w}\nu_{LT} + 2G_{LT}\right)$$

$$f_{crw} = \frac{\pi^2 (0.5)^2}{6\left(10^2\right)}\left(\sqrt{2.6 \times 0.8 \times 10^{12}} + \left(0.33 \times 0.8 \times 10^6\right) + \left(2 \times 0.425 \times 10^6\right)\right)$$

$$f_{crw} = 10{,}512\,\text{psi}$$

The minimum buckling strength (f_{cr}) is obtained from the local flange buckling (e.2.3). For φ_c (for local buckling) = 0.8 and λ (for 1.2D + 1.6L) = 0.8. The nominal buckling strength is:

$$P_{cr} = \lambda \varphi_c f_{cr} A_g = (0.8 \times 0.8)(4,250)(9.50) = 25,840 \, lbs$$

f. Check for the minimum nominal strength for severability and strength limit states:

The minimum nominal longitudinal strength (P_l) of all elements of a cross section must be higher than the nominal axial strength due to buckling and the design strength with factored loads.

Serviceability limit state check:

From (d) and (e.1):

$$P_s \leq \varphi_0 \frac{\pi^2 E_L}{\left(\dfrac{KL_e}{r}\right)^2} A_g \leq 0.3 F_L^c A_g$$

$$P_s = 13.5 \, kips \leq (P_{cr})_{Zallow} = 35.9 \, kips \leq 0.3 F_L^c A_g = 85.5 \, kips$$

Strength limit state check:

From (d) and (e.2.3):

$$P_u \leq \lambda \varphi_c F_{cr} A_g \leq 0.7 \lambda F_L^c A_g$$

$$P_u = 19.4 \, kips \leq \lambda \varphi_c F_{cr} A_g = 25.8 \, kips \leq 0.7 \lambda F_L^c A_g = 159.6 \, kips$$

It is found that the column ($10 \times 5 \times 1/2''$) under axial compression satisfies the allowable material and buckling strength limit states.

g. Check for global torsional buckling strength:

The global torsional buckling strength is determined from Eq. (10.16.2) as:

Polar moment of inertia $I_p = 143.3 + 10.51 = 153.8 \, in^4$, $k_\omega = 1.0$

$$C_\omega = \frac{I_z d^2}{4} = \frac{10.51(10^2)}{4} = 262.8 \, in^6$$

Weak axis:

$$\left(f_{cr}^{GTB}\right)_Z = \frac{1}{I_p}\left(CG_{LT} + \frac{\pi^2 E_L C_\omega}{(k_\omega L)^2}\right)$$

$$\left(f_{cr}^{GTB}\right)_Z = \frac{1}{153.8}\left((0.792 \times 0.425 \times 10^6) + \frac{\pi^2 (2.6 \times 10^6 \times 262.8)}{(12 \times 5)^2}\right) = 14,368 \, psi$$

$$\varphi_c \lambda \left(f_{cr}^{GTB}\right)_Z = (0.7 \times 0.8)14,368 = 8,046 \, psi$$

Strong axis:

$$\left(f_{cr}^{GTB}\right)_T = \frac{1}{I_p}\left(JG_{LT} + \frac{\pi^2 E_L C_\omega}{(k_\omega L)^2}\right)$$

$$\left(f_{cr}^{GTB}\right)_Z = \frac{1}{153.8}\left((0.792 \times 0.425 \times 10^6) + \frac{\pi^2 (2.6 \times 10^6 \times 262.8)}{(12 \times 5)^2}\right) = 14,368 \, psi$$

$$\varphi_c \lambda \left(f_{cr}^{GTB} \right)_Z = (0.7 \times 0.8) 14{,}368 = 8{,}046 \, \text{psi}$$

The global torsional buckling strength (2,931 psi) is higher than the design compressive strength (1,579 psi). The pultruded column (10 × 5 × 1/2″ I-shaped section) based on LRFD satisfies strength and serviceability limit states.

Example 10.9

Using the ASD method, design a pultruded FRP column under 2,500 lbs of dead load including column self-weight and 10,000 lbs of live load. Assume: the effective lengths $(KL)_T$ about the strong axis and $(KL)_Z$ about the weak axis are 10 ft long and 5 ft long, respectively.

Solution

a. Determine service load and axial compression strength:
   Total service load (P_s) = (2.5 kips) + (7.5 kips) = 10.0 kips. Induced compression stress (f_s) is:

$$P_s = 10{,}000 \, \text{lbs} \quad \text{and} \quad f_s = \frac{P_s}{A_g} = \frac{10{,}000}{9.50} = 1{,}053 \, \text{psi}$$

b. Trial section with the slenderness and deformation limits:
   The slenderness ratio of an axial compression member is determined from Eq. (10.5) as: (Assuming: pinned condition $K = 1.0$)

Strong axis:

$$\left(\frac{KL_e}{r} \right)_T = \frac{(10 \times 12)}{\sqrt{\dfrac{I_T}{A}}} \leq 300; \quad \text{or} \quad \sqrt{\frac{A}{I_T}} \leq 2.50$$

Weak axis:

$$\left(\frac{KL_e}{r} \right)_Z = \frac{(5 \times 12)}{\sqrt{\dfrac{I_Z}{A}}} \leq 300; \quad \text{or} \quad \sqrt{\frac{A}{I_Z}} \leq 5.0$$

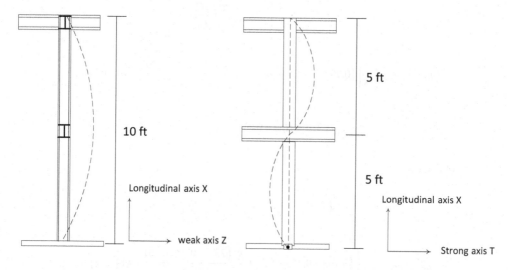

10 ft

Longitudinal axis X

weak axis Z

5 ft

5 ft

Longitudinal axis X

Strong axis T

EXAMPLE FIGURE 10.9 Pultruded FRP column with different effective length in each axis.

Select 10 × 5 × 1/2" I-shaped section:

Gross area $(A) = 9.50$ in². Major and minor radius of gyration (r) is 3.88" and 1.05", respectively. Moment of inertia about the strong axis $(I_T) = 143.3$ in⁴ and Moment of inertia about the weak axis $(I_z) = 10.51$ in⁴. Torsional constant $(C) = 0.792$ in⁴. Longitudinal modulus $(E_L) = 2.6$ (10⁶) psi, Transverse modulus $(E_T) = 0.8$ (10⁶) psi and Shear modulus $(G_{LT}) = 0.425$ (10⁶) psi.

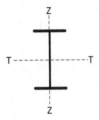

$$\frac{(KL)_T}{(KL)_z} = \frac{10}{5} = 2 \quad < \left(\frac{r_T}{r_z} = \frac{3.88}{1.05} = 3.70 \right)$$

Thus, $(KL)_z$ (the weak axis) controls when the ratio of (r_T/r_z) is higher than $(KL)_T/(KL)_z$.

Weak axis:

At serviceability load: $P_S = 10.0$ kips

$$\sqrt{\frac{A}{I_z}} = \sqrt{\frac{9.50}{10.51}} = 0.951 \leq 5.0$$

The FRP I-shaped column (10 × 5 × 1/2") is selected for this design.

c. Determine the allowable material strength:

For material strength from Eq. (10.6):

$$P_L = 0.3F_L^c A_g = 0.3(30,000)(9.50) = 85,500\,\text{lbs} = 85.5\,\text{kips}$$

d. Determine the allowable buckling strength:

(δ_0/L) initial out of straightness fraction is 0.05 in/ft (manufacturers).

From Eqs. (10.23 and 10.24):

$$\left(P_{cr}\right)_{Tallow} = \varphi_0 \frac{\pi^2 E_L}{\left(\frac{KL_{eT}}{r_T}\right)^2} A_g$$

where φ_0, as per Eq. (10.24) is:

$$\varphi_0 = 1 - 125\frac{\delta_0}{L} = 1 - 125(0.05) = 0.48$$

$$\left(P_{cr}\right)_{T,allow} = 0.48\left(\frac{\pi^2\left(2.6\times10^6\right)}{\left(\frac{10\times12}{3.88}\right)^2}\right)\times(9.50) = 122,331\,\text{lbs}$$

$$(P_{cr})_{Zallow} = \varphi_0 \frac{\pi^2 E_L}{\left(\dfrac{KL_{eZ}}{r_Z}\right)^2} A_g$$

$$(P_{cr})_{Z,\,allow} = 0.48 \left(\frac{\pi^2\left(2.6\times10^6\right)}{\left(\dfrac{5\times12}{1.05}\right)^2}\right) \times (9.50) = 35,835\,\text{lbs}$$

The minimum allowable material strength P_L of all elements comprising the cross section must be higher than the allowable strength under buckling and the design strength under working load.

From (c) and (d):

$$P_s \le \varphi_0 \frac{\pi^2 E_L}{\left(\dfrac{KL_e}{r}\right)^2} A_g \le 0.3 F_L^c A_g$$

$$P_s = 10\,\text{kips} \le (P_{cr})_{allow} = 35.8\,\text{kips} \le 0.3 F_L^c A_g = 85.5\,\text{kips}$$

The FRP I-shaped column under the axial compression load (10 kips) satisfies the allowable material and buckling strength limits.

e. Check for the allowable global torsional buckling strength:

Polar moment of inertia $I_p = 143.3 + 10.51 = 153.8$ in^4, $k_\omega = 1.0$

The global torsional buckling strength is determined from Eq. (10.16.2) as:

$$C_\omega = \frac{I_z d^2}{4} = \frac{10.51\left(10^2\right)}{4} = 262.8\,\text{in}^6$$

Weak axis:

$$\left(f_{cr}^{GTB}\right)_Z = \frac{1}{I_p}\left(CG_{LT} + \frac{\pi^2 E_L C_\omega}{(k_\omega L)^2}\right)$$

$$\left(f_{cr}^{GTB}\right)_Z = \frac{1}{153.8}\left((0.792\times0.425\times10^6) + \frac{\pi^2\left(2.6\times262.8\times10^6\right)}{(12\times5)^2}\right) = 14,368\,\text{psi}$$

$$\left(f_{cr}^{GTB}\right)_{Zallow} = \frac{\left(f_{cr}^{GTB}\right)_Z}{F.S.} = \frac{14,368}{3} = 4,789\,\text{psi}$$

Strong axis:

$$\left(f_{cr}^{GTB}\right)_{Zallow} = \frac{\left(f_{cr}^{GTB}\right)_Z}{F.S.} = \frac{14,368}{3} = 4,789\,\text{psi}$$

$$\left(f_{cr}^{GTB}\right)_{T} = \frac{1}{153.8}\left(\left(0.792\times0.425\times10^{6}\right)+\frac{\pi^{2}\left(2.6\times262.8\times10^{6}\right)}{\left(12\times10\right)^{2}}\right)=5,234\,\text{psi}$$

$$\left(f_{cr}^{GTB}\right)_{Tallow} = \frac{\left(f_{cr}^{GTB}\right)_{T}}{F.S.}=\frac{5,234}{3}=1,745\,\text{psi}$$

The allowable global torsional buckling strength (1,745 psi) is higher than the design compressive strength (1,053 psi) for the column.

f. Check for the allowable local buckling strength:

Based on the ASD method, the F.S. for compression members is taken as 3.

f.1) Determine the allowable local flange buckling from Eq. (10.17.3):

$$f_{crf} = \frac{G_{LT}}{\left(\dfrac{b_{f}}{2t_{f}}\right)^{2}}=\frac{\left(0.425\times10^{6}\right)}{\left(\dfrac{10}{2\times0.5}\right)^{2}}=4,250\,\text{psi}$$

$$\left(f_{crf}\right)_{allow}=\frac{F_{crf}}{F.S.}=\frac{4,250}{3}=1,416\,\text{psi}$$

f.2) Determine the allowable web flange buckling from Eq. (10.17.4):

$$f_{crw} = \frac{\pi^{2}t_{w}^{2}}{6h^{2}}\left(\sqrt{E_{L,w}E_{T,w}}+E_{T,w}v_{LT}+2G_{LT}\right)$$

$$f_{crw} = \frac{\pi^{2}\left(0.5\right)^{2}}{6\left(10^{2}\right)}\left(\sqrt{2.6\times0.8\times10^{12}}+\left(0.33\times0.8\times10^{6}\right)+\left(2\times0.425\times10^{6}\right)\right)$$

$$f_{crw} = 10,512\,\text{psi}$$

$$\left(f_{crw}\right)_{allow}=\frac{f_{crw}}{F.S.}=\frac{10,512}{3}=3,504\,\text{psi}$$

The minimum allowable local buckling strength (1,416 psi) is higher than the design compressive strength (1,053 psi) for the column. The pultruded column (10 × 5 × 1/2" I-shaped section) based on ASD satisfies both the strength and serviceability limit states.

EXERCISES

Problem 10.1: A cooling tower, 8′ high column (8″ × 8″ × ½″) made of E-glass fiber and polyester was loaded with 10 kips of axial dead load and 8 kips of live load. Find instantaneous axial shortening of the column and also its shortening after 15 years.

Problem 10.2: A box section (b & h with wall thickness t) is subjected to concentric axial compression load. Find the buckling load and the critical concentric axial load.

Problem 10.3: Using LRFD method, design a pultruded FRP wide flange 10 ft long column having dead load including column self-weight of 6,000 lbs and live load of 9,000 lbs. Given: Pultruded FRP column properties (manufacturer): Longitudinal compression strength (F_c) = 30,000 psi, Longitudinal modulus (E_L) = 2.6 (10^6) psi, Transverse modulus (E_T) = 0.8 (10^6) psi and Shear modulus (G_{LT}) = 0.425 (10^6) psi.

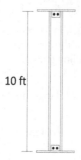

10 ft

PROBLEM FIGURE 10.3   Simply supported FRP column.

Problem 10.4: Determine the nominal strength of a pultruded FRP 10 × 10 × 1/2″ wide flange column that has an unbraced length of 16 and 8 ft with respect to the strong axis and weak axis, respectively as shown in Problem Figure 10.4 (Assuming the pultruded FRP wide flange column, pin-supported at both ends and pin – braced at the mid-height).

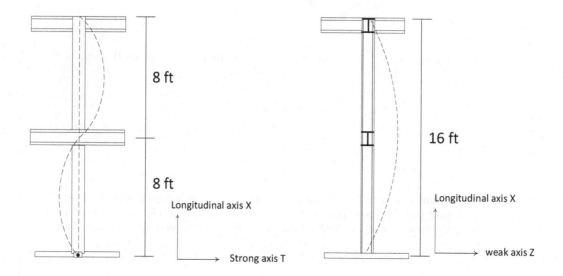

PROBLEM FIGURE 10.4   Pultruded FRP column with unbraced length.

Problem 10.5: Using the LRFD method, design a pultruded FRP column under 3,000 lbs of dead load including column self-weight and 12,000 lbs of live load. Assume: the effective lengths $(KL)_T$ about the strong axis and $(KL)_Z$ about the weak axis are 10 ft long and 5 ft long, respectively.

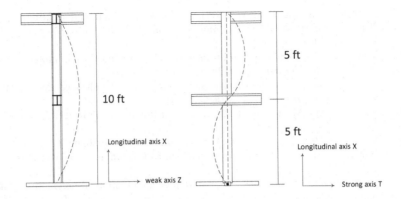

PROBLEM FIGURE 10.5   Pultruded FRP column with different effective length in each axis.

Problem 10.6: A polyester/E-glass $6'' \times 6'' \times 3/8''$ pultruded angle under tension is attached to a gusset plate using two numbers of $3/4''$ diameter stainless steel bolts through one leg. Find the nominal concentric tensile load using the ASD method when the load is acting concentrically. The hole is drilled $1/8''$ larger than the bolt diameter.

REFERENCES AND SELECTED BIOGRAPHY

ACMA, Pre-Standard for Load & Resistance Factor Design (LRFD) of Pultruded Fiber Reinforced Polymer (FRP) Structures, Arlington, VA, 2021.

ANSI/AISC 360–05, An American National Standard, Specification for Structural Steel Buildings, AISC, Chicago, IL, 2005.

ASCE, Structural Plastics Design Manual, Alexander Bell Drive, Reston, VA, 1984.

ASCE/SEI7–10, Minimum Design Loads for Buildings and Other Structures, Alexander Bell Drive, Reston, VA, 2010.

Bank, L.C., *Composites for Construction: Structural Design with FRP Materials*, John Wiley & Sons, Inc., Hoboken, NJ, 2006.

Bank, L.C., Gentry, T.R., and Nadipelli, M., Local buckling of pultruded FRP beams: analysis and design, *Journal of Reinforced Plastics and Composites*, 15(3), 1996, pp. 283–294.

Barbero, E.J., *Introduction to Composite Materials Design*, 2nd edition, CRC Press, Boca Raton, FL, 2011.

Bedford Reinforced Plastics, *Design Guide*, Bedford Reinforced Plastics, Inc., Bedford, PA, 2012.

Boresi, A.P., and Schmidt, R.J., *Advanced Mechanics of Materials*, 6th edition, John Wiley and Sons, Inc., New York, NY, 2003.

Clarke, J.L., *Structural Design of Polymeric Composites – EUROCOMP Design Code and Handbook*, E & FN Spon, London, UK, 1996.

Creative Pultrutions, *The Pultex Pultrusion Design Manual of Standard and Custom Fiber Reinforced Structural Profiles*, 5, revision (2), Creative Pultrutions Inc., Alum Bank, PA, 2004.

Fiberline, *Design Manual*, Fiberline, Kolding, Denmark, 2003.

GangaRao, H.V.S., and Blanford, M.M., Critical buckling strength prediction of pultruded glass fiber reinforced polymeric composite columns, *Journal of Composite Materials*, 48(29), 2014, pp. 3685–3702.

Johnston, B.G., The Structural Stability Research Council, *The Guide to Stability Design Criteria for Metal Structures*, 3rd edition, 1976, p. 616.

Kollar, L.P., and Pluzsik, A., Analysis of thin walled composite beams with arbitrary layup, *Journal of Reinforced Plastics and Composites*, 21, 2002, pp. 1423–1465.

Kollar, L.P., and Springer, G.S., *Mechanics of Composite Structures*, Cambridge University Press, New York, NY, 2003.

Mottram, J.T., Brown, N.D., and Anderson, D., Physical testing for concentrically loaded columns of pultruded glass fibre reinforced plastic profile, *Structures and Buildings*, 156(2), 2003, pp. 205–219.

Nagaraj, V., and GangaRao, H.V.S., Static behavior of pultruded GFRP beams, *Journal of Composites for Construction*, 1(3), 1997, pp. 120–129.

Prachasaree, W., Limkatanyu, S., Kaewjuea, W., and GangaRao, H.V.S., Simplified buckling-strength determination of pultruded FRP structural beams, *Practice Periodical on Structural Design and Construction, ASCE*, 24(2), 2019, p. 04018036.

Roberts, T.M., Influence of shear deformation on buckling of pultruded fiber reinforced plastic profiles, *Journal of Composites for Construction*, 6(2), 2002, pp. 241–248.

Roberts, T.M., and Al-Ubaidi, H., Flexural and torsional properties of pultruded fiber reinforced plastic I-profiles, *Journal of Composites for Construction*, 6(1), 2002, pp. 28–34.

Strongwell Corporation, Design Manual: EXTERN and Other Proprietary Pultruded Products, Bristol, VA, 2010.

Timoshenko, S., *Strength of Materials, Part II*, D. Van Nostrand Co, Inc., 1956, pp. 172–173.

Tomblin, J., and Barbero, E., Local buckling experiments on FRP columns, *Thin-Walled Structures*, 18(2), 1994, pp. 97–116.

Zureick, A., and Scott, D., Short-term behavior and design of fiber-reinforced polymeric slender members under axial compression, *Journal of Composites for Construction, ASCE*, 1(4), 1997, pp. 140–149.

Zureick, A., and Steffen, R., Behavior and design of concentrically loaded pultruded angle struts, *Journal of Structural Engineering, ASCE*, 126(3), 2000, pp. 406–416.

11 Design of Connections for FRP Members

Connections, typically, are the Achilles heels in any structural system assembled with structural members. Pultruded FRP structural members are assembled with bolted connections and/or adhesive bonding. In the design of any structural system, the configuration of structural construction systems must be selected before the design of a structural component. A connection must be designed to transfer forces from one member to another through a set of flat plates, angles, channels, or other forms of connectors, which connect different members in the same plane or even in orthogonal or other planes. To prevent catastrophic structural failures, a good design dictates that a member failure must precede the connection failure.

11.1 CONNECTIONS

The behavior of a connection is complex since several forces through different components come together at and near a connecting mechanism. The frequent causes of connection failure are: (1) the members may warp at a connection and deform the plane sections due to noncollinearity of forces, (2) uneven load distribution through a connection, resulting in bending and torsional moment transfer through a connecting mechanism, (3) local restraint preventing deformation necessary for adequate stress distribution, thus resulting in stress risers, and (4) connection indeterminacy inducing excess (beyond design values) stresses and deformations of the material and fasteners, resulting in stress concentrations and complex stress distribution. Thus, the proper representation of theoretical equations relating to precise connection responses is difficult to simulate even for isotropic members, and especially more difficult with FRP composites due to their anisotropic behavior.

11.2 SCOPE

In this chapter, a connection design methodology is presented based on the design provisions proposed in the pre-standard (ACMA, 2021). In addition, EUROCOMP (1996) and CROW-CUR Recommendations (96, 2019) are suggested as guides for additional design considerations. The fasteners of FRP members can connect one member to another (Figure 11.1), including connecting elements such as gusset plates. The bolt connectors in any design are typically made of steel. The steel connectors (bolts) are designed as per ANSI/AISC specifications. FRP bolt design is not discussed in this chapter because of uncertain structural data on the resistance of FRP bolt threads. For connections of FRP members with steel bolts, for example, FRP may be slipping in a hole drilled to receive connector bolts and in the direction of the applied loads until the FRP bears against the bolt, which is known as a bearing type connection. Bearing-type connections (bearing failure mode) are preferred for pultruded FRP members due to certain complexities associated with other failure modes, such as shear-out, cleavage, or tension failures (see Figure 11.2 for failure modes). The following details and information do not apply to bolted connections with more than three bolts in a row, i.e., parallel to the direction of the load and/or with three or more bolts in a single line with the main load acting perpendicular to this line of bolting at a connection. A brief discussion and design guidance are provided for connections with adhesive bonding. Riveted connections are not discussed herein due to their infrequent use in FRP structural systems.

DOI: 10.1201/9781003196754-11

11.3 CONNECTION

11.3.1 MECHANICAL CONNECTIONS

The mechanical fasteners (steel bolts) are typically used to connect pultruded FRP structural components as shown in Figure 11.1. In general, stress concentration is induced around a bolted location and also around a joint due to abrupt stiffness variations. The stress concentration is accentuated further due to reduction in material cross section (holes created for structural members). The potential action of concentrated force acting around the holes further accentuates the stress concentration-related responses.

The joint strength is affected by the connections (or holes) in each direction of different structural members connecting at a joint. When mechanical connections of FRP structural members are under several loads and moments, the failure modes of mechanical fasteners are typically given in one of the several failure modes (Figure 11.2).

- **Bearing failure:** This type of failure is local, in the region of (bolt) force transfer. The FRP structural members crush locally, in the applied load direction but do not cause catastrophic failure (Figure 11.2a).
- **Net–tension failure:** when a specimen is narrow in comparison with the bolt diameter, then the failure may be attributed to crack propagating transverse to applied load direction (Figure 11.2b).
- **Cleavage failure:** This type of failure mode is known as the shear-out failure that can also be sometimes characterized by a single plane "cleavage" failure, where the apparent laminate transverse tensile strength is less than the corresponding in-plane shear strength (Figure 11.2c).
- **Shear-out failure:** This type of failure is predominant in the fiber direction of unidirectional laminates. The section of structural members parallel to the applied load direction pushes past the remaining structural members (Figure 11.2d), meeting at a joint.
- **Combined failure:** This type of failure is a combination of two different failure modes, i.e., net-tension and shear-out. The failure mode is usually found on the edge near a corner, transitioning from shear to tension (Figure 11.2e).

Note: Failure modes of mechanically fastened connections of FRP members are different from the failure modes of conventional structural members such as steel due to mismatch of strength and stiffness of FRP versus steel and for other reasons.

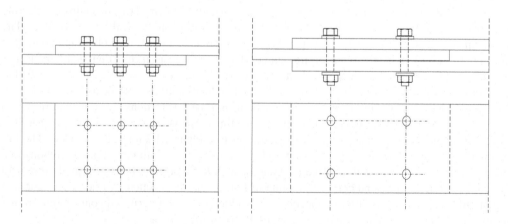

FIGURE 11.1 Bolted connections.

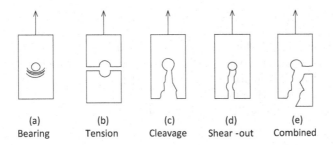

(a)   (b)   (c)   (d)   (e)
Bearing   Tension   Cleavage   Shear -out   Combined

FIGURE 11.2 Failure modes of mechanical fasteners: (a) bearing, (b) tension, (c) cleavage, (d) shear-out, and (e) combined.

The differences in failure modes are due to several reasons. These are (1) higher notch sensitivity of FRP structural members when compared to conventional steel structural members, resulting in higher stress concentration around the perimeter of a hole, (2) higher brittleness of FRP structural members which results in higher stress concentration around the perimeter of a hole, and (3) influence of parameters such as fiber volume fraction, resin properties, voids, stacking sequence, etc. Herein, notch sensitivity, C, is defined by Hart-Smith (1978) as the ratio of effective stress concentration factor, k_{tc} at the ultimate failure of the composite minus 1.0 divided by the stress concentration factor of an isotropic material, k_{te} minus 1.0, or $C = (k_{tc}-1)/(k_{te}-1)$ where k_{te} is approximately equal to member width divided by the product of the hole diameter and number of bolts in a row along the load direction k_{tc} can be found from the literature as studied by Hassan (1994). In 1994, Hassan studied the C-factor for different composite joints with different fiber architectures and arrived at specific numbers for C-factors for multi-bolt connections, not single bolts.

The adjusted C-factors for 0° fibers (in load direction) composites are (Hassan, 1994; Mosallam, 2011):

1. $C = 0.22$ for two bolts in fiber direction;
2. $C = 0.4$ for two bolts in a row (perpendicular to fiber direction);
3. $C = 0.16$ for three bolts in fiber direction;
4. $C = 0.50$ for three bolts in a row (perpendicular to fiber direction); and
5. $C = 0.30$ for four bolts (two bolts in a row for two rows).

In general, connections of pultruded FRP structural parts are designed and constructed as simple framing joints that are unrestrained and pin-connected. The beam reactions are transferred to the supports (columns and/or beams) through the beam web. A pair of angles (L-shape) is the most common type of force transfer for FRP connections. A detail of a bolted framing connection with a pair of angles is shown in Figure 11.3.

Sometimes, the top or bottom flange or both the top and bottom flanges must be shaped when the pultruded FRP beams are at the same depth. For example, the connection of a structural part with a

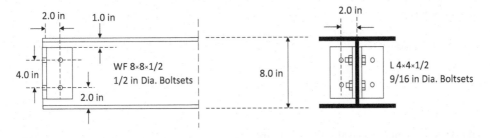

FIGURE 11.3 Typical bolted framing connection with a pair of angles.

cut back of the FRP flanges is illustrated in Figure 11.4. However, this type of connection should be avoided if possible because it will reduce the beam capacity and potentially initiate the web failure, with shear-out or bearing type failure. The type of failure depends upon the web fabric architecture.

The most common connection types, (beam-beam, beam-column, and beam rested on a column, etc.) as in Figure 11.5 are customarily used in framed structures with FRP WF and I shape beams. Most connection details of FRP framed structures provided by the manufacturers are specifically designed to suit a certain type of FRP structural system.

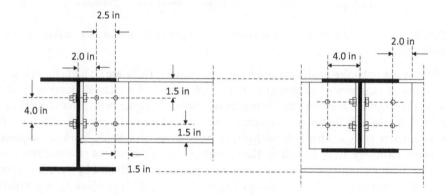

FIGURE 11.4 Typical bolted framing connection with a pair of angles and shaped flange.

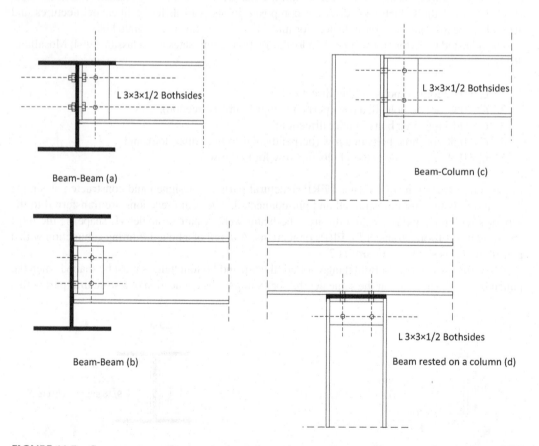

FIGURE 11.5 Common connection types as beam-beam, beam-column, and beam rested on a column (BRP Inc., 2012).

11.3.2 Adhesive (Bonded) Connections

Bonded connections with adhesives (as shown in Figure 11.6) are efficient because of no reductions due to stress concentrations from holes or fiber breakages around holes drilled to accommodate fasteners as mechanical connections. However, stress concentrations do exist because of a change in stiffness at a joint due to the addition of bonded plates at a joint. In practice, the adhesively bonded connections have been limited because of the absence of (1) long-term behavior of adhesives under varying environmental conditions, (2) accurate analysis methods to evaluate stress and strain distributions around joints and along the length of an adherend, and (3) reliable materials properties of adhesives including field installation difficulties (Da Silva et al., 2011). In addition, failure modes of bonded connections are influenced significantly by defects and voids, thickness variations, environmental effects, and processing variations of adhesives. It is equally important to note that failure of bounded connections may depend on deficiencies in surface preparation.

Three different failure modes of bonded connections are:

- **Tensile rupture of structural members:** when a bonded connection with adhesive is very strong and does not fail in the adhesive, then the parts of structural members at fully sustained tension loads (outside the joint locations) will be ruptured as shown in Figure 11.7a due to stress concentration effects at corners, with a sudden change in stiffness at/or near the corner locations.
- **Interface shear rupture:** this failure mode can occur in both tensile-shear (in-plane) and torsion-shear (out-of-plane) when the bonded connector resistance is limited by the shear strength of the adhesive, which is illustrated in Figure 11.7b.
- **Peel (or cleavage) failure:** It is the weakest failure mode related to adhesives. It is possible to find delamination of fibrous composite adherends as shown in Figure 11.7b (Hart-Smith, 1987). It should be noted that the cleavage failure is often found on thick sections, while the peel failure is found when thin sections are connected.

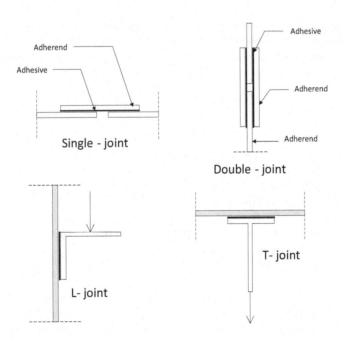

FIGURE 11.6 Common adhesively bonded joints.

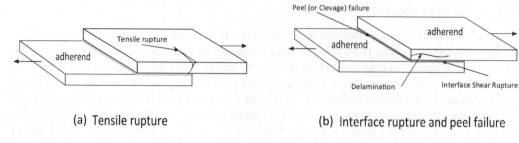

(a) Tensile rupture   (b) Interface rupture and peel failure

FIGURE 11.7   Failure modes of adhesively bonded joints: (a) tensile rupture and (b) interface rupture and peel failure.

11.4   DESIGN METHODOLOGY

As before (Chapters 7–10), both the allowable stress and load-resistance factor design methods are highlighted for connection designs as we did for member designs. The design variations herein are with reference to failure modes, resistance factors, stress concentration, durability considerations, and strain distribution complexities around a connector.

In the allowable stress design (ASD), the design allowable strength of bolts is determined using the minimum nominal strength with safety factor. The design of connections must be satisfied by making sure that:

$$R_a < R_u/\Omega \tag{11.1}$$

where

R_a = required strength
R_u = ultimate strength
Ω = factor of safety.

In the LRFD method, the design strength of bolts is determined based on the strength of its component with adjustment factors for end use conditions. The minimum nominal strength is considered from all possible critical failure paths. The design of connections must be ensured by sufficient strengths (Section 11.7) as:

$$R_u \leq \lambda\varphi R_n C_\Delta C_M C_T \tag{11.2}$$

where

R_u = ultimate connection strength due to factored loads, i. e, end of service life of a structure
R_n = nominal connection strength
φ = either the resistance factor (φ_b) for steel bolt or resistance factor (φ_c) for FRP connections
    with strength expression (Table 11.1)
λ = time effect factor (Table 7.2)
C_Δ = geometry factor which takes into account the connection geometry
C_M = moisture condition factor (Table 11.3)
C_T = temperature condition factor (Table 11.3).

For bolted connections, the reduction factors and partial safety factors according to the pre-standard (ACMA, 2021) and EUROCOMP (1996), respectively, are summarized in Tables 11.1 and 11.2. These reduction factors are used and referred to as different strength modes of bolts and bolted connections.

TABLE 11.1

Reduction Factors for Different Strength Modes of Bolted Connections (ACMA, 2021)

Strength Mode		Bolted Connections	
		Two or Three Rows	
Tension of bolts	φ for steel bolt	0.75	0.75
Shear of bolts			
Tension and shear of bolts			
Tension (through the thickness) R_{tt}	φ for FRP connections	0.50	0.45
Perpendicular pin bearing R_{br}		0.80*	0.45
Net tension R_{nt}		0.45	0.50, 0°–5°
			0.45, 5°–90°
Shear-out between rows R_{sh}		0.50	0.45
Cleavage for a single row R_{cl}		0.50	–
Block shear for multi-rows R_{bs}		–	0.45
Shear for clip angle $R_{sh,sp}$	φ for FRP connections	0.70	0.75

Note: The reduction factors will be refined as additional data is made available from laboratory testing or field evaluations.

* Not provided by ACMA (2021).

TABLE 11.2

γ_m Partial Safety Factors for Materials (EUROCOMP, 1996)

Factor	Type		Value
$\gamma_{m,1}$	Experimental test		1.15
Laminate properties	Laminate theory		1.5
$\gamma_{m,2}$	Fully cured		1.1
Manufacturing process (pultrusion)	Non-fully cured		1.7
$\gamma_{m,2}$	Fully cured		1.4
Manufacturing process (hand lay-ups)	Non-fully cured		2.0
$\gamma_{m,3}$	Short-term loading	$0 < T < 25°C$ and $T_g > 80°C$	1.0
Environment and loading duration		$25 < T < 50°C$ and $T_g > 90°C$	
	Long-term loading	$0 < T < 25°C$ and $T_g > 80°C$	2.5
		$25 < T < 50°C$ and $T_g > 90°C$	

Note: T_g = glass transition temperature of pultruded FRP materials. The material partial safety factors in Table 11.2 are in the denominator for Eq. (7.4.1) in Section 7.2.6. The product of the material partial factors ($\gamma_{m,1}\gamma_{m,2}\gamma_{m,3}$) is the reciprocal of the reduction factor (φ).

11.4.1 Geometry Factor C_Δ

The connection geometry factor (pitch, edge distance, and bolt spacing) C_Δ is used to account for the effect of the connection geometry. The geometry factor (C_Δ) according to the pre-standard (ACMA, 2021) is equal to 1, unless the pitch spacing, s, is less than specified in Table 11.5, then $C_\Delta = s/s_{min}$.

TABLE 11.3

Moisture Condition (C_M) and Temperature (C_T) factor (ACMA, 2021)

Materials	Properties	Moisture C_M	Temperature C_T
Vinyl ester	Strength	0.75	$1.7-0.008T$
	Modulus	0.90	$1.5-0.006T$
Polyester	Strength	0.75	$1.9-0.010T$
	Modulus	0.90	$1.7-0.008T$

Note: Factors given in Tables 11.1 through 11.3 may vary from one code to another and may even be modified with additional field data; so a designer is alerted to use appropriate factors depending on the type of design to be adopted for a structure or depending upon the geographical location of a structure.

11.4.2 MOISTURE CONDITION C_M AND TEMPERATURE C_T FACTOR

The moisture condition and temperature factors are taken into account for the environmental effects. The moisture condition factor is used for sustained in-service moisture and the temperature factor is considered for a sustained in-service temperature higher than 90°F but less than T_g −40°F. These moisture conditions and temperature factors are based on the applicable adjustment factors for the reference strength of a structural member as given in the pre-standard (ACMA, 2021). Both moisture conditions and temperature factors are given in Table 11.3.

11.5 HIGH-STRENGTH BOLTS

Typically, bolts are simple pins inserted in drilled holes of members in a structural system which would prevent the movement of connected members in the perpendicular direction to the direction of the bolt length. A bolt assembly consists of a bolt and nut with a potential need for a washer. For high-strength bolts, heavy hexagonal nuts are widely used for structural bolting. Bolts, nuts, and washers for pultruded FRP members meet the specification in accordance with ASTM standards as given below:

- **ASTM A307:** Standard Specification for Carbon Steel Bolts, Studs, and Threaded Rod 60 ksi Tensile Strength (ASTM A307, 2014)
- **ASTM A325:** Standard Specification for Structural Bolts, Steel, Heat Treated, 120/105 ksi Minimum Tensile Strength (ASTM A325, 2014)
- **ASTM F593:** Standard Specification for Stainless Steel Bolts, Hex Cap Screws, and Studs (ASTM F593, 2013)
- **ASTM A563:** Standard Specification for Carbon and Alloy Steel Nuts (ASTM A563, 2015)
- **ASTM F594:** Standard Specification for Stainless Steel Nuts (ASTM F594, 2015)
- **ASTM A436:** Standard Specification for Austenitic Gray Iron Castings (ASTM, 2015)
- **ASTM F844:** Standard Specification for Washers, Steel, Plain (Flat), Unhardened for General Use (ASTM F844, 2013).

Diameters of bolts range from 3/8″ (9.53 mm) to 1 inch (25.4″). In terms of determining bolt length, the end of a bolt must be extended beyond or at least in flush with the outer face of the nut when properly installed. The length of a bolt shank with thread, which is in bearing with FRP materials, should not exceed one-third of the thickness of the plate component (ACMA, 2021).

TABLE 11.4
Nominal Strength of Bolts

	Nominal Strength Per Unit Area F_n (ksi)		
Load Descriptions	**ASTM A307**	**ASTM A325**	**ASTM F593**
Tension F_{nt} (static)	60	90	60
Shear F_{nv} (threads excluded from shear plane)	48	68	48

Note: For shear F_{nv} of bolts ASTM A325, threads are excluded from the shear plane. The nominal strength of bolts ASTM F593 is the lowest strength for stainless steel in Alloy Group 1 (304) and Alloy Group 2 (316).

The mechanical properties of bolts are generally determined by applying loading to a full-size bolt. The nominal strengths of A307, A325, and F593 high-strength bolts used for the connections of pultruded FRP members are given in Table 11.4.

11.6 BOLT SPACING AND EDGE DISTANCES

The following connections do not allow more than three rows of three bolts per row. As shown in Figure 11.8, the pitch spacing (s) is the distance from the center of a bolt hole to an adjacent bolt hole in the same bolt row and in the connector load direction. The bolt gage (g or g_s) is the distance between two consecutive bolt holes perpendicular to the direction of connection loads.

- The staggered distance (l_s) is the distance between two consecutive bolt holes on adjacent gage lines.
- The end distance (e_1) is defined as the distance from the center of a bolt hole to the adjacent edge in the direction in which the bolt bears. The end distance must be sufficient to ensure adequate bearing capacity.
- The edge distance (e_2) is defined as the distance from the center of a bolt hole to the adjacent edge at right angles to the loading direction.

Typically, the pitch spacing (s) between centers of bolts is not less than 4.0 times the nominal diameter (d) of the bolt to ensure the full effectiveness of bolts. The maximum pitch spacing between bolt holes is 12 times the maximum thickness of the pultruded FRP components bolted together (ACMA, 2021). In addition, the minimum requirement for bolt spacings is summarized in Table 11.5. These bolt spacings vary from one FRP specification guide to another. The spacings may have to be revised after collecting additional experimental data on mechanical joints of FRP shapes under static loads.

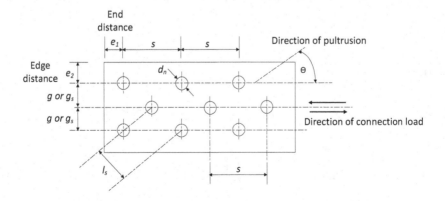

FIGURE 11.8 Bolt spacing and edge distances.

TABLE 11.5

Minimum Requirements for Bolt Spacing (Refer to Figure 11.8, ACMA, 2021)

Spacing	Case		Minimum Required Spacing (Nominal Diameter of Bolt d)
S	All		$4d$
g_s	All		$2d$
G	All		$4d$
e_1	Single row of bolts	Under tension	$4d$
		Under compression	$2d$
	Two or three bolt rows	Under tension	$2d$
		Under compression	$2d$
e_2	All		$1.5d$
l_s	All		$2.8d$

Note: Minimum e_1 may be reduced to $2d$ when the connected member has a perpendicular element attached to the end that the connection load is acting towards. When minimum spacing of "s" cannot meet the requirement, the connection strength will be reduced according to the geometry factor (C_Δ) defined in Section 11.4.1.

11.7 NOMINAL STRENGTH OF BOLTED CONNECTIONS

In general, the aspects of shear, bearing, and tension according to both the connected materials and fasteners need to be considered to determine the strength of a connection. For bolt connections, when the bolt is loaded in shear, the bolt tends to shear off along the contact plane of the connected materials as shown in Figure 11.9. Sometimes, the bolt in a connection is adequate to transfer the load in shear from one member to another, but the connection may still fail unless the connected materials are able to transfer the (shear) load into the bolts. The bolts are assumed to be acting under uniform bearing pressure around the perimeter of bolt holes as shown in Figure 11.10, i.e., no shear lag in the bearing plane. The bearing area is also assumed to be a rectangular area, which may be debatable, but a good design practice.

The stress of a bolt in single shear is:

$$f_{sh} = \frac{P}{A_b} \tag{11.3}$$

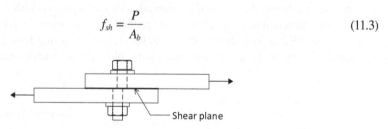

FIGURE 11.9   Bolt in single shear.

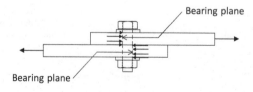

FIGURE 11.10   Bolt in bearing.

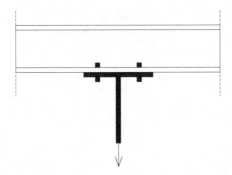

FIGURE 11.11   Bolts in pure tension (hanger type).

The stress of a bolt in bearing is:

$$f_{sh} = \frac{P}{d_n t} \qquad (11.4)$$

where
  P = load transmitted by a bolt
  A_b = nominal cross-sectional area of a bolt
  d_n = nominal diameter of a bolt
  t = thickness of the connected part.

The bolts in a connection may be loaded in pure tension (a hanger type connection as shown in Figure 11.11). The bolt stress in tension is determined as:

$$f_t = \frac{P_t}{A_b} \qquad (11.5)$$

where
  P_t = tension load transmitted by a bolt
  A_b = nominal cross-sectional area of a bolt.

11.7.1   Nominal Strength of Single Row Bolted Connections

For FRP materials, the nominal connection strength is taken as the minimum of six possible critical failure strengths as given below:

1. Bolt shear strength (R_{bt}) in Table 11.6
2. Tension (through the thickness) strength (R_{tt}) in Table 11.6
3. Pin-bearing strength (R_{br}) in Table 11.6
4. Net tension strength (R_{nt}) in Table 11.7
5. Shear-out strength (R_{sh}) in Table 11.6
6. Cleavage strength (R_{cl}) in Table 11.6.

It should be noted that the semi-empirical pin-bearing strength (R_{br}), net tension strength (R_{nt}), shear-out strength (R_{sh}), and cleavage strength (R_{cl}) as given in Table 11.6 are for connections in double-lap shear (Figure 11.12) when a connection force component is in the plane of the connection to be designed as per this approach.

Note: When the connection is for single-lap shear, the strength obtained from Table 11.6 must be reduced by a multiplier of 0.6 (ACMA, 2021).

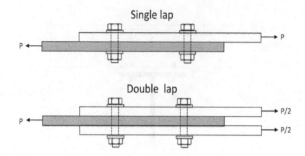

FIGURE 11.12 Single and double-lap connections.

- When there is a perpendicular element (such as the flanges of I-shaped, box profiles, and channels) of FRP at the end of the pultruded member or connecting component of FRP material then, the connection strengths under shear-out (R_{sh}) and cleavage (R_{cl}) need not be determined (ACMA, 2021).
- When there are two perpendicular elements (such as the flanges of I-shaped, box profiles, and channels) having their planes aligned with the connection force then, the connection strengths of net tension strength (R_{nt}) need not be determined (ACMA, 2021).
- When there is only one perpendicular element (such as leg angle) then the net tension strength is determined from equations as given in Tables 11.7 and 11.9 (ACMA, 2021).

All possible nominal strengths excluding pin bearing strength (R_{br}) are applied to connections with a single row of bolts and constant thickness of the FRP member or connecting component. Connection strength is computed based on the assumption that each bolt in a single row of bolts (not more than three bolts) is equally loaded. For single-row bolt connections, the connection aligned in the bolt axis (a row of bolts) is perpendicular to the connection force.

TABLE 11.6
Nominal Strength of Single Bolted Connection (ACMA, 2021)

Strength	Equations Based on the Pre-Standard (ACMA, 2021)		Requirement
Tension of bolts	$R_{bt} = F_{nt}A_b$	(11.6)	
Shear of bolts	$R_{bt} = F_{nv}A_b$	(11.7)	
Combined tension and shear of bolts	$R_{bt} = F_{nt}^t A_b$	(11.8)	
	where, $\quad F_{nt}^t = 1.3F_{nt} - \dfrac{F_{nt}}{\varphi_b F_{nv}} f_v \le F_{nt}$	(11.9)	
Tension per bolt (through the thickness)	$R_{tt} = \text{Min}\left[0.5\pi d_w t F_{sh,tt}, 0.4\pi d_w t F_{sh,int} \right]$	(11.10)	
Perpendicular pin bearing per bolt	$R_{br} = td\xi F_\theta^{br}$ $R_{br} = 0.5tdF_\theta^{br}$ If, on one of the two sides to a bolted connection, there is no washer and no nut (there will be a washer and nut or bolt head on the other side).	(11.11)	$\xi = 1$ if fastener threads are excluded from bearing surface and 0.6 otherwise $F_\theta^{br} = F_L^{br}$ When θ is $\le 5°$ $F_\theta^{br} = F_T^{br}$ When θ is $> 5°–90°$

(Continued)

TABLE 11.6 (*Continued*)
Nominal Strength of Single Bolted Connection (ACMA, 2021)

Strength	Equations Based on the Pre-Standard (ACMA, 2021)		Requirement
Shear-out per bolt	$R_{sh} = 1.4\left(e_1 - \dfrac{d_n}{2}\right)tF_{sh}$	(11.12)	In a line that is parallel to the direction of the connection force $e_1 < 4d = e_{1,min}$
Cleavage per bolt	$R_{cl} = \text{Min}\begin{pmatrix} 0.15\begin{bmatrix} (2e_2 - d_n)F'_L \\ +2e_1F_{sh} \end{bmatrix}t_c \\ R_{br}\left(\dfrac{10}{9} - \dfrac{4d_n}{9e_1}\right)^2 \end{pmatrix}$	(11.13)	A single bolt, centrally positioned, $e_1 < 4d = e_{1,min}$ Or, $R_{cl} = R_{br}$ for $e_1 \geq 4d$
	$R_{cl} = K_{cl}0.5\begin{bmatrix} (e_2 - 0.5g - d_n)F'_L \\ +2e_1F_{sh} \end{bmatrix}t_c$	(11.14)	A row of bolts (with the maximum number of bolts in the row set to three) at uniform gage spacing g

where F_n = nominal tensile strength F_{nt}, or nominal shear strength F_{nv} of steel bolt is either from Table 11.4 or from ASTM F593 with $F_{nv} = 0.66F_{nt}$ as threads are excluded from the shear plane, A_b = nominal unthreaded body area of bolt, F_{tnt} = nominal tensile strength modified to include the effect of shear stress, f_v = required shear stress, d_w = nominal diameter of the washer, and t = thickness of FRP material resisting through the thickness tension.

$F_{sh,tt}$ = shear strength in the through the thickness plane of the FRP material taken to be the in-plane shear strength, $F_{sh,int}$ = interlaminar shear strength of the FRP material, θ = angle of loading, the orientation between the direction of the connection force and the direction of pultrusion or the principal direction of the FRP material. F^{br}_L and F^{br}_T = pin-bearing strength in the longitudinal and transverse direction of FRP, respectively, t_c = minimum thickness of the connected members, F_{sh} = in-plane shear of FRP material appropriate to the mode of failure, F^t_L = tensile strength of the FRP material in the longitudinal direction.

TABLE 11.7
Net Tension Strength for Single Row Bolted Connection (ACMA, 2021)

Strength	Equations Based on the Pre-Standard (ACMA, 2021)		Requirement
Net tension strength	The connection force is between 0° and 5° to the longitudinal direction of FRP material perpendicular to a single row of bolts		For a single row bolt connection ($n = 1$ and $S_{pr} = w/d$) For $e_1/w \leq 1$: $\Theta = 1.5 - 0.5(w/e_1)$
	$R_{nt} = \dfrac{1}{K_{nt,L}}(w - nd_n)tF'_L$	(11.15.1)	For $e_1/w \geq 1$: $\Theta = 1.0$ For a shape: $C_L = 0.5$ and for plate: $C_L = 0.4$
	$K_{nt,L} = C_L\left(S_{pr} - 1.5\dfrac{(S_{pr} - 1)}{(S_{pr} + 1)}\Theta\right) + 1$	(11.15.2)	For a shape or plate: $C_T = 0.5$ $w = e_3 + e_4$
	The connection force is from 5° to 90° to the longitudinal direction of FRP material perpendicular to a single row of bolts		$e_3 = e_4 = e_2$, for a connection with two side edges having a side distance e_2
	$R_{nt} = \dfrac{1}{K_{nt,T}}(w - nd_n)tF'_T$	(11.16.1)	$e_3 = e_2$, $e_4 = 2e_{2,min}$, for a connection with one side edge having a side distance e_2 and with the other side distance $> 2e_{2,min}$
	$K_{nt,T} = C_T\left(S_{pr} - 1.5\dfrac{(S_{pr} - 1)}{(S_{pr} + 1)}\Theta\right) + 1$	(11.16.2)	$e_3 = e_4 = 2e_{2,min}$, for a connection having its two side edges with side distance $> 2e_{2,min}$

(Continued)

TABLE 11.7 (*Continued*)

Net Tension Strength for Single Row Bolted Connection (ACMA, 2021)

Strength	Equations Based on the Pre-Standard (ACMA, 2021)	Requirement
	The connection force is between 0° and 5° to the longitudinal direction of FRP material perpendicular to a single row of bolts	For a single row of bolts with constant gage spacing across the effective width ($n = 2$ or 3 and $g_{max}/d = 5$, $e_{1,max} = 2e_{1,min}$
	$R_{nt} = \dfrac{1}{K_{nt,L}}(w - nd_n)tF'_L$ (11.17.1)	and $e_{2,max} = 2e_{2,min}$) For a shape: $C_L = 0.5$ and for plate: $C_L =$
	$K_{nt,L} = 0.65C_L$ (11.17.2)	0.4
	The connection force is from 5° to 90° to the longitudinal direction of FRP material perpendicular to a single row of bolts	For a shape or plate: $C_T = 0.5$ $w = e_3 + e_4 + (n-1)g$ $e_3 = e_4 = e_2$, for a connection with two side edges having a side distance e_2
	$R_{nt} = \dfrac{1}{K_{nt,T}}(w - nd_n)tF'_T$ (11.18.1)	$e_3 = e_2$, $e_4 = 2e_{2,min}$, for a connection with one side edge having a side distance e_2
	$K_{nt,T} = 0.65C_T$ (11.18.2)	and with the other side distance $> 2e_{2,min}$ $e_3 = e_4 = 2e_{2,min}$, for a connection having its two side edges with side distance $> 2e_{2,min}$

where t = minimum thickness of the connected members, d = bolt diameter, d_n = nominal hole diameter, n = number of bolts across the effective width, $n = 1$–3, F'_L and F'_T = tensile strength in the longitudinal and transverse direction of FRP material, respectively, and w = effective width.

11.7.2 NOMINAL STRENGTH OF BOLTED CONNECTIONS WITH TWO OR THREE ROWS OF BOLTS

The nominal connection strength of the two- or three-row bolted connections is taken as the minimum of seven possible critical failure strengths as:

1. Bolt tension strength (R_{bt}) in Table 11.6
2. Tension (through the thickness) strength (R_{tt}) in Table 11.6
3. Perpendicular pin-bearing strength (R_{br}) in Table 11.6
4. Net tension strength at first bolt row ($R_{nt,f}$) in Table 11.9
5. Shear-out strength (R_{sh}) in Table 11.10
6. Block shear strength for the concentric load (R_{bs}) in Table 11.10
7. Block shear strength for eccentric in-plane load ($R_{bs,e}$) in Table 11.10.

Block shear failure occurs when one or more segments (or blocks) of FRP plate material tears out from the end of a tension member or a gusset plate. Typically, failure segments are rectangular in shape due to both the shear along the longitudinal axis and tension in the perpendicular direction to the longitudinal axis.

   Example: Block shear segments on the tension member are illustrated as shown in Figure 11.13. Thus, block shear capacity may be defined as the summation of both parts: (1) tension rupture strength provided by the side of the block perpendicular to the load causing tension and (2) shear rupture strength provided by the side of the block parallel to the load causing tension.

Bolt strength (R_{bt}), through the thickness tension strength (R_{tt}) and pin-bearing strength (R_{br}) must be considered for connections with a single row of bolts and having a constant thickness of the FRP member or connecting component. The nominal strength of the multi-row connection is assumed to be attained from one of these three modes.

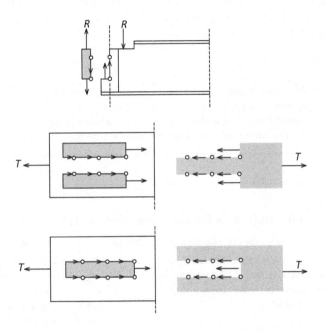

FIGURE 11.13   Block shear failure of web and block shear rupture with failure blocks.

Typically, the nominal strength of the multi-row connection is the summation of equal strength contributions from each of the bolts with the minimum of possible critical failure strengths. Load distribution per bolt row as a proportion of the connection force transmitted through bearing is given in Table 11.8 (ACMA, 2021). Load distribution in Table 11.8 is based on each row having the same number of bolts, up to three bolts. Each bolt in a particular row is assumed to bear equally the portion of the load resisted by that row.

Moreover, when the multi-row connection is for single-lap shear, semi-empirical pin-bearing strength (R_{br}) in Table 11.6, net tension strength (R_{nt}) in Table 11.9, shear-out strength (R_{sh}) in Table 11.10, and cleavage strength (R_{cl}) in Table 11.6 must be reduced by a multiplier of 0.6 (ACMA, 2021).

- When there is a perpendicular element (such as the flanges of I-shaped, box profiles, and channels) of FRP at the end of the pultruded member or connecting component of FRP material then, the connection strengths of shear-out strength (R_{sh}) and cleavage strength (R_{cl}) need not be determined (ACMA, 2021).
- When there are two perpendicular elements (such as the flanges of I-shaped, box profiles and channels) having their planes aligned with the connection force, then the connection strength of net tension strength ($R_{nt, f}$) need not be determined (ACMA, 2021).
- When there is only one perpendicular element (such as leg angle or T-profiles) then the net tension strength is determined from equations as given in Tables 11.8 and 11.9 (ACMA, 2021).

TABLE 11.8

Load Distributions for Multi-Bolted Connections

Materials	No. of Rows (n)	Proportion of Load at		
		First Row L_{br}	Second Row	Third Row
FRP/FRP	2	0.5	0.5	–
FRP/Steel	2	0.6	0.4	–
FRP/FRP	3	0.4	0.2	0.4
FRP/Steel	3	0.5	0.3	0.2

Note: Pin-bearing strength (R_{br}), net tension strength at first bolt row ($R_{nt,f}$), shear-out strength (R_{sh}), block shear strength for concentric load (R_{bs}) and block shear strength for eccentric load ($R_{bs,e}$) are for multi-row bolted connections in double-lap shear when a connection force component is in the plane of the connection.

TABLE 11.9

Net Tension Strength for Multi-Row Bolted Connection (ACMA, 2021)

Strength	Equations Based on the Pre-Standard (ACMA, 2021)		Requirement
Net tension strength at first bolt row	The connection force is between 0° and 5° to the longitudinal direction of FRP material and perpendicular to the bolt rows, with constant pitch spacing. $R_{nt,f} = 0.2wtF'_L$	(11.19)	For a single row bolt connection ($n = 1$ and $S_{pr} = w/d$) $w = e_3 + e_4$ $e_3 = e_4 = e_2$, for a connection with two side edges having a side distance e_2
	The connection force is from 5° to 90° to the longitudinal direction of FRP material and perpendicular to the bolt rows, with constant spacing (s). $R_{nt,f} = 0.2wtF'_T$	(11.20)	$e_3 = e_2$, $e_4 = 2e_{2,min}$, for a connection with one side edge having a side distance e_2 and with the other side distance $> 2e_{2,min}$ $e_3 = e_4 = 2e_{2,min}$, for a connection having its two side edges with side distance $> 2e_{2,min}$
			For rows of bolts with constant gage spacing across the effective width ($n = 2$ or 3 and $S_{pr} = g/d$) $w = e_3 + e_4 + (n-1)g$ $e_3 = e_4 = e_2$, for a connection with two side edges having a side distance e_2 $e_3 = e_2$, $e_4 = 2e_{2,min}$, for a connection with one side edge having a side distance e_2 and with the other side distance $> 2e_{2,min}$ $e_3 = e_4 = 2e_{2,min}$, for a connection having its two side edges with side distance $> 2e_{2,min}$

where t = minimum thickness of the connected members, $n = 1$–3, F'_L and F'_T = tensile strength in the longitudinal and transverse direction of FRP material, respectively, w = effective width, d = bolt hole diameter

TABLE 11.10

Shear Out and Block Shear Strength for Multi-Row Bolted Connection (ACMA, 2021)

Strength	Equations Based on the Pre-Standard (ACMA, 2021)		Requirement
Shear-out between rows of bolts	$R_{sh} = 1.4\left(e_1 - \dfrac{d_n}{2} + s\right)tF_{sh}$	(11.21.1)	For two rows of bolts separated by pitch spacing (s)
	$R_{sh} = 2(n-1)stF_{sh}$	(11.21.2)	For three rows of bolts separated by pitch spacing (s)
Block shear	$R_{bs} = 0.5\left(A_{ns}F_{sh} + A_{nt}F_L^t\right)$	(11.22.1)	The connection force is concentric to the group of bolts, tensile and parallel to the direction of FRP material.
	$R_{bs,e} = 0.5\left(A_{ns}F_{sh} + 0.5A_{nt}F_L^t\right)$	(11.22.2)	For a bolt group subject to eccentric in-plane loading
Shear of clip angle	$R_{sh,sp} = l_{sp}t_{sp}F_{sh}$	(11.23)	Nominal strength at the knee of the clip

where t = minimum thickness of the connected members, d_n = nominal hole diameter, F_L^t = tensile strength in the longitudinal direction of FRP material, F_{sh} = in-plane shear strength of FRP material appropriate to the shear-out failure, w = effective width, L_{br} = proportion of the connection force taken in bearing at the first bolt row, A_{ns} = net area subjected to shear, A_{nt} = net area subjected to tension, l_{sp} = depth of shear plane at the fillet radius of the leg angle profile and t_{sp} = minimum thickness of FRP material.

A_{nt} = the greater of (1) the maximum of the sectional area in any cross section perpendicular to the member axis, or (2) $t\,(n\,d_n - \Sigma b_s)$, n = number of holes extending in any diagonal or zig-zag line progressively across the number or part of the number (n_{max} = 3), b_s = the lesser of ($s^2/4g_s$) or $0.65g_s$.

11.7.3 Nominal Strength of Bolted Connections (EUROCOMP, 1996)

The ultimate bearing, tensile and shear-out strengths of the laminated composites, as presented in EUROCOMP (1996), are given below:

The ultimate bearing strength ($\sigma_{r,k}$) is:

$$\sigma_{r,k} = E_r \varepsilon_{c,ult} \tag{11.24.1}$$

where the radial elastic modulus (E_r) is:

$$\frac{1}{E_r} = \frac{1}{E_L}Cos^4\varphi + \left(\frac{1}{G_{LT}} - \frac{2v_{LT}}{E_L}\right)Cos^2\varphi Sin^2\varphi + \frac{1}{E_T}Sin^4\varphi \tag{11.24.2}$$

where

$\varepsilon_{c,ult}$ = ultimate strain of FRP member in compression
E_L = elastic modulus of the longitudinal direction
E_T = elastic modulus of the transverse direction
G_{LT} = shear modulus.

The ultimate tensile strength ($\sigma_{\varphi,k}$) is:

$$\sigma_{\varphi,k} = E_\varphi \varepsilon_{\varphi,ult} \tag{11.24.3}$$

where the circumferential elastic modulus (E_φ) is:

$$\frac{1}{E_\varphi} = \frac{1}{E_L} Sin^4\varphi + \left(\frac{1}{G_{LT}} - \frac{2v_{LT}}{E_L} \right) Sin^2\varphi Cos^2\varphi + \frac{1}{E_T} Cos^4\varphi \qquad (11.24.4)$$

The ultimate shear-out strength ($\tau_{sn,k}$) is:

$$\tau_{sn,k} = G_{sn}\varepsilon_{sh,ult} \qquad (11.24.5)$$

where the radial shear modulus (G_{sn}) is:

$$\frac{1}{G_{sn}} = \frac{1}{G_{LT}} + \left(\frac{1+v_{LT}}{E_L} - \frac{1+v_{TL}}{E_T} - \frac{1}{G_{LT}} \right) Sin^2 2\varphi \qquad (11.24.6)$$

where

φ = direction in polar coordinate as shown in Figure 11.14

$\varepsilon_{sh,ult}$ = ultimate shear strain of FRP member

v_{LT} = Poisson's ratio of FRP for a plane with a normal along the longitudinal and a parallel along the transverse direction

v_{TL} = Poisson's ratio of FRP for a plane with a normal along the transverse and parallel along the longitudinal direction.

Note: Ultimate strain should be based on experiments carried out on the laminated composites used for testing and evaluation. These values can vary depending on the type of resin used for making the laminated composites, and the data will be provided by the manufacturer.

The ultimate bearing, tensile and shear stresses (σ_r, σ_φ, and τ_{sn}) induced by applied loads must incorporate stress distribution around the bolt hole, effective stress concentration, and notch sensitivity factor. The normalized stress distribution factor (K_r, K_t, and K_s) for bearing, tensile and shear stress (σ_r, σ_φ, and τ_{sn}) can be determined by referring to the stress charts provided (by EUROCOMP Design Code, 1996-Figures 5.13–5.18). Thus, the induced bearing, tensile and shear stresses incorporating those effects ($\sigma_{r,s}$, $\sigma_{\varphi,s}$, and $\tau_{sn,s}$) are defined as:

$$\sigma_{r,s} = K_r\sigma_r \qquad (11.24.7)$$

$$\sigma_{\varphi,s} = K_t\sigma_\varphi \qquad (11.24.8)$$

$$\tau_{ns,s} = K_s\tau_{ns} \qquad (11.24.9)$$

In addition, the normalized stress distribution factors are modified to account for non-uniform stress distribution through the thickness of the laminate due to bolt bending and loss of lateral restraint.

When the induced stresses ($\sigma_{r,s}$, $\sigma_{\varphi,s}$, and $\tau_{sn,s}$) of laminated composites under in-plane loading are greater than the limit stresses (the allowable stresses in ASD methodology and the adjusted stresses ($\sigma_{r,k}$, $\sigma_{\varphi,k}$, and $\tau_{sn,k}$) in the LRFD methodology), the connection design is not safe and must be redesigned to satisfy limit stress and serviceability. If the above check does not satisfy the limit stresses, then the changes of the geometric parameter are recommended in the form of:

1. increase the fastener diameter
2. increase the thickness of the laminated plate
3. increase the number of fasteners within the connecting mechanism
4. combination of the above (1 thru 3).

For bearing failure, the failure needs to be evaluated around the loaded part of the circumference of a hole at a sufficient number of locations. Bearing failure occurs at any one of the selected hole boundary points when the radial compressive stress ($\sigma_{r,\,s}$) reaches the characteristic compressive strength ($\sigma_{r,\,k}$) of the laminate edge (EUROCOMP, 1996) (Figure 11.14b).

In addition, the net-section failure due to net tension needs to be evaluated at a sufficient number of locations around the circumference of a hole. Failure at any of the evaluated points is assumed to have occurred when the tangential stress ($\sigma_{\varphi,\,s}$) at the hole edge reaches the characteristic strength ($\sigma_{\varphi,\,k}$) of the laminate in the direction considered as shown in Figure 11.14a. Both the tensile and compressive net-section failures are to be considered. Hence, the characteristic strengths (in tension and compression) are required to evaluate net-section failure (EUROCOMP, 1996).

Shear-out failure needs to be evaluated along the shear-out planes as shown in Figure 11.14c. It is assumed that failure occurs when the maximum shear stress (τ_{sn}) along the shear-out plane reaches the characteristic shear strength ($\tau_{sn,\,k}$) of the laminated composite (EUROCOMP, 1996).

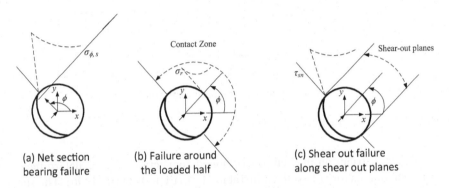

(a) Net section bearing failure   (b) Failure around the loaded half   (c) Shear out failure along shear out planes

FIGURE 11.14 Evaluation of failure around the hole circumference (EUROCOMP, 1996): (a) net section bearing failure, (b) failure around the loaded half, and (c) shear-out failure along shear-out planes.

The compliance check for bearing strength is:
For bearing strength:

$$\varphi_F C_m \sigma_{r,s} \leq \frac{1}{\gamma_m} \sigma_{r,k} A_G \tag{11.25.1}$$

For tensile strength:

$$\varphi_F C_m \sigma_{\varphi,s} \leq \frac{1}{\gamma_m} \sigma_{\varphi,k} A_G \tag{11.25.2}$$

For shear-out strength:

$$\varphi_F C_m \tau_{sn,s} \leq \frac{1}{\gamma_m} \tau_{sn,k} A_G \tag{11.25.3}$$

where

φ_F and γ_m = load factors and reduction factors, respectively
A_G = aging factor.

The factor (C_m) is a product of ($C_{m,1}$ and $C_{m,3}$). C_m accounts for lap joint types ($C_{m,1}$) and the ratio of the center to center distance between holes and the diameter of both holes. The factor ($C_{m,1}$) is taken as 2 and 1.2 for single-lap joints and double-lap joints with symmetric loads, respectively. The factor ($C_{m,3}$) is given in Figure 11.15 (EUROCOMP, 1996).

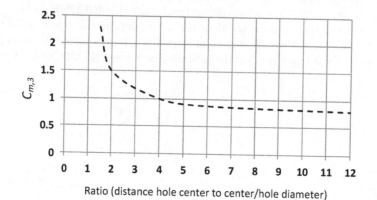

FIGURE 11.15   Factor $C_{m,3}$ (EUROCOMP, 1996).

The shear load distribution in each row of bolted connections can be determined by using the shear distribution factor (μ) as given in Table 11.11 (EUROCOMP, 1996). The shear distribution factor is different for each bolt row. Thus, the distribution of shear load (F_x^{DR}) of each fastener row is obtained from a resultant between the average shear load (F_x^R) and the distribution factor (μ).

$$F_x^{DR} = \mu F_x^R \qquad (11.26.1)$$

To calculate the shear load distribution of a bolt in each row, the total unfactored uniaxial load (F^{xtotal}) or applied unidirectional load is divided by the number of fastener rows (n^R). Then, the average shear load (F_x^R) of each bolt row is assumed to be equal to:

$$F_x^R = \frac{F_x^{Total}}{n^R} \qquad (11.26.2)$$

Substituting Eq. (11.26.1) into (11.26.2), the shear load distribution on each bolt row is given below:

$$F_x^{DR} = \frac{\mu F_x^{Total}}{n^R} \qquad (11.26.3)$$

The average shear load on a bolt (F_x^F) can be calculated as the average shear load (F_x^{DR}) divided by the number of bolts on the same row (n^F).

$$F_x^F = \frac{F_x^{DR}}{n^F} = \frac{\mu F_x^{Total}}{n^R n^F} \qquad (11.26.4)$$

TABLE 11.11
μ – Shear Load Distribution Factor of μ- Bolt Row
(EUROCOMP, 1996)

Numbers of Rows	Materials	Distribution of Shear Load			
		Row 1	Row 2	Row 3	Row 4
1	Glass FRP/glass FRP	1.0	–	–	–
1	Glass FRP/metal	1.0	–	–	–
2	Glass FRP/glass FRP	1.0	1.0	–	–
2	Glass FRP/metal	1.15	0.85	–	–
3	Glass FRP/glass FRP	1.1	0.8	1.1	–
3	Glass FRP/metal	1.5	0.85	0.65	–
4	Glass FRP/glass FRP	1.2	0.8	0.8	1.2
4	Glass FRP/metal	1.7	1.0	0.8	0.6
>4	Not recommended	–	–	–	–

The terms of by-pass load distribution are referred to the situation in which the load coming onto a joint is taken wholly by the composite laminate and there is no contribution from the bolts in resisting this load. Six load cases of bearing and by-pass load distribution are (Figure 11.16):

1. The effective width is equal to $4d$ when there is a bearing load acting on the fastener, resulting in tension.
2. The effective width is equal to $4d$ when there is a bearing load acting on the fastener, resulting in compression.
3. The effective width is greater than $4d$ when there is a by-pass load, i.e., the load is taken by the plate rather than the fastener.
4. The effective width is greater than $4d$ when there is a by-pass load, but it is compressive.
5. Shear load acting on a fastener as shown in Figure 11.16.
6. Shear load acting on a fastener is the same as the load case 5 but in the opposite direction.

Note: d = hole diameter

Resolve the resultant force in each bolt due to the concentric uniaxial load (F_x), into each of the six load cases as mentioned before. The by-pass load distribution is determined from the equilibrium conditions. In addition, the individual bolts are sized to determine the by-pass load distribution. By using the free body diagram and equilibrium condition of each bolt, the fraction of the resultant force (F_x) on each bolt can be determined.

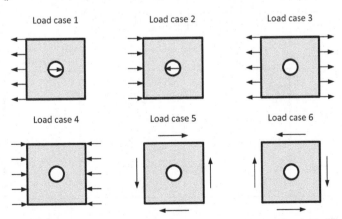

FIGURE 11.16 Load cases for bearing and by –pass load distribution. (Adapted from EUROCOMP, 1996.)

Example 11.1

Using the LRFD method, find the tensile capacity for the connection shown in Example Figure 11.1. Single-lap joint with 7/8″ diameter A325 (bearing-type connection) in standard holes is used to connect pultruded FRP vinyl ester plates (12 × 1/2″).

The time effect factor $\lambda = 0.8$ for $W = 1.2D + 1.6L$, Geometry factor $C_\Delta = 1.0$ for pitch spacing $(s) \geq s_{min}$, Moisture condition factor $C_M = 0.75$ for vinyl ester material and temperature condition factor $C_T = 0.74$ for strength ($T = 120°F$).

Solution

a. Shear per bolt:
   The nominal shear strength of bolts A325 is 68 ksi (Table 11.4). The reduction factor (ϕ_b) for bolt tension is 0.75. The nominal shear strength is determined from Eq. (11.7) as:

$$R_{bt} = F_{nv} A_b = 68(0.601) = 40.9 \text{ kips}$$
$$\lambda \phi_b R_{bt} C_\Delta C_M C_T = (0.8 \times 0.75)40.9(1.0 \times 0.75 \times 0.74) = 13.62 \text{ kips}$$

b. Tension per bolt:
   The nominal tensile strength of bolts A325 is 90 ksi (static tension F_{nt}, Table 11.4). The reduction factor (ϕ_b) for bolt tension is 0.75. The nominal tensile strength is determined from Eq. (11.6) as:

$$R_{bt} = F_{nt}^l A_b = 90(0.601) = 54.1 \text{ kips}$$
$$\lambda \phi_b R_{bt} C_\Delta C_M C_T = (0.8 \times 0.75)54.1(1.0 \times 0.75 \times 0.74) = 18.02 \text{ kips}$$

c. Combined tension and shear per bolt:
   The reduction factor (ϕ_b) for combined tension and shear strengths is 0.75. The nominal tensile strength including shear stress effects is determined from Eqs. (11.8 and 11.9) as:

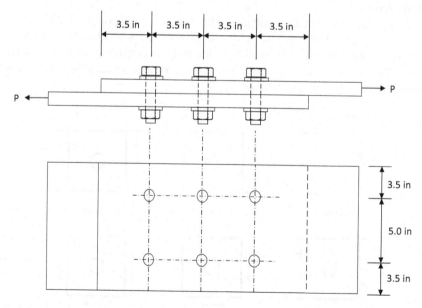

EXAMPLE FIGURE 11.1   Single-lap joint connection.

$$F_{nt}' = 1.3F_{nt} - \frac{F_{nt}}{\varphi_b F_{nv}} f_v \leq F_{nt} = 90 \text{ kips}, \quad \text{where } f_v = \text{required shear stress}$$

$$R_{bt} = F_{nt}' A_b = 90(0.601) = 54.1 \text{ kips}$$

$$\lambda_b R_{bt} C_\Delta C_M C_T = (0.8 \times 0.75)(54.1)(1.0 \times 0.75 \times 0.74) = 18.02 \text{ kips}$$

Note: Tensile strength per bolt (through the thickness) is not considered in this example, because there is no induced tensile strength through the thickness for connections with prying action along with a force component aligned with the bolt axis.

d. Perpendicular pin bearing strength per bolt:

The characteristic pin bearing strength in the longitudinal direction of FRP is 32 ksi (manufacturer). The nominal hole dia. is taken as 1/16" larger than the nominal bolt diameter (7/8"). Fastener threads are excluded, so $\xi = 1.0$.

The orientation between the direction of the connection force and the strong direction of the FRP plates is 0°. The reduction factor (ϕ_c) for the perpendicular pin-bearing strength per bolts is 0.8 (Table 11.1). The perpendicular pin-bearing strength per bolt is determined from Eq. (11.11) as:

$$R_{br} = td\xi F_\theta^{br} = \left(\frac{7}{8} + \frac{1}{16}\right)\left(\frac{1}{2}\right)(1.0)32 = 15 \text{ kips}$$

$$\lambda \varphi_c R_{br} C_\Delta C_M C_T = (0.8 \times 0.8)(15)(1.0 \times 0.75 \times 0.74) = 5.33 \text{ kips}$$

The multi-row connection is for single-lap shear and its strength must be reduced by a multiplier of 0.6 (ACMA, 2021).

$$(0.6)(\lambda \varphi_c R_{br} C_\Delta C_M C_T) = 0.6(5.33) \text{ kips} = 3.20 \text{ kips}$$

e. Net tension strength at first bolt row (refer to Table 11.9):

Effective width (w) is determined as: $w = e_3 + e_4 + (n-1)g = 2(3.5) + (2-1)5 = 12''$, where $e_3 = e_4 = e_2$ for a connection with side edges having a side distance.

Number of bolts across the effective width, $n = 3$, number of bolt rows = 2 with FRP/FRP materials connected.

The characteristic tensile strength in the longitudinal direction (F_L') of FRP is 24 ksi (manufacturer). The reduction factor (φ) for the net tension at the first bolt row is 0.50 (Table 11.1).

From Eq. (11.19), net tension strength is (Table 11.9):

$$R_{nt,f} = 0.2wtF_L' = 0.2(12)(0.5)(24) = 28.8 \text{ kips}$$

$$\lambda \varphi_c R_{nt,f} C_\Delta C_M C_T = (0.8 \times 0.50)28.8(1.0 \times 0.75 \times 0.74) = 6.4 \text{ kips}$$

The multi-row connection is for single-lap shear and its strength must be reduced by a multiplier of 0.6.

$$(0.6)(\lambda \varphi_c R_{nt,f} C_\Delta C_M C_T) = 0.6(6.4) \text{ kips} = 3.84 \text{ kips}$$

f. Shear out between rows of bolts:

For two rows of bolts, the shear- out strength per line of bolts is determined as:

The characteristic in-plane shear strength to the shear-out failure (F_{sh}) of FRP is 6 ksi (manufacturer). The reduction factor (φ) for the shear out between bolt rows is 0.45, as given in Table 11.1.

From Table 11.10 and Eq. (11.21.1):

$$R_{sh} = 1.4\left(e_1 - \frac{d_n}{2} + s\right)tF_{sh} = 1.4\left(3.5 - \frac{\left(\frac{7}{8} + \frac{1}{16}\right)}{2} + 3.5\right)(0.5 \times 6) = 27.4 \text{ kips}$$

$$\lambda\varphi_c R_{sh}C_\Delta C_M C_T = (0.8 \times 0.45)27.4(1.0 \times 0.75 \times 0.74) = 5.47 \text{ kips}$$

The multi-row connection is for single-lap shear and its strength must be reduced by a multiplier of 0.6 (ACMA, 2021).

$$(0.6)(\lambda\varphi_c R_{sh}C_\Delta C_M C_T) = 0.6(5.47) = 3.28 \text{ kips}$$

g. Block shear:

The connection force is concentric to the group of bolts, tensile and parallel to the direction of FRP materials. The reduction factor (ϕ_c) for the block shear is 0.45. The block shear strength for the multi-bolted connection is determined from Table 11.10 and Eq. (11.22.1):

Net area subjected to shear (A_{ns}) = 0.5 × 2 [(3.5 × 3)–(7/8 +1/16)(2.5)] = 8.16 in²

Net area subjected to tension (A_{nt}) = 0.5[5–(7/8 +1/16)] = 2.03 in²

The characteristic in-plane shear strength to the shear-out failure (F_{sh}) is 6 ksi (manufacturer).

The characteristic tensile strength in the longitudinal direction (F^t_L) is 24 ksi (manufacturer).

$$R_{bs} = 0.5\left(A_{ns}F_{sh} + A_{nt}F^t_L\right) = 0.5\left((8.16 \times 6) + (2.03 \times 24)\right) = 48.8 \text{ kips}$$

$$\lambda_c\varphi_c R_{bs}C_\Delta C_M C_T = (0.8 \times 0.45)48.8(1.0 \times 0.75 \times 0.74) = 9.75 \text{ kips}$$

The multi-row connection is for single-lap shear and its strength must be reduced by a multiplier of 0.6 (ACMA, 2021).

$$(0.6)(\lambda\varphi_c R_{bs}C_\Delta C_M C_T) = 0.6(9.75) = 5.85 \text{ kips}$$

h. Determine the tensile capacity of connection:

The nominal strength of the multi-row connection is the sum of the strengths from each of the bolts. Load distribution per bolt row proportioned to the connection force transmitted through bearing is given in Table 11.8.

h.1) Shear per bolt:

Maximum proportion of load in the first row (L_{br} = 0.5):

Each bolt: 0.5P/3 = 13.62 kips

Ultimate tensile capacity P = 6(13.62) = 81.72 kips

h.2) Tension per bolt:

Maximum proportion of load in the first row (L_{br} = 0.5):

Each bolt: 0.5P/3 = 18.02 kips

Ultimate tensile capacity P = 6(18.02) = 108.12 kips

h.3) Combined tension and shear of bolts:

Maximum proportion of load in the first row (L_{br} = 0.5):

Each bolt: 0.5P/3 = 20.4 kips

Ultimate tensile capacity P = 6(20.4) = 122.4 kips

h.4) Perpendicular pin bearing strength per bolt:

Maximum proportion of load in the first row (L_{br} = 0.5):

Each bolt: 0.5P/3 = 3.20 kips

Ultimate tensile capacity P = 6(3.20) = 19.2 kips

h.5) Net tensile strength at first bolt row:

Maximum proportion of load in the first row (L_{br} = 0.5):

First bolt row: $0.5P = 3.84$ kips
Ultimate tensile capacity $P = 2(3.84) = 7.68$ kips
h.6) Shear out between rows of bolts:
Maximum proportion of load in the first row ($L_{br} = 0.5$):
First bolt row: $0.5P = 3.28$ kips
Ultimate tensile capacity $P = 2(3.28) = 6.56$ kips
h.7) Block shear:
The block shear strength: $P = 5.85$ kips
Ultimate tensile capacity $P = 5.85$ kips

Summary:
The ultimate tensile capacity of the connection is 5.85 kips due to block shear failure.
Alternate method:
("Design Guide for FRP Composite Connections" American Society of Civil Engineers (ASCE), A. Mosallam, 2011). A comprehensive treatment of this alternate approach can be found in Mosallam (2011).

Using the alternate method, find the tensile capacity for the connection shown in Example Figure 11.1. Single-lap joint with 7/8" diameter A325 (bearing type connection) in standard holes is used to connect pultruded FRP vinyl ester plates (12 × 1/2").

Solution

The tensile capacity of a mechanical connection is determined as:

a. Ultimate net-section tension failure load:
   The characteristic tensile strength in the longitudinal direction (F_{tl}) of FRP is 24 ksi (manufacturer). The ultimate net-tension failure load is (Mosallam, 2011):

$$P_{ult}^{net-tension} = \eta(twF_{tu}) = 0.39(0.5)12(24) = 55.7 \text{ kip}$$

and,

$$\eta = \frac{1}{1+C(k_{te}-1)}\left(1-\frac{nd}{w}\right) = \frac{1}{1+0.3\left(\dfrac{12}{3(0.9375)}-1\right)}\left(1-\frac{3(0.9375)}{12}\right) = 0.39$$

where η = joint efficiency = (strength of notched member)/(strength of unnotched member)
   $k_{te} \approx \dfrac{w}{nd}$ = stress concentration factor of isotropic material of a joint $\cong 1.0$ for isotropic
materials at full plastic deformation (Hart-Smith, 1978)
   w = member width
   t = member thickness
   n = number of bolts in a row along the loading direction
   d = hole diameter
   $C = 0.3 = (k_{tc}-1)/(k_{te}-1)$ (Refer to Section 11.3.1 and Hassan, 1994)
b. Cleavage failure load:
   The characteristic pin bearing strength in the longitudinal direction of FRP is 32 ksi (manufacturer). The ultimate cleavage failure load, $P_{ult}^{cleavage}$, is determined as:

$$P_{ult}^{cleavage} = F_{br}t\frac{d^2}{d_{bolt}}\eta = 32(0.5)\frac{0.94^2}{0.875}0.39 = 6.2 \text{ kip}$$

and,

$$\eta_{cleavage} = \frac{1}{1+C(k_{te}-1)}\left(1-\frac{nd}{w}\right)\Psi = 0.39$$

where

$$\Psi = \left(\frac{6}{5} - \frac{3d}{5e} \right)^v = 1, \text{since } v = 0 \text{ for case } 5$$

$v = 2, 1, 0$ for cases 2 & 4, 1 & 3, and 5 (refer to Section 11.3.1)

c. Bearing failure load:

The ultimate bearing failure load is determined by (Mosallam, 2011):

$$P_{ult}^{cleavage} = F_{br}t\frac{d^2}{d_{bolt}} = 32(0.5)\frac{0.94^2}{0.875} = 16.07 \text{ kip}$$

The ultimate tensile capacity of the joint is 6.2 kip.

$$P_{ult}^{cleavage} < P_{ult}^{bearing} < P_{ult}^{net\text{-}tension}$$

Note: The difference between results computed from the LRFD and simplified method (Mosallam, 2011) is about 5.6% (5.85 kip vs. 6.2 kip).

Example 11.2

Using the ASD method, find the tensile capacity for the connection shown in Example Figure 11.2. Single-lap joint with 7/8″ diameter A325 (bearing type connection) in standard holes is used to connect pultruded FRP vinyl ester plates (12 × 1/2″).

The time effect factor $\lambda = 0.8$ for $W = 1.2D+1.6L$, Geometry factor $C_\Delta = 1.0$ for pitch spacing $(s) \geq s_{min}$, Moisture condition factor $C_M = 0.75$ for vinyl ester material and temperature condition factor $C_T = 0.74$ for strength ($T = 120°F$).

Solution

a. Shear per bolt:

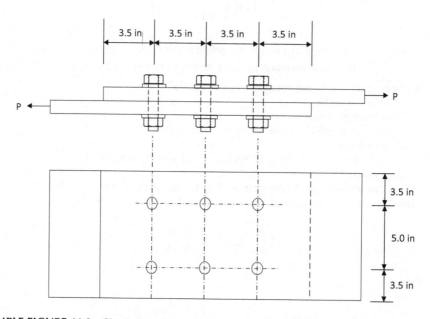

EXAMPLE FIGURE 11.2 Single-lap joint connection.

The nominal shear strength (F_{nv}) of bolts A325 is 68 ksi (Table 11.4). Based on the ASD method, the safety factor for a connection is 4.0. The allowable shear strength is determined from Eq. (11.7) as:

$$\left(F_v\right)_{allow} = \frac{F_{nv}}{F.S.} = \frac{68}{4} = 17\,\text{ksi}$$

$$\left(R_{bt}\right)_{allow} = \left(F_v\right)_{allow} A_b = 17(0.601) = 10.21\,\text{kips}$$

b. Tension per bolt:

The nominal tensile strength (F_{nt}) of bolts A325 is 90 ksi (Table 11.4). Based on the ASD method, the safety factor of a connection is 4.0. The allowable tensile strength is determined from Eq. (11.6) as:

$$\left(F_{n_t}\right)_{allow} = \frac{F_{nt}}{F.S.} = \frac{90}{4} = 22.5\,\text{ksi}$$

$$\left(R_{bt}\right)_{allow} = \left(F_{n_t}\right)_{allow} A_b = 22.5(0.601) = 13.52\,\text{kips}$$

c. Combined tension and shear per bolt:

Based on the ASD method, the safety factor of a connection is 4.0. The nominal tensile strength including shear stress effects is determined from Eqs. (11.8 and 11.9) as:

$$F_{nt}^t = 1.3F_{nt} - \frac{F_{nt}}{\varphi_b F_{nv}} f_v = 90\,\text{kips} \leq F_{nt}$$

$$\left(F_{nt}^t\right)_{allow} = \frac{F_{nt}}{F.S.} = \frac{90}{4} = 22.5\,\text{ksi}$$

$$\left(R_{bt}\right)_{allow} = \left(F_{nt}^t\right)_{allow} A_b = 22.5(0.601) = 13.52\ \text{kips}$$

Note: Tensile strength per bolt (through the thickness) is not considered in this computation. The tensile strength through the thickness is for connections with prying action along with the connection force component aligned with the axis of bolting.

d. Perpendicular pin bearing strength per bolt:

The characteristic pin bearing strength in the longitudinal direction of FRP is 32 ksi (manufacturer). The nominal hole diameter is taken as 1/16" larger than the nominal bolt diameter (7/8"). Fastener threads are excluded, so $\xi = 1.0$.

The orientation between the direction of the connection force and the strong direction of the FRP plates is 0°. Based on the ASD method, the safety factor of a connection is 4.0. The allowable perpendicular pin-bearing strength per bolt is determined from Eq. (11.11) as:

$$R_{br} = dt\xi F_\theta^{br} = \left(\frac{7}{8} + \frac{1}{16}\right)\left(\frac{1}{2}\right)(1.0)32 = 15\,\text{kips}$$

Note: Washer and nut are provided on both sides of a bolted connection.

$$\left(R_{br}\right)_{allow} = \frac{dt\xi F_\theta^{br}}{F.S.} = \frac{15}{4} = 3.75\,\text{kips}$$

The multi-row connection is for single-lap shear and its strength must be reduced by a multiplier of 0.6 (ACMA, 2021).

$$\left(R_{br}\right)_{allow} = 0.6(3.75) = 2.25\ \text{kips}$$

e. Net tensile strength at first bolt row (refer to Table 11.9):

The effective width (w) is determined as: $w = e_3 + e_4 + (n-1)g = 2(3.5) + (2-1)5 = 12''$, where $e_3 = e_4 = e_2$ for a connection with side edges having a side distance.

Number of bolts across the effective width, $n = 3$, Number of bolt rows = 2 with FRP/FRP materials connected. The characteristic tensile strength in the longitudinal direction (F_{tL}) of FRP is 24 ksi (manufacturer). Based on the ASD method, the safety factor of a connection is 4.0.

From Eq. (11.19):

$$R_{nt,f} = 0.2wtF_L^t = 0.2(12)(0.5)(24) = 28.8 \, \text{kips}$$

$$\left(R_{nt,f}\right)_{allow} = \frac{R_{nt,f}}{F.S.} = \frac{28.8}{4} = 7.2 \, \text{kips}$$

The multi-row connection is for single-lap shear and its strength must be reduced by a multiplier of 0.6 (ACMA, 2021).

$$\left(R_{nt,f}\right)_{allow} = 0.6(7.2) = 4.32 \, \text{kips}$$

f. Shear out between rows of bolts:

For two rows of bolts, the shear-out strength per line of bolts is determined as:

The characteristic in-plane shear strength to the shear-out failure (F_{sh}) of FRP is 6 ksi (manufacturers). Based on the ASD method, the safety factor of a connection is 4.0.

From Table 11.10 and Eq. (11.21.1):

$$R_{sh} = 1.4\left(e_1 - \frac{d_n}{2} + s\right)tF_{sh} = 1.4\left(3.5 - \frac{(7/8 + 1/16)}{2} + 3.5\right)(6 \times 0.5) = 27.4 \, \text{kips}$$

$$\left(R_{sh}\right)_{allow} = \frac{R_{sh}}{F.S.} = \frac{27.4}{4} = 6.86 \, \text{kips}$$

The multi-row connection is for single-lap shear and its strength must be reduced by a multiplier of 0.6 (ACMA, 2021).

$$\left(R_{sh}\right)_{allow} = 0.6(6.86) = 4.11 \, \text{kips}$$

g. Block shear:

The connection force is concentric to the group of bolts. The force is in tension and parallel to the direction of FRP materials. Based on the ASD method, the safety factor of a connection is 4.0. The block shear strength for the multi-bolted connection is determined from Table 11.10 and Eq. (11.22.1) as:

The characteristic in-plane shear strength to the shear-out failure (F_{sh}) of FRP is 6 ksi. The characteristic tensile strength in the longitudinal direction (F_L^t) of FRP is 24 ksi. Net area subjected to shear (A_{ns}) = 2 × 0.5 [(3.5 × 3)−(7/8 +1/16)(2.5)] = 8.16 in² Net area subjected to tension (A_{nt}) = 0.5[5−(7/8 +1/16)] = 2.03 in²

$$R_{bs} = 0.5\left(A_{ns}F_{sh} + A_{nt}F_L^t\right) = 0.5\left((8.16 \times 6) + (2.03 \times 24)\right) = 48.8 \, \text{kips}$$

$$\left(R_{bs}\right)_{allow} = \frac{R_{bs}}{F.S.} = \frac{48.8}{4} = 12.2 \, \text{kips}$$

The multi-row connection is for single-lap shear and its strength must be reduced by a multiplier of 0.6 (ACMA, 2021).

$$(R_{bs})_{allow} = 0.6(12.2) = 7.33 \text{ kips}$$

h. Determine the tensile capacity of connection:

The nominal strength of the multi-row connection is the sum of the strengths from each of the bolts. Load distribution per bolt row proportioned to the connection force transmitted through bearing is given in Table 11.8.

h.1) Bolt shear:

Maximum proportion of load in the first row ($L_{br} = 0.5$):

Each bolt per row: $0.5P/3 = 10.2$ kips

Tensile service capacity $P = 6(10.2) = 61.2$ kips

h.2) Bolt tension:

Maximum proportion of load in the first row ($L_{br} = 0.5$):

Each bolt per row: $0.5P/3 = 13.5$ kips

Tensile service capacity $P = 6(13.5) = 81.0$ kips

h.3) Combined tension and shear of bolts:

Maximum proportion of load in the first row ($L_{br} = 0.5$):

Each bolt per row: $0.5P/3 = 13.5$ kips

Tensile service capacity $P = 6(13.5) = 81.1$ kips

h.4) Perpendicular pin bearing strength per bolt:

Maximum proportion of load in the first row ($L_{br} = 0.5$):

Each bolt per row: $0.5P/3 = 2.25$ kips

Tensile service capacity $P = 6(2.25) = 13.5$ kips

h.5) Net tensile strength at first bolt row:

Maximum proportion of load in the first row ($L_{br} = 0.5$):

First bolt row: $0.5P = 4.32$ kips

Tensile service capacity $P = 2(4.32) = 8.64$ kips

h.6) Shear out between rows of bolts:

Maximum proportion of load in the first row ($L_{br} = 0.5$):

First bolt row: $0.5P = 4.11$ kips

Tensile service capacity $P = 2(4.11) = 8.22$ kips

h.7) Block shear:

The allowable block shear strength: $P = 7.33$ kips

Tensile service capacity $P = 7.33$ kips.

Summary:

The allowable tensile capacity of the connection is 7.33 kips which is controlled by block shear strength, with F.S. of 4.0. For long-term serviceability, the knockdown factor (0.8) for environmental effects may be applied to the allowable tensile capacity (safety factor = $0.8 \times (1/4) = 0.20$ or 20% failure strain). Thus, the allowable tensile capacity of the connection is 5.86 kips for long-term serviceability.

Alternate method:

("Design Guide for FRP Composite Connections" American Society of Civil Engineers (ASCE), A. Mosallam, 2011). A comprehensive treatment of this alternate approach can be found in Mosallam (2011).

Using the ASD method, find the tensile capacity for the connection shown in Example Figure 11.2. Single-lap joint with 7/8" diameter A325 (bearing type connection) in standard holes is used to connect pultruded FRP vinyl ester plates (12 × 1/2").

Solution

The tensile capacity of a mechanical connection is determined as:

a. Ultimate net-section tension failure load:

The ultimate net-tension failure load is determined as:

$$P_{ult}^{net-tension} = \eta(twF_{tu}) = 0.39(0.5)12(24) = 55.7 \text{ kip}$$

and,

$$\eta = \frac{1}{1+C(k_{te}-1)}\left(1-\frac{nd}{w}\right) = 0.39$$

where η = joint efficiency = (strength of notched member)/(strength of unnotched member)
  $k_{te} \approx \dfrac{w}{nd}$ = stress concentration factor of isotropic material of a joint (Hart-smith, 1978)
  w = member width
  t = member thickness
  n = number of bolts in a row along the loading direction
  d = hole diameter
  C = 0.3, assuming fiber orientation is along the pultrusion direction (refer to Section 11.3.1)
  b. Cleavage failure load:
    The characteristic pin bearing strength in the longitudinal direction of FRP is 32 ksi (manufacturer). The ultimate cleavage failure load is determined as:

$$P_{ult}^{cleavage} = F_{br}t\frac{d^2}{d_{bolt}}\eta = 32(0.5)\frac{0.94^2}{0.875}0.39 = 6.2\,\text{kip}$$

and

$$\eta_{cleavage} = \frac{1}{1+C(k_{te}-1)}\left(1-\frac{nd}{w}\right)\Psi = 0.39$$

where

$$\Psi = \left(\frac{6}{5}-\frac{3d}{5e}\right)^\nu = 1$$

$$\nu = 0$$

  c. Bearing failure load:
    The ultimate bearing failure load is:

$$P_{ult}^{bearing} = F_{br}t\frac{d^2}{d_{bolt}} = 32(0.5)\frac{0.94^2}{0.875} = 16.07\,\text{kip}$$

The ultimate tensile capacity of the joint is 6.2 kip

$$P_{ult}^{cleavage} < P_{ult}^{bearing} < P_{ult}^{net\text{-}tension}$$

Note: The difference between results from the ASD with F.S. of 4.0 and the simplified method (Mosallam, 2011) results in about 15% (7.33 kip vs. 6.2 kip), variation.

Example 11.3

Design a double-lap joint under axial factored tension 10.0 kips using the LRFD method and the minimum requirements of the bolted connection for bolt spacing. The double-lap joint with 3/4″ diameter A325 bolt is fit in standard holes of FRP plates (12 × 1/2″).

Solution

From Figure 11.8 and Table 11.5, the minimum requirement of the bolted connection is accepted as $e_1 = 2d$, $e_2 = 1.5d$, $s = 4d$, $g = 4d$. The multi-row bolted connection is assumed to be $n = 2$ rows.

a. Bolt capacity:

a.1) Shear per bolt:

The nominal shear strength of bolts A325 is 68 ksi (Table 11.4). The reduction factor (ϕ_b) for bolt tension is 0.75. The nominal shear strength is determined from Eq. (11.7) as:

$$R_{bt} = F_{nv}A_b = 68(0.442) = 30.1\,\text{kips}$$
$$\lambda\varphi_b R_{bt}C_\Delta C_M C_T = (0.8 \times 0.75)30.1(1.0 \times 0.75 \times 0.74) = 10.02\,\text{kips}$$

a.2) Tension per bolt:

The nominal tensile strength of bolts A325 is 90 ksi (Table 11.4). The reduction factor (ϕ_b) for bolt tension is 0.75. The nominal tensile strength is determined from Eq. (11.6) as:

$$R_{bt} = F_{nt}A_b = 90(0.442) = 39.8\,\text{kips}$$
$$\lambda\varphi_b R_{bt}C_\Delta C_M C_T = (0.8 \times 0.75)39.8(1.0 \times 0.75 \times 0.74) = 13.25\,\text{kips}$$

a.3) Combined tension and shear of bolts per bolt:

The reduction factor (ϕ_b) for combined tension and shear strengths is 0.75. The nominal tensile strength including shear stress effects is determined from Eqs. (11.8 and 11.9) as:

$$F_{nt}' = 1.3F_{nt} - \frac{F_{nt}}{\varphi_b F_{nv}} f_v \le F_{nt} = 90 \ \text{kips}$$
$$R_{bt} = F_{nt}'A_b = 90(0.442) = 39.8\,\text{kips} \qquad \left(\text{Table}\,11.4\right)$$
$$\lambda\varphi_b R_{bt}C_\Delta C_M C_T = (0.8 \times 0.75)39.8(1.0 \times 0.75 \times 0.74) = 13.25\,\text{kips}$$

Note: Tensile strength per bolt (through the thickness) is not considered in this. The check for tensile strength through the thickness is for connections with prying action along with a connection force component aligned with the axis of bolting.

a.4) Perpendicular pin bearing strength per bolt:

The characteristic pin bearing strength in the longitudinal direction of FRP is 32 ksi (manufacturer). The nominal hole diameter is taken as 1/16″ larger than the nominal bolt diameter (3/4″). Fastener threads are excluded so $\xi = 1.0$.

The orientation between the direction of the connection force and the strong direction of the FRP plates is 0°. The reduction factor (ϕ_c) for the perpendicular pin-bearing strength per bolts is 0.8. The perpendicular pin-bearing strength per bolt is determined from Eq. (11.11) as:

$$R_{br} = dt\xi F_\theta^{br} = \left(\frac{3}{4}+\frac{1}{16}\right)\left(\frac{1}{2}\right)(1.0)32 = 13\,\text{kips}$$
$$\lambda\varphi_c R_{br}C_\Delta C_M C_T = (0.8 \times 0.8)13(1.0 \times 0.75 \times 0.74) = 4.62\,\text{kips}$$

b. Determine the load distribution per bolt row and the number of bolts per row for the double-lap connection:

The nominal strength of the multi-row connection is the sum of equal strength contributions from each of the bolts. Load distribution per bolt row as a proportion of the connection force transmitted through bearing is given in Table 11.7 (ACMA, 2021).

- To determine the number of bolts, the minimum nominal bolt capacity (of the connection) under perpendicular pin bearing strength is used (4.62 kip from a.4, as shown above).
- For FRP-FRP material bolted connections with two rows, the maximum distributed load for each row is 10.0 (0.5) = 5.0 kips.
- Number of bolts in each row = 5.0 kips/(0.8 × 5.23 kips) = 1.2 using two bolts per row

Minimum requirements for connection geometry as per the pre-standard (ACMA, 2021) must be satisfied as follows:

From Figure 11.8 and Table 11.5:

- Spacing (s) = 3.0″≥4d = 4(3/4″)
- Gage (g) = 6.0″≥4d = 4(3/4″)
- End distance (e_1) = 1.5″≥2d = 2(3/4″)
- Edge distance (e_2) = 1 1/8″≥1.5d = 1.5(3/4″)

The design configuration of the connection is shown in Example Figure 11.3.

c. Check for net tension strength at the first bolt row:

The characteristic tensile strength in the longitudinal direction (F^t_L) of FRP is 24 ksi (manufacturer). As per (ACMA, 2021), rows of bolts with constant gage spacing across the effective width are $n = 2$. The reduction factor (φ) for the net tension at the first (bolt) row is 0.50.

For $w = e_3 + e_4 + (n-1)\,g$, where $e_3 = e_4 = e_2$ for a connection with side edges having a side distance. The effective width (w) is determined from Eq. (11.19):

For $n = 2$ and $w = 2(1.5d) + (2-1)4d = 7d$

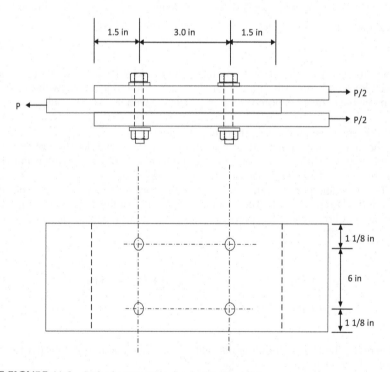

EXAMPLE FIGURE 11.3   Bolted connection for double-lap joint.

$$R_{nt,f} = 0.2wtF_L^t$$
$$R_{nt,f} = 0.2(7)(0.75)(0.5)24 = 12.6\,\text{kips}$$
$$\lambda\varphi_c R_{nt,f}C_\Delta C_M C_T = (0.8\times0.5)12.6(1.0\times0.75\times0.74) = 2.80\,\text{kips}$$

The maximum proportion of load in the first row (L_{br}) is 0.5.
First bolt row: $0.5P = 2.80$ kips
The nominal net tensile strength ($P = 2(2.80) = 5.60$ kips) is less than the required design factored tension load ($P = 10$ kips). Hence, the design is unsatisfactory, and a new bolt arrangement will be needed.

d. Check for shear out between rows of bolts:

The characteristic in-plane shear strength to the shear-out failure (F_{sh}) is 8 ksi (manufacturer). The reduction factor (ϕ_c) for the shear out between rows of bolts is 0.45. For two rows of bolts, the shear-out strength per line of bolts is determined from Table 11.10 and Eq. (11.21.1) as:

Number of bolts across the effective width $n = 2$

$$R_{sh} = 1.4\left(e_1 - \frac{d_n}{2} + s\right)tF_{sh} = 1.4\left(1.5 - 0.5\left(\frac{3}{4} + \frac{1}{16}\right) + 3\right)(0.5\times8) = 22.9\,\text{kips}$$
$$\lambda\varphi_c R_{sh}C_\Delta C_M C_T = (0.8\times0.45)22.9(1.0\times0.75\times0.74) = 4.07\,\text{kip}$$

Maximum proportion of load in the first row ($L_{br} = 0.5$):
First bolt row: $0.5P = 4.07$ kips
The nominal shear-out capacity ($P = 2(4.07) = 8.14$ kips) is less than the required factored design tension load ($P = 10.0$ kips). Hence, the design is unsatisfactory.

e. Check for block shear capacity:

The connection force is concentric to the group of bolts, tensile and parallel to the direction of FRP materials. The block shear strength for the multi-bolted connection is determined from Table 11.10 and Eq. (11.22.1) as:

The characteristic in-plane shear strength to the shear-out failure (F_{sh}) is 8 ksi.
The characteristic tensile strength in the longitudinal direction (F_L^t) is 24 ksi.
Net area subjected to shear (A_{ns}) = $0.5 \times 2 \times [(4.5)-(3/4 +1/16)(1.5)] = 3.28$ in²
Net area subjected to tension (A_{nt}) = $0.5[(6)-(3/4 +1/16)] = 2.59$ in²

$$R_{bs} = 0.5\left(A_{ns}F_{sh} + A_{nt}F_L^t\right) = 0.5\left((3.28\times8)+(2.59\times24)\right) = 44.2\,\text{kips}$$
$$\lambda\varphi_c R_{bs}C_\Delta C_M C_T = (0.8\times0.45)44.2(1.0\times0.75\times0.74) = 8.83\,\text{kips}$$

The nominal block shear capacity ($P = 8.83$ kips) is less than the required factored design tension load ($P = 10$ kips). Hence, the design is unsatisfactory.

Example 11.4

Design a double-lap joint under axial service tension load 5.0 kips using the ASD method and the minimum requirement of the bolted connection for bolt spacing. The double-lap joint with 3/4″ diameter A325 bolt is fit in standard holes of FRP plates (12 × 1/2″).

Solution

From Figure 11.8 and Table 11.5, the minimum requirement of the bolted connection is accepted as $e_1 = 2d$, $e_2 = 1.5d$, $s = 4d$, $g = 4d$. The multi-row bolted connection is assumed to be $n = 2$ rows.

a. Bolt capacity:

  a.1) Shear per bolt:

    The nominal shear strength of bolts A325 is 68 ksi (Table 11.4). Based on the ASD method, the safety factor of a connection is 4.0. The allowable shear strength is determined from Eq. (11.7) as:

$$F_v = \frac{F_{nv}}{F.S.} = \frac{68}{4} = 17\,\text{ksi}$$

$$R_{bt} = F_{nv} A_b = 17(0.442) = 7.51\,\text{kips}$$

  a.2) Tension per bolt:

    The nominal strength of bolts A325 is 90 ksi (Table 11.4). Based on the ASD method, the safety factor of a connection is 4.0. The allowable tensile strength is determined from Eq. (11.6) as:

$$\left(F_{nt}\right)_{allow} = \frac{F_{nt}}{F.S.} = \frac{90}{4} = 22.5\,\text{ksi}$$

$$\left(R_{bt}\right)_{allow} = \left(F_{nt}\right)_{allow} A_b = 22.5(0.442) = 9.95\ \text{kips}$$

  a.3) Combined tension and shear of bolts per bolt:

    Based on the ASD method, the safety factor of a connection is 4.0. The nominal tensile strength including shear stress effects is determined from Eqs. (11.8 and 11.9) as:

$$F_{nt}^t = 1.3F_{nt} - \frac{F_{nt}}{\varphi_b F_{nv}} f_v = 90\,\text{kips} \le F_{nt}$$

$$\left(F_{nt}^t\right)_{allow} = \frac{F_{nt}}{F.S.} = \frac{90}{4} = 22.5\,\text{ksi}$$

$$\left(R_{bt}\right)_{allow} = \left(F_{nt}^t\right)_{allow} A_b = 22.5(0.442) = 9.95\,\text{kips}$$

Tensile strength per bolt (through the thickness) is not considered in this. The tensile strength through the thickness is for the connections with prying action and a connection force component aligned with the axis of bolting.

  a.4) Perpendicular pin bearing strength per bolt:

    The characteristic pin bearing strength in the longitudinal direction of FRP is 32 ksi (manufacturers). The nominal hole diameter is taken as 1/16″ larger than the nominal bolt diameter (3/4″).

    The orientation between the direction of the connection force and the strong direction of the FRP plates is 0°. Based on the ASD method, the safety factor of a connection is 4.0. The perpendicular pin-bearing strength per bolt is determined from Eq. (11.11) as:

$$R_{br} = dt\xi F_\theta^{br} = \left(\frac{3}{4} + \frac{1}{16}\right)\left(\frac{1}{2}\right)(1.0)32 = 13\,\text{kips}$$

$$\left(R_{br}\right)_{allow} = \frac{dt\xi F_\theta^{br}}{F.S.} = \frac{13}{4} = 3.25\,\text{kips}$$

b. Determine the load distribution per bolt row and number of bolts per row for the connection:

The nominal strength of the multi-row connection is the sum of equal strength contributions from each of the bolts. Load distribution per bolt row as a proportion of the connection force transmitted through bearing is given in Table 11.6.

- For FRP-FRP material bolted connections with two rows, the maximum distributed load for each row is 5.0(0.5) = 2.5 kips.
- The knock down factor for long term serviceability is taken as 0.8 (total safety factor = 1/4 × 0.8 = 0.2 or 20% failure strain). To determine the number of bolts, the allowable perpendicular pin bearing strength (a.4) with knock down factor for long-term serviceability is 3.25(0.8) = 2.6 kips.
- Number of bolts each row = 2.5 kips/(2.6 kips) = 0.96 using two bolts per row
- The minimum requirements for connection geometry as per the pre-standard (ACMA, 2021) must be satisfied as follows:

From Figure 11.8 and Table 11.5:

- Spacing (s) = 3.0″≥4d = 4(3/4″)
- Gage (g) = 6.0″≥4d = 4(3/4″)
- End distance (e_1) = 1.5″≥2d = 2(3/4″)
- Edge distance (e_2) = 1 1/8″≥1.5d = 1.5(3/4″)

The design configuration of the connection is shown in Example Figure 11.4.

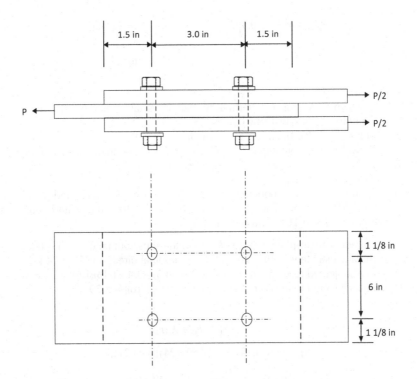

EXAMPLE FIGURE 11.4 Bolted connection for double-lap joint.

c. Check for net tension strength at the first bolt row:

The characteristic tensile strength in the longitudinal direction (F^t_L) of FRP is 24 ksi (manufacturer). As per (ACMA, 2021), rows of bolts with constant gage spacing across the effective width are $n = 2$.

For $w = e_3 + e_4 + (n-1) g$, where $e_3 = e_4 = e_2$ for a connection with side edges having a side distance. The effective width (w) is determined from Eq. (11.19) as:

For $n = 2$ and $w = 2(1.5d) + (2-1)4d = 7d$

$$R_{nt,f} = 0.2 wt F^t_L = 0.2(7 \times 0.75)(0.75)24 = 12.6 \, \text{kips}$$

$$\left(R_{nt,f}\right)_{allow} = \frac{R_{nt,f}}{F.S.} = \frac{12.6}{4} = 3.14 \, \text{kips}$$

The knock down factor for long term serviceability is taken as 0.8 (total safety factor = $1/4 \times 0.8 = 0.2$ or 20% failure strain).

Maximum load in the first row ($L_{br} = 0.5$) and First bolt row: $0.5P = 0.8(3.14)$, $P = 5.02$ kips:

The allowable net tensile strength ($P = 5.02$ kips) is higher than the required service tension load ($P = 5$ kips). Hence, the design is satisfactory.

d. Check for shear out between rows of bolts:

The characteristic in-plane shear strength to the shear-out failure (F_{sh}) of FRP is 8 ksi. For two rows of bolts, the shear-out strength per line of bolts is determined from Table 11.10 and Eq. (11.21.1) as:

Number of bolts across the effective width $n = 2$

$$R_{sh} = 1.4\left(e_1 - \frac{d_n}{2} + s\right) t F_{sh}$$

$$R_{sh} = 1.4\left(1.5 - 0.5\left(\frac{3}{4} + \frac{1}{16}\right) + 3\right)(0.5 \times 8) = 22.9 \, \text{kips}$$

$$\left(R_{sh}\right)_{allow} = \frac{R_{sh}}{F.S.} = \frac{22.9}{4} = 5.73 \, \text{kips}$$

The knockdown factor for long term serviceability is taken as 0.8 with the maximum proportion of load in the first row (L_{br}) as 0.5:

First bolt row: $0.5P = 0.8(5.73)$, $P = 9.16$ kips

The allowable shear-out capacity ($P = 9.16$ kips) is higher than the service design tension load ($P = 5.0$ kips). Hence, the design is satisfactory.

e. Check for block shear capacity:

The connection force is concentric to the group of bolts, tensile and parallel to the direction of FRP materials. The block shear strength for the multi-bolted connection is determined from Table 11.10 and Eq. (11.22.1) as:

The characteristic in-plane shear strength to the shear-out failure (F_{sh}) is 8 ksi.
The characteristic tensile strength in the longitudinal direction (F^t_L) is 24 ksi.
Net area subjected to shear (A_{ns}) = $0.5 \times 2 \times [(4.5)-(3/4+1/16)(1.5)] = 3.28 \, \text{in}^2$
Net area subjected to tension (A_{nt}) = $0.5[(6)-(3/4+1/16)] = 2.59 \, \text{in}^2$

$$R_{bs} = 0.5\left(A_{ns} F_{sh} + A_{nt} F^t_L\right)$$

$$R_{bs} = 0.5\left((3.28 \times 8) + (2.59 \times 24)\right) = 44.2 \, \text{kips}$$

$$\left(R_{bs}\right)_{allow} = \frac{R_{bs}}{F.S.} = \frac{44.2}{4} = 11.05 \, \text{kips}$$

The knock down factor for long term serviceability is taken as 0.8 (total safety factor = 1/4 × 0.8 = 0.2 ~ 20% failure strain).

The allowable block shear capacity (P = 11.05 × 0.8 = 8.84 kips) is higher than the service design tension load (P = 5.0 kips). Hence, the design is satisfactory.

Example 11.5

Using the LRFD method, determine the load-carrying capacity of the framed connection as shown in Example Figure 11.5. The bolts are 3/4″ dia. A325 (bearing-type connection) in standard holes. The frame connection is used to connect pultruded FRP WF 12 × 12 × 1/2″ and WF 8 × 8 × 1/2″ with 2Ls 4 × 4 × 1/2″

a. For WF 8 × 8 × 1/2″
b. For WF 12 × 12 × 1/2″
c. For 2Ls 4 × 4 × 1/2″

Solution

The load capacity of the simple frame connection is obtained from a minimum capacity of three components: a) WF 8 × 8 × 1/2″, b) WF 12 × 12 × 1/2″, and c) 2Ls 4 × 4 × 1/2″.

a. The connection is a bolted connection through the web of the pultruded FRP WF 8 × 8 × 1/2″:

The bolted connection is considered a double shear connection.

a.1) Shear per bolt:

The nominal strength of bolts A325 is 68 ksi (Table 11.4). The reduction factor (ϕ_b) for tensile strength of bolts is 0.75. The nominal shear strength is determined from Eq. (11.7) as:

$$R_{bt} = F_{nv}A_b = 68(0.442) = 30.1\,\text{kips}$$

$$\lambda\varphi_b R_{bt} C_\Delta C_M C_T = (0.8 \times 0.75)30.1(1.0 \times 0.75 \times 0.74) = 10.02\,\text{kips}$$

a.2) Tension per bolt:

The nominal strength of bolts A325 is 90 ksi (Table 11.4). The reduction factor (ϕ_b) for tensile strength of bolts is 0.75. The nominal tensile strength is determined from Eq. (11.6) as:

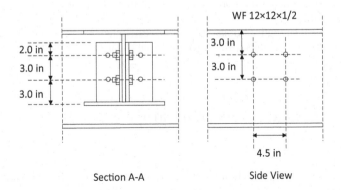

EXAMPLE FIGURE 11.5 Simple frame connection.

$$R_{bt} = F_{nt} A_b = 90(0.442) = 39.8 \, \text{kips}$$

$$\lambda \varphi_b R_{bt} C_\Delta C_M C_T = (0.8 \times 0.75) 39.8 (1.0 \times 0.75 \times 0.74) = 13.25 \, \text{kips}$$

a.3) Combined tension and shear of bolts per bolt:

The reduction factor (ϕ_b) for combined tensile and shear strength of bolts is 0.75. The nominal tensile strength including the shear stress effect is determined from Eqs. (11.8 and 11.9) as:

$$F_{nt}^t = 1.3 F_{nt} - \frac{F_{nt}}{\varphi_b F_{nv}} f_v \leq F_{nt} = 90 \, \text{kips}$$

$$R_{bt} = F_{nt}^t A_b = 90(0.442) = 39.8 \, \text{kips}$$

$$\lambda \varphi_b R_{bt} C_\Delta C_M C_T = (0.8 \times 0.75) 39.8 (1.0 \times 0.85 \times 0.74) = 15.01 \, \text{kips}$$

Note: Tensile strength per bolt (through the thickness) is not considered in this example. The tensile strength through the thickness is for the connections with prying action along with connection force component aligned with the axis of bolting.

a.4) Perpendicular pin bearing strength per bolt:

The characteristic pin bearing strengths in both the longitudinal and transverse directions are 32 and 16 ksi, respectively (manufacturers). Fastener threads are excluded so $\xi = 1.0$.

The reduction factor (ϕ_b) for perpendicular pin bearing strength of bolts is 0.8. The nominal hole diameter is taken as 1/16″ larger than the nominal bolt diameter (3/4″). The orientation between the direction of the connection force and the strong direction of the FRP plates is 90°. The perpendicular pin-bearing strength is determined from Eq. (11.11):

$$R_{br} = dt\xi F_\theta^{br} = \left(\frac{3}{4} + \frac{1}{16}\right)\left(\frac{1}{2}\right)(1.0)32 = 13 \, \text{kips}$$

$$\lambda \varphi_c R_{br} C_\Delta C_M C_T = (0.8 \times 0.8) 13 (1.0 \times 0.75 \times 0.74) = 4.62 \, \text{kips}$$

The net tensile strength and shear-out strength need not be determined, since there are two perpendicular elements (the flanges) of an FRP wide flange beam at the end member or connecting component of FRP material. The net tensile strength and shear-out strength given herein are not considered in this example.

b. The connection is bolted through the web of the pultruded FRP WF 12 × 12 × 1/2″:

The bolted connection is considered to be a single shear connection.

b.1) Bolt shear:

The nominal strength of bolts A325 is 68 ksi (Table 11.4). The reduction factor (ϕ_b) for tensile strength of bolts is 0.75. The nominal shear strength is determined from Eq. (11.7) as:

$$R_{bt} = F_{nv} A_b = 68(0.442) = 30.1 \, \text{kips}$$

$$\lambda \varphi_b R_{bt} C_\Delta C_M C_T = (0.8 \times 0.75) 30.1 (1.0 \times 0.75 \times 0.74) = 10.02 \, \text{kips}$$

b.2) Bolt tension:

The nominal strength of bolts A325 is 90 ksi (Table 11.4). The reduction factor (ϕ_b) for tensile strength of bolts is 0.75. The nominal tensile strength is determined from Eq. (11.6) as:

$$R_{bt} = F_{nt}A_b = 90(0.442) = 39.8\,\text{kips}$$

$$\lambda\varphi_b R_{bt}C_\Delta C_M C_T = (0.8 \times 0.75)39.8(1.0 \times 0.75 \times 0.74) = 13.25\,\text{kips}$$

b.3) Combined tension and shear of bolts:

The reduction factor (ϕ_b) for combined tension and shear strength of bolts is 0.75. The nominal tensile strength including shear stress effects is determined from Eqs. (11.8 and 11.9) as:

$$F_{nt}^t = 1.3F_{nt} - \frac{F_{nt}}{\varphi_b F_{nv}}f_v \le F_{nt} = 90\,\text{kips}$$

$$R_{bt} = F_{nt}^t A_b = 90(0.442) = 39.8\,\text{kips}$$

$$\lambda\varphi_b R_{bt}C_\Delta C_M C_T = (0.8 \times 0.75)39.8(1.0 \times 0.75 \times 0.74) = 13.25\,\text{kips}$$

Tensile strength per bolt (through the thickness) is not considered in this example. The tensile strength through the thickness is the connections with prying action and a connection force component aligned with the axis of bolting.

b.4) Perpendicular pin bearing strength per bolt:

The characteristic pin bearing strengths in both the longitudinal and transverse directions are 32 and 16 ksi, respectively (manufacturer). Fastener threads are excluded so $\xi = 1.0$.

The reduction factor for perpendicular pin bearing strength of bolts ϕ_b is 0.8. The nominal hole diameter is taken as 1/16″ larger than the nominal bolt diameter (3/4″). The orientation between the direction of the connection force and the strong direction of the FRP plates is 90°. The perpendicular pin-bearing strength is determined from Eq. (11.11) as:

$$R_{br} = dt\xi F_\theta^{br} = \left(\frac{3}{4} + \frac{1}{16}\right)\left(\frac{1}{2}\right)(1.0)16 = 6.5\,\text{kips}$$

The strength is obtained from the pre-standard (ACMA, 2021) for double-lap shear connection. In this example, the multi-row connection is for single-lap shear. Therefore, the strength must be reduced by a multiplier of 0.6.

$$\lambda\varphi_c R_{br}C_\Delta C_M C_T = 0.6(0.8 \times 0.8)6.50(1.0 \times 0.75 \times 0.74) = 1.39\,\text{kips}$$

The net tensile strength and shear-out strength need not be determined, since there are two perpendicular elements (the flanges) of an FRP wide flange beam at the end member or connecting component of FRP material. The net tensile strength and shear-out strength given herein are not considered in this example.

c. The connection is bolted through the web of the clip angle 2Ls 4 × 4 × 1/2″:

c.1) Shear strength of clip angle (frame type)

Depth of shear plane at the fillet radius of the leg angle profile (l_{sp}) = 4″ (manufacturers)

The reduction factor (ϕ_c) for shear strength of a clip angle is 0.75. The characteristic in-plane shear strength to the shear-out failure (F_{sh}) is 6 ksi.

The nominal shear strength at the knee of the clip angle is determined from Eq. (11.23) as:

$$\varphi_c R_{sh,sp} = 0.75 l_{sp} t_{sp} F_{sh} = 0.75(4 \times 0.5)6 = 9.0\,\text{kips}$$

d. Determine the nominal capacity of the simple frame connection
     The minimum strength is found in the case of perpendicular pin bearing strength per
     bolt (1.39 kips per bolt from b.4) for FRP wide flange beam (12 × 12 × 1/2"). The design
     load carrying capacity is determined to be (1.39 × 4 = 5.56) kips.

Example 11.6

Using the ASD method with F.S. = 2 and the knockdown factor for long-term serviceability and
other environmental issues = 0.4, determine the load-carrying capacity of the framed connection
as shown in Example Figure 11.6. The bolts are 3/4" diameter A325 (bearing type connection) in
standard holes. The frame connection is used to connect pultruded FRP WF 12 × 12 × 1/2" and
WF 8 × 8 × 1/2" with 2Ls 4 × 4 × 1/2".

   a) For WF 8 × 8 × 1/2"
   b) For WF 12 × 12 × 1/2"
   c) For 2Ls 4 × 4 × 1/2"

Solution

The load-carrying capacity of a simple frame connection is obtained from the minimum capacity
of three different beam types as (a) WF 8 × 8 × 1/2", (b) WF 12 × 12 × 1/2", and (c) 2Ls 4 × 4 × 1/2".
The connection is bolted through the web of an FRP wide flange beam (8 × 8 × 1/2"). The bolted
connection is considered a double shear connection.

   a. Determine bolt strength
      a.1) Shear per bolt:
         The nominal strength of bolts A325 is 68 ksi (Table 11.4). Based on the ASD method,
      the safety factor of joint and connection is 2.0. The allowable shear strength is deter-
      mined from Eq. (11.7) as:

$$F_v = \frac{F_{nv}}{F.S.} = \frac{68}{2} = 34\,\text{ksi}$$
$$R_{bt} = F_v A_b = 34(0.442) = 15.02\,\text{kips}$$

      a.2) Tension per bolt:
         The nominal strength of bolts A325 is 90 ksi (Table 11.4). Based on the ASD method,
      the safety factor of a connection is 2.0. The allowable tensile strength is determined from
      Eq. (11.6) as:

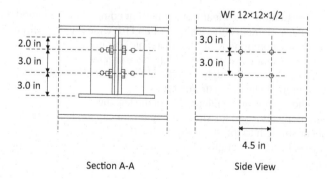

Section A-A   Side View

EXAMPLE FIGURE 11.6 Simple frame connection.

$$\left(F_{n_t}\right)_{allow} = \frac{F_{nt}}{F.S.} = \frac{90}{2} = 45\,\text{ksi}$$

$$\left(R_{bt}\right)_{allow} = \left(F_{n_t}\right)_{allow} A_b = 45(0.442) = 19.90\,\text{kips}$$

a.3) Combined tension and shear of bolts per bolt:

Based on the ASD method, the safety factor of joint and connection is 2.0. The nominal tensile strength including shear stress effects is determined from Eqs. (11.8 and 11.9) as:

$$F_{nt}^t = 1.3F_{nt} - \frac{F_{nt}}{\varphi_b F_{nv}} f_v \le F_{nt} = 90\,\text{kips}$$

$$\left(F_{nt}^t\right)_{allow} = \frac{F_{nt}}{F.S.} = \frac{90}{2} = 45\,\text{ksi}$$

$$\left(R_{bt}\right)_{allow} = \left(F_{nt}^t\right)_{allow} A_b = 45(0.442) = 19.90\,\text{kips}$$

Note: Tensile strength per bolt (through the thickness) is not considered in this example. The tensile strength through the thickness is for the connections with prying action along with the connection force component aligned with the axis of bolting.

a.4) Perpendicular pin bearing strength per bolt:

The characteristic pin bearing strengths in both the longitudinal and transverse directions are 32 and 16 ksi, respectively (manufacturers). Fastener threads are excluded so $\xi = 1.0$.

The nominal hole diameter is taken as 1/16″ larger than the nominal bolt diameter 3/4″ The orientation between the direction of the connection force and the strong direction of the FRP plates is 90°. The perpendicular pin-bearing strength per bolt is determined from Eq. (11.11) as:

$$R_{br} = dt\xi F_{\theta}^{br} = \left(\frac{3}{4} + \frac{1}{16}\right)\left(\frac{1}{2}\right)(1.0)16 = 6.5\,\text{kips}$$

$$\left(R_{br}\right)_{allow} = \frac{dt\xi F_{\theta}^{br}}{F.S.} = \frac{6.5}{2} = 3.25\,\text{kips}$$

The net tension strength and shear-out strength need not be determined, since there are two perpendicular elements (the flanges) of an FRP wide flange beam at the end member or connecting component of FRP material. The net tension strength and shear-out strength given herein are not considered in this example.

b. The connection is bolted through the web of the FRP WF 12 × 12 × 1/2″:

The bolted connection is considered a single shear connection.

b.1) Shear per bolt:

The nominal strength of bolts A325 is 68 ksi (Table 11.4). Based on the ASD method, the safety factor of a connection is 2.0. The allowable shear strength is determined from Eq. (11.7) as:

$$F_v = \frac{F_{nv}}{F.S.} = \frac{68}{2} = 34\,\text{ksi}$$

$$R_{bt} = F_v A_b = 34(0.442) = 15.02\,\text{kips}$$

b.2) Tension per bolt:

The nominal strength of bolts A325 is 90 ksi (Table 11.4). Based on the ASD method, the safety factor of a connection is 2.0. The allowable tensile strength is determined from Eq. (11.6) as:

$$\left(F_{n_t}\right)_{allow} = \frac{F_{nt}}{F.S.} = \frac{90}{2} = 45\,\text{ksi}$$

$$\left(R_{bt}\right)_{allow} = \left(F_{n_t}\right)_{allow} A_b = 45(0.442) = 19.90\,\text{kips}$$

b.3) Combined tension and shear of bolts:

Based on the ASD method, the safety factor of a connection is 2.0. The nominal tensile strength including shear stress effects is determined from Eqs. (11.8 and 11.9) as:

$$F_{nt}^t = 1.3F_{nt} - \frac{F_{nt}}{\varphi_b F_{nv}} f_v = 90\,\text{kips} \le F_{nt}$$

$$\left(F_{nt}^t\right)_{allow} = \frac{F_{nt}}{F.S.} = \frac{90}{2} = 45\,\text{ksi}$$

$$\left(R_{bt}\right)_{allow} = \left(F_{nt}^t\right)_{allow} A_b = 45(0.442) = 19.90\,\text{kips}$$

Tensile strength per bolt (through the thickness) is not considered in this example. The tensile strength through thickness is for the connections with prying action and a connection force component aligned with the axis of bolting.

b.4) Perpendicular pin bearing strength per bolt:

The characteristic pin bearing strengths in both the longitudinal and transverse directions are 32 and 16 ksi, respectively (manufacturer). Fastener threads are excluded so $\xi = 1.0$.

The nominal hole diameter is taken as 1/16″ larger than the nominal bolt diameter (3/4″). The orientation between the direction of the connection force and the strong direction of the FRP plates is 90°. The perpendicular pin-bearing strength per bolt is determined from Eq. (11.11) as:

$$R_{br} = dt\xi F_\theta^{br} = \left(\frac{3}{4} + \frac{1}{16}\right)\left(\frac{1}{2}\right)(1.0)16 = 6.5\,\text{kips}$$

The multi-row connection is for a single-lap shear connection, therefore the pin-bearing strength (b.4), must be reduced by a multiplier of 0.6 (ACMA, 2021). Based on the ASD method, the safety factor of a connection is 2.0, which is adopted hereunder to get $\left(R_{br}\right)_{allow}$.

$$\left(R_{br}\right)_{allow} = 0.6\left(\frac{dtF_\theta^{br}}{F.S.}\right) = 0.6\left(\frac{6.5}{2}\right) = 1.95\,\text{kips}$$

The net tensile strength and shear-out strength need not be determined, since there are two perpendicular elements (the flanges) of an FRP wide flange beam at the end member or connecting component of FRP material. The net tensile strength and shear-out strength given herein are not considered in this example.

c. The connection is bolted through the web of the clip angle 2Ls 4 × 4 × 1/2″.

c.1) Shear strength of clip angle (frame type)

Based on the ASD method, the safety factor of a connection is 2.0. The nominal shear strength at the knee of a clip angle is determined from Eq. (11.23) as:

$$R_{sh,sp} = l_{sp} t_{sp} F_{sh} = (4 \times 0.5)6 = 12 \text{ kips}$$

$$\left(R_{sh,sp}\right)_{allow} = \frac{R_{sh,sp}}{F.S.} = \frac{12}{2} = 6 \text{ kips}$$

d. Determine the allowable capacity of the simple frame connection

The minimum strength is found in the case of perpendicular pin bearing strength per bolt (1.95 kips per bolt from b.4) for pultruded FRP WF 12 × 12 × 1/2". For long-term serviceability, the load-carrying capacity is reduced by using the knockdown factor (0.4), The allowable strength including the long-term serviceability factor is (0.4 × 1.95 = 0.78) kips. The design load carrying capacity is determined to be (0.78 × 4 = 3.12) kips. For long-term serviceability, the knockdown factor (0.4) for environmental effects may be applied to the allowable tensile capacity (safety factor = 0.4 × (1/2) = 0.20~or 20% failure strain).

Example 11.7

Design a semi-rigid beam-to-column connection shown in Example Figure 11.7.1 to resist a moment of 15.5 ft-kips and a shear load of 4.5 kips. Use 1/2-in-diameter, A325N high-strength steel bolts in standard holes (in practice add 1/16" to bolt hole to fit ½" diameter bolt). Design this connection using allowable stress design (ASD) methodology. For this problem assumed allowable FRP design values are: bending stress is 30,000 psi, bearing capacity ($F_{bearing}$) is 34,500 psi, and shear stress is 7,000 psi and 60% rigidity in the connection (ASCE Manual, "Design guide for FRP composite connections", Mosallam, 2011).

Solution

a. Reduction of fixed end moment (semi-rigid approximation)

The fixed end induced moment of 15.5 ft-kips is reduced to 9.30 ft-kips due to 60% rigidity in the connection.

b. Design with A325N high-strength bolts

b.1) Bolt Shear: Connected Leg (double-shear at beam leg):

Allowable shear strength in one bolt: $F_v = 10$ ksi (AISC, 2016)

Assume $d_{bolt} = 1/2$ in; area of one bolt: $A_{bolt} = 0.196$ in^2.

Allowable shear per bolt (for beam only) = $2 A_{bolt} F_v = 2 \times 0.196 \times 10 = 3.92$ kips

Number of bolts required: $n = 4.5/3.92 = 1.15$

Use a minimum of two bolts.

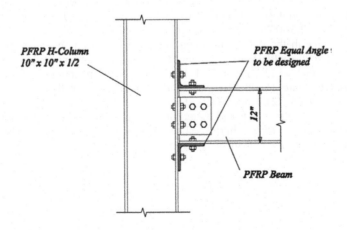

EXAMPLE FIGURE 11.7.1 Connection details.

b.2) Bolt Shear: Outstanding Leg (single – shear at column flange):

F_v = 10 ksi (AISC, 2016)

Assume d_{bolt} = 1/2 in; area of one bolt: A_{bolt} = 0.196 in².

Allowable shear per bolt = $A_{bolt} F_v$ = 10 × 0.196 = 1.96 kips

Number of bolts required: n = 4.5/1.96 = 2.3.

Use a minimum of four bolts, two bolts per row, in column flange (total four bolts), as shown in Example Figure 11.7.1.

c. Bolt bearing (in the web of the beam)

Assume thickness of the web = 1/2 in.

Using a safety factor F.S. = 2.5,

$F_{bearing}$ = 34,500/2.5 = 13,800 psi

Bolt bearing area = thickness of web × bolt diameter = 0.5 in × 0.5 in = 0.25 in²

The allowable bearing stress per bolt = 13,800/4 = 3,450 psi

Required number of bolts is 4.5/3.45 = 1.30

Use two 1/2-in diameter bolts on top and bottom of the I-beam (12″ × 10″ × ½″).

d. Size of the Pultruded FRP angle connecting Beam Web to Column

Check shear on the gross area of FRP angle:

Using a safety factor F.S. = 2.5,

F_v = 7,000/2.5 = 2,800 psi = 2.8 ksi

Assume 1.5″ clearance at the top and bottom

Max angle length = height of the beam–2 × (beam flange thickness)–2 × (clearance)

Therefore,

Maximum FRP angle length = (12–2 × 0.5–2 × 1.5) = 8″

Angle thickness = 4.5/(8 × 2.8)≈0.20. Try 1/2-in thick pultruded FRP angle.

e. Number of bolts through the beam flanges and the top and bottom angles

The capacity of one bolt in shear is 9.3 kips (AISC, 2016).

The minimum capacity of one bolt in bearing = 32.6 kips (AISC, 2016).

The force in each flange of the beam is calculated as follows:

T = C = moment/(height of the beam) = (9.30 × 12)/12 = 9.30 kips

The number of bolts required is 9.3/9.3 = 1.0. Use four bolts (Example Figure 11.7.2), use at least two bolts on the top flange and two bolts on the bottom flange of the beam connecting the top and bottom angles to the beam for symmetry.

f. Bolts through the column flanges

The capacity of one bolt in tension = 19.44 kips (AISC, 2016).

Therefore, number of bolts = 9.30/19.44 = 0.48 (use a minimum of two bolts).

Use a 9″ × 9″ × 3/4″ FRP angle which has a length of one inch less than the column flange width (9 in). Use pultruded FRP equal-leg angle to accommodate the required number of bolts (refer to Figure 11.7.2).

Find the contraflexure point located at the top angle. The distance from the C.G. of the bolts in tension (top angle bolts with force T) to the top of the horizontal leg of the top angle, is 2+2.5–0.75 = 3.75 in, where angle thickness is 3/4″.

Tensile force in the bolts connecting the top and bottom angles to the column is found by using the actual vertical distance between tension-compression couple:

T = C = (9.3 × 12)/21.0 = 5.31 kips (distance from centroid of the connecting (Example Figure 11.7.2) bolt angle and column flange at the top to the bottom bolt = 4.5+12+4.5 = 21″), and M = 5.31 × (3.75/2) = 9.96 in-kip.

Required thickness (t) is found by recalling the section modulus (S) of a rectangular shape is $bt^2/6$ and S = M/F_b, where F_b = bending stress, the required thickness of the FRP is calculated using the following expression:

$$t = \sqrt{\frac{6M}{F_b b}} = \sqrt{\frac{6(9.96)}{12(9)}} = 0.74″$$

where F_b = 30,000/2.5 = 12,000 psi and b = 9 in

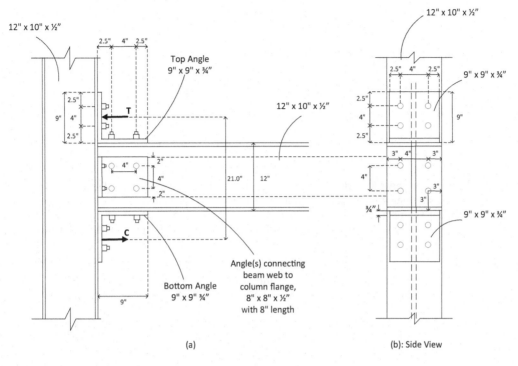

(a)

(b): Side View

EXAMPLE FIGURE 11.7.2 Analysis of the top cleat angle.

The required thickness $t = 0.74$ in $< 0.75''$ thickness for angle as originally suggested for this design.

g. Shear-out and bearing strengths compliance check

g.1) Taking into consideration the web of the beam and assuming that the bolts in the web of the beam are taking 40% of the total shear load:

First, assume the number of steel bolts = 4.

g.1.1) Determine shear load distribution (EUROCOMP, 1996):

The shear load distribution in each row of bolts can be determined by using the shear distribution factor (μ) as given in Table 11.11.

The shear load factors for a two-row joint for rows 1 and 2 are given:

- Distribution factor (μ) for row 1 = 1.15
- Distribution factor (μ) for row 2 = 0.85
- Numbers of bolt row (n^R) = 2
- Number of bolts on the row 1 (n^f) = 2
- Number of bolts on the row 2 (n^f) = 2.

Substituting the distribution factor (μ), number of fastener rows (n^R), and number of fasteners in a row (n^f) into Eq. (11.26.3) average shear loads on a row and a fastener are given as (EUROCOMP, 1996):

$$F_x^{DR} = \frac{\mu F_x^{Total}}{n^R}$$

The average shear loads on row 1:

$$F_x^{DR} = \frac{\mu F_x^{Total}}{n^R} = \frac{1.15 F_x^{Total}}{2} = 0.57 F_x^{Total}$$

The average shear loads on row 2:

$$F_x^{DR} = \frac{\mu F_x^{Total}}{n^R} = \frac{0.85 F_x^{Total}}{2} = 0.43 F_x^{Total}$$

From Eq. (11.26.4), the average shear loads on a fastener $\left(F_x^F\right)$ of row 1:

$$F_x^F = \frac{F_x^{DR}}{n^F} = \frac{\mu F_x^{Total}}{n^R n^F} = \frac{1.15 F_x^{Total}}{2 \times 2} = 0.285 F_x^{Total}$$

From Eq. (11.26.4), the average shear loads on a fastener $\left(F_x^F\right)$ of row 2:

$$F_x^F = \frac{F_x^{DR}}{n^F} = \frac{\mu F_x^{Total}}{n^R n^F} = \frac{0.85 F_x^{Total}}{2 \times 2} = 0.215 F_x^{Total}$$

The average shear load distribution in each bolt is illustrated in Example Figure 11.7.3.
g.1.2) Determine bearing load distribution for individual bolts in a row:

From Example Figure 11.7.3, the average shear load on each bolt in row 1 can be classified into load case 1 as shown in Example Figure 11.7.4. In addition, the average shear load of each bolt in row 2 can be classified into load cases 1 and 3 as shown in Example Figure 11.7.5.

- For load case 1, there is a bearing load acting on the bolt and the nature of loading is tensile. In this case, the effective width over which the bearing load is active is equal to $4d$.
- For load case 3, there is a by-pass load, i.e., the load is taken by the plate rather than the bolt. In this case, the effective width over which the bearing load is active is greater than $4d$.

g.1.3) Determine normalized stress distribution around the bolt hole, effective stress concentration, and notch sensitivity factor:

The normalized stress distribution factors of quasi-isotropic composite material for bearing (K_r) and shear-out (K_s) can be determined by referring to the stress charts provided (by EUROCOMP Design Code, 1996-Figures 5.13–5.18).

- For load case 1: $K_r = 1.1$ and $K_s = 0.59$ (dominant)
- For load case 3: $K_r = 1.1$ and $K_s = 0.80$

g.1.4) Calculate the induced shear-out strength in FRP material (Taking into consideration the web of the beam and assuming that the bolts in the web of the beam are taking 40% of the total shear load:

First, assume the number of steel bolts = 4 (EUROCOMP Design Code, 1996)

$$\tau_{sh,s} = K_s \tau_{ns} = K_s \frac{F_x^F}{A} = 0.59 \frac{0.285 F_x^{Total}}{t d_{bolt}} = 0.59 \frac{0.285 \times 4.5 \times 0.4^*}{0.75 \times 0.5} = 0.807 \, \text{ksi}$$

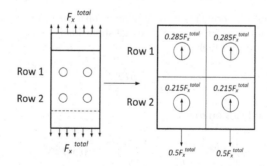

EXAMPLE FIGURE 11.7.3   Average shear load distribution in each bolt (EUROCOMP, 1996).

Load Case 1

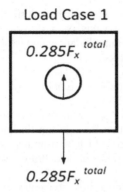

EXAMPLE FIGURE 11.7.4 Load distribution of row 1.

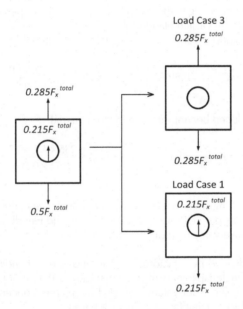

EXAMPLE FIGURE 11.7.5 Load distribution of row 2 (EUROCOMP, 1996).

*Bolts in the beam web are taking 40% of the total shear load (refer to g.1)

g.1.5) Calculate the induced bearing strength in FRP material:

$$\sigma_{r,s} = K_r \sigma_r = K_r \frac{F_x^F}{A} = 1.1 \frac{0.285 F_x^{Total}}{td_{bolt}} = 1.1 \frac{0.285 \times 4.5 \times 0.4}{0.75 \times 0.5} = 1.504 \, \text{ksi}$$

g.1.6) Compute the resisting shear-out strength of the composite material:

$\tau_{sh,\, k} = 24.07$ ksi (as per EUROCOMP Design Code and Handbook, 1996)

Dividing the above value by a safety factor of 2.5, we have $\tau_{sh,\, k} = 9.63$ ksi.

The factor (C_m) is a product of $(C_{m,1}$ and $C_{m,3})$. This factor accounts for lap joint types $(C_{m,1})$ and the ratio of the distance between center to center distance of holes and diameter of both holes. The factor $(C_{m,1})$ is taken as 2 for single-lap joints and 1.2 for double-lap joints with symmetric loads. The factor $(C_{m,3})$ is given in Figure 11.15 (EUROCOMP, 1996).

$C_{m,3}$ is taken as 1.0 all the time because the bolt spacing is maintained at least $4 \times d_{bolt}$.

After dividing the shear-out value by a stress concentration factor of $C_{m,1} = 1.2$ (double-lap joint) and multiplying it by an aging factor of 0.4, shear-out strength of the composite material: $\tau_{sh, k} = 3.21$ ksi.

Induced shear-out stress: $\tau_{sh, s} = 0.807$ ksi $\leq \tau_{sh, k} = 3.21$ ksi; hence acceptable g.1.7). Calculate resisting bearing strength of FRP material:

For bearing, we have $\sigma_{r, k} = 38$ ksi.

Dividing this stress value by a safety factor of 2.5 produces $\sigma_{r, k} = 15.2$ ksi.

Nonuniform stress distribution through the thickness of the laminate due to non-symmetry, bolt bending, and loss of lateral restraint, we will divide by a factor $C_m = 1.2$ (EUROCOMP Design Code and Handbook, 1996), resulting in further reduction of $\sigma_{r, k}$ ($\sigma_{r, k} = 12.7$ ksi).

Furthermore, $\sigma_{r, k}$ is divided by a stress concentration factor of 1.6, and multiplying by an aging factor of 0.4, we have in bearing, $\sigma_{r, k} = 3.16$ ksi.

$\sigma_{r, s} = 1.504$ ksi $\leq \sigma_{r, k} = 3.16$ ksi; hence acceptable.

g.2) Assume the top and bottom angles attached to the column flange (single shear) with four bolts, each resists 30% of the transverse shear and the remaining 40% is resisted by beam web angles connecting to the column flange. The above percentages are based on the EUROCOMP (1996) Design Handbook.

g.2.1) Calculate the induced shear-out strength in FRP material:

$$\tau_{sh,s} = K_s \tau_{ns} = K_s \frac{F_x^F}{A} = 0.59 \frac{0.285 F_x^{Total}}{t d_{bolt}} = 0.59 \frac{0.285 \times 4.5 \times 0.3}{0.75 \times 0.5} = 0.605 \, \text{ksi}$$

g.2.2) Calculate induced bearing strength in FRP material:

$$\sigma_{r,s} = K_r \sigma_r = K_r \frac{F_x^F}{A} = 1.1 \frac{0.285 F_x^{Total}}{t d_{bolt}} = 1.1 \frac{0.285 \times 4.5 \times 0.3}{0.75 \times 0.5} = 1.128 \, \text{ksi}$$

g.2.3) Calculate resisting shear-out strength of the composite material:

$\tau_{sh, k} = 24.07$ ksi (as per EUROCOMP Design Code and Handbook, 1996).

Dividing the above value by a safety factor of 2.5, we have:

$\tau_{sh, k} = 9.63$ ksi.

$\tau_{sh, k}$ is further divided by a factor $C_m = 2$ to account for bending effects due to the eccentricity in the single-lap joint configuration (EUROCOMP, 1996). Thus, $\tau_{sh, k} = 4.82$ ksi.

Furthermore, the shear-out value is divided by a stress concentration factor of 1.2, and multiplying it by an aging factor of 0.4, we have:

$\tau_{sh, k} = 1.61$ ksi

(induced) $\tau_{sh, s} = 0.605$ ksi $< \tau_{sh, k} = 1.61$ ksi (allowable); hence, the proposed shear-out strength design is acceptable.

g.2.4) Calculate resisting bearing strength of FRP material:

In bearing, we have $\sigma_{r, k} = 38,000$ psi.

Dividing by a safety factor of 2.5, we have $\sigma_{r, k} = 15,200$ psi.

The factor (C_m) is a product of ($C_{m,1}$ and $C_{m,3}$). This factor accounts for lap joint types ($C_{m,1}$) and the ratio of the distance between holes center to center and diameter of both holes. The factor ($C_{m,1}$) is taken as 2 for single-lap joints and 1.2 for double-lap joints with symmetric loads. The factor ($C_{m,3}$) is given in Figure 11.15 (EUROCOMP, 1996). $C_{m,3}$ is taken as 1.0 all the time because the bolt is maintained at least $4 \times d_{bolt}$.

Dividing by a factor $C_m = 2$ to account for bending effects due to eccentricity in the single-lap joint; we have $\sigma_{r, k} = 7,600$ psi.

Dividing by a stress concentration factor of 1.6 and multiplying by an aging factor of 0.4, $\sigma_{r, k} = 1.9$ ksi.

(induced) $\sigma_{r, s} = 1.128$ ksi $\leq \sigma_{sr, k} = 1.9$ ksi (allowable); hence acceptable.

Note: The above computations for shear-out and bearing strengths will hold good for the bottom angle, as well, in case of loading reversals.

h. **Check maximum force in bolt group** (for bolts in the web when subjected to a combination of induced moment and shear)

The torsional moment is $M = 9.30$ ft-kip $= 111.6$ in-kip (eccentricity of shear load: $e = \dfrac{M}{V}$).

The polar moment of inertia is calculated as:

Note: The four middle holes bear 0.285 of the transferred loads, whereas each set of away holes (group of four holes at top and bottom) bears 0.215/4 of the loads.

$\Sigma x^2 = 12(2^2) = 48$ in^2
$\Sigma y^2 = 4(11.5^2) + 4(7.5^2) + 4(2^2) = 770$ in^2
$\Sigma x^2 + \Sigma y^2 = 818$ in^2

The vertical and horizontal components of the stresses on the connectors are as follows:

$$f_x = \frac{\alpha M y}{A\left(\sum x^2 + \sum y^2\right)}$$

$$f_y = \frac{\alpha M x}{A\left(\sum x^2 + \sum y^2\right)}$$

where M = torsional moment

x and y = the position of the C.G. of the bolt group

A = area of one bolt

$\alpha = \dfrac{0.215}{0.285} \approx 0.75$, a ratio indicating the portion of the stresses distributed on the connectors (EUROCOMP, 1996)

Example Figure 11.7.6 illustrates immediate-hole and away-hole bearing capacities of 0.285 and 0.215, respectively.

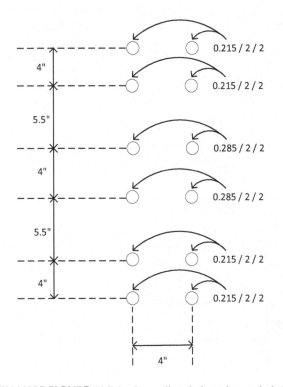

0.215 / 2 / 2

0.215 / 2 / 2

0.285 / 2 / 2 C.G. in horizontal direction (x) = 2"

0.285 / 2 / 2 C.G. in vertical direction (y) = 11.5"

0.215 / 2 / 2

0.215 / 2 / 2

4"

5.5"

4"

5.5"

4"

4"

EXAMPLE FIGURE 11.7.6 Immediate-hole and away-hole bearing capacities.

$$f_x = \frac{0.75(111.6)(11.5)}{0.196(818)} = 6.0 \, \text{ksi}$$

$$f_y = \frac{0.75(111.6)(2)}{0.196(818)} = 1.04 \, \text{ksi}$$

Assuming the shear load (P) along the "y" axis is equally shared by all the connectors, the direct shear stress is as follows:

$$f_S = \frac{P}{\Sigma A}$$

$$f_s = \frac{4.5}{12(0.196)} = 1.91 \, \text{ksi}$$

The total resultant stress is:

$$f = \sqrt{(f_y + f_s)^2 + f_x^2}$$

$$f = \sqrt{(1.91 + 1.04)^2 + 6.0^2} = 6.69 \, \text{ksi}$$

The induced strength in shear-out due to this load is:

$$f \times K_s = 6.69 \times 0.59 = 3.94 \, \text{ksi}$$

This induced strength of 3.94 ksi is lower than the allowable shear strength of 4.0 ksi as given by the manufacturer.

i. Checking block-shear in the web of the beam (refer to Example Figure 11.7.7)

The capacity of the connection based on web tear-out (block shear) is calculated as follows:

The thickness of the web is 0.5 in.
Block shear load $P_{block} = A_v \times \tau_{sh} + A_t \times \sigma_n$
The shear and tension areas of the web tear out are found by: (2 angles in web) × (net length of plane) × (thickness)
Shear area $A_v = 2[6-1.5\,(0.5)] \times 0.5 = 5.25 \, \text{in}^2$
Tension area $A_t = 2[2-0.5\,(0.5)] \times 0.5 = 1.75 \, \text{in}^2$
Allowable value of stress in shear = [7,000/(2.5 × 1.2)] × 0.4 = 933.3 psi
Allowable value of stress in tension = [76,250/(2.5 × 1.2)] × 0.4 = 10,166.67 psi

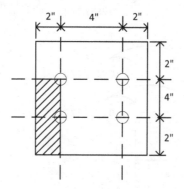

EXAMPLE FIGURE 11.7.7   Web tear-out (block shear).

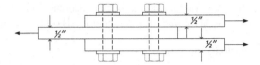

EXAMPLE FIGURE 11.7.8 Double-lap shear of beam-web connection.

The allowable block shear load $P_{block} = [5.25 \times 933.3 + 1.75 \times 10,166.67] = 22.7$ kips $> 0.4 \times 4.5 = 1.8$ kips.

Note: The factor 1.2 in the above stress calculation is attributed to stress concentration due to holes in double-lab shear. A designer should modify the stress concentration factor of 1.2 to a higher value if warranted. Web of WF $12'' \times 10'' \times \frac{1}{2}''$, as shown in Example Figure 11.7.8, need not be checked for shear because the beam shear is evaluated as a part of the beam design.

(SI unit) Example 11.8

Using the LRFD method and EUROCOMP, design a composite plate sandwiched between two steel plates (Example Figure 11.8.1), in a double-lap configuration when the joint is subjected to a unit static load (F_x) of 10 kN. The minimum spacing, edge distance, and side distance are $4d_h$, $4d_b$, and $4d_b$, respectively.

Given: lap length details for connection: Overlap length (L) = 153.6 mm, Width (W) = 153.6 mm, Thickness (t) = 7 mm, Edge distance (e) = 50.8 mm, Side distance (s) = 50.8 mm, Spacing in x direction (W_x) = 52 mm, Spacing in y direction (W_y) = 52 mm, diameter of bolt (d_b) = 12.7 mm, diameter of hole (d_h) = 13 mm, and Unfactored uniaxial design load (F_x)= 10 kN. The ultimate strengths as obtained from experiments are given as: $\sigma_{r,\,k}$ = 400 MPa, $\sigma_{\phi,\,k}$ = 525 MPa and $\tau_{sn,\,k}$ = 166 MPa. The resistance factors (γ_m) for bearing, net-tension, and shear-out stress are given to be 1.25, 1.25, and 1.35 respectively. Aging factor (A_G) = 0.4.

Solution

a. Determine shear load distribution:

The shear load distribution in each row of bolts can be determined by using the shear distribution factor (μ) as given in Table 11.11. The shear load factors for a two-row joint for rows 1 and 2 are given:

- Distribution factor (μ) for row 1 = 1.15
- Distribution factor (μ) for row 2 = 0.85
- Numbers of bolt row (n^R) = 2
- Number of bolts on the row 1 (n^f) = 2
- Number of bolts on the row 2 (n^f) = 2.

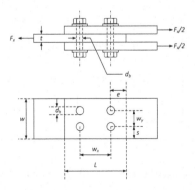

EXAMPLE FIGURE 11.8.1 Double-lap connection under axial load.

Substituting the distribution factor (μ), number of fastener rows (n^R), and number of fastener in a row (n^f) into Eq. (11.26.3), average shear loads on each bolt row and a fastener are given as:

The average shear loads on row 1:

$$F_x^{DR} = \frac{\mu F_x^{Total}}{n^R} = \frac{1.15 F_x^{Total}}{2} = 0.57 F_x^{Total}$$

The average shear loads on row 2:

$$F_x^{DR} = \frac{\mu F_x^{Total}}{n^R} = \frac{0.85 F_x^{Total}}{2} = 0.43 F_x^{Total}$$

From Eq. (11.26.4), the average shear loads on a fastener (F^{xF}) of row 1:

$$F_x^F = \frac{F_x^{DR}}{n^F} = \frac{\mu F_x^{Total}}{n^R n^F} = \frac{1.15 F_x^{Total}}{2 \times 2} = 0.285 F_x^{Total}$$

From Eq. (11.26.4), the average shear loads on a fastener (F_x^f) of row 2:

$$F_x^F = \frac{F_x^{DR}}{n^F} = \frac{\mu F_x^{Total}}{n^R n^F} = \frac{1.15 F_x^{Total}}{2 \times 2} = 0.285 F_x^{Total}$$

The average shear load distribution in each both is illustrated in Example Figure 11.8.2.

b. Determine bearing and by-pass load distribution for individual bolts in a row:

From Example Figure 11.8.2, the average shear load on each bolt in row 1 can be classified into load case 1 as shown in Example Figure 11.8.3. In addition, the average shear load of each bolt in row 2 can be classified into load cases 1 and 3 as shown in Example Figure 11.8.4.

- For load case 1, there is a bearing load acting on the bolt and the nature of loading is tensile. In this case, the effective width is equal to 4d.
- For load case 3, there is a by-pass load, i.e., the load is taken by the plate rather than the bolt. In this case, the effective width is greater than 4d.

c. Determine normalized stress distribution around the bolt hole, effective stress concentration, and notch sensitivity factor:

The normalized stress distribution factors of quasi-isotropic composite material for net-tension (K_t), bearing (K_r), and shear-out (K_s) are determined from stress charts (Figures 5.13–5.18, pp. 172–176) provided in EUROCOMP Design Code (1996).

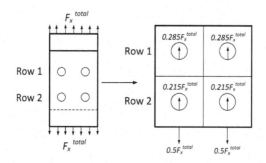

EXAMPLE FIGURE 11.8.2   Average shear load distribution in each bolt (EUROCOMP, 1996).

Load Case 1

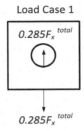

$0.285F_x{}^{total}$

$0.285F_x{}^{total}$

EXAMPLE FIGURE 11.8.3   Load distribution of row 1 (EUROCOMP, 1996).

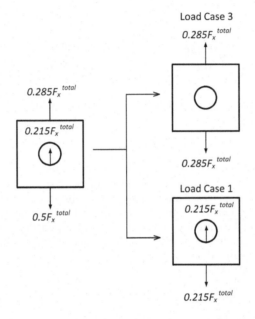

Load Case 3

$0.285F_x{}^{total}$

$0.285F_x{}^{total}$

$0.215F_x{}^{total}$

$0.285F_x{}^{total}$

$0.5F_x{}^{total}$

Load Case 1

$0.215F_x{}^{total}$

$0.215F_x{}^{total}$

EXAMPLE FIGURE 11.8.4   Load distribution of row 2 (EUROCOMP, 1996).

- For load case 1: $K_r = 1.1$, $K_t = 0.65$, $K_s = 0.59$
- For load case 3: $K_r = 1.1$, $K_t = -0.40$, $K_s = 0.80$

d. Determine the induced bearing, tensile and shear stress under applied load:

To find bearing tensile and shear stresses ($\sigma_{r,\,s}$, $\sigma_{\phi,\,s}$ and $\tau_{sn,\,s}$), stress distribution factors are used only for load case 1 as it is the most critical loading condition for the bolts in the connection. Then, the bearing, tensile and shear stresses ($\sigma_{r,\,s}$, $\sigma_{\phi,\,s}$ and $\tau_{sn,\,s}$) are determined as:

Substituting the average shear loads on a bolt ($F^x{}_F$) of row 1 into Eq. (11.24.7), bearing stress due to the total design load ($F_x{}^f$) is given as:

$$\sigma_{r,s} = K_r \sigma_r = K_r \frac{F_x^F}{A}$$

$$\sigma_{r,s} = 1.1 \frac{0.285 F_x^{Total}}{d_h t} = 1.1 \frac{0.285 (10,000\,\text{N})}{13(7)\,\text{mm}^2} = 34.5\,\text{MPa}$$

Substituting the average shear loads on a bolt (F_x^F) of row 1 into Eq. (11.24.8), net-section tensile stress due to the total design load (F_x^F) is given as:

$$\sigma_{\varphi,s} = K_t \sigma_\varphi = K_r \frac{F_{xx}^F}{A}$$

$$\sigma_{\varphi,s} = 0.65 \frac{0.285 F_x^{Total}}{d_h \times t} = 0.65 \frac{0.285 (10,000\,\text{N})}{13(7)\,\text{mm}^2} = 20.3\,\text{MPa}$$

Substituting the average shear loads on a bolt (F_x^F) of row 1 into Eq. (11.24.9), shear stress due to the total design load (F_x^F) is given as:

$$\tau_{ns,s} = K_s \tau_{ns} = K_s \frac{F_x^F}{A}$$

$$\tau_{ns,s} = 0.59 \frac{0.285 F_x^{Total}}{d_h \times t} = 0.59 \frac{0.285 (10,000\,\text{N})}{13(7)\,\text{mm}^2} = 18.4\,\text{MPa}$$

For the LRFD approach, the load factor (ϕ_F) of 1.5 is applied to the above bearing, tensile, and shear stresses of ($\sigma_{r,\,s}$, $\sigma_{\phi,\,s}$ and $\tau_{sn,\,s}$).
- Bearing: $\phi_F \sigma_{r,\,s} = 1.5(34.5) = 51.8\,\text{MPa}$
- Net-tension: $\phi_F \sigma_{\varphi,\,s} = 1.5(20.3) = 30.5\,\text{MPa}$
- Shear-out: $\phi_F \tau_{sn,\,s} = 1.5(18.4) = 27.6\,\text{MPa}$

e. Determine the ultimate strengths of laminated composites:

The ultimate bearing, tensile and shear-out strengths ($\sigma_{r,\,k}$, $\sigma_{\varphi,\,k}$ and $\tau_{sn,\,k}$) of laminated composites can be generally determined from Eqs. (11.24.1–11.24.6) as per EUROCOMP (1996). However, in the current example, experimental strength values are used to compare with the design stresses in bearing, tensile, and shear. The ultimate strengths as obtained from experiments are given as:
- Bearing: $\sigma_{r,\,k} = 400\,\text{MPa}$
- Net-tension: $\sigma_{\varphi,\,k} = 525\,\text{MPa}$
- Shear-out: $\tau_{sn,\,k} = 166\,\text{MPa}$

f. Perform compliance check:

The resistance factors (γ_m) for bearing, net-tension, and shear-out stress are given as 1.25, 1.25, and 1.35, respectively. In addition, the aging factor ($A_G = 0.4$) is also introduced to account for the strength reduction of composite materials due to physical aging and chemical aging.

The factor ($C_{m,1}$) is taken as 1.2 for double-lap joints with symmetric loads. From Figure 11.15, the factor ($C_{m,3}$) for the ratio of the center to center hole distance to hole diameter as 4 is 1.0. Thus, the factor (C_m) is determined as:

$$C_m = C_{m,1} C_{m,3} = 1.2(1.0) = 1.2$$

The compliance check performed is given in Eqs. (11.25.1–11.25.3) as:

For bearing strength:

$$\varphi_F C_m \sigma_{\varphi,s} \leq \frac{1}{\gamma_m} \sigma_{\varphi,k} A_G$$

$$1.5(1.2)20.3 \leq \left(\frac{525}{1.25} \right) 0.4$$

$$36.5\,\text{MPa} \leq 168\,\text{MPa}$$

For tensile strength:

$$\varphi_F C_m \tau_{sn,s} \leq \frac{1}{\gamma_m} \tau_{sn,k} A_G$$

$$1.5(1.2)18.4 \leq \left(\frac{166}{1.35}\right)0.4$$

$$33.1\,\text{MPa} \leq 49.2\,\text{MPa}$$

For shear-out strength:

$$\varphi_F C_m \tau_{sn,s} \leq \frac{1}{\gamma_m} \tau_{sn,k} A_G$$

$$1.5(1.2)18.4 \leq \left(\frac{166}{1.35}\right)0.4$$

$$33.1\,\text{MPa} \leq 49.2\,\text{MPa}$$

The bearing, tensile, and shear-out capacities are higher than the design stresses. Hence, the design is satisfactory.

Alternate method:

("Design Guide for FRP Composite Connections" American Society of Civil Engineers (ASCE), A. Mosallam, 2011). A comprehensive treatment of this alternate approach can be found in Mosallam (2011).

Using the LRFD method and EUROCOMP, design a composite plate sandwiched between two steel plates (Example Figure 11.8.1), in a double-lap configuration when the joint is subjected to a unit static load (F_x) of 10 kN. The minimum spacing, edge distance, and side distance are $4d_h$, $4d_b$, and $4d_b$, respectively.

Solution

The tensile capacity of a mechanical connection is determined as:

a. Ultimate net-section tension failure load:
   The ultimate net-tension failure load is determined as:

$$P_{ult}^{net-tension} = \eta(t w F_{tu}) = 0.34(7)153.6(525) = 189.7\,\text{kN}$$

and

$$\eta = \frac{1}{1+C(k_{te}-1)}\left(1-\frac{nd}{w}\right) = 0.34$$

where η = joint efficiency = (strength of notched member)/(strength of unnotched member defined in Section 11.3)

$k_{te} \approx \dfrac{w}{nd}$ = stress concentration factor of isotropic material of a joint (Hart-smith, 1978)

w = member width
n = number of bolts in a row along loading direction
d = hole diameter
$C = 0.3$, assuming fiber orientation is along the pultrusion direction (refer to Section 11.3.1)

b. Cleavage failure load:
   The ultimate cleavage failure load is:

$$P_{ult}^{cleavage} = F_{br}t\frac{d^2}{d_{bolt}}\eta = 400(7)\frac{13^2}{12.7}0.34 = 12.7\,kN$$

and

$$\eta = \frac{1}{1+C\left(k_{te}-1\right)}\left(1-\frac{nd}{w}\right)\Psi = 0.34$$

where

$$\Psi = \left(\frac{6}{5}-\frac{3d}{5e}\right)^{\nu}$$

$\nu = 0$

c. Bearing failure load:

$$P_{ult}^{bearing} = F_{br}t\frac{d^2}{d_{bolt}} = 400\ (7)\ \frac{13^2}{12.7} = 37.3\,kip$$

The ultimate tensile capacity of the joint is 6.2 kip

$$P_{ult}^{cleavage} < P_{ult}^{bearing} < P_{ult}^{net-tension}$$

Note: The applied load (F^x = 10 kN) is lower than the ultimate tensile capacity of the joint (12.7 kN).

(SI unit) Example 11.9

Using the LRFD method and EUROCOMP, design a three-row six bolt joint (Example Figure 11.9.1), in a double-lap configuration that is subjected to a unit static load (F_x) of 30 kN.

Given: lap length details for connection: Width (W) = 153.6 mm, Thickness (t) = 16 mm, diameter of bolt (d_b) = 15.88 mm, diameter of hole (d_h) = 16 mm, and Unfactored uniaxial design load

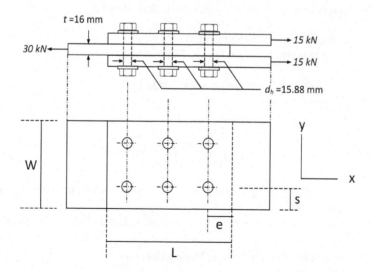

EXAMPLE FIGURE 11.9.1 Double-lap connection under axial load.

(F_x)= 30 kN. The experimental ultimate strengths are: $\sigma_{r,\,k}$ = 400 MPa, $\sigma_{\varphi,\,k}$ = 525 MPa and $\tau_{sn,\,k}$ = 166 MPa. The resistance factors (γ_m) for bearing, net-tension, and shear-out stress are given to be 1.25, 1.25, and 1.35 respectively. Aging factor (A_G) = 0.4.

Solution

Try: connection configuration: Edge distance (e) = 50.8 mm, Side distance (s) = 50.8 mm, Spacing in x direction (W_x) = 52 mm, Spacing in y direction (W_y) = 52 mm, and Overlap length (L) = 205.6 mm.

a. Determine shear load distribution:

The shear load distribution in each row of bolts can be determined by using the shear distribution factor (μ) as given in Table 11.11. The shear load factors for a two-row joint for row 1, 2 and 3 is given:
- Distribution factor (μ) for row 1 = 1.15
- Distribution factor (μ) for row 2 = 0.85
- Distribution factor (μ) for row 3 = 0.65
- Numbers of bolt row (n^R) = 3
- Number of bolts on the row 1 (n^F) = 2
- Number of bolts on the row 2 (n^F) = 2
- Number of bolts on the row 3 (n^F) = 2.

Substituting the distribution factor (μ), number of fastener rows (n_R), and number of fastener in a row (n_F) into Eq. (11.26.3), the average shear loads in a row per a fastener are given as:

The average shear loads in row 1:

$$F_x^{DR} = \frac{\mu F_x^{Total}}{n^R} = \frac{1.5 F_x^{Total}}{3} = 0.5 F_x^{Total}$$

The average shear loads in row 2:

$$F_x^{DR} = \frac{\mu F_x^{Total}}{n^R} = \frac{0.85 F_x^{Total}}{3} = 0.283 F_x^{Total}$$

The average shear loads in row 3:

$$F_x^{DR} = \frac{\mu F_x^{Total}}{n^R} = \frac{0.65 F_x^{Total}}{3} = 0.217 F_x^{Total}$$

From Eq. (11.26.4), the average shear loads per fastener (F_{xF}) of row 1:

$$F_x^F = \frac{F_x^{DR}}{n^F} = \frac{\mu F_x^{Total}}{n^R n^F} = \frac{1.5 F_x^{Total}}{3 \times 2} = 0.25 F_x^{Total}$$

From Eq. (11.26.4), the average shear loads per fastener (F_{xF}) of row 2:

$$F_x^F = \frac{F_x^{DR}}{n^F} = \frac{\mu F_x^{Total}}{n^R n^F} = \frac{0.85 F_x^{Total}}{3 \times 2} = 0.14 F_x^{Total}$$

From Eq. (11.26.4), the average shear loads per fastener (F_{xF}) of row 3:

$$F_x^F = \frac{F_x^{DR}}{n^F} = \frac{\mu F_x^{Total}}{n^R n^F} = \frac{0.65 F_x^{Total}}{3 \times 2} = 0.11 F_x^{Total}$$

The average shear load distribution in each bolt is illustrated in Example Figure 11.9.2.

b. Determine bearing and by-pass load distribution for individual bolts in a row:

From Example Figure 11.9.2, the average shear load of each bolt in row 1 can be classified into load case 1 as shown in Example Figure 11.9.3. In addition, the average shear load of each bolt in rows 2 and 3 can be classified into load cases 1 and 3 as shown in Example Figure 11.9.4.

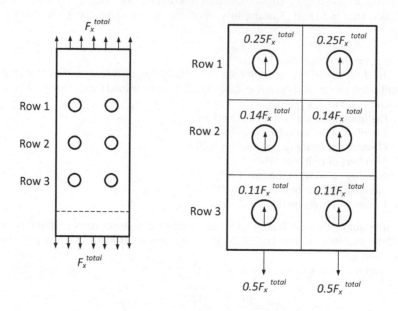

EXAMPLE FIGURE 11.9.2   Average shear load distribution in each bolt (EUROCOMP, 1996).

Load Case 1

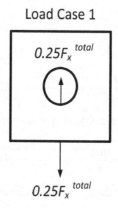

EXAMPLE FIGURE 11.9.3   Load distribution of row 1 (EUROCOMP, 1996).

- For load case 1, there is a bearing load acting on the bolt and the nature of loading is tensile. In this case, the effective width is equal to $4d$.
- For load case 3, there is a by-pass load, i.e., the load is taken by the plate rather than the bolt. In this case, the effective width is greater than $4d$.

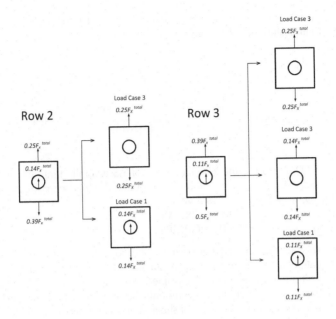

EXAMPLE FIGURE 11.9.4   Load distribution of row 2 and 3 (EUROCOMP, 1996).

c. Determine normalized stress distribution around the bolt hole, effective stress concentration, and notch sensitivity factor:

The normalized stress distribution factors for net-tension (K_r), bearing (K_t), and shear-out (K_s) are determined from stress charts provided in EUROCOMP Design Code (1996) (Figures 5.13–5.18).

- For load case 1: $K_r = 1.35$, $K_t = 0.65$, $K_s = 0.59$
- For load case 3: $K_r = 1.35$, $K_t = -0.40$, $K_s = 0.80$.

d. Determine induced bearing, tensile, and shear stress under applied load:

Further in calculating bearing tensile and shear stress $(\sigma_{r,s}, \sigma_{\varphi,s},$ and $\tau_{sn,s})$, stress distribution factors are used only for load case 1 as it is the most critical loading condition for the bolts in the connection. Then, the bearing, tensile, and shear stresses $(\sigma_{r,s}, \sigma_{\varphi,s},$ and $\tau_{sn,s})$ are determined as:

Substituting the average shear loads on a bolt (F_x^F) of row 1 into Eq. (11.24.7), bearing stress due to the total design load (F_x^F) is given as:

$$\sigma_{r,s} = K_r \sigma_r = K_r \frac{F_x^F}{A}$$

$$\sigma_{r,s} = 1.35 \frac{0.285 F_x^{Total}}{d_h t} = 1.35 \frac{0.25(30,000\,\text{N})}{16(16)\,\text{mm}^2} = 39.6\,\text{MPa}$$

Substituting the average shear loads on a bolt (F_x^F) of row 1 into Eq. (11.24.8), net-section tensile stress due to the total design load (F_x^F) is given as:

$$\sigma_{\varphi,s} = K_t \sigma_\varphi = K_r \frac{F_{xx}^F}{A}$$

$$\sigma_{\varphi,s} = 0.65 \frac{0.285_x^{Total}}{d_h \times t} = 0.65 \frac{0.25(30,000\,\text{N})}{16(16)\,\text{mm}^2} = 19.1\,\text{MPa}$$

Substituting the average shear loads on a bolt (F_x^f) of row 1 into Eq. (11.24.9), shear stress due to the total design load (F_x^f) is given as:

$$\tau_{ns,s} = K_s \tau_{ns} = K_s \frac{F_x^F}{A}$$

$$\tau_{ns,s} = 0.59 \frac{0.25 F_x^{Total}}{d_h \times t} = 0.59 \frac{0.25(30,000\,\text{N})}{16(16)\,\text{mm}^2} = 17.3\,\text{MPa}$$

For the LRFD approach, the load factor (φ_F) of 1.5 is applied into the above bearing, tensile and shear stresses ($\sigma_{r,s}$, $\sigma_{\varphi,s}$ and $\tau_{sn,s}$) to obtain the load-carrying capacity.
- Bearing: $\varphi_F \sigma_{r,s} = 1.5(39.6) = 59.4\,\text{MPa}$
- Net-tension: $\varphi_F \sigma_{\varphi,s} = 1.5(19.1) = 28.7\,\text{MPa}$
- Shear-out: $\varphi_F \tau_{sn,s} = 1.5(17.3) = 26.0\,\text{MPa}$.

e. Determine the ultimate strengths of laminated composites:

The ultimate bearing, tensile and shear-out strengths ($\sigma_{r,k}$, $\sigma_{\varphi,k}$ and $\tau_{sn,k}$) of the laminated composites can be generally determined from Eqs. (11.24.1–11.24.6). However, in the current example, the results of experimental strength provided are used to compare to the design above bearing, tensile and shear stress. The ultimate strengths as obtained from experiments are given as:
- Bearing: $\sigma_{r,k} = 400\,\text{MPa}$
- Net-tension: $\sigma_{\varphi,k} = 525\,\text{MPa}$
- Shear-out: $\tau_{sn,k} = 166\,\text{MPa}$.

f. Perform compliance check:

The resistance factors (γ_m) for bearing, net-tension, and shear-out stress are given to be 1.25, 1.25, and 1.35 respectively. In addition, the aging factor ($A_G = 0.4$) is also introduced to account for the strength reduction of the composite materials due to physical aging and chemical aging.

The factor ($C_{m,1}$) is taken as 1.2 for double-lap joints with symmetric loads. From Figure 11.15, the factor ($C_{m,3}$) for the ratio of the center to center hole distance to hole diameter as 3.25 is 1.1. Thus, the factor (C_m) is determined as:

$$C_m = C_{m,1} C_{m,3} = 1.2(1.1) = 1.32$$

The compliance check performed is given in Eqs. (11.25.1–11.25.3) as:
For bearing strength:

$$\varphi_F C_m \sigma_{r,s} \le \frac{1}{\gamma_m} \sigma_{r,k} A_G$$

$$1.5(1.32)39.6 \le \left(\frac{400}{1.25}\right)0.4$$

$$78.4\,\text{MPa} \le 128\,\text{MPa}$$

For tensile strength:

$$\varphi_F C_m \sigma_{\varphi,s} \le \frac{1}{\gamma_m} \sigma_{\varphi,k} A_G$$

$$1.5(1.32)19.1 \le \left(\frac{525}{1.25}\right)0.4$$

$$37.8\,\text{MPa} \le 168\,\text{MPa}$$

For shear-out strength:

$$\varphi_F C_m \tau_{sn,s} \le \frac{1}{\gamma_m} \tau_{sn,k} A_G$$

$$1.5(1.32)17.3 \le \left(\frac{166}{1.35}\right)0.4$$

$$34.3\,\text{MPa} \le 49.2\,\text{MPa}$$

The bearing, tensile and shear-out capacities are higher than the design stresses. Hence, the design is satisfactory.

11.8 NOMINAL STRENGTH OF ADHESIVE CONNECTIONS

Adhesive connections are gaining acceptance in infrastructural applications of polymer composite components. Their superior performance to mechanical connections is attributed to relatively smoother stress transfer, and stress distribution from one member to another at the connection location when compared with mechanical connections. However, mechanical connections such as steel and FRP bolts provide higher localized stress concentration in joints near the bolt holes. The stress distribution in adhesive connections can provide more uniform stress distribution and better damping and fatigue resistance over mechanical connections. However, special care must be provided in terms of surface preparation before bonding and external loads cannot be applied soon after bonding because of adhesive curing time requirements.

For infrastructural applications, an adhesive in a bonded joint must resist substantial loads, while transferring those loads into structural components. Typically, the shear strength of adhesive varies from 5 (polyurethane) to 50 MPa (epoxy) (Da Silva et al., 2011). However, there are several significant factors affecting the efficacy of an adhesive connection. These factors are (1) stress concentration at the tip of a bonded connection, (2) connection efficiency (defined as the ratio of joint strength to unjointed strength of the same material and configuration), and (3) connection stiffness. Stress concentration over a small area will increase peeling and cleavage stress, and result in inefficient adhesive connection configuration.

The primary design objective here is to develop a reliable adhesive joint and evaluate for joint integrity under shear and peel stresses with emphasis on peel stress concentration at the tip of an adhesive joint. Shear stresses are induced by unequal straining of the adherends, whereas peel stress exists due to eccentricities of applied loads, which are found normally at the free end of a lap joint. Stress concentrations are affected by: (1) thickness of the adherends, (2) overlap length and thickness of the bonding material including its modulus, and (3) stacking sequence of the laminates of different stiffnesses (GangaRao and Palakamshetty, 2001) and others.

In addition, adherend thickness, joint geometry, effective adhesive length, fiber volume fraction and fiber orientation of members can affect the joint efficiency. The reliability of an adhesive joint not only depends on the abovementioned factors but is also dominated by joint stiffness. The factors controlling joint stiffness include the adhesive properties and thickness, inter-laminar shear modulus of the adherends and stress distribution in the adhesive area. For a comprehensive treatment on the design of adhesive connections, the reader is referred to the EUROCOMP design code entitled "Structural Design of Polymer Composites" (1996).

Humidity and temperature conditions may limit preparation, usability, resistance, and durability of adhesive connections. The strength of adhesive joints will decrease under the effect of environment moisture absorbed by the polymer adhesives. It was found that adhesive joints are weakened by prolonged exposure under high relative humidity, say 80%–100%, and under temperatures close to the glass transition temperature. Adhesive joints can withstand well under moderate relative humidity, say 50% or less, for long-term serviceability (Brewis et al., 1980a, 1980b). In addition, adhesive joints tend to weaken rapidly when stressed under high relative humidity and increasing temperatures, exceeding 90°F. For more information, the reader is directed to refer to the work by Kinloch (1983).

Failure Modes:

The potential failure modes of adhesive joints depend on different load conditions such as tension, torsional or axial shear, edge peel forces, and peel forces coupled with bending forces varying across a joint. Such complex external forces can result in cleavage failures (Figure 11.2c). In adhesive joint design, joint shear force transfer from one member to another is preferred to attain maximum strength. Even though tensile force transfer is the most efficient approach, it is difficult to attain in practice without inducing small amounts of bending or peel stresses. Potential joint force transfer through peeling or cleavaging should be minimized, which is produced through out-of-plane loads or eccentric in-plane loads between adherends.

As stated earlier, the failure modes in adhesive joints depend on: (1) mechanical properties of both the adhesives and adherends including their thickness, (2) fiber orientation of adherends, thickness of adhesive ranging around 1/64″ to 1/8″ (0.4–3 mm) with smaller thickness as the preferred target, (3) magnitude and eccentricity of forces in lap joints inducing bending across a joint, and (4) lap length. It is preferable to maintain the joint fiber orientation parallel to the direction of load which is the highest on a joint response. Similarly, thermal expansion/contraction coefficients of an adhesive and adherend can play a major role in force transfer between members and potentially result in thermal failures in joints, due to excess stress inducement in the adhesive. In any adhesive joint, surface preparation and surface treatment are extremely critical. Based on the authors' experiences, the final cleaning of a joint with acetone would leave a very fine/thin film which may interfere with the bonding characteristics of adhesives with adherends. Similarly, application of proper primer and adequate pressure during curing needs to be carefully monitored while constructing an adhesive joint.

Design approach:

Joint design with structural adhesive is very complex and even uncertain in attaining requisite design strength because of long-term reliability of adhesive strength under varying environmental conditions and consistency in field application of adhesives. However, recent advances in the application of adhesives in aerospace structures have mostly alleviated the above shortcomings. Typically, adhesive joint designs are based on either analytical or numerical approaches, while occasional testing is preformed to validate designs. In the analysis and design of a joint, perfect bonding is assumed between the adhesive and adherend. A joint is designed to be at least as strong as the weakest member meeting at that joint. Typically, through-thickness strength of the adherend at the joint is checked against the induced loadings. Herein, the joint design equations are given for relatively simple plate-to-plate configurations. Therefore, while designing joints, the designer should adopt a few clever techniques by breaking the joint down into a series of individual plate-to-plate joints such as the single- or double-lap joints. It is impossible to suggest a particular adhesive for field applications because of the availability of a wide range of adhesives in the market. Therefore, a designer should consult adhesive manufacturers to establish the characteristic mechanical properties before proceeding with the design. Similarly, curing process parameters including curing initiation and time of cure need to be understood properly for safe designs.

Adherends are assumed to be stiff and do not deform as much as the adhesives (Figure 11.17). An exaggerated schematic view of adhesive deformation is shown in Figure 11.14. The induced shear stress is also assumed to be uniformly distributed within the adhesive.

The shear stress within adhesive (f^{sa}) per unit width is:

$$f_{sa} = \frac{P_k}{2c} \quad \left(\text{per unit width of the joint}\right) \tag{11.27}$$

where

P_k = applied load per unit width
$2c$ = over-lap length

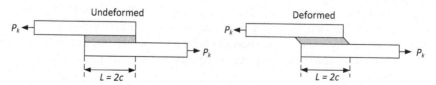

FIGURE 11.17   Undeformed and deformed adhesive joint.

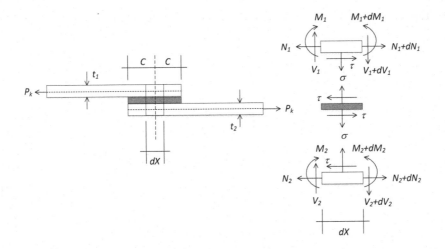

FIGURE 11.18 Free body diagram of a single-lap joint element.

The single-lap joint is shown in Figure 11.17. This model is to be an adhesive–beam model with a two-parameter elastic medium and Euler beams (Goland and Reissner, 1944). The equilibrium equations of adherends, as per Figures 11.17 and 11.18 are:

$$\frac{dN_1}{dx}+\tau=0 \quad \frac{dV_1}{dx}+\sigma=0 \quad \frac{dM_1}{dx}+\frac{t_1}{2}\tau-V_1=0 \quad (11.28.1)$$

$$\frac{dN_2}{dx}-\tau=0 \quad \frac{dV_1}{dx}+\sigma=0 \quad \frac{dM_2}{dx}+\frac{t_2}{2}\tau-V_2=0 \quad (11.28.2)$$

where

τ = shear stress

σ = peel stress

t_1 and t_2 are the thicknesses of the adherends

For balanced joints ($t_1 = t_2$), the shear stress (τ) is:

$$\tau = \frac{G_a}{t_a}\left[\left(u_2-u_1\right)+\left(\frac{t_1 dw_1}{2dx}+\frac{t_2 dw_2}{2dx}\right)\right] \quad (11.29.1)$$

where shear strain $=\left(\dfrac{u_2-u_1}{t_a}\right)$ from relative horizontal displacement between the centers of top and

bottom adherends; $\dfrac{dw}{dx}$ is the slope of the deflected shape which is multiplied by half the thickness of the adherend to establish approximate shear displacement.

The peel stress (σ) is defined as:

$$\sigma = \frac{E_a}{t_a}\left(w_2-w_1\right) \quad (11.29.2)$$

where

G_a, E_a = adhesive shear and longitudinal modulus, respectively

t_a = adhesive thickness

t_1 and t_2 = thicknesses of the adherends

u = horizontal displacement

w = vertical displacement.

Note: The above equations are valid under the assumption that γ and τ are uniform along a joint which implies that the shear force resultant is distributed linearly in the upper and lower adherends.

From Eqs. (11.28 and 11.29), the governing equations of shear and peel stresses are (Oplinger, 1998):

$$\frac{d^3\tau}{dx^3} - \beta_c^2 \frac{d\tau}{dx} = 0 \tag{11.30.1}$$

$$\frac{d^4\sigma}{dx^4} - 4\beta_\sigma^4\sigma = 0 \tag{11.30.2}$$

$$\beta_c = \alpha_a \sqrt{\frac{8G_a}{A_1 t_a t}} \tag{11.30.3}$$

$$\beta_\sigma = \frac{1}{\sqrt{2}}\left(\sqrt[4]{\frac{2E_a}{D_1 t_a}}\right) \tag{11.30.4}$$

$$\alpha_a = 0.25(1+\alpha_k) = 0.25\left(1 + \frac{A_1 t_1^2}{4D_1}\right) \tag{11.30.5}$$

where
  D_1 = bending stiffness of adherends
  A_1 = extension (axial) stiffness
  t_1 = thickness of adherends.

11.8.1 Lap Length of Single and Double-Lap Joints

To determine the lap length of joints, joint notations are referred to and presented in Figure 11.16. The lap length parameter (β/t) of single and double-lap connections is obtained from (EUROCOMP, 1996):

The half-lap length (c) of both the single- and double–lap joints can be approximated as (EUROCOMP, 1996):

$$\frac{\beta}{t} = \sqrt{\frac{8G_a}{Et_a t}} \tag{11.31}$$

where
  G_a = adhesive shear modulus
  E = adherend modulus
  t = minimum adherend thickness (unit: m)
  t_a = adhesive layer thickness (unit: m).

Case: Single-lap joint ($100 \le \beta/t \le 1{,}000$)

$$c(\text{unit: mm}) = 25{,}230\big/\left(\beta/t\right) \tag{11.32.1}$$

Case: Double-lap joint ($100 \le \beta/t \le 1{,}000$)

$$c(\text{unit: mm}) = 10{,}790\big/\left(\beta/t\right) \tag{11.32.2}$$

where β/t = lap length parameter in Eq. (11.31). Equations (11.32.1 and 11.32.2) are obtained from the curve fitting of Figure 5.37, given in EUROCOMP design code (1996).

11.8.2 Shear Strength of Adhesive Joint

The differential equations in Eqs. (11.30.1 and 11.30.2) solved for certain type of boundary conditions give a solution of the maximum shear stress. For design purpose, the nominal shear strength based on EUROCOMP code is provided as (EUROCOMP, 1996):

Case: Single-lap joint

$$\tau_{max} = \frac{\sigma}{8}(1+3k)\sqrt{\frac{8G_a t}{Et_a}} \qquad (11.33.1)$$

$$\text{where,} \quad \sigma = \frac{P_k \varphi_f}{t} \qquad (11.33.2)$$

$$k = \frac{Cosh(u_2 c)Sinh(u_1 L)}{Sinh(u_1 L)Cosh(u_2 c) + 2\sqrt{2}Cosh(u_1 L)Sinh(u_2 c)} \qquad (11.33.3)$$

$$u_2 = \frac{1}{t}\sqrt{\frac{3\sigma}{2E}(1-v^2)} \quad \frac{u_1}{u_2} = 2\sqrt{2} \qquad (11.33.4)$$

where

P_k = the characteristic load per unit width
φ_f = load factor
v = adherend Poisson's ratio
k = bending moment factor
L = lap length
t_0 and t_i = the outer and inner adherend thickness, respectively
t = minimum adherend thickness (unit: m)
t_a = adhesive layer thickness (unit: m)
E = adherend modulus
G_a = adhesive shear modulus.

The joint notations are referred in Figure 11.19.

Case: Double-lap joint

$$\tau_{max} = \frac{\lambda P_k \varphi_f}{4t}\left(\frac{Cosh(\lambda c)}{Sinh\lambda c} + \Omega\frac{Sinh(\lambda c)}{Cosh(\lambda c)}\right) \qquad (11.34.1)$$

$$\lambda^2 = \frac{G_a}{t_a}\left(\frac{1}{E_0 t_0} + \frac{2}{E_i t_i}\right) \qquad (11.34.2)$$

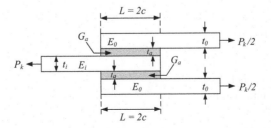

FIGURE 11.19 Joint notations of adhesive lap joint.

$$\Omega = \text{Max}\left(\frac{1-\psi}{1+\psi}, \frac{1+\psi}{1-\psi}\right)$$ (11.34.3)

$$\psi = \frac{E_i t_i}{2 E_0 t_0}$$ (11.34.4)

Note: Equations (11.34.1 through 11.34.4) are identical to Eq. (5.12) of EUROCOMP Design Code "Structural Design of Polymer Composites" (1996).
where

P_k = the characteristic load per unit width
φ_f = load factor
ν = adherend Poisson's ratio
k = bending moment factor
t_0 and t_i = outer and inner adherend thickness, respectively
t = minimum adherend thickness (unit: m)
t_a = adhesive layer thickness (unit: m)
E_0 and E_i = outer and inner adherend stiffness, respectively
G_a = adhesive shear modulus
τ_{max} = maximum shear stress.

The joint notations are given in Figure 11.19.

11.8.3   PEEL STRENGTH OF ADHESIVE JOINT

The maximum adhesive peel stresses for single and double-lap joints are (EUROCOMP, 1996):

Case: Single-lap joint ($\lambda > 2.5$)
The maximum adhesive peel stress for the single-lap and single-strap connections is:

$$\lambda = \frac{c}{t}\left(\frac{6E_a t}{E t_a}\right)^{0.25} > 2.5$$ (11.35.1)

$$\frac{\sigma_{max}}{\sigma} = \left(\frac{k}{2} + k_e \frac{t}{c}\right)\sqrt{\frac{6E_a t}{E t_a}}$$ (11.35.2)

$$\sigma = \frac{P_k \varphi_f}{t}$$ (11.35.3)

$$k_e = \frac{kc}{t}\sqrt{\frac{3(1-v^2)\sigma}{E}}$$ (11.35.4)

where

k_e = rotation factor
E_a = adhesive elastic modulus
E = adherend modulus
c = half lap length in mm (SI) units
t = minimum adherend thickness
P_k = applied load per unit width
ν = adherend Poisson's ratio
φ_f = load factor.

Case: Double-lap joint

The maximum adhesive peel stress for the double-lap and double-strap connections is determined as:

$$\sigma_{max} = \tau_{max} \left(\frac{3E_{ea}\left(1-v^2\right)t_0}{E_0 t_a} \right)^{0.25} \tag{11.36.1}$$

$$E_{ea} = \left(\frac{1-v_a}{\left(1+v_a\right)\left(1-2v_a\right)} \right) E_a \tag{11.36.2}$$

where

E_{ea} = effective transverse modulus of the adhesive
E_0 = outer adherend tensile modulus
E_a = outer adhesive tensile modulus
v = Poisson's ratio of the outer adherend
t_0 = adherend thickness
t_a = adhesive layer thickness
v_a = adhesive Poisson's ratio
τ_{max} = maximum shear stress in Eq. (11.34.1).

Note: Equation (11.33.1) is identical to Eqs. (5.11) and (11.33 and 11.35) are identical to Eqs. (5.14 and 5.17) of EUROCOMP Design Code "Structural Design of Polymer Composites" (1996).

11.9   DESIGN RECOMMENDATIONS FOR ADHESIVELY BONDED JOINTS

The recommendations for a reliable bonded connection design are presented as follows:

- Designers must avoid adhesive failures. The bonded connection should be designed to fail in the adherends (cohesive failure) before bond failure of an adhesive (Hart-smith, 1987).
- The bonded connection should be designed for the primary stress in shear or compression while tensile, cleavage, and peel loads should be minimized for this type of connection. In addition, the bond layer thickness is limited if possible by varying from 0.125 to 0.39 mm (0.005″–0.0015″) due to defects and processes of the bond materials (Oplinger, 1998), but the bond layer thickness should not exceed 3 mm to attain reasonable (not optimal) adhesive joint strength.
- The adherend stiffnesses should be designed and nearly equal since adhesively bonded joints are adversely influenced by unequal adherend stiffness. Use of tapering end of adherends and the ductile adhesive can reduce the peel stresses and shear stress.
- The use of symmetric joint configurations such as double-lap is more preferable than the use of single-lap connection, Which would reduce the effect of eccentric loading and consequent bending moment transfer from one adherend to another.
- The use of adhesive fillets and tapering the adherend ends may significantly increase the load-bearing capacity of the connection by reducing stress concentration at the ends of the overlap. In addition, the use of a peel ply on the bond surface during manufacturing of the adherend is recommended to prevent surface contamination (EUROCOMP, 1996).

Example 11.10

(SI unit): using the LRFD method, design a single-lap joint as shown in Example Figure 11.10. The joint is loaded under an in-plane tensile load per unit width (P_k) of 0.1 kN/m. (Given: load factor $\varphi_f = 1.35$).

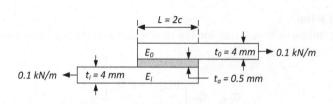

EXAMPLE FIGURE 11.10 Single-lap of adhesive joint.

Mechanical properties of adhesive (brittle epoxy): Elastic modulus (E_a) = 950 MPa, Shear modulus (G_a) = 371 MPa, Tensile strength = 58.9 MPa and Shear strength = 21.5 MPa, allowable tensile stress = 30 MPa and allowable shear stress = 11 MPa.

Mechanical properties of adherend: Elastic modulus (E_0 and E_i) = 15 GPa, Normal stress ($\sigma^{z,\,k}$) = 24 MPa, and Poisson's ratio (ν) = 0.3.

Solution

a. Determine parameter (β/t):
   The parameter (β/t) is determined from Eq. (11.31) as:

$$\frac{\beta}{t} = \sqrt{\frac{8G_a}{Et_at}} = \sqrt{\frac{8(371)}{15(10^3)(0.0005 \times 0.004)}} = 315\frac{1}{mm}$$

b. Determine half-lap length (c):
   The half-lap is determined from Eq. (11.32.1) as:

$$c = 25,230/(\beta/t) = 25,230/315 = 80\,mm$$

Thus, the half lap length (c) = 80 mm is provided, thus the lap length ($L = 2c$) is 160 mm.
c. Determine the maximum induced shear stress in the adhesive:
   The maximum induced shear stress in adhesive is determined from Eqs. (11.33.1–11.33.4) as:

$$\sigma = \frac{P_k\varphi_f}{t} = \frac{1 \times 100 \times 1.35}{(0.004)} = 0.034\,MPa$$

$$u_2 = \frac{1}{t}\sqrt{\frac{3\sigma}{2E}(1-v^2)} = \frac{1}{(0.004)}\sqrt{\frac{3 \times 0.034}{2 \times 15(10^3)}(1-0.3^2)} = 0.438\,m^{-1}$$

$$u_1 = 2\sqrt{2}(u_2) = (2\sqrt{2})0.438 = 1.24\,m^{-1}$$

$$k = \frac{Cosh(u_2c)Sinh(u_1L)}{Sinh(u_1L)Cosh(u_2c) + 2\sqrt{2}Cosh(u_1L)Sinh(u_2c)}$$

$$k = \frac{Cosh(0.0351)Sinh(0.1983)}{Sinh(0.1983)Cosh(0.099) + 2\sqrt{2}Cosh(0.1983)Sinh(0.0351)} = 0.662$$

$$\tau_{max} = \frac{\sigma}{8}(1+3k)\sqrt{\frac{8G_a}{E_0t_at}}$$

$$\tau_{max} = \frac{0.034}{8}(1+3 \times 0.623)\sqrt{\frac{8 \times 371}{15 \times 10^3(0.004 \times 0.0005)}} = 3.96\,MPa$$

d. Check for adjusted shear stress
   The adjusted shear stress (τ^{adj}) in the adhesive ($\gamma_f = 0.25$) is:

$$\tau_{adj} = \gamma_f(21.5) = 0.25(21.5) = 5.38\,MPa$$

$$\tau_{allowable} = 11\,MPa$$

The maximum induced shear stress (3.96 MPa) is less than the adjusted shear stress (5.38 MPa) and the allowable shear stress (11 MPa). Hence, the design is satisfactory.

e. Determine maximum adhesive peel stress:

The maximum adhesive peel stress is determined from Eqs. (11.35.1–11.35.4) as:

$$\lambda = \frac{c}{t}\left(\frac{6E_a t}{Et_a}\right)^{0.25} = \frac{80}{4}\left(\frac{6 \times 950 \times 4}{15 \times 1,000 \times 0.5}\right)^{0.25} = 26.4 > 2.5$$

$$\sigma = \frac{P_k \varphi_f}{t} = \frac{1 \times 100 \times 1.35}{(0.004)} = 0.034\,\text{MPa}$$

$$k_e = \frac{kc}{t}\sqrt{\frac{3(1-v^2)\sigma}{E}} = \frac{0.662 \times 0.080}{0.004}\sqrt{\frac{3(1-0.3^2)0.034}{15(10^3)}} = 0.0329$$

$$\frac{\sigma_{max}}{\sigma} = \left(\frac{k}{2}+k_e\frac{t}{c}\right)\sqrt{\frac{6E_a t}{Et_a}} = \left(\frac{0.662}{2}+\frac{0.0329 \times 4}{80}\right)\sqrt{\frac{6(950)4}{15(10^3)0.5}} = 0.580$$

$$\sigma_{max} = 0.580(0.034) = 0.02\,\text{MPa}$$

f. Check for adjusted peel stress

The adjusted adhesive peel stress in the adhesive ($\gamma_f = 0.25$) is:

$$\sigma_{adj} = \gamma_f(58.9) = 0.25(58.9) = 14.7\,\text{MPa}$$

The maximum peel stress (0.02 MPa) is less than the adjusted peel stress (14.7 MPa) and the allowable tensile stress (30 MPa). Hence, the design is satisfactory.

g. Check for adjusted adherend through through-thickness tensile stress

The adjusted adherend through-thickness tensile stress ($\gamma_f = 0.25$) is:

$$\sigma_{z,adj} = \gamma_f(\sigma_{z,k}) = 0.25(24) = 6.0\,\text{MPa}$$

The maximum peel stress (0.02 MPa) is less than the adjusted adherend through-thickness tensile stress (6.0 MPa). Hence, the design is satisfactory.

Example 11.11

(SI unit): using the LRFD method, design a double-lap joint as shown in Example Figure 11.11. The 15 cm wide joint is loaded under an in-plane tensile load (P_k) of 3 kN. (Using load factor $\varphi_f = 1.35$)

Adhesive mechanical properties (brittle epoxy): Elastic modulus (E_a) = 950 MPa, Shear modulus (G_a) = 371 MPa, Tensile strength = 58.9 MPa and Shear strength = 21.5 MPa, Allowable tensile stress ($\sigma_{0,k}$) = 30 MPa, and Allowable shear stress ($\tau_{0,k}$) = 11 MPa

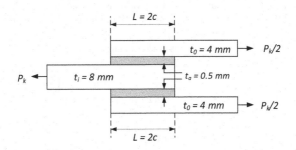

EXAMPLE FIGURE 11.11   Double lap of adhesive joint.

Adherend mechanical properties: Elastic modulus (E^0 and E^i) = 15 GPa, Normal stress ($\sigma^{z, k}$) = 24 MPa, and Poisson's ration (ν) = 0.3.

Solution

a. Determine parameter (β/t):

The parameter (β/t) is determined from Eq. (11.28) as:

$$\frac{\beta}{t} = \sqrt{\frac{8G_a}{Et_a t}} = \sqrt{\frac{8(371)}{15(10^3)(0.0005 \times 0.004)}} = 315 \frac{1}{\text{mm}}$$

b. Determine half-lap length (c):

The half-lap length is determined from Eq. (11.32.2) as:

$$c = 10{,}790 / (\beta/t) = 10{,}790 / 315 = 35\,\text{mm}$$

Thus, the half lap length (c) = 35 mm is provided, thus the lap length ($L = 2c$) is 70 mm.

c. Determine maximum shear stress:

The maximum shear stress is determined from Eqs. (11.34.1–11.34.4) as:

$$\lambda = \sqrt{\frac{G_a}{t_a}\left(\frac{1}{E_0 t_0} + \frac{2}{E_i t_i}\right)} = \sqrt{\frac{371(10^3)}{0.0005}\left(\frac{1}{15 \times 10^6 \times 0.004} + \frac{2}{15 \times 10^6 \times 0.008}\right)} = 157$$

$$\psi = \frac{E_i t_i}{2E_0 t_0} = \frac{15 \times 0.008}{2 \times 15 \times 0.004} = 1$$

where $E_{0,i}$ = outer and inner adherend modulus

$$\Omega = \text{the greater of } \frac{1-\psi}{1+\psi} \text{ or } \frac{1+\psi}{1-\psi} \quad \text{Thus,} \Omega = 0$$

$$\tau_{max} = \frac{\lambda P_k \varphi_f}{4}\left(\frac{Cosh(\lambda c)}{Sinh(\lambda c)} + \Omega \frac{Sinh(\lambda c)}{Cosh(\lambda c)}\right) = \frac{157 \times 3{,}000 \times 1.35}{4 \times 0.15}\left(\frac{Cosh(157 \times 0.035)}{Sinh(157 \times 0.035)}\right) = 1.06\,\text{MPa}$$

d. Check for adjusted shear stress:

The adjusted shear stress (τ_{adj}) is determined as follows:

$$\tau_{adj} = \gamma_f (21.5) = 0.25(21.5) = 5.38\,\text{MPa}$$

$$\tau_{allowable} = 11\,\text{MPa}$$

The maximum induced shear stress (1.06 MPa) is less than the adjusted shear stress (5.38 MPa) and the allowable shear stress (11 MPa). Hence, the design is satisfactory.

e. Determine maximum adhesive peel stress:

The maximum adhesive peel stress is determined from Eqs. (11.36.1 and 11.36.2) as:

$$E_{ea} = \left(\frac{1-v_a}{(1+v_a)(1-2v_a)}\right)E_a = \left(\frac{1-0.3}{(1+0.3)(1-2(0.3))}\right)15 = 15.2\,\text{GPa}$$

$$\sigma_{max} = \tau_{max}\left(\frac{3E_{ea}(1-v^2)t_0}{E_0 t_a}\right)^{0.25} = 1.06\left(\frac{3 \times 15.2 \times (1-0.3^2) \times 4}{15 \times 8}\right) = 1.47\,\text{MPa}$$

f. Check for adjusted peel stress

The adjusted adhesive peel stress in the adhesive ($\gamma_f = 0.25$) is:

$$\sigma_{adj} = \gamma_f(58.9) = 0.25(58.9) = 14.7\,\text{MPa}$$

The maximum peel stress (1.47 MPa) is less than the adjusted peel stress (14.7 MPa) and the allowable tensile stress (30 MPa). Hence, the design is satisfactory.

g. Check for adjusted adherend through-thickness tensile stress

The adjusted adherend through-thickness tensile stress ($\gamma_f = 0.25$) is:

$$\sigma_{z,adj} = \gamma_f(\sigma_{z,k}) = 0.25(24) = 6.0\,\text{MPa}$$

The maximum peel stress (1.47 MPa) is much less than the adjusted adherend through-thickness tensile stress (6.0 MPa). Hence, the design is satisfactory.

EXERCISES

Problem 11.1: Using LRFD method, find the tensile capacity for the connection shown in Problem Figure 11.1. Single-lap joint with 7/8″ diameter A325 (bearing type connection) in standard holes is used to connect pultruded FRP vinyl ester plates (12 × 1/2″). The time effect factor $\lambda = 0.8$ for $W = 1.2D + 1.6L$, Geometry factor $C_\Delta = 1.0$ for pitch spacing $(s) \geq s_{min}$, Moisture condition factor $C_M = 0.85$ for vinyl ester material and Temperature condition factor $C_T = 0.74$ for strength $(T = 120°F)$.

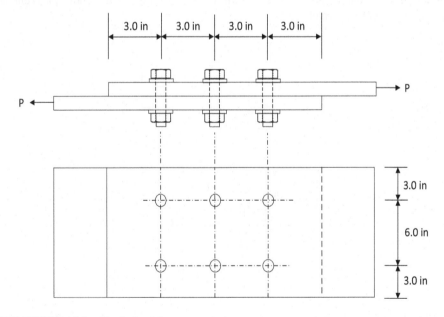

PROBLEM FIGURE 11.1 Single-lap joint connection.

Problem 11.2: Design a double-lap joint under axial factored tension 12.5 kips using the LRFD method and the minimum requirements of the bolted connection for bolt spacing. The double-lap joint with 3/4″ diameter A325 bolt is fit in standard holes of FRP plates (12 × 1/2″).

Problem 11.3 Using the LRFD method, determine the load-carrying capacity of the framed connection as shown in Problem Figure 11.3. The bolts are 3/4″ dia. A325 (bearing type connection) in standard holes. The frame connection is used to connect pultruded FRP WF 12 × 12 × 1/2″ and WF 10 × 10 × 3/8″ with 2Ls 4 × 4 × 1/2″. (1) For WF 10 × 10 × 3/8″ (2) For WF 12 × 12 × 1/2″ (3) For 2Ls 4 × 4 × 1/2″

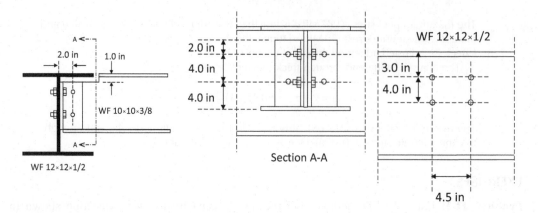

PROBLEM FIGURE 11.3   Simple-frame connection.

Problem 11.4: Using the LRFD method and EUROCOMP, design a composite plate sandwiched between two steel plates (Problem Figure 11.4), in a double-lap configuration when the joint is subjected to a unit static load (F_x) of 12.5 kN. The minimum spacing, edge distance, and end distance are $4d_h$, $4d_b$, and $4d_b$, respectively.

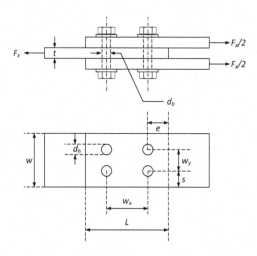

PROBLEM FIGURE 11.4   Double-lap connection under axial load.

Problem 11.5: Using both ASD & LRFD methods, find the maximum shear force that can be transferred by a clip angle (6″ × 6″ × ¼″ × 10″ long pultruded) assuming four numbers of ¾″ diameter steel bolts are used in each leg assuming only shear through the angles and bearing at the bolt holes (no eccentricity).

Note: Recommended end distance to bolt diameter is 3.0, plate width to bold diameter is 5.0, longitudinal and transverse spacing to bolt diameter is 4.0, bolt diameter to plate thickness is 1.0. Use properties from the Creative Pultrusion Design Manual.

Problem 11.6: Using the ADS method, design a 50′ long beam under 0.2 k/ft uniform load with a joint at the midspan. The composite beam is a box section with outer dimensions of 12″ × 18″ and a wall thickness of 1″. The bending modulus is 4.5×10^6 psi. The joint at the center of this 50′ long beam (with two 25′ long pieces of FRP tubes) is designed with an 18″ long FRP tube of 10″ × 16″ with ¾″ wall thickness. Design the size and number of bolts and check for deflection limit state at the beam center. Neglect the additional stiffness offered by the 18″ long FRP tube whose properties are the same as those of the main beam (12″ × 18″ × 1″).

Problem 11.7: Using the ASD method, design a semi-rigid beam-column connection with 80% rigidity and identical to the one shown in Example Figure 11.7.1 except the moment resistance is 12 ft-kips and the shear load acting on the beam is 6 kips. Use ½″ diameter high-strength steel bolts (A325N) and drilling holes that are 1/16″ larger than the bolt diameter. Assume bending, bearing and shear stress, respectively, are 32, 38, and 8 ksi.

REFERENCES AND SELECTED BIOGRAPHY

American Society for Testing and Materials, Standard Specification for Structural Bolts, Steel, Heat Treated, 120/105 ksi Minimum Tensile Strength, ASTM – A325, ASTM International, West Conshohocken, PA.

American Society for Testing and Materials, Standard Specification for Washers, Steel, Plain (Flat), Unhardened for General Use, ASTM F844, ASTM International, West Conshohocken, PA, 2013.

American Society for Testing and Materials, Standard Specification for Stainless Steel Bolts, Hex Cap Screws, and Studs, ASTM – F593, ASTM International, West Conshohocken, PA, 2013.

American Society for Testing and Materials, Standard Specification for Carbon Steel Bolts, Studs, and Threaded Rod 60000 PSI Tensile Strength, ASTM – A307, ASTM International, West Conshohocken, PA, 2014.

American Society for Testing and Materials, Standard Specification for Carbon and Alloy Steel Nuts, ASTM A563, ASTM International, West Conshohocken, PA, 2015.

American Society for Testing and Materials, Standard Specification for Stainless Steel Nuts, ASTM F594, ASTM International, West Conshohocken, PA, 2015.

American Society for Testing and Materials, Standard Specification for Austenitic Gray Iron Castings, ASTM A436, ASTM International, West Conshohocken, PA, 2015.

ANSI/AISC 360-16, An American National Standard, Specification for Structural Steel Buildings, AISC, Chicago, IL, 2016.

ASCE, Structural Plastics Design Manual, Alexander Bell Drive, Reston, VA, 1984.

ASCE/ACMA, Pre-Standard for Load & Resistance Factor Design (LRFD) of Pultruded Fiber Reinforced Polymer (FRP) Structures, Arlington, VA, 2021.

ASCE/SEI7–10, Minimum Design Loads for Buildings and Other Structures, Alexander Bell Drive, Reston, VA, 2010.

Bank, L.C., *Composites for Construction: Structural Design with FRP Materials*, Hoboken, NJ, John Wiley & Sons, Inc., 2006.

Bedford Reinforced Plastics, *Design Guide*, Bedford, PA, Bedford Reinforced Plastics, Inc., 2012.

Brewis, D.M., Comyn, J., and Tegg, J.L., The durability of some epoxide adhesive-bonded joints on exposure to moist warm air, *International Journal of Adhesion and Adhesive*, 1(1), 1980a, pp. 35–39.

Brewis, D.M., Comyn, J., and Tegg, J.L., The uptake of water vapor by an epoxide adhesive formed from the diglycidyl ether of bisphenol-A and di(1-aminoproyl-3-ethoxy) ether, *Polymer*, 1, 1980b, p. 134.

Clarke, J.L., *Structural Design of Polymeric Composites – EUROCOMP Design Code and Handbook*, London, UK, E & FN Spon, 1996.

Creative Pultrutions, *The Pultex Pultrusion Design Manual of Standard and Custom Fiber Reinforced Structural Profiles*, 5, revision (2), Alum Bank, PA, Creative Pultrutions Inc, 2004.

CROW-CUR Recommendations 96:2019, Fibre-Reinforced Polymers in Building and Civil Engineering Structures, CROW, November 2019, ISBN-978-90-6628-6733.

Da Silva, F.M.L., Ochsner, A., and Adams, D.R., *Handbook of Adhesion Technology*, Berlin Heidelberg, Springer-Verlag, 2011.

Fiberline, *Design Manual*, Kolding, Denmark, Fiberline, 2003.

GangaRao, H., and Palakamshetty, S., Design of adhesively bonded FRP joints, *Modern Plastics*, 78(3), 2001, pp. 79–82.

Goland, M., and Reissner, E., The stresses in cemented joints, *Journal of Applied Mechanics*, 11, 1944, pp. A17–A27.

Hart-Smith, L.J., Mechanically fastened joints for advanced composites: Phenomenological considerations and simple analysis, *Proc of the 4th Conference on Fibrous Composites in Structural Design*, Plenum Press, New York, NY, 1978.

Hart-Smith, L.J., Design of adhesively bonded joints, In Chapter 7 in *Jointing Fiber Reinforced Plastics*, Edited by Matthews, F.L., Elsevier, London, UK, 1987.

Hassan, N.K., Evaluation of Multi-bolted Connections for GFRP Members, Ph.D. Dissertation, Department of CE, University of Manitoba, Winnipeg, CA, 1994.

Kinloch, A.J., *Durability of Structural Adhesives*, London, UK, Applied Science Publishers, 1983.

Mosallam, A.S., *Design Guide for FRP Composite Connections*, ASCE, Reston, VA, 2011.

Oplinger, D.W., Mechanical fastening and adhesive bonding, In *Handbook of Composites*, Edited by Peters, S.T., Chapman & Hall, London, 1988, pp. 610–666.

Strongwell Corporation, Design Manual: EXTERN and Other Proprietary Pultruded Products, Bristol, VA, 2010.

12 Design of Combined Loads for FRP Members

All structural systems in practice are made of two- or three-dimensional frames or trusses or arches, and perhaps even some combination of the above. Those systems are assembled by joining structural components in different forms and are subjected to axial, bending, shear, and/or torsional loads. Typically, structural components in a system are under combined loads, such as axial plus bending loads, or axial plus bending loads coupled with torsional loads, or other load combinations including shear. The behavior of a component under combined loading is different from its behavior under one loading condition only, such as axial load or bending load. This chapter deals with interactive structural responses under different loads acting simultaneously on FRP composite components. The FRP component responses are unique under combined load conditions. However, FRP components are designed for single dominant load condition (say, axial or bending only), if the magnitude of one of the loads is relatively small, i.e., approximately less than 15% of the total effect from all the external loads combined.

In Chapter 9, lateral torsional buckling, stress concentration, and bending response issues are elaborated for FRP components under bending loads only. Similarly, local stability issues are dealt with in Chapter 10 for components under compression. This implies that a component design can be accentuated by instability aspects of design under combined bending and torsional moments and compression forces. Furthermore, since our focus herein is on the first-order analysis, secondary bending moment (product of axial compression and transverse deflection) effects induced under transverse deflections are neglected. FRP components under bending and tension are usually governed by material rupture, which is limited by the FRP rupture strain of the order of around 2% (~20,000 microstrains) for glass composites. Limit states under combined stresses may not lead to proper designs (cross-sectional dimensions) unless buckling issues are addressed adequately. Therefore, this chapter addresses design issues based on interaction equations, which describe the behavior of a component under combined loads.

12.1 MEMBERS UNDER COMBINED LOADS

Pultruded FRP members are often subjected to combined loads such as flexure coupled with tension or compression acting about the longitudinal axis. For example, a pultruded FRP beam under combined axial compression and bending is known as a beam-column. In general, the bending moment in a beam-column may be induced from (1) transverse loading perpendicular to the longitudinal axis of a beam and concentric axial compression (or tension) loading parallel to the beam axis, (2) joint moment induced due to noncolinear forces (axial, torsion) acting at a joint, (3) eccentric axial load at one or both ends of a member, and (4) others such as out-of-straightness of a member. A few of these loading scenarios are shown in Figure 12.1.

End moments in a pultruded beam are not induced excessively because the rigid frame connections of pultruded members are not rigid enough to develop adequate moment transfer from one member to another. In practice, the pultruded FRP beam-column members are part of braced frame structures. Thus, most of the pultruded FRP beam-column members are due to the combined loads between concentrated axial and transverse loads with potentially small joint movements (case 1 in Figure 12.1).

The interaction equations based on cases without joint movements of braced frame structures are presented herein.

DOI: 10.1201/9781003196754-12

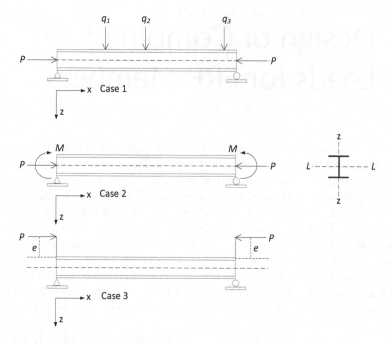

FIGURE 12.1 Examples of beam-column members.

12.2 INTERACTION OF COMBINED LOADS

The interactions of combined loads are presented herein for singly and doubly symmetric pultruded FRP members. As mentioned above, the interaction of combined loads is used for pultruded FRP members under transverse and concentric axial loads. Occasionally, transverse loads do not pass through a symmetrical plane of the cross section (or the shear center of cross sections). Then, torsional moment will be induced and must be considered for the analysis and design (Qureshi et al., 2016) of an FRP member. For example, pultruded FRP members are under combined transverse and concentric axial loads and by neglecting the secondary moment, an approximate interaction equation for stresses under the combined loads can be expressed as:

$$f_{max} = f_a + f_{bT} + f_{bZ} \qquad (12.1.1)$$

$$\frac{f_a}{f_{max}} + \frac{f_{bT}}{f_{max}} + \frac{f_{bZ}}{f_{max}} = 1 \qquad (12.1.2)$$

where
f_a = axial stress
f_{bT} and f_{bZ} = bending about the strong and weak centroidal axes (x and y) of a cross section, respectively
f_{max} = nominal strength.

Using the same methodology in the design of conventional engineering structural components under combined load effects, the pre-standard for LRFD (ACMA, 2021) provides provisions for the design of pultruded FRP members under combined loads with different interaction equations for different types of cross sections. These interaction equations are given in Table 12.1 and are similar to those in the Steel Construction Manual (ANSI/AISC 360–05).

12.2.1 NOMINAL STRENGTH

Through the LRFD method, the available axial P_c and flexural M_c strengths are determined using the basic strength (P_n, M_n, or T_n) with adjustment factors for (1) end use conditions, (2) member strength in structural assemblies, and (3) member stiffnesses as given in Section 2.4.4 of the pre-standard document (ACMA, 2021).

$$P_c = \lambda \varphi_c P_n \tag{12.2.1}$$

$$M_c = \lambda \varphi_b M_n \tag{12.2.2}$$

$$T_c = \lambda \varphi_T T_n \tag{12.2.3}$$

where

P_c = available axial (compression or tensile) strength

M_c and T_c = available flexural and torsional strengths, respectively

P_n, M_n, and T_n = nominal strengths of axial force, bending, and torsion with adjustment factors, respectively

λ = time effect factor

φ_c, φ_b, and φ_T = resistance factors for axial, flexural and torsional strengths, respectively.

To use the allowable stress design (ASD) approach, the design allowable strength is determined by the minimum nominal strength with a safety factor. The available allowable strength of members must be satisfied as:

$$R_a = R_u/\Omega \tag{12.4}$$

where

R_a = available allowable strength (P_a or M_a or T_a)

R_u = ultimate strength (P_u or M_u or T_u)

Ω = factor of safety (F.S.).

Interactions through the ASD method as given in Table 12.2 are identical to the interaction expressions for the LRFD method, which are based on AISC-ASDS (HI-3). Equations in Table 12.1 are verified with limited experimental data generated for FRP sections under combined loads (Qureshi et al., 2016).

TABLE 12.1

Interaction of FRP Members for LRFD Method (ACMA, 2021)

Section	Combined Loads	Equations Based on the Pre-Standard (ACMA, 2021)
Doubly and singly symmetric members	Flexure and compression (both axes) Flexure and tension (both axes)	$\dfrac{P_u}{P_c} + \dfrac{M_{uT}}{M_{cT}} + \dfrac{M_{uZ}}{M_{cZ}} \le 1.0$ (12.3.1)
	Only strong axis flexure and compression Only strong axis flexure and tension	$\dfrac{P_u}{P_c} + \dfrac{M_{uT}}{M_{cT}} \le 1.0$ (12.3.2)
▢ Rectangular hollow tubes ⊥ Open doubly symmetric shapes	Axial, flexure and torsion	$\dfrac{P_u}{P_c} + \dfrac{M_u}{M_c} + \left(\dfrac{T_u}{T_c}\right)^2 \le 1.0$ (12.3.3)

TABLE 12.2
Interaction of FRP Members for ASD Method Based on AISC – ASDS (H1-3)

Section	Combined Loads	Equations Based on AISC – ASDS (H1-3)
Doubly and singly symmetric members	Flexure and compression (both axes) Flexure and tension (both axes)	$\dfrac{P_s}{P_a} + \dfrac{M_{sx}}{M_{ax}} + \dfrac{M_{sy}}{M_{ay}} \le 1.0$ (12.5.1)
	Only strong axis flexure and compression Only strong axis flexure and tension	$\dfrac{P_s}{P_a} + \dfrac{M_{sx}}{M_{ax}} \le 1.0$ (12.5.2)
Rectangular hollow tubes Open doubly symmetric shapes	Axial, flexure and torsion	$\dfrac{P_s}{P_a} + \dfrac{M_s}{M_a} + \left(\dfrac{T_s}{T_a}\right)^2 \le 1.0$ (12.6)

12.3   DEFLECTION LIMITS

Deflection of a member under combined service loads for transverse loading along the span length with the concentric axial load P may be determined by considering additional moment $(P\delta)$ due to the concentric axial load. The bending moment along the member length is:

$$M_{max} = M\left(\text{transverse load}\right) + \left(P\delta\right) \tag{12.7.1}$$

By neglecting the shear effects, deflection of a member under combined service loads (transverse loading along the span length with the concentric axially applied load P) can be written in terms of transverse deflection (δ_0) with the amplification factor.

$$\delta = \delta_0 \frac{1}{1 - \left(\dfrac{P}{P_E}\right)} \tag{12.7.2}$$

where
  δ = deflection of a structural member under the combined transverse and concentric axial loads
  P = concentric axial load
  P_E = the critical Euler's strength under axial loading
  δ_0 = transverse deflection under transverse (bending) load on a section.

Example 12.1

Using the LRFD method, design a small pultruded FRP purlin for a simple span of 8 ft. 10 lbs/ft of Dead load and 30 lbs/ft of live load are assumed to be acting on the member. The pultruded FRP beam is placed on a slope of 1:3 as shown in Example Figure 12.1.

Given:

Pultruded beam properties (manufacturers manual):

  Longitudinal tensile strength (F_t) = 30,000 psi, Shear strength (F_v) = 4,500 psi, Longitudinal modulus (E_L) = 2.6 (10^6) psi, Transverse modulus (E_T) = 0.8 (10^6) psi, and Shear modulus (G_{LT}) = 0.425 (10^6) psi.

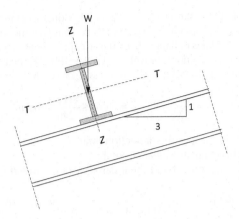

EXAMPLE FIGURE 12.1   Simply supported pultruded FRP I beam under bi-axial bending.

Solution

a. Determine factored loads:
   Total service load (W) = (10 lbs/ft) + (30 lbs/ft) = 40 lbs/ft
   Bending @ strong axis (T): W Cos 18.4° = 40 Cos 18.4° =38.0 lbs/ft
   Bending @ weak axis (Z): W Sin 18.4° = 40 Sin 18.4° = 12.63 lbs/ft
   Total factored load (W_u) = 1.2 (10 lbs/ft) +1.6 (30 lbs/ft) = 60 lbs/ft
   Bending @ strong axis (T): W_u Cos 18.4° = 60 Cos 18.4° = 56.9 lbs/ft
   Bending @ weak axis (Z): W_u Sin 18.4° = 60 Sin 18.4° = 18.94 lbs/ft.

b. Trial section with the deformation limit:
   Trial I-shaped section is evaluated using the serviceability limit (the deflection limit L/240) including and shear deformation response. The shear rigidity $k(GA)$ can be approximated with $(G_{LT})_{web} A_{web}$, i.e., where $k \cong 1.0$.
   The pultruded FRP beam deflection is:
   From Eq. (9.6.4):

$$w_{max} = \frac{5WL^4}{384(E_L I_T)} + \frac{WL^2}{8k(G_{LT}A)} = \frac{5(38.0/12)(8 \times 12)^4}{384(2.6 \times 10^6)I_T} + \frac{(38.0/12)(8 \times 12)^2}{8(0.425 \times 10^6)A_{web}}$$

$$w_{max} = \frac{1.35}{I_T} + \frac{0.008}{A_{web}} \leq \frac{8(12)}{240} = 0.4\,in$$

Select 6 × 3 × 3/8″ I-shaped beam section:

3 in

6 in

6×3×3/8″

Given:

Self-weight = 3.39 lbs/ft, Area (A) = 4.23 in², Web area (A_w) = 1.97 in², Sectional modulus about the strong axis (S_T) = 7.43 in³, Sectional modulus about the weak axis (S_Z) = 1.14 in³,

Moment of inertia about the strong axis $(I_T) = 22.3$ in^4, Moment of inertia about the weak axis $(I_Z) = 1.71$ in^4, Torsional Constant $(C) = 0.198$ in^4. Longitudinal modulus of flange $(E_{Lf}) = 2.6$ (10^6) psi, Transverse modulus of flange $(E_{Tf}) = 0.8$ (10^6) psi, Longitudinal modulus of web (E_{Lw}) $= 2.6$ (10^6) psi, Transverse modulus of web $(E_{Tw}) = 0.8$ (10^6) psi, Shear modulus $(G_{LT}) = 0.425$ (10^6) psi, Poisson's ratio $(\nu_{LT}) = 0.33$, Web thickness $(t_w) = 0.375''$, and flange width $(b_f) = 3''$.

$$w_{max} = \frac{1.35}{22.3} + \frac{0.008}{1.97} = 0.065 \leq \frac{8(12)}{240} = 0.4 \,\text{in}$$

The pultruded FRP beam satisfies the deflection limit.

c. Determine design bending and shear strength:

The design bending moment and shear capacity of pultruded FRP beam are:

$$M_{uT} = \frac{W_{uT}L^2}{8} = \frac{56.9(8^2)}{8} = 455.2 \,\text{lbs} - \text{ft} \quad \text{and} \quad V_{uT} = \frac{W_{uT}L}{2} = \frac{56.9(8)}{2} = 227.8 \,\text{lbs}$$

$$M_{uZ} = \frac{W_{uZ}L^2}{8} = \frac{18.94(8^2)}{8} = 151.2 \,\text{lbs} - \text{ft} \quad \text{and} \quad V_{uZ} = \frac{W_{uZ}L}{2} = \frac{18.94(8)}{2} = 75.8 \,\text{lbs}$$

Induced flexural and average web shear stresses are determined as:

$$f_{buT} = \frac{M_{uT}}{S_T} = \frac{455.2(12)}{7.43} = 735 \,\text{psi} \quad \text{and} \quad f_{vuT} = \frac{V}{A_{web}} = \frac{227.8}{1.97} = 116 \,\text{psi}$$

$$f_{buZ} = \frac{M_{uZ}}{S_Z} = \frac{151.2(12)}{1.17} = 1,551 \,\text{psi} \quad \text{and} \quad f_{vuZ} = \frac{V}{A_{web}} = \frac{75.8}{(4.23 - 1.97)} = 33.5 \,\text{psi}$$

d. Bending about a strong axis:

d.1) Determine nominal material strength:

The resistance of the FRP I-shaped beam is determined as:

$\varphi = 0.65$ for material rupture and time factor λ for $1.2D + 1.6L = 0.8$

$$\varphi\lambda F_t = (0.65 \times 0.8)30,000 = 15,600 \,\text{psi} \quad \text{and} \quad \varphi\lambda F_v = (0.65 \times 0.8)4,500 = 2,340 \,\text{psi}$$

$$\varphi\lambda F_t = 15,600 \,\text{psi} > f_{bu} = 735 \,\text{psi} \quad \text{and} \quad \varphi\lambda F_v = 2,340 \,\text{psi} > f_{vu} = 116 \,\text{psi}$$

It is found that the FRP I-shaped beam satisfies the material strength limit.

d.2) Determine lateral torsional buckling strength:

C_b (moment modification factor) = 1.0 for a simply supported uniformly distributed beam under constant torque

For $d \geq 6''$ then $L_{b,\,eff} = 0.95L_b$

The lateral torsional buckling strength is determined from Eq. (9.13) as:

$$C_\omega = \frac{I_z d^2}{4} = \frac{(1.71)(6^2)}{4} = 15.39 \,\text{in}^6$$

$$f_n^{LTB} = \frac{C_b \pi}{S_T L_{b,eff}} \sqrt{E_{T,f} I_z G_{LT} C + \frac{\pi^2 E_{T,f}^2 I_z C_\omega}{L_{b,eff}^2}}$$

$$f_n^{LTB} = \frac{\pi}{7.43(0.95 \times 8 \times 12)} \sqrt{\begin{array}{c}(0.8 \times 10^6)(1.71)(0.425 \times 10^6)(0.198) \\[6pt] + \dfrac{\pi^2 (0.8 \times 10^6)^2 (1.71)(15.39)}{(0.95 \times 8 \times 12)^2}\end{array}}$$

$$f_n^{LTB} = 1,704 \,\text{psi}$$

For $\varphi = 0.55$ and time factor λ for $1.2D + 1.6L = 0.8$, the lateral torsional buckling strength is:

$$\varphi\lambda f_n^{LTB} = (0.55\times0.8)1,704 = 750\,\text{psi}$$

$$f_{buT} = 735\,\text{psi} < \varphi\lambda f_n^{LTB} = 750\,\text{psi}$$

The cross section of the FRP I-shaped beam is sufficient to prevent global lateral torsional buckling.

d.3) Determine local buckling strength:

d.3.1) The local flange buckling strength is determined from Eq. (9.22) as:

$$f_{crf}^{local} = \frac{G_{LT} + 0.25\sqrt{E_{L,f}E_{T,f}}}{\left(\dfrac{b_f}{2t_f}\right)^2}$$

$$f_{crf}^{local} = \frac{\left(0.425\times10^6\right) + 0.25\sqrt{\left(2.6\times10^6\right)\left(0.8\times10^6\right)}}{\left(\dfrac{3}{2\times0.375}\right)^2} = 49,097\,\text{psi}$$

For $\varphi = 0.8$ and λ (for $1.2D + 1.6L$) = 0.8, the local flange buckling strength is determined as:

$$\varphi\lambda\left(f_{crf}^{local}\right) = (0.8\times0.8)49,097 = 31,422\,\text{psi}$$

$$f_{buT} = 735\,\text{psi} < \varphi\lambda\left(f_{crf}^{local}\right) = 31,422\,\text{psi}$$

Hence, the cross section of the FRP I-shaped beam is sufficient to prevent local flange buckling.

d.3.2) The local web buckling strength is determined from Eq. (9.23) as:

$$f_{crw}^{local} = \frac{15\sqrt{E_{L,f}E_{T,f}}}{\left(\dfrac{d}{t_w}\right)^2}$$

$$f_{crw}^{local} = \frac{15\sqrt{\left(2.6\times10^6\right)\left(0.8\times10^6\right)}}{\left(\dfrac{6}{0.375}\right)^2} = 84,505\,\text{psi}$$

For $\varphi = 0.8$ and λ (for $1.2D + 1.6L$) = 0.8, the local web buckling strength is:

$$\varphi\lambda\left(f_{crw}^{local}\right) = (0.8\times0.8)84,505 = 54,083\,\text{psi}$$

The local web buckling strength (54,083 psi) obtained from the pre-standard (ACMA, 2021) is higher than the material strength (30,000 psi) of the FRP beam. The cross section of the FRP I-shaped beam provides adequate resistance to local web buckling.

d.4) Check for web shear buckling strength:

Web shear buckling strength from Eqs. (9.35.1–9.35.3):

$$\beta = \frac{G_{LT}}{\sqrt{E_{L,w}E_{T,w}}} = \frac{\left(0.425\times10^6\right)}{\sqrt{\left(2.6\times10^6\right)\left(0.8\times10^6\right)}} = 0.295$$

$$k = 2.67 + 1.59\beta = 2.67 + 1.59(0.295) = 3.14$$

$$f_{cr} = \frac{k\sqrt[4]{E_{L,w}\left(E_{T,w}\right)^3}}{\left(\dfrac{d_w}{t_w}\right)^2} = \frac{3.14\sqrt[4]{\left(2.6\times10^6\right)\left(0.8\times10^6\right)^3}}{\left(\dfrac{6-0.375-0.375}{0.375}\right)^2} = 17,200\,\text{psi}$$

The compressive web shear buckling strength (with the resistance factor $\varphi = 0.8$ and time factor λ for $1.2D + 1.6L$ is 0.8) is determined as:

$$\varphi\lambda\left(f_{cr}^{local}\right) = (0.8\times0.8)17,200 = 11,008\,\text{psi}$$

$$f_{vuT} = 116\,\text{psi} < \varphi\lambda\left(f_{cr}^{local}\right) = 11,008\,\text{psi}$$

The web shear buckling strength (11,008 psi) is higher than the design shear strength (116 psi). The nominal available strength under bending about the strong axis is obtained from the lateral torsional buckling strength (750 psi, Section d.2).

e. Bending about a weak axis:
   e.1) Determine nominal material strength:
      The resistance of the pultruded FRP I-shaped beam is determined as:
      $\varphi = 0.65$ for material rupture and time factor λ for $1.2D + 1.6L = 0.8$

$$\varphi\lambda F_t = (0.65\times0.8)30,000 = 15,600\,\text{psi} \quad\text{and}\quad \varphi\lambda F_v = (0.65\times0.8)4,500 = 2,340\,\text{psi}$$

$$\varphi\lambda F_t = 15,600\,\text{psi} > f_{buZ} = 1,551\,\text{psi} \quad\text{and}\quad \varphi\lambda F_v = 2,340\,\text{psi} > f_{vuZ} = 33.5\,\text{psi}$$

The FRP I-shaped beam satisfies the material strength limit.
   e.2) Determine lateral torsional buckling strength:
      e.2.1) using the complex expression:
         The lateral torsional buckling strength is:
         C_b (moment modification factor) = 1.0 for a simply supported uniformly distributed beam, under constant torque.
         For $d \geq 6''$ then $L_{b,\,eff} = 0.95L_b$,
         From Eq. (9.13):

$$C_\omega = \frac{I_z d^2}{4} = \frac{(1.71)\left(6^2\right)}{4} = 15.39\,\text{in}^6$$

$$f_n^{LTB} = \frac{C_b\pi}{S_z L_{b,eff}}\sqrt{E_{T,f}I_Z G_{LT}C + \frac{\pi^2 E_{T,f}^2 I_Z C_\omega}{L_{b,eff}^2}}$$

$$f_n^{LTB} = \frac{\pi}{1.14(0.95\times8\times12)}\sqrt{\begin{array}{l}\left(0.8\times10^6\right)(1.71)\left(0.425\times10^6\right)(0.198) \\[4pt] +\dfrac{\pi^2\left(0.8\times10^6\right)^2(1.71)(15.39)}{(0.95\times8\times12)^2}\end{array}}$$

$$f_n^{LTB} = 11,107\,\text{psi}$$

(**Alternative method**): e.2.2) using the simplified expression from Eq. (9.19):

$$f_n^{LTB} = \frac{\pi E_{T,f}}{2 S_Z L_{b,eff}} \sqrt{2 I_Z C + \left(\frac{\pi h I_Z}{L_{b,eff}} \right)^2}$$

$$f_n^{LTB} = \frac{\pi \left(0.8 \times 10^6 \right)}{2 \left(1.14 \times 0.95 \times 8 \times 12 \right)} \sqrt{\left(2 \times 1.71 \times 0.198 \right) + \left(\frac{\pi \times 6 \times 1.71}{0.95 \times 8 \times 12} \right)^2} = 10{,}825 \, \text{psi}$$

Note: The difference in the lateral torsional strength $\left(f_n^{LTB} \right)$ between the complex and simplified expressions is about 2.5%.

For $\varphi = 0.55$ and time factor λ for $1.2D + 1.6L = 0.8$, the lateral torsional buckling strength is:

$$\varphi \lambda f_n^{LTB} = (0.55 \times 0.8) 11{,}107 = 4{,}887 \, \text{psi}$$

$$f_{buZ} = 1{,}551 \, \text{psi} < \varphi \lambda f_n^{LTB} = 4{,}887 \, \text{psi}$$

The cross section of the FRP I-shaped beam is sufficient to prevent global lateral torsional buckling.
e.3) Determine local buckling strength:
    e.3.1) Local flange buckling:
        The local flange buckling strength is determined from Eq. (9.31) as:

$$f_{crf}^{local} = \frac{4 t_f^2}{b_f^2} G_{LT} = \frac{4 (0.375)^2}{3^2} \left(0.425 \times 10^6 \right) = 26{,}563 \, \text{psi}$$

For $\varphi = 0.8$ and λ (for $1.2D + 1.6L$) $= 0.8$, the flange local buckling strength is determined as:

$$\varphi \lambda \left(f_{crf}^{local} \right) = (0.8 \times 0.8) 26{,}563 = 17{,}000 \, \text{psi}$$

$$f_{buZ} = 1{,}551 \, \text{psi} < \varphi \lambda \left(f_{crf}^{local} \right) = 17{,}000 \, \text{psi}$$

Hence, the cross section of the FRP I-shaped beam is sufficient to prevent compressive flange local buckling.
e.4) Check for web shear buckling strength about a weak axis:
    The web shear buckling strength is:

$$\beta = \frac{G_{LT}}{\sqrt{E_{L,w} E_{T,w}}} = \frac{\left(0.425 \times 10^6 \right)}{\sqrt{\left(2.6 \times 10^6 \right) \left(0.8 \times 10^6 \right)}} = 0.295$$

$$k = 2.67 + 1.59 \beta = 2.67 + 1.59 (0.295) = 3.14$$

$$f_{cr} = k\frac{\sqrt[4]{E_{L,w}\left(E_{T,w}\right)^3}}{\left(\dfrac{d_w}{t_w}\right)^2} = 3.14\frac{\sqrt[4]{\left(2.6\times10^6\right)\left(0.8\times10^6\right)^3}}{\left(\dfrac{3}{2\times0.375}\right)^2} = 210,703\,\text{psi}$$

For $\varphi = 0.8$ and λ (for $1.2D + 1.6L$) = 0.8, the compressive web shear buckling strength is determined as:

$$\varphi\lambda\left(f_{cr}^{local}\right) = (0.8\times0.8)210,703 = 134,850\,\text{psi}$$

$$f_{vuZ} = 38.5\,\text{psi} < \varphi\lambda\left(f_{cr}^{local}\right) = 134,850\,\text{psi}$$

Web shear buckling strength is higher than the design shear load of the FRP beam. The nominal available strength due to bending about the weak axis is obtained from the lateral torsional buckling strength (4,887 psi).

f. Resistance of the pultruded FRP member under biaxial bending:
    The interaction of the FRP I-shaped beam is determined from Eq. (12.3.1) as:

$$\frac{P_u}{P_c} + \frac{M_{uT}}{M_{cT}} + \frac{M_{uZ}}{M_{cZ}} = 0 + \frac{(455.2\times12)}{750(7.43)} + \frac{(151.2\times12)}{4,887(1.14)} = 0.98 + 0.33 = 1.31 \le 1.0$$

The design of I-shaped beam ($6 \times 3 \times 3/8''$) is not adequate for strength and deflection limits under biaxial bending. The above process needs to be repeated with a larger section so that the interaction of the FRP I-shaped beam is satisfactory.

Example 12.2

Using the ASD method, design a small pultruded FRP purlin for a simple span of 8 ft with 10 lbs/ft of dead load and 30 lbs/ft of live load acting on the member. The pultruded FRP beam is placed on a slope of 1:3 as shown in Example Figure 12.2.

Given: Pultruded beam properties as per manufacturers:

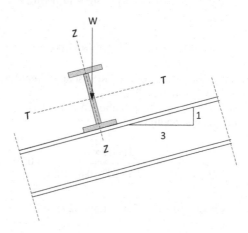

EXAMPLE FIGURE 12.2   Simply supported pultruded FRP I beam under bi-axial bending.

Longitudinal tensile strength (F_t) = 30,000 psi, Shear strength (F_v) = 4,500 psi, Longitudinal modulus (E_L) = 2.6 (10^6) psi, Transverse modulus (E_T) = 0.8 (10^6) psi, and Shear modulus (G_{LT}) = 0.425 (10^6) psi.

Solution

a. Determine service loads:

Total service load (W) = (10 lbs/ft) + (30 lbs/ft) = 40 lbs/ft

Bending @ strong axis (T): W Cos 18.4° = 40 Cos 18.4° =38.0 lbs/ft

Bending @ weak axis (Z): W Sin 18.4° = 40 Sin 18.4° = 12.63 lbs/ft

b. Trial section with the deformation limit:

Trial I-shaped section is evaluated using the serviceability limit (the deflection limit L/240) including the shear deformation response. The shear rigidity $k(GA)$ can be approximated with $(G_{LT})_{web} A_{web}$.

The pultruded FRP beam deflection is determined from Eq. (9.6.4) as:

$$w_{max} = \frac{5WL^4}{384(E_L I_T)} + \frac{WL^2}{8k(G_{LT}A)} = \frac{5(38.0/12)(8\times12)^4}{384(2.6\times10^6)I} + \frac{(38.0/12)(8\times12)^2}{8(0.425\times10^6)A_{web}}$$

$$w_{max} = \frac{1.35}{I_T} + \frac{0.008}{A_{web}} \leq \frac{8(12)}{240} = 0.4\,\text{in}$$

Select 6 × 3 × 3/8" I-shaped beam section:

Given: Self-weight = 3.39 lbs/ft, Area (A) = 4.23 in², Web area (A_w) = 1.97 in², Sectional modulus about the strong axis (S_T) = 7.43 in³, Sectional modulus about the weak axis (S_Z) = 1.14 in³, Moment of inertia about the strong axis (I_T) = 22.3 in⁴, Moment of inertia about the weak axis (I_Z) = 1.71 in⁴, and Torsional constant (C) = 0.198 in⁴.

Longitudinal modulus of flange (E_{Lf}) = 2.6 (10^6) psi, Transverse modulus of flange (E_{Tf}) = 0.8 (10^6) psi, Longitudinal modulus of web (E_{Lw}) = 2.6 (10^6) psi, Transverse modulus of web (E_{Tw}) = 0.8 (10^6) psi, Shear modulus (G_{LT}) = 0.425 (10^6) psi, Poisson's ratio (v_{LT}) = 0.33, Web thickness (t_w) = 0.375", and flange width (b_f) = 3".

$$w_{max} = \frac{1.35}{22.3} + \frac{0.008}{1.97} = 0.065 \leq \frac{8(12)}{240} = 0.4\,\text{in}$$

The pultruded FRP beam satisfies the deflection limit state, as shown above.

c. Determine design bending and shear strength:

The design bending moment and shear of the FRP beam are:

$$M_{ST} = \frac{W_{ST}L^2}{8} = \frac{38.0(8^2)}{8} = 304\,\text{lbs} - \text{ft} \quad \text{and} \quad V_{ST} = \frac{W_{ST}L}{2} = \frac{38.0(8)}{2} = 152\,\text{lbs}$$

$$M_{SZ} = \frac{W_{SZ}L^2}{8} = \frac{12.63(8^2)}{8} = 101\,\text{lbs} - \text{ft} \quad \text{and} \quad V_{SZ} = \frac{W_{SZ}L}{2} = \frac{12.63(8)}{2} = 50.5\,\text{lbs}$$

Induced flexural and average web shear stresses are determined as:

$$f_{bST} = \frac{M_{ST}}{S_T} = \frac{304(12)}{7.43} = 491\,\text{psi} \quad \text{and} \quad f_{vST} = \frac{V}{A_{web}} = \frac{152}{1.97} = 77.16\,\text{psi}$$

$$f_{bSZ} = \frac{M_{SZ}}{S_Z} = \frac{101(12)}{1.17} = 1,036\,\text{psi} \quad \text{and} \quad f_{vSZ} = \frac{V}{A_{web}} = \frac{50.5}{(4.23-1.97)} = 22.4\,\text{psi}$$

d. Bending about a strong axis:
   d.1) Determine the allowable material strength:
     The resistance of the FRP I-shaped beam is determined as:

$$F_{at} = \frac{F_t}{F.S.} = \frac{30,000}{2.5} = 12,000 \, \text{psi} \quad \text{and} \quad F_{av} = \frac{F_v}{F.S.} = \frac{4,500}{2.5} = 1,800 \, \text{psi}$$

$$F_{at} = 12,000 \, \text{psi} > f_{bST} = 491 \, \text{psi} \quad \text{and} \quad F_{av} = 1,800 \, \text{psi} > f_{vST} = 77.2 \, \text{psi}$$

     The FRP I-shaped beam satisfies the material strength limit.
   d.2) Determine the allowable lateral torsional buckling strength:
     d.2.1)Using the complex expression:
       C_b (moment modification factor) = 1.0 for a simply supported uniformly loaded beam, under constant torque.
       For $d \geq 6''$ then $L_{b,\,eff} = 0.95 L_b$
       The lateral torsional buckling strength is determined from Eq. (9.13) as:

$$C_\omega = \frac{I_z d^2}{4} = \frac{(1.71)(6^2)}{4} = 15.39 \, \text{in}^6$$

$$f_n^{LTB} = \frac{C_b \pi}{S_T L_{b,eff}} \sqrt{E_{T,f} I_z G_{LT} C + \frac{\pi^2 E_{T,f}^2 I_z C_\omega}{L_{b,eff}^2}}$$

$$f_n^{LTB} = \frac{\pi}{7.43(0.95 \times 8 \times 12)} \sqrt{\begin{array}{l}\left(0.8 \times 10^6\right)(1.71)\left(0.425 \times 10^6\right)(0.198) \\[2mm] + \dfrac{\pi^2\left(0.8 \times 10^6\right)^2 (1.71)(15.39)}{(0.95 \times 8 \times 12)^2}\end{array}}$$

$$f_n^{LTB} = 1,704 \, \text{psi}$$

(Alternative method): d.2.2) using the simplified expression from Eq. (9.19):

$$f_n^{LTB} = \frac{\pi E_{T,f}}{2 S_T L_{b,eff}} \sqrt{2 I_z C + \left(\frac{\pi h I_z}{L_{b,eff}}\right)^2}$$

$$f_n^{LTB} = \frac{\pi\left(0.8 \times 10^6\right)}{2\left(7.43 \times 0.95 \times 8 \times 12\right)} \sqrt{(2 \times 1.71 \times 0.198) + \left(\frac{\pi 6 \times 1.71}{0.95 \times 8 \times 12}\right)^2} = 1,661 \, \text{psi}$$

Note: The difference in the lateral torsional strength $\left(f_n^{LTB}\right)$ between the complex and simplified expressions is about 2.5%.

For F.S. = 2.5, the allowable lateral–torsional buckling strength is:

$$F_a^{LTB} = \frac{f_n^{LTB}}{F.S.} = \frac{1,704}{2.5} = 682 \, \text{psi}$$

$$f_{bST} = 491 \, \text{psi} < F_a^{LTB} = 682 \, \text{psi}$$

The cross section of the FRP I-shaped beam is sufficient to prevent global lateral–torsion buckling.

d.3) Determine the allowable local buckling strength:

d.3.1) The local flange buckling strength is determined from Eq. (9.22) as:

$$f_{crf}^{local} = \frac{G_{LT} + 0.25\sqrt{E_{L,f}E_{T,f}}}{\left(\dfrac{b_f}{2t_f}\right)^2}$$

$$f_{crf}^{local} = \frac{\left(0.425 \times 10^6\right) + 0.25\sqrt{\left(2.6 \times 10^6\right)\left(0.8 \times 10^6\right)}}{\left(\dfrac{3}{2 \times 0.375}\right)^2} = 49,097\,\text{psi}$$

For F.S. = 2.5, the allowable local flange buckling is determined as:

$$F_{crf}^{local} = \frac{\left(f_{crf}^{local}\right)}{F.S.} = \frac{49,097}{2.5} = 19,639\,\text{psi} \qquad f_{bST} = 491\,\text{psi} < F_{crf}^{local} = 19,639\,\text{psi}$$

Hence, the cross section of the proposed FRP I-shaped beam is sufficient to prevent compressive local flange buckling.

d.3.2) The local web buckling strength is determined from Eq. (9.23) as:

$$f_{crw}^{local} = \frac{15\sqrt{E_{L,f}E_{T,f}}}{\left(\dfrac{d}{t_w}\right)^2}$$

$$f_{crw}^{local} = \frac{15\sqrt{\left(2.6 \times 10^6\right)\left(0.8 \times 10^6\right)}}{\left(\dfrac{6}{0.375}\right)^2} = 84,505\,\text{psi}$$

$$F_{crw}^{local} = \frac{\left(f_{crw}^{local}\right)}{F.S.} = \frac{84,505}{2.5} = 33,802\,\text{psi}$$

The local web buckling strength (33,802 psi) obtained from the pre-standard (ACMA, 2021) is higher than the material strength (30,000 psi). The cross section of the FRP I-shaped beam provides adequate resistance to local web bucking.

d.4) Determine the allowable web shear buckling strength:

Web shear buckling strength from Eqs. (9.35.1–9.35.3):

$$\beta = \frac{G_{LT}}{\sqrt{E_{L,w}E_{T,w}}} = \frac{\left(0.425 \times 10^6\right)}{\sqrt{\left(2.6 \times 10^6\right)\left(0.8 \times 10^6\right)}} = 0.295$$

$$k = 2.67 + 1.59\beta = 2.67 + 1.59(0.295) = 3.14$$

$$f_{cr} = k\frac{\sqrt[4]{E_{L,w}\left(E_{T,w}\right)^3}}{\left(\dfrac{d_w}{t_w}\right)^2} = 3.14\frac{\sqrt[4]{\left(2.6 \times 10^6\right)\left(0.8 \times 10^6\right)^3}}{\left(\dfrac{6 - 0.375 - 0.375}{0.375}\right)^2} = 17,200\,\text{psi}$$

The allowable web shear buckling strength is:

$$F_{cr}^{local} = \frac{\left(f_{cr}^{local}\right)}{F.S.} = \frac{17,200}{2.5} = 6,880 \, \text{psi}$$

$$f_{vST} = 77.16 \, \text{psi} < \left(F_{cr}^{local}\right) = 6,880 \, \text{psi}$$

The web shear buckling strength (6,880 psi) is higher than the design shear strength (77.16 psi) of the FRP beam. The nominal available strength under bending about the strong axis is obtained from the allowable lateral torsional buckling strength (682 psi in Section d.2).

e. Bending about a weak axis:
   e.1) Determine the allowable material strength:
      The resistance of the pultruded FRP I beam is determined as:

$$F_{at} = \frac{F_t}{F.S.} = \frac{30,000}{2.5} = 12,000 \, \text{psi} \quad \text{and} \quad F_{av} = \frac{F_v}{F.S.} = \frac{4,500}{2.5} = 1,800 \, \text{psi}$$

$$F_{at} = 12,000 \, \text{psi} > f_{bSZ} = 1,036 \, \text{psi} \quad \text{and} \quad F_{av} = 1,800 \, \text{psi} > f_{vSZ} = 22.4 \, \text{psi}$$

The proposed FRP I-shape satisfies the material strength limit.
   e.2) Determine the allowable lateral torsional buckling strength:
      The lateral torsional buckling strength is determined from Eq. (9.13) as:
      C_b (moment modification factor) = 1.0 for a simply supported uniformly beam.
      For $d \geq 6''$ then $L_{b,\,eff} = 0.95 L_b$,

$$C_\omega = \frac{I_z d^2}{4} = \frac{(1.71)\left(6^2\right)}{4} = 15.39 \, \text{in}^6$$

$$f_n^{LTB} = \frac{C_b \pi}{S_z L_{b,eff}} \sqrt{E_{T,f} I_Z G_{LT} C + \frac{\pi^2 E_{T,f}^2 I_Z C_\omega}{L_{b,eff}^2}}$$

$$f_n^{LTB} = \frac{\pi}{1.14(0.95 \times 8 \times 12)} \sqrt{\begin{array}{c} \left(0.8 \times 10^6\right)(1.71)\left(0.425 \times 10^6\right)(0.198) \\ \\ + \dfrac{\pi^2 \left(0.8 \times 10^6\right)^2 (1.71)(15.39)}{(0.95 \times 8 \times 12)^2} \end{array}}$$

$$f_n^{LTB} = 11,107 \, \text{psi}$$

(Alternative method): e.2.2) using the simplified expression from Eq. (9.19):

$$f_n^{LTB} = \frac{\pi E_{T,f}}{2 S_Z L_{b,eff}} \sqrt{2 I_Z C + \left(\frac{\pi h I_Z}{L_{b,eff}}\right)^2}$$

$$f_n^{LTB} = \frac{\pi\left(0.8 \times 10^6\right)}{2(1.14 \times 0.95 \times 8 \times 12)} \sqrt{(2 \times 1.71 \times 0.198) + \left(\frac{\pi 6 \times 1.71}{0.95 \times 8 \times 12}\right)^2} = 10,825 \, \text{psi}$$

Note: The difference in the lateral torsional strength $\left(f_n^{LTB}\right)$ between the complex and simplified expressions is about 2.5%.

For F.S. = 2.5, the allowable lateral–torsional buckling strength is determined as:

$$F_a^{LTB} = \frac{f_n^{LTB}}{F.S.} = \frac{11,107}{2.5} = 4,443 \, \text{psi}$$

$$f_{bSZ} = 1,036 \, \text{psi} < F_a^{LTB} = 4,443 \, \text{psi}$$

The cross section of the FRP I-shaped beam is sufficient to prevent global lateral torsional buckling.

e.3) Determine the allowable local buckling strength:

Local flange buckling:

From Eq. (9.31):

$$f_{crf}^{local} = \frac{4t_f^2}{b_f^2} G_{LT} = \frac{4(0.375)^2}{3^2} (0.425 \times 10^6) = 26,563 \, \text{psi}$$

For F.S. = 2.5, the allowable compressive local flange buckling strength is:

$$F_{crf}^{local} = \frac{\left(f_{crf}^{local}\right)}{F.S.} = \frac{26,563}{2.5} = 10,625 \, \text{psi}$$

$$f_{bSZ} = 1,036 \, \text{psi} < F_{crf}^{local} = 10,625 \, \text{psi}$$

The cross section of the proposed FRP I-section is sufficient to prevent compressive local flange buckling.

e.4) Determine the allowable web shear buckling strength:

The web shear buckling strength is:

$$\beta = \frac{G_{LT}}{\sqrt{E_{L,w} E_{T,w}}} = \frac{\left(0.425 \times 10^6\right)}{\sqrt{\left(2.6 \times 10^6\right)\left(0.8 \times 10^6\right)}} = 0.295$$

$$k = 2.67 + 1.59\beta = 2.67 + 1.59(0.295) = 3.14$$

$$f_{cr} = \frac{k\sqrt[4]{E_{L,w}\left(E_{T,w}\right)^3}}{\left(\dfrac{d_w}{t_w}\right)^2} = \frac{3.14\sqrt[4]{\left(2.6 \times 10^6\right)\left(0.8 \times 10^6\right)^3}}{\left(\dfrac{3}{2 \times 0.375}\right)^2} = 210,703 \, \text{psi}$$

Using a F.S. of 2.5, the allowable web shear buckling strength is:

$$F_{cr}^{local} = \frac{\left(f_{cr}^{local}\right)}{F.S.} = \frac{210,703}{2.5} = 84,281 \, \text{psi}$$

$$f_{vSZ} = 22.4 \, \text{psi} < \left(F_{cr}^{local}\right) = 84,281 \, \text{psi}$$

The web shear buckling strength is higher than the design shear strength of the FRP beam. The nominal available strength under bending about the weak axis is obtained from the allowable lateral torsional strength (4,443 psi).

f. Determine the resistance of the pultruded FRP member under biaxial bending:
The interaction of the proposed FRP I section is determined from Eq. (12.3.1) as:

$$\frac{P_s}{P_a} + \frac{M_{sT}}{M_{aT}} + \frac{M_{sZ}}{M_{aZ}} = 0 + \frac{(304 \times 12)}{(682 \times 7.43)} + \frac{(101 \times 12)}{(4,443 \times 1.14)} = 0.720 + 0.239 = 0.959 \leq 1.0$$

The proposed I-shaped beam (6 × 3 × 3/8″) is adequate for strength and deflection limits under biaxial bending.

12.4 STRENGTH LIMITS

The required strengths under moments for frames, beam-columns, connections, and connecting members under combined axial and bending loads must be considered for their influence under induced lateral displacements of a structural system. For example, a beam-column member is subjected to both the primary and secondary bending moments (primary moment – first-order bending moment induced by end moments or transverse loadings without axial loads and secondary moment – additional moment induced by interaction of axial loads coupled with induced lateral displacements). To determine the required strength, two different methods are elaborated hereunder:

1. Using a second-order elastic analysis of the structural system as a whole or for an individual member with appropriate factored loads: maximum bending moment obtained from the second-order analysis is the required bending strength.
2. Using a first-order elastic analysis of a member or an approximate method: In such an event, the bending moment from the first-order elastic analysis is adjusted by moment amplification factors accounting for the second-order effects. The approximate relations for the axial compression and bending (required strengths) are presented as follows (ACMA, 2021):

$$M_u = B_1 M_{nt} + B_2 M_{lt} \tag{12.8.1}$$

$$P_u = P_{nt} + P_{lt} \tag{12.8.2}$$

where P_{nt} and M_{nt} are the first-order elastic axial and flexural strengths in a member assuming there is no lateral translation whether the structural member is braced or not. P_{lt} and M_{lt} are required as the first order elastic axial and flexural strengths in a member as a result of lateral translation of a structural member.

In practice, first-order elastic analysis with moment amplification factors is preferred due to its simplicity and straightforwardness with reasonable accuracy.

For example, as shown in Figure 12.2, a symmetric frame is under symmetric gravity loadings (W) and lateral load H. To obtain maximum non-sway moment (M_{nt}) first-order elastic analysis of the symmetric frame is conducted under factored gravity loads (W). At the same time, maximum sway moment (M_{lt}) is also determined using the first-order elastic analysis method for the same symmetric frame with lateral loads only. Both the non-sway and sway moments will be modified by appropriate moment amplification factors, (B_1 and B_2) for a non-sway mode and sway mode, respectively.

For a general frame and loading as shown in Figure 12.3, the first-order elastic analysis of non-sway frame is first analyzed to obtain the maximum non-sway moment (M_{nt}) and lateral restraint (R). The non-sway frame is idealized by providing each lateral support consistently against lateral displacements.

In the next step, the first-order elastic analysis of the original frame with lateral restraint (R) in the displacement direction (opposite to the provided support reaction) is applied to attain the maximum

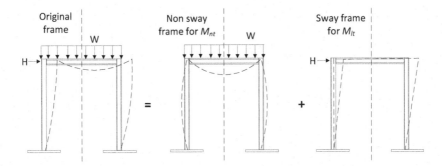

FIGURE 12.2   Symmetric frame with symmetric gravity loads and lateral load.

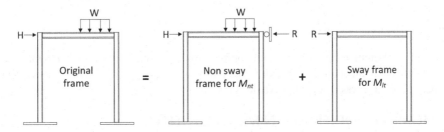

FIGURE 12.3   Symmetric frame with asymmetric gravity loads and lateral load.

sway moment (M_{lt}). Finally, appropriate moment amplification factors (B_1 and B_2 as given in Eqs. (12.9.1 and 12.9.3)) for non-sway and sway modes are used to modify the maximum non-sway and sway moments, respectively.

12.5   MOMENT AMPLIFICATION FACTOR B_1 (EFFECT OF MEMBER CURVATURE)

The moment amplification factor B_1 of the first order no sway moment relates the curvature and (P-δ) effect of the member. B_1 is important when a column is slender with a high axial load ratio. Typically, the moment amplification factor B_1 is equal to or higher than unity. B_1 is applied for both the braced and unbraced structural systems (ACMA, 2021).

$$B_1 = \frac{C_m}{\left(1 - \dfrac{P_u}{P_e}\right)} \geq 1.0 \tag{12.9.1}$$

where C_m = equivalent moment factor based on first-order elastic analysis, assuming no translation of the frame
  P_u = required axial compression load (factored load) from both the no lateral and lateral translation cases
  P_e = elastic buckling load in the plane of bending for members with zero sway

$$C_m = 0.6 - 0.4\left(\frac{M_1}{M_2}\right) \tag{12.9.2}$$

where M_1/M_2 (Figure 12.4) is the ratio of smaller $|M_1|$ to larger $|M_2|$ moments at the ends of that portion of the member that is unbraced in the plane of bending (ACMA, 2021). The moment ratio is assigned to be positive when the member is bent in double curvature and negative if the member is bent in single curvature.

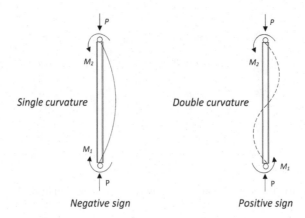

FIGURE 12.4 Sign convention of M_1/M_2.

For beam-column members under axial compression at the ends and transverse loads along the span length, the equivalent moment factor is equal to unity ($C_m = 1.0$). C_m is always less than or equal to 1; hence, $C_m = 1.0$ gives a conservative design value.

12.6 MOMENT MODIFICATION FACTOR B_2 (EFFECT OF LATERAL DISPLACEMENT)

The moment amplification factor B_2 of the first-order moment without sway is associated with the lateral displacement of a member known as P – delta (Δ) effect. Modification factor B_1 is applied for both the braced and unbraced structural system. For multistory frames, the moment amplification factor B_2 is significant when a group of columns in the same story is slender. The importance of B_2 factor increases with higher levels of gravity loads.

$$B_2 = \frac{1}{1 - \left(\dfrac{\sum P_u}{\sum H}\right)\dfrac{\Delta_l}{L}}$$

where
$\sum P_u$ = required axial compression load (factored load) from cases with and without lateral translation cases of all columns in a story
$\sum H$ = summation of all story lateral loads corresponding to lateral drift (lateral displacement between stories)
Δ_l = lateral inter-story displacement
L = story height.

Example 12.3

Using the LRFD method, an FRP WF 8 × 8 × 3/8″ beam under combined service loads of 70 lbs/ ft uniform load (moment about the strong axis (T) and concentric axial compression dead and live loads of 1.2 and 3.5 kips, respectively. The member is a part of a braced frame in TT and ZZ planes. Check the adequacy of this member.

The nominal stiffness and strength are assumed to be as per Section 2.4.2 in the pre-standard (ACMA, 2021). Therefore, the adjustment factors will not be applied in this example.

Given: wide flange beam properties (Manufacturers): Longitudinal tensile strength (F_t) = 30,000 psi, Longitudinal compression strength (F_c) = 30,000 psi, Shear strength (F_v) = 4,500 psi, Longitudinal modulus (E_L) = 2.6 (10^6) psi, Transverse modulus (E_T) = 0.8 (10^6) psi, Shear modulus (G_{LT}) = 0.425 (10^6) psi, and Poisson's ratio (ν_{LT}) = 0.33.

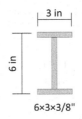

3 in

6 in

6×3×3/8"

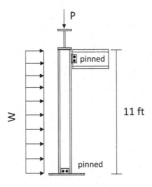

P

pinned

11 ft

W

pinned

EXAMPLE FIGURE 12.3 Simply supported pultruded FRP beam.

Given: WF 8 × 8 × 3/8″ section: Self-weight = 6.97 lbs/ft, Area (A) = 8.73 in², Web area (A_w) = 2.72 in², Sectional modulus about the strong axis (S_T) = 24.8 in³, Moment of inertia about the strong axis (I_T) = 99.1 in⁴, Moment of inertia about the weak axis (I_Z) = 32.0 in⁴, Torsional constant (C) = 0.409 in⁴. Web thickness (t_w) = 0.375 inch and width of flange (b_f) = 8 inch

Solution

The interaction of a pultruded FRP member under bending about the strong axis and compression is determined from Eq. (12.3.2) as:

$$\frac{P_u}{P_c} + \frac{M_{uT}}{M_{cT}} \le 1.0$$

a. Determine the available axial strength P_c:
   The available axial strength P_c of the interaction relations is obtained from a minimum value of the critical local and global strengths. These are based on: (a.1) material strength P_L, (a.2) flexural and local buckling strength, and (a.3) global torsional buckling strength.

a.1) Determine material strength:
   For nominal material strength limit state (λ = 0.8 for 1.2D + 1.6L as per ACMA, 2021):
   From Eq. (10.7):

$$P_E = 0.7\lambda F_L^c A_g = 0.7(0.8)(30,000)(8.73) = 146,664\,\text{lbs} \cong 147\,\text{kips}$$

a.2) Determine flexural and local buckling strength, where $K_T = K_Z = 1$
   a.2.1) Flexural buckling about the strong axis from Eq. (10.18.1):

$$F_{crT} = \frac{\pi^2 E_L}{\left(\dfrac{K_T L_T}{r_T}\right)^2} = \frac{\pi^2\left(2.6\times10^6\right)}{\left(\dfrac{11\times12}{3.37}\right)^2} = 16,726\,\text{psi} \cong 16.8\,\text{ksi}$$

a.2.2) Flexural buckling about the weak axis from Eq. (10.18.2):

$$F_{crZ} = \frac{\pi^2 E_L}{\left(\dfrac{K_Z L_Z}{r_Z}\right)^2} = \frac{\pi^2 \left(2.6 \times 10^6\right)}{\left(\dfrac{11 \times 12}{1.69}\right)^2} = 4,206\,\text{psi} \cong 4.21\,\text{ksi}$$

Euler buckling load P_e in the plane of bending is: $P_e = 4,206 \times 8.73 = 36,718$ psi $\cong 36.8$ kips

a.2.3) Local flange buckling from Eq. (10.18.3):

$$F_{crf} = \frac{G_{LT}}{\left(\dfrac{b_f}{2t_f}\right)^2} = \frac{\left(0.425 \times 10^6\right)}{\left(\dfrac{8}{2 \times 0.375}\right)^2} = 3,735\,\text{psi} \cong 3.74\,\text{ksi}$$

a.2.4) Local web buckling from Eq. (10.18.4):

$$F_{crw} = \frac{\pi^2 t_w^2}{6h^2}\left(\sqrt{E_{L,w} E_{T,w}} + E_{T,w} \nu_{LT} + 2G_{LT}\right)$$

$$F_{crw} = \frac{\pi^2 (0.375)^2}{6(8^2)}\left(\sqrt{\left(2.6 \times 0.8 \times 10^{12}\right)} + \left(0.33 \times 0.8 \times 10^6\right) + \left(2 \times 0.425 \times 10^6\right)\right)$$

$$F_{crw} = 9,246\,\text{psi} \cong 9.25\,\text{ksi}$$

For $\varphi_c = 0.8$ and λ (for $1.2D + 1.6L$) = 0.8, the minimum buckling strength is obtained from the local flange buckling (F_{crf}). The available buckling strength is determined as:

$$P_{cr} = \lambda \varphi_c F_{cr} A_g = (0.8 \times 0.8)(3,735)(8.73) = 20,868\,\text{lbs} \cong 20.9\,\text{kips}$$

a.3) Determine global torsional buckling strength:

$(C_\omega = I_z h^2/4 = 32\ (8^2)/4 = 512\ \text{in}^6$ for wide flange). Polar moment of inertia $I_p = 99.1 + 32.0 = 131.1\ \text{in}^4$, $k_\omega = 1.0$. The global torsional buckling load is determined from Eq. (10.16.2) as:

$$f_{cr}^{GTB} = \frac{1}{I_p}\left(CG_{LT} + \frac{\pi^2 E_L C_\omega}{\left(k_\omega L\right)^2}\right)$$

$$f_{cr}^{GTB} = \frac{1}{131.1}\left(\left(0.409 \times 0.425 \times 10^6\right) + \frac{\pi^2 \left(2.6 \times 512 \times 10^6\right)}{\left(12 \times 11\right)^2}\right) = 7,078\,\text{psi}$$

$$\varphi \lambda f_{cr}^{GTB} = (0.7 \times 0.8)7,078 = 3,963\,\text{psi}$$

$$P_{cr}^{GTB} = A_g \left(\varphi \lambda f_{cr}^{GTB}\right) = 8.73(3,963) = 34,600\,\text{lbs} = 34.6\,\text{kips}$$

The minimum available axial strength is obtained from the local flange buckling strength ($P_c = 20.9$ kips).

b. Determine nominal bending strength M_a:

The nominal bending strength M_a of the interaction relation is taken as the minimum value of the critical local and global strengths as (b.1) material flexural strength, (b.2) lateral torsional buckling strength, (b.3) local buckling strength, and (b.4) web shear buckling strength.

b.1) Determine material strength:

The resistance of the pultruded FRP WF section is determined as:

$\varphi = 0.65$ and λ (for $1.2D + 1.6L$) $= 0.8$

$$F_b = \varphi \lambda F_t = (0.65 \times 0.8)30,000 = 15,600 \, \text{psi}$$

$$M_b = S_T F_b = \frac{24.8(15,600)}{12 \times 1,000} = 32.2 \, \text{kips.ft}$$

b.2) Determine lateral torsional buckling strength:

b.2.1) Using the complex expression from Eq. (9.13):

C_b (moment modification factor) $= 1.0$ for a simply supported uniformly beam (For $d \geq 6''$ then $L_{b, \, eff} = 0.95L_b$).

$$C_\omega = \frac{I_z d^2}{4} = \frac{32.0(8^2)}{4} = 512 \, \text{in}^6$$

$$f_n^{LTB} = \frac{C_b \pi}{S_T L_b} \sqrt{E_{T,f} I_z G_{LT} C + \frac{\pi^2 E_{T,f}^2 I_z C_\omega}{L_b^2}}$$

$$f_n^{LTB} = \frac{\pi}{24.8(0.95 \times 11 \times 12)} \sqrt{\left(0.8 \times 10^6\right)(32.0)\left(0.425 \times 10^6\right)(0.409) + \frac{\pi^2\left(0.8 \times 10^6\right)^2(32.0)(512)}{(0.95 \times 11 \times 12)^2}}$$

$$f_n^{LTB} = 3,371 \, \text{psi}$$

(Alternative method) b.2.2) Using the simplified expression from Eq. (9.19):

$$f_n^{LTB} = \frac{\pi E_{T,f}}{2 S_T L_{b,eff}} \sqrt{2 I_z C + \left(\frac{\pi h I_z}{L_{b,eff}}\right)^2}$$

$$f_n^{LTB} = \frac{\pi\left(0.8 \times 10^6\right)}{2(24.8 \times 0.95 \times 11 \times 12)} \sqrt{(2 \times 32.0 \times 0.409) + \left(\frac{\pi 8 \times 32.0}{0.95 \times 11 \times 12}\right)^2} = 3,315 \, \text{psi}$$

Note: The difference in the lateral torsional strength $\left(f_n^{LTB}\right)$ between the complex and simplified expressions is about 1.7%.

For $\varphi = 0.55$ and λ (for $1.2D + 1.6L$) = 0.8, the lateral torsional buckling strength is:

$$F_b^{LTB} = \varphi \lambda f_n^{LTB} = (0.55 \times 0.8)3,371 = 1,483\,\text{psi}$$

$$M_b^{LTB} = S_T F_b^{LTB} = 24.8(1,483) = 36,784 \text{ lbs.in} = 3.07\,\text{kip.ft}$$

b.3) Determine local buckling strength:
   b.3.1) The local flange buckling strength is determined from Eq. (9.22) as:

$$f_{crf}^{local} = \frac{G_{LT} + 0.25\sqrt{E_{L,f}E_{T,f}}}{\left(\dfrac{b_f}{2t_f}\right)^2}$$

$$f_{crf}^{local} = \frac{\left(0.425 \times 10^6\right) + 0.25\sqrt{\left(2.6 \times 10^6\right)\left(0.8 \times 10^6\right)}}{\left(\dfrac{8}{2 \times 0.375}\right)^2} = 6,904\,\text{psi}$$

For $\varphi = 0.8$ and λ (for $1.2D + 1.6L$) = 0.8), the local flange buckling strength is:

$$F_{crf}^{local} = \varphi \lambda \left(f_{crf}^{local}\right) = (0.8 \times 0.8)6,904 = 4,419\,\text{psi}$$

$$M_{crf}^{local} = S_T F_{crf}^{local} = \frac{24.8(4,419)}{12 \times 1,000} = 9.13\,\text{kip.ft}$$

b.3.2) The local web buckling strength is determined from Eq. (9.23) as:

$$f_{crw}^{local} = \frac{15\sqrt{E_{L,f}E_{T,f}}}{\left(\dfrac{d}{t_w}\right)^2}$$

$$f_{crw}^{local} = \frac{15\sqrt{\left(2.6 \times 10^6\right)\left(0.8 \times 10^6\right)}}{\left(\dfrac{8}{0.375}\right)^2} = 47,534\,\text{psi}$$

For $\varphi = 0.8$ and λ (for $1.2D + 1.6L$) = 0.8, the local web buckling strength is determined as:

$$F_{crw}^{local} = \varphi \lambda \left(f_{crw}^{local}\right) = (0.8 \times 0.8)47,534 = 30,422\,\text{psi}$$

$$M_{crw}^{local} = S_T F_{crw}^{local} = \frac{(24.8)30,422}{12 \times 1,000} = 62.9\,\text{kip.ft}$$

b.4) Determine web shear buckling strength:
   Web shear buckling strength from Eqs. (9.35.1–9.35.3):

$$\beta = \frac{G_{LT}}{\sqrt{E_{L,w}E_{T,w}}} = \frac{\left(0.425 \times 10^6\right)}{\sqrt{\left(2.6 \times 10^6\right)\left(0.8 \times 10^6\right)}} = 0.295$$

$$k = 2.67 + 1.59\beta = 2.67 + 1.59(0.295) = 3.14$$

$$f_{cr} = k\frac{\sqrt[4]{E_{L,w}\left(E_{T,w}\right)^3}}{\left(\dfrac{d_w}{t_w}\right)^2} = 3.14\frac{\sqrt[4]{\left(2.6\times10^6\right)\left(0.8\times10^6\right)^3}}{\left(\dfrac{8-0.375-0.375}{0.375}\right)^2} = 9,019\,\text{psi}$$

For $\varphi = 0.8$ and λ (for $1.2D + 1.6L$) = 0.8, the web shear buckling strength is determined as:

$$F_{cr} = \varphi\lambda\left(f_{cr}^{local}\right) = (0.8\times0.8)9,019 = 5,772\,\text{psi}$$

$$M_{cr} = S_T F_{cr} = \frac{(24.8)5,772}{12\times1,000} = 11.9\,\text{kip.ft}$$

The nominal moment strength is obtained from the global lateral torsional buckling strength ($M_c = 3.07$ kips.ft from b.2.1 and b.2.2).

c. Determine factored bending moment about the strong axis and axial compression load:
Factored load: $W = 1.6$ (70 lbs/ft) = 112 lbs/ft

$$M_{nt} = \frac{1}{8}WL^2 = \left(\frac{0.112\,\text{kips/ft}}{8}\right)(11)^2 = 1.69\,\text{kip.ft}$$

Factored column loads: dead load of 1.2 kips and live load of 3.5 kips
From Eq. (12.8.2):

$$P_{nt} = 1.2(1.2) + 1.6(3.5) = 7.04\,\text{kips}$$

$$P_u = P_{nt} + P_{lt} = 7.04 + 0 = 7.04\,\text{kips}$$

$C_m = 1.0$ for compression members subjected to transverse loading between points of support, Euler buckling load P_e in the plane of bending is 36.8 kips (from a.2.2).
From Eqs. (12.8.1 and 12.9.1):

$$B_1 = \frac{C_m}{\left(1-\dfrac{P_u}{P_e}\right)} = \frac{1}{\left(1-\dfrac{7.04}{36.8}\right)} = 1.234$$

$$M_u = B_1 M_{nt} + B_2 M_{lt} = 1.234(1.69) + 0 = 2.09\,\text{kip.ft}$$

The interaction of the wide flange member is determined from Eq. (12.3.2) as:

$$\frac{P_u}{P_c} + \frac{M_u}{M_b^{LTB}} = \frac{7.04}{20.9} + \frac{2.09}{3.07} = 1.02 \leq 1.0$$

The FRP wide flange member (8 × 8 × 3/8″) is adequate for the combined loads for the strength limit stage. It should be noted that the combined resistance ratio is about 2% above 1.00 and considered safe based on the conventional wisdom of practicing design engineers; however, the section size or other materials can be increased to bring the combined resistance ratio below or equal to 1.00.

Example 12.4

Using the ASD method, FRP WF 8 × 8 × 3/8″ beam under combined service loads as a uniform lateral load W of 70 lbs/ft (moment about the strong axis (T)) and concentric axial compression dead and live loads of 1.2 and 3.5 kips, respectively. The member is a part of a braced frame in TT and ZZ planes. Check the adequacy of this member.

Given: wide flange beam properties (Manufacturers): Longitudinal tensile strength (F_t) = 30,000 psi, Longitudinal compression strength (F_c) = 30,000 psi, Shear strength (F_v) = 4,500 psi, Longitudinal modulus (E_L) = 2.6 (10^6) psi, Transverse modulus (E_T) = 0.8 (10^6) psi, Shear modulus (G_{LT}) = 0.425 (10^6) psi, and Poisson's ratio (ν_{LT}) = 0.33.

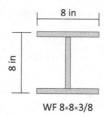

WF 8×8×3/8

Given: WF 8 × 8 × 3/8″ section: Self-weight = 6.97 lbs/ft, Area (A) = 8.73 in^2, Web area (A_w) = 2.72 in^2, Sectional modulus about the strong axis (S_T) = 24.8 in^3, Moment of inertia about the strong axis (I_T) = 99.1 in^4, Moment of inertia about the weak axis (I_Z) = 32.0 in^4, Torsional constant (C) = 0.409 in^4, Web thickness (t_w) = 0.375 inch, and width of flange (b_f) = 8 inch.

Solution

The interaction of a pultruded FRP member under bending about a strong axis and compression is determined from Eq. (12.3.2):

$$\frac{P_u}{P_c} + \frac{M_{uT}}{M_{cT}} \leq 1.0$$

a. Determine allowable axial strength P_a:

The allowable axial strength P_a of the interaction relations is obtained from the minimum of the critical local and global strengths by considering $K_T = 1.0$. These are based on: (a.1) material strength P_L, (a.2) flexural buckling strength, (a.3) global torsional buckling strength, and (a.4) local buckling strength.

a.1) Determine material strength:
From Eq. (10.6):

$$P_L = 0.3F_L^c A_g = 0.3(30,000)(8.73) = 78,570 \, \text{lbs} \cong 78.6 \, \text{kips}$$

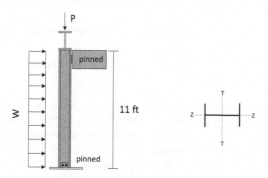

EXAMPLE FIGURE 12.4 Simply supported pultruded FRP beam.

a.2) Determine allowable buckling strength:
(δ_0/L) initial out-of-straightness fraction is 0.05 in/ft (manufacturers) from Eqs. (10.24 and 10.25):

$$\varphi_0 = 1 - 125\frac{\delta_0}{L} = 1 - 125\left(\frac{0.05}{12}\right) = 0.48$$

$$\left(P_{cr}\right)_{allow} = \varphi_0 \frac{\pi^2 E_L}{\left(\dfrac{KL_e}{r}\right)^2} A_g = 0.48\left(\frac{\pi^2\left(2.6\times10^6\right)}{\left(\dfrac{11\times12}{1.69}\right)^2}\right)\times(8.73) = 17,626\,\text{lbs}$$

$$= 17.63\,\text{kips}$$

a.3) Determine the allowable global torsional buckling strength:
($C_\omega = I_z h^2/4 = 32\,(8^2)/4 = 512\,\text{in}^6$ for wide flange). $I_p = 99.1 + 32.0 = 131.1\,\text{in}^4$, $k_\omega = 1.0$. From Eq. (10.16.2):

$$f_{cr}^{GTB} = \frac{1}{I_p}\left(CG_{LT} + \frac{\pi^2 E_L C_\omega}{\left(k_\omega L\right)^2}\right)$$

$$f_{cr}^{GTB} = \frac{1}{131.1}\left(\left(0.409\times0.425\times10^6\right) + \frac{\pi^2\left(2.6\times512\times10^6\right)}{\left(12\times11\right)^2}\right) = 7,078\,\text{psi}$$

$$\left(f_{cr}^{GTB}\right)_{allow} = \frac{f_{cr}^{GTB}}{F.S.} = \frac{7,078}{2.5} = 2,831\,\text{psi}$$

$$\left(P_{cr}^{GTB}\right)_{allow} = A_g\left(f_{cr}^{GTB}\right)_{allow} = 8.73(2,831) = 24,716\,\text{lbs} \cong 24.7\,\text{kips}$$

a.4) Determine the allowable local buckling strength:
Based on the ASD method, the F.S. is taken as 2.5.

a.4.1) Local flange buckling from Eq. (10.18.3):

$$F_{crf} = \frac{G_{LT}}{\left(\dfrac{b_f}{2t_f}\right)^2} = \frac{\left(0.425\times10^6\right)}{\left(\dfrac{8}{2\times0.375}\right)^2} = 3,735\,\text{psi}$$

$$\left(F_{crf}\right)_{allow} = \frac{F_{crf}}{F.S.} = \frac{3,735}{2.5} = 1,494\,\text{psi}$$

$$\left(P_{crf}\right)_{allow} = A_g\left(F_{crf}\right)_{allow} = 8.73\times1,494 = 13,043\,\text{lbs} \cong 13.0\,\text{kips}$$

a.4.2) Local web buckling from Eq. (10.18.4):

$$F_{crw} = \frac{\pi^2 t_w^2}{6h^2}\left(\sqrt{E_{L,w}E_{T,w}} + E_{T,w}\nu_{LT} + 2G_{LT}\right)$$

$$F_{crw} = \frac{\pi^2\left(0.375\right)^2}{6\left(8^2\right)}\left(\sqrt{2.6\times0.8\times10^{12}} + \left(0.33\times0.8\times10^6\right) + \left(2\times0.425\times10^6\right)\right)$$

$$F_{crw} = 9,246 \, psi$$

$$\left(F_{crw}\right)_{allow} = \frac{F_{crw}}{F.S.} = \frac{9,246}{2.5} = 3,698 \, psi$$

$$\left(P_{crw}\right)_{allow} = A_g \left(F_{crw}\right)_{allow} = 8.73 \times 3,698 = 32,288 \, psi \cong 32.3 \, kips$$

The allowable axial strength is obtained as a minumum of (a.1–a.4), which corresponds to the allowable local buckling strength (P_a = 13.0 kips).

b. Determine allowable bending strength M_a:

The allowable bending strength M_a is obtained from the minimum of the critical local and global strengths. These are based on: (b.1) material flexural strength, (b.2) lateral torsional buckling strength, (b.3) local buckling strength, and (b.4) web shear buckling strength.

b.1) Determine allowable material strength:

(F.S. = 2.5 for allowable flexural strength)

$$\left(f_b\right)_{allow} = \frac{F_t}{2.5} = \frac{30,000}{2.5} = 12,000 \, psi$$

$$M_a = S_T \left(f_b\right)_{allow} = \frac{24.8(12,000)}{12 \times 1,000} = 24.8 \, kip.ft$$

b.2) Determine the allowable lateral torsional buckling strength:

b.2.1) Using the complex expression from Eq. (9.13):

C_b (moment modification factor) = 1.0, (For $d \geq 6''$ then $L_{b,\,eff} = 0.95L_b$).

$$C_\omega = \frac{I_z d^2}{4} = \frac{32.0\left(8^2\right)}{4} = 512 \, in^6$$

$$f_n^{LTB} = \frac{C_b \pi}{S_T L_b} \sqrt{E_{T,f} I_z G_{LT} C + \frac{\pi^2 E_{T,f}^2 I_z C_\omega}{L_b^2}}$$

$$f_n^{LTB} = \frac{\pi}{24.8(0.95 \times 11 \times 12)} \sqrt{\begin{array}{c} \left(0.8 \times 10^6\right)(32.0)\left(0.425 \times 10^6\right)(0.409) \\[2mm] + \dfrac{\pi^2 \left(0.8 \times 10^6\right)^2 (32.0)(512)}{\left(0.95 \times 11 \times 12\right)^2} \end{array}}$$

$$f_n^{LTB} = 3,371 \, psi$$

Allowable lateral torsional buckling stress (F.S. = 2.5) is determined as:

$$\left(f_n^{LTB}\right)_{allow} = \frac{f_n^{LTB}}{2.5} = \frac{3,371}{2.5} = 1,348 \, psi$$

$$M_a = S_T \left(f_n^{LTB}\right)_{allow} = (24.8)1,348 = 33,440 \, lbs.in = 2.79 \, kip.ft$$

(Alternative method): b.2.2) using the simplified expression from Eq. (9.19):

$$f_n^{LTB} = \frac{\pi E_{T,f}}{2 S_T L_{b,eff}} \sqrt{2 I_z C + \left(\frac{\pi h I_z}{L_{b,eff}} \right)^2}$$

$$f_n^{LTB} = \frac{\pi \left(0.8 \times 10^6 \right)}{2 \left(24.8 \times 0.95 \times 11 \times 12 \right)} \sqrt{\left(2 \times 32.0 \times 0.409 \right) + \left(\frac{\pi \times 8 \times 32.0}{0.95 \times 11 \times 12} \right)^2} = 3{,}315 \, \text{psi}$$

Note: The difference in the lateral torsional strength $\left(f_n^{LTB} \right)$ between the complex and simplified expressions is about 1.7%.

    b.3) Determine the allowable local buckling strength:
        b.3.1) The local flange buckling strength is determined from Eq. (9.22) as:

$$f_{crf}^{local} = \frac{G_{LT} + 0.25 \sqrt{E_{L,f} E_{T,f}}}{\left(\dfrac{b_f}{2 t_f} \right)^2}$$

$$f_{crf}^{local} = \frac{\left(0.425 \times 10^6 \right) + 0.25 \sqrt{\left(2.6 \times 10^6 \right) \left(0.8 \times 10^6 \right)}}{\left(\dfrac{8}{2 \times 0.375} \right)^2} = 6{,}904 \, \text{psi}$$

Allowable local flange buckling stress ($F.S. = 2.5$) is determined as:

$$\left(f_{crf}^{local} \right)_{allow} = \frac{f_{crf}^{local}}{2.5} = \frac{6{,}904}{2.5} = 2{,}762 \, \text{psi}$$

$$M_a = S_T \left(f_{crf}^{local} \right)_{allow} = \frac{(24.8) \, 2{,}762}{12 \times 1{,}000} = 5.71 \, \text{kip.ft}$$

        b.3.2) The local web buckling strength is determined from Eq. (9.23) as:

$$f_{crw}^{local} = \frac{15 \sqrt{E_{L,f} E_{T,f}}}{\left(\dfrac{d}{t_w} \right)^2}$$

$$f_{crw}^{local} = \frac{15 \sqrt{\left(2.6 \times 10^6 \right) \left(0.8 \times 10^6 \right)}}{\left(\dfrac{8}{0.375} \right)^2} = 47{,}534 \, \text{psi}$$

Allowable local web buckling stress (with $F.S. = 2.5$) is determined as:

$$\left(f_{crw}^{local} \right)_{allow} = \frac{47{,}534}{2.5} = 19{,}014 \, \text{psi}$$

$$M_a = S_T \left(f_{crw}^{local} \right)_{allow} = \frac{(24.8)19{,}014}{12 \times 1{,}000} = 39.3 \, \text{kip.ft}$$

b.4) Allowable web shear buckling strength:
   Web shear buckling strength from Eqs. (9.35.1–9.35.3):

$$\beta = \frac{G_{LT}}{\sqrt{E_{L,w}E_{T,w}}} = \frac{\left(0.425 \times 10^6\right)}{\sqrt{\left(2.6 \times 10^6\right)\left(0.8 \times 10^6\right)}} = 0.295$$

$$k = 2.67 + 1.59\beta = 2.67 + 1.59(0.295) = 3.14$$

$$f_{cr} = k \frac{\sqrt[4]{E_{L,w}\left(E_{T,w}\right)^3}}{\left(\dfrac{d_w}{t_w}\right)^2} = 3.14 \frac{\sqrt[4]{\left(2.6 \times 10^6\right)\left(0.8 \times 10^6\right)^3}}{\left(\dfrac{8 - 0.375 - 0.375}{0.375}\right)^2} = 9{,}019 \, \text{psi}$$

Allowable web shear buckling stress (F.S. = 2.5) is determined as:

$$\left(f_{cr}^{local} \right)_{allow} = \frac{9{,}019}{2.5} = 3{,}608 \, \text{psi}$$

$$M_a = S_T \left(f_{cr}^{local} \right)_{allow} = \frac{(24.8)3{,}608}{12 \times 1{,}000} = 7.46 \, \text{kip.ft}$$

The minimum allowable bending strength is obtained from the allowable lateral torsional buckling strength, $M_a = 2.79$ kips.ft (from b.2.1).

c. Determine service bending moment about the strong axis and axial compression load:
    From (12.8.2):

$$M_{nts} = \frac{1}{8}WL^2 = \left(\frac{0.07 \, \text{kips/ft}}{8} \right)(11)^2 = 1.06 \, \text{kip.ft}$$

$$P_{nts} = 4.7 \, \text{kips}$$

$$P_s = P_{nts} + P_{lts} = 4.7 + 0 = 4.7 \, \text{kips}$$

$C_m = 1.0$ for compression members subjected to transverse loading between points of support. Allowable Euler's buckling load $(P_{cr})_{allow}$ in the plane of bending is 17.63 kips (from a.2).
From Eqs. (12.8.1 and 12.9.1):

$$B_1 = \frac{C_m}{\left(1 - \dfrac{P_s}{(P_{cr})_{allow}}\right)} = \frac{1}{\left(1 - \dfrac{4.7}{17.63}\right)} = 1.364$$

$$M_{nts} = 1.06 \, \text{kip.ft}$$

$$M_s = B_1 M_{nts} + B_2 M_{lts} = 1.364(1.06) + 0 = 1.45 \, \text{kip.ft}$$

The interaction of the FRP wide flange member is determined from Eq. (12.5.2) as:

$$\frac{P_s}{P_a} + \frac{M_s}{M_a} = \frac{4.7}{13.0} + \frac{1.45}{2.79} = 0.881 \le 1.0$$

M_a is controlled by stresses induced under lateral torsional buckling; P_a is controlled by local flange buckling stress. At service loads, the FRP wide flange section (WF 8 × 8 × 3/8") is adequate for the combined loads.

Example 12.5

From Example 12.3, if the FRP section (WF 8 × 8 × 3/8") is under combined service loads (uniform lateral load W of 70 lbs/ft causing moment about the strong axis and concentric axial compression load including self-weight P of 3.2 kips), determine the maximum deflection of the members at service loads.

Solution

By neglecting the shear effect, the deflection under combined loads (transverse loading along the span length with the concentric axial load P) is obtained by multiplying deflection due to bending (δ_o) with the amplification factor. The maximum deflection at the mid-span is:

From Eqs. (10.18.1 and 12.7.2):

$$P_E = \frac{\pi^2 E_L}{\left(\dfrac{K_T L_e}{r}\right)^2} A_g = \left(\frac{\pi^2 (2.6 \times 10^6)}{\left(\dfrac{11 \times 12}{1.69}\right)^2}\right) \times (8.73) = 36{,}720 \text{ lbs} = 36.7 \text{ kips}$$

$$\delta = \delta_0 \frac{1}{1 - \left(\dfrac{P}{P_E}\right)} = \left(\frac{5WL^4}{384(E_L I_T)}\right) \frac{1}{1 - \left(\dfrac{P}{P_E}\right)}$$

$$\delta = \left(\frac{5(70/12)(11 \times 12)^4}{384(2.6 \times 99.1 \times 10^6)}\right) \left(\frac{1}{1 - \left(\dfrac{3.2}{36.7}\right)}\right) = 0.089(1.10) = 0.098''$$

$$\delta = 0.098'' < \frac{L}{240} = \frac{11 \times 12}{240} = 0.55''$$

Example 12.6

Using the LRFD method, check the adequacy of a pultruded column (WF 10 × 5 × 1/2") for 10 ft $(KL)_T$ and 5 ft $(KL)_Z$ about the strong and weak axes, respectively, under axial loads and uniform lateral load, given as (1) dead load including a column self-weight of 1,000 lbs, (2) a live load of 6,000 lbs, and (3) a uniform lateral load of 50 lbs/ft (moment about the strong axis T).

Given: I-shaped 10 × 5 × 1/2" section

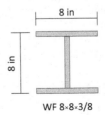

WF 8×8×3/8

Rectangular hollow tubes

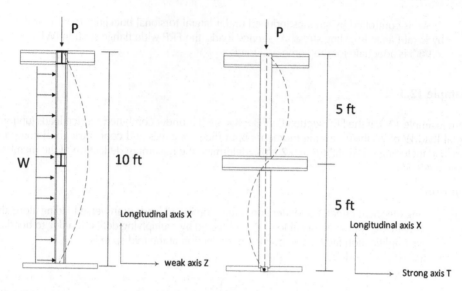

EXAMPLE FIGURE 12.6 Pultruded FRP WF column $10 \times 5 \times 1/2''$ under combined loads.

Gross area $(A) = 9.50$ in². The major and minor radius of gyration (r) is $3.88''$ and $1.05''$, respectively, Moment of inertia about the strong axis $(I_T) = 143.3$ in⁴, Moment of inertia about the weak axis $(I_Z) = 10.51$ in⁴, Section modus about the weak axis $(S_Z) = 4.20$ in³, and Torsional constant (C) = 0.792 in⁴. Section modulus about the strong axis $(S_T) = 28.7$ in³.

Solution

The interaction of a pultruded FRP member under strong axis flexure and compression is determined from Eq. (12.3.2) as:

$$\frac{P_s}{P_a} + \frac{M_{sT}}{M_{aT}} \leq 1.0$$

a. Determine the available axial strength P_c:

The available axial strength P_c is obtained from the minimum value of the critical local and global strengths. These are (a.1) material strength P_L, (a.2) flexural and local buckling strength, and (a.3) global torsional buckling strength.

a.1) Determine nominal material strength:

For nominal material strength at strength limit state from Eq. (10.7):
$\lambda = 0.8$ (for $1.2D + 1.6L$):

$$P_E = 0.7\lambda F_L^c A_g = 0.7(0.8)(30,000)(9.50) = 159,600 \, \text{lbs} = 159.6 \, \text{kip}$$

a.2) Determine nominal buckling strength:

For strength limit state:

The critical buckling strength (F_{cr}) with $K_T = 1.0$, is taken as the minimum of the buckling strengths obtained from different failure modes. These are (1) flexural buckling about the strong axis, (2) flexural buckling about the weak axis, (3) local flange buckling, and (4) local web buckling.

a.2.1) Flexural buckling about the strong axis from Eq. (10.18.1):

$$F_{crT} = \frac{\pi^2 E_L}{\left(\dfrac{K_T L_T}{r_T}\right)^2} = \frac{\pi^2 \left(2.6 \times 10^6\right)}{\left(\dfrac{10 \times 12}{3.88}\right)^2} = 26,827 \, \text{psi}$$

a.2.2)Flexural buckling about the weak axis from Eq. (10.18.2):

$$F_{crZ} = \frac{\pi^2 E_L}{\left(\dfrac{K_Z L_Z}{r_Z}\right)^2} = \frac{\pi^2 \left(2.6 \times 10^6\right)}{\left(\dfrac{5 \times 12}{1.05}\right)^2} = 7,859\,\text{psi}$$

a.2.3)Local flange buckling from Eq. (10.18.3):

$$F_{crf} = \frac{G_{LT}}{\left(\dfrac{b_f}{2t_f}\right)^2} = \frac{\left(0.425 \times 10^6\right)}{\left(\dfrac{10}{2 \times 0.5}\right)^2} = 4,250\,\text{psi}$$

a.2.4) Local web buckling from Eq. (10.18.4):

$$F_{crw} = \frac{\pi^2 t_w^2}{6h^2}\left(\sqrt{E_{L,w} E_{T,w}} + E_{T,w} v_{LT} + 2G_{LT}\right)$$

$$F_{crw} = \frac{\pi^2 (0.5)^2}{6\left(10^2\right)}\left(\sqrt{2.6 \times 0.8 \times 10^{12}} + \left(0.33 \times 0.8 \times 10^6\right) + \left(2 \times 0.425 \times 10^6\right)\right)$$

$$F_{crw} = 10,512\,\text{psi}$$

The buckling strength (F_{cr}) corresponds to local flange buckling (F_{crf}). For $\varphi_c = 0.8$ and λ for $(1.2D + 1.6L) = 0.8$, the nominal buckling strength is determined as:

$$P_{cr} = \lambda \varphi_c F_{cr} A_g = (0.8 \times 0.8)(4,250)(9.50) = 25,840\,\text{lbs}$$

a.3) Determine the global torsional buckling strength:
The global torsional buckling load is determined as:
Polar moment of inertia $I_p = 143.3 + 10.51 = 153.8$ in[4], $k_\omega = 1.0$

$$C_\omega = \frac{I_z d^2}{4} = \frac{10.51\left(10^2\right)}{4} = 262.8\,\text{in}^6$$

a.3.1)about the weak axis from Eq. (10.16.2):

$$\left(f_{cr}^{GTB}\right)_Z = \frac{1}{I_p}\left(CG_{LT} + \frac{\pi^2 E_L C_\omega}{\left(k_\omega L\right)^2}\right)$$

$$\left(f_{cr}^{GTB}\right)_Z = \frac{1}{153.8}\left(\left(0.792 \times 0.425 \times 10^6\right) + \frac{\pi^2 \left(262.8 \times 2.6 \times 10^6\right)}{\left(12 \times 5\right)^2}\right) = 14,368\,\text{psi}$$

$$\varphi\lambda\left(f_{cr}^{GTB}\right)_Z = (0.7 \times 0.8)14,368 = 8,046\,\text{psi}$$

a.3.2)about the strong axis from Eq. (10.16.2):

$$\left(f_{cr}^{GTB}\right)_T = \frac{1}{I_p}\left(CG_{LT} + \frac{\pi^2 E_L C_\omega}{\left(k_\omega L\right)^2}\right)$$

$$\left(f_{cr}^{GTB}\right)_T = \frac{1}{153.8}\left(\left(0.792 \times 0.425 \times 10^6\right) + \frac{\pi^2 \left(2.6 \times 262.8 \times 10^6\right)}{\left(12 \times 10\right)^2}\right) = 5,234\,\text{psi}$$

$$\varphi\lambda\left(f_{cr}^{GTB}\right)_T = (0.7 \times 0.8)5,234 = 2,931\,\text{psi}$$

$$P_{cr} = \varphi\lambda\left(f_{cr}^{GTB}\right)_T A_g = 2,931 \times 9.50 = 27,845\,\text{lbs}$$

The minimum available axial strength corresponds to local flange buckling strength ($P_{cr} = 25.8$ kips from a.2.3).

b. Determine the nominal bending strength M_c:

The available bending strength M_c is obtained from the minimum of the critical local and global strengths. These are (b.1) material flexural strength, (b.2) lateral torsional buckling strength, (b.3) local buckling strength, and (b.4) web shear buckling strength.

b.1) Determine material strength:

The resistance of the FRP wide flange column is:
$\varphi = 0.65$ (for material rupture) and λ (for $1.2D + 1.6L$) = 0.8

$$F_b = \varphi\lambda F_t = (0.65 \times 0.8)30,000 = 15,600\,\text{psi} \qquad M_b = S_T F_b = \frac{28.7(15,600)}{12 \times 1,000} = 37.31\,\text{kip.ft}$$

b.2) Determine lateral torsional buckling strength:

C_b (moment modification factor) = 1.0 for a simply supported uniformly beam. Polar moment of inertia $I_p = 143.3 + 10.51 = 153.8$ in^4, $k_\omega = 1.0$. (For $d \geq 6''$ then $L_{b, eff} = 0.95L_b$.) From Eq. (9.13):

$$C_\omega = \frac{I_z d^2}{4} = \frac{10.51\left(10^2\right)}{4} = 262.8\,\text{in}^6$$

b.2.1) about the strong axis with $L_b = 10$ ft:

$$f_n^{LTB} = \frac{C_b \pi}{S_T L_{b,eff}}\sqrt{E_{T,f} I_z G_{LT} C + \frac{\pi^2 E_{T,f}^2 I_z C_\omega}{L_{b,eff}^2}}$$

$$f_n^{LTB} = \frac{\pi}{28.7(0.95 \times 10 \times 12)}\sqrt{\frac{\left(0.8 \times 10^6\right)(10.51)\left(0.425 \times 10^6\right)(0.792)}{} + \frac{\pi^2\left(0.8 \times 10^6\right)^2(10.51)(262.8)}{(0.95 \times 12 \times 10)^2}}$$

$$f_n^{LTB} = 1,961\,\text{psi}$$

(**Alternative method**): b.2.2) about the strong axis with $L_b = 10$ ft using the simplified expression from Eq. (9.19):

$$f_n^{LTB} = \frac{\pi E_{T,f}}{2S_T L_{b,eff}}\sqrt{2I_z C + \left(\frac{\pi h I_z}{L_{b,eff}}\right)^2}$$

$$f_n^{LTB} = \frac{\pi\left(0.8 \times 10^6\right)}{2(28.7 \times 0.95 \times 10 \times 12)}\sqrt{(2 \times 10.51 \times 0.792) + \left(\frac{\pi \times 10 \times 10.51}{0.95 \times 10 \times 12}\right)^2} = 1,922\,\text{psi}$$

Note: The difference in the lateral torsional strength $\left(f_n^{LTB}\right)$ between the complex and simplified expressions is less than 1%.

For $\varphi = 0.55$ and λ (for $1.2D + 1.6L$) = 0.8, the lateral torsional buckling strength is:

$$F_b^{LTB} = \varphi\lambda f_n^{LTB} = (0.55 \times 0.8)1,961 = 862.8\,\text{psi}$$

$$M_b^{LTB} = S_T F_b^{LTB} = 28.7(862.8) = 24,764\,\text{lbs.in} = 2.06\,\text{kip.ft}$$

b.3) Determine local buckling strength:
    b.3.1) The local flange buckling strength is determined from Eq. (9.22) as:

$$f_{crf}^{local} = \frac{G_{LT} + 0.25\sqrt{E_{L,f}E_{T,f}}}{\left(\dfrac{b_f}{2t_f}\right)^2}$$

$$f_{crf}^{local} = \frac{\left(0.425 \times 10^6\right) + 0.25\sqrt{\left(2.6 \times 10^6\right)\left(0.8 \times 10^6\right)}}{\left(\dfrac{5}{2 \times 0.5}\right)^2} = 31,422\,\text{psi}$$

For $\varphi = 0.8$ and λ (for $1.2D + 1.6L$) = 0.8, the local flange buckling strength is determined as:

$$F_{crf}^{local} = \varphi\lambda\left(f_{crf}^{local}\right) = (0.8 \times 0.8)31,422 = 20,110\,\text{psi}$$

$$M_{crf}^{local} = S_T F_{crf}^{local} = \frac{24.8(20,110)}{12 \times 1,000} = 48.1\,\text{kip.ft}$$

b.3.2) The local web buckling strength is determined from Eq. (9.23) as:

$$f_{crw}^{local} = \frac{15\sqrt{E_{L,f}E_{T,f}}}{\left(\dfrac{d}{t_w}\right)^2}$$

$$f_{crw}^{local} = \frac{15\sqrt{\left(2.6 \times 10^6\right)\left(0.8 \times 10^6\right)}}{\left(\dfrac{10}{0.5}\right)^2} = 54,083\,\text{psi}$$

For $\varphi = 0.8$ and λ (for $1.2D + 1.6L$) = 0.8, the local web buckling strength is determined as:

$$F_{crw}^{local} = \varphi\lambda\left(f_{crw}^{local}\right) = (0.8 \times 0.8)54,083 = 34,613\,\text{psi}$$

$$M_{crw}^{local} = S_T F_{crw}^{local} = \frac{(24.8)34,613}{12 \times 1,000} = 82.8\,\text{kip.ft}$$

b.4) Determine web shear buckling strength:
    Web shear buckling strength from Eqs. (9.35.1–9.35.3):

$$\beta = \frac{G_{LT}}{\sqrt{E_{L,w}E_{T,w}}} = \frac{\left(0.425 \times 10^6\right)}{\sqrt{\left(2.6 \times 10^6\right)\left(0.8 \times 10^6\right)}} = 0.295$$

$$k = 2.67 + 1.59\beta = 2.67 + 1.59(0.295) = 3.14$$

$$f_{cr} = k \frac{\sqrt[4]{E_{L,w}\left(E_{T,w}\right)^3}}{\left(\dfrac{d_w}{t_w}\right)^2} = 3.14 \frac{\sqrt[4]{\left(2.6 \times 10^6\right)\left(0.8 \times 10^6\right)^3}}{\left(\dfrac{10-0.5-0.5}{0.5}\right)^2} = 10,405\,\text{psi}$$

For $\varphi = 0.8$ and λ (for $1.2D + 1.6L$) = 0.8, the web shear buckling strength is determined as:

$$F_{cr} = \varphi\lambda\left(f_{cr}^{local}\right) = (0.8 \times 0.8)10,405 = 6,659\,\text{psi}$$

$$M_{cr} = S_T F_{cr} = \frac{(24.8)6,659}{12 \times 1,000} = 15.9\,\text{kip.ft}$$

The nominal moment strength is obtained from the global lateral torsional buckling strength ($M_c = 2.06$ kips.ft from b.2.1 and b.2.2).

c. Determine the factored bending moment about the strong axis and axial compression load:
Factored load: $W = 1.6$ (50 lbs/ft) = 80 lbs/ft

$$M_{nt} = \frac{1}{8}WL^2 = \left(\frac{0.08\,\text{kips/ft}}{8}\right)(10)^2 = 1.0\,\text{kip.ft}$$

Factored column loads: dead load of 1.0 kips and live load of 6.0 kips
From (12.8.2):

$$P_{nt} = 1.2(1.0) + 1.6(6.0) = 10.8\,\text{kips}$$

$$P_u = P_{nt} + P_{lt} = 10.8 + 0 = 10.8\,\text{kips}$$

$C_m = 1.0$ for compression members subjected to transverse loading between points of support, Euler buckling load P_e in the plane of bending is determined from (a.2.2) flexural buckling @ weak axis:
From Eqs. (10.18.1) and (12.9.1):

$$P_{crZ} = \varphi\frac{\pi^2 E_L}{\left(\dfrac{K_Z L_Z}{r_Z}\right)^2}A_g = 0.7\left(\frac{\pi^2\left(2.6 \times 10^6\right)}{\left(\dfrac{5 \times 12}{1.05}\right)^2}\right)\frac{9.50}{1,000} = 52.3\ \text{kips}$$

$$B_1 = \frac{C_m}{\left(1 - \dfrac{P_u}{P_e}\right)} = \frac{1}{\left(1 - \dfrac{10.8}{52.3}\right)} = 1.26$$

$$M_u = B_1 M_{nt} + B_2 M_{lt} = (1.26 \times 1.0) + 0 = 1.26\,\text{kip.ft}$$

The interaction of the pultruded FRP I-shaped beam is determined from (12.3.2) as:

$$\frac{P_u}{P_c} + \frac{M_u}{M_{cT}} = \frac{10.8}{25.8} + \frac{1.26}{2.06} = 0.419 + 0.61 = 1.03 \le 1.0$$

The pultruded FRP (WF 10 × 5 × 1/2") section is adequate for combined loads from the strength viewpoint. It should be noted that the combined resistance ratio is about 3% above 1.00 and considered safe based on the conventional wisdom of practicing design engineers; however, the section size or other materials can be increased to bring the combined resistance ratio below or equal to 1.00.

Example 12.7

Using the ASD method, check the adequacy of a pultruded column (WF 10 × 5 × 1/2″) for 10 ft $(KL)_T$ and 5 ft $(KL)_Z$ about the strong axis and weak axis, respectively, under axial loads and uniform lateral load, given as (1) dead load including a column self-weight of 1,000 lbs, (2) a live load of 6,000 lbs, and (3) a uniform lateral load of 50 lbs/ft (moment about the strong axis T).
I-shaped 10 × 5 × 1/2″ section

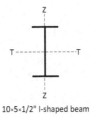

10×5×1/2″ I-shaped beam

Given: Gross area (A) = 9.50 in². The major and minor radius of gyration (r) is 3.88″ and 1.05″, respectively, Moment of inertia about the strong axis (I_T) = 143.3 in⁴, Moment of inertia about the weak axis (I_Z) = 10.51 in⁴, Section modus about the weak axis (S_Z) = 4.20 in³, and Torsional constant (C) = is 0.792 in⁴.

Solution

The interaction of a pultruded FRP member under only strong axis flexure and compression is determined from Eq. (12.3.2) as:

$$\frac{P_s}{P_a} + \frac{M_{sT}}{M_{aT}} \le 1.0$$

a. Determine the available axial strength P_c:
   The available axial strength P_c is obtained from the minimum value of the critical local and global strengths. These are (a.1) material strength P_L, (a.2) flexural and local buckling strength, and (a.3) global torsional buckling strength.

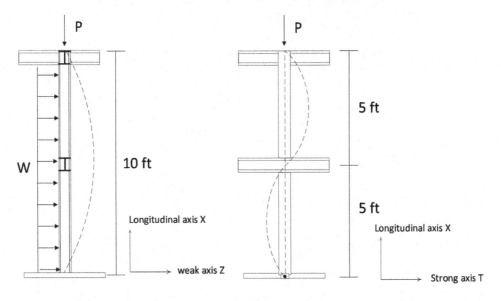

EXAMPLE FIGURE 12.7   Pultruded FRP WF column 10 × 5 × 1/2″ under combined loads.

a.1) Determine allowable material strength:
   For nominal material strength at serviceability limit state from Eq. (10.6):

$$P_L = 0.3 F_L^c A_g = 0.3(30,000)(9.50) = 88,500 \, \text{lbs} = 88.5 \, \text{kips}$$

a.2) Determine allowable buckling strength:
   (δ_0/L) initial out of straightness fraction is 0.05 in/ft (manufacturers).
   From Eq. (10.25) is:

$$\varphi_0 = 1 - 125 \frac{\delta_0}{L}$$

$$\varphi_0 = 1 - 125 \left(\frac{0.05}{12} \right) = 0.48$$

a.2.1) flexural buckling @ the strong axis from Eq. (10.18.1):

$$\left(P_{crT} \right)_{allow} = \varphi_0 \frac{\pi^2 E_L}{\left(\dfrac{K_T L_T}{r_T} \right)^2} A_g$$

$$\left(P_{crT} \right)_{allow} = 0.48 \left(\frac{\pi^2 \left(2.6 \times 10^6 \right)}{\left(\dfrac{10 \times 12}{3.88} \right)^2} \right) \times (9.50) = 122,431 \, \text{lbs} \cong 122 \, \text{kips}$$

a.2.2) flexural buckling @ the weak axis from Eq. (10.18.2):

$$\left(P_{crZ} \right)_{allow} = \varphi_0 \frac{\pi^2 E_L}{\left(\dfrac{K_Z L_Z}{r_Z} \right)^2} A_g$$

$$\left(P_{crZ} \right)_{allow} = 0.48 \left(\frac{\pi^2 \left(2.6 \times 10^6 \right)}{\left(\dfrac{5 \times 12}{1.05} \right)^2} \right) \times (9.50) = 35,864 \, \text{lbs} \cong 35.9 \, \text{kips}$$

a.3) Determine allowable global torsional buckling strength:
   a.3.1) about the weak axis from Eq. (10.16.2):
      $k_\omega = 1.0$. Polar moment of inertia $I_p = 153.8 \, \text{in}^4$. Based on the ASD method, the F.S. is 2.5.

$$C_\omega = \frac{I_z d^2}{4} = \frac{10.51 \left(10^2 \right)}{4} = 262.8 \, \text{in}^6$$

$$\left(f_{cr}^{GTB} \right)_Z = \frac{1}{I_p} \left(C G_{LT} + \frac{\pi^2 E_L C_\omega}{\left(k_\omega L \right)^2} \right)$$

$$\left(f_{cr}^{GTB} \right)_Z = \frac{1}{153.8} \left(\left(0.792 \times 0.425 \times 10^6 \right) + \frac{\pi^2 \left(2.6 \times 262.8 \times 10^6 \right)}{\left(12 \times 5 \right)^2} \right) = 14,368 \, \text{psi}$$

$$\left(f_{crZ}^{GTB} \right)_{allow} = \frac{\left(f_{cr}^{GTB} \right)_Z}{F.S.} = \frac{14,368}{2.5} = 5,747 \, \text{psi}$$

a.3.2) about the strong axis from Eq. (10.16.2):

$$\left(f_{cr}^{GTB}\right)_T = \frac{1}{I_p}\left(CG_{LT} + \frac{\pi^2 E_L C_\omega}{\left(k_\omega L\right)^2}\right)$$

$$\left(f_{cr}^{GTB}\right)_T = \frac{1}{153.8}\left(\left(0.792 \times 0.425 \times 10^6\right) + \frac{\pi^2\left(2.6 \times 262.8 \times 10^6\right)}{\left(12 \times 10\right)^2}\right) = 5,234 \text{ psi}$$

$$\left(f_{crT}^{GTB}\right)_{allow} = \frac{\left(f_{cr}^{GTB}\right)_T}{F.S.} = \frac{5,234}{2.5} = 2,094 \text{ psi}$$

$$\left(P_{cr}\right)_{allow} = \left(f_{crT}^{GTB}\right)A_g = 2,094 \times 9.50 = 19,889 \text{ lbs} \cong 19.89 \text{ kips}$$

Allowable torsional buckling strength is obtained from the global torsional buckling strength about the strong axis ($(P_{cr})_{allow}$ = 19.89 kips).

a.4) Determine the allowable local buckling strength:

Based on ASD method, the F.S. for compression members is 2.5.

a.4.1) Allowable local flange buckling is determined from Eq. (10.18.3):

$$F_{crf} = \frac{G_{LT}}{\left(\dfrac{b_f}{2t_f}\right)^2} = \frac{\left(0.425 \times 10^6\right)}{\left(\dfrac{10}{2 \times 0.5}\right)^2} = 4,250 \text{ psi}$$

$$\left(F_{crf}\right)_{allow} = \frac{F_{crf}}{F.S.} = \frac{4,250}{2.5} = 1,700 \text{ psi}$$

$$\left(P_{crf}\right)_{allow} = A_g\left(F_{crf}\right)_{allow} = 9.50 \times 1,700 \text{ lbs} = 16.15 \text{ kips}$$

a.4.2) Allowable web flange buckling is determined from Eq. (10.18.4):

$$F_{crw} = \frac{\pi^2 t_w^2}{6h^2}\left(\sqrt{E_{L,w}E_{T,w}} + E_{T,w}\nu_{LT} + 2G_{LT}\right)$$

$$F_{crw} = \frac{\pi^2\left(0.5\right)^2}{6\left(10^2\right)}\left(\sqrt{2.6 \times 0.8 \times 10^{12}} + \left(0.33 \times 0.8 \times 10^6\right) + \left(2 \times 0.425 \times 10^6\right)\right) = 10,512 \text{ psi}$$

$$\left(F_{crw}\right)_{allow} = \frac{F_{crw}}{F.S.} = \frac{10,512}{2.5} = 4,205 \text{ psi}$$

$$\left(P_{crw}\right)_{allow} = A_g\left(F_{crw}\right)_{allow} = 9.50 \times 4,205 \text{ psi} = 39.9 \text{ kips}$$

The allowable local buckling strength is obtained from the allowable local flange buckling strength ($(P_{crf})_{allow}$ = 16.15 kips).

b. Determine the allowable bending strength M_a:

The allowable bending strength M_a is obtained from the minimum of the critical local and global strengths. These are (b.1) material flexural strength, (b.2) lateral torsional buckling strength, (b.3) local buckling strength, and (b.4) web shear buckling strength.

b.1) Allowable material strength:

The resistance of the pultruded FRP I-shaped beam is determined as:

($F.S.$ = 2.5 for allowable flexural strength)

$$\left(f_b\right)_{allow} = \frac{F_t}{2.5} = \frac{30,000}{2.5} = 12,000 \text{ psi}$$

$$M_a = S_T \left(f_b\right)_{allow} = \frac{28.7(12,000)}{12 \times 1,000} = 28.7 \, \text{kip.ft}$$

b.2) Allowable lateral torsional buckling strength:

C_b (moment modification factor) = 1.0 for a simply supported uniformly beam. Polar moment of inertia $I_p = 143.3 + 10.51 = 153.8 \, \text{in}^4$, $k_\omega = 1.0$. The lateral torsional buckling strength is determined as:

b.2.1) about the strong axis with $L_b = 10 \, \text{ft}$; Eq. (9.13) gives:

$$C_\omega = \frac{I_z d^2}{4} = \frac{10.51\left(10^2\right)}{4} = 262.8 \, \text{in}^6$$

$$f_n^{LTB} = \frac{C_b \pi}{S_T L_{b,eff}} \sqrt{E_{T,f} I_z G_{LT} C + \frac{\pi^2 E_{T,f}^2 I_z C_\omega}{L_{b,eff}^2}}$$

$$f_n^{LTB} = \frac{\pi}{28.7(0.95 \times 10 \times 12)} \sqrt{\begin{array}{l}\left(0.8 \times 10^6\right)(10.51)\left(0.425 \times 10^6\right)(0.792) \\[2mm] + \dfrac{\pi^2 \left(0.8 \times 10^6\right)^2 (10.51)(262.8)}{(0.95 \times 12 \times 10)^2}\end{array}}$$

$$f_n^{LTB} = 1,961 \, \text{psi}$$

(**Alternative method**): b.2.2) about the strong axis with $L_b = 10 \, \text{ft}$ using the simplified expression from Eq. (9.18):

$$f_n^{LTB} = \frac{\pi E_{T,f}}{2 S_T L_{b,eff}} \sqrt{2 I_z C + \left(\frac{\pi h I_z}{L_{b,eff}}\right)^2}$$

$$f_n^{LTB} = \frac{\pi\left(0.8 \times 10^6\right)}{2(28.7 \times 0.95 \times 10 \times 12)} \sqrt{(2 \times 10.51 \times 0.792) + \left(\frac{\pi \times 10 \times 10.51}{0.95 \times 10 \times 12}\right)^2} = 1,922 \, \text{psi}$$

Note: The difference in the lateral torsional strength $\left(f_n^{LTB} \text{ for}\right)$ between the complex and simplified expressions is less than 1%.

The allowable lateral torsional buckling strength ($F.S. = 2.5$) is determined as:

$$F_b^{LTB} = \frac{f_n^{LTB}}{F.S.} = \frac{1,961}{2.5} = 785 \, \text{psi}$$

$$\left(M_b^{LTB}\right)_{allow} = S_T F_b^{LTB} = 28.7(785) = 22,517 \, \text{lbs.in} \cong 1.88 \, \text{kip.ft}$$

b.3) Allowable Local buckling strength:

b.3.1) The local flange buckling strength is determined from Eq. (9.22) as:

$$f_{crf}^{local} = \frac{G_{LT} + 0.25\sqrt{E_{L,f}E_{T,f}}}{\left(\dfrac{b_f}{2t_f}\right)^2}$$

$$f_{crf}^{local} = \frac{\left(0.425\times10^6\right) + 0.25\sqrt{\left(2.6\times10^6\right)\left(0.8\times10^6\right)}}{\left(\dfrac{5}{2\times0.5}\right)^2} = 31,422\,\text{psi}$$

The allowable local flange buckling strength ($F.S. = 2.5$) is:

$$F_{crf}^{local} = \frac{\left(f_{crf}^{local}\right)}{F.S.} = \frac{31,422}{2.5} = 12,569\,\text{psi}$$

$$\left(M_{crf}^{local}\right)_{allow} = S_T F_{crf}^{local} = \frac{28.7(12,569)}{12\times1,000} = 30.1\,\text{kip.ft}$$

b.3.2) The local web buckling strength is determined from Eq. (9.23) as:

$$f_{crw}^{local} = \frac{15\sqrt{E_{L,f}E_{T,f}}}{\left(\dfrac{d}{t_w}\right)^2}$$

$$f_{crw}^{local} = \frac{15\sqrt{\left(2.6\times10^6\right)\left(0.8\times10^6\right)}}{\left(\dfrac{10}{0.5}\right)^2} = 54,083\,\text{psi}$$

The allowable local web buckling strength ($F.S. = 2.5$) is determined as:

$$\left(F_{cr}\right)_{allow} = \frac{\left(f_{cr}^{local}\right)}{F.S.} = \frac{54,083}{2.5} = 21,633\,\text{psi}$$

$$\left(M_{cr}\right)_{allow} = S_T F_{cr} = \frac{(28.7)21,633}{12\times1,000} = 51.7\,\text{kip.ft}$$

b.4) Allowable web shear buckling strength:

Web shear buckling strength from Eqs. (9.35.1–9.35.3):

$$\beta = \frac{G_{LT}}{\sqrt{E_{L,w}E_{T,w}}} = \frac{\left(0.425\times10^6\right)}{\sqrt{\left(2.6\times10^6\right)\left(0.8\times10^6\right)}} = 0.295$$

$$k = 2.67 + 1.59\beta = 2.67 + 1.59(0.295) = 3.14$$

$$f_{cr} = k\frac{\sqrt[4]{E_{L,w}\left(E_{T,w}\right)^3}}{\left(\dfrac{d_w}{t_w}\right)^2} = 3.14\frac{\sqrt[4]{\left(2.6\times10^6\right)\left(0.8\times10^6\right)^3}}{\left(\dfrac{10-0.5-0.5}{0.5}\right)^2} = 10,405\,\text{psi}$$

Allowable web shear buckling strength ($F.S. = 2.5$) is:

$$\left(F_{cr}\right)_{allow} = \frac{\left(f_{cr}^{local}\right)}{F.S.} = \frac{10,405}{2.5} = 4,162\,\text{psi}$$

$$\left(M_{cr}\right)_{allow} = S_T F_{cr} = \frac{(28.7)4,162}{12 \times 1,000} = 9.95\,\text{kip.ft}$$

Allowable moment strength is obtained from the global lateral torsional buckling strength ($M_c = 1.88$ kips.ft).

c. Determine the factored bending moment about the strong axis and axial compression load:

Service uniform load: $W = 50$ lbs/ft

$$M_{nt} = \frac{1}{8} WL^2 = \left(\frac{0.05\,\text{kips/ft}}{8}\right)(10)^2 = 0.625\,\text{kip.ft}$$

Service column loads: dead load of 1.0 kips and live load of 6.0 kips
From Eq. (12.8.2):

$$P_{nt} = 1.0 + 6.0 = 7.0\,\text{kip}$$

$$P_s = P_{nt} + P_{lt} = 7.0 + 0 = 7.0\,\text{kip}$$

$C_m = 1.0$ for compression members subjected to transverse loading between points of support, Euler buckling load P_e in the plane of bending $= 35.9$ kips (about the weak axis in a.2.2).

From Eq. (12.9.1)

$$B_1 = \frac{C_m}{\left(1 - \dfrac{P_s}{P_e}\right)} = \frac{1}{\left(1 - \dfrac{7.0}{35.9}\right)} = 1.242$$

$$M_s = B_1 M_{nt} + B_2 M_{lt} = (1.242 \times 0.625) + 0 = 0.776\,\text{kip.ft}$$

The interaction of the FRP member is determined from Eq. (12.5.2) as:

$$\frac{P_s}{P_a} + \frac{M_s}{M_{aT}} = \frac{7.0}{16.15} + \frac{0.776}{1.88} = 0.433 + 0.413 = 0.846 \le 1.0$$

The pultruded FRP member (WF $10 \times 5 \times 1/2''$) is adequate for the combined loads.

Example 12.8

Using the LRFD method, design a column supported cantilever highway sign as in Example Figure 12.8.1. Total post height $= 23$ ft.

Based on AASHTO standard, design wind velocity $= 90$ miles/h, Base pressure: Truss $= 0.05$ ksi, and Base pressure: Sign $= 0.04$ ksi, Ice load $= 3$ lbs/ft^2. (Assume 1/6 of the truss plane in the first bay is comprised of structural elements and the center of gravity for truss and sign is at 0.6 (truss span).).

Solution

a. Trial section: FRP circular post:

Circular hallow 16″ section: Gross area $(A) = 24.3$ in², Wall thickness $(t) = 0.5″$, Radius of gyration (r) is 5.48″, Moment of inertia $(I) = 732$ in⁴, Section of modulus $(S_T) = 91.5$ in³, Polar moment of inertia $= 1,464$ in⁴, Weight $= 20.5$ lbs/ft., and in-plane shear strength $F_{LT}^v = 6,500$ psi

b. Determine design load for post:

b.1) Dead loads on structural members:

Dead load of truss and sign $(P_{t+s}) = 1.5$ kips

Dead load of post $(P_p) = (20.5$ lbs/ft$) (23)/1,000 = 0.47$ kips

b.2) Ice load on structural members:

Ice load on sign (P_{ice}) (ice on truss and post negligible) $= 2$ $(3$ lb/ft²$)$ $(10 \times 8) = 0.48$ kips

Loads on the column are presented in Example Figure 12.8.2.

b.3) Design wind load (based on AASHTO):

Post height is less than 30 ft, then design wind velocity is 90 miles/h.

Assume 1/6 of truss plane in the first bay is comprised of structural elements and center of gravity for truss and sign $= 0.6$ (truss span)

– Base pressure: Truss $= 0.05$ ksi

Design wind pressure on truss and post $= 0.05$ $(V^2)/10^4 = 0.05(90^2)/10^4 = 0.0405$ kips/ft²

– Base pressure: Sign $= 0.04$ ksi

Design wind pressure on sign $= 0.05$ $(V^2)/10^4 = 0.04(90^2)/10^4 = 0.0324$ kips/ft²
Wind loads on structural members are presented in Example Figure 12.8.3.

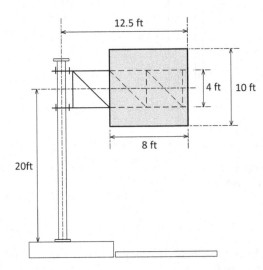

EXAMPLE FIGURE 12.8.1 Column supported cantilever highway sign.

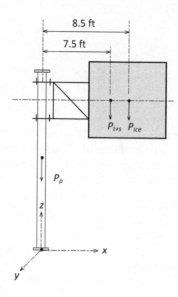

EXAMPLE FIGURE 12.8.2 Loads on post in *z*-direction.

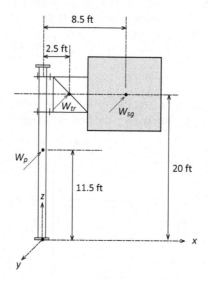

EXAMPLE FIGURE 12.8.3 Wind loads on structural members.

c. Determine moments for post:

c.1) Moment about *x*-axis:
- Moment due to wind load (W_{tr}) on truss about *x*-axis = 0.0405 (4 × 4 × 1/6) 20 = 2.16 kips.ft
- Moment due to wind load (W_{sg}) on sign about *x*-axis = 0.0324 (10 × 8) 20 = 51.8 kips.ft
- Moment due to wind load (W_p) on post about *x*-axis = 0.0405 (23 × (16″/12″)) (23/2) = 14.3 kips.ft

c.2) Moment about *y*-axis:
Assume center of gravity for truss and sign = 0.6 (truss span)
- Moment due to (P_{t+s}) about *y*-axis = (P_{t+s})(0.6 × 12.5) = 1.5 kips (7.5 ft) = 11.25 kips.ft
- Moment due to (P_{ice}) about *y*-axis = (P_{ice}) (12.5−4) = 0.48 kips (8.5 ft) = 4.08 kips.ft

 c.3) Moment about z-axis:
- Moment due to wind load (W_{tr}) on truss about z-axis = 0.0405 (4 × 4 × 1/6) (12.5−8−(4/2)) = 0.27 kips.ft
- Moment due to wind load (W_{sg}) on sign about z-axis = 0.0324 (10 × 8) (12.5−(8/2)) = 22.0 kips.ft

d. Determine shear for the post:
- Shear force due to wind load on truss = 0.0405 (4 × 4 × 1/6) = 0.108 kips
- Shear force due to wind load on sign = 0.0324 (10 × 8) = 2.59 kips
- Shear force due to wind load on post = 0.0405 (23 × (16″/12″)) = 1.24 kips

e. Determine factored load and moment combination:

 Load combination 4: 1.2D + 1.6W + 0.5L + 0.5 (L_r or S or R):

 where D = dead load, L = live load due to occupancy, L_r = roof live load, S is snow load, R = load due to initial rain water or ice exclusive of the ponding condition, W = wind load.

 Factored axial load (P_u) = 1.2 ($P_{t+s} + P_p$) + 0.5 (P_{ice})
- Factored dead load of truss and sign (P_{t+s}) = 1.2 (1.5) = 1.8 kips
- Factored dead load of post (P_p) = 1.2 (0.47) = 0.56 kips
- Factor ice load on sign (P_{ice}) = 0.5 (0.48) = 0.24 kips

 Factored axial load (P_u) = 1.8 + 0.56 + 0.24 = 2.6 kips

 Factored shear load (V_u) = 1.0 (V due to wind load)
- Factored shear load due to wind load on truss = 1.0 (0.108) = 0.108 kips
- Factored shear load due to wind load on sign = 1.0 (2.59) = 2.59 kips
- Factored shear load due to wind load on post = 1.0 (1.24) = 1.24 kips

 Factored shear load (V_u) = 0.108 + 2.59 + 1.24 = 3.94 kips

 Factored load (M_{ux}) = 1.0 (M due to wind load about x-axis)
- Factored moment due to wind load on truss about x-axis = 1.0 (2.16) = 2.16 kips.ft
- Factored moment due to wind load on sign about x-axis = 1.0 (51.8) = 51.8 kips.ft
- Factored moment due to wind load on post about x-axis = 1.0 (14.3) = 14.3 kips.ft

 Factored load (M_{ux}) = 2.16 + 51.8 + 14.3 = 68.3 kips.ft

 Factored load (M_{uy}) = 1.2 (M due to dead load about y-axis) + 0.5 (M due to live load about y-axis)
- Factored moment due to (P_{t+s}) about y-axis = 1.2 (11.25) = 13.5 kips.ft
- Factored moment due to (P_{ice}) about y-axis = 0.5 (4.08) = 2.04 kips.ft

 Factored load (M_{ux}) = 13.5 + 2.04 = 15.5 kips.ft

 Factored load (T_{uz}) = 1.0 (M due to wind load about z-axis)
- Moment due to wind load on truss about z-axis = 1.0 (0.27) = 0.27 kips.ft
- Moment due to wind load on sign about z-axis = 1.0 (22.0) = 22.0 kips.ft

 Factored load (T_{uz}) = 0.27 + 22.0 = 22.3 kips.ft

 f. Determine the available axial strength P_c:

 The available axial strength P_c of the interaction relations is obtained from a minimum value of the critical local and global strengths. These are based on: (f.1) material strength P_L, (f.2) flexural buckling strength, (f.3) local buckling strength, and (f.4) global torsional buckling strength.

f.1) Determine material strength from Eq. (10.7):

 For nominal material strength limit state (λ = 1.0 for 1.2D + 1.6W + 0.5L + 0.5 (L_r or S or R):

$$P_E = 0.7\lambda F_L^c A_g = 0.7(1.0)(30,000)(24.3) = 510,300 \text{ lbs} = 510 \text{ kips}$$

f.2) Determine flexural buckling from Eq. (10.22.1):

 (From Figure 10.4, K = 2.1)

$$F_{crT} = \frac{\pi^2 E_L}{\left(\frac{KL_T}{r}\right)^2} = \frac{\pi^2 \left(2.6 \times 10^6\right)}{\left(\frac{2.1 \times 20 \times 12}{5.48}\right)^2} = 3{,}036\,\text{psi} = 3.04\,\text{ksi}$$

Euler buckling load P_e in the plane of bending is: $P_e = 3.04 \times 24.3 = 73.9$ kips

f.3) Determine local buckling from Eq. (10.22.2):

$$F_{cr} = \frac{2t}{D}\sqrt{\frac{2}{3}G_{LT}\sqrt{E_L E_T}} \le \frac{2t}{D}\sqrt{\frac{E_L E_T}{3}}$$

$$F_{cr} = \frac{2(0.5)}{16}\sqrt{\frac{2}{3}\left(0.425 \times 10^6\right)\sqrt{\left(2.6 \times 0.8 \times 10^{12}\right)}} \le \frac{2(0.5)}{16}\sqrt{\frac{\left(2.6 \times 0.8 \times 10^{12}\right)}{3}}$$

$$F_{cr} = 0.04\left(10^6\right)\text{psi} \le 0.052\left(10^6\right)\text{psi}$$

For $\varphi_c = 0.8$ and λ (for $1.2D + 1.6L$) = 1.0, the minimum buckling strength is obtained from the global buckling (F_{crf}). The available flexural buckling strength including environmental factors is determined as:

$$P_{cr} = \lambda \varphi_c F_{cr} A_g = (0.8 \times 1.0)(3.04)(24.3) = 59.1\,\text{kips}$$

f.4) Determine global torsional buckling strength:
$C_\omega = Id^2/4 = 46{,}848$ in⁶. $C = 4(A^2)/(2\pi r/t) = 23.5$ in⁴, $I_p = 1{,}464$ in⁴, $k_\omega = 1.0$.
The global torsional buckling load is determined from Eq. (10.16.2) as:

$$f_{cr}^{GTB} = \frac{1}{I_p}\left(CG_{LT} + \frac{\pi^2 E_L C_\omega}{\left(k_\omega L\right)^2}\right)$$

$$f_{cr}^{GTB} = \frac{1}{1{,}464}\left(\left(23.5 \times 0.425 \times 10^6\right) + \frac{\pi^2 \left(2.6 \times 46{,}848 \times 10^6\right)}{\left(12 \times 20\right)^2}\right) = 21{,}090\,\text{psi}$$

$$\varphi_c \lambda f_{cr}^{GTB} = (0.7 \times 1)21{,}090 = 14{,}763\,\text{psi}$$

$$P_{cr}^{GTB} = A_g\left(\varphi_c \lambda f_{cr}^{GTB}\right) = 24.3(14{,}763) = 358{,}735\,\text{lbs} = 358\,\text{kips}$$

Thus, the minimum axial strength is obtained from the global flexural buckling strength ($P_c = 59.1$ kips).

g. Determine nominal bending strength M_a:
The nominal bending strength M_a is obtained from the material flexural strength. The global lateral torsional buckling, local buckling, and web shear buckling can be ignored for the circular tube section in this design.
The resistance of the FRP circular tube section is determined as:
$\varphi = 0.65$ and $\lambda = 1.0$ for $1.2D + 1.6W + 0.5L + 0.5$ (L_r or S or R)

$$F_b = \varphi \lambda F_t = (0.65 \times 1.0)30{,}000 = 19{,}500\,\text{psi}$$

$$M_b = S_T F_b = \frac{91.5}{12}\left(\frac{19{,}500}{1{,}000}\right) = 148.7\,\text{kip.ft}$$

h. Determine nominal torsional strength T_a:

The nominal torsional strength of pultruded FRP structural members is influenced by the torsional buckling or torsional rupture strength. The minimum strengths corresponding to torsional buckling (stiffness control) and torsional rupture (material control) are defined as the nominal torsional strength of the cross section.

For strength control:

Note: torsional constant (C) in Table 9.4 = polar moment of inertia:

In-plane shear strength $F_{LT}^v = 6,500$ psi

$$T_n = F_n C$$

$$T_n = \left(\frac{6,500\,\text{psi}}{16/2}\right)\left(\frac{1}{1,000}\right)\left(\frac{1,464\,\text{in}^4}{12}\right) = 99.1\,\text{kip.ft}$$

Note: γ in Eq. (9.38.1) is shearing strain per unit length and the shear strength (stress) F_v must be divided by radius $(d_w/2) = R$.

$$\tau = \frac{R}{C}T_n = (\gamma_{LT}G_{LT}) \quad \text{and} \quad T_n = (\gamma_{LT}G_{LT})\frac{C}{R} = \left(\frac{F_v}{R}\right)C \quad \text{thus,} \quad \gamma = \left(\frac{\gamma_{LT}}{R}\right)$$

For stiffness (buckling) control from Eq. (9.39.4):

$$F_{cr1} = \frac{0.2\left(E_T^c\right)^{5/8}\left(E_L^c\right)^{3/8}}{\left(\dfrac{R}{t}\right)^{3/2}} \leq F_{LT}^v$$

$$F_{cr1} = \frac{0.2\left(0.8\left(10^6\right)\right)^{5/8}\left(2.6\left(10^6\right)\right)^{3/8}}{\left(\dfrac{8}{0.5}\right)^{3/2}} = 3,890 < 6,500\,\text{psi}$$

For stiffness (buckling) control from Eq. (9.39.5):

$$F_{cr2} = \frac{0.7\left(E_T^c\right)^{5/8}\left(E_L^c\right)^{3/8}}{\left(\dfrac{R}{t}\right)^{5/4}\sqrt{\dfrac{L}{R}}} \leq F_{LT}^v$$

$$F_{cr2} = \frac{0.7\left(0.8\left(10^6\right)\right)^{5/8}\left(2.6\left(10^6\right)\right)^{3/8}}{\left(\dfrac{8}{0.5}\right)^{5/4}\sqrt{\dfrac{20\times12}{8}}} = 4,971 < 6,500\,\text{psi}$$

Thus, the critical strength (F_{cr}) for stiffness control is 3,890 psi.

The nominal torsional strength is obtained from the torsional stiffness which is under buckling control.

For $\varphi = 0.7$ and $\lambda = 1.0$ for $1.2D + 1.6W + 0.5L + 0.5$ (L_r or S or R), the torsional strength is determined from Eq. (9.46.2) and Table 9.4 as:

$$\bar{C} = \frac{\pi t}{2}(2R - t)^2 = \frac{\pi(0.5)}{2}\left((2\times8) - 0.5\right)^2 = 188.8\,\text{in}^3$$

$$T_n = F_{cr}\bar{C} = \frac{3,890}{1,000}\left(\frac{188.8}{12}\right) = 61.2\,\text{kip.ft}$$

$$\varphi\lambda T_n = (0.7\times1.0)61.2 = 42.8 \text{ kip.ft}$$

i. Determine nominal material shear strength:

The resistance of the post circular section is determined using $\varphi = 0.65$ for material rupture and $\lambda = 1.0$ for $1.2D + 1.6W + 0.5L + 0.5$ (L_r or S or R),

$$\varphi\lambda F_v = (0.65\times1.0)6,500 = 4,225 \text{ psi}$$

$$(\varphi\lambda F_v)A = (4.23 \text{ ksi})(24.3) = 102.7 \text{ kips}$$

The shear strength (102.7 kips) of the post circular section is much high than the factored shear load (V_u)

j. Check for the interaction of combined loads:

The interaction of the FRP circular tube under combined loads is determined from Eq. (12.3.3) as:

$$\frac{P_u}{P_c} + \frac{M_u}{M_c} + \left(\frac{T_u}{T_c}\right)^2 \le 1.0$$

$$\frac{P_u}{P_c} + \frac{M_{ux}}{M_c} + \frac{M_{uy}}{M_c} + \left(\frac{T_{uz}}{T_c}\right)^2 \le 1.0$$

$$\frac{2.6}{59.1} + \frac{68.3}{148.7} + \frac{15.5}{148.7} + \left(\frac{22.3}{42.8}\right)^2 = 0.879 \le 1.0$$

The pultruded FRP circular tube column (dia.16″) is adequate for the combined loads as the strength state.

EXERCISES

Problem 12.1: Using the LRFD method, design a small pultruded FRP purlin for a simple span of 10 ft. 12.5 lbs/ft of Dead load and 35 lbs/ft of live load are assumed to be acting on the member. The pultruded FRP beam is placed on a slope of 1:3 as shown in Problem Figure 12.1. Given: Pultruded beam properties (manufacturers): Longitudinal tensile strength (F_t) = 30,000 psi, Shear strength (F_v) = 4,500 psi, Longitudinal modulus (E_L) = 2.6 (10^6) psi, Transverse modulus (E_T) = 0.8 (10^6) psi, and Shear modulus (G_{LT}) = 0.425 (10^6) psi.

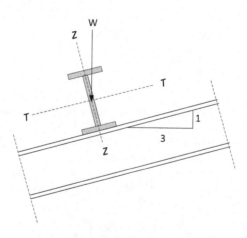

PROBLEM FIGURE 12.1   Simply supported pultruded FRP I beam under bi-axial bending.

Problem 12.2: Using the LRFD method, an FRP WF 8 × 8 × 1/2″ beam under combined service loads of 65 lbs/ft uniform load (moment about the strong axis (T) and concentric axial compression dead and live loads of 1.5 and 3.75 kips, respectively. The member is a part of a braced frame in TT and ZZ planes. Check the adequacy of this member. Given: wide flange beam properties (Manufacturers): Longitudinal tensile strength (F_t) = 30,000 psi, Longitudinal compression strength (F_c) = 30,000 psi, Shear strength (F_v) = 4,500 psi, Longitudinal modulus (E_L) = 2.6 (10^6) psi, Transverse modulus (E_T) = 0.8 (10^6) psi, Shear modulus (G_{LT}) = 0.425 (10^6) psi, and Poisson's ratio (ν_{LT}) = 0.33.

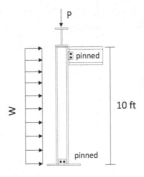

PROBLEM FIGURE 12.2   Simply supported pultruded FRP beam.

Problem 12.3: Using the LRFD method, check the adequacy of a pultruded column (WF 12 × 6 × 1/2″) for 10 ft $(KL)_T$ and 5 ft $(KL)_Z$ about strong and weak axes, respectively, under axial loads and uniform lateral load, given as (1) dead load including column self-weight of 1,200 lbs, (2) live load of 6,500 lbs, and (3) a uniform lateral load of 65 lbs/ft (moment about the strong axis T).

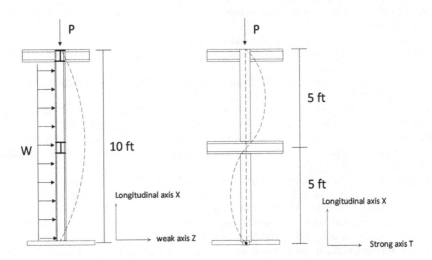

PROBLEM FIGURE 12.3   Pultruded FRP WF column 12 × 6 × 1/2″ under combined loads.

Problem 12.4: Using the LRFD method, design a circular shaft that is 20′ long and is simply supported on the ends and subjected to 200 lb/ft of uniform transverse (vertical) loading and 600 ft-lbs of torsional load. Assume the shaft is a 12″ diameter FRP tube with 1″ wall thickness which is made of vinylester/E-glass composite with 60% fiber volume fraction.

Problem 12.5: Using the LRFD method, design a column supported cantilever highway sign as in Problem Figure 12.5. Total post height = 23 ft. Based on AASHTO standard, design wind velocity = 90 miles/h, Base pressure: Truss = 0.05 ksi, and Base pressure: Sign = 0.04 ksi, Ice load = 3 lbs/ft².

(Assume 1/6 of the truss plane in the first bay is composed of structural elements and the center of gravity for truss and sign is at 0.6 (truss span).)

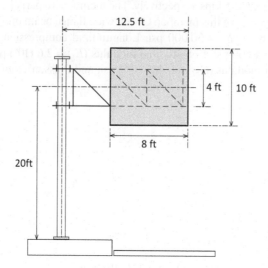

PROBLEM FIGURE 12.5   Column supported cantilever highway sign.

REFERENCES AND SELECTED BIOGRAPHY

ACMA, Pre-Standard for Load & Resistance Factor Design (LRFD) of Pultruded Fiber Reinforced Polymer (FRP) Structures, Arlington, VA, 2021.

ANSI/AISC 360–05, An American National Standard, Specification for Structural Steel Buildings, AISC, Chicago, IL, 2005.

ASCE, Structural Plastics Design Manual, Alexander Bell Drive, Reston, VA, 1984.

ASCE/SEI7–10, Minimum Design Loads for Buildings and Other Structures, Alexander Bell Drive, Reston, VA, 2010.

Bank, L.C., *Composites for Construction: Structural Design with FRP Materials*, John Wiley & Sons, Inc., Hoboken, NJ, 2006.

Barbero, E.J., and Turk, M., Experimental investigation of beam-column behavior of pultruded structural shapes, *Journal of Reinforced Plastics and Composites*, 19(3), 2000, pp. 249–265.

Bedford Reinforced Plastics, *Design Guide*, Bedford Reinforced Plastics, Inc., Bedford, PA, 2012.

Clarke, J.L., *Structural Design of Polymeric Composites – EUROCOMP Design Code and Handbook*, E & FN Spon, London, UK, 1996.

Creative Pultrutions, *The Pultex Pultrusion Design Manual of Standard and Custom Fiber Reinforced Structural Profiles*, 5, revision (2), Creative Pultrutions Inc., Alum Bank, PA, 2004.

Fiberline, *Design Manual*, Fiberline, Kolding, Denmark, 2003.

Johnston, B.G., The Structural Stability Research Council, *The* Guide *to Stability Design Criteria for Metal Structures*, 3rd edition, John Wiley and Sons, New York, NY, 1976.

Kollar, L.P., and Springer, G.S., *Mechanics of Composite Structures*, Cambridge University Press, New York, NY, 2003.

Qureshi, M.A., GangaRao, H.V.S., Hayat, N., and Majjigapu, P., Response of closed glass fiber reinforced polymer sections under combined bending and torsion, *Journal of Composite Materials*, 51(2) 2016, pp. 1–20.

Strongwall Corporation, Design Manual: EXTERN and Other Proprietary Pultruded Products, Bristol, VA, 2010.

Appendix A
Classification of Laminated Composite Stacking Sequence

The coefficients of stiffness and compliance matrices of laminated composites are derived in Section 4.3. The stiffness characteristics of laminated composites are given by those laminate stiffness (*ABD*) and compliance (*abd*) matrices in which the fiber type, fiber orientation, and laminated stacking sequence play a major role. In this section, influences of laminated stacking sequence are discussed. The differences in laminated stacking sequence of the same types of fiber layers often lead to varying stiffness of the composite laminates including the cases with different types of fiber layers. Described in sections A.1 through A.9 are the standard laminated stacking sequences, which include (1) symmetric laminate, (2) balanced laminate, (3) symmetric balanced laminate, (4) cross-ply laminate, (5) symmetric cross-ply laminate, (6) angle-ply laminate, (7) symmetric angle-ply laminate, and (8) quasi-isotropic laminate.

Note: Even though different stacking sequences are needed for complex applications, our focus in this textbook is limited to symmetric and balanced laminates only. Therefore, different stacking sequences that are used in the industry are discussed below.

A.1 SYMMETRIC LAMINATES

When every layer with specific thickness, specific material properties, and specific fiber orientation on one side of a reference surface is identical to each layer on the opposite side with identical distance from the reference surface (mid-plane), then that composite laminate is classified as a symmetrical laminate (Figure A.1). In other words, the plies above the mid-plane are a mirror image of plies below.

Two important properties of symmetric laminates have to be kept in mind: (1) in-plane and out-of-plane coupling stiffness $[B_{ij}]$ is equal to zero, and (2) hygrothermal moment resultants (M^T and M^M) are also zero. However, for unsymmetrical laminates, the in-plane and out-of-plane couplings $[B_{ij}]$ of laminates are not equal to zero.

Let us consider symmetric laminates with n layers:

From Eq. (4.14.2):

$$B_{ij} = \frac{1}{2} \sum_{n}^{k=1} \bar{Q}_{ij}^k \left(z_k^2 - z_{k-1}^2 \right) \tag{A.1.1}$$

$$B_{ij} = \frac{1}{2} \left(\begin{array}{c} \bar{Q}_{ij}^1 \left(z_1^2 - z_0^2 \right) + \bar{Q}_{ij}^2 \left(z_2^2 - z_1^2 \right) + \bar{Q}_{ij}^3 \left(z_3^2 - z_2^2 \right) + \cdots + \bar{Q}_{ij}^{n-2} \left(z_{n-2}^2 - z_{n-3}^2 \right) \\ + \bar{Q}_{ij}^{n-1} \left(z_{n-1}^2 - z_{n-2}^2 \right) + \bar{Q}_{ij}^n \left(z_n^2 - z_{n-1}^2 \right) \end{array} \right) \tag{A.1.2}$$

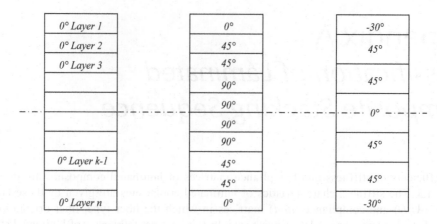

FIGURE A.1 Symmetric laminates: (a) $[0°]_n$, (b) $[0°/45°_2/90°_2]_s$, and (c) $[-30°/45°_2/0°/45°_2/-30°]$.

where the transformation of the reduced stiffness of orthotropic materials is defined in Eq. (3.31.5) as:

$$
\left[\bar{Q}_{ij}^R\right] =
\begin{bmatrix}
\bar{Q}_{11} & \bar{Q}_{12} & \bar{Q}_{16} \\
\bar{Q}_{12} & \bar{Q}_{22} & \bar{Q}_{26} \\
\bar{Q}_{16} & \bar{Q}_{26} & \bar{Q}_{66}
\end{bmatrix}
=
\begin{bmatrix}
c^2 & s^2 & -2cs \\
s^2 & c^2 & 2cs \\
cs & -cs & c^2 - s^2
\end{bmatrix}
\begin{bmatrix}
\dfrac{E_{11}}{1 - v_{12}v_{21}} & \dfrac{v_{12}E_{22}}{1 - v_{12}v_{21}} & 0 \\[2ex]
v_{21}E_{11} & \dfrac{E_{22}}{1 - v_{12}v_{21}} & 0 \\[2ex]
1 - v_{12}v_{21} & 0 & G_{12}
\end{bmatrix}
$$

$$
\times
\begin{bmatrix}
c^2 & s^2 & cs \\
s^2 & c^2 & -cs \\
-2cs & 2cs & c^2 - s^2
\end{bmatrix}
$$

$$(A.1.3)$$

Note: more information on the reduced stiffness is provided in Chapters 3 and 4.

For symmetric laminates, the transformation of reduced stiffness of laminated plies and the distances from the reference surface of each layer are given below:

$$
\bar{Q}_{ij}^1 = \bar{Q}_{ij}^n, \quad \bar{Q}_{ij}^2 = \bar{Q}_{ij}^{n-1}, \quad \bar{Q}_{ij}^3 = \bar{Q}_{ij}^{n-2}, \dots \quad \bar{Q}_{ij}^{\left(\frac{n}{2}\right)} = \bar{Q}_{ij}^{\left(\frac{n}{2}\right)+1}
$$

$$(A.2)$$

$$
\left(z_1^2 - z_0^2\right) = -\left(z_n^2 - z_{n-1}^2\right), \left(z_2^2 - z_1^2\right) = -\left(z_{n-1}^2 - z_{n-21}^2\right), \left(z_3^2 - z_2^2\right)
$$

$$
= -\left(z_{n-2}^2 - z_{n-3}^2\right), \dots, \left(z_{\left(\frac{n}{2}\right)}^2 - z_{\left(\frac{n}{2}\right)-1}^2\right) = -\left(z_{\left(\frac{n}{2}\right)+1}^2 - z_{\left(\frac{n}{2}\right)}^2\right)
$$

$$(A.3)$$

Substituting Eqs. (A.2 and A.3) into (A.1) gives:

$$
B_{ij} = 0
$$

$$(A.4)$$

Thus, it can be concluded that the in-plane and out-of-plane couplings $[B_{ij}]$ are equal to zero for symmetric laminates. In addition, hygrothermal moment resultants are equal to zero for symmetric laminates. The laminate stiffness matrix in Eq. (4.23.2) is simplified as follows:

$$
\begin{bmatrix} N \\ M \end{bmatrix} = \begin{bmatrix} A & 0 \\ 0 & D \end{bmatrix} \begin{bmatrix} \varepsilon^o \\ \kappa^o \end{bmatrix} - \begin{bmatrix} N^T \\ M^T \end{bmatrix} - \begin{bmatrix} N^M \\ M^M \end{bmatrix}
\tag{A.5}
$$

In the case of general symmetric laminates, elements of the laminate stiffness matrix are not zero except for $[B_{ij}]$. Therefore, characteristics of general symmetric laminates are anisotropic for both the in-plane and bending behaviors. However, in-plane force and bending moment resultants can be decoupled into two independent sets of laminate constitutive relations because of the absence of $[B_{ij}]$. The abbreviated form of symmetric laminated compliance is:

$$
\begin{bmatrix} \varepsilon^o \\ \kappa^o \end{bmatrix} = \begin{bmatrix} a & 0 \\ 0 & d \end{bmatrix} \begin{bmatrix} N + N^T + N^M \\ M + M^T + M^M \end{bmatrix}
\tag{A.6.1}
$$

$$
\text{where,} \quad \begin{bmatrix} a_{11} & a_{12} & a_{16} \\ a_{12} & a_{22} & a_{26} \\ a_{16} & a_{26} & a_{66} \end{bmatrix} = \begin{bmatrix} A_{11} & A_{12} & A_{16} \\ A_{12} & A_{22} & A_{26} \\ A_{16} & A_{26} & A_{66} \end{bmatrix}^{-1}
\tag{A.6.2}
$$

$$
\begin{bmatrix} d_{11} & d_{12} & d_{16} \\ d_{12} & d_{22} & d_{26} \\ d_{16} & d_{26} & d_{66} \end{bmatrix} = \begin{bmatrix} D_{11} & D_{12} & D_{16} \\ D_{12} & D_{22} & D_{26} \\ D_{16} & D_{26} & D_{66} \end{bmatrix}^{-1}
\tag{A.6.3}
$$

A.2 BALANCED LAMINATES

The laminates are classified as balanced (typically, with reference to in-plane deformation) when every layer with specific thickness, properties, and fiber orientation is identical to another layer with opposite fiber orientation. Note that, another layer with opposite fiber orientation does not have to be located adjacent to a specific layer or exactly at the same distance on the opposite side of the reference surface. Individual layers (only 0° and 90°) without pairs of the opposite fiber orientation can be added to the balanced laminates, and orthotropic in a global coordinate system (i.e., [0°/ 45°$_2$/−45°$_2$/0°], [+30°/−30°/+30°], [−30°/0°/30°], etc.) as shown in Figure A.2, and still qualify as balanced laminates. In other words, balanced laminates have $-\theta$ ply of the same density as $+\theta$ in the stacking configuration; at some location in the laminate thickness.

Characteristics of balanced laminates are given as elements (A_{16} and A_{26}) of in-plane stiffness $[A_{ij}]$ and are always equal to zero. The elements (A_{16} and A_{26}) of the laminated stiffness matrix are established from the transformation of reduced stiffness ($\bar{Q}_{16}^k$ and $\bar{Q}_{26}^k$), as shown in Chapter 4.

If the sum of ($\bar{Q}_{16}^k$ and $\bar{Q}_{26}^k$) for all layers is equal to zero, then the elements (A_{16} and A_{26}) are also equal to zero. The transformation of reduced stiffness of ($\bar{Q}_{16}$ and $\bar{Q}_{26}$) is an odd function, [i.e. $f(-x) = -f(x)$]. For pairs of opposite fiber orientation (θ and $-\theta$), the results of both ($\bar{Q}_{16}$ and $\bar{Q}_{26}$) are different in sign convention with the same absolute values. Thus, for multiple layers of pairs of the opposite fiber orientation, the sum of those pairs is equal to zero.

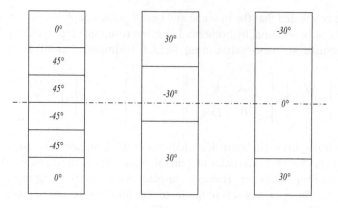

FIGURE A.2   Balanced laminates: (a) [0°/45°$_2$/−45°$_2$/0°], (b) [+30°/-30°/+30°], and (c) [−30°/0°/30°].

$$\left(\bar{Q}_{16}\right)_{\theta} = -\left(\bar{Q}_{16}\right)_{-\theta}$$

(A.7.1)

$$\left(\bar{Q}_{26}\right)_{\theta} = -\left(\bar{Q}_{26}\right)_{-\theta}$$

(A.7.2)

$$A_{16} = \sum_{k=1}^{n} \bar{Q}_{16}^{k} h_k = 0$$

(A.8.1)

$$A_{26} = \sum_{k=1}^{n} \bar{Q}_{26}^{k} h_k = 0$$

(A.8.2)

For multiple layers of pairs of opposite fiber orientations, the sum of those pairs in terms of coefficients of hygrothermal shear expansions (α_{xy} and β_{xy}) is equal to zero. Fiber orientation ($\theta = 0°$ and 90°) provides zero for both (α_{xy} and β_{xy}).

From the definition of hygrothermal force and moment resultants in Eqs. (4.20.2 and 4.22.1), it can be concluded that shear force and twisting moment resultants ($\bar{Q}_{16}, \bar{Q}_{26}$) in the absence of hygrothermal strains are equal to zero, for balanced laminates.

$$
\begin{bmatrix}
N_{xx} \\
N_{yy} \\
N_{xy} \\
M_{xx} \\
M_{yy} \\
M_{xy}
\end{bmatrix}
=
\begin{bmatrix}
A_{11} & A_{12} & 0 & B_{11} & B_{12} & B_{16} \\
A_{12} & A_{22} & 0 & B_{12} & B_{22} & B_{26} \\
0 & 0 & A_{66} & B_{16} & B_{26} & B_{66} \\
B_{11} & B_{12} & B_{16} & D_{11} & D_{12} & D_{16} \\
B_{12} & B_{22} & B_{26} & D_{12} & D_{22} & D_{26} \\
B_{16} & B_{26} & B_{66} & D_{16} & D_{26} & D_{66}
\end{bmatrix}
\begin{bmatrix}
\varepsilon_{xx}^{o} \\
\varepsilon_{yy}^{0} \\
\gamma_{xy}^{o} \\
\kappa_{xx}^{o} \\
\kappa_{yy}^{o} \\
\kappa_{xy}^{o}
\end{bmatrix}
-
\begin{bmatrix}
N_{xx}^{T} \\
N_{yy}^{T} \\
0 \\
M_{xx}^{T} \\
M_{yy}^{T} \\
0
\end{bmatrix}
-
\begin{bmatrix}
N_{xx}^{M} \\
N_{yy}^{M} \\
0 \\
M_{xx}^{M} \\
M_{yy}^{M} \\
0
\end{bmatrix}
$$

(A.9.1)

Inversion of Eq. (A.9.1) gives the laminate compliance matrix of laminated composites.

$$
\begin{bmatrix} \varepsilon_{xx}^{o} \\ \varepsilon_{yy}^{o} \\ \varepsilon_{xy}^{o} \\ \kappa_{xx}^{o} \\ \kappa_{yy}^{o} \\ \kappa_{xy}^{o} \end{bmatrix} = \begin{bmatrix} a_{11} & a_{12} & 0 & b_{11} & b_{12} & b_{16} \\ a_{12} & a_{22} & 0 & b_{12} & b_{22} & b_{26} \\ 0 & 0 & a_{66} & b_{16} & b_{26} & b_{66} \\ b_{11} & b_{12} & b_{16} & d_{11} & d_{12} & d_{16} \\ b_{12} & b_{22} & b_{26} & d_{12} & d_{22} & d_{26} \\ b_{16} & b_{26} & b_{66} & d_{16} & d_{26} & d_{66} \end{bmatrix} \begin{bmatrix} N_{xx} + N_{xx}^{T} + N_{xx}^{M} \\ N_{yy} + N_{yy}^{T} + N_{yy}^{M} \\ N_{xy} \\ M_{xx} + M_{xx}^{T} + M_{xx}^{M} \\ M_{yy} + M_{yy}^{T} + M_{yy}^{M} \\ M_{xy} \end{bmatrix} \tag{A.9.2}
$$

A.3   BALANCED - SYMMETRIC LAMINATES

In this laminate type, characteristics of both symmetric and balanced laminates are combined. (i.e., $[0°/45°_2/-45°_2/-45°_2/45°_2/0°]$, $[-30°/30°]_s$, $[-30°/0°/30°]_s$ as shown in Figure A.3). The in-plane and out-of-plane couplings $[B_{ij}]$ and elements (A_{16} and A_{26}) of the in-plane laminate stiffness $[A_{ij}]$ are taken to be zero. In addition, hygrothermal in-plane shear force and moment resultants are equal to zero. The characteristics of balanced symmetric laminates are orthotropic for in-plane behavior, but anisotropic properties can still exist on the bending behavior.

$$
\begin{bmatrix} N_{xx} \\ N_{yy} \\ N_{xy} \\ M_{xx} \\ M_{yy} \\ M_{xy} \end{bmatrix} = \begin{bmatrix} A_{11} & A_{12} & 0 & 0 & 0 & 0 \\ A_{12} & A_{22} & 0 & 0 & 0 & 0 \\ 0 & 0 & A_{66} & 0 & 0 & 0 \\ 0 & 0 & 0 & D_{11} & D_{12} & D_{16} \\ 0 & 0 & 0 & D_{12} & D_{22} & D_{26} \\ 0 & 0 & 0 & D_{16} & D_{26} & D_{66} \end{bmatrix} \begin{bmatrix} \varepsilon_{xx}^{o} \\ \varepsilon_{yy}^{o} \\ \gamma_{xy}^{o} \\ \kappa_{xx}^{o} \\ \kappa_{yy}^{o} \\ \kappa_{xy}^{o} \end{bmatrix} - \begin{bmatrix} N_{xx}^{T} \\ N_{yy}^{T} \\ 0 \\ 0 \\ 0 \\ 0 \end{bmatrix} - \begin{bmatrix} N_{xx}^{M} \\ N_{yy}^{M} \\ 0 \\ 0 \\ 0 \\ 0 \end{bmatrix}
$$

$$\tag{A.10.1}$$

The laminate compliance matrix (abd) of symmetric balanced laminates is obtained by inverting Eq. (A.10.2).

$$
\begin{bmatrix} \varepsilon_{xx}^{o} \\ \varepsilon_{yy}^{o} \\ \varepsilon_{xy}^{o} \\ \kappa_{xx}^{o} \\ \kappa_{yy}^{o} \\ \kappa_{xy}^{o} \end{bmatrix} = \begin{bmatrix} a_{11} & a_{12} & 0 & 0 & 0 & 0 \\ a_{12} & a_{22} & 0 & 0 & 0 & 0 \\ 0 & 0 & a_{66} & 0 & 0 & 0 \\ 0 & 0 & 0 & d_{11} & d_{12} & d_{16} \\ 0 & 0 & 0 & d_{12} & d_{22} & d_{26} \\ 0 & 0 & 0 & d_{16} & d_{26} & d_{66} \end{bmatrix} \begin{bmatrix} N_{xx} + N_{xx}^{T} + N_{xx}^{M} \\ N_{yy} + N_{yy}^{T} + N_{yy}^{M} \\ N_{xy} \\ M_{xx} \\ M_{yy} \\ M_{xy} \end{bmatrix} \tag{A.10.2}
$$

$$
\text{where,} \quad \begin{bmatrix} a_{11} & a_{12} & 0 \\ a_{12} & a_{22} & 0 \\ 0 & 0 & a_{66} \end{bmatrix} = \begin{bmatrix} A_{11} & A_{12} & 0 \\ A_{12} & A_{22} & 0 \\ 0 & 0 & A_{66} \end{bmatrix}^{-1} \tag{A.10.3}
$$

and

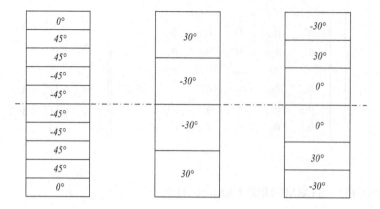

FIGURE A.3 Balanced - symmetric laminates: (a) $[0°/45°_2/-45°_2]_s$, (b) $[-30°/30°]_s$, and (c) $[-30°/0°/30°]_s$.

$$\begin{bmatrix} d_{11} & d_{12} & d_{16} \\ d_{12} & d_{22} & d_{26} \\ d_{16} & d_{26} & d_{66} \end{bmatrix} = \begin{bmatrix} D_{11} & D_{12} & D_{16} \\ D_{12} & D_{22} & D_{26} \\ D_{16} & D_{26} & D_{66} \end{bmatrix}^{-1} \tag{A.10.4}$$

A.4 ANTISYMMETRIC CROSS-PLY LAMINATES

Laminates are classified as antisymmetric cross-ply when laminated layers are oriented either 0° or 90° without symmetrical lay-ups about the reference surface (i.e., [0°/90°], [0°/90°/0°/90°], [90°/0°/90°/90°], etc., as shown in Figure A.4). The important characteristic of antisymmetric cross-ply laminates is given as (1) elements (A_{16}, A_{26}, B_{16}, B_{26} D_{16}, and D_{26}) of the laminated stiffness are equal to zero and (2) hygrothermal in-plane shear force and twisting moment resultants are also equal to zero.

It is clear that if ($\bar{Q}_{16}^k$ and $\bar{Q}_{26}^k$) of every layer are equal to zero, then (A_{16}, A_{26}, B_{16}, B_{26} D_{16}, and D_{26}) are also equal to zero, since fiber orientations of antisymmetric cross-ply laminates are aligned in either 0° or 90°. In addition, shear coefficients of hygrothermal expansions are equal to zero when fiber orientations are aligned with either 0° or 90°. It can be concluded that the shear force and twisting moment resultants in the absence of hygrothermal strain are equal to zero as in the balanced laminates. Therefore, the laminate stiffness matrix (ABD) can be simplified as follows:

$$\begin{bmatrix} N_{xx} \\ N_{yy} \\ N_{xy} \\ M_{xx} \\ M_{yy} \\ M_{xy} \end{bmatrix} = \begin{bmatrix} A_{11} & A_{12} & 0 & B_{11} & B_{12} & 0 \\ A_{12} & A_{22} & 0 & B_{12} & B_{22} & 0 \\ 0 & 0 & A_{66} & 0 & 0 & B_{66} \\ B_{11} & B_{12} & 0 & D_{11} & D_{12} & 0 \\ B_{12} & B_{22} & 0 & D_{12} & D_{22} & 0 \\ 0 & 0 & B_{66} & 0 & 0 & D_{66} \end{bmatrix} \begin{bmatrix} \varepsilon_{xx}^o \\ \varepsilon_{yy}^o \\ \gamma_{xy}^o \\ \kappa_{xx}^o \\ \kappa_{yy}^o \\ \kappa_{xy}^0 \end{bmatrix} - \begin{bmatrix} N_{xx}^T \\ N_{yy}^T \\ 0 \\ M_{xx}^T \\ M_{yy}^T \\ 0 \end{bmatrix} - \begin{bmatrix} N_{xx}^M \\ N_{yy}^M \\ 0 \\ M_{xx}^M \\ M_{yy}^M \\ 0 \end{bmatrix}$$

$$\tag{A.11.1}$$

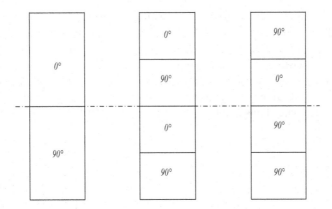

FIGURE A.4 Antisymmetric cross-ply laminate: (a) [0°/90°], (b) [0°/90°/0°/90°], and (c) [90°/0°/90°/90°].

The laminate compliance matrix (*abd*) of antisymmetric cross-ply laminates is also obtained from inversion of Eq. (A.11.1).

$$
\begin{bmatrix}
\varepsilon_{xx}^{o} \\
\varepsilon_{yy}^{o} \\
\varepsilon_{xy}^{o} \\
\kappa_{xx}^{o} \\
\kappa_{yy}^{o} \\
\kappa_{xy}^{o}
\end{bmatrix}
=
\begin{bmatrix}
a_{11} & a_{12} & 0 & b_{11} & b_{12} & 0 \\
a_{12} & a_{22} & 0 & b_{12} & b_{22} & 0 \\
0 & 0 & a_{66} & 0 & 0 & b_{66} \\
b_{11} & b_{12} & 0 & d_{11} & d_{12} & 0 \\
b_{12} & b_{22} & 0 & d_{12} & d_{22} & 0 \\
0 & 0 & b_{66} & 0 & 0 & d_{66}
\end{bmatrix}
\begin{bmatrix}
N_{xx} + N_{xx}^{T} + N_{xx}^{M} \\
N_{yy} + N_{yy}^{T} + N_{yy}^{M} \\
N_{xy} \\
M_{xx} + M_{xx}^{T} + M_{xx}^{M} \\
M_{yy} + M_{yy}^{T} + M_{yy}^{M} \\
M_{xy}
\end{bmatrix}
\tag{A.11.2}
$$

For special case of antisymmetric cross-ply laminates, if the stacking sequence of laminates is arranged in forms of $[0°/90°]_n$ antisymmetric cross-ply laminated stiffness in Eq. (A.11.1) still can be simplified as follows: $A_{11}=A_{22}$, $B_{11}=-B_{22}$, $D_{11}=D_{22}$ and $A_{16}=A_{26}=B_{12}=B_{16}=B_{26}=B_{66}=D_{16}=D_{26}=0$. Thus, the laminate stiffness (*ABD*) of antisymmetric cross-ply laminates $[0°/90°]_n$ is:

$$
\begin{bmatrix}
N_{xx} \\
N_{yy} \\
N_{xy} \\
M_{xx} \\
M_{yy} \\
M_{xy}
\end{bmatrix}
=
\begin{bmatrix}
A_{11} & A_{12} & 0 & B_{11} & 0 & 0 \\
A_{12} & A_{11} & 0 & 0 & -B_{11} & 0 \\
0 & 0 & A_{66} & 0 & 0 & 0 \\
B_{11} & 0 & 0 & D_{11} & D_{12} & 0 \\
0 & -B_{11} & 0 & D_{12} & D_{11} & 0 \\
0 & 0 & 0 & 0 & 0 & D_{66}
\end{bmatrix}
\begin{bmatrix}
\varepsilon_{xx}^{o} \\
\varepsilon_{yy}^{o} \\
\gamma_{xy}^{o} \\
\kappa_{xx}^{o} \\
\kappa_{yy}^{o} \\
\kappa_{xy}^{o}
\end{bmatrix}
-
\begin{bmatrix}
N_{xx}^{T} \\
N_{yy}^{T} \\
0 \\
M_{xx}^{T} \\
M_{yy}^{T} \\
0
\end{bmatrix}
-
\begin{bmatrix}
N_{xx}^{M} \\
N_{yy}^{M} \\
0 \\
M_{xx}^{M} \\
M_{yy}^{M} \\
0
\end{bmatrix}
$$

$$(A.11.3)$$

A.5 SYMMETRIC CROSS-PLY LAMINATES

Laminates from plies of fiber orientation in either 0° or 90°, satisfying the condition of symmetric laminates are classified to be symmetric cross-ply laminates. (i.e., $[0°/90°_2/90°_2/0°]$, $[0°/90°/0°]_s$, $[90°/0°]_s$ as shown in Figure A.5). The characteristics of symmetric cross-ply laminates are obtained by combining antisymmetric cross-ply and symmetric laminates. The in-plane and out-of-plane couplings $[B_{ij}]$ and all hygrothermal moment resultants are equal to zero because of symmetric lay-up. The elements (A_{16}, A_{26}, B_{16}, B_{26} D_{16}, and D_{26}) of the laminate stiffness (ABD), hygrothermal in-plane shear force and twisting moment resultants are equal to zero because of cross-ply laminates.

The characteristics of symmetric cross-ply laminates are orthotropic for both the in-plane and bending behaviors. In addition, the symmetric cross-ply laminate is sometimes known as "specially orthotropic" laminate. Thus, the laminate stiffness (ABD) of symmetric cross-ply laminates is given as:

$$
\begin{bmatrix} N_{xx} \\ N_{yy} \\ N_{xy} \\ M_{xx} \\ M_{yy} \\ M_{xy} \end{bmatrix} = \begin{bmatrix} A_{11} & A_{12} & 0 & 0 & 0 & 0 \\ A_{12} & A_{22} & 0 & 0 & 0 & 0 \\ 0 & 0 & A_{66} & 0 & 0 & 0 \\ 0 & 0 & 0 & D_{11} & D_{12} & 0 \\ 0 & 0 & 0 & D_{12} & D_{22} & 0 \\ 0 & 0 & 0 & 0 & 0 & D_{66} \end{bmatrix} \begin{bmatrix} \varepsilon_{xx}^{o} \\ \varepsilon_{yy}^{o} \\ \gamma_{xy}^{o} \\ \kappa_{xx}^{o} \\ \kappa_{yy}^{o} \\ \kappa_{xy}^{o} \end{bmatrix} - \begin{bmatrix} N_{xx}^{T} \\ N_{yy}^{T} \\ 0 \\ 0 \\ 0 \\ 0 \end{bmatrix} - \begin{bmatrix} N_{xx}^{M} \\ N_{yy}^{M} \\ 0 \\ 0 \\ 0 \\ 0 \end{bmatrix}
$$

$$\text{(A.12.1)}$$

The laminate compliance matrix (abd) of symmetric cross-ply laminates is obtained from inversion of Eq. (A.12.1).

$$
\begin{bmatrix} \varepsilon_{xx}^{o} \\ \varepsilon_{yyy}^{o} \\ \varepsilon_{xy}^{\circ} \\ \kappa_{xx}^{o} \\ \kappa_{yy}^{o} \\ \kappa_{xy}^{o} \end{bmatrix} = \begin{bmatrix} a_{11} & a_{12} & 0 & 0 & 0 & 0 \\ a_{12} & a_{22} & 0 & 0 & 0 & 0 \\ 0 & 0 & a_{66} & 0 & 0 & 0 \\ 0 & 0 & 0 & d_{11} & d_{12} & 0 \\ 0 & 0 & 0 & d_{12} & d_{22} & 0 \\ 0 & 0 & 0 & 0 & 0 & d_{66} \end{bmatrix} \begin{bmatrix} N_{xx} + N_{xx}^{T} + N_{xx}^{M} \\ N_{yy} + N_{yy}^{T} + N_{yy}^{M} \\ N_{xy} \\ M_{xx} \\ M_{yy} \\ M_{xy} \end{bmatrix} \quad \text{(A.12.2)}
$$

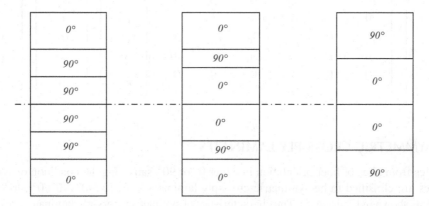

FIGURE A.5   Symmetric cross-ply laminate: (a) $[0°/90°_2]_s$, (b) $[0°/90°/0°]_s$, and (c) $[90°/0°]_s$.

A.6 ANTISYMMETRIC ANGLE-PLY LAMINATES

The laminates are classified as antisymmetric angle-ply when fiber orientations of all layers are aligned in either θ and $-\theta$ where fiber orientation (θ) is specified for all layers. In addition, numbers of θ layers are necessarily equal to numbers of $-\theta$ layers (i.e., [30°/−30°], [60°/60°/−60°/−60°], [75°/75°/−75°/−75°], etc., as shown in Figure A.6). It should be noted that if antisymmetric angle-ply laminates are one of the balanced laminates, then the characteristics (A_{16}, A_{26}, N_{xy}^T, and $N_{xy}^M = 0$) of balanced laminates also dominate the antisymmetric angle-ply laminates. Elements (B_{11}, B_{22}, and B_{66}) of the in-plane and out-of-plane couplings [B_{ij}] and hygrothermal in-plane moment resultants are equal to zero.

$$
\begin{bmatrix} N_{xx} \\ N_{yy} \\ N_{xy} \\ M_{xx} \\ M_{yy} \\ M_{xy} \end{bmatrix} = \begin{bmatrix} A_{11} & A_{12} & 0 & 0 & 0 & B_{16} \\ A_{12} & A_{22} & 0 & 0 & 0 & B_{26} \\ 0 & 0 & A_{66} & B_{12} & B_{26} & 0 \\ 0 & 0 & B_{16} & D_{11} & D_{12} & D_{16} \\ 0 & 0 & B_{26} & D_{12} & D_{22} & D_{26} \\ B_{16} & B_{26} & 0 & D_{16} & D_{26} & D_{66} \end{bmatrix} \begin{bmatrix} \varepsilon_{xx}^o \\ \varepsilon_{yy}^o \\ \gamma_{xy}^0 \\ \kappa_{xx}^o \\ \kappa_{yy}^o \\ \kappa_{xy}^o \end{bmatrix} - \begin{bmatrix} N_{xx}^T \\ N_{yy}^T \\ 0 \\ 0 \\ 0 \\ M_{xy}^T \end{bmatrix} - \begin{bmatrix} N_{xx}^M \\ N_{yy}^M \\ 0 \\ 0 \\ 0 \\ M_{xy}^M \end{bmatrix}
$$

$$(A.13.1)$$

Inversion of Eq. (A.13.1) gives the laminate compliance matrix (abd) of antisymmetric angle-ply laminates.

$$
\begin{bmatrix} \varepsilon_{xx}^o \\ \varepsilon_{yy}^o \\ \varepsilon_{xy}^o \\ \kappa_{xx}^o \\ \kappa_{yy}^o \\ \kappa_{xy}^o \end{bmatrix} = \begin{bmatrix} a_{11} & a_{12} & 0 & 0 & 0 & b_{16} \\ a_{12} & a_{22} & 0 & 0 & 0 & b_{26} \\ 0 & 0 & a_{66} & b_{16} & b_{26} & 0 \\ 0 & 0 & b_{16} & d_{11} & d_{12} & d_{16} \\ 0 & 0 & b_{26} & d_{12} & d_{22} & d_{26} \\ b_{16} & b_{26} & 0 & d_{16} & d_{26} & d_{66} \end{bmatrix} \begin{bmatrix} N_{xx} + N_{xx}^T + N_{xx}^M \\ N_{yy} + N_{yy}^T + N_{yy}^M \\ N_{xy} \\ M_{xx} \\ M_{yy} \\ M_{xy} + M_{xy}^T + M_{xy}^M \end{bmatrix}
$$

$$(A.13.2)$$

For the special case of antisymmetric angle-ply laminates, if numbers of layers increase, then elements (D_{16} and D_{26}) of the laminate stiffness matrix (ABD) decrease and (D_{16} and D_{26}) approach zero when numbers of layers are close to infinity. Thus, the laminate stiffness (ABD) of antisymmetric angle-ply laminates with an infinite number of layers is simplified as:

$$
\begin{bmatrix} N_{xx} \\ N_{yy} \\ N_{xy} \\ M_{xx} \\ M_{yy} \\ M_{xy} \end{bmatrix} = \begin{bmatrix} A_{11} & A_{12} & 0 & 0 & 0 & B_{16} \\ A_{12} & A_{22} & 0 & 0 & 0 & B_{26} \\ 0 & 0 & A_{66} & B_{12} & B_{26} & 0 \\ 0 & 0 & B_{16} & D_{11} & D_{12} & 0 \\ 0 & 0 & B_{26} & D_{12} & D_{22} & 0 \\ B_{16} & B_{26} & 0 & 0 & 0 & D_{66} \end{bmatrix} \begin{bmatrix} \varepsilon_{xx}^o \\ \varepsilon_{yy}^o \\ \gamma_{xy}^o \\ \kappa_{xx}^o \\ \kappa_{yy}^o \\ \kappa_{xy}^o \end{bmatrix} - \begin{bmatrix} N_{xx}^T \\ N_{yy}^T \\ 0 \\ 0 \\ 0 \\ M_{xy}^T \end{bmatrix} - \begin{bmatrix} N_{xx}^M \\ N_{yy}^M \\ 0 \\ 0 \\ 0 \\ M_{xy}^M \end{bmatrix}
$$

$$(A.13.3)$$

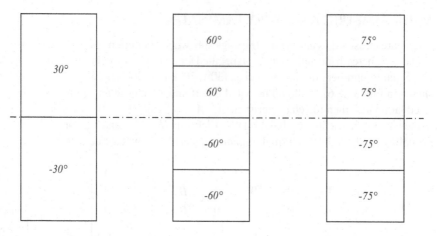

FIGURE A.6 Antisymmetric angle-ply laminate: (a) [30°/−30°], (b) [60°/60°/−60°/−60°], and (c) [75°/75°/−75°/−75°].

Inversion of Eq. (A.13.3) gives the laminate compliance matrix (*abd*) of antisymmetric angle-ply laminates, which is:

$$
\begin{bmatrix}
\varepsilon_{xx}^o \\
\varepsilon_{yy}^o \\
\varepsilon_{xy}^o \\
\kappa_{xx}^o \\
\kappa_{yy}^o \\
\kappa_{xy}^o
\end{bmatrix}
=
\begin{bmatrix}
a_{11} & a_{12} & 0 & 0 & 0 & b_{16} \\
a_{12} & a_{22} & 0 & 0 & 0 & b_{26} \\
0 & 0 & a_{66} & b_{16} & b_{26} & 0 \\
0 & 0 & b_{16} & d_{11} & d_{12} & 0 \\
0 & 0 & b_{26} & d_{12} & d_{22} & 0 \\
b_{16} & b_{26} & 0 & 0 & 0 & d_{66}
\end{bmatrix}
\begin{bmatrix}
N_{xx} + N_{xx}^T + N_{xx}^M \\
N_{yy} + N_{yy}^T + N_{yy}^M \\
N_{xy} \\
M_{xx} \\
M_{yy} \\
M_{xy} + M_{xy}^T + M_{xy}^M
\end{bmatrix}
\qquad \text{(A.13.4)}
$$

A.7 SYMMETRIC ANGLE-PLY LAMINATES

In this case, fiber orientations of all layers are aligned at either θ or $-\theta$ and satisfied the criterion of symmetric laminates (i.e., [30°/−30°/−30°/30°], [60°/60°/−60°/−60°], etc., as shown in Figure A.7). The laminate stiffness (*ABD*) of antisymmetric angle-ply laminates in Eq. (A.13.1) is simplified by taking the in-plane and out-of-plane couplings [B_{ij}] and also taking the hygrothermal moment resultants as zero.

In a special case, number of layers approach infinity, magnitudes of the elements (D_{16} and D_{26}) decrease and can be approximated to zero. The laminate stiffness (*ABD*) and compliance (*abd*) are the same as in the case of symmetrical cross-ply or especially orthotropic laminates. These are given in Eqs. (A.13.3 and A.13.4).

A.8 QUASI-ISOTROPIC LAMINATES

The characteristics of quasi-isotropic laminates are only isotropic under in-plane loadings, whereas the behavior under other loadings as transverse and out-of-plane shear is different from that of isotropic laminates. In general, the quasi-isotropic laminates are one-of-a-kind balanced laminates (A_{16}, A_{26}, N_{xy}^T, and $N_{xy}^M = 0$) (i.e., [0°/45°/−45°/90°]$_s$, [0°/60°/−60°]$_s$ as shown in Figure A.8.

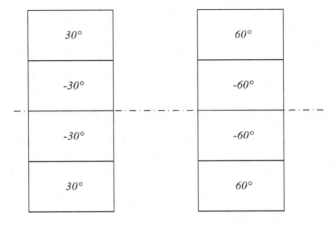

FIGURE A.7 Symmetric angle-ply laminate: (a) [30°/–30°]$_s$ and (b) [60°/60°]$_s$.

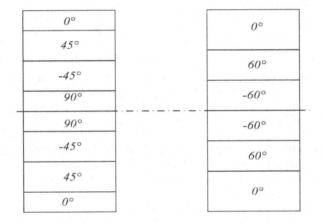

FIGURE A.8 Quasi-isotropic laminate: (a) [0°/45°/–45°/90°]$_s$ and (b) [0°/60°/–60°]$_s$.

The laminates are classified as quasi-isotropic laminates when they satisfy: (1) at least three layers of different fiber orientations, (2) numbers of layers in different fiber orientations are equal, (3) each layer is made from the same material and thickness, and (4) fiber orientations of different layers are equal to ($i\pi/n$) degrees where n is numbers of layers, i is a sequence of layers (e.g., if three different orientated layers are formed to be quasi-isotropic then, fiber orientation of layer second is ($2\pi/3$) degrees). Thus, the in-plane laminated stiffness [A_{ij}] can be simplified as:

$$A_{11} = A_{22} \tag{A.14.1}$$

$$A_{66} = 0.5\left(A_{11} - A_{12}\right) \tag{A.14.2}$$

$$A_{16} = A_{26} = 0 \tag{A.14.3}$$

$$N_{xy}^T = N_{xy}^M = M_{xy}^T = M_{xy}^M = 0 \tag{A.14.4}$$

The laminate stiffness (*ABD*) of quasi-isotropic laminates is presented below:

$$
\begin{Bmatrix}
N_{xx} \\
N_{yy} \\
N_{xy} \\
M_{xx} \\
M_{yy} \\
M_{xy}
\end{Bmatrix}
=
\begin{bmatrix}
A_{11} & A_{12} & 0 & B_{11} & B_{12} & B_{16} \\
A_{12} & A_{11} & 0 & B_{12} & B_{22} & B_{26} \\
0 & 0 & \dfrac{A_{11}-A_{12}}{2} & B_{16} & B_{26} & B_{66} \\
B_{11} & B_{12} & B_{16} & D_{11} & D_{12} & D_{16} \\
B_{12} & B_{22} & B_{26} & D_{12} & D_{22} & D_{26} \\
B_{16} & B_{26} & B_{66} & D_{16} & D_{26} & D_{66}
\end{bmatrix}
\begin{bmatrix}
\varepsilon_{xx}^{o} \\
\varepsilon_{yy}^{o} \\
\gamma_{xy}^{o} \\
\kappa_{xx}^{o} \\
\kappa_{yy}^{o} \\
\kappa_{xy}^{0}
\end{bmatrix}
-
\begin{bmatrix}
N_{xx}^{T} \\
N_{yy}^{T} \\
0 \\
M_{xx}^{T} \\
M_{yy}^{T} \\
0
\end{bmatrix}
-
\begin{bmatrix}
N_{xx}^{M} \\
N_{yy}^{M} \\
0 \\
M_{xx}^{M} \\
M_{yy}^{M} \\
0
\end{bmatrix}
$$

A.14.5)

Inversion of Eq. (A.14.5) gives the laminated compliance matrix (*abd*) of laminated composites as:

$$
\begin{bmatrix}
\varepsilon_{xx}^{o} \\
\varepsilon_{yy}^{o} \\
\varepsilon_{xy}^{o} \\
\kappa_{xx}^{o} \\
\kappa_{yy}^{o} \\
\kappa_{xy}^{o}
\end{bmatrix}
=
\begin{bmatrix}
a_{11} & a_{12} & 0 & b_{11} & b_{12} & b_{16} \\
a_{12} & a_{11} & 0 & b_{12} & b_{22} & b_{26} \\
0 & 0 & 2(a_{11}-a_{12}) & b_{16} & b_{26} & b_{66} \\
b_{11} & b_{12} & b_{16} & d_{11} & d_{12} & d_{16} \\
b_{12} & b_{22} & b_{26} & d_{12} & d_{22} & d_{26} \\
b_{16} & b_{26} & b_{66} & d_{16} & d_{26} & d_{66}
\end{bmatrix}
\begin{bmatrix}
N_{xx}+N_{xx}^{T}+N_{xx}^{M} \\
N_{yy}+N_{yy}^{T}+N_{yy}^{M} \\
N_{xy} \\
M_{xx}+M_{xx}^{T}+M_{xx}^{M} \\
M_{yy}+M_{yy}^{T}+M_{yy}^{M} \\
M_{xy}
\end{bmatrix}
$$

(A.14.6)

If the quasi-isotropic laminates are also symmetric then, the in-plane and out-of-plane couplings [B_{ij}] are zero. The laminate stiffness matrix (*ABD*) can be presented as:

$$
\begin{bmatrix}
N_{xx} \\
N_{yy} \\
N_{xy} \\
M_{xx} \\
M_{yy} \\
M_{xy}
\end{bmatrix}
=
\begin{bmatrix}
A_{11} & A_{12} & 0 & 0 & 0 & 0 \\
A_{12} & A_{22} & 0 & 0 & 0 & 0 \\
0 & 0 & \dfrac{A_{11}-A_{12}}{2} & 0 & 0 & 0 \\
0 & 0 & 0 & D_{11} & D_{12} & D_{16} \\
0 & 0 & 0 & D_{12} & D_{22} & D_{26} \\
0 & 0 & 0 & D_{16} & D_{26} & D_{66}
\end{bmatrix}
\begin{bmatrix}
\varepsilon_{xx}^{o} \\
\varepsilon_{yy}^{o} \\
\gamma_{xy}^{o} \\
\kappa_{xx}^{o} \\
\kappa_{yy}^{o} \\
\kappa_{xy}^{o}
\end{bmatrix}
-
\begin{bmatrix}
N_{xx}^{T} \\
N_{yy}^{T} \\
0 \\
0 \\
0 \\
0
\end{bmatrix}
-
\begin{bmatrix}
N_{xx}^{M} \\
N_{yy}^{M} \\
0 \\
0 \\
0 \\
0
\end{bmatrix}
$$

(A.14.7)

The laminate compliance matrix (*abc*) of quasi-isotropic laminates is obtained from inversion of Eq. (A.14.7).

$$
\begin{bmatrix}
\varepsilon_{xx}^{o} \\
\varepsilon_{yy}^{o} \\
\varepsilon_{xy}^{o} \\
\kappa_{xx}^{o} \\
\kappa_{yy}^{o} \\
\kappa_{xy}^{o}
\end{bmatrix}
=
\begin{bmatrix}
a_{11} & a_{12} & 0 & 0 & 0 & 0 \\
a_{12} & a_{22} & 0 & 0 & 0 & 0 \\
0 & 0 & 2(a_{11}-a_{12}) & 0 & 0 & 0 \\
0 & 0 & 0 & d_{11} & d_{12} & d_{16} \\
0 & 0 & 0 & d_{12} & d_{22} & d_{26} \\
0 & 0 & 0 & d_{16} & d_{26} & d_{66}
\end{bmatrix}
\begin{bmatrix}
N_{xx}+N_{xx}^{T}+N_{xx}^{M} \\
N_{yy}+N_{yy}^{T}+N_{yy}^{M} \\
N_{xy} \\
M_{xx} \\
M_{yy} \\
M_{xy}
\end{bmatrix}
$$

(A.14.8)

A.9 ISOTROPIC LAMINATES

Isotropic laminates are made from isotropic layers that may be different or the same in terms of material properties. Since properties of each layer are independent of the direction due to isotropy, the transformation of the reduced stiffness matrix is equal to the reduced stiffness matrix in the principal material coordinate system $\left(\left[\bar{Q}_{ij}\right]=\left[Q_{ij}\right]\right)$.

Thus, the laminate stiffness (ABD) is also independent of direction for the in-plane stiffness $[A_{ij}]$, in-plane and out-of-plane couplings $[B_{ij}]$, and bending–twisting stiffness $[D_{ij}]$. It should be noted that the in-plane and out-of-plane couplings $[B_{ij}]$ still exist. Therefore, the characteristic of isotropic laminates is identical to anisotropic properties. The elements of the laminated stiffness can be simplified as:

$$A_{11} = A_{22} \quad B_{11} = B_{22} \quad D_{11} = D_{22} \tag{A.15.1}$$

$$A_{66} = 0.5\left(A_{11} - A_{12}\right) \quad B_{66} = 0.5\left(B_{11} - B_{12}\right) \quad D_{66} = 0.5\left(D_{11} - D_{12}\right) \tag{A.15.2}$$

$$A_{16} = A_{26} = B_{16} = B_{26} = D_{16} = D_{26} = 0 \tag{A.15.3}$$

$$N_{xy}^T = N_{xy}^M = M_{xy}^T = M_{xy}^M = 0 \tag{A.15.4}$$

The laminate stiffness (ABD) of quasi-isotropic laminates is presented below:

$$
\begin{bmatrix} N_{xx} \\ N_{yy} \\ N_{xy} \\ M_{xx} \\ M_{yy} \\ M_{xy} \end{bmatrix} =
\begin{bmatrix}
A_{11} & A_{12} & 0 & B_{11} & B_{12} & 0 \\
A_{12} & A_{22} & 0 & B_{12} & B_{22} & 0 \\
0 & 0 & \dfrac{A_{11}-A_{12}}{2} & 0 & 0 & \dfrac{B_{11}-B_{12}}{2} \\
B_{11} & B_{12} & 0 & D_{11} & D_{12} & 0 \\
B_{12} & B_{22} & 0 & D_{12} & D_{22} & 0 \\
0 & 0 & \dfrac{B_{11}-B_{12}}{2} & 0 & 0 & \dfrac{D_{11}-D_{12}}{2}
\end{bmatrix}
\begin{bmatrix} \varepsilon_{xx}^o \\ \varepsilon_{yy}^o \\ \gamma_{xy}^o \\ \kappa_{xx}^o \\ \kappa_{yy}^o \\ \kappa_{xy}^o \end{bmatrix} -
\begin{bmatrix} N_{xx}^T \\ N_{yy}^T \\ 0 \\ M_{xx}^T \\ M_{yy}^T \\ 0 \end{bmatrix} -
\begin{bmatrix} N_{xx}^M \\ N_{yy}^M \\ 0 \\ M_{xx}^M \\ M_{yy}^M \\ 0 \end{bmatrix}
$$

$$\tag{A.15.5}$$

Inversion of Eq. (A.15.5) gives the laminated compliance matrix of isotropic laminates.

$$
\begin{bmatrix} \varepsilon_{xx}^o \\ \varepsilon_{yy}^0 \\ \varepsilon_{xy}^0 \\ \kappa_{xx}^o \\ \kappa_{yy}^{ox} \\ \kappa_{xy}^0 \end{bmatrix} =
\begin{bmatrix}
a_{11} & a_{12} & 0 & b_{11} & b_{12} & 0 \\
a_{12} & a_{22} & 0 & b_{12} & b_{22} & 0 \\
0 & 0 & 2\left(a_{11}-a_{12}\right) & 0 & 0 & 2\left(b_{11}-b_{12}\right) \\
b_{11} & b_{12} & 0 & d_{11} & d_{12} & 0 \\
b_{12} & b_{11} & 0 & d_{12} & d_{22} & 0 \\
0 & 0 & 2\left(b_{11}-b_{12}\right) & 0 & 0 & 2\left(d_{11}-d_{12}\right)
\end{bmatrix}
\begin{bmatrix} N_{xx}+N_{xx}^T+N_{xx}^M \\ N_{yy}+N_{yy}^T+N_{yy}^M \\ N_{xy} \\ M_{xx}+M_{xy}^T+M_{xy}^M \\ M_{yy}+M_{xy}^T+M_{xy}^M \\ M_{xy} \end{bmatrix}
$$

$$\tag{A.15.6}$$

If laminates are symmetric, then the laminate stiffness (ABD) in Eq. (A.15.5) is simplified by taking $[B_{ij}]$ and hygrothermal moment resultants to be zero. The stiffness of symmetric isotropic laminates is presented as:

$$
\begin{bmatrix} N_{xx} \\ N_{yy} \\ N_{xy} \\ M_{xx} \\ M_{yy} \\ M_{xy} \end{bmatrix} = \begin{bmatrix} A_{11} & A_{12} & 0 & 0 & 0 & 0 \\ A_{12} & A_{22} & 0 & 0 & 0 & 0 \\ 0 & 0 & \dfrac{A_{11}-A_{12}}{2} & 0 & 0 & 0 \\ 0 & 0 & 0 & D_{11} & D_{12} & 0 \\ 0 & 0 & 0 & D_{12} & D_{22} & 0 \\ 0 & 0 & 0 & 0 & 0 & \dfrac{D_{11}-D_{12}}{2} \end{bmatrix} \begin{bmatrix} \varepsilon_{xx}^{o} \\ \varepsilon_{yy}^{o} \\ \gamma_{xy}^{o} \\ \kappa_{xx}^{\circ} \\ \kappa_{yy}^{o} \\ \kappa_{xy}^{o} \end{bmatrix} - \begin{bmatrix} N_{xx}^{T} \\ N_{yy}^{T} \\ 0 \\ 0 \\ 0 \\ 0 \end{bmatrix} - \begin{bmatrix} N_{xx}^{M} \\ N_{yy}^{M} \\ 0 \\ 0 \\ 0 \\ 0 \end{bmatrix}
$$

(A.15.7)

Inversion of Eq. (A.15.7) gives the laminated compliance matrix (abd) of symmetric isotropic laminates as:

$$
\begin{bmatrix} \varepsilon_{xx}^{o} \\ \varepsilon_{yy}^{o} \\ \varepsilon_{xy}^{o} \\ \kappa_{xx}^{\circ} \\ \kappa_{yy}^{o} \\ \kappa_{xy}^{o} \end{bmatrix} = \begin{bmatrix} a_{11} & a_{12} & 0 & 0 & 0 & 0 \\ a_{12} & a_{22} & 0 & 0 & 0 & 0 \\ 0 & 0 & 2(a_{11}-a_{12}) & 0 & 0 & 0 \\ 0 & 0 & 0 & d_{11} & d_{12} & 0 \\ 0 & 0 & 0 & d_{12} & d_{22} & 0 \\ 0 & 0 & 0 & 0 & 0 & 2(d_{11}-d_{12}) \end{bmatrix} \begin{bmatrix} N_{xx}+N_{xx}^{T}+N_{xx}^{M} \\ N_{yy}+N_{yy}^{T}+N_{yy}^{M} \\ N_{xy} \\ M_{xx} \\ M_{yy} \\ M_{xy} \end{bmatrix}
$$

(A.15.8)

A.10 SINGLE ISOTROPIC LAMINATES

When laminates are made of one kind of homogenous and isotropic materials, then the laminates can be considered isotropic over the total laminated thickness. The laminate stiffness (ABD) is simplified to be isotropic. The in-plane and out-of-plane couplings $[B_{ij}]$ will not exist in this case. Also, elements ($A_{16}, A_{26}, D_{16},$ and D_{26}) are equal to zero. Thus, the in-plane, bending, and twisting of laminates are independent, and the laminate stiffness matrix (ABD) is given below:

$$
\begin{bmatrix} N_{xx} \\ N_{yy} \\ N_{xy} \\ M_{xx} \\ M_{yy} \\ M_{xy} \end{bmatrix} = \begin{bmatrix} A & vA & 0 & 0 & 0 & 0 \\ vA & A & 0 & 0 & 0 & 0 \\ 0 & 0 & \dfrac{1-v}{2}A & 0 & 0 & 0 \\ 0 & 0 & 0 & D & vD & 0 \\ 0 & 0 & 0 & vD & D & 0 \\ 0 & 0 & 0 & 0 & 0 & \dfrac{1-v}{2}D \end{bmatrix} \begin{bmatrix} \varepsilon_{xx}^{\circ} \\ \varepsilon_{yy}^{\circ} \\ \gamma_{xy}^{o} \\ \kappa_{xx}^{o} \\ \kappa_{yy}^{o} \\ \kappa_{xy}^{o} \end{bmatrix} - \begin{bmatrix} N_{xx}^{T} \\ N_{yy}^{T} \\ 0 \\ 0 \\ 0 \\ 0 \end{bmatrix} - \begin{bmatrix} N_{xx}^{M} \\ N_{yy}^{M} \\ 0 \\ 0 \\ 0 \\ 0 \end{bmatrix}
$$

(A.16.1)

Inversion of Eq. (A.16.1) gives laminate compliance (*abd*) of single isotropic laminate.

$$
\begin{bmatrix} \varepsilon_{xx}^{o} \\ \varepsilon_{yy}^{0} \\ \gamma_{xy}^{0} \end{bmatrix} = \begin{bmatrix} a & -va & 0 \\ -va & a & 0 \\ 0 & 0 & 2(1+v)a \end{bmatrix} \begin{bmatrix} N_{xx} + N_{xx}^{T} + N_{xx}^{M} \\ N_{yy} + N_{yy}^{T} + N_{yy}^{M} \\ N_{xy} \end{bmatrix} \tag{A.16.2}
$$

$$
\begin{bmatrix} \kappa_{xx}^{o} \\ \kappa_{yy}^{o} \\ \kappa_{xy}^{o} \end{bmatrix} = \begin{bmatrix} d & -vd & 0 \\ -vd & d & 0 \\ 0 & 0 & 2(1+v)d \end{bmatrix} \begin{bmatrix} M_{xx} \\ M_{yy} \\ M_{xy} \end{bmatrix} \tag{A.16.3}
$$

$$
a = \frac{1}{Eh} \tag{A.16.4}
$$

$$
d = \frac{12}{Eh^{3}} \tag{A.16.5}
$$

where
  E = elastic modulus (Young's modulus)
  v = Poisson's ratio
  h = total thickness of laminates.

Appendix B
Durability of FRP Composites under Environmental Conditions

B.1 INTRODUCTION

Lack of understanding of durability performance for FRP composites is a major technical barrier toward high volume use in infrastructure systems (Karbhari et al., 2003; Micelli et al., 2015; Shi et al., 2011). The durability response is identified typically in terms of chemical, physical, and mechanical aging as a function of moisture, temperature, pH, sustained stress, ultraviolet (UV) radiation, and material constituents. This section focuses on the durability response of FRP composites including (1) composite laminates of different fibers and fabrics bonded with compatible resins; (2) structural shapes and systems, such as FRP bars, bridges deck, etc.; and (3) wraps and strips strengthening or reinforcing steel-reinforced members and other hybrids. Long-term predictive models are reviewed and strength reduction factors for the design of FRP composite structures are presented from the research work of the authors and others. The aging response of FRP composites is based on deterioration mechanisms, observed through many research findings.

B.2 MOISTURE

Although FRP composites, other than carbon composites, are not susceptible to electrochemical corrosion as with the conventional materials, debonding and strength reductions under moisture uptake still play a major role in the durability response. Depending on the polymer chemistry, the moisture absorption rate varies and leads to reversible or irreversible physical, thermal, mechanical, and/or chemical changes (Bisby, 2006). As explained below, moisture absorption in FRPs is extremely complex, requiring a comprehensive understanding of its influence on the thermomechanical behavior of FRP composites, especially with reference to interlaminar bond forces under freeze–thaw (FT) cycling and wet–dry cycling. In addition, the moisture response phenomena of FRP composites under corrosive media with various diffusion models are explained below.

B.2.1 IMMERSION IN CORROSIVE MEDIA

Water absorption of FRP laminates, in general (except aramid), has an adverse effect on fatigue due to hydrolysis of the resin system; however, there is no major influence on fibers or fabrics, i.e., the influence of fiber type, volume fraction, and fabric architecture are not influencing the composite laminates exposed to moisture; however, matrix influenced bond between reinforcement layers is affected.

For example, glass-carbon hybrid laminates showed excellent durability, similar to carbon laminates (Nakada and Miyano, 2009). Such excellent performance is mainly attributed to constituents least affected by water other than electrochemical reaction. Glass-aramid laminates showed continued degradation with time since moisture uptake of aramid fiber is significantly higher than that of carbon fibers (Jones, 2001). The results revealed that unidirectional laminates had higher degradation than laminates with bidirectional fabrics because debonding along the longitudinal fibers was prone to occur in unidirectional laminates when water molecules penetrated the interface between fiber and matrix along the fiber length direction.

B.2.1.1 Different Resins

The tensile strength of glass/epoxy composites decreased by about ~20% after exposure for 1 year to seawater at room temperature and by ~30% after 1 year of exposure at 65°C. These reductions are attributed to hydrolysis and fiber/matrix separation at the interface (Mourad et al., 2012; Eldridge and Fam, 2014). However, the modulus of glass/polyurethane composite did not change significantly. This behavior was attributed to the highly heterogeneous materials consisting of stiff regions and soft regions due to the nonhomogeneous moisture distribution. For glass/epoxy composites under immersions for 3, 6, and 12 months at room temperature and 65°C, both the tensile strength and modulus show an initial increase due to post-curing with no significant degradation. Although nonhomogeneous water distribution creates localized defects and lowers the strength of glass/epoxy composites in the initial conditioning phase, the relaxation of residual stresses can potentially improve failure strength.

Similar observations were made by Eldridge and Fam (2014). For glass/vinyl ester composites exposed to seawater at room temperature, the flexural modulus decreased 18.5% between the 3- and 6-month periods (Hammami and Al-Ghuilani, 2004). Vijay and GangaRao (1999) conducted screening tests for medium reactivity isophthalic unsaturated polyester (MUPE), high reactivity isophthalic unsaturated polyester (HUPE), isocyanurate vinyl ester (IVE), and urethane-modified vinyl ester (UMVE). The results showed that GFRP bars with UMVE exhibited the lowest vulnerability under different environmental conditions, while MUPE exhibited less vulnerability to different conditioning schemes compared with HUPE. Mounts (2007) compared the durability of urethane and UMVE GFRP in an alkaline environment and showed the peak moisture content of urethane GFRP was almost five times as much as that of UMVE-based GFRP composites. This large variation may be responsible for urethane GFRP to be more severely aged than the UMVE-based GFRP composites.

B.2.1.2 Fiber Volume

Lower fiber volume fraction usually results in FRP composites that are fairly sensitive to moisture uptake, and significantly affect the deterioration process over time both at the level of bulk resin and the fiber–matrix interface (Mourad et al., 2012). Some researchers reported that composite specimens with higher fiber volume fraction absorbed more water compared with the specimens of lower fiber volume fraction (Bian et al., 2012; Hammami and Al-Ghuilani, 2004). This is because relatively high fiber content may prevent the matrix from fully bonding with the fiber, resulting in a high void ratio. Thus, microcracks appear at the interface between the fiber and the matrix. Such microcracks, under the combined action of water diffusion and hydrogen ion exchange, accelerate the degradation process. Moisture uptake and desorption response of high fiber volume fraction of pultruded carbon/epoxy composites in deionized water were observed from previous studies (Karbhari and Xian, 2009). It is shown that specimens with thickness/length ratio lower than 0.045 do not reach equilibrium within 60 days and show a good correlation with Langmuir dual-mode diffusion response (Karbhari and Xian, 2009). The phenomenon was attributed to hygrothermal history, including uptake and desorption of moisture, resulting in a significantly higher rate of moisture uptake on re-immersion.

Kumosa et al. (2004) showed that moisture uptake of modified polyester and vinyl ester composites exposed to relative humidity (RH) of 80% at 50°C, followed the Fickian law up to certain time span of immersion. Herein, the behavior of the epoxy-based composites demonstrated to be anomalous. It has been shown that not all FRP composites exhibit the same moisture uptake following the single-phase Fickian diffusion model. Glass/epoxy composites, notched composites (Buck et al., 2001), and composites immersed in liquid or at high-temperature saturate rapidly with time. The rapid saturation could be from void content in the resin, quality of the fiber matrix interfaces, relaxation of resin in the presence of moisture through hydrolysis and elevated temperature, water molecules binding to the molecular structure of the resin, etc.

B.3 pH

When composites are exposed to reactive environments, it is desirable to have no susceptible linkages between the resin and fiber interface. If partial linkage breakdown is unavoidable, then it is desirable to have a high concentration of linkages. For example, in GFRPs, two types of chemical bonds (siloxane linkage between glass and coupling agent, and within coupling agent ester linkage for polymer resins and in anhydride-hardened epoxies) are susceptible to bond breakage leading to higher rates of bond degradation (Vijay, 1999).

B.3.1 ACID SOLUTION

Acid attack on resin causes multiple cracks, resulting in resin flaking and fiber damage. Marru et al. (2014) revealed that the performance of glass-reinforced epoxy (GRE) pipes in extreme alkaline medium (10% NaOH) was better than that in an extreme acid medium (3% H_2SO_4). Further, degradation of composite pipes followed the Arrhenius principle (temperature-dependent reaction rates). Large strength decrease in FRP composites occurs in high density of the acid solution. Nakayama et al. (2004) observed that after 2,500 h in 2%, 5%, and 10% sulfuric acid solution at 80°C, the flexural strength losses of GFRP (E-glass/unsaturated polyester) were about 40%, 60%, and 80%, respectively. Moreover, the GFRP failed at the end of 1,500 h in a 20% sulfuric acid solution. Hammami and Al-Ghuilani (2004) compared the durability of glass vinyl ester composites exposed to 2% and 5% nitric acid and concluded that GFRP laminates decrease in flexural strength drastically with increasing acid concentration and immersion time. Corrosive fluid causes matrix expansion and leaching, resulting in pits. After longer immersion time, pits may form blisters and eventually swelling of resin and collapse. The bond strength deteriorated with the acid concentration is shown by Kajorncheappunngam (1999, 2002). The decrease in tensile strength of glass reinforced epoxy in acid (1-M HCl) was more than that in alkaline medium (5-M NaOH), and the effect of saline water was less severe than that of acid or alkaline medium.

B.3.2 ALKALINE SOLUTION

Fiber cracking is found to be the main culprit in material property losses in acid-aged GFRP composites, whereas matrix cracking and fiber/matrix debonding are the major reasons for strength decay in GFRP composites aged in caustic soda and distilled water (Kajorncheappunngam, 1999). Alkaline hydrolysis tends to occur when OH reacts with ester bonds, which are the weakest locations in the chemical structure of the polymer (Chin et al., 1997). Alkaline solution not only attacks the matrix but also glass fibers. The chemical reaction of silica in glass and alkaline solution causes hydroxylation and dissolution, followed by notching, which is caused by the formation of calcium hydroxide crystals on the glass surface (Yilmaz and Glasser, 1991). Moreover, at pH exceeding 9, caustic soda can attack the backbone of the glass molecules through the following chemical reactions (Sonawala and Spontak, 1996):

$$Si - O - Si + NaOH \rightarrow Si - O - Na + Si - OH \tag{B.1}$$

Many researchers have demonstrated the tensile strength of GFRP decreased dramatically in high concentrations of alkaline solution with time, whereas the modulus was not significantly influenced by alkaline solution. The results show that a pH of 12.5 has the most effect on the GFRP composites compared to pH 10, 7, and 2.5. However, the alkaline solution did not affect the modulus of elasticity of the GFRP laminates. Moisture absorption of GFRP bars was strongly dependent on tap water, saltwater, and alkaline conditioning. Alkaline conditioning doubled in moisture gain by weight as compared to tap water and saltwater conditioning (Vijay, 1999). For sand-coated bars, maximum strength reductions in salt and alkaline conditioning at room temperature were 18.5% and 32.2%, respectively, over a 15-month duration. For ribbed glass composite bars, maximum strength reductions in salt and alkaline conditioning at room temperature were 24.5% and 30%, respectively, over a 30-month duration. However, carbon laminates have superb durability qualities and are unaffected even after a long period of exposure to pH solutions (Saadatmanesh et al., 2010).

Many of the conditioned glass bars showed an increase in stiffness associated with a reduction in failure stress, which implied that the bars were more brittle than the unconditioned bars. This is because the modulus of a glassy polymer is relatively insensitive to changes in molecular weight (Martin et al., 1972). Significant reduction is observed only when the measurements are made at temperatures approaching the polymer glass transition temperature (T_g where polymer changes from brittle sate to ductile with temperature increase), (De'Neve and Shanahan, 1993), i.e., above 160°F (~65°C). The strength degradation of GFRP composites is more severe in an extreme acid medium (5% H2SO4/1-M HCl) than in an extreme alkaline medium (10% NaOH/5-M NaOH), whereas a medium with a pH of 12.5 has the most deterioration effect on GFRP composites. No significant degradation of stiffness of GFRP composites was observed. Although the data based on the Arrhenius model agreed well with those obtained from short-term (~600days) experiments, it is valid up to a certain amount of conditioning time and temperature (GangaRao et al., 2020). Any additional long-term response data from field would be helpful to refine current prediction models (Zhou et al., 2011).

B.4 TEMPERATURE

Although fibers are temperature-resistant and can retain most of their strength and stiffness at temperatures below their process temperatures, most polymer matrices are susceptible to high temperature (Fitzer, 1998; Sauder et al., 2004; ASTM E119, 2011; Williams et al., 2008). Once the temperature exceeds or even comes around 20°C below the glass transition temperature (T_g) in-service, the strength of resin decreases, which begins to affect shear properties of composites. Moreover, the fiber matrix interface becomes susceptible to aggressive reactions under higher temperatures, leading to matrix degradation (Ray and Rathore, 2014). On the other hand, cold temperature and FT cycling may affect the durability of FRP composites through differential thermal expansion between the polymer matrix and fiber components or between concrete and FRP, or steel and FRP. This could result in damage to the interface between FRP composites and other materials such as concrete and steel (Bisby, 2006).

B.4.1 HIGH TEMPERATURE

The tensile properties of CFRP sheets, hybrid carbon/glass fiber-reinforced polymer (C/GFRP) sheets, and hybrid carbon/basalt fiber-reinforced polymer (C/BFRP) sheets were evaluated at a temperature ranging from 16°C to 200°C (Cao et al., 2009). The tensile strength of carbon fibers in different FRP sheets decreased significantly when the temperature increased from 16°C to 55°C and remained almost stable by retaining 67% of its original strength after the polymer exceeds its T_g (55°C). Meanwhile, at a temperature range of 16°C–200°C, the hybridization of carbon fibers with glass or basalt fibers exhibited the same variation in the tensile strength as CFRP sheets and reduced scatter. The performance of GFRP bars (E-glass/vinyl ester) and GFRP reinforced concrete elements under three different elevated temperatures (100°C, 200°C, and 300°C) for three different periods (1, 2, and 3 h) revealed no significant losses. Losses in tensile strength were proportional to the level of temperature and exposure time. The bars with concrete cover showed higher residual tensile strength compared to their counterparts without cover, especially at 100°C for a period of 1 h. The stress–strain relationships of GFRP and CFRP bars remained almost linear at elevated temperatures not exceeding (T_g, −20°C) until failure. The elastic modulus remained almost unchanged until 300°C–350°C but decreased dramatically beyond 400°C.

As shown in Figure B.1, three failure modes were observed depending on the temperature range (25°C–300°C) and the FRP laminate type: (1) within the temperature range of 100°C–150°C, the

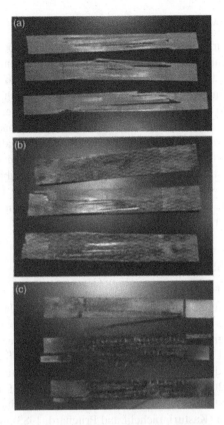

FIGURE B.1 Failure modes of CFRP, GFRP and C/GFRP composites subjected to different temperature range for 45 min (Hawileh et al., 2015): (a) within 100°C–150°C, (b) within 200°C–250°C, and (c) at 300°C.

CFRP, GFRP, and C/GFRP specimens have failed similarly to those specimens tested at room temperature by brittle fiber ruptures at different positions along the specimen gauge length; (2) within the temperature range of 200°C–250°C, the epoxy adhesives had been softened and the specimens failed primarily by partial loss of adhesion followed by sheet splitting, and (3) the epoxy adhesives were burned and the specimens failed by rupture of the fibers above 300°C (Hawileh et al., 2015).

Different failure modes suggest the degradation of thermal stability of epoxy resin with increasing temperature, resulting in the debonding of the FRP/matrix interface. Once the temperature exceeds the ignition point of resin, then the resin cannot protect the fiber surfaces, lowering their mechanical behavior and shortening their useful life. In particular, the heat resistance of FRP bars made with carbon fiber and phenolic matrix resin is almost the same as that of steel bars (Sumida and Mutsuyoshi, 2008).

B.4.2 HYGROTHERMAL

The service temperature of a composite may strongly affect the water absorption rate. An increase in temperature causes a higher level of water diffusion; hence, the onset of saturation showed advancement in the number of hours of exposure to saturation. All specimens showed an increase in moisture uptake by increasing temperature (Santhosh et al., 2012). After 100 h exposure, the moisture absorption values were found to be 0.26% and 0.29% for glass/isopolyester, 0.21% and 0.24% for carbon/isopolyester, and 0.18% and 0.20% glass/isopolyester/gel coat at 60°C and 70°C, and 95% RH at the end of 100 h, respectively. Micrographs of fractured specimens of composites confirmed that matrix/fiber bonding sustained after 100 h of exposure at 60°C and 70°C.

B.4.3 LOW TEMPERATURE AND FT (FREEZE-THAW) CYCLING

FRPs under lower temperature (below 0°C) exhibit mechanical property changes along with additional microcracks in FRP materials. This can further result in the increase of water absorption at higher temperatures, leading to an increased matrix plasticization and hydrolysis. The expansion of frozen water results in debonding and transverse microcrack growth. Rupture elongation showed similar degradation trends to the tensile strength. For both basalt fiber reinforced polymer (BFRP) composites and carbon basalt fiber hybrid FRP sheets, FT cycling had a negligible effect on their tensile properties, which means the hybridization of carbon and basalt fibers can contribute to the stability of the tensile properties of FRP sheet under FT environments (Shi et al., 2014). However, both the shear and flexural strengths of saturated GFRP samples subjected to lower temperatures (between −60°C and 0°C) were not affected appreciably, and the tensile strength (Robert and Benmokrane, 2010; Wu et al., 2006; Wu and Yan, 2011), and flexural modulus of elasticity appeared to be stable at a temperature ranging from −40°C to 50°C.

B.5 SUSTAINED STRESS

For structural applications, FRP components must be able to resist some level of sustained load for a prolonged duration without exhibiting excess creep and/or relaxation. In combination with environmental attacks such as moisture, aggressive mediums (acid, alkali, and salt), temperature, and sustained stress can aggravate strength loss, interfacial degradations, and brittleness in FRP composites (GangaRao et al., 2007; GangaRao and Liang, 2014). Tensile stresses from external loads can enhance moisture diffusion by increasing the rate of moisture uptake and moisture saturation content (Wang et al., 2015). However, for synergistic effects of bending strain and moisture, the tensile and compressive stresses from bottom to top of a composite section under bending have opposing effects on diffusion (Kasturiarachchi and Pritchard, 1983). Moreover, elevated temperatures (e.g., 60°C) combined with lower bending strains (i.e., 30% bending) result in a decrease in apparent weight gain before failure, associated with the occurrence of material leeching-out through desorption (Helbling and Karbhari, 2008; Ghorbel and Valentin, 1993; Harper and Naeem, 1989) and/or with chemical breakdown through the hydrolysis of polymer groups (Chin et al., 2001; Weitsman, 1998).

The study of Miller (2014) indicated that even after the 2,000-h duration with 30% sustained tensile stress to failure at 60°C, both the glass/polyester and glass/vinyl ester samples only exhibited a 10%–15% reduction in tensile strength which is consistent with the data (Debaiky et al., 2006; Vijay et al., 2001). Similarly, from 6% to 10% decrease in elastic modulus was observed by several researchers. Meanwhile, acoustic emission analyses indicated that there was limited activity at lower stress, especially levels of stress less than 30% of the ultimate value. Thus, if a low in-service load (less than 25%–30% of ultimate strength of FRP) is anticipated, sustained loading conditioning may not be necessary for durability assessment of commercially available pultruded FRP composites. Similar behavior of FRP under low in-service load has also been observed by Helbling et al. (2006) in which E-glass/vinyl ester composites exposed at 40°C under 30% of the tensile strain capacity show moderate (~20%–30%) degradation in tensile strength. However, for specimens exposed at 40°C with 45% of the ultimate tensile strain capacity, and 60°C with both 30% and 45% of the tensile strain capacity showed pronounced decreases in tensile strength. This phenomenon is attributed to severe fiber interface degradation. A corrosive medium coupled with sustained stress tends to aggravate the degradation of FRP strength. Vijay (1999) showed that the reductions of tensile strength of GFRP bars subjected to 27% applied stress were of the order of 1.5 times those of the unstressed bars at the same age and alkaline conditioning. Stress corrosion of E-glass fibers is known to generate weak zones that reduce the fiber strength and enhance crack propagation, especially in conjunction with moisture (Helbling et al., 2006).

B.5.1 Creep

The static fatigue (creep) of FRP can compromise structural durability. Carbon and glass fibers are not generally expected to creep significantly in the absence of other degradation mechanisms. However, polymer materials used as matrices for FRPs are known to experience significant creep and/or relaxation under sustained load. For example, studies by Ascione et al. (2012) indicated that the longitudinal deformation of polyester resin increased 150% and 271%, respectively, under the sustained load of 8.3% and 13.64% of ultimate tensile strength at the end of 547 days. On the contrary, deformations in glass fibers and GFRP specimens were almost negligible under 10%–15% sustained stress of the failure stress. Similarly, creep behavior of pultruded GFRP specimens subjected to constant traction (Bottoni et al., 2014) (15% of the tensile strength) and shear stresses for shear creep revealed that they are larger than the creep under traction, which confirms that the viscoelastic properties of the pultruded composites related to normal or shear loadings are independent due to the directional arrangement of fibers, and the Findley model has been used with success for creep predictions.

Also, creep behavior of FRP composites under varying temperatures, moisture contents, and loads revealed that unidirectional pultruded E-glass/vinyl ester composites at room temperature and dry conditions have a threshold value of 50%, above which the tertiary creep approaches faster or narrows the secondary phase, as shown in Figure B.2a (Batra, 2009). The rate of creep increased rapidly with an increase in moisture diffusion and at higher temperatures, as shown in Figure B.2b (Batra, 2009). The creep rupture due to higher stress in short time is attributed to slip between the fiber and matrix. Meanwhile, polyurethane samples performed better when compared to vinyl ester laminates. It is observed that higher loading of 80% can lead to creep rupture or failure in a very short time. GangaRao and Liang (2014) showed that the bending energy expended by GFRP samples was found to be a function of (1) induced sustained strain level as a percent of static failure strain, (2) time to failure under varying sustained strain levels, and (3) mechanical properties of the GFRP composite material.

B.6 UV RADIATION

Exposure to UV radiation causes photochemical damage of FRP composites near the exposed surface, leading to discoloration and reduction in molecular weight that results in the degradation of composites. The surface flaws can lead to stress concentrations and aggravate resin erosion under environmental attacks (GangaRao and Kumar, 1995). Moreover, effects of UV radiation appear to be exacerbated by other factors such as moisture, chemical media, elevated temperature, and thermal cycling (Bisby, 2006).

B.6.1 Independent Effect

Homam and Sheikh (2000) investigated the durability of carbon/epoxy and glass/epoxy composites exposed to UV-A lamp radiation at 156 watt/m^2 and 38°C. For the 1,200–4,800 h of UV radiation, the tensile strength and stiffness of the exposed FRP specimens remained slightly higher than those of the control specimens, and the strain at rupture was reduced slightly. At the same time, the lap shear strengths of GFRP and CFRP specimens were not affected by the exposure. Several studies have shown that the matrix-dominated transverse properties can undergo severe deterioration due to chain scission induced by photo-oxidation from UV radiation.

B.6.2 Coupled Effect

When exposed sequentially to UV radiation followed by condensation, the CFRP/epoxy composite specimens initially lost weight during the UV radiation cycle and subsequently gained weight during the condensation cycle (Kumar et al., 2002). The actual change in weight was a simple

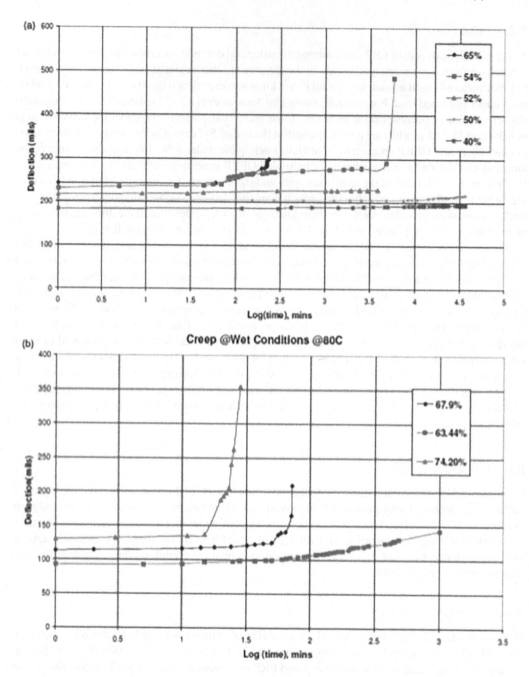

FIGURE B.2 Deflection-time behaviors of three-point bending creep with constant loads at (a) dry conditions and room temperature and (b) wet conditions and 80°C (Batra, 2009).

time-shifted linear superposition of results obtained for individual exposure conditions. However, when the specimens were cyclically exposed to both UV radiation and condensation, the specimens exhibited continuous weight loss at a steady rate throughout the exposure duration. After 1,000 h of cyclic exposure to UV radiation and condensation (6 h cycle), an average of 1.25% decrease in specimen weight was observed. Micrographic observation confirmed that UV radiation and condensation

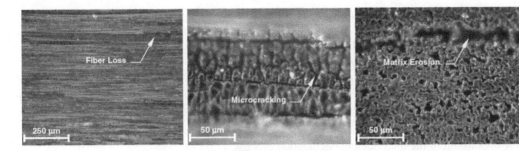

FIGURE B.3 Optical micrographs of CFRP/epoxy composite specimens after 1,000 h of cyclic exposure to both UV radiation and condensation (Kumar et al., 2002).

operate in a synergistic manner leading to extensive matrix erosion, matrix microcracking, fiber debonding, fiber loss, and void formation (Figure B.3). The transverse tensile strength of the CFRP/ epoxy composite specimens decreased 21% and 29% after exposure to 1,000 h of UV radiation followed by 1,000 h of condensation at 50°C, and cyclic exposure to both 1,000 h of UV radiation and condensation. The moduli for either of the subsequent or cyclical exposures were the same as those of unconditioned specimens. This indicates that there are some synergistic effects that govern the changes in matrix properties when the specimens are exposed to a combination of both UV radiation and condensation (Kumar et al., 2002).

It can be seen that the UV radiation does not affect much the fiber-dominated longitudinal properties of CFRP/epoxy composites, whereas it degrades the matrix-dominated transverse and shear properties. The combined exposure of UV radiation, high-temperature cycles, and high-humidity cycles had more detrimental environmental effect on FRP than single exposure effect. On the other hand, in a limited manner, FT cycles can compensate, any decrease in mechanical properties of FRP under the combined effects of UV radiation, high-temperature cycles, high-humidity cycles along with FT cycles, because of continued post-curing of the resin.

B.7 INTERLAMINAR SHEAR STRENGTH

GangaRao et al. (2020) have focused their efforts on the Interlaminar Shear Strength (ILSS) as a measure of the durability of composites because researchers (Lorenzo, 2018; Barker, 2019) have shown that ILSS degradation rates are higher than those in bending and tension, under identical accelerated aging conditions of a polymer composite. Additional ILSS failure, even under bending, is the controlling factor at ultimate bending strength levels (Bazli et al., 2020). The above inferences were made based on extensive testing and evaluation of coupons in both the longitudinal and transverse directions, under pH solutions of 3, 7, and 13, and temperatures ranging from −10°F, +72°F, and +160°F, which are well below the glass transition temperature of the resin. The ILSS testing was carried out from 1 to 450 days.

B.7.1 CORRELATION BETWEEN ACCELERATED VERSUS NATURAL AGING

The aging data of FRP composites correlated by applying the Arrhenius model (time-temperature superposition principle) has been adopted by many researchers. The model represents an exponential decay for the data based on accelerated testing under laboratory conditions as stated in the previous paragraph. It is important to note that the overall environmental degradation under natural aging, i.e., outside environment, is represented in the model by considering only the reductions through temperature variations. These lab-based accelerated aging reduction rates, over 1.5 years, correlated with the limited available data under natural aging (over ~2 years) to establish the strength reduction factors, accounting for environmental, physical, and chemical aging of FRP composites.

The core concept of the Arrhenius model is based on the assumption that the strength degradation is independent of any effect on material strength under temperatures ranging from (−10°F to 160°F); however, dependent on pH variations and FT effects. The degradation rate, k, which is the inverse of time, is written as:

$$k = Ae^{\left(-Ea/RT\right)} \tag{B.2}$$

where

A = Constant of the material and degradation process,
E_a = Activation energy,
R = Universal gas constant, and
T = Temperature in Kelvin units.

The experimental results revealed that glass composites with vinyl ester resin exhibited severe ILSS degradation (~60%) within the first 30–40 days (Barker, 2019), followed by gradual loss of ILSS till 150 days, before stabilizing to a very slow rate. Therefore, ILSS loss was correlated using the Arrhenius model separately for the first 150 days of accelerated testing and different degradation rates starting at 150 and ending at 500 days. The alkaline environment leads to maximum damage to glass fiber-reinforced composites due to alkaline-silica reactions (ASRs). This was observed from the slope of Arrhenius plots where E_a (energy of the reaction material), i.e., silica in glass fiber, undergoing ASR.

B.7.2 STRENGTH RETENTIONS RESISTANCE UNDER DIFFERENT ENVIRONMENTS

The reduction factors for ILSS under longer-term exposure to various environments (pH, FT, exposure time, and others), also known as "resistance" or "knockdown" factors, have to be established as accurately as possible for design purposes. Thus, the structural safety of in-service structural systems is not only assured but also a guaranteed minimum number of years of service life can be projected. Based on the 12-year field data of naturally aged vinyl ester-glass composites, resistance factors for ILSS up to 100 years of service life are projected using the Arrhenius model. These ILSS reduction factors (also known as resistance factors) are incorporated in the load and resistance factor design (LRFD) approach, which is commonly used in infrastructure design projects.

B.7.3 SUMMARY

The Arrhenius model and time-temperature superposition principles were discussed extensively by GangaRao et al. (2007 and 2020) and Wang et al. (2015) to determine the resistance factors in the LRFD approach for infrastructure systems made of polymer composites. Both the laboratory and field data revealed that ILSS is more sensitive and significant to reductions than tensile or bending strength. The dissolution of silica in glass fibers weakened the glass fiber-vinyl ester interface, resulting in ILSS reductions under high pH (~12). However, the resin gets affected under low pH, but not as significant as silica reaction under high pH. GFRP composites exposed to neutral and acidic environments are also under freezing and room temperatures are not overly sensitive to chemical reactions and show degradation after 150 days of accelerated aging. In a neutral environment (ph = 7), for example, the ILSS reduces by 30% over 3 years of service and will retain ~30% strength after 100 years of service life.

B.8 CONCLUSIONS

A review on the durability of fiber-reinforced polymer composites under environments (Wang et al., 2015) concluded that moisture attacks including pH variations weaken the FRP composite strength more than the stiffness. The extreme acid environment tends to be more aggressive than extreme alkaline condition toward FRP composites except for glass fiber composites under high pH (>10). Alkaline conditioning has more detrimental effects on the tensile strength of GFRP composites than tap water and saltwater. Although Fickian law is popular in describing moisture diffusion in polymers, a linear superposition of phenomenologically independent contributions from Fickian diffusion and polymeric relaxation can be applied to composites immersed in liquid at high temperature.

The tensile strength of FRP composites decreases significantly once the temperature exceeds T_g, whereas the elastic modulus remains relatively unaffected until the exposure temperature increases to the ignition point of resin. Moreover, temperature strongly affects the moisture absorption rate of FRP composites, and hygrothermal effect can decrease T_g of FRPs due to its plasticization effect. Limited FT cycles cause insignificant changes in tensile and flexural properties of FRP composites. Sustained loading tends to accelerate the degradation of mechanical properties and damage growth of FRP composites when combined with environmental conditions. Due to its viscoelastic nature, polymer matrix accounts for a large part of creep of FRP composites.

UV radiation degrades the matrix-dominated transverse properties more than the fiber-dominated longitudinal properties of composites. The coupled exposure of UV radiation with high-temperature cycles and/or high humidity cycles had more detrimental effects on FRP than single exposure. However, the coupled exposure of UV radiation with FT cycles can compensate partially for the decrease in mechanical properties of FRP due to post-curing.

Note: The contents of this Appendix B are primarily a summary of the technical paper published by GangaRao along with three other authors (Wang et al., 2015).

REFERENCES SELECTED BIOGRAPHY

Agarwal, S., GangaRao, H., Liang, R., and Gupta, R., Physical and chemical aging of thermoset composites, November 2019, WVU Report Submitted to EPRI.

Ascione, L., Berardi, V.P., and D'Aponte, A., Creep phenomena in FRP materials, *Mechanics Research Communications*, 43, 2012, pp. 15–21.

ASTM E119, 2011. Standard test methods for fire tests of building construction and materials, 2012.

Barker, W.T., Short beam shear strength evaluations of GFRP composites – Correlation through accelerating and natural aging, M.S. Thesis, Submitted to the College of Engineering and Mineral Resources, West Virginia University, 2019.

Batra, S., Creep rupture and life prediction of polymer composites, Master's Thesis, Submitted to the College of Engineering and Mineral Resources, West Virginia University, 2009.

Bazli, M., Jafari, A., Ashrafi, H., Zhao, X.L., Bai, Y., and Raman, R.S., Effects of UV radiation, moisture and elevated temperature on mechanical properties of GFRP pultruded profiles, *Construction Building Materials*, 231, 2020, Article 117137.

Bian, L., Xiao, J., Zeng, J, and Xing, S., Effects of seawater immersion on water absorption and mechanical properties of GFRP composites, *Journal of Composite Materials*, 46(25), 2012, pp. 3151–3162.

Bisby, L.A., ISIS Educational Modules 8 – Durability of FRP composites for construction. A Canadian Network of Centers of Excellence, www.isiscanada.com/ publications/educational-modules/, accessed 19 October 2015, 2006.

Bottoni, M., Mazzotti, C., and Savoia, M., Creep tests on GFRP pultruded specimens subjected to traction or shear, *Composite Structures*, 108, 2014, pp. 514–523.

Buck, S.E., Lischer, D.W., and Nemat-Nasser, S., Mechanical and microstructural properties of notched E-glass/vinyl ester composite materials subjected to the environment and a sustained load, *Materials Science and Engineering – A*, 317(1–2), 2001, pp. 128–134.

Cao, S., Zhis, W.U, and Wang, X., Tensile properties of CFRP and hybrid FRP composites at elevated temperatures, *Journal of Composite Materials*, 43(4), 2009, pp. 315–330.

Chin, J.W., Aouadi, K., Haight, M.R., Hughes, W.L., and Nguyen, T., Effects of water, salt solution and simulated concrete pore solution on the properties of composite matrix resins used in civil engineering applications, *Polymer Composites*, 22(2), 2001, pp. 282–297.

Chin, J.W., Nguyen, T., and Aouadi, K., Effects of environmental exposure on fiber reinforced plastic (FRP) materials used in construction, *Journal of Composites, Technology and Research*, 19(4), 1997, pp. 205–213.

Debaiky, A.S., Nkurunziza, G., Benmokrane, B., and Cousin, P., Residual tensile properties of GFRP reinforcing bars after loading in severe environments, *Journal of Composites for Construction*, 10(5), 2006, pp. 370–380.

De'Neve, B., and Shanahan, M.E.R., Water absorption by an epoxy resin and its effect on the mechanical properties and infra-red spectra, *Polymer*, 34(24), 1993, pp. 5099–5105.

Dittenber, D., GangaRao, H., and Liang, R., Durability and life cycle performance of pultruded and infused FRPs for infrastructure – Baseline aging effects on mechanical properties, NSF # 1230351, 2012.

Eldridge, A., and Fam, A., Environmental aging effect on tensile properties of GFRP made of furfuryl alcohol bioresin compared to epoxy, *Journal of Composites for Construction*, 18(5), 2014, p. 04014010.

Fitzer, E., Composites for high temperature, *Pure and Applied Chemistry*, 60(3), 1998, pp. 287–302.

GangaRao, H.V.S., Barker, W., and Allan, M., Degradation mechanism of glass fiber/vinyl ester-based composite materials under accelerated and natural aging, *Construction and Building Materials Journal*, 256, 2020, p.119462.

GangaRao, H., and Kumar, S., Design and fatigue response of concrete bridge decks reinforced with FRP rebar. In *Proceedings 2nd International RILEM Symposium (FRPPRCS-2)*, Non-metallic (FRP) Reinforcement for Concrete Structures, Gent, Belgium, 1995.

GangaRao, H., and Liang, R., Creep response characterization and life prediction of GFRP composites under bending. Research and application of composite materials in infrastructure. In *Proceedings of the Workshop on Global Innovations in Infrastructure with Advanced Composites*, Nanjing, China, 30–31 October 2014, pp. 48–58.

GangaRao, H., Taly, N., and Vijay, P.V., *Reinforced Concrete Design with FRP Composites*, Boca Raton, FL, Taylor & Francis Group, 2007.

Ghorbel, I., and Valentin, D., Hydrothermal effects on the physico-chemical properties of pure and glass fiber reinforced polyester and vinyl ester resins, *Polymer Composites*, 14(4), 1993, pp. 324–334.

Hammami, A., and Al-Ghuilani, N., Durability and environmental degradation of glass-vinyl ester composites, *Polymer Composites*, 25(6), 2004, pp. 609–616.

Han, S.O., and Drzal, L.T., Water absorption effects on hydrophilic polymer matrix of carboxyl functionalized glucose resin and epoxy resin, *European Polymer Journal*, 39(9), 2003, pp. 1791–1799.

Harper, J.F., and Naeem, M., The moisture absorption of glass fibre reinforced vinyl ester and polyester composites, *Materials & Design*, 10(6), 1989, pp. 297–300.

Hawileh, R.A., Abu-Obeidah, A., and Abdalla, J.A., Temperature effect on the mechanical properties of carbon, glass and carbon-glass FRP laminates, *Construction and Building Materials*, 75, 2015, pp. 342–348.

Helbling, C., Abanilla, M., Lee, L., and Karbhari, V.M., Issues of variability and durability under synergistic exposure conditions related to advanced polymer composites in civil infrastructure, *Composites Part A – Applied Science and Manufacturing*, 37(8), 2006, pp. 1102–1110.

Helbling, C.S., and Karbhari, V.M., Investigation of the sorption and tensile response of pultruded E-glass/vinyl ester composites subjected to hygrothermal exposure and sustained strain, *Journal of Reinforced Plastics and Composites*, 27(6), 2008, pp. 613–638.

Homam, S.M., and Sheikh, S.A., Durability of fiber reinforced polymers used in concrete structures. In *Proceedings 3rd International Conference on Advanced Composite Materials in Bridges and Structures*, Ottawa, Ontario, Canada, August 2000, pp. 751–758.

Jones, R.H., *Environmental Effects on Engineered Materials*, New York, NY, CRC Press, 2001, p. 415.

Kajorncheappunngam, S., The effects of environmental aging on the durability of glass/epoxy composites, PhD Thesis, Submitted to the College of Engineering and Mineral Resources, West Virginia University, 1999.

Kajorncheappunngam, S., Gupta, R.K., and GangaRao, H., Effect of aging environment on degradation of glass-reinforced epoxy, *Journal of Composites for Construction*, 6(1), 2002, pp. 61–69.

Karbhari, V.M., Chin, J.W., Hunston, D., et al., Durability gap analysis for fiber-reinforced polymer composites in civil infrastructure, *Journal of Composites for Construction*, 7(3), 2003, pp. 238–247.

Karbhari, V.M., and Xian, G., Hygrothermal effects on high V_F pultruded unidirectional carbon/epoxy composites – Moisture uptake, *Composites Part B – Engineering*, 40(1), 2009, pp. 41–49.

Kasturiarachchi, K.A., and Pritchard, G., Water absorption of glass/epoxy laminates under bending stresses, *Composites*, 14(3), 1983, pp. 244–250.

Kumar, B.G., Singh, R.P., and Nakamura, T., Degradation of carbon fiber-reinforced epoxy composites by ultraviolet radiation and condensation, *Journal of Composite Materials*, 36(24), 2002, pp. 2713–2733.

Kumosa, L., Benedikt, B., Armentrout, D., and Kumosa, M., Moisture absorption properties of unidirectional glass/polymer composites used in composite (non-ceramic) insulators, *Composites Part A – Applied Science and Manufacturing*, 35(9), 2004, pp. 1049–1063.

Lorenzo, M., Durability of GFRP composites under harsh environments – Effect of pH and temperature, M.S. Thesis, Submitted to the College of Engineering and Mineral Resources, West Virginia University, 2018.

Marouani, S., Curtil, L., and Hamelin, P., Ageing of carbon/epoxy and carbon/vinyl ester composites used in the reinforcement and/or the repair of civil engineering structures, *Composites Part B – Engineering*, 43(4), 2012, pp. 2020–2030.

Marru, P., Latane, V., Puja, C., Vikas, K., Kumar, P., and Neogi, S., Lifetime estimation of glass reinforced epoxy pipes in acidic and alkaline environment using accelerated test methodology, *Fibers and Polymers*, 15(9), 2014, pp. 1935–1940.

Martin, J.R., Johnson, J.F., and Cooper, A.R., Mechanical properties of polymers – The influence of molecular weight and molecular weight distribution, *Journal of Macromolecular Science – Reviews in Macromolecular Chemistry*, 8(1), 1972, pp. 57–199.

Micelli, F., Mazzotta, R., Leone, M., et al., Review study on the durability of FRP-confined concrete, *Journal of Composites for Construction*, 19(3), 2015, p. 04014056.

Miller, B.L., Synergistic effect of mechanical loading and environmental conditions on the degradation of pultruded glass fiber-reinforced polymers (GFRP), MS Thesis, North Carolina State University, 2014.

Mounts, J.L., Durability of glass-fiber-reinforced polymer composites in an alkaline environment, Master Thesis, Submitted to the College of Engineering and Mineral Resources, West Virginia University, 2007.

Mourad, A.H.I., Magid, B.M.A., El-Maaddawy, T., and Grami, M.E., Effect of seawater and warm environment on glass/epoxy and glass/polyurethane composites, *Applied Composite Materials*, 17(5), 2012, pp. 557–573.

Nakada, M., and Miyano, Y., Accelerated testing for long-term fatigue strength of various FRP laminates for marine use, *Composites Science and Technology*, 69(6), 2009, pp. 805–813.

Nakayama, M., Hosokawa, Y., Muraoka, Y., and Katayama, T., Life prediction under sulfuric acid environment of FRP using X-ray analysis microscope, *Journal of Materials Processing Technology*, 155, 2004, pp. 1558–1563.

Ray, B.C., and Rathore, D., Durability and integrity studies of environmentally conditioned interfaces in fibrous polymeric composites – Critical concepts and comments, *Advances in Colloid Interface Science*, 209, 2014, pp. 68–83.

Robert, M., and Benmokrane, B., Behavior of GFRP reinforcing bars subjected to extreme temperatures, *Journal of Composites for Construction*, 14(4), 2010, pp. 353–360.

Saadatmanesh, H., Tavakkolizadeh, M., and Mostofinejad, D., Environmental effects on mechanical properties of wet lay-up fiber-reinforced polymer, *ACI Materials Journal*, 107(3), 2010, pp. 267–274.

Santhosh, K., Muniraju, M., Shivakumar, N.D., and Raguraman, M., Hygrothermal durability and failure modes of FRP for marine applications, *Journal of Composite Materials*, 46(15), 2012, pp. 1889–1896.

Sauder, C., Lamon, J., and Pailler, R., The tensile behavior of carbon fibers at high temperatures up to 2400 C, *Carbon*, 42(4), 2004, pp. 715–725.

Shi, J., Zhu, H., and Wu, Z., Durability of wet lay-up FRP composites and their epoxy reins in alkaline environment. In *4th International Symposium on Innovation & Sustainability of Structures in Civil Engineering*, Xiamen, China, 28–30 October 2011, pp. 1129–1135.

Shi, J.W., Zhu, H., Wu, G., and Wu, Z.S., Tensile behavior of FRP and hybrid FRP sheets in freeze–thaw cycling environments, *Composites Part B – Engineering*, 60, 2014, pp. 239–247.

Sonawala, S.P., and Spontak, R.J., Degradation kinetics of glass-reinforced polyesters in chemical environments, *Journal of Materials Science*, 31(18), 1996, pp. 4745–4756.

Sumida, A., and Mutsuyoshi, H., Mechanical properties of newly developed heat-resistant FRP bars, *Journal of Advanced Concrete Technology*, 6(1), 2008, pp. 157–170.

Vijay, P.V., Aging and design of concrete members reinforced with GFRP bars, PhD Thesis, Submitted to the College of Engineering and Mineral Resources, West Virginia University, 1999.

Vijay, P.V., and GangaRao, H., Accelerated and natural weathering of glass fiber reinforced plastic bars. In *Fourth International Symposium, Fiber Reinforced Polymer Reinforcement for Reinforced Concrete Structure ACI*, Michigan, 1999, pp. 605–614.

Wang, J., GangaRao, H., Liang, R., Zhou, D., Liu, W., and Fang, Y., Durability of glass fiber-reinforced polymer composites under the combined effects of moisture and sustained loads, *Journal of Reinforced Plastics and Composites*, 34(21), 2015, pp. 1739–1754.

Weitsman, Y.J., Effects of fluids on polymeric composites – A review, Mechanical and Aerospace Engineering and Engineering Science, The University of Tennessee Report MA ES98-5.0-CM, August 1998.

Williams, B., Kodur, V., Green, M.F., and Bisby, L., Fire endurance of fiber-reinforced polymer strengthened concrete T-beams, *ACI Structural Journal*, 205(1) 2008, pp. 60–67.

Wu, H.C., Fu, G., Gibson, R.F., Yan, A., Warnemuende, K., and Anumandla, V., Durability of FRP composite bridge deck materials under freeze–thaw and low temperature conditions, *Journal of Bridge Engineering*, 11(4), 2006, pp. 443–451.

Wu, H., and Yan, A., Time-dependent deterioration of FRP bridge deck under freeze/thaw conditions, *Composites Part B – Engineering*, 42(5), 2011, pp. 1226–1232.

Yilmaz, V.T., and Glasser, F.P., Reaction of alkali-resistant glass fibers with cement. Part I – Review, assessment and microscopy, *Glass Technology*, 32(3), 1991, pp. 91–98.

Zhou, J., Chen, X., and Chen, S., Durability and service life prediction of GFRP bars embedded in concrete under acid environment, *Nuclear Engineering and Design*, 241(10), 2011, pp. 4095–4102.

Index

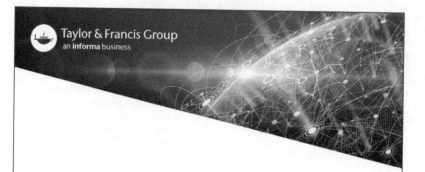